15th International Conference on Quality in Research (QIR 2017)

IOP Conference Series: Materials Science and Engineering
Volume 316

Bali, Indonesia
24 – 27 July 2017

ISBN: 978-1-5108-6063-6
ISSN: 1757-8981

Printed from e-media with permission by:

Curran Associates, Inc.
57 Morehouse Lane
Red Hook, NY 12571

proceedings.com

Some format issues inherent in the e-media version may also appear in this print version.

Printed by Curran Associates, Inc. (2018)

For permission requests, please contact the Institute of Physics at the address below.

Institute of Physics
Dirac House, Temple Back
Bristol BS1 6BE UK

Phone: 44 1 17 929 7481
Fax: 44 1 17 920 0979

techtracking@iop.org

Additional copies of this publication are available from:

Curran Associates, Inc.
57 Morehouse Lane
Red Hook, NY 12571 USA
Phone: 845-758-0400
Fax: 845-758-2633
Email: curran@proceedings.com
Web: www.proceedings.com

Table of contents

Volume 316

Quality in Research: International Symposium on Materials, Metallurgy, and Chemical Engineering

24–27 July 2017, Bali, Indonesia

Accepted papers received: 7 February 2018
Published online: 28 March 2018

Preface

Papers

Intl. Conf. on Saving Energy in Refrigeration and Air Conditioning (ICSERA)

Intl. Symp. on Materials and Metallurgy

The 3rd Biannual Meeting on Bioprocess Engineering

IOP Conf. Series: Materials Science and Engineering **316** (2018) 011001 doi:10.1088/1757-899X/316/1/011001

THE 15th INTERNATIONAL CONFERENCE

PREFACE

WELCOME FROM THE RECTOR OF UNIVERSITAS INDONESIA

It is both a pleasure and honor for me to welcome you all to the 15th International Conference on QiR (Quality in Research) 2017 in Nusa Dua, Bali, Indonesia.
Universitas Indonesia strives to be one of the leading research universities and the most outstanding academic institution in the world. UI is distinctive among research universities in its commitment to the academic invention and research activities through various scientific programs. QiR 2017 is our main academic conference in the field of engineering and technology which has been successfully held for the last two decades. It is our hope that this world class scientific program would showcase our scientists and researchers achievements and provide forums for scientific exchanges in their respective fields.

The theme this year of 'Science, Technology and Innovation for Sustainable World', is very relevant with the fact that the globalization today results in very competitive atmosphere in all aspects. However, this flourishing competition should consider the harmony and balance between human needs and the environment quality for creating favorable sustainable future. Scientists and researchers, hand in hand with industrial experts are creating and developing new sustainable technologies that enable us to make products and services more efficient, design better buildings, produce safer cars, keep people healthier and building smarter cities.

I extend my sincere thanks to the Faculty of Engineering Universitas Indonesia, supporting parties and institutions for their participation and contributions in QiR 2017. I would also thank our colleagues from Universitas Udayana and Politeknik Negeri Bali for their gracious support and hospitality. Additionally, I extend a hearty thank you to the members of the organizing committees for dedicating their valuable time so that each one of us enjoys an exceptional conference program over the next several days. May we have a successful, stimulating, fruitful and rewarding conference.

Prof. Dr. Ir. Muhammad Anis, M.Met.
Rector
Universitas Indonesia

IOP Conf. Series: Materials Science and Engineering **316** (2018) 011001 doi:10.1088/1757-899X/316/1/011001

PREFACE

WELCOME FROM THE DEAN OF FACULTY OF ENGINEERING UNIVERSITAS INDONESIA

Welcome to the 15th International Conference on QiR (Quality in Research) 2017. The Faculty of Engineering Universitas Indonesia is delighted to host our flagship international academic event this year back in Bali, Indonesia. This two-day, biennial conference is presented together with our co-hosts Universitas Udayana and Politeknik Negeri Bali with the hope that this would be able to provide an international media for exchange of the knowledge, experience and research as well as the review of progress and discussion on the state of the art and future trend of prospective collaboration and networking in broad field of science, technology and innovation.

The main theme for this year conference, "Science, Technology and Innovation for Sustainable World" is consistent with the mission of our faculty to be a leading institution with the initiatives that responds to local, national and global societal needs. In that context, the Faculty of Engineering Universitas Indonesia is performing state-of-the arts research and development in engineering and architecture areas which results in technology and innovation which contribute to sustainable development at both national and global level. QiR 2017 provides platforms and forums to disseminate our scientific achievements and exchange information with our counterparts from Indonesia and all over the world. This event will allow for further research and education collaborations between Universitas Indonesia and its partners worldwide.

I would like to express my deepest appreciation to our sponsors, supported parties and various contributors for their never ending supports of this conference. I would also like to convey my gratitude to all of our distinguished speakers for making the time to share their knowledge with us. To our fellow researchers and/or practitioners from Indonesia and overseas, welcome and enjoy your stay in this Nusa Dua, Bali. I would also like to invite all participants in expressing our appreciation to all members of the QiR 2017 organizing committee for their hard work in making this conference success.

Prof. Dr. Ir. Dedi Priadi, DEA
Dean Faculty of Engineering
Universitas Indonesia

PREFACE

WELCOME FROM THE QiR 2015 ORGANIZING COMMITTEE

On behalf of the organizing committee, it is a great pleasure for us to welcome you to the 15th International Conference on Quality in Research (QiR) 2017 to be held in Bali, Indonesia on July, 24 - 27, 2017. This biennial event is co-organized with the Faculty of Engineering Universitas Udayana and Politeknik Negeri Bali.

The main theme for this year conference is "Science, Technology and Innovation for Sustainable World". Under this theme the conference focuses on the innovative research and contribution in science and technology toward achieving sustainable world. In line with this theme, it is our utmost pleasure to hold the QiR 2017 in conjunction with the 6th IEEE-International Conference on Advanced Logistics and Transport (ICALT), the 2nd International Symposium on Biomedical Engineering (ISBE 2017), International Conference in Saving Energy in Refrigeration and Air Conditioning (ICSERA) and the 3rd Biannual Meeting on Bioprocess Engineering.

The QiR 2017 brings together national and international academicians, researchers, executives, government, industrial and business officials, practitioners and leaders to present and discuss a vast range of engineering, architectural designs and community development based on green and smart technology. It is our hope and aim that this conference would be able to provide an international media for exchange of the knowledge, experience and research as well as the review of progress and discussion on the state of the art and future trend of prospective collaboration and networking in broad field of science, technology and innovation. Furthermore, QiR 2017 benefits industry sector, since it would create a close contact between and among the audiences. The audiences mostly come from different job and activities: therefore this is a great potential and opportunity to meet each other, creating fruitful discussions and broaden business relationship.

QiR has been growing, since its first event two decades ago, into our flagship academic event with international reputation. This year, we have received almost 1000 submissions from more than 26 countries. Along with our events in conjunction, more than 500 oral and poster presentations is scheduled with expected 700 participants gather in the event.
On behalf of QiR 2017 committee, we would like to thank all of our speakers, participants, contributors, partners and professional associations for their generous contributions. We also would like to acknowledge the support from our International Advisory Board members and distinguished reviewers. Last but not least, a special thanks to our local co-organizer, Universitas Udayana and Politeknik Negeri Bali.
We wish all of you a productive and rewarding conference, also a pleasant and memorable stay in Nusa Dua, Bali, Indonesia.
Thank you and we hope to see you again in QiR 2019.

Ardiyansyah, Ph.D.
General Chair of QiR 2017 Organizing Committee

IOP Conf. Series: Materials Science and Engineering **316** (2018) 011002 doi:10.1088/1757-899X/316/1/011002

CONFERENCE ORGANIZER

ADVISOR
- Prof. Prof. Dr. Ir. Dedi Priadi, DEA.,
- Dr. Ir. Muhamad Asvial, M.Eng.
- Ir. Hendri DS Budiono, M.Eng
- Dr. Badrul Munir, ST., M.Eng.Sc
- Jos Istiyanto, S.T., M.T., Ph.D.
- Dr. Ir. Wiwik Rahayu,DEA.
- Prof. Dr. Akhmad Herman Yuwono, M.Phil., Eng.

GENERAL CHAIR
Ardiyansyah, PhD., Universitas Indonesia

CO-CHAIR
Dr. Eny Kusrini, Universitas Indonesia

INTERNATIONAL ADVISORY BOARD
- Prof. Muhammad Anis, Universitas Indonesia
- Prof. Rosari Saleh, Universitas Indonesia
- Prof. Dedi Priadi, Universitas Indonesia
- Prof. Hiroshi Murase, Nagoya University, Japan
- Prof. Manabu Tanaka, Director of JWRI
- Prof. Kazuhiro Ito, Professor at JWRI
- Assoc. Prof. Yosuke Kawahito, Associate Professor at JWRI
- Prof. Afshin Ghajar, Oklahoma State University
- Prof. Josaphat Tetuko Sri Sumantyo, Chiba University
- Prof. Pega Hrnjak, University of Illinois at Urbana Champaign
- Prof. Greet Vanden Berghe, KU Leuven
- Prof. Joong Kee Lee, KIST, Korea
- Prof. Pekka Leviäkangas, University of Oulu
- Prof. Marie-Anne Guerry, Vrije Universiteit Brussel
- Prof. Rainer Leisten, University of Duisburg Essen
- Prof. Hamid Ullah - Universiti Teknologi Brunei

STEERING COMMITTEE
- Dr. Tri Tjahjono, Universitas Indonesia
- Prof. Yulianto S. Nugroho, Universitas Indonesia
- Prof. Benyamin K., Universitas Indonesia
- Prof. Winarto, Universitas Indonesia
- Dr. Ing. Dalhar Susanto, Universitas Indonesia
- Prof. Widodo Wahyu Purwanto, Universitas Indonesia
- Prof. Isti Surjandari Prajitno, Universitas Indonesia
- Prof. Suardana, Universitas Udayana
- I Made Rajendra, M.Eng, Politeknik Negeri Bali

SCIENTIFIC PUBLICATION PARTNER
- Dr. Nyoman Suwartha
- Dr. Mohammed Ali Berawi

TECHNICAL PROGRAM COMMITTEE
- Dr. Cindy Rianti Priadi
- Sugeng Supriadi, Ph.D
- Dr. Basari
- Chairul Hudaya, Ph.D
- Wahyuaji N. Putra, MT
- Dr.-Ing. Yulia Nurliani Lukito

CONFERENCE ORGANIZER

- Dr. Bambang Heru Susanto
- Komarudin, Ph.D
- I Dewa Gede Ary Subagia, PhD
- Dr. Wayan Nata Septiadi
- Dr. I Nyoman Suamir

Secretariat and Registration
Herra Astasusmini, SE
Agnes Sagita Nauli, S.I.A.
Indah Sari Dewi

Treasurer
- Evy Surpiningsih, S.Pd., MM
- Nuri Nugraini, Amd

Programme and Protocol
Tikka Anggraeni, M.Si.

Design and Documentation
- Rengga Wibisono, S.Sos.
- Muhammad Badi

Web and Information System
- I Gede Dharma Nugraha, S.T., M.T.
- Boma Anantasatya Adhi, ST., M.T.
- Ruki Harwahyu, S.T., M.T., M.Eng.
- Ardiansyah, ST., M.Eng.
- Gunawan Heri Saputra, Amd

Exhibition and Sponsorship
- Dr. Ir. Nahry., MT.
- Dr. Adi Surjosatyo, M.Eng.
- Dr. Muhammad Suryanegara
- Kemas Ridwan Kurniawan, ST., MSc., PhD.
- Dr. Tania Surya Utami, S.T., M.T.
- Ir. Erlinda Muslim, MEE.

Venue and Facilities
- Jumiardi, S.Ars
- Hadi Mulyadi

Meal
- Yunita Dewi Hapsari
- Indri Feriani

Conference Organizing Committee :
Faculty of Engineering Universitas Indonesia
Dekanat Building 3th Floor Kampus UI, Depok 16424, Indonesia
Phone : +62-21- 7863503, Fax : +62-21 – 7270050
Email : qir@eng.ui.ac.id,
Website : http://qir.eng.ui.ac.id
www.eng.ui.ac.id

ACKNOWLEDGEMENT

The 15th International Conference on QiR (Quality in Research) Organizing Committee wishes to express its gratitude and appreciation to :

Prof. Dr. Ir. Muhammad Anis M.Met., Rector of Universitas Indonesia
for consenting to be the guest of honour

All invited speakers session, moderators and conference speakers, for their participation.
All conference Sponsors, Supporters, Exhibitors and advertisers for their generous support.
All participants and others who have in one way or another contributed towards the success of this conference.

PLATINUM SPONSOR

SILVER SPONSOR

SUPPORTED BY

IOP Conf. Series: Materials Science and Engineering **316** (2018) 011004 doi:10.1088/1757-899X/316/1/011004

Peer review statement

All papers published in this volume of *IOP Conference Series: Materials Science and Engineering* have been peer reviewed through processes administered by the proceedings Editors. Reviews were conducted by expert referees to the professional and scientific standards expected of a proceedings journal published by IOP Publishing.

IOP Conf. Series: Materials Science and Engineering **316** (2018) 012001 doi:10.1088/1757-899X/316/1/012001

Studies of the water adsorption on Lampung's natural zeolite of Indonesia for cooling application

D A Wulandari[1,2]*, Nasruddin[1], Lemington[3]

[1] Department of Mechanical Engineering, Universitas Indonesia, Kampus Baru UI, Depok 16424, Indonesia

[2] Department of Mechanical Engineering, Universitas Negeri Jakarta, jl. Rawamangun Muka, Jakarta Timur 13220, Indonesia

[3] Institute of Refrigeration and Cryogenics, Shanghai Jiao Tong University, 800 Dongchuan Road, Shanghai, 200240, P.R. China

Email: dyah.arum51@ui.ac.id; nasruddin@eng.ui.ac.id; lemington@live.com

Abstract. Part of minerals that originally formed from volcanic rock and ash layers reacting further with alkaline groundwater is called natural zeolite, where its sources are not always available in all countries. Indonesia is located in the ring of fire which have a huge sources of zeolite, one of the area is Lampung, South Sumatra. Natural zeolite has been considered as one of potential heat adsorbent medium which can contribute to the energy consumption and reduce air pollution in the using of cooling application. The characteristic of this Lampung natural zeolite such as adsorption kinetics, adsorption water uptake, and adsorption capacity were test with ASAP 2020 system. Sorption kinetics by this experiment of zeolite samples were carried out in a constant temperature and humidity chamber. The chamber can supply constant air condition with deviations of ±0.5 °C for temperature and ±3% for relative humidity. The data based on rate of adsorption and the defined working condition was set as 20°C and 70% RH. Pore volume is a significant parameter for determining the limitation of water uptake, which can describe the saturated condition of zeolite. Sorption isotherm models used to describe sorption phenomena are commonly deduced from the Polanyi potential theory were investigated. The water adsorption quantity increased with the increase of relative pressure. To sum up, this pure zeolite has a less heat and mass transfer performance so its need to be activated before using in cooling application to get their great potential and by being coated in a desiccant heat exchanger systems.

1. Introduction

The use of natural zeolite from Indonesia as an adsorbent for cooling application can be a huge alternative energy resources for energy conservation and environmentally protection. Natural zeolites are scattered in several regions in Indonesia on more than twenty places in Indonesia among others in Sumatra, Java and East Nusa Tenggara because it's spread on the land near the volcanic mountains [1] with lots of mordenit and klinoptilolit type [2]. More research show that a lot of it using for other needs.

Nowadays, widely utilization of natural zeolite in Indonesia have been applied in agriculture, livestock, fisheries and some industries [3]. However, there is a research about natural zeolite in Indonesia as a adsorber in CO2 capture modified with Na+ to achieve the higher of adsorption capacity was conducted [4]. Besides, the research of chromium (VI) waste adsorption by zeolite has done by isotherm adsorption method to determine maximum capacity of zeolite to chromium (VI) waste [5]. Subsequent work by [6] and [6] show about conversion of palm oil into bio gasoline using Ni/ZA or natural zeolite catalyst and Cr/ZA catalyst on the conversion reaction of waste cooking oil into liquid fuel. In addition, the dehydration process on the bioethanol purification using natural zeolite result the purity more than 99% but has lower yield and much ethanol can be absorbed by natural zeolite from Gunung Kidul which could be seen from the addition of adsorbent weight by

20% [7] and [8] also as a molecular sieve [9]. This kind of natural zeolite used for the phase purification of methane in biogas which can be converted into electrical energy was doing by [10]. In medical science it can be used as a raw material of antiseptic [11]. Another field by hydrocracking of waste tires has been done too [12].

Thus, there is limited study about the use of natural zeolite from Indonesia using for cooling application. In this paper, an Indonesian natural zeolite from Lampung were under testing and some key parameters were studied. First about the sorption kinetics and then the sorption isotherms data were measured and investigated.

2. Methodology / experimental

The material used in this paper mainly included natural zeolite from Lampung, Indonesia. This source being selected because the sufficient value of total surface area, total pores volume and relatively small pores distance which can result high pore density and adsorbing rate [13].

2.1. Measurement of sorption kinetics

Sorption kinetics of the samples were being tested in humidity chamber and a constant temperature with the reference of natural zeolite. This humidity chamber could deliver constant air condition by the temperature deviations of ±0.5°C and relative humidity deviation of ±3%. The setup temperature and humidity of defined working condition were 20°C and 70%RH respectively. The samples were dried in the oven by the temperature of 100°C as for pretreatment. Recorded the weights of the samples at setting intervals by an electronic balance, the accuracy was 0.001 g.

2.2. Measurement of sorption isotherms

The samples were carried out in an improved Accelerated Surface Area and Porosimetry System (ASAP 2020) system at 20°C to get their water sorption isotherms data. This system apply the principle of static volumetric technique to acquire nitrogen adsorption/desorption isotherms at temperature -196°C. The error from this experimental of the ASAP apparatus is particularly caused by pressure and temperature transducers. The fidelity of pressure transducer is about 0.1% while temperature is ±0.02°C.

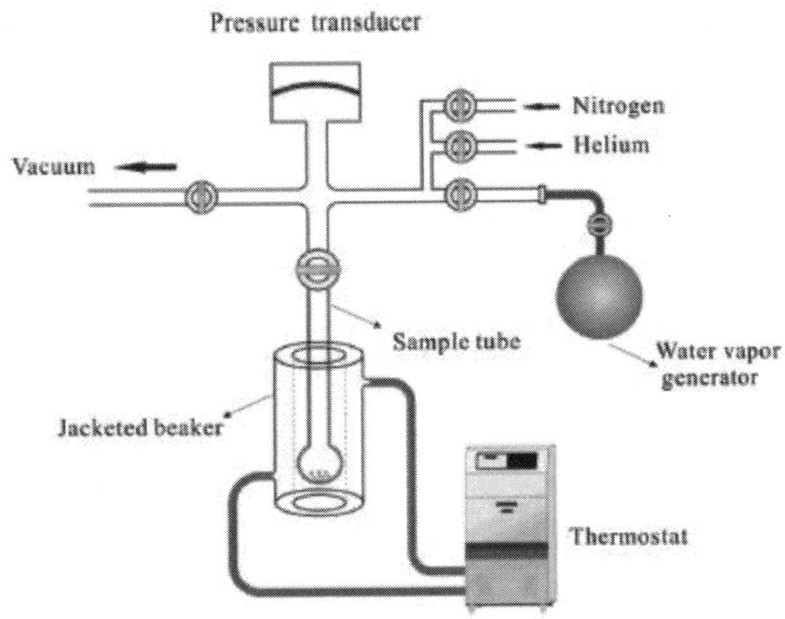

Figure 1. The schematic diagram of the improved ASAP 2020 system [14]

Fig. 1 represent the schematic diagram of the improved ASAP 2020 system. There are many items such as a jacketed beaker, a water vapor generator and a thermostatic water bath in that ASAP 2020 apparatus. The jacketed beaker and the thermostatic bath are integrated together to control the temperature of the sample tube with the accuracy of the thermostat is ±0.05°C. By utilizing the improved ASAP 2020 machine, sensitive water sorption isotherms can be examined and sorption temperature of a test sample can also be exactly managed.

3. Results and discussion

3.1. Sorption kinetics

The natural zeolite's dynamic sorption amounts on water vapor were find out in a thermo-humidistat chamber under the test condition of temperature 20°C and 70%RH. The water sorption quantity was indicated as per water uptake on per unit mass of natural zeolite (i.e. g g^{-1}).

As represented in Fig. 2, for all the composition of relative humidity, the dynamic water sorption increases harshly at the initial stage of sorption. Then it grows gradually until the sorption equilibrium is obtained. Contrast with the lowest relative humidity, the highest relative humidity not only produced higher dynamic sorption quantities, but also had faster sorption rates.

The reason for this is that water sorption by higher relative humidity having more physical sorption. The released sorption heat influenced the time needed to reach sorption equilibrium.

IOP Conf. Series: Materials Science and Engineering **316** (2018) 012001 doi:10.1088/1757-899X/316/1/012001

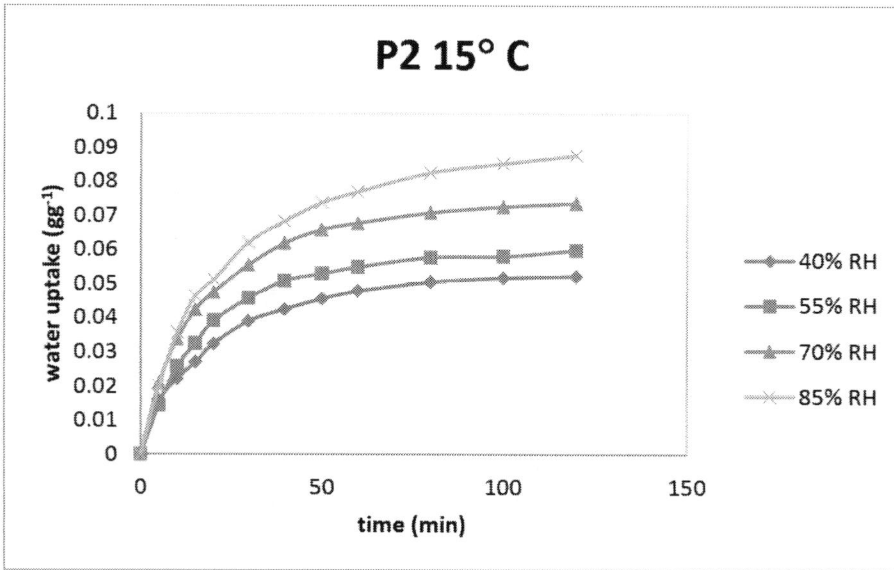

Figure 2. Water sorption kinetic curves of natural zeolite from Lampung, Indonesia at 15°C with different relative humidity

3.2. Sorption isotherms

The water sorption isotherms behavior of natural zeolite were conducted by an improved ASAP 2020 system at 20°C. The isotherms were provided in Fig. 3. It can be noticed that, water sorption capacity of Lampung's natural zeolite fluently grew with the enhanced of relative pressure.

With the increase of relative pressure, the more liquid absorption process occurred. Sorption isotherm models to describe sorption phenomena are commonly sum up from the Polanyi potential theory:

$$\varepsilon = -RTln\left(\frac{P}{Po}\right) \quad \ldots\ldots (1)$$

where,

ε is sorption potential (kJ kg^{-1}),

T is the sorption temperature (K),

P is the vapor pressure (Pa),

Po is the saturated vapor pressure at sorption temperature T (Pa).

IOP Conf. Series: Materials Science and Engineering **316** (2018) 012001 doi:10.1088/1757-899X/316/1/012001

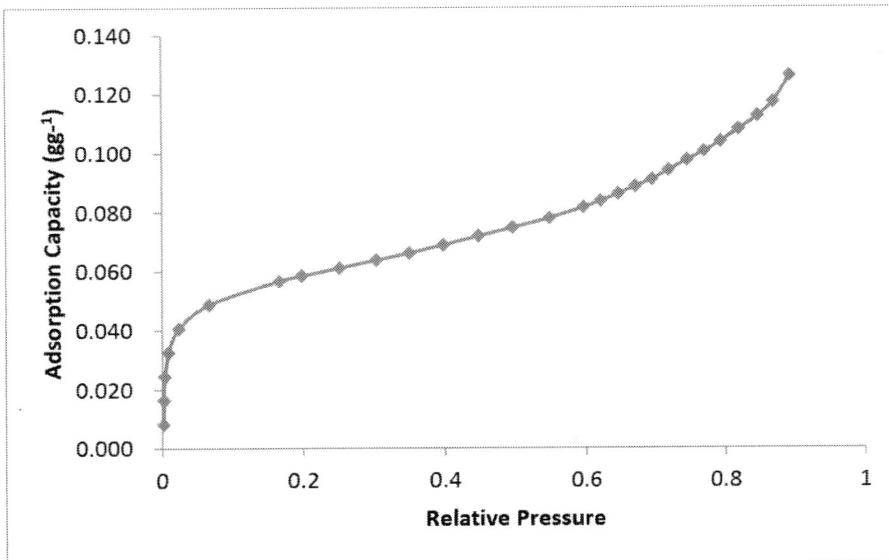

Figure 3. Water sorption isotherms of Lampung's natural zeolite at 20°C by an ASAP 2020 apparatus.

By using the sorption potential, pressure and temperature can be merge into a single parameter. The correlation curves which represent the relation between adsorbed water and sorption potential can be achieved, as shown in Fig. 4.

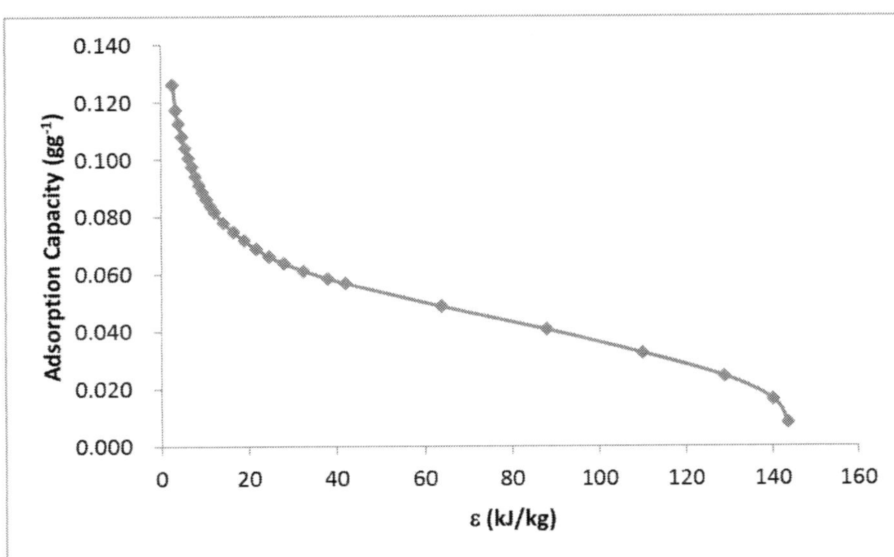

Figure 4. Correlation curves between adsorbed water and sorption potential of Lampung's natural zeolite.

4. Conclusion

Investigation on sorption kinetics (i.e. dynamic sorption quantity and rate coefficient) indicated that Lampung's natural zeolite having denote higher dynamic sorption quantities and faster sorption rates.

Sorption isotherms were acquired by an improved ASAP 2020 system to explore water sorption behavior of Lampung's natural zeolite and the experimental data had good agreement with calculated data.

5. Acknowledgement

The author would like to thank Prof. Wang Ruzhu for his assistance, guidance and experience during the experiment at Institute of Refrigeration and Cryogenics, Key Laboratory of Power Mechanical Engineering, Shanghai Jiao Tong University, Shanghai, China and the Indonesian Ministry of Research and Higher Education for the scholarship (BPPDN).
This article's publication is supported by the United States Agency for International Development (USAID) through the Sustainable Higher Education Research Alliance (SHERA) Program for Universitas Indonesia's Scientific Modeling, Application, Research and Training for City-centered Innovation and Technology (SMART CITY) Project, Grant #AID-497-A-1600004, Sub Grant #IIE-00000078-UI-1.

6. References

[1] K. Kusdarto, "POTENCY OF ZEOLITE IN INDONESIA," *Jurnal Zeolit Indonesia,* vol. 7, no. 2, pp. 78-87, 2008.
[2] D. Y. Lestari, "Kajian modifikasi dan karakterisasi zeolit alam dari berbagai negara," in *Prosiding Seminar Nasional Kimia dan Pendidikan Kimia*, 2010.
[3] L. Thamzil and H. Zamroni, "Penggunaan Zeolit dalam Bidang Industri dan Lingkungan," *Jurnal Zeolit Indonesia,* vol. 1, 2010.
[4] R. Adriany, "Pemanfaatan Zeolit Alam Termodifi kasi Kation Na+ untuk Penangkapan CO2," *Lembaran Publikasi Minyak dan Gas Bumi,* vol. 46, no. 3, 2012.
[5] M. Handayani and E. Sulistiyono, "Uji Persamaan Langmuir dan Freundlich Pada Penyerapan Limbah Chrom (Vi) Oleh Zeolit," in *Prosiding Seminar Nasional Sains*, 2009.
[6] P. E. Rahayu, S. Priatmoko, and S. Kadarwati, "Konversi Minyak Sawit Menjadi Biogasoline Menggunakan Katalis Ni/Zeolit Alam," *Indonesian Journal of Chemical Science,* vol. 2, no. 2, 2013.
[7] S. K. Wahono, A. Kristiani, S. Tursiloadi, and H. Abimanyu, "Characterization and utilization of gunungkidul natural zeolite for bioethanol dehydration," *Energy Procedia,* vol. 47, pp. 263-267, 2014.
[8] S. K. Wahono, R. Maryana, and D. Pratiwi, "Modification of Gunungkidul Natural Zeolite as Bioethanol Dehydrating Agents," *Energy Procedia,* vol. 65, pp. 116-120, 2015.
[9] K. Khaidir, D. Setyaningsih, and H. Haerudin, "MODIFICATION OF NATURAL ZEOLITE AS MOLECULAR SIEVE MATERIAL ON BIOETHANOL DEHYDRATION," *Jurnal Zeolit Indonesia,* vol. 8, no. 2, pp. 97-105, 2009.

[10] S. K. Wahono, R. Maryana, M. Kismurtono, M. Kismurtono, and K. KNisa, "Modifikasi Zeolit Lokal Gunungkidul Sebagai Upaya Peningkatan Performa Biogas Untuk Pembangkit Listrik," 2010.

[11] L. M. Estiaty, "ENGINEERING OF ZEOLITE MINERAL WITH WET IMPREGNATION INHIBITOR METAL METHOD AS RAW MATERIAL OF ANTISEPTIC BY CONTINOUS FLOW METHOD," *Jurnal Zeolit Indonesia,* vol. 9, no. 2, pp. 61-70, 2010.

[12] A. B. Majid, "Karakterisasi dan Uji Aktivitas Katalitik Zeolit Alam Indonesia pada Hidrorengkah Ban Bekas dengan Preparasi Sederhana Characterization of Indonesian Natural Zeolite and Their Activity on Hydrocracking of Waste Tires with Simple Preparation."

[13] Y. I. Supriyatna, D. C. Birawidha, and S. Sumardi, "Lampung Zeolite Utilization as Gas Emission Adsorbant on Charcoal Making Process," *International Journal of Civil & Environmental Engineering IJCEE-IJENS Vol,* vol. 11, pp. 116902-7575, 2011.

[14] X. Zheng, T. Ge, Y. Jiang, and R. Wang, "Experimental study on silica gel-LiCl composite desiccants for desiccant coated heat exchanger," *International journal of refrigeration,* vol. 51, pp. 24-32, 2015.

IOP Conf. Series: Materials Science and Engineering **316** (2018) 012002 doi:10.1088/1757-899X/316/1/012002

Characteristics Air Flow in Room Chamber Test Refrigerator Household Energy Consumption with Inlet Flow Variation

Edy Susanto, M. Idrus Alhamid[1], Nasruddin, Budihardjo

Faculty of Engineering, Universitas Indonesia
Kampus Baru UI Depok, Depok 16424, Indonesia

E-mail: mamak@eng.ui.ac.id,

Abstract. Room Chamber is the most important in making a good Testing Laboratory. In this study, the 2-D modeling conducted to assess the effect placed the inlet on designing a test chamber room energy consumption of household refrigerators. Where the geometry room chamber is rectangular and approaching the enclosure conditions. Inlet varied over the side parallel to the outlet and compared to the inlet where the bottom is made. The purpose of this study was to determine and define the characteristics of the airflow in the room chamber using CFD simulation. CFD method is used to obtain flow characteristics in detail, in the form of vector flow velocity and temperature distribution inside the chamber room. The result found that the position of the inlet parallel to the outlet causes air flow cannot move freely to the side of the floor, even flow of air moves up toward the outlet. While by making the inlet is below, the air can move freely from the bottom up to the side of the chamber room wall as well as to help uniform flow.

1. Introduction

Energy has become an interesting issue to be discussed. Indonesia as one of the developing countries will continue to turn into a developed country, is in need of energy to drive the economic growth rate. But the reality of Indonesia is now starting to face major problems, namely energy crisis. Addition of population and economic growth are not able to be coupled with energy generation growth will infrastructure. It led the government to continue to think in order to get out of this problem. In 2009, the government has issued Government Regulation no.70 on the Conservation of Energy which aims to preserve the continuity of existing resources through prudent use of resources for achieving a balance between development, equity and environmental development. Energy conservation efforts directed to improve the equitable and sustainable development. In connection with the use of technologies that will be developed production and use of more energy efficient in terms of technical, economic and environmental health. Realistic measures of energy conservation programs that create an energy policy that relates on energy efficiency standards and labeling. Energy efficiency standards and labeling for household appliances is one of the popular strategies to save energy and become a place of learning for the community or the consumer to be able to use energy wisely [1]. However, this rule cannot be applied because one of the problems is the lack Testing Laboratory, particularly Fridge Household Consumption Test. [2] says that in applying energy efficiency standards and labels it is necessary to perform several stages, one of which is a set of test procedures. The test procedure is a method or regulation performed in the laboratory testing to measure performance and evaluates the energy use of various brands and models in connection with the characteristics of such products [3].

[1] mamak@eng.ui.ac.id,

IOP Conf. Series: Materials Science and Engineering **316** (2018) 012002 doi:10.1088/1757-899X/316/1/012002

Room Chamber is the most important in making a good Testing Laboratory. Terms of conditions room chamber to test household refrigerator energy consumption becomes a major factor to the lack Testing Laboratory in Indonesia. Based on the ISO 15502: 2008 states that the temperature and velocity of air in the room must be kept constant chamber ranges from 32 ^{0} C and 0.25 m / s [4]. Therefore, it is necessary to design appropriate in its design.

The purpose of this study was to determine and define the characteristics of the airflow in the room chamber using CFD simulation. The use of numerical computational simulation methods can shorten the time and cost of a cheap in analyzing the phenomenon of fluid flow [5, 6]. The condition of the cube-shaped room chamber where the small inlet and outlet make it like an enclosure. The heat transfer that occurs close to natural convection, because of the slow fluid movement and flow of fluid flow due to differences in density caused by the difference in temperature.

2. Numerical Methods

Two models of the inlet flow were varied to determine and define the characteristics of the air flow in accordance with the required standards. CFD method is used to simulate fluid flow in the room chamber. In a settlement with numerical methods, computational modeling with 2-D flow approach, necessary steps are: The first stage of the modeling and geometry, mesh manufacture and provision of boundary condition definition. Where the geometry room chamber 2-D is 224 mm long, 240 mm high, 20 mm wide outlet channel. The second phase is a treatment that consists of setup conditions and finds solutions. Stage three is a post processing, consisting of obtaining results (graphics, vector, contours) and analyze them.

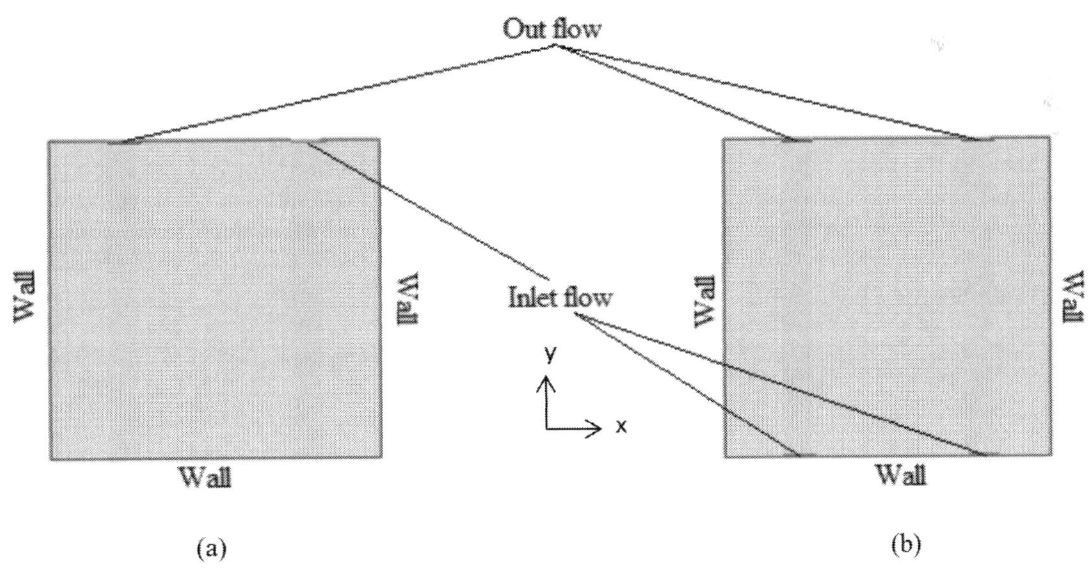

Figure 1. a. Boundary condition with the inlet and outlet aligned on top
b. Boundary condition with the inlet at the bottom and the outlet at the top

Make mesh generation using this type of meshing folder quadrilateral, meshing on the surface of the inlet and outlet are made more stringent property so that the observed changes in the fluid, whereas at greater distances from the inlet and outlet of the meshing is made more tenuous to save computer memory and speed up the convergence as the figure 1. And on the assumptions above, the general equation that can be used are:

- Conservation of mass:

$$\frac{\partial \rho}{\partial t} + div(\rho u) = 0 \tag{1}$$

- Momentum equation:

$$\frac{\partial(\rho u)}{\partial t} + div(\rho uu) = -\frac{\partial p}{\partial x} + div(\mu\ grad\ u) + S_x \tag{2.a}$$

$$\frac{\partial(\rho v)}{\partial t} + div(\rho vu) = -\frac{\partial p}{\partial y} + div(\mu\ grad\ u) + S_y \tag{2.b}$$

- Energy equation:

$$u\frac{\partial T}{\partial t} + v\frac{\partial T}{\partial y} = \left(\frac{k}{\rho cp}\right)\left(\frac{\partial^2 T}{\partial x^2} + \frac{\partial^2 T}{\partial y^2}\right) + S_h \tag{3}$$

3. Results and Discussion

3.1. Vector and Contours Velocity in Inlet and Outlet Inline

Figure 2 shows the numerical results of the velocity vector in which the phenomenon of flow in through the inlet channel flow into the chamber with a room large enough speed of about 1.03 m / sec at the surface of the fluid inlet. While after moving to reach about 1.7 m from floor room chamber movement direction vector flow has begun to change, until at a distance of 1.1 m from the floor surface velocity vector starting to reverse. However, it can be seen also in part downstream moves down closer to the floor surface and not a bit vector flow actually turned toward the outlet (outlet flow). Phenomena such as the flow of this happened because of the inability to continue to flow down to the bottom towards the floor surface due to the effect buoyancy. Effect buoyancy or buoyancy forces caused by the temperature difference resulting density of the fluid is small. The result is a fluid will move turned toward the outlet or move upwards.

IOP Conf. Series: Materials Science and Engineering **316** (2018) 012002 doi:10.1088/1757-899X/316/1/012002

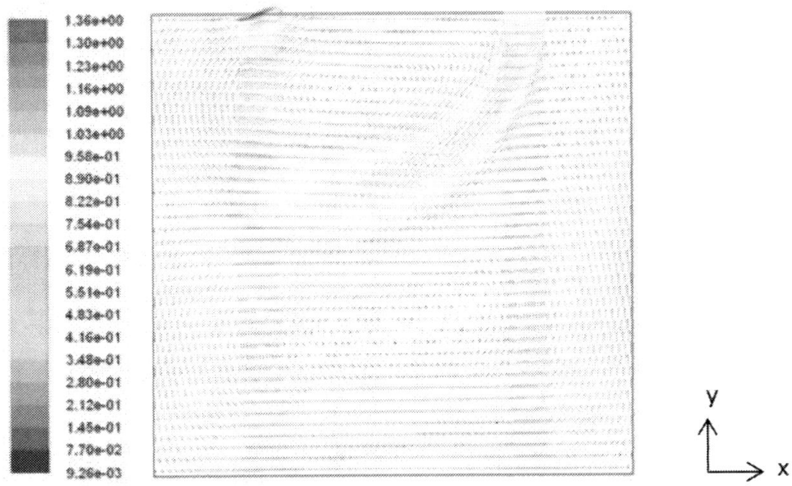

Figure 2. Velocity vector room chamber with inlet and outlet parallel

While the figure 3, we can see more clearly the flow phenomena. Figure 3 is the result of a numeric contour velocity, which is clearly visible color degradation. This color degradation represents a change of pace flowing fluid. Contour velocity yellow are highly concentrated in the upper zone and the color began to fade along with the reduced speed of the fluid. There is one phenomenon under contour velocity that occurs contour velocity yellow blue. Where the contour lines forming the opal. This indicates that under contour velocity yellow vortex flow occurs at a flow rate that is not too large. But this is more or less contour causes the incoming air flow through the inlet turned toward the outlet.

Figure 3. Contour velocity room chamber with inlet and outlet parallel

3.2. Vector Contours Velocity and Temperature at Inlet in the down position

Figure 4 shows the results of the numerical velocity vector in which the position is under the inlet and outlet located on the side. The figures show that the flow pattern tends to move around freely. The movement upward flow near occurs the wall to the left and continue moving clockwise around the

IOP Conf. Series: Materials Science and Engineering **316** (2018) 012002 doi:10.1088/1757-899X/316/1/012002

walls of the chamber room. This occurs because the temperature on the lower side is higher than the upper side so that the fluid air to move freely upwards. But when air fluid approaching the upper side has a lower temperature causes the air density increases so that the fluid moves down the air back which is gravitation style effect on the condition of the enclosure. The movement of the fluid flow such as this would speed up and simplify the deployment temperature. With the rapid spread of the temperatures accelerate the uniform conditions at room chamber.

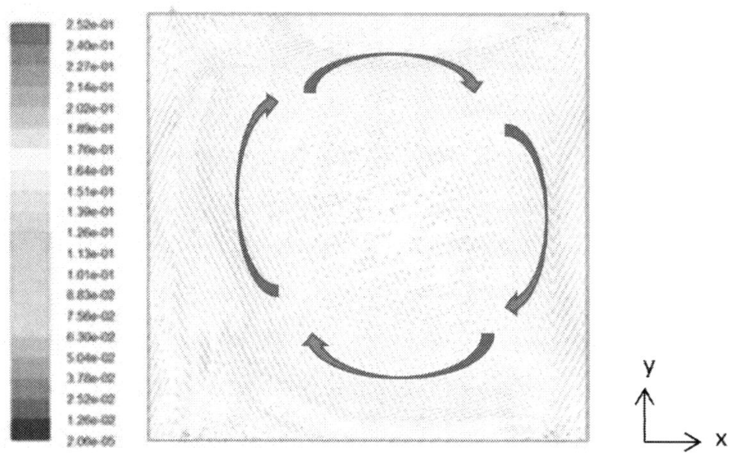

Figure 4. The vector velocity room chamber where the inlet at the bottom.

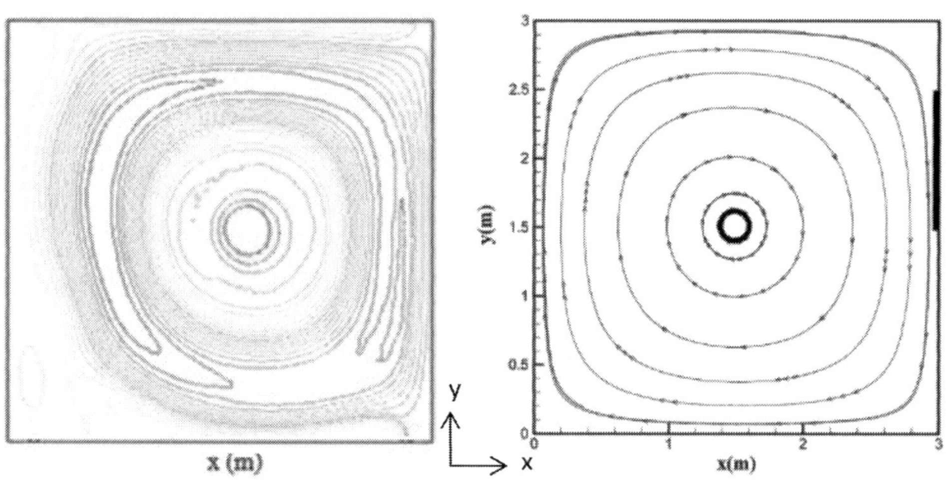

Present work Khorasanizadeh et.al

Figure 5. Stream lines to the inlet flow temperature below

Figure 5 is clearly visible from the similarity between the flow phenomena that has been done by Khorasanizadeh et al with that being done now. Conducted a study on the performance of numerical 2D floor heating by comparing with centralized heating systems. Include analysis the following phenomena airflow and temperature distribution [7].

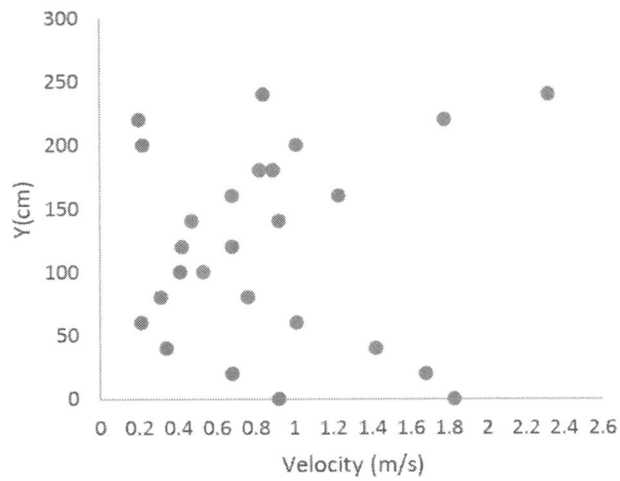

Figure 6. Comparison of inlet velocity distribution and outlet parallel with inlet is below at x = 112 cm

From figure. 6 it can be seen that the flow velocity distribution for inlet and outlet conditions is parallel to a random pattern, in contrast to the inlet condition on the underside of the floor showing a more directional flow pattern.

4. Conclusion

In designing a room chamber, should conduct a study or numerical analysis beforehand. Because method this can save time and cost of manufacture. Apart from that, the phenomena flow that occurs in the room chamber can be easily analyzed. This will make it easier to determine the channel system, both the inlet and the outlet. From the above analysis, it can be concluded that by positioning the inlet temperature is below makes the distribution more quickly uniform. Although the air flow rate in the room chamber has not reached the desired target of 0.25 m / sec, so the next development is needed.

5. Acknowledgments

The authors would like to thanks the Ministry of Research, Technology and Higher Education, The Republic of Indonesia through “Hibah Infrastruktur Laboratorium 2012”.

This article’s publication is supported by the United States Agency for International Development (USAID) through the Sustainable Higher Education Research Alliance (SHERA) Program for Universitas Indonesia’s Scientific Modeling, Application, Research and Training for City-centered Innovation and Technology (SMART CITY) Project, Grant #AID-497-A-1600004, Sub Grant #IIE-00000078-UI-1.

6. References

[1] T. M. I. Mahlia and R. Saidur, "A review on test procedure, energy efficiency standards and energy labels for room air conditioners and refrigerator–freezers," *Renewable and Sustainable Energy Reviews,* vol. 14, pp. 1888-1900, 2010.

[2] T. M. I. Mahlia, *et al.*, "Theory of energy efficiency standards and labels," *Energy Conversion and Management,* vol. 43, pp. 743-761, 2002.

[3] A. K. Meier and J. E. Hill, "Energy test procedures for appliances," *Energy and buildings,* vol. 26, pp. 23-33, 1997/01/01 1997.
[4] BSNi, "Lemari Pendingin untuk Rumah Tangga – Karakteristik dan Metode Uji," ed, 2008.
[5] A. H. György Bicsák, Dr. Árpád Veress, "Numerical Simulation of Combustion Processes in a Gas Turbine," *AIP Conference Proceedings,* vol. 1493, November 2012.
[6] F. di Mare, *et al.*, "Large eddy simulation of a model gas turbine combustor," *Combustion and Flame,* vol. 137, pp. 278-294, 2004.
[7] H. Khorasanizadeh, *et al.*, "Numerical study of air flow and heat transfer in a two-dimensional enclosure with floor heating," *Energy and buildings,* vol. 78, pp. 98-104, 2014.

IOP Conf. Series: Materials Science and Engineering **316** (2018) 012003 doi:10.1088/1757-899X/316/1/012003

The floating houses of Sintang City: space, resources and political nexus

M S Lubis[1], T Y Harjoko[1], D Susanto[1]

[1]Department of Architecture, Faculty of Engineering, Universitas Indonesia, Kampus Baru UI, Depok 16424, Indonesia.

gotty49@gmail.com

Abstract. Sintang is a city in the interior of West Kalimantan, located at the confluence of two large rivers, Kapuas River and Melawi River. Growing from the river's edge, the present Sintang still maintains its riverine culture in the form of floating houses ('rumah lanting'), as its legacy of the past, when rivers played important role as main transportation routes for political and economic activities. This paper explores several concepts related to power, space, and resources, to examine the complexity of life of lanting dwellers, conducted at two locations of lanting houses that spread around Sintang urban area: Pasar Sungai Durian and Pasar Inpres. The results show that there are two main factors affecting the life as well as the spatiality of lanting dwellers: First, the relationship with river's ecology and river's behavior, and second, the linkages with the activities on the mainland, where both of them are strongly influenced by political and economic considerations. The impact of changes in the river's ecology and behavior on the lanting dwellers has shown how government policies affected the rivers and thus, the lanting communities. The key contribution of this study is a clear depiction of the political and power relations in the context of spatiality and resources on the riverine culture, especially the lanting communities along Kapuas and Melawi River in Sintang City, which tends to reveal the marginalization of the lanting community among the other urban community. So far, there has been no serious and detailed studies about the politics of space in relation to nature and resources and local livelihoods in the area. This paper wishes to explore the challenges to strive for a livable and just place for river community in the future.

1. Introduction

Sintang is a city in the interior of West Kalimantan, which is located at the confluence of two large rivers, Sungai Kapuas and Sungai Melawi. Growing from the river's edge, the present Sintang still maintains its riverine culture in the form of floating houses ('rumah lanting'), as its legacy of the past, when rivers played an important role as main transportation routes for political and economic activities. As stated by Evers (2015), "Considerations of security from floods, wild animals and human enemies have shaped housing and settlement patterns around Borneo and, indeed, in many other parts of Southeast Asia. This has been a common practice among various Austronesian groups throughout Southeast Asia".

In the decade of 1940s to at least 1970s, river transportation still played an important role for the people in inland region of West Kalimantan, i.e. Sanggau, Sekadau, Sintang, and Kapuas Hulu. At that time, people used to board motorboats and navigate to Pontianak. Floating villages became real transit points for passing motorboats, and were used for purchases of fuel, as well as other necessities such as

IOP Conf. Series: Materials Science and Engineering **316** (2018) 012003 doi:10.1088/1757-899X/316/1/012003

rice, sugar, etc. Motorboats used to depart from the upstream, usually carrying produce from the inland region such as rubber, ironwood, and bengkirai wood. The produce was brought to Pontianak and traded for basic groceries in the Chinese market. Those were then brought back into the inland cities and villages to be traded or exchanged for the local community's produce (Heidhues, 2003).

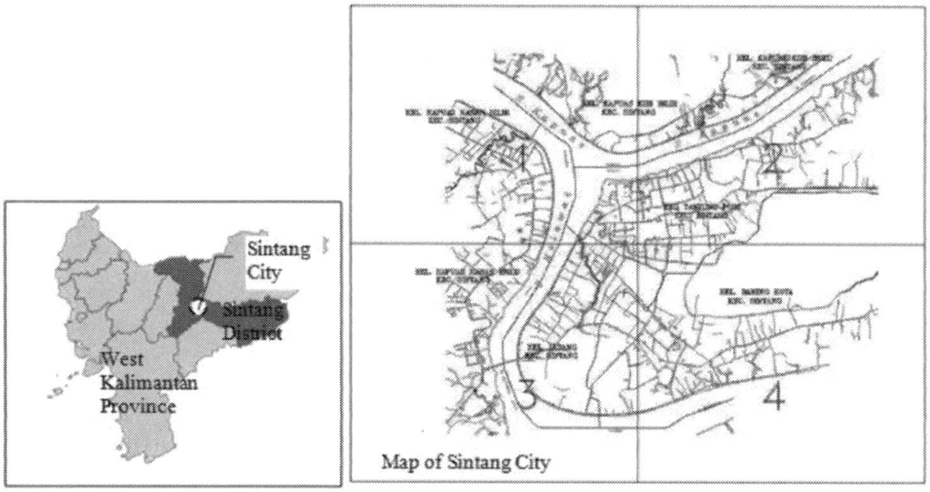

Figure 1. Map of Sintang City

After the construction of roads in the inland Borneo (around 1950s), the conditions of the river deteriorated. There were many illegal gold mining activities along the riverbanks; the waste was habitually and directly dumped into the water of the river. That made the river contaminated by the mine waste, including mercury (Cleary, 1992). The river that was once used as the source of clean water and fish became gradually 'unreliable', due to the high level of pollutants and sediments. Until now, all types of waste are increasingly polluting the Kapuas River, ranging from household waste, from pesticides and plantation fertilizers, to mercury.

Figure 2. Floating houses in Kapuas River

2. Methodology

This paper explores concepts related to power, space, and resources, in order to examine the complexity of life of lanting dwellers, in terms of relations within and between the communities; the inter-relationship of humans, nature and resources, and the consequent implications on the spatiality. This study was conducted at two locations of lanting houses that spread around Sintang urban area: Sungai Durian and Pasar Inpres. In this study, I used ethnographic methods, especially the spatial ethnography, which was also supported by the participant observation, in-depth interviews, and literature studies.

3. Space, resources, and political nexus from past to present

3.1. Sintang in the past

The arrival of the colonialists in Indonesia gave birth to the story of colonialism and the struggle for independence. It also created the cities built in order to serve the interests of the colonizers. At that time, Sintang was a kingdom located in the interior of West Kalimantan, which had the largest territory in the region . It was the old kingdom known to have existed since the pre-Islamic period and had a strong influence both on the upstream and downstream of the Kapuas River. In the upstream, the kingdom was respected for its political power, while in the downstream it became an engine of the economic development.

The establishment of the territories in West Borneo, especially along the Kapuas river, had foundations in the economic 'pockets' or river trading spots, which were generally located at the junctions of two rivers, typically in the form of boat houses and floating houses. These trading pockets had the opportunity and then control the economy of local people who cultivated and sold their produce to the traders. (Listiana, 2012).

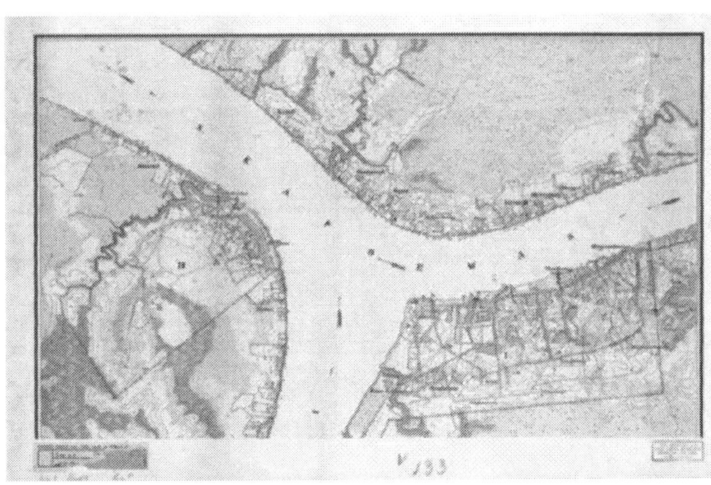

Figure 3. Sintang Map, ca.1888 (source: KITLV, Netherlands)

Furthermore, the growing trading centers would then engage in the political practice of conquest and expansion into the surrounding regions. In this phase, the kingdom has become the main ruler. The process of production and marketing was determined by the royal family and their bureaucrats through traditional bureaucratic and institutional networks, while the people were the only provider of labor and had no bargaining power to determine the size and value of production. It is known that the kingdom of Sintang had applied the concept of land ownership influenced by Hinduism, which states that the landowner is the king *(vorstendomein)*. Because families and bureaucrats of the kingdom assisted the king, he gave them plots of lands called *'apanage'* or *'lungguh'* (Pranoto, 2010 in Listiana, 2012).

Regarding the territory of the kingdom, there are many conflicting pieces of information and opinions. Before Dutch intervention, the traditional kingdoms especially in western Kalimantan had no definite boundaries. The frequent occurrences of disputes between the royals could be identified as evidence of the vagueness of territorial boundaries. The claims of the royal territory were usually based on the statement of submission coming from small kingdoms in the upper stream, in the form of tribute to the kingdom. The territory of the royal power might have been referred to as a tax collection system to the districts of the Dayak community. The arrival of the Dutch colonialists to Sintang changed the politics and government system of the Sintang Kingdom. The Dutch managed to change the political status and territory of Sintang kingdom gradually, through the contractual agreements between the kingdom and the colonial government. Through the contract signed on November 24, 1823, the Governor-General Baron van der Capellen 'legally' ratified the transfer of the Sintang Kingdom to the Dutch power (Sjamsuddin, 2015).

The kingdom's glory was actually depended mostly on the trading and commercial activities along the Kapuas watershed. All commercial ships from the upstream had to pass through Sintang in order to navigate downstream. Also, the attempt to conquest the small kingdom in Selimbau were undertaken in order to control the trading routes leading to Sarawak (then under the British colonialists) in the north.

Being located at the confluence of two large rivers had made Sintang an important trading center with abundance of natural resources that could be found in the surrounding area. The rivers were not only important as transportation routes, but were also providing abundant amount of fish for people's livelihood (Enthoven, 1905). On land, the Dayaks used to extract bird's nests from the crack in the rocks of Mount Kelam, selling them to the kingdom, at a low price. Wax collected (by hundreds of loads) was a very important commodity during that time. The soil was also capable of producing a good quality rice, meeting the needs of the entire inland settlements during the harvest season. Rattan was easily found in the forest and along the Melawi's riverbank. Best quality gold was mined at various spots, as well as iron ore, grindstone and magnet (Heidhues, 2003).

Figure 4. Floating hospital in Sintang, ca.1920(source: KITLV, Netherlands)

3.2. From early Independence to The New Order era

In the early era of the Republic of Indonesia's independence, the socio-environmental conditions of Kalimantan have not showed drastic changes, compared with the previous period of time. Through the

IOP Conf. Series: Materials Science and Engineering **316** (2018) 012003 doi:10.1088/1757-899X/316/1/012003

Government Regulation no. 64 of 1957 issued by President Soekarno, the central government decentralized forest governance outside of Java, placing it under control of the provincial governments. The decentralization policy in the forestry sector during Soekarno was supported and shaped by the strategy of political economy based on the ideas of socialism and patriotism. This strategy was essentially against the Western investment. Therefore, in the Soekarno era, both foreign and local businessmen could not exploit the forest resources. This was in total contrast to the New Order era (New Order of Suharto) policies.

In the New Order era, Indonesia lost 40 million hectares of forest because they were converted into the palm oil plantations in the so-called transmigration areas, because of HPH and HTI concessions, agricultural expansion, and illegal logging practices. The damage to the environment of the forest and the river obviously affects people's lives, especially when it comes to the lanting dwellers. Illegal logging that utilizes the river for rafting timber down the stream; gold mining in the upper part of Kapuas River, and the 'clearance' of massive areas of land to give way to plantations, have been causing negative intrusion into people's lives, as well as suffer and insecurity to the communities along the river. In 1983, the head of Sintang District issued a Local Regulation (Peraturan Daerah) prohibiting people to live in the floating houses along the Kapuas River in the Sintang City area. As a result, the number of floating houses had decreased drastically, although they never disappeared completely.

3.3. Post-Suharto Era

The Asian financial crises and consequent collapse of several Asian economies in 1997, followed by rapid political upheavals, eventually forced President Suharto to step down from the presidency in May 1998. Since then, there has been an opportunity to voice opposition to forest destruction and environmental exploitation. Some positive changes in terms of policies and practices, related to the forest issues, begin to emerge. But the economic crisis also created 'new demands' that have encouraged even greater forest and environmental destruction. This is exacerbated by an ambitious but chaotic decentralization process and its implementation. (Chidley, 2002).

3.4. Sintang today

Today, floating settlements are still surviving along several rivers of Kalimantan, especially in the rural areas. In Sintang, floating settlements are concentrated in two different areas: Sungai Durian and Pasar Inpres. Besides living in their floating homes, the lanting dwellers are also running their businesses, such as mini fuel-stations, low-cost motels, cafes and grocery shops. In Sungai Durian, formerly known as the Pasar Cina (Chinese market), the majority of the residents of floating houses areof Chinese origin, and the rest are Javanese, Malay and Dayak.

One of the owners of the floating houses I interviewed is Mr. H (48), the third-generation owner of the house that he inherited from his grandfather; the first owner. Mr. H himself was born in lanting house, but unlike his predecessors who lived permanently'on the water', Mr. H also owns a property on the land, which he is using for both residential and business activities. Life in the floating houses for already several generations forged a very close relationship between the communities and the river, but the current land-oriented urbanization process has led the lanting communities to live their lives in two interconnected spaces: the river and the land. Today, most of the Chinese lanting residents have their homes on the shore. However, they are still maintaining floating houses as their family heritage.

In Pasar Inpres, most of lanting dwellers are immigrants from surrounding villages seeking better jobs or continuing their education in Sintang city. Most of them are Dayaks who came from the areas of Ambalau and Serawai. They decided to stay in lanting mainly for two reasons: First, they could build or rent the lanting house at the lower cost than those on the land. Second, living on the water allows them to have easy access to water transportation, and their home villages are mostly accessible by boats.

Figure 5. Floating houses as foreground of urban scenery in Sintang

Unlike in the past, today there are not many Malays who are still living in floating homes. The availability of large quantities of land at relatively low prices, particularly in the floodplains of Kampung Raja across the river, has made Malays to choose to live on the land, above the surface in their stilt houses, still maintain the character of local traditional rural living. This is consistent with the ideas of Evers and Korff (2002) that Malays tend to choose to live in kampongs, even though these kampongs are administratively located within the city. The life of Malay people tends to circulate predominantly around the palace and the mosque, rather than in the entire city. That is in contrast to the Chinese way of life. Although most Chinese immigrants who arrived in Indonesia came from the rural areas in southern China, their lifestyle is urban in its essence. The concept of densely populated neighborhood settlements located in the city centers is typical of the Chinese communities in almost all Indonesian cities, and Sintang is not an exception. In other words, the areas of the city that have a distinctive urban character are usually the areas inhabited by the people of Chinese origin, while the areas populated by Malays are still retaining the character of the kampong (local traditional village).

4. Conclusion

Capacity of the river (as space) and resources (commodities) to shape particular economic and social mechanisms has made them synonymous with the 'instruments of power'. These elements (space, resources, power) are also forming a mutually constitutive relationship. Floating villages and riverine settlements are what could be described as clear manifestation of spaces that are constructed and reconstructed periodically. This is based on various purposes and motivations, and involving various actors and political considerations. The origin of the growthof floating settlement in Sintang was the desire of the authorities to control the territory of their power along the river networks.Eventually these settlements grew into the economic pockets, while being generally located in the junctions of the rivers. Their trading and commercial activities, as well as the availability of vital natural resources in the mainland of Kalimantan such as forest products and mining products,have been playing an important role indetermining both direction and nature of Sintang's economy. However, the land-oriented urbanization process has made an impact on the riverine settlements, with many serious consequences.Communitiesbegan livinga semi-urban existence in two different (although interconnected) spaces: the river and the land. The conditions of the rivers deteriorated, becoming increasingly polluted. The government's one-sidedpolicy, geared towards the conventional land-based development,has become one of the mainfuture challenges, even obstacles, to both hopes and desiresto build a livable and just living place for the communities of the floating villages.

5. Acknowledgement

The research was funded by Ministry of Research, Technology and Higher Education of the Republic of Indonesia, under the Domestic Graduate Scholarship (BPP-DN) scheme. We would like to express our gratitude to all those who have contributed to this paper: all colleagues and individuals supplying us with the valuable information supporting this research, in both Sintang and Pontianak. We would like to thank residents of floating houses in Tanjung Puri and Sungai Durian, Sintang City, as well as all our colleagues from the Department of Architecture and Faculty of Engineering, at the University of Indonesia.

References

[1] Chidley L and Marr C (ed.) 2002 Forests, People and Rights: a down to earth special report. June 2002.
[2] Cleary M and Eaton P 1992 Borneo Change and Development. Oxford University Press. Singapore.
[3] Enthoven J J K 1905 Sejarah dan Geografi Daerah Sungai Kapuas Kalimantan Barat. Institut Dayakologi. Pontianak.
[4] Evers H D 2015 Kampung Air: Water Settlements on the island of Borneo. JMBRAS, VOL. 88, Part 1 2015 pp. 1–xx
[5] Evers H D and Korff R 2002 Urbanisme di Asia Tenggara. Jakarta: Yayasan Obor Indonesia.
[6] Heidhues M S 2003 Penambang Emas, Petani, dan Pedagang di "Distrik Tionghoa" Kalimantan Barat. Yayasan Nabil Press. Jakarta.
[7] Kusno A 2007 Space, Power and Identity: Patches of the Postcolonial Past, Present and Future Jakarta. *Journal of Comparative Cultural Studies in Architecture* 1/2007 pp 37-42
[8] Listiana D 2012 Tanah Sintang Masa Kolonial: telaah sederhana tentang perubahan status pemerintahan dan wilayah kekuasaan. STAIN Pontianak Press. Pontianak.
[9] Sjamsuddin H 2015 Kerajaan Sintang: perlawanan dan perubahan di Kalimantan Barat 1822-1942. Penerbit Ombak. Yogyakarta.
[10] Tolo E Y S 2013 Sejarah Ekonomi Politik Tata Kelola Hutan di Indonesia. Harian Indo Progress.

The concept of sustainable prefab modular housing made of natural fiber reinforced polymer (NFRP)

E Setyowati[1*], E E Pandelaki[2]

[1]Bachelor Program of Architecture, Architecture Department, Faculty of Engineering, Diponegoro University, Jl. Prof. Sudarto, SH, Tembalang, Semarang, 50275, Indonesia
[2]Master Degree Program of Architecture, Architecture Department, Faculty of Engineering, Diponegoro University, Jl. Prof. Sudarto, SH, Tembalang, Semarang, 50275, Indonesia.
*Corresponding author: ernisetyowati@arsitektur.undip.ac.id

Abstract. This research aims to formulate the concept of public housing based on research results on natural fiber reinforced polymer (FRP) material which has been done in the road map of research. Research output is the public housing design and specifications of FRP made of water hyacinths and coconut fiber. Method used is descriptive review of the concept based on references and material test which consists of density, water absorption, modulus of rupture (MOR), tensile strength, absorption coefficient and Sound Transmission Loss (STL). The entire tests of material were carried out in the laboratory of materials and construction, while the acoustic tests carried out using the impedance tubes method. The test results concluded that the FRP material may have a density between 0.2481 – 0.2777 g/cm^3, the absorption coefficient is average of 0.450 – 0.900, the Modulus of Elasticity is between 4061 – 15193 kg/cm^2, while the average of sound transmission loss is 52 – 59 dB. Furthermore, that the concept of public housing must be able to be the embryo of the concept of environment-friendly and low emissions housing.

1. Introduction

Technological developments increasingly rapidly identified one of them with the progress of the industry where material synthetic fibers began to be replaced by natural fibre (Pritchard,G, 2015). Building materials made of natural local fibers has been widely observed and implemented in construction. Natural fibers often used are the stem of banana, flax fiber, fiber of pineapple, coconut fibres and wood fiber. The past two year research have produced waffle acoustical material made of coconut fibers and sawdust which were produced using hot press machine with waffles matras. This Panel has a good acoustic performance, because the absorption coefficient have ranged between 0.6-0.8 and the Sound Transmission Loss is between 32-54 dB.

Road map of the research began from the innovation of materials made of waste and forest plantations (Setyowati, E, et.al, 2015a; Setyowati, E, et.al, 2015b). Coconut fibers and sawdust processed into composite absorber for the implementation of the housing in the noisy region (Setyowati, E, 2013; Setyowati, E and Sadwikasari, A.F, 2013; Setyowati, E and Trilistyo, H, 2013; Setyowati,E, 2015). The novelty in this research is the natural fiber composition system and technology with synthetic polymer that can be used for building wall panel. These Building materials are certainly very light compared to conventional wall panels that are often used in building construction. With its lightweight performance, then it secure the building against earthquake hazards. In addition, the cost for structural

components can be saved due to its performance. Climate change, pollution and environmental deterioration due to industry sparking much research related to how to make innovative, creative materials, reduce emissions and sustainable (Yan, L.,et.al,2016). In the automotive industry, the matural fiber is becoming increasingly popular and it replacing carbon (carbon fibre) and glass fiber (glass fibre). In the building industry, the use of natural fibers in concrete or polymer replace content of steel in concrete and it become a major step in the development of the sustainable building construction industries (Wei, J.Q and Meyer,C, 2015; Saba, N, et.al, 2015; Yan, L, et.al, 2014).
As mentioned previously, that the industry had already begun abandon advanced synthetic fibers like glass fiber (glass fibre) and carbon fiber (carbon fibre) in manufacturing material. Even the car industry like Opel, Ford and Daimler-Chrysler have already used natural fibers on its component (Pritchard,G, 2015). In the building industry, natural fibre reinforced polymer had already been introduced, but haven't been popular yet. Hemp and some other bio fiber have a high hemiselulose content, so do the water hyacinth and coconut fibres (Staib,G., et.al, 2008; Kang, S, et.al, 2001). The following figure shows the concept of modular housing with wall panel made of Natural Fibre reinforced polymer (NFRP).

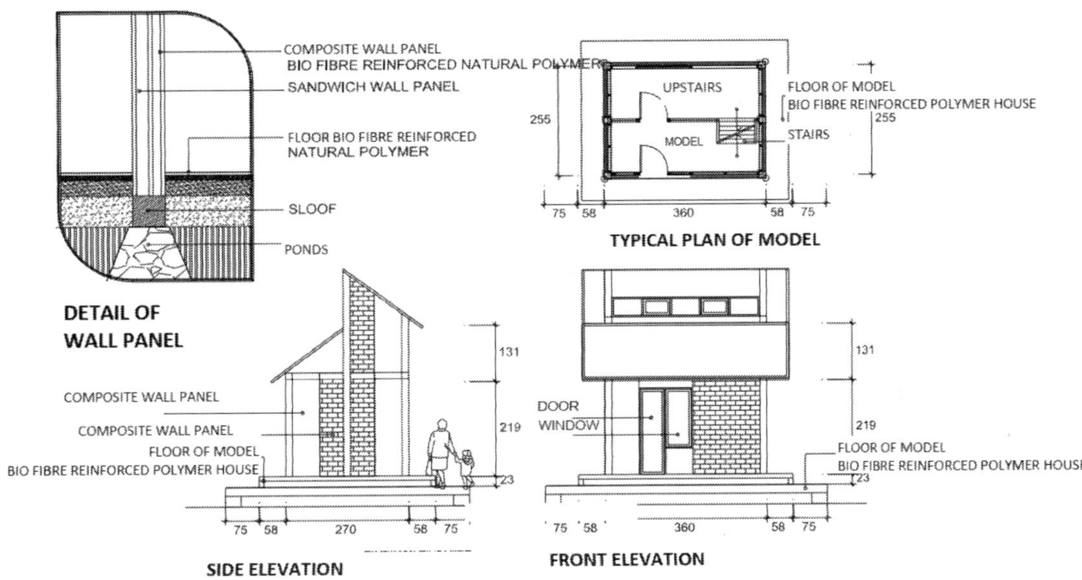

Figure 1. The Concept of Modular housing made of NFRP

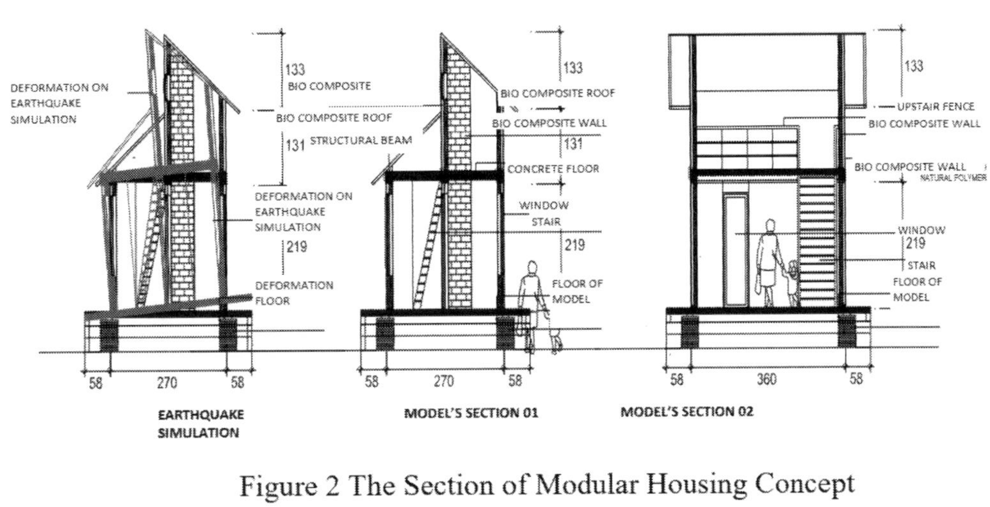

Figure 2 The Section of Modular Housing Concept

Figure 1 shows the concept of modular housing with wall panel made of natural fibre reinforced polymer (NFRP). Natural fibers used are water hyacinth and coconut fibers. Wall Panels made of NFRP used to baffle interior and exterior wall panel. Because it is made from polymer, then the wall panel is highly resistant to weather constraints. Figure 2 shows the elasticity of modular housing in the earthquake vibrations withstand that will probably arise. Due to the lightweight performance, then the modular housing model is reliable as earthquake resistant housing. Observing the process of manufacturing wall panel made of natural fiber reinforced polymer (NFRP), then on the next parts are methodology, results and discussions.

2. Methodology of Experimental Research

In this research, the concept of modular housing from the NFRP was based on the availability of natural resources in abundance, ease the process of manufacture material and technology that are beneficial to the development of housing in Indonesia. To apply the material made of natural fibre, then the NFRP needs to get a series of test materials which are carried out at the materials and constructionlaboratory

2.1. Density Test

Example of test measuring 10 cm x 10 cm material with electric scales weighed and recorded as the initial weight (m). Length (p), width (l) and thickness (t). Then the volume is calculated using the formula as below:

$$V = p \times l \times t \tag{1}$$

V is the volume (cm^3), p is the length (cm), l is the width (cm) and t is the thickness (cm). After obtaining the value of the volume, then the magnitude of the density can be obtained by the formula:

$$\rho = \frac{m}{V} \tag{2}$$

with ρ is density (g/cm^3), m is the weight of the initial sample test (g) and V is the volume of the test of sample (cm^3)

2.2. Modulus of Elasticity Test

Modulus of Elasticity test used UTM Instron MOE. This test used one point loading, the point of imposition on the middle of the length of sample. Loading executed until the limit of the elastic point of sample. The length of the span is 15 times the nominal thickness, but not less than 7.5 cm. MOE values obtained using the formula

$$MOE = \frac{\Delta PL^3}{4\Delta ybh^3} \tag{3}$$

with MOE is Modulus of Elasticity (kg/cm^2), P is the difference in the burden under the limit of proportion (kg), L is the length of the span (cm), y is the change in deflection under the limit of proportion (cm), b is the width of the test sample (cm), and h is the test sample thickness (cm)

2.3. Modulus of Rupture

The static bending strength or Modulus of Rupture (MOR) is one of the very important mechanical properties. The strength of bending fracture or Modulus of Rupture (MOR) is the mechanical properties of wood related to the strength of the wood that is the size of the wood's ability to withstand loads or style beyond the work and tend to change the shape and size of the wood. Modulus of Rupture (MOR) is calculated from the maximum load (the load at the time of broken) in the test using the pliable constancy the same test for MOE (Bowyer et al. 2003).

2.4. Absorption Coefficient Test

The Absorption coefficients of materials can be observed and thoroughly calculated using impedance tubes. The sound absorption coefficient (α_0) was calculated by comparing the sound pressure that falls on the surface of the material and reflects. Testing against the absorption on a material refers to standard of JIS a 1405 1963. This coefficient can be calculated using the following equation (ASTM E 413)

$$\alpha_0 = \frac{4}{n + \left(\frac{1}{n}\right) + 2} \quad (4)$$

with α_0 is the Sound absorption (dB) and n is the ratio of the standing wave. Standing wave ratio (n) is calculated by substituting it value prisoner/sound attenuation, while for determining equations, the standing wave ratio are used by the following equation (ASTM D 790):

$$n = 10^{\left(\frac{L}{20}\right)} \quad (5)$$

Where n is the ratio of the standing wave and L is the difference in sound intensity level. On the process of the absorption Coefficient Test (α), the lab has used the device namely the Impedance tube equipped with 1 unit of microphone to transmit sound frequencies in the low frequency range up to high frequency (ASTM D 790).

2.5. Sound Transmission Loss Test

Unlike the absorption coefficient Test, the transmissions loss (STL) used the impedance tube equipped with four microphones that have the sensitivity to high frequency sounds with. STL voice transmission value partition is defined as the ratio between logarithm of transmitted sound pressure (Wt) and sound pressure comes on the surface material/material partitions (Wi). In general the STL definition above can be formulated as (ASTM E 2611-09):

$$TL = 10\log\frac{W_t}{W_i} \quad (6)$$

$$TL = 10\log\frac{1}{r} \quad (7)$$

The value of r is the transmission coefficient of sound material in question, namely the value of the ratio between the powerful voice that is transferred through a material with a sound that falls on the surface of the material. Based on the ASTM E 413-2004, measurement of STL in the tubes is carried out using impedance frequency which have range between 125 Hz up to 4000 Hz in 1/3 octave filter (ASTM E 2611-09).

3. Results and Discussion

This research focused on the innovation of natural fibre material processed into material Fiber Reinforced Polymer (FRP). To be able to produce materials FRP as intended, then this study uses polymer (polyesther) mixed with fibers of nature: water hyacinth and coconut fibers. The catalyst used was methyl ethyl ketone peroxide composition material comparison: natural fibres: polyesther catalyst = 200 ml: 200 ml: 100 ml: 20 ml respectively. The results of the Test weight and density is as follows:

Table 1. Density Test

Sample	thickness (mm)	weight (gram)	Ø (mm)	T (gram)	Volume (cm^3)	density
A	15	11	29	1.5	39.6111	0.2777
B	30	19	29	2.8	73.9407	0.2570
C	15	10	29	1.5	39.6111	0.2525
D	30	19	29	2.9	76.5815	0.2481

with A and B are 15 mm and 30 mm thickness of water hyacinth composite respectively; C and D are 15 mm and 30 mm thickness of coconut fibre composite respectively. From the 4 samples tested, then it can be inferred that the FRP material density is 0.2481 – 0.2777 g/cm^3. Based on the category of composite material, then this material can be concluded the Low-density Composite. Because the value of the density of the material affected by the dimensions of the components forming the material. According to T.M. Maloney (1998), that based on its density, particle board can be divided into three groups (Maloney, t. m., 1998):

a) *Low density particleboard*,. i.e. the particle board that has a density less than 0.59 g/cm^3.
b) *Medium density particleboard*, i.e. the particle board that has a density between 0.59 – 0.8 g/cm^3.
c) *High density particleboard*, i.e. the particle board that has a density more than 0.8 g/cm^3.

To test the Modulus of Elasticity and the Modulus of Rupture is obtained the following results:

Table 2. Modulus of Elasticity and Modulus of Rupture

Natural Fibre Reinforced Polymer	MOE (kg/cm^3)	MOR (kg/cm^3)
15 mm thickness water hyacinth	4,061	29
30 mm thickness water hyacinth	12,280	126
15 mm thickness coconut fibre	4,534	59
30 mm thickness coconut fibre	15,174	138

The nature of stiffness is a measure of the ability of an object to resist changing incurred as a result of loading. This trait is expressed with a Modulus of Elasticity (MOE) and applies only to the limit of proportion (Bowyer et al. 2003). Based on the testing that has been done is obtained value ranges between MOE 4,061 – 10,238 kg/cm^2 on particle board also 0.4 g/cm^3,

IOP Conf. Series: Materials Science and Engineering **316** (2018) 012004 doi:10.1088/1757-899X/316/1/012004

while MOE values between 12,280 – 15,193 kg/cm^2 on the NFRP water hyacinths 0.6 g/cm^3 density. Increased density can increase MOE about 46.18 – 235%. This is in accordance with statement of Bowyer et al. 2003, that the higher the level of density of the material produced, the higher nature of power supple. Meanwhile, the larger the particle size the more increases the power supple.
Modulus of Rupture (MOR) or broken modulus is the ability of the Board to hold a load of bending up to maximum limit (broken constancy). Based on the testing that has been done is obtained a value of MOR ranges between 29 – 74 kg/cm^2 on the material also 0.4 g/cm^3, while the value of MOR between 116 – 167 kg/cm^2 on material density 0.6 g/cm^3. Increased density can increase MOR about 125 – 300%. This is in accordance with statement of Bowyer et al. (2003) that the higher level of density of the material, it will be the higher nature of supple..

On the process of the absorption Coefficient Test (α) laboratory is using the device Impedance tube equipped with 1 unit of microphone to transmit sound frequencies in the range of low to high. The results of absorption coefficient tests are:

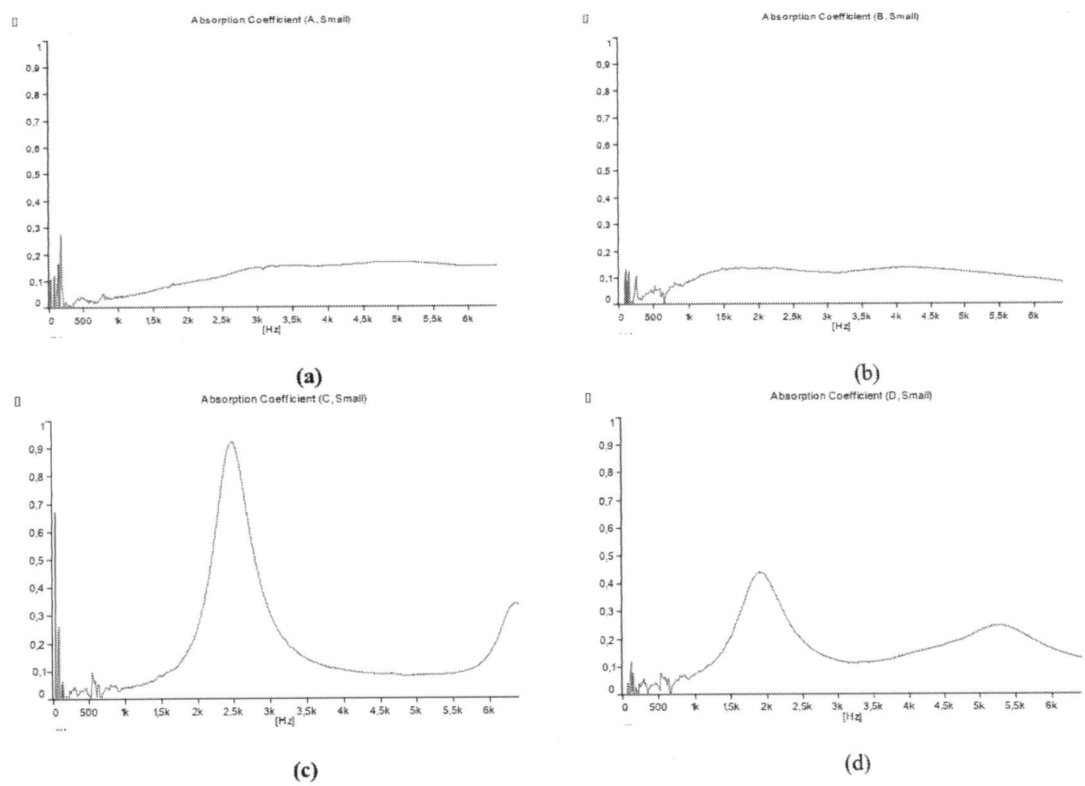

Figure 3. The results of Absorption Coefficient Tests

In contrast to materials FRP (composite) water hyacinth, then materials FRP (composite) coconut fibers have a better acoustic performance. Coconut fibers in composite materials with 15 mm thick, it looks that the material has a coefficient of absorption between 0.500 – 0.900. This material can also be categorized into the absorber is good, especially in the 2000 Hz frequency – 3250 Hz. highest Coefficient (peak) recorded at 2500 Hz frequency (2.5 k).

IOP Conf. Series: Materials Science and Engineering **316** (2018) 012004 doi:10.1088/1757-899X/316/1/012004

Materials FRP (composite) coconut fibers with thickness 30 mm also has a better acoustic performance than the material water hyacinth, although not as good as FRP (Composite thick coconut fibers 15 mm. Material has a coefficient of absorption between 0.050 – 0.450. Two thousand Hz is the frequency of the peak absorption coefficient of the material performance of FRP coconut fibers. From the graph, it can be concluded that the coconut fibre composite material thickness 30 mm has the ability as the absorber is primarily on the frequency-frequency: 1500 – 2500 Hz and 4500 – 5500 Hz. In contrast with the methods implemented in Test coefficient of absorption, then test the Sound Transmission Loss (STL) using impedance tube equipped with a microphone that has 4 units voice sensitivity is high. The results of Sound Transmission Loss Tests are as below:

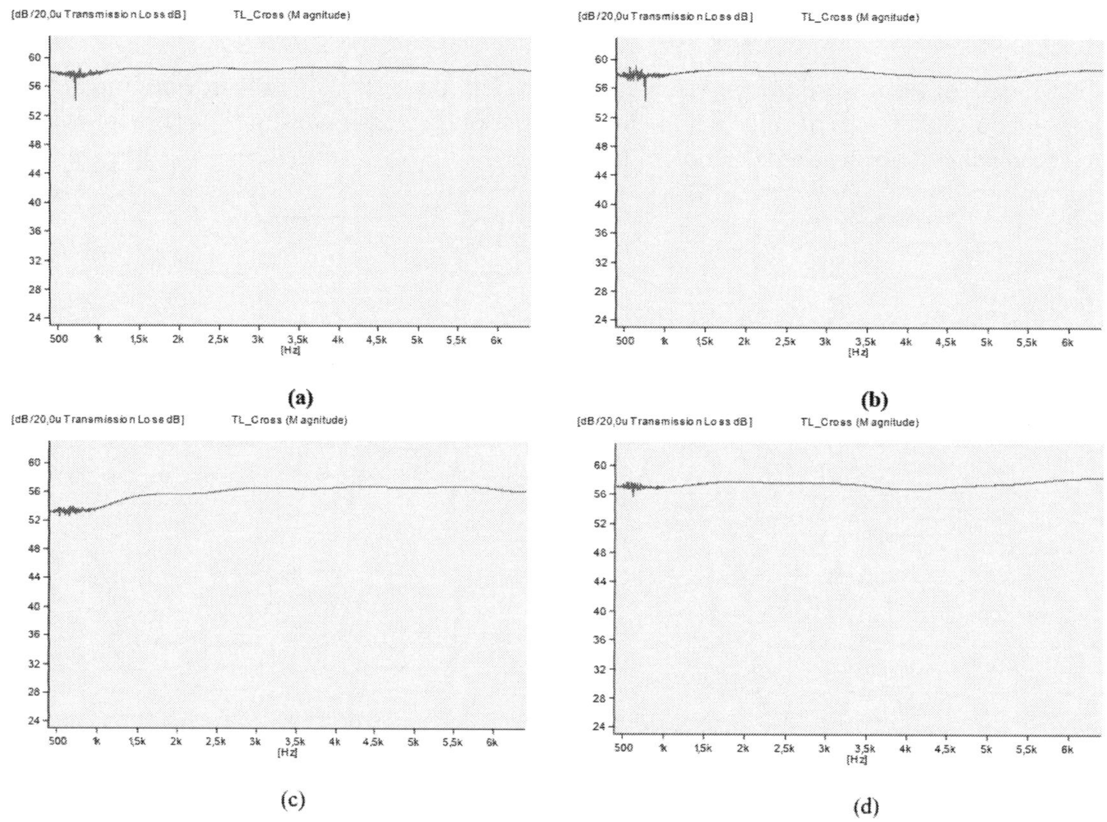

Figure 4. The Results of Sound Transmission Loss Tests

Test results showed the STL an average rating of 59 dB, it means that the material can absorp sound coming towards the surface up to 59 dB (fig. 4a). Similar to the results, the composite of water hyacinth with 15 mm thickness has Transmission Loss which is quite high, i.e. in the range of 57-59 dB comparing the 3.0 cm material. The conclusion is that the material could be used as absorber (fig. 4b).

Coconut fibers composite STL performance was lower compared with the FRP water hyacinth. Test results showed the value of STL ranged between 52 – 56 dB (figure 4 c). Performance of FRP material STL (composite) 30 mm thick coconut fibers resemble the performance shown by FRP materials water hyacinth. The value of the STL this material ranges between 56 – 58 dB. The conclusion is that this material can be used as a reliable sound reducing on the building (figure 4 d). The buildings in an area prone to such noisy airport region, industry, railroads and highways are advised to use this material as an attempt to reduce the sound on the components of the wall of the building.

4. Conclusion
Natural fibre material has quite a good performance in the development of materials for housing low income communities. Because the material is easy to find, then this would be very appropriate material for housing in Indonesia. NFRP material from water hyacinth and coconut fibers proved to have a good acoustic power, so it can be used for housing in the area is prone to noise. The power supple and its elasticity can withstand vibrations or earthquakes in areas prone to disasters.

5. Acknowledgement
This research is financed by the budget of the Research Grant from Diponegoro University Number: 275-084/UN7.5.1/PG/2017. The funders do not have any intervention on the data and the content of the manuscript. Authors thank all private sectors who have contributed in this research and gave the valuable opinion, advices and discourses.

References

[1[Pritchard,G, (2015), Plants moved up the Reinforced Plastics agenda, The ninth conference on wood and biofibre plastic composites

[2] Setyowati, E, Satyapratama, A, Atmadja, S.T, Hardiman, G (2015), Manufacture of Acoustical One Side Waffle Panel Made of Natural Resources with Hydraulic Hot Press Machine, J. Teknologi.

[3] Setyowati, E, Hardiman, G., Atmadja, S.T. (2015), Green Materials Comparation of Sawdust and Coconut Fibre Acoustical Waffle Panel, J. Applied Mechanics and Materials, Vol. 747, pp. 221-225.

[4] Setyowati, E. (2013), Sustainable Master Plan Design in Residential Area Near Airport, The 1st Annual International Conference Proceedings on Architecture and Civil Engineering, pp. 405-411, 18-19 March 2013, Singapore.

[5] Setyowati, E. and Sadwikasari, A.F. (2013), Building Material Composition Influence to Sound Transmission Loss Reduction, J. Advanced Materials Research, Vol. 789, pp. 242-247, terindeks SCOPUS.

[6] Setyowati, E. and Trilistyo, H. (2013), Climate Assessment of Orientation Design in the Housing Master Plan Close to The Airport, GSTF Journal of Engineering Technology, Vol.2 (1), p. 158-164.

[7] Setyowati, E. (2015), Strategi Kontrol Kebisingan Melalui Master Plan Perumahan, Monograf, CV. Tigamedia Pratama, Semarang, Indonesia

[8] Yan, L., Chouw, N., Huang, L., Kasal, B., 2016, Effect of alkali treatment on microstructure and mechanical properties of coir fibres, coir fibre reinforced-polymer composites and reinforced-ementitious composites, Contrusction and Building Material, Vol. 112 (2016) pp. 168–182

[9] Wei,J.Q and Meyer, C., (2015), Degradation mechanisms of natural fibre in the matrix of cement composites, Cement concrete Research, 73, p. 1-16.

[10] Saba, N, Paridah, M.T., Jawaid, M, (2015), Mechanical properties of kenaf fibre reinforced polymer composite: a review, Construction and Building Material, 76, p. 87-96.

[11] Yan, L., Chouw,N., Jayaraman, K., Flax fibre and its composites – A review,

Composite. B Vol. 56 (2014) pp. 296–317.

[12] Staib, G., Dorrhofer, A., Rosenthal, M., (2008), Component and Systems – Modular Construction – Design Structure New Technology – 1st Edition, Institute fur Internationale Architektur – Dokumentation GmbH & Co. KG, Munchen.
[13] Kang, S., Hong, S.I., Choe, C.R., Park, M., Rim, S., Kim, J., (2001), Preparation an characterization of epoxy composites filled with functionalized nano silica particles obtained via sol-gel process, J. Polymer, 42, 879-887.
[14] Bowyer JL, Haygreen JG, dan Shmulsky R. 2003. Forest Product and Wood Science. Iowa: The Iowa State University Press.
[15] Japanese Standard Association (JSA). 1963. JIS A 1405. Methods of Test for Sound Absorption of Acoustical Material by the Tube Method.
[16] American Society for Testing and Materials. 2004. ASTM E 413: Classification for Rating Sound Insulation.
[17] ASTM D790, Standard Test Methods for Flexural Properties of Unreinforced and Reinforced Plastics and Electrical Insulating Materials, American Society for Testing and Materials, Philadelphia, Pennsylvania, 2010.
[18] American Society for Testing and Materials : Standard Test Method for Measurement of Normal Incidence Sound Transmission of acoustical Material Based on The Transfer Matrixs Method, ASTM E 2611-09.
[19] Maloney, T.M., (1998), Modern Particleboard and dry Process Fibreboard Manufacturing, Miller Freeman Inc, San Fransisco.

Thermal Effectiveness of Wall Indoor Fountain in Warm Humid Climate

J A P Seputra

University of Atma Jaya Yogyakarta, Jl. Babarsari 44, Yogyakarta, 55281, Indonesia

Abstract. Nowadays, many buildings wield indoor water features such as waterfalls, fountains, and water curtains to improve their aesthetical value. Despite the provision of air cooling due to water evaporation, this feature also has adverse effect if applied in warm humid climate since evaporation might increase air humidity beyond the comfort level. Yet, there are no specific researches intended to measure water feature's effect upon its thermal condition. In response, this research examines the influence of evaporative cooling on indoor wall fountain toward occupant's thermal comfort in warm humid climate. To achieve this goal, case study is established in Waroeng Steak Restaurant's dining room in Surakarta-Indonesia. In addition, SNI 03-6572-2001 with comfort range of 20.5–27.1°C and 40-60% of relative humidity is utilized as thermal criterion. Furthermore, *Computational Fluid Dynamics* (CFD) is employed to process the data and derive conclusions. Research variables are; feature's height, obstructions, and fan types. As results, Two Bumps Model (ToB) is appropriate when employs natural ventilation. However, if the room is mechanically ventilated, Three Bumps Model (TeB) becomes the best choice. Moreover, application of adaptive ventilation is required to maintain thermal balance.

Keywords: Computational Fluid Dynamics; Evaporative Cooling; Indoor Water Fountain; Room Ventilation; Thermal Comfort

1. Introduction

According to Al-Obaidi et al, warm humid region as Indonesia has uncomfortable climatic condition because it receives huge amount of solar radiation along with long period of clear day resulting in high temperature and humidity all year [1]. Therefore, the effective passive design technique might be used to improve indoor thermal comfort is 'comfort ventilation' [2;3;4]. However, naturally ventilated building always has problems with low wind speed which often occurs in equatorial belt countries. In response, addition of ancillary tools to increase wind speed by using ceiling fans are necessary to provide cooling in naturally ventilated buildings [5].

Another comfort strategy which follows ventilation is the evaporative cooling. Analogous with sweat on human's skin, evaporation happens when wind blows on sweat's surface and transform the substance's liquid form into gaseous form. During this process, huge amount of heat, known as heat of vaporization, is absorbed and produces cooling effect toward its surroundings. Although there are many findings stated that evaporative cooling would not be effective in humid environment and could only be applied in a condition which does not require rigorous comfort's requisite [6;7], in fact, people in humid climates have acclimatized to be accustomed to 45-80% relative humidity and wind speed of 0.2-0.5m/s [8;9;10]. To assess the limitation above, a research was conducted in Thailand by examining occupant's thermal sensation due to indoor wall fountain inside a conditioned room. This research proves that the occupants feel 0.1–0.3°C cooler when subjected to water feature [11].

Moreover, air cooling by evaporative mist is becoming popular in Japan to cool cities as a solution of Urban Heat Island (UHI). Although Japan has a humid climate, its effective temperature shows that increased humidity is not significant compared to the improvement of thermal comfort due to evaporated water mist [12]. Due to those contentions and the dearth of academic records on evaporative cooling in naturally ventilated indoor space, this research is aimed to figure out the effect of indoor water feature toward occupant's thermal comfort in warm humid climate. Indoor wall fountain is chosen as the most popular water feature installed inside building in Indonesia.

Over the last decades, numerical studies aided by computer software have been implemented by researchers. For instance, Prianto (2012), employs computer simulation to test balcony effect and room furniture layout in naturally ventilated building toward its thermal comfort condition [13]. Almhafdy et al. (2015), used *Computational Fluid Dynamics* (CFD) to figure out the performance of natural ventilation occurred at courtyard by experimenting on aspect ratios and cantilevered roofs upon wind speed [14]. Furthermore, a numerical validation on CFD was conducted by developing evaporative cooling system inside a greenhouse. The result was satisfactory and concluded that correct amount of ventilation rate is the main factor to improve the efficiency of evaporative cooling system [15]. Therefore, this research also harnesses CFD to figure out the effectiveness of indoor wall fountain in warm humid climate based on two thermal indices, i.e. air temperature and relative humidity.

2. Methodology

Research is divided into four main stages; validation, simulation, analysis, and conclusion. Validation is initialized by developing a case study in Waroeng Steak Restaurant's dining room in Nusukan, Surakarta. Field measurements are done by retrieving the data of air temperature and relative humidity inside and outside the dining room by two HOBO loggers. Each device is set to store the data within 5 minutes interval and activated simultaneously by built-in timer. Data recording is undertaken for 2 weeks to store dynamism of occupant activities and weather conditions. Field results are then compared with the results produced by identical CFD model to figure out the level of agreement.

CFD Simulation is commenced after the validation produces error below 5%. Research variables developed are; feature's height (low, mid, tall), number of obstructions (one, two, three), and types of fans (forward, backward, ceiling). Each of them are simulated sequentially to obtain best combination of wall indoor fountain components. Thermal performance is measured by investigating the cooling effect and distribution pattern of indoor air temperature and relative humidity.

Analysis stage consists of graphical examination to find the balance of air temperature and relative humidity. Distribution patterns are measured by *evenness factor,* expressed as a fraction of 0 to 1, 1 represents complete mixing and good uniformity inside the room whereas 0 shows no mixing at all.

SNI 03-6572-2001 is used as criterion in Indonesia since this standard has been contextually adapted for warm humid climate [16]. This standard is commonly used as environmental benchmark, such as a research in Dormitory Building of Universitas Sam Ratu Langi Indonesia, it was conducted to measure comfort, ventilation, heat and lighting to confirm its energy efficient design [17]. SNI was also utilized to analyze the role of natural ventilation on seven stories building of Economy and Bussiness Faculty Universitas Brawijaya Indonesia for reducing electric consumption and determine the dimension of fenestration as a recommendation [18]. Furthermore, SNI was the basis of a building assessment in Jember Fashion Carnival Center to achieve thermal comfort for the occupants [19].

Table 1. Comfort criteria of SNI 03-6572-2001

Dry bulb Temperature	Relative Humidity	Air Movement	Surface Radiation
Cool, 20.5-22.8°C	Advisable, 40-50%	0.1m/s at 25°C	
Neutral, 22.8-25.8°C	Tight, 55-60%	0.2m/s at 26.8°C	
Warm,25.8-27.1°C		0.25m/s at 26.9°C	Similar with DBT
		0.3m/s at 27.1°C	
		0.35m/s at 27.2°C	

Lastly, the conclusion stage points out important results and findings. These outcomes would be beneficial to provide guidelines for interior designers or room planners to effectively incorporate water features inside building in warm humid climate. Identified research weaknesses and suggestions for future research are also presented in this section

2.1. Equation of Water Pools' Evaporation Rate

Evaporation rate from a pool of water due to the direct contact with dry air on its surface is expressed by equation below [20],

$gh = \Theta A (xs - x)$

where,
- gh = amount of evaporated water per hour (kg/h),
- Θ = (25 + 19 v) = evaporation coefficient (kg/m^2h),
- v = velocity of air above the water surface (m/s),
- A = water surface area (m^2),
- xs = humidity ratio saturated air at the same temperature as the water surface (kg/kg) (kg H_2O in kg Dry Air)
- x = humidity ratio air (kg/kg) (kg H_2O in kg Dry Air)

Therefore, evaporation rate is affected by air velocity on the water surface, contrast between air and water's humidity ratio, and water surface exposure. This implies that faster wind, drier air, and greater exposure of water to the adjacent air will bolster evaporation as portrayed by Figure 1 below.

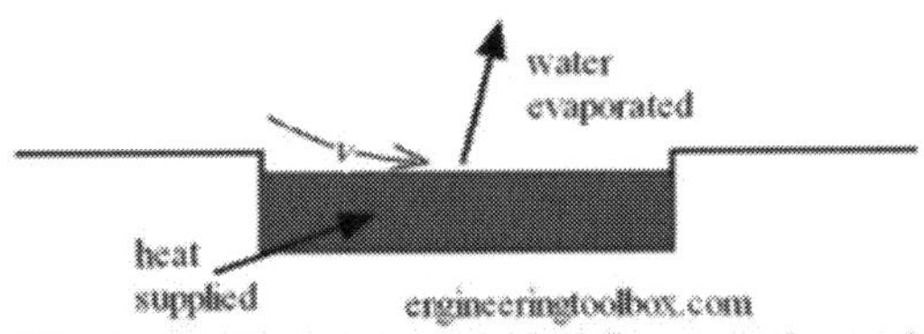

Figure 1. Water evaporation from static pool

3. Results and Discussion

Research presents the utilization of *Computational Fluid Dynamics* (CFD) in transient mode due to the time dependent case. Accuracy of CFD program is determined by its ability in incorporating heat transfer between two phase of fluids, i.e water and air. This phase change phenomenon or evaporation is exploited throughly in this paper. However, validation on CFD in simulating the case is necessary to figure out CFD's degree of agreements with the real environment.

3.1. CFD Validation

To achieve good level of certainty in using CFD, a validation study is established in Waroeng Steak Restaurant's dining room, Jalan Adi Sumarmo No. 133 Nusukan – Surakarta (see Figure 2). Research is focused on the wall fountain installed inside the dining room as shown in Figure 3 below.

Figure 2. Dining Room of "Waroeng Steak".

Figure 3. Indoor Wall Fountain.

Field measurement is conducted to figure out the thermal effect of indoor wall fountain inside the dining room toward its immediate surroundings. Two HOBO data loggers are placed at different locations as depicted by Figure 4 and Figure 5 below. The first device is installed near the fountain to

IOP Conf. Series: Materials Science and Engineering **316** (2018) 012005 doi:10.1088/1757-899X/316/1/012005

measure its thermal effect and the second is tied on nearby door to record initial air temperature and relative humidity before being affected by the water fountain.

Figure 4. Measurement tool near the fountain.

Figure 5. Measurement tool on entrance's wall.

Outdoor air temperature of 33°C and relative humidity of 55% are recorded as the prevailing condition. Water fountain's flow rate is 0.24 m^3/s and the incoming wind speed is 0.1 m/s. Figure 6 and Figure 7 below illustrate CFD domain and point probes aimed to extract the values of air temperature and relative humidity.

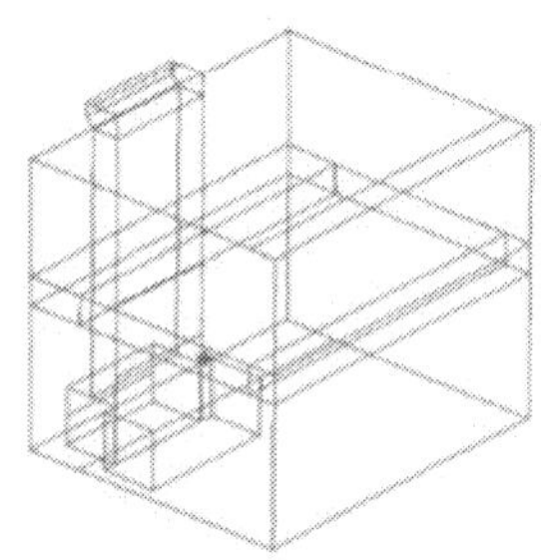

Figure 6. CFD model and probe (perspective).

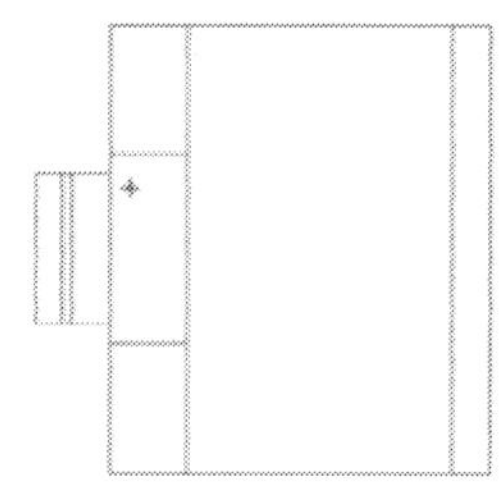

Figure 7. CFD model and probe (plan).

Values taken from field measurement are 32.9°C and 58.4%, whereas simulation shows 31.85°C and 60% respectively. Therefore, they have good agreement since the deviations or errors do not exceed 5% in each parameters, i.e. 3.34% and 2.68%.

3.2. CFD Simulation and Results

To obtain numerical values of air temperature and relative humidity, sets of line probes are modeled as depicted by Figure 8 and Figure 9 below.

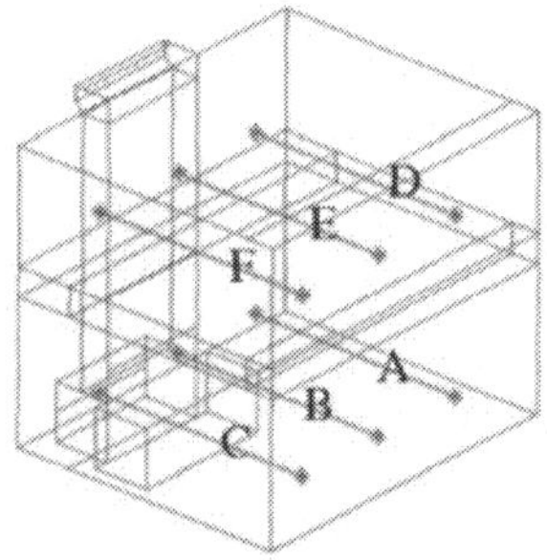

Figure 8. CFD model and line probes (perspective).

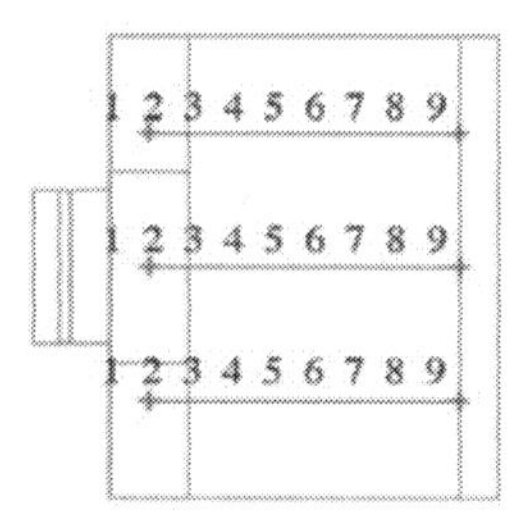

Figure 9. CFD model and line probes (plan).

First, simulation is begun by creating three kinds of falling distance model; tall, mid, and low height without obstacles. After simulation finished, CFD produces images of propagating air temperature and corresponding relative humidity as shown in Table 2 below.

Table 2. Examples of CFD results depicting air temperature and relative humidity contours.

Tall Model	Mid Model	Low Model
	Air Temperature Contours	
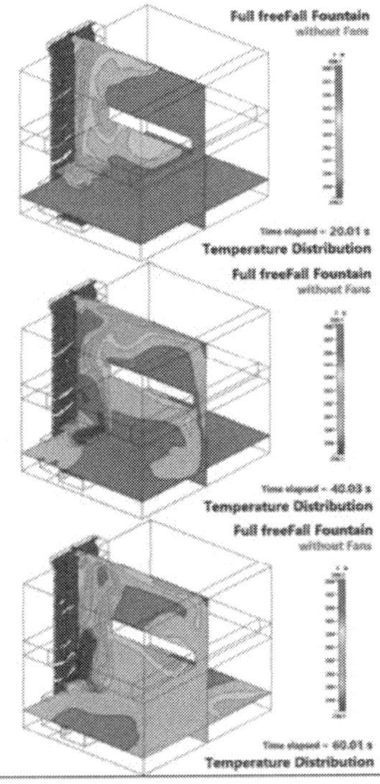	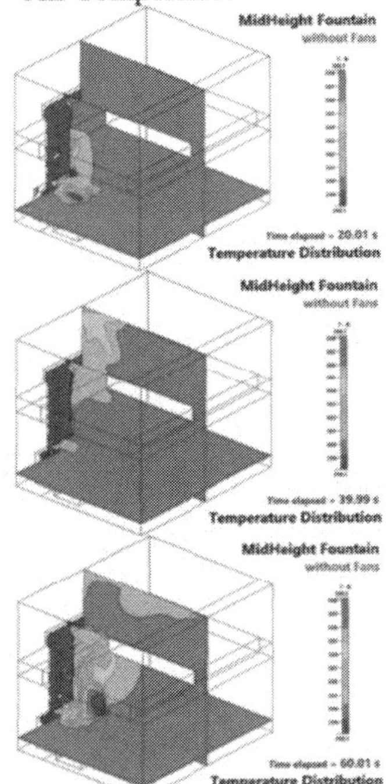	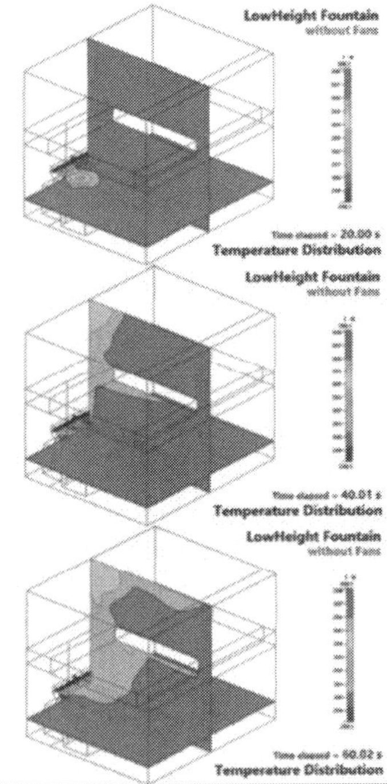
	Relative Humidity Contours	
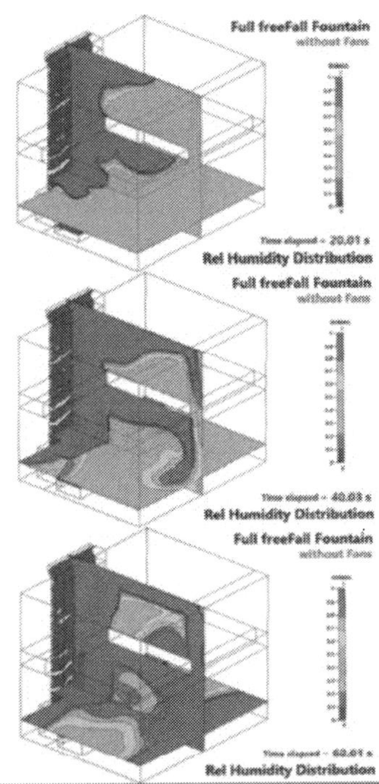	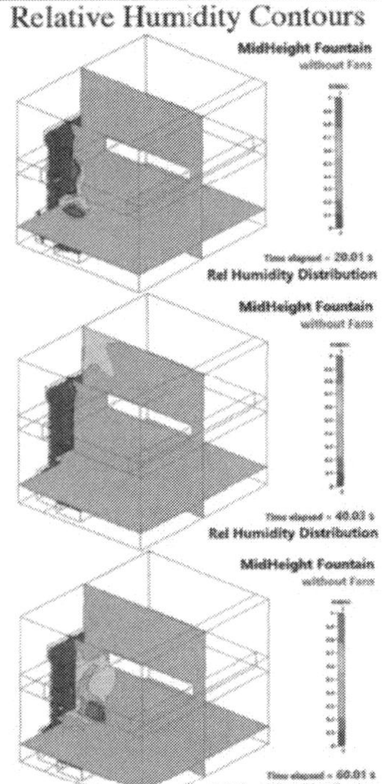	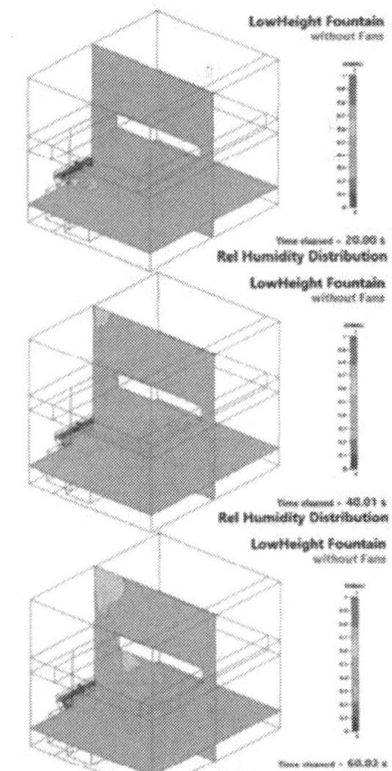

Since showing all the contour images as above is too inefficient, research combines the calculation of air temperature and relative humidity into graphs as portrayed in Table 3 below. Fluctuating lines indicate uneven air temperature and humidity distribution while straight lines show better uniformity.

IOP Conf. Series: Materials Science and Engineering **316** (2018) 012005 doi:10.1088/1757-899X/316/1/012005

Table 3. Comparisons of air temperature and relative humidity on height experiments.

Tall Model	Mid Model	Low Model
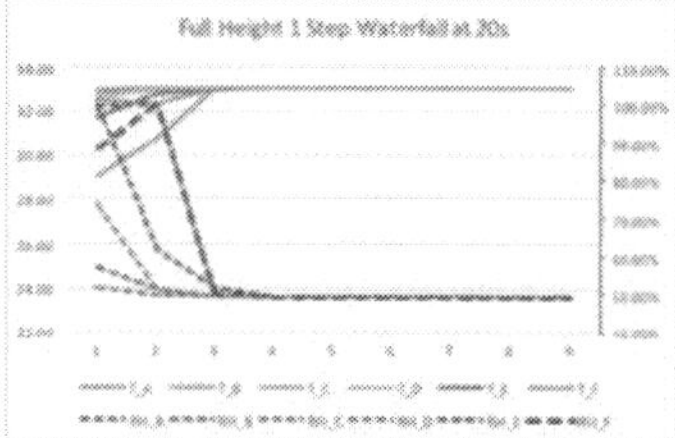	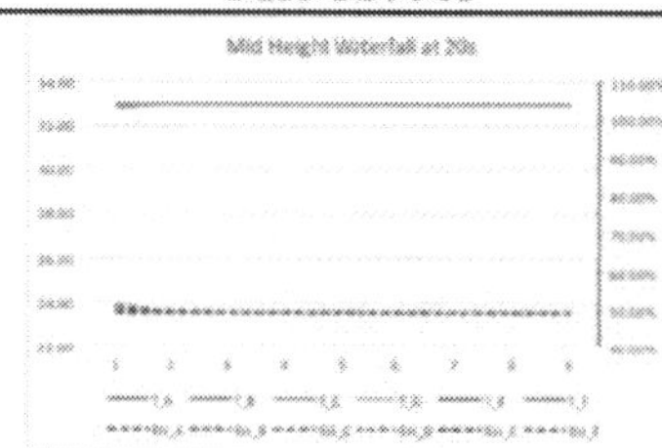	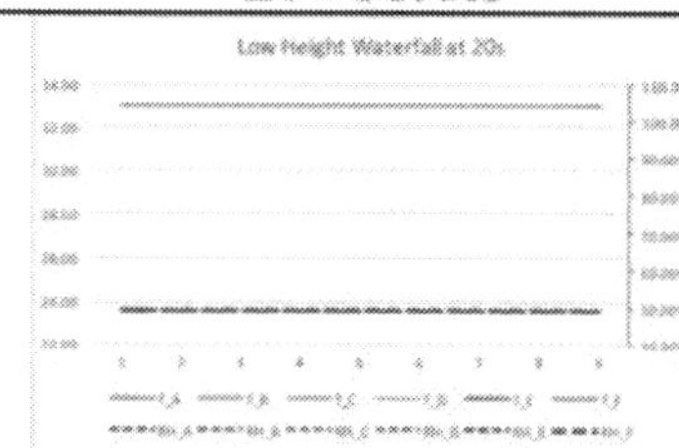
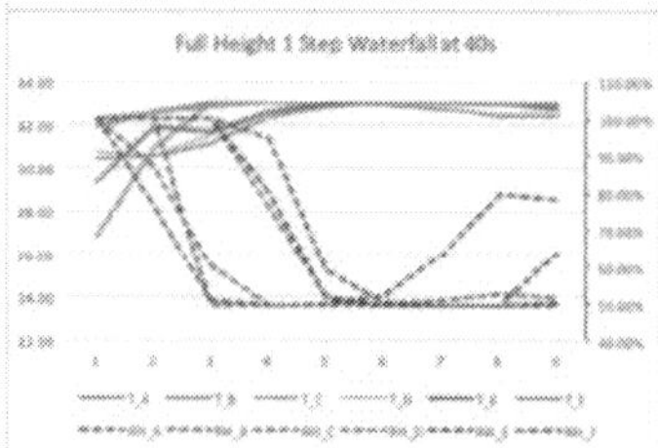	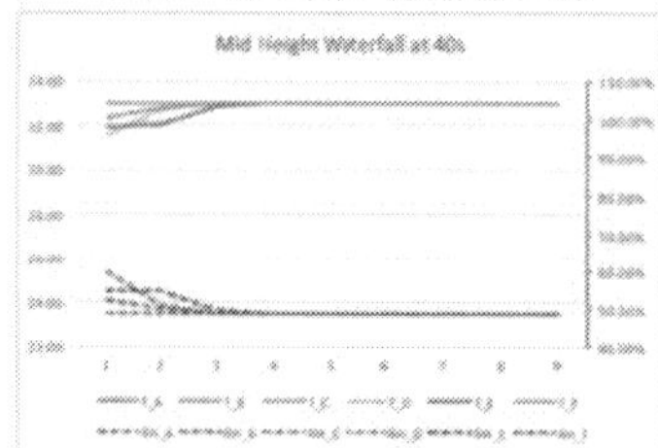	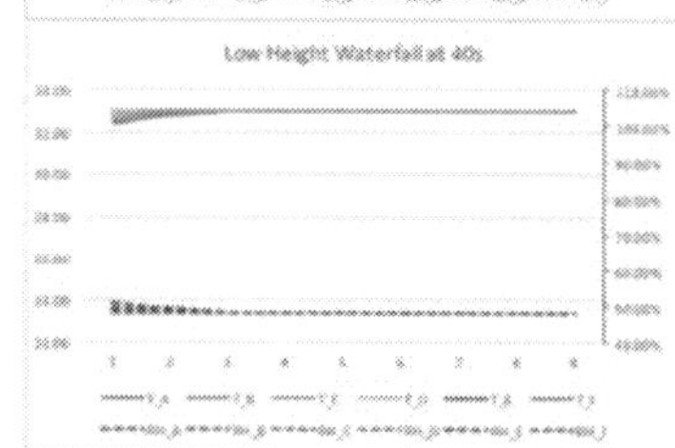
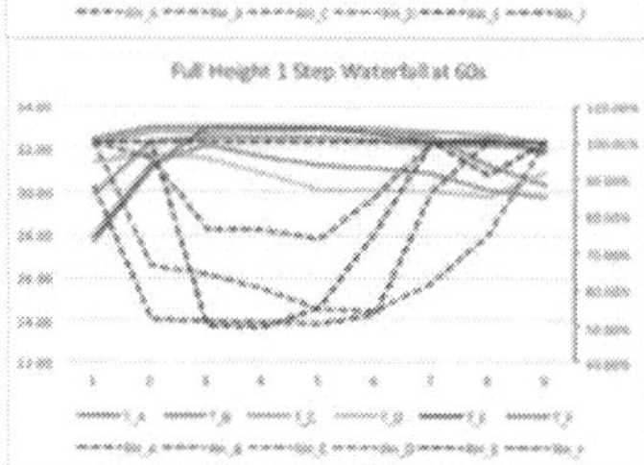	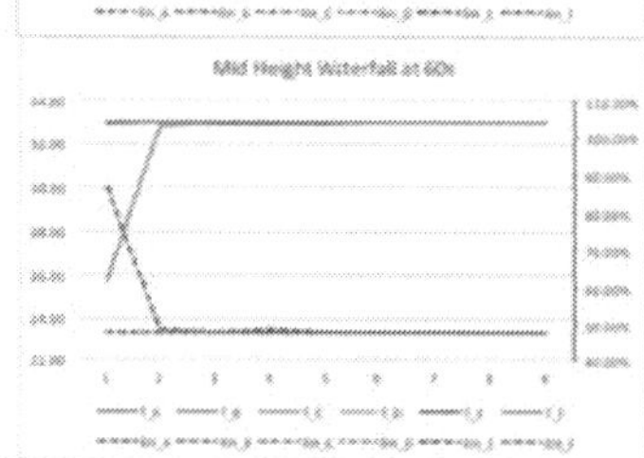	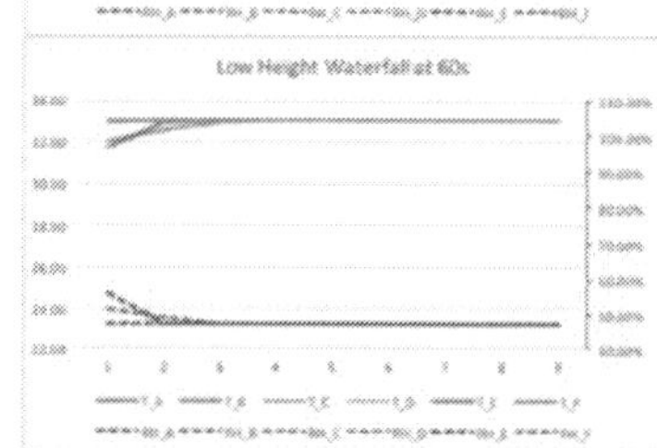

Tall Model possesses greatest potential in cooling the room but provides poor distribution. Next, Two and Three Bumps Model are simulated and the results are shown in Table 4 below.

Table 4. Comparisons of air temperature and relative humidity on bump experiments.

One Bump (OeB)	Two Bumps (ToB)	Three Bumps (TeB)
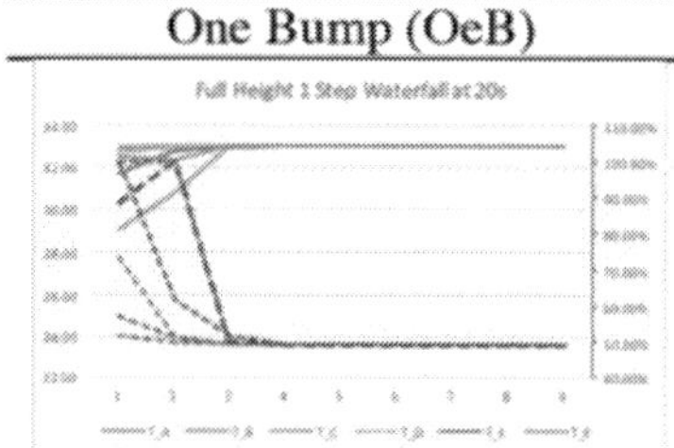	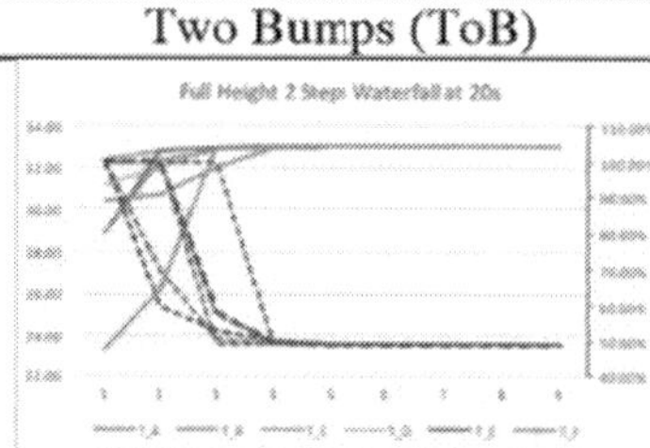	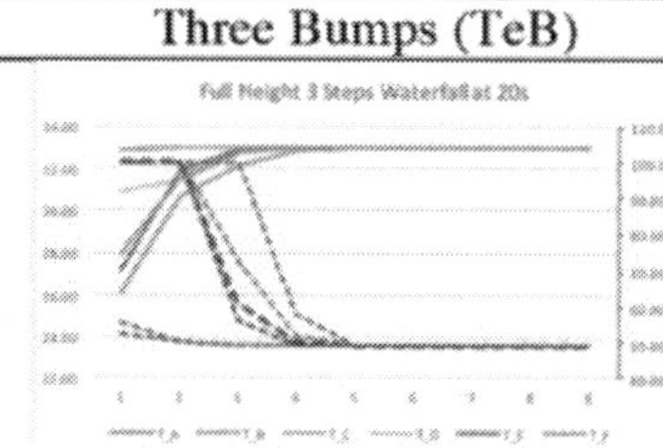
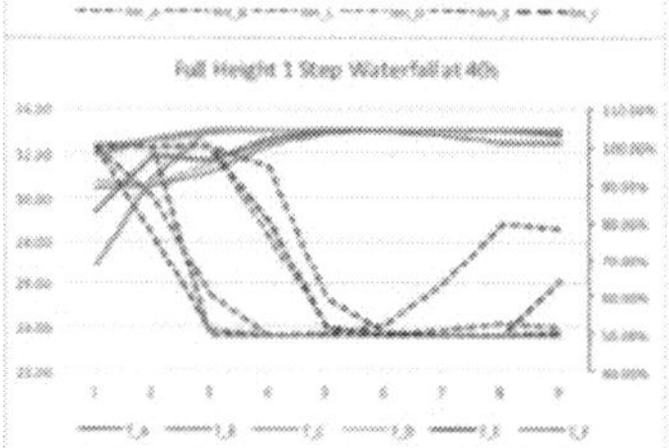	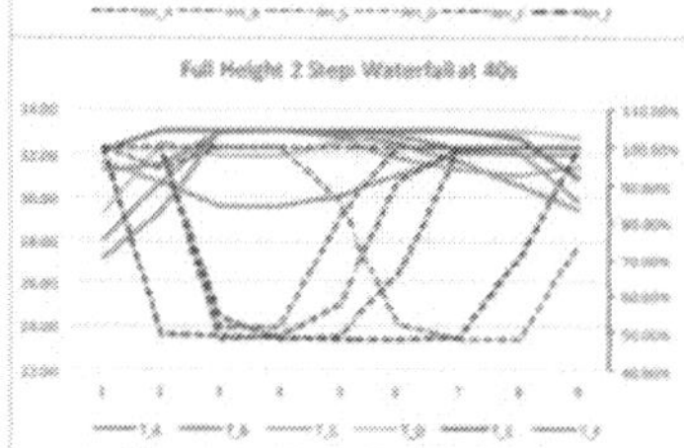	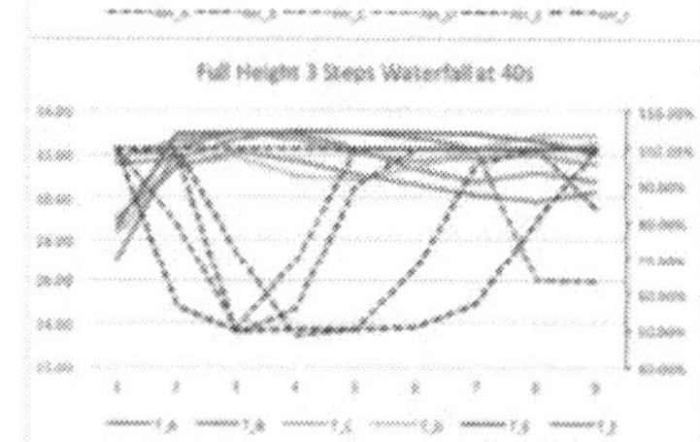
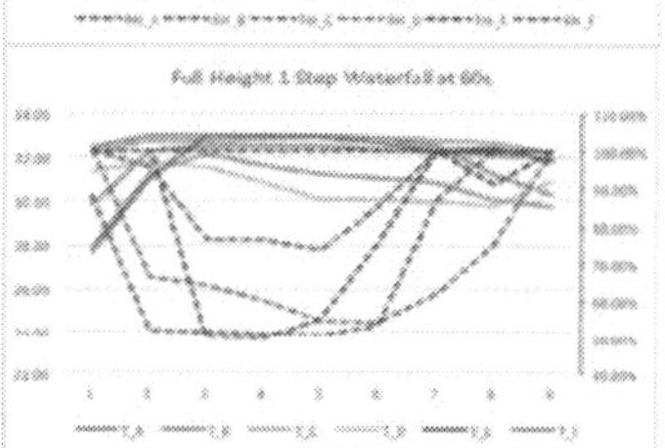	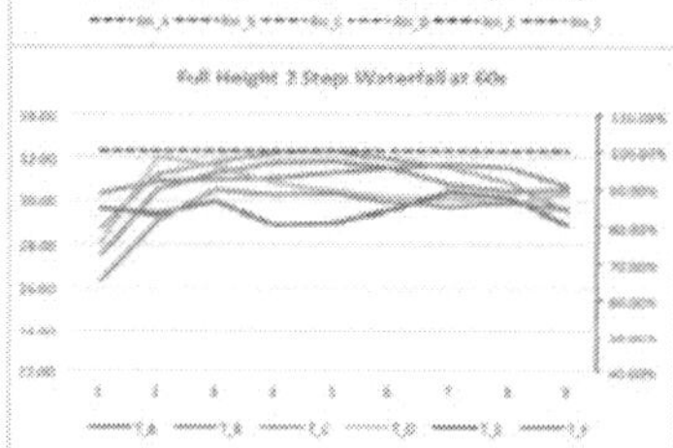	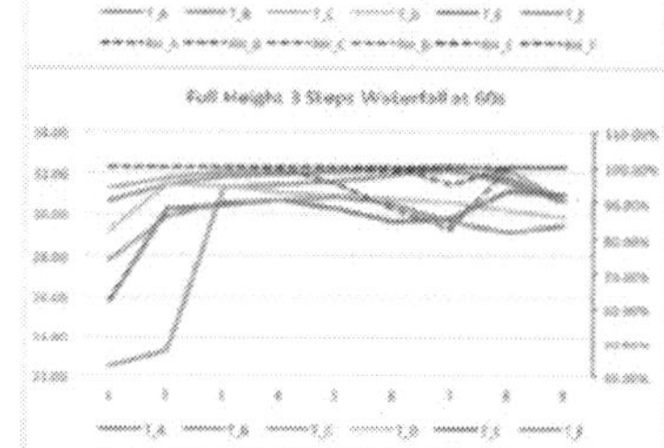

Two Bumps (ToB) Model possesses higher evaporation rate due to faster waterfalls and better uniformity. After that, fans are turned on to enhance wind speed. However, since there are no clear performance differences between two and three bumps model, both of them are simulated again. Fans are created in two types, i.e. two walls and one ceiling mounted fans, and specified as Table 5 below.

Table 5. Fan's technical specifications

Fan's Type	Radius	Hub Radius	Speed	Thickness	Blade angle
Wall	0.4 m	0.1 m	1500 rpm	0.2 m	14 deg
Ceiling	0.8 m	0.1 m	350 rpm	0.2 m	14 deg

Fans are arranged symmetrically as shown by Table 6 below, i.e. Forward Fans blow air to the room, Backward Fans direct air to the fountain, and Ceiling Fans send the air down to the workplanes.

Table 6. Fans installation inside the examined room along with the trajectories.

Forward Fans (FFans)	Backward Fans (BFans)	Ceiling Fans (CFans)
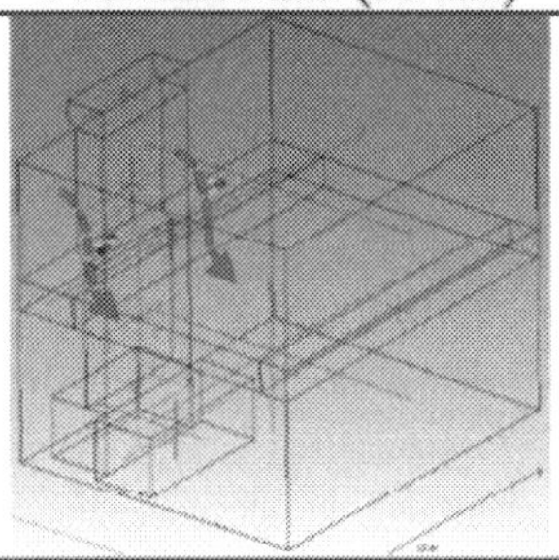	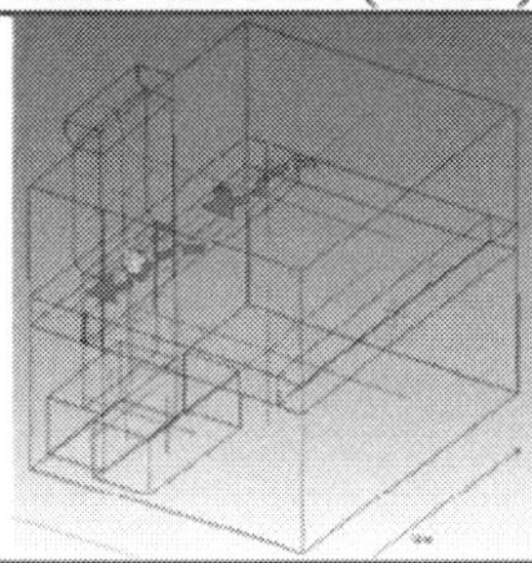	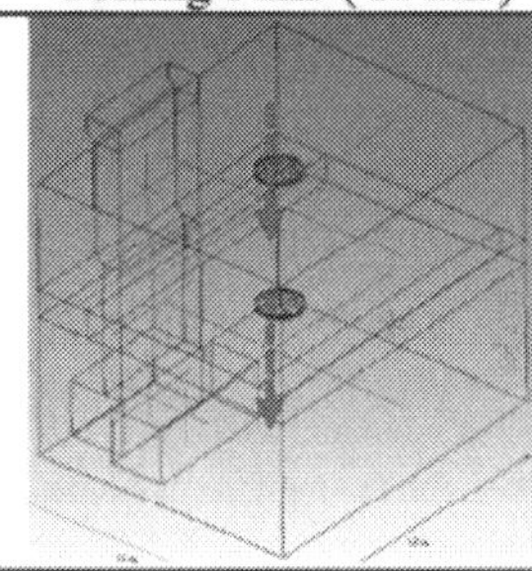

Two Bumps Model (ToB) is simulated with mechanical assistance by turning on the fans. The results are shown in Table 7 below.

Table 7. Comparisons of air temperature and relative humidity on fan experiments in ToB.

FFans	BFans	CFans
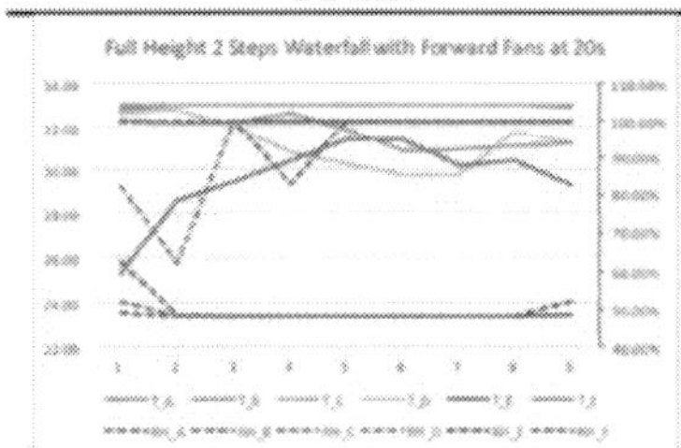	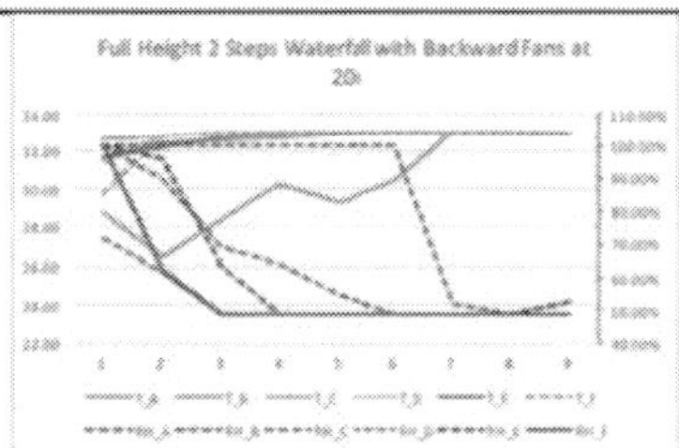	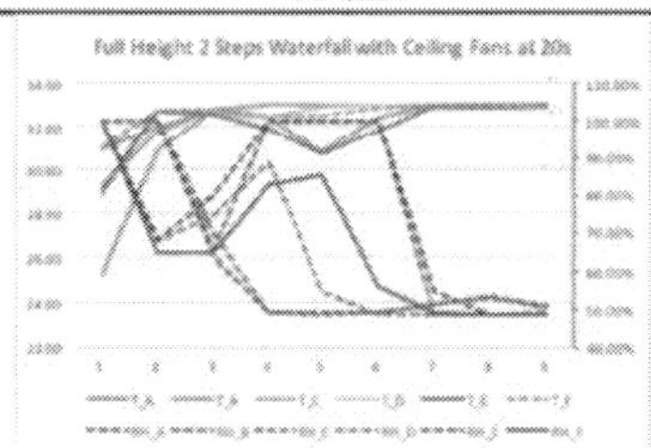
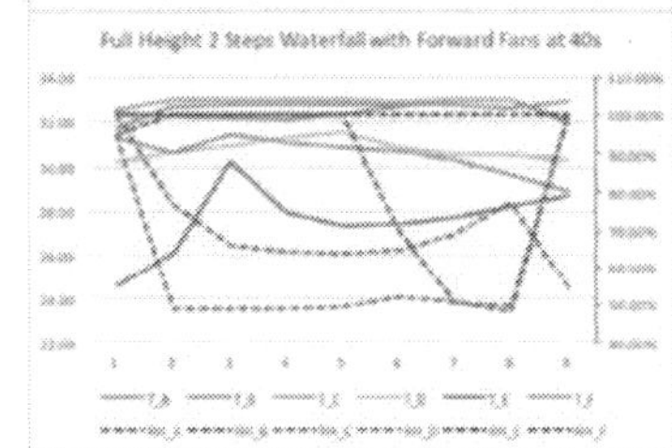	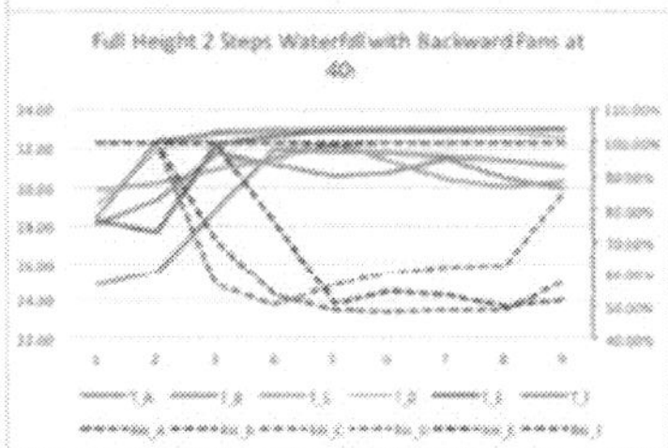	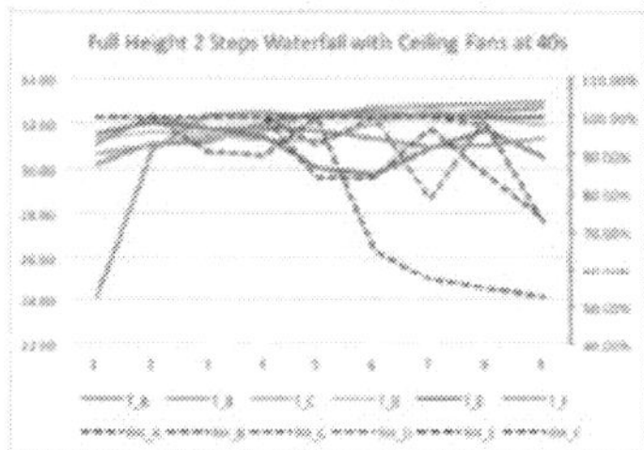
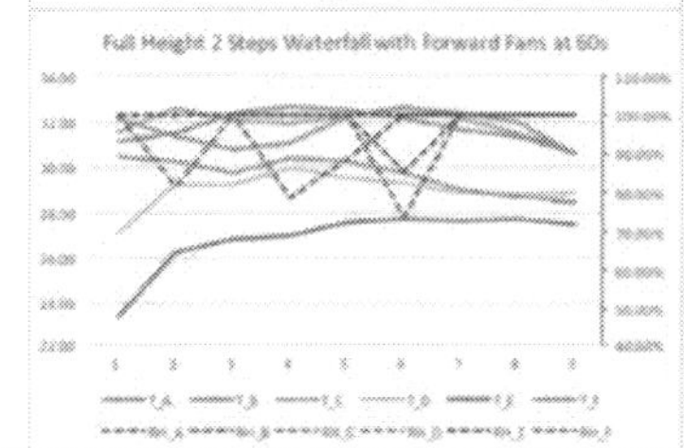	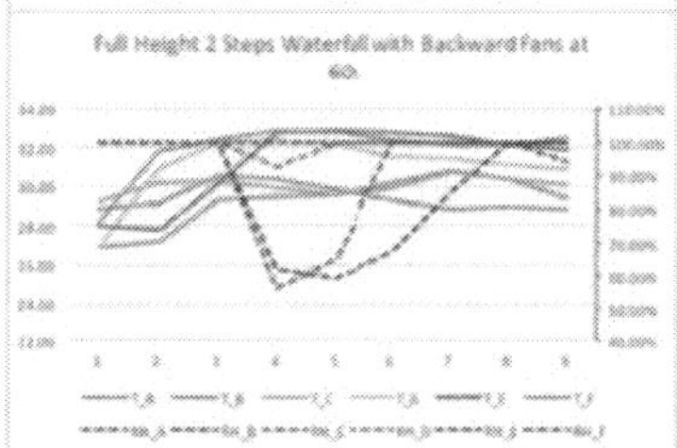	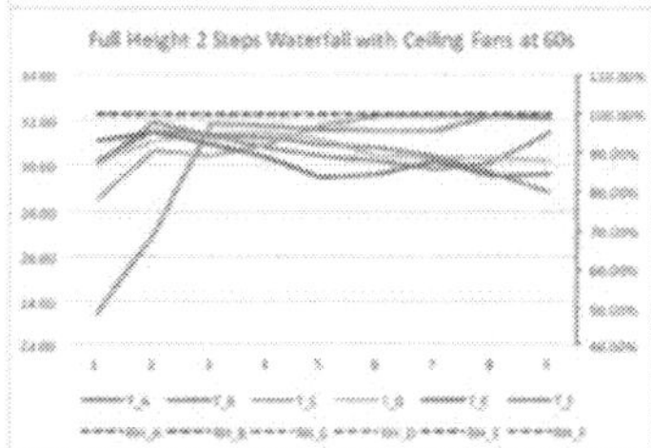

Better result is produced by ceiling fans due to the closely stacked and straightest lines. Lastly, examination is completed by simulating fans for Three Bumps Model (TeB).

Table 8. Comparisons of air temperature and relative humidity on fan experiments in TeB.

FFans	BFans	CFans

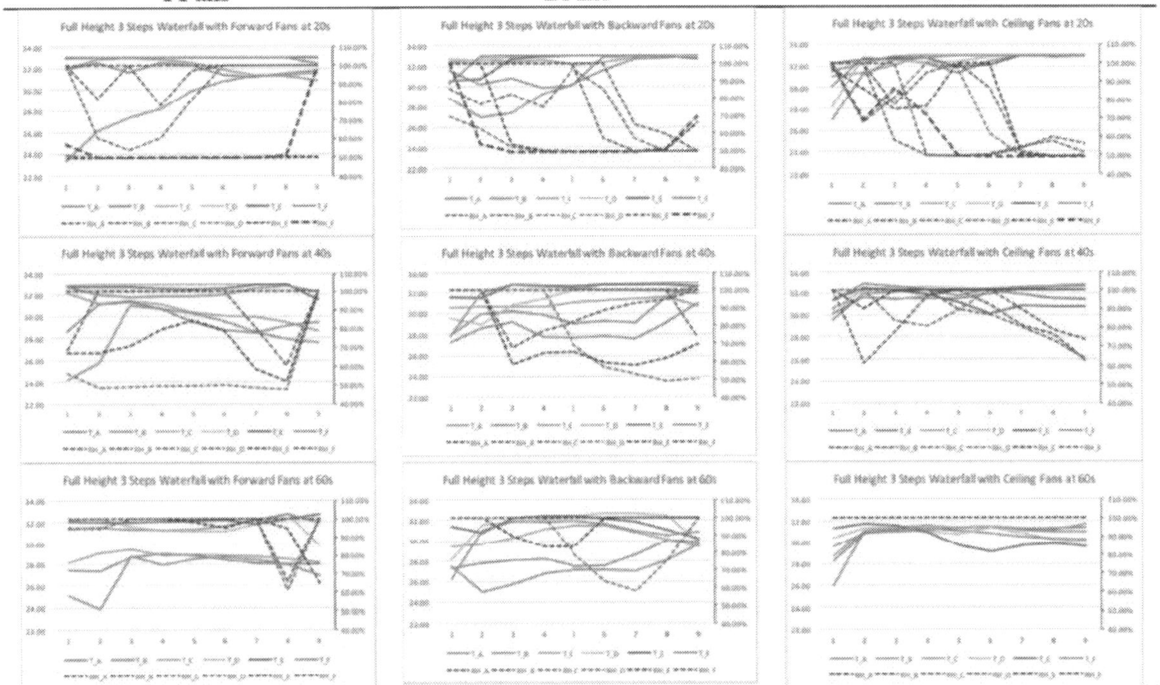

Greater oscillations of air temperature and relative humidity occurred at the beginning of simulation as shown by Table 8 above. It indicates the dominance of airflow exerted by fans over the evaporation process. However, as the time goes by 60 seconds all the thermal conditions have been mixed up and produce less diversities, especially on ceiling fans' case.

4. Conclusions

To identify the optimum combinations simulated by CFD, numerical values of air temperature and relative humidity are extracted and averaged in the Table 9 below. Evenness is added at the rightmost colomn as a ratio of minimum to its average value after 60 seconds to determine air uniformity level, value of 1 represents a perfect mixing and value of 0 means no mixing at all.

Table 9. Averaged results of each simulated model by recorded time

Model Type	20 seconds		40 seconds		60 seconds		Evenness at 60 s
	Avg Tmp (°C)	Avg RH (%)	Avg Tmp (°C)	Avg RH (%)	Avg Tmp (°C)	Avg RH (%)	
Tall	32.82	54.95%	31.71	85.20%	32.38	67.04%	0.94
Mid	32.99	49.38%	32.91	49.06%	32.85	48.56%	0.78
Low	33.00	49.17%	32.97	48.37%	32.95	47.58%	0.96
ToB_NoFans	32.34	60.60%	31.72	81.66%	30.39	100.00%	0.87
TeB_NoFans	32.48	59.00%	31.69	85.70%	30.55	99.25%	0.74
ToB_FFans	31.92	73.19%	31.13	85.98%	30.13	98.30%	0.78
ToB_BFans	32.35	62.38%	31.23	84.15%	30.49	95.95%	0.88
ToB_CFans	32.25	69.07%	31.59	94.02%	30.62	100.00%	0.77
TeB_FFans	31.95	72.35%	31.02	86.45%	29.96	97.58%	0.80

TeB_BFans	32.08	68.60%	30.86	88.69%	29.97	96.87%	0.83
TeB_CFans	32.38	70.54%	31.82	92.99%	30.75	100.00%	0.85

According to the results above, there is not even a single case manages to reach air temperature below 27.1°C since the outside air itself was 33°C, far higher than the expected SNI standard. As response, research is reoriented to find the lowest temperature possible and use the upper limit of 80% relative humidity taken from Chow, et.al. (2010) as requisite to determine the best configuration. Consequently, the highlighted numbers are the cases which fulfill the requirements above.

By applying those new rules, ToB is better than TeB since it produces cooler average air temperature (31.48°C to 31.57°C) and provides better evenness ratio (0.87 to 0.74). Meanwhile, FFans Model reduces air temperature better than the other models by increasing air velocity on water surface using negative pressure. However, if the fans are activated, TeB reduces air temperature more than ToB (31.2°C to 31.3°C) and provides slightly better uniformity (0.83 to 0.81).

Nevertheless, all simulated cases are having problem with high relative humidity (>80%) after 20 seconds. It suggests that the ventilation rate should be increased to replace old stale air after 20 seconds to lower the humidity hence maintain indoor thermal comfort much longer. Still, a fluctuating outdoor air condition will be a real challenge to solve this problem. Thus, implementation of sophisticated technology such as thermal sensors and actuators is necessary. Additionally, future study to examine adaptive ventilation technology as well as subjective responses from local residents is required to figure out whole thermal effectiveness of evaporating water feature toward indoor thermal comfort in warm humid climate.

References

[1] Al-Obaidi K M, Ismail M and Abdul Rahman A M 2014 *Passive Cooling Techniques Through Reflective and Radiative Roofs in Tropical Houses in Southeast Asia*: A Literature Review, Frontiers of Architectural Research 3 pp 283-297.

[2] Koenigsberger O H, Ingersoll T G, Mayhew A and Szokolay S V 1973 *Manual of Tropical Housing and Building* Part 1: Climatic Design. Longmans London UK pp 119-130.

[3] Givoni B 1994 *Passive and Low Energy Cooling of Buildings* Van Nostrand Reinhold, New York, USA.

[4] Nandita A and Ashish K 1983 *Passive Cooling Techniques for The Architect in Baroda India* (*Proc. 2nd Int. Conf. PLEA Conf.*) ed Yannas S (Pergamon Press) pp 725-721.

[5] Vijayalaxmi J 2009 The thermal comfort of a naturally ventilated house resulting from the evaporative cooling of a ceiling fan in the hot-humid climate of chennai, india *Int. J. of Ventilation* **8:2** pp 135-144.

[6] Baca M, et al 2011 Evaporative cooling efficiency according to climate conditions *Procedia Engineering* **21** pp 283 – 290.

[7] Peng J, et al 2015 Application on evaporative cool technique in garment workshop in hot-humid area *Procedia Engineering* **121** pp 2029 – 36.

[8] Khedari J, Yamtraipat N and Hirunlabh J 2000 Thailand ventilation comfort chart *Energy and Building* **32** pp 45-49.

[9] Chow T T, et al 2010 Thermal sensation of hong kong people with increased air velocity, temperature and humidity in air-conditioned environment *Building and Environment* **45** pp 77-83.

[10] Nicol J F and Humphrey M A 2002 Adaptive thermal comfort and sustain thermal standard for buildings *Energy and Buildings* **34** pp 63-72.

[11] Sudprasert S, Kasorn K and Khedari J 2015 Thermal sensation of thai students in an air conditioned space with a pond type water source and air velocity step change *Int. J. of Ventilation* **14:1** pp 91-108.

[12] Farnhama C, et al 2011 Study of mist-cooling for semi-enclosed spaces in osaka, japan *Procedia Environmental Sciences* vol 4 pp 228–238.

[13] Prianto E 2002 Alternatif disain arsitektur daerah tropis lembab dengan pendekatan kenyamanan thermal *Dimensi Teknik Arsitektur* vol 30 no 1 pp 85 – 94.

[14] Almhafdy A, et al 2015 Thermal performance analysis of courtyards in a hot humid climate using computational fluid dynamics cfd method *Procedia - Social and Behavioral Sciences* vol 170 pp 474–483.

[15] Sapounas A A, et al 2008 Aspects of cfd modelling of a fan and pad evaporative cooling system in greenhouses *Int. J. of Ventilation* **6:4** pp 379-388.

[16] Badan Standarisasi Nasional 2001 SNI 03-6572-2001 *Tata Cara Perancangan Sistem Ventilasi dan Pengkondisian Udara pada Bangunan Gedung, Indonesia.*

[17] Sangkertadi 2011 Evaluation of indoor comfort and building energy performance of students residence at sam ratulangi university in manado, indonesia *Architecture & Environment* vol 10 no 1 pp 97-115.

[18] Fathony D A, Sufianto H and Yatnawijaya B 2015 Optimalisasi penghawaan alami pada bangunan pendidikan berlantai banyak studi kasus gedung f feb ub *Jurnal Mahasiswa Jurusan Arsitektur Universitas Brawijaya* vol 3 no 3.

[19] Kusumawardani N, Thojib J and Martiningrum I 2015 Sistem ventilasi alami sebagai dasar perancangan jfc center di kabupaten jember *Jurnal Mahasiswa Jurusan Arsitektur Universitas Brawijaya* vol 3 no 1.

[20] http://www.engineeringtoolbox.com/evaporation-water-surface-d_690.html

IOP Conf. Series: Materials Science and Engineering **316** (2018) 012006 doi:10.1088/1757-899X/316/1/012006

The Crux of Minimalist Architecture: A Local Strategy of Housing Design in Jakarta or a Break Free from Traditions?

Yulia Nurliani Lukito[1] and Bella Previta Handoko[1]

[1] Department of Architecture, Faculty of Engineering, Universitas Indonesia, Depok 16424, Indonesia

Email: bella.previta@gmail.com

Abstract. During the 1950s, the idea of Minimalism presents itself as one of the response of the search of universal language in art and architecture. This particular style, which was started as an art movement, has received many critics in the relation to the loss of art but nevertheless Minimalism has spread all over the world and influenced many disciplines, including architecture. In minimalist architecture, elements of design convey simplicity, basic geometrical forms, with no decoration, and the use of white color, modern materials and clean spaces. The "less is more" movement in architecture, which can be seen in the works of Mies van der Rohe and also in the International Style that celebrates materiality and rationality, is also understood as Minimalism. Moreover, an important historical connection to minimalist architecture is the relationship to popular representations of how the upscale modern family lived. Recently, the idea of minimalist architecture appears in Indonesia as a preferable housing style. Adapting minimalist architecture to be suitable for a tropical climate can be done partly by modifying the forms and the microclimate such as using passive system approach or additional equipment that creates comfort in the building. This paper investigates the idea of minimalist architecture in Jakarta, Indonesia, and how the idea is widely used for housing. Some questions related to this study are investigating whether minimalist architecture in Jakarta shares the same principles with minimalist architecture in its earlier time or it is only a trend in housing design. Not only this study analyzes the moment when the idea of Minimalism develops in the history of modern architecture but also some important characteristics of minimalist architecture in different era and space. In addition, this study also discusses how minimalist architecture that happens in Jakarta becomes a way of dealing with both modern and local conditions, including a break free from traditions.

1. Introduction

The International Style is known as a major architectural style emerged in the 1920s and the 1930s and this Style is a result of dissatisfaction of eclectic buildings of a mix of decorative elements from different architectural periods and styles. This modern style is the dominant tendency in Western architecture and the idea of finding a universal language in architecture that has spread worldwide. The term is first used by Hitchcock and Johnson for the architectural exhibition of Modern Architecture: International Exhibition for the Museum of Modern Art (MoMA), New York, held in 1932. Since the International Style emphasizes more on architectural style and aesthetics and lacks social aspects of the modern movement promoted in Europe, architects are able to use this universal language in different countries. The idea of the International Style spreads widely after the World War II that leaves destruction and the need to rebuild cities as soon as possible and to fulfill the need to

IOP Conf. Series: Materials Science and Engineering **316** (2018) 012006 doi:10.1088/1757-899X/316/1/012006

build houses and buildings. In addition to this, the growth of industries has caused the production of new building materials in large-scale and architects start to design buildings by using mass-produced materials.

According to William Jordy [1], the characteristics of International Style are:

- Abstraction; physics of an object is simplified by reducing the ornaments to get the main essentials of the object.
- Neutral color; white or black are classified as neutral color, because it doesn't show any emotion.
- Exposed materials; materials like steel, concrete, and glass is used because those are industrial products and do not need complicated finishing.
- Simple geometric form; cube is used mostly it is a neutral form and using cube means reducing the possibility of historical traces.

Minimalism is also interpreted as diffused boundaries or smudged borders, which means that sometimes the artist or the architect blur the gap between two objects in order to create a unifying effect [2]. Moreover, Hitchcock and Johnson [3] also state that the most common characteristics of the International Style are rectilinear form with almost no ornamentation, open interior spaces and the use of modern materials such as reinforced concrete, glass and steel.

In its earlier stage, the International Style offers two patterns that are seamless and membranous walls. Architects make some modifications to those patterns as a result of their own interpretation of the style. During the 1920s and the 1930s, the use of new materials such as glass, concrete and steel were still limited. As an example of the application of the International Style's principles is the Farnsworth House designed by Mies van der Rohe in 1949. The design of the house is simple with a cube-shape, uplifted piles as the foundations and glass walls (see Figure 1). The house is divided into three segments with eight piles as its main structure. The distance between two columns was probably related to the possible wide span of the materials. The house is known for using new technology and materials for both structural and non-structural parts, including concrete and glass. The glass walls also gave support to the whole structure since they stood from the floor to the ceiling. For the stairs, the architect used plain horizontal steps from concrete, and marble for the floor. The architect painted the wall with white color. Mies van der Rohe implemented characteristic of the International Style into his design and in this way he created his design identity.

Figure 1. Mies van der Rohe's Farnsworth House

Unfortunately, this building has received some critics from the owner, Dr. Edith Farnsworth. One of the reasons is because the ventilation was really bad. The whole building had only one opening, which was the door, that made the air circulation inside the house very bad. Moreover, the owner could not have the privacy because of the glass walls. The owner was afraid that someone watched her from outside the building. "...The truth is that in this house with its four walls of glass I feel like a prowling animal, always on the alert. I am always restless. Even in the evening. I feel like a sentinel on guard day and night.... Mies talks about his 'free space': but his space is very fixed... Any arrangement of furniture becomes a major problem, because the house is transparent, like an X-ray." [4]. In short, the owner felt that the house failed to give her a privacy.

Although the Farnsworth House does not function well as a weekend home for Dr. Edith Farnsworth, technologically, this house is one of the most successful example of the International Style applied for a building with a good relationship between inside and outside and in maximizing the use of modern building materials. In 1950s, people started to accept modern industrial products for buildings, including standardized measurements from fabrications, and exposed materials. However, this acceptance also marks a new problematic phenomenon of losing traditional meaning of a house. Some other critics for the International Style is that after the 1937 this style has more individual expression of the architect but not of the owner, moreover, the building cannot adapt well to the surroundings because the house is designed following the architect's personal tastes and preferences.

In relation to the physical condition of a house, new materials such as glass, concrete and steel prevent the building to cope with its surroundings and its local context, and this condition causes the building to become unable to respond well to the site and local conditions. Like the case of the Farnsworth House, the use of modern materials also makes the building unable to fulfill the owner's needs. As the result of those critics and failures of the International Style, some architects started to find a more personal approach to the site. The work of Le Corbusier after the 1937, for instance, started to put the site and the psychological comfort of the space for the users as more important than the technological consideration [1]. However, the most important critic of the International Style, which is its inability to adapt to the surroundings and local conditions, nevertheless occurs.

The search of the universal architectural language still continues, including during the 1960s with the idea of Minimalism, known as the climax of modern art and architecture before a shift into postmodernism [5][6]. The term Minimalism is used for an exhibition at Museum of Modern Art in 1960. Minimalism is brave enough to come out and said that art should not be framed [5]. Minimalism is typified by artworks composed of simple geometric shapes based on the square and the rectangle. Minimalism is an extension of the idea that art should have its own reality and not be an imitation of some other thing such as a landscape or a person. With this highly purified form of beauty, minimalist avoided overt symbolism and emotional content, but instead called attention to the materiality of the works as well as order, simplicity and harmony.

In architecture, minimalist idea glorifies a style in which things come down to its simplest, most necessary elements [7]. This kind of architecture was widely accepted as part of new interpretations of modern architecture and the dynamics of reductive and clean elements. Moreover, in minimalist architecture, there are some beliefs that the knowledge of materials – including the understanding of the limits and the way those materials work – as well as the lived experience of the viewer could create a harmony in architecture. The objective of minimalism is not about an object or a building but the client, the person who is going to use the space or see the building. This is the reason why minimalist architecture should be considered as an architectural style, especially after the fall of the International Style to create an adaptive design language, an opportunity to use products of industry and the needs of housing. Minimalism in architecture, is usually understood as rooted in the works of De Stijl, its rectangular forms and primary colors, as well as the designs by the pioneer of modern architecture, Mies van der Rohe, a German architect who laid the ground work for less is more. The designs are honest, in terms of being exactly what they look like and delivering nothing more than they promise [8].

In Jakarta, Minimalism in architecture became a trend especially since early 2000's. Minimalist architecture also typically uses basic geometric shapes, white colors, open-plan spatial arrangements, straight elements, flat roofs and modern materials such as steel, concrete, and glass. However, does minimalist architecture also engage the idea of minimalism in art and architecture just like in the West? And how does the use of this style contribute to the understanding of space?

The object of this study is a residential building designed by the architect Andra Matin in 2009 located on Jalan Bangka VIII, South Jakarta. Andra Matin is known for simple and modern designs that attempts to show a dynamic interplay of internal and external boundaries. The concept of the building is very similar to other Andra Matin's residential works such as I&L House and IH House with dominant geometric form, floating platform and open plan. The concrete structure holds the floors - open but non private spaces on the lower level and private spaces on the upper level – creates a floating effect to the whole building. The architect puts the service areas such as kitchen together with the swimming pool on the lower level in order to create more private spaces for the family on the upper floor. Andra Matin's identity in his residential also includes the idea of contrast – between inside and outside – and hierarchy of spaces – such as public and private [9][10].

Figure 2. Andra Matin's Residential Work Comparison

The owner of the house has changed for three times and there are some changes made to the functions of the rooms although the façades stay the same. The land is approximately 1,600 square meter with two-stories building. The current family who live in the house since 2013 consist of seven people (a father, a mother, a daughter, a son, a grandmother, a servant, and a security). The father spends most of his time working outside the house while the mother mostly stays at home as a housewife. The daughter studies at the Universitas Indonesia while the son studies at ITB Bandung. The grandmother, the servant and the security spend most of their time in the house. The family have changed some functions of the room, for instance, the former fitness room on the lower level is now an audio room used to for Playstation and karaoke, and the study room is now a bedroom for the grandmother. It is the intention of this study elaborates how the design of the house is influenced by the idea of minimalism, which arguably has been adapted by the architect and, and analyze how the family adapts to the house's design.

2. Methods

Since this paper focus on how a building reflects the principles of Minimalism, it is necessary to have literature surveys on minimalism especially in architecture. In order to get information of the house, we conducted field research and did an interview for the family members who live in the house and studied Andra Matin's design from secondary sources. We then analyze how the minimalist thinking affects the design of the house and how the inhabitants live inside the house.

3. Results & Discussion

Minimalist architecture becomes a trend in Jakarta, especially for middle class people because it gives simple and good expressions for the house and modern lifestyle for the owner. Moreover, minimalist architecture is sometimes associated with a modern and high class architecture since it looks different from the surrounding buildings. A clean and geometric form with flat roof does not actually refer to common tropical house. A typical tropical house is known for having some openings for cross ventilations, as well as dominant and extended roofs. Minimalist architecture should be adaptable to the needs of its inhabitants' needs and local contexts since the original ideas are actually transported from the West. Below are some discussions of the characteristics of minimalist architecture from the house in Bangka, South Jakarta.

3.1. Physical Conditions of the House

Figure 3. The glass walls of the house.

The walls of the building are approximately ninety percent made from glass with only few openings, since the house is air-conditioned. The number of openings on each level are only four but they are relatively small. In minimalist architecture, a glass wall can be used to make a boundary, and at the same time gives a transparency and a connection between inside and outside. In the house, a glass sometimes creates a boundary to separate two activities that has no problem with transparency, for instance in living room. A glass wall is used to separate the interior and the exterior, and replace the function of a solid wall. This separation occurs in some walls facing the pool and some walls facing the entrance. The glass walls absorb the sunlight and illuminate the whole house during the day. However, without air conditions, the house is very hot and the air circulation is very bad even though all of the windows are open. At some windows, the owner use curtains to reduce the heat. Depicted in Figure 3 are some glass walls and areas in the house with bad air circulation, especially during the day, because of the lack of openings.

For materials, the architect uses concrete columns and walls as structure and boundaries. Solid materials like concrete gives more privacy therefore the concrete walls are used for dividing the interior of the house. For example, there is a concrete wall to separate the guest area and the living room to create boundaries and privacy. In this way, the family can freely use the living room although there is a guest in the guest room. Approximately ninety percent of the façades use glass materials.

Figure 4. Concrete walls separates the guest and the family areas.

3.2. Functionality of the House

The main building is used by the family to do their daily activities. Only the grandmother stays most of the time in the pavilion – which is used to be a study room – since she has some physical constrains. In this way, someone can assist her whenever she need a help.

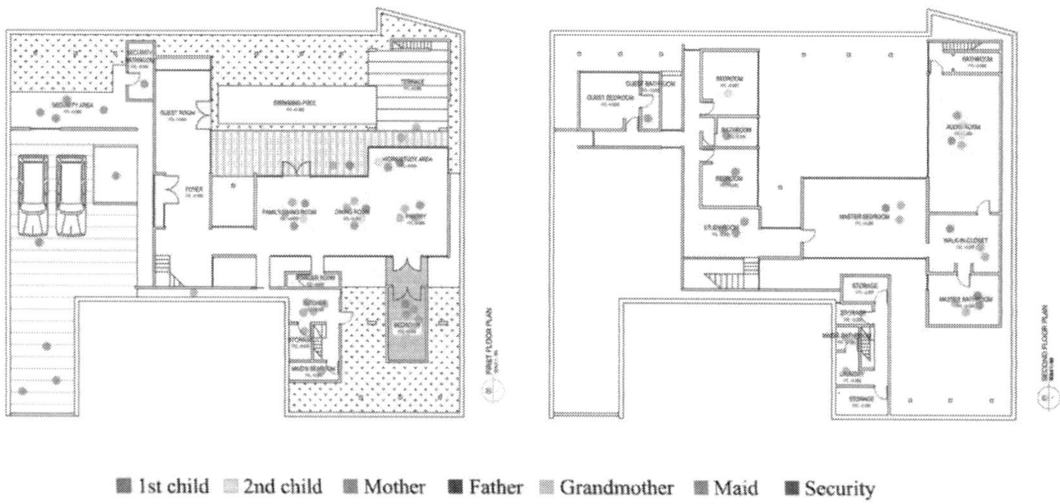

Figure 5. Activity and Density Diagram of the House.

As shown in Figure 5, the mother has various activities and occupies many areas in the house. The father usually has less activities in the house comparing to the mother. After coming back from his office, the father usually spends sometimes sitting next to the pool. The first child usually does her activities next to her mother. She also uses her bedroom or the study area located on the upper floor. The second child spends most of his time outside for his study at a university. When at home, he spends his time with the family in the living room or spending his time alone in his own bedroom. The maid spends most of her time during the day in the kitchen, sometimes cleans the whole house, and she stays in her room during the night. The security occupies his space near the main gate. He sometimes accesses the kitchen though the pathway.

The house implements some principles of minimalist architecture through the idea of geometric form and open space in the rooms. As a result, one can feel that the house form is a combination of some boxes and planar surfaces. The house form is an abstraction of geometric form and living conditions such as an open space to support family activities. The use of neutral color and exposed materials, especially concrete and glass, give simplicity and create boundaries between different activities and inside-outside. The compact design of the bedrooms on the upper level and open spaces on the lower level forces the inhabitants to gather and care of each other.

The house has different owners from the first it was built until now. Some functions of the rooms also change. This indicates that the owners did some adaptation to the house although in some cases the house may not support the changes very well.

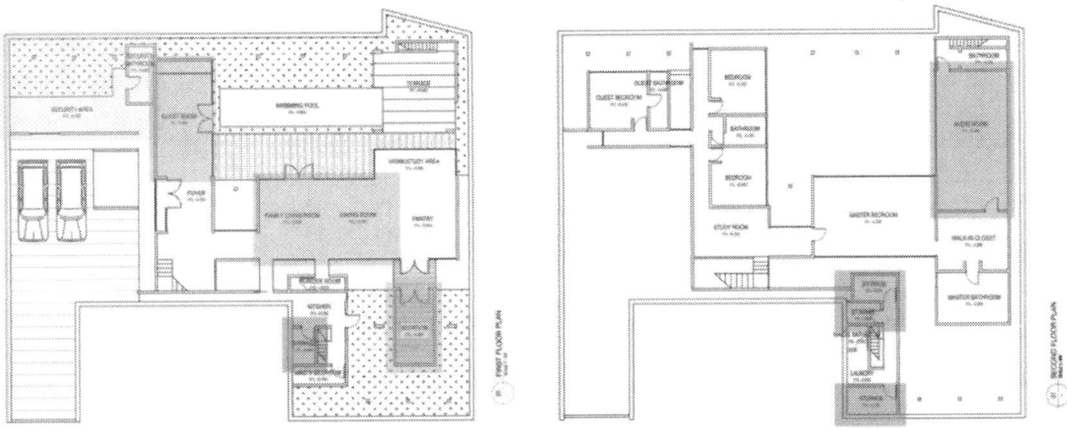

Figure 6. Changes in the room functions includes the security area, the guest room and the grandmother's room.

Figure 6 shows how some changes in the function of the room. The guest room of the late owner is changed into the security area since the location is not in the main building that indicates hierarchy and different level of relationship with the family. The security area faces the main gate that makes this room very strategic to see who is coming to the house. The former family room is now a guest room while the former study area is now the grandmother's room. However, since the study room is made mostly by glass walls, this room does not give good privacy. The owner then put curtains all over the room. The former fitness room, with the glass wall looks straight to the pool, is now an audio room. This room was changed because the whole family like to sing and the children like to play electronic games. Unfortunately, because the only access to the room is through the master bedroom, the room is rarely used especially by children.

From our interview, the owner decided to buy the house because the location of the house is strategic. The location of the house is not on the main street, which gives more privacy for the owner. The family seems to be happy to live in the house especially with Andra Matin's signature of minimalist house. The glass façade of the house does not bother the family like Dr. Edith Fansworth felt about her house since the site of the house is hidden from the main street. Moreover, open spaces resulted from the glass walls on the lower level of the house supports the gathering activities of the family. The separation of service area and a pavilion for the security also enhances the privacy of the family. Unfortunately, there are some disadvantages of the house. The glass wall on the upper floor makes the air circulation becomes very bad and the house becomes very hot during the day. The audio room, which was designed at first as a fitness room but now is completely very solid and close, cannot be accessed from outside the master bedroom. Thus the occupancy of the room is very low. A study room which is now converted into the grandmother's bedroom gives only little privacy to the occupant.

4. Conclusion

The house on Jalan Bangka, South Jakarta, does implements some principles of minimalist architecture such as using exposed industrial materials, neutral colors, and a simple cube form. The house design is most likely following the taste of the architect since his design always uses industrial materials, neutral colors, and a simple cube form. It seems that in designing the house, what comes first is the idealism of the architect then the consideration of owner's need and lived experience in the house. Although the building can adapt well to the needs of the current owner, yet it still produces some disadvantages of design, especially in relation to adaptation to local context and privacy. As a way to fulfill the need of the family, the current owner of the house need to adjust the function of the rooms according to the needs of the family.

IOP Conf. Series: Materials Science and Engineering **316** (2018) 012006 doi:10.1088/1757-899X/316/1/012006

In discussion in this study has shown how the idea of minimalism is used for house design, with the application of general principle of geometric form, open layout, the use of concrete and glass material and white color. However, the use of minimalist architecture has no connection to the concept of abstracting art and architecture to the simplest and purest form but to follow a trend and give modern identity to the house, away to common practice of tropical house. With the use of minimalist architecture, the house cannot actually give a certain comfort for tropical climate without the support of air conditioner and in the case of privacy with the support of curtain. The owner can tolerate the inability of the house to support their needs so they change some part of the house for their convenience. In short, minimalist architecture glorifies a style in which it all comes down to its simplest element but unfortunately sometimes does not adapt to local contexts. This is the crux of minimalism. No matter how ideal the architect design the building, nor how far the architect wants the building to get away from traditions, in the end, it always comes back to the people who use it.

5. Acknowledgement

The research for this paper is conducted partially with the support of PITTA program University of Indonesia. We thank the owner of the house in Bangka, Ms. Tharra Ayuriany and her family for allowing us to use the house as our case study.
This article's publication is supported by the United States Agency for International Development (USAID) through the Sustainable Higher Education Research Alliance (SHERA) Program for Universitas Indonesia's Scientific Modeling, Application, Research and Training for City-centered Innovation and Technology (SMART CITY) Project, Grant #AID-497-A-1600004, Sub Grant #IIE-00000078-UI-1.

References

[1] Jordy W 1965 The International Style in the 1930s *J. of the Society of Architectural Historians,* vol 24 no 1 10-14.
[2] Obendorf H 2009 *Minimalism: designing simplicity* (Dordrecht: Springer)
[3] Hitchcock H Johnson P 1932 *The International Style* (New York: W. W. Norton & Company)
[4] Wendl N 2015 Sex and Real Estate, Reconsidered: What Was the True Story Behind Mies van der Rohe's Farnsworth House? *Arch Daily.*
[5] Foster, H. 1996. *The Crux of Minimalism,* (Cambridge: MIT Press)
[6] Krauss E 1986 *The Originality of the Avant Garde and Other Modernist Myths* (Cambridge: The MIT Press)
[7] Cuito A 2002 *Minimalist Lofts* (Barcelona: Paco Asensio)
[8] Levine N 1986 *Abstraction and representation in Modern Architecture: the International Style and Frank Lloyd Wright* (New Haven: Yale University Press)
[9] Hartanto R and Sopandi S 2015 A cinematic sandwich: Andra Matin's house *Uncube Magazine.*
[10] Muynck B 2013 Andra Matin interview *Mark Magazine* 44

Degradation Behaviour of Gamma Irradiated Poly(Acrylic Acid)-graft-Chitosan Superabsorbent Hydrogel

Dhena Ria Barleany[1], Alpin Ilhami[1], Dea Yusuf Yudanto[1], Erizal[2]

[1]Department of Chemical Engineering, Faculty of Engineering, University of Sultan Ageng Tirtayasa, Banten, Indonesia
[2]Centre for Application of Isotopes and Radiation, Jakarta, Indonesia

Email: dbarleany@yahoo.com

Abstract. A series of superabsorbent hydrogels were prepared from chitosan and partially neutralized acrylic acid at room temperature by gamma irradiation technique. The effect of irradiation and chitosan addition to the degradation behaviour of polymer were investigated. The gel content, swelling capacity, Equillibrium Degree of Swelling (EDS), Fourier Transform Infra Red (FTIR), and Scanning Electron Microscopy (SEM) study were also performed. Natural degradation in soil and thermal degradation by using of TGA analysis were observed. The variation of chitosan compositions were 0.5, 1, 1.5, and 2 g and the total irradiation doses were 5, 10, 15, and 20 kGy. The highest water capacity of 583.3 g water/g dry hydrogel was resulted from 5 kGy total irradiation dose and 0,5 g addition of chitosan. From the thermal degradation evaluation by using of TGA analysis showed that irradiation dose did not give a significant influence to the degradation rate. The rate of thermal degradation was ranged between 2.42 to 2.55 mg/min. In the natural test of degradation behaviour by using of soil medium, the hydrogel product with chitosan addition was found to have better degradability compared with the poly(acrylic acid) polymer without chitosan.

1. Introduction

Absorbent hygiene products such as baby care products, feminine hygiene products and adult incontinence products are an essential part of everyday life. Baby care products include baby diapers, training pants, dry and wet baby wipes. Feminine hygiene products encompass sanitary pads, tampons, panty shields, interlabial devices, wipes and cosmetic removal pads. Adult incontinence products cover adult incontinence pads, adult pant diapers, underpads, personal/medicated wipes, and nursing pads. Amongst these products, baby diapers, sanitary pads and adult incontinence pads are the three mostly used disposable absorbent hygiene products [1].

Disposable diapers are absorbent products for personal hygiene designed to absorb and retain urine and faeces from adults or from babies with incontinence problems [2]. The global market for Baby Disposable Diapers is forecast to reach US$33,4 billion by the year 2017 [3]. The usage rate of nappies/diapers/pants in Indonesia is fairly low compared to other countries, as it is still developing. Per capita use of nappies/diapers/pants in volume terms in 2016 in Indonesia amongst children aged 0-36 months was 295 units per child per year, lower than 1,244 units of North America and 963 units in Western Europe. This is because some children in rural areas still use traditional nappies/diapers which can be washed and reused [4].

In modern societies, disposable diapers constitute a significant percentage of municipal solid wastes. They have been traditionally landfilled or incinerated as only limited recycling processes are being implemented in some parts of the world [5]. Diapers made from several material like synthetic pulp, polychlorine dibenzodioxins, superabsorbent gel of polyacrylic acid, and plastic. The introduction of superabsorbent polymers (polyacrylates) as a component of the diaper core in the 1980s was the most important improvement. Unlike traditional cellulose materials, polyacrylates have the ability to retain liquid in the diaper core, keeping it away from the baby's skin, even under pressure [6]. However, the use of synthetic materials in this product causes some problems to the degradation process in soil. Synthetic material has resistant property to microorganism. Resistant property from synthetic material must be reduced or eliminated in order to avoid enviroment problem [7]. Technological progress and the need for alternative eco-friendly and inexpensive resources have put forward the use of natural polymers in superabsorbents. Unfortunately, the design of the hydrogels by using natural polymers usually would be biocompatible and biodegradable, but they have inferior mechanical properties compared to petroleum based polymers [8]. In recent years, modification of synthetic polymers with the natural polymers have been utilized for fabricating eco-friendly, super water absorbency, and great mechanical properties of superabsorbent hydrogels using grafting method [9].

Superabsorbents prepared with polysaccharides have attracted great attention because of their abundant resources, low production cost, and biodegradability [10, 11]. Polysaccharides are among the favourites with the main examples being starch [12], chitosan and guar gum [13]. Chitosan is a highly deacetylated derivative of chitin, shows interesting chemical, physical and biological properties [8]. Therefore, chitosan acts as a desirable backbone to graft poly(acrylic acid) forming a superabsorbent material [14]. The superabsorbent resins grafted with chitosan can absorb aqueous solutions up to hundreds of times their own dry weight [15] and should have the antibacterial activities [16]. Some synthetic materials were grafted onto chitosan with the aim to enhance the thermal and structural properties of the polymers. Two different deacetylation degree (DD%) and molecular weight of chitosan were subjected to grafting reactions with poly (n-butyl acrylate) to prepare natural/synthetic hybrid materials, then the products grafting percentage and efficiency were studied against acrylate/chitosan ratio and DD% [17]. A preparation of a porous carboxymethyl chitosan grafted poly (acrylic acid) (CMCTS-g-PAA) superabsorbent polymers with excellent properties was developed [18]. Synthesis and characterization of chitosan hydrogels, in macro- and nano-size, grafted with N-vinylcaprolactam (NVCL) using gamma radiation was recently evaluated, and the effect of dose and monomer concentration in the grafting process was studied [19]. The previous study [20] observed the effect of neutralization degree to the swelling properties of a novel poly (acrylic acid)-graft-chitosan superabsorbent hydrogel. In this study, partially neutralized acrylic acid was grafted onto chitosan backbone and the effect of some reaction conditions to the swelling capacity and degradation behaviour of grafted polymers were discussed.

This paper aims to determine the conditions that result good mechanical strength, superabsorbent property, and degradability. This paper reports the preparation of (acrylic acid)-graft-chitosan superabsorbent hydrogels through ϒ-irradiation cross-linking procedures, besides the evaluation of superabsorbent properties, mechanical properties, and biodegradability of hydrogels, as well as the influence of irradiation dose and chitosan composition in the reactant mixture on the properties of hydrogels. The data resulted from this research is very important and useful to predict the optimum condition in the synthesis of poly (acrylic acid)-graft-chitosan superabsorbent hydrogel with great performances in some criterias.

2. Materials and Methods

2.1. Materials

Pharmaceutical grade of chitosan with deacetillation degree of 85% used for backbones of polymer chains on the grafting process, was purchased from PT. Biotech Surindo Indonesia. Analytical grade of acrylic acid (Merck, Darmstard, Germany), and industrial grade of potassium hydroxide (NaOH) for neutrallization of acrylic acid before irradiation procedure, were purchased from Cv. Cahaya Kimia Indonesia. Distilled water was used for preparation of the hydrogel and swelling measurement.Chitosan food grade from PT. Biotech Surindo with a degree of deacetylation (DD) = 87.2% and a viscosity of 37.10 cps, glacial acetic acid from Merck, technical glycerol from Bratachem, red ginger oleoresin from PT. Lansida, and ethanol 96% from Sigma-Aldrich.Some text.

2.2. Methods

2.2.1. Preparation of poly(acrylic acid)-graft-chitosan . 85 mL of distilled water was poured into beaker glass, then 15 mL of acrylic acid solution and 5.6 g of potassium hydroxide was added until potassium acrylate was formed. Chitosan was added to this solution and mixed well with magnetic stirrer. After homogen, the reactant mixture solution was packed in a small plastic container then irradiated with the variation of ϒ-irradiation total doses of 5 kGy, 10 kGy, 15 kGy and 20 kGy to form poly(acrylic acid)-graft-chitosan). The hydrogels were dried by putting them into oven with the temperatur was maintained at 60°C over 24 hours. The dried hydrogels were crushed into powder size by using of a blender.

2.2.2. Gel Content Measurement. An amount of 0.01 g of hydrogel powder was measured and packed in the tea bag. Tea bag with the hydrogel powder inside was then immerse into water for 24 hours at 30°C. After this process, the tea bag then was put in the oven for 24 hours at 60°C and weighed. Gel content was calculated using the equation (1).

$$gel\,content(\%) = (\frac{W_1}{W_0})x\,100 \tag{1}$$

where W_1= weight of dried gel after soaking (g) and W_0 = initial weight of gel (g)

2.2.3. Swelling Study. The swelling capacity of gel was determined according to a tea bag method [12]. Powdered hydrogel (0.1 g) were immersed in distilled water (100 mL) at room temperature. At 30 seconds time intervals, the samples were removed from the swelling medium and blotted on a filter prior to weighing to remove excess surface moisture. The swelling ratio was calculated according to the following expression :

$$swelling\ capacity\left(\frac{g\ water}{g\ dry\ hydrogel}\right) = \frac{W_t - W_0}{W_0} \tag{2}$$

where W_t= weight of the swollen gel at time t (g) and W_0 is the initial weight of the dried gel (g). The studies were carried out for 240 seconds.

2.2.4. TGA. An accurately of hydrogel powder sample (10 mg) was put into a tiny cup, then characterized by TGA (Thermogravimetric Analysis) at temperature range of 30°C to 500°C. The result then appeared as a thermogram.

2.2.5. Biodegradation Study. Biodegradable property of poly(acrylic acid)-graft-chitosan was studied by using soil. An amount of 50 g of soil and 2 g of hydrogel were prepared and blend in a glass cup. Samples were taken and weighed after a regular time interval of 2 days. The percentage of weight loss after particular time interval was determined by using of equation (3).

IOP Conf. Series: Materials Science and Engineering **316** (2018) 012007 doi:10.1088/1757-899X/316/1/012007

$$weight\ loss\ (\%) = \left(\frac{W_2 - W_1}{W_1} \right) x100 \quad (3)$$

where W_1 is the initial weight of the sample and W_2 is the weight after 2 days.

3. Results and Discussion

In this research, superabsorbent hydrogel for diaper application was made from acrylic acid, potassium hydroxide and chitosan with gamma (γ) irradiation of Co-60 as an iniatior or crosslinking agent [21].

(a) (b)

Figure 1. Superabsorbent hydrogel (a) variation of chitosan (b) variation of gamma irradiation dose with 0,5 g of chitosan

From figure 1 (a), the mass variation of chitosan used in the reactant affects the hydrogel colour. Figure 1 (b) shows how ϒ-ray of radiation makes the hydrogel texture difference. Hydrogel with high dose of radiation has a slight white colour due to crosslinking process between chitosan and neutralized poly(acrylic acid). In low dose of ϒ-ray radiation, chitosan acts as filler because of the unperfect crosslinking reaction.

3.1. Swelling Capacity

Swelling value shows the ratio between mass of water or solution absorbed by hydrogel to mass of dry hydrogel. Swelling test can show the hydrogel performance. On swelling test, hydrogel powder was immersed into water with variation of time from 30 seconds to 240 seconds with 30 seconds of time interval.

ϒ-irradiation technique has been performed to form crosslinking reaction between acrylic acid with potassium acrylate, and potassium acrylate with chitosan. Weight variation of chitosan in this research are 0, 0.5 g, 1 g, 1.5 g; and 2 g; and the irradiation dose was varied of 5 kGy, 10 kGy, 15 kGy and 20 kGy with a constant 50% netralization of acrylid acid by using of potassium hydroxide.

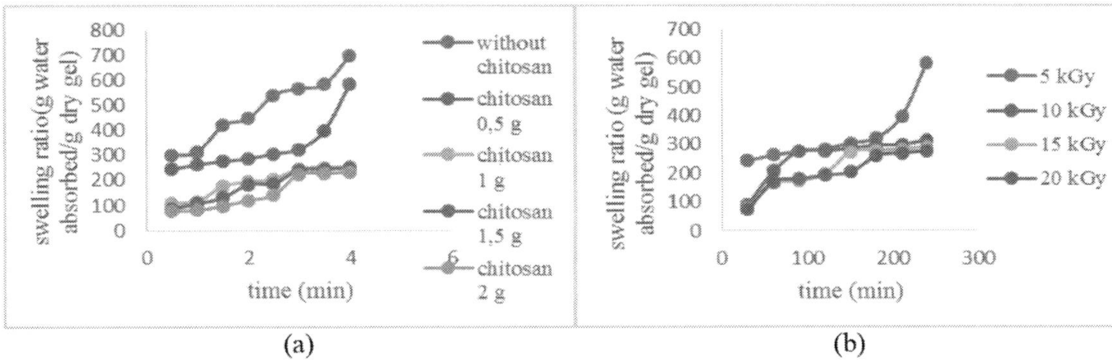

Figure 2 Swelling capacity of superabsorbents with (a) variation of chitosan weight (b) variation of gamma radiation dose

In figure 2 (a) shows that the swelling of hydrogel decreases from 2 g of chitosan to 0 g of chitosan, meanwhile in figure 2 (b) the swelling of hydrogel increases from 20 kGy to 5 kGy with variation of time, 30 seconds to 240 seconds. Hydrogel with 240 seconds of time immersion has the greatest swelling capacity. At a longer time, the amount of water absorbed in the gel matrix would be higher . However, when it was reached a saturation condition, the gel won't absorb more water. More chitosan presents in the reactant cause the decreasing of water swelling capacity. Like other natural polymers, at constant total dose of radiation, there will be excess chitosan in the matrix which is not crosslinked properly and fill spaces between the crosslinked-network [21].

In figure 2 (b), total dose of gamma radiation has responsible for swelling capacity. Higher dose of radiation causes the decreasing of hydrogel absorption capacity. 5 kGy of radiation result 583.3 g water/g dry hydrogel of swelling, and for 20 kGy of radiation generate 277 g water/g dry hydrogel of swelling. Total dose of radiation affects density of crosslinking reaction and absorption capacity. High density of the bond indicates small pore in the network, this may cause the difficulty for water to be absorbed inside [22].

3.2. Equilibrium Degree of Swelling (EDS)

Equilibrium degree of swelling (EDS) is one of important parameters that shows swelling ratio in maximum condition. The maximum condition occures because of the hydrophilic group in hydrogel which has the hydrogen bond with water, or the hydrophilic group in hydrogel reaches the equilibrium condition. In general, every hydrogel has different value of EDS, depending on the component. Figure 3 presents the effect of dose of radiation that has measured for 24 hours at room temperature. Higher dose of radiation causes the increasing of density of the network, and as a result, the EDS value decreases.

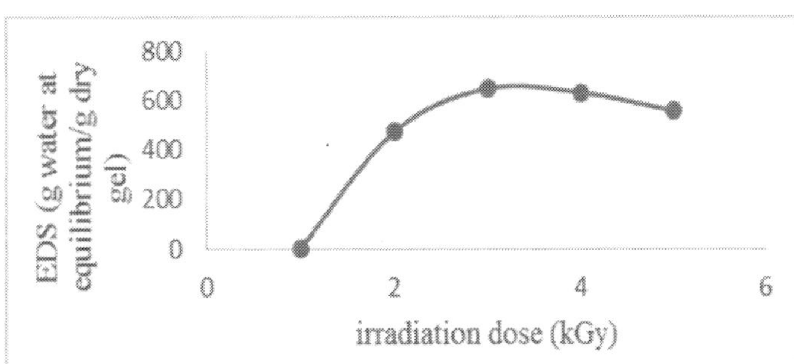

Figure 3. EDS value at constant chitosan composition (0.5 g)

3.3. Gel Content Analysis

Gel fraction is a parameter that used to evaluate the efficiency in the synthesis process of hydrogel depending on total dose of ϒ-radiation. Gel fraction has been tested on variation of radiation doses. Figure 4 presents how the radiation gives effect to the gel fraction. This figure shows that the gel content increases during the increasing of total dose of radiation [23].

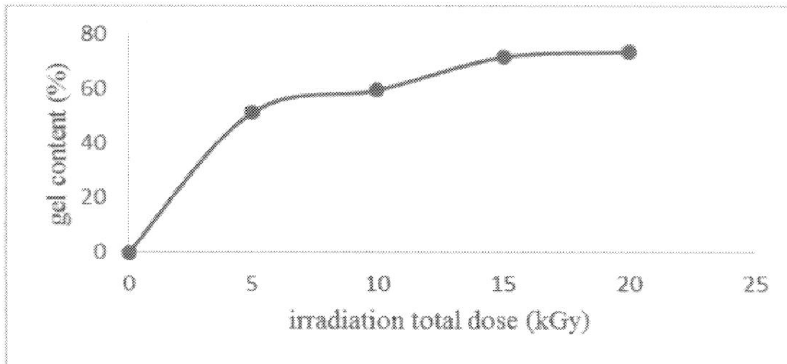

Figure 4. Gel content (%) at variation of ϒ- radiation total dose

3.4. Functional Group Analysis with FTIR

Infrared spectroscopy or *Frourier Transform Infrared* (FTIR) is the instrument that is used to find the functional groups of organic, polymer or coating compunds. FTIR is able to analyze material in many forms such as liquid, solid, fiber, paste, powder, or another form. Functional groups of modificated hydrogel were confirmed by FTIR spectroscopy. Functional groups of chitosan was confirmed by the FTIR spectra study as shown in Figure 5. In figure 5 (a), the hydroxyl group appears at 3475.88 cm^{-1} that is bothered by weak amina groups ($-NH_2$) at 3700 cm^{-1}. CO at 1015.57 cm^{-1} indicates the bonding between CO and CN. In figure 5 (b), acrylid acid in the form of carboxylic acid, the peak at 1725.40 cm^{-1} shows appearance of carbonyl groups (C=O) in carboxylic compound and hydroxyl groups at 2876.9 cm^{-1} are wider. This enlargement can be affected by dymer on the hydrogen bond. Spectra of poly (acrylic acid-co-potassium acrylate) at 3454.51 cm^{-1} shows the existing of hydroxyl groups. Peak of carbonyl groups at 1712.79 cm^{-1} in derivative of carboxylic compound are a result of partial neutralization of acrylic acid by potassium hydroxide. Peak at 2951,09 cm^{-1} shows the bonding of C-H. Spectra of modificated hydrogel from figure 5 (d) shows that hydroxil groups appears at 3450.65 cm^{-1}, C-H at 2881.65 cm^{-1}, 1519.19 cm^{-1} for C-C (aryl groups), and 1168.86 cm^{-1} for -C-O- and -C-N-.

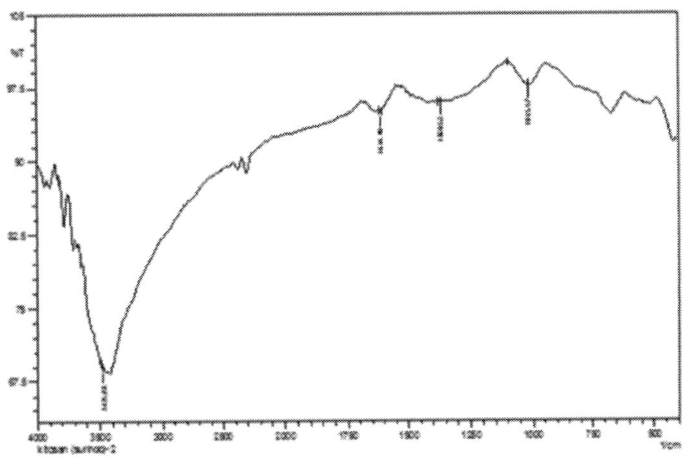

a

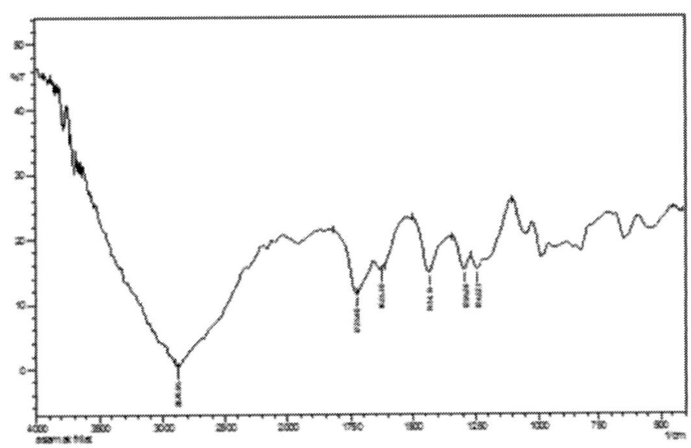

b

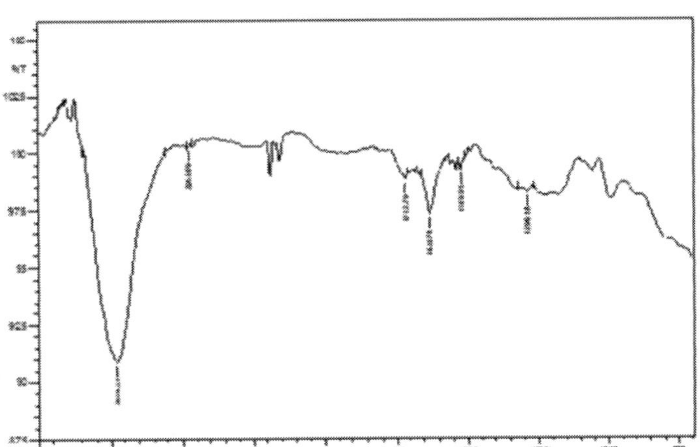

c

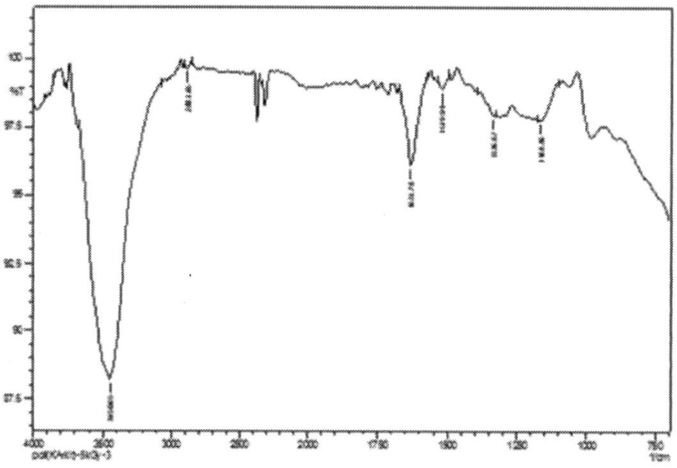

d

Figure 5. FTIR spectra of (a) Chitosan (b) acrylic acid (c) poly (acrylic acid) and (d) poly (acrylic acid)-graft-chitosan

3.5. Thermal Degradation

The thermal decomposition of poly(acrylic acid)-graft-chitosan was analyzed by using of TGA as shown in figure 6. High dose of radiation results the increasing of melting point temperature of the hydrogel. Melting point of the hydrogel obtained from this research for 5 kGy, 10 kGy, 15 kGy, 20 kGy variation of total dose of radiation are respectively 302.88°C, 304.32°C, 319.98°C and 350.17°C. The synthesis of hydrogel using the lowest dose (5 kGy) affects the unproper reaction during the radiation. In another word, low dose of radiation produce the hydrogel with low density of crosslink, so the hydrogel becomes more amorphous.

Derivative graph of TGA in figure 6 (a), (b) and (c) show more than one peak of endotherm decomposition, this are indicate that chitosan in the hydrogels result unstable chain during the reaction, so that the products becomes sensitive to thermal charge and easier to degrade. From the thermogram, can be concluded that rate of heating of the hydrogels from 30°C to 500°C is 10°C/minute as shown in table 1.

From table 1, the fastest rate of hydrogel degradation in this research was obtained at 10 kGy of radiation and the slowest was at 20 kGy. 10 kGy is the best dose of ϒ-irradiation to synthesize hydrogel which results fastest thermal degradation. Lower and higher dose of radiation did not give good performance on the crosslink reaction. Dose of radiation which was too low brought weaker chain of the network. In another hand, too high dose of radiation caused too high chain density and resulted bad performance on swelling.

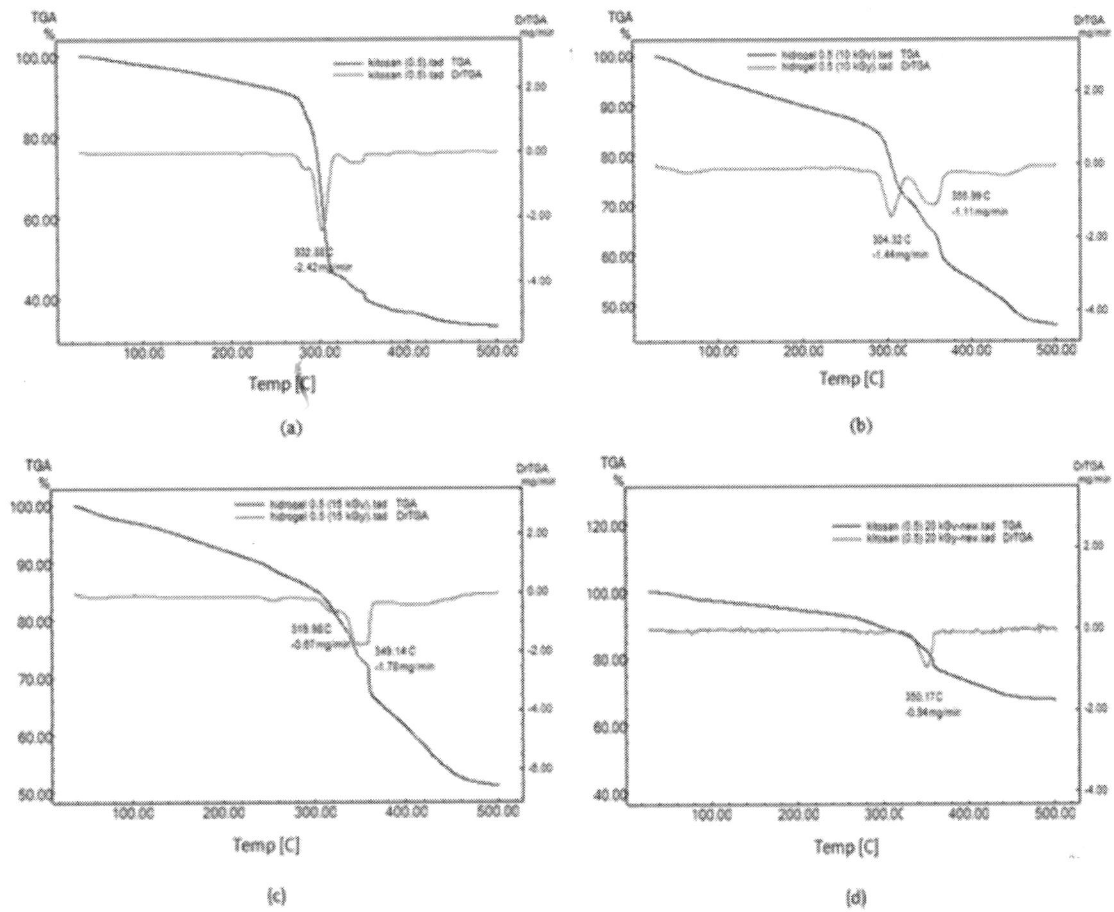

Figure 6. TGA thermogram of hydrogel prodused by varied total dose of radiation (a) 5 kGy (b) 10 kGy (c) 15 kGy (d) 20 kGy

Table 1. Rate of thermal decomposition respect to the grafted hydrogel

Radiation dose (kGy)	Degradation rate (mg/min)
5	2.42
10	2.55
15	2.45
20	0.94

3.6. Biodegradation Study

Biodegradation is the degradation caused by biological activity like bacteria and enzyme. The most simple method for biodegradation is by measuring the loss of mass [24]. Weight loss of the hydrogel can be measured by weighing the initial weight and the weight of material after degradation procedure. Figure 7 shows the weight loss of hydrogel for some time intervals. Based on the data can be explained that the longer contact time between hydrogel and soil, the weight loss of hydrogel due to biodegradation of microorganism becomes higher. Effect of the chitosan existence in the grafted polymer is shown in figure 7. In this figure, the addition of chitosan causes the increasing of biodegradable ability of the hydrogel. Mircoorganism needs more time to decompose hydrogel produced from pure synthetic material than hydrogel produced from synthetic material with the addition of chitosan as natural material. Like any other natural materials, chitosan is biodegradable. Microorganism activity can easily degrade chitosan in the soil. Chitosan based grafted polymer can be a solution to produce an eco-friendly superabsorbent resin for diaper application with great performance.

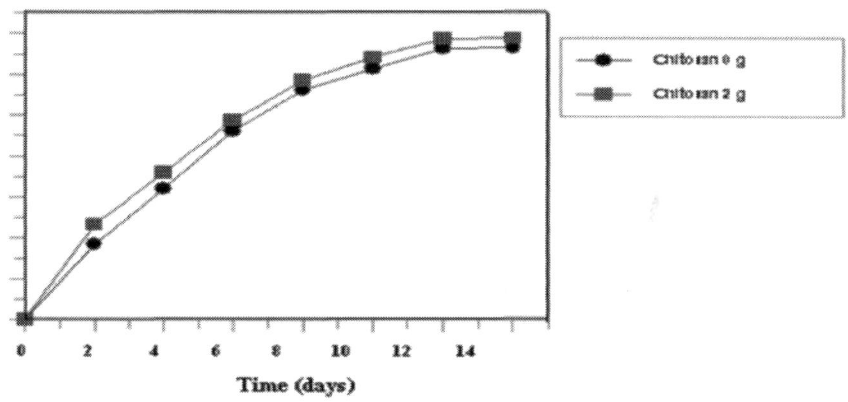

Figure 7. Biodegradation of poly(acrylic acid)-graft-chitosan at varied using of chitosan

Hydrogel with 2 g of chitosan has 13.8% rate of biodegradation, faster than hydrogel without chitosan which only reaches 13.3%. Hydrogel with more chitosan is easy to digest by microorganism in the soil. Microorganism activity will break the chemical bond [25]. The difference weight loss between hydrogel with chitosan and hydrogel without chitosan were not significant because some species of microorganisms in the soil can not decompose chitosan [7].

3.7. SEM Study

The surface morphology of the hydrogel was characterized by SEM at magnifications of 750x. Pore size of modified hydrogel is ranged from 10 μm to 25 μm and was confirmed by SEM as shown in figure 8. Hydrogel without chitosan with same composition and neutralization condition of acrylic acid has 38-39 μm of pore size. Hydrogel poly(acrylic acid)-graft-chitosan has a smaller pore than poly(acrylic acid) without chitosan. In some parts of the matrix found that chitosan filled in some pores of hydrogel and caused reduction of the hydrogel pore size.

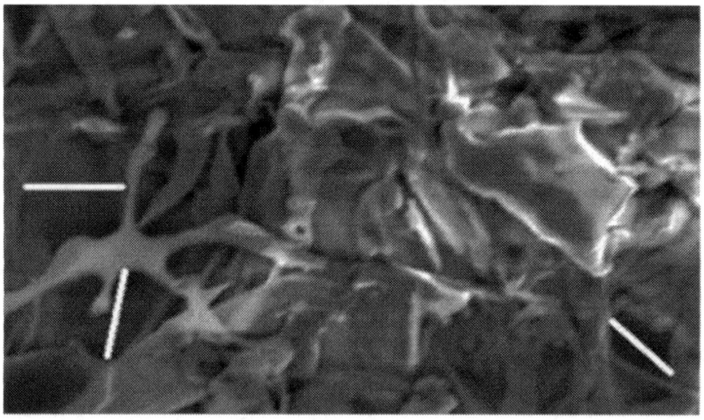

Figure 8. SEM micrograph of poly(acrylic acid)-graft-chitosan

4. Conclusion

The best product of poly (acrylic acid)-graft-chitosan superabsorbent hydrogel obtained from this study was produced with addition of 0.5 g of chitosan and synthesize using 5 kGy total dose of ϒ-ray radiation which has reached 583.3 g water/g dry hydrogel of swelling capacity. This product has the biggest absorption capacity than the other compositions and doses. In addition, rate of thermal degradation of this superabsorbent hydrogel is 2.42 mg/min, and the biodegradation ability is better than product which irradiated with total dose of 10 kGy, 15 kGy and 20 kGy.

5. References

[1] Das D, 2014 *Composite Non-Woven Materials* (United Kingdom: Woodhead Publishing Limited) p 74-88
[2] Torrijos M, Sousbie P, Rouez M, Lemunier M, Lessard Y, Galtier L, Simao A and Steyer J P 2014 *Waste Management* **34** 669-75
[3] Global Industry Analysts Inc. (GIA) 2017 Baby Disposable Diapers- Global Startegic Business Report, [online] Available at: http://www.prweb.com/releases/baby_disposable_diapers/premium_designer_diapers/prweb8973346.htm [Accessed 30 Mar. 2017]
[4] Euromonitor International 2017 Nappies/Diapers/Pants in Indonesia, [online] Available at: http://www.euromonitor.com/nappies-diapers-pants [Accessed 30 Mar. 2017].
[5] Colón J, Mestre-Montserrat M, Puig-Ventosa I and Sánchez A 2013 *Waste Management* **33** 1097-103
[6] Kosemund K, Schlatter H, Ochsenhirt J L, Krause E L, Marsman D S and Erasala G N 2009 *Regulatory Toxicology and Pharmacology* **53** 81-9
[7] Pradhan A K, Rana P K and Sahoo P K 2014 *International Journal of Biological Macromolecules* **74** 620-6
[8] Spagnol C, Rogdrigues F H A, Pereira A G B P, Fajardo A R Rubira A F and Muniz E C 2012 *Cellulose* **19 (4)** 1225-37
[9] Singh V, Kumar P and Sanghi R 2012 *Progress in Polymer Science* **37 (2)** 340-64
[10] Likhitha M, Sailaja R R N, Priyambika V S and Ravibabu M V 2014 *International Journal of Biological Macromolecules* **65** 500-8
[11] Huang Z, Liu S, Fang G and Zhang B 2013 *Carbohydrate Polymers* **92** 2314-20
[12] Barleany D R, Ulfiyani F, Istiqomah S, Heriyanto H, Rahmayetty and Erizal 2015 *AIP Conference Proceeding 1699 2nd International Conference of Chemical and Material Engineering (ICCME) 2015* (Semarang 29-30 September Indonesia) 040008-1–040008-10 doi: 10.1063/1.4938323
[13] Glampedaki P and Duschk V 2015 *Functional Finishes for Textiles* (United Kingdom: Woodhead Publishing Series in Textiles) p 283-302
[14] Rusu A G, Popa M, Ibanescu C, Danu M and Verestiuc L 2016 *Materials Letters* **164** 320-4
[15] Erizal, Abbas B, Sukaryo S G and Barleany D R 2015 *Indo. J. Chem.* **15 (3)** 281-7
[16] Barleany D R, Alim I P, Rizkiyah N, Lusi U T, Heriyanto H and Erizal 2016 *Proceeding of International Conference on Technology, Innovation, and Society (ICTIS) 2016* (Padang 20-21 July Indonesia)
[17] Anbinder P, Macchi C, Amalvy J and Somoza A 2016 *Carbohydrate Polymers* **145** 86-94
[18] Chen Y, Zhang Y, Wang F, Meng W, Yang X, Li P, Jiang J, Tan H and Zheng Y 2016 *Materials Science and Engineering C* **63** 18-29
[19] Cruz A, García-Uriostegui L, Ortega A, Isoshima T and Burillo G 2017 *Carbohydrate Polymers* **155** 303-12
[20] Barleany D R, Lestari R S D, Yulvianti M, Susanto T R, Shalina and Erizal 2017 *International Journal on Advanced Science Engineering and Information Technology* **7 (2)** 702-8
[21] Abd. Alla S G, Sen M and El-Naggar A W M 2012 *Carbohydrate Polymers* **89 (2)** 478-85
[22] Hemvichian K, Chantawong A and Suwanmala P 2014 *Radiation Physics and Chemistry* **103**

IOP Conf. Series: Materials Science and Engineering **316** (2018) 012007 doi:10.1088/1757-899X/316/1/012007

167-71.
[23] Demeter M, Virgolici M, Vancea C, Scaresorenau A, Kaya M G A and Meltzer V 2017 *Radiation Physics and Chemistry* **131** 51-9
[24] Saruchi, Kaith B S, Jindal R and Kumar V 2015 *Polymer Degradation and Stability* **115** 24-31
[25] Mittal H, Mishra S B, Mishra A K, Kaith B S, Jindal R and Kalia S 2013 *Carbohydrate Polymers* **98 (1)** 397-404

IOP Conf. Series: Materials Science and Engineering **316** (2018) 012008 doi:10.1088/1757-899X/316/1/012008

THE USE OF MULTI-REACTOR CASCADE PLASMA ELECTROLYSIS FOR LINEAR ALKYLBENZENE SULFONATE DEGRADATION

Nelson Saksono[1,a], Ibrahim[1], Zainah[1], Trisutanti Budikania[2]

[1]*Department of Chemical Engineering, Universitas Indonesia, Depok 16242, Indonesia*
[2]*Politeknik AKA, Bogor, Indonesia*
Phone: +62217863516, Fax: +62217863515
[a)] *Corresponding Author: nelson@che.ui.ac.id*

Abstract

Plasma electrolysis is a method that can produce large amounts of hydroxyl radicals to degrade organic waste. The purpose of this study is to improve the effectiveness of Linear alkylbenzene sulfonate (LAS) degradation by using multi-reactor cascade plasma electrolysis. The reactor which operated in circulation system, using 3 reactors series flow and 6 L of LAS with initial concentration of 100 ppm. The results show that the LAS degradation can be improved multi-reactor cascade plasma electrolysis. The greatest LAS degradation is achieved up to 81.91% with energy consumption of 2227.34 kJ/mmol that is obtained during 120 minutes by using 600 Volt, 0.03 M of KOH, and 0.5 cm of the anode depth.

Keywords: LAS degradation; plasma electrolysis; hydroxyl radical

Introduction

Linear alkylbenzene sulfonate (LAS) is an anionic surfactant alkyl chain containing about 10-14 carbon atoms which is bound to the aromatic rings sulfonated [1]. LAS is one type of organic compounds which commonly found in detergent content at this time. Its use is fairly widespread among industries and households alike. Wastewater that contained LAS will flow into the river, so that the river will become polluted. This condition can cause some problems (e.g. shallowing waters, inhibiting oxygen transfer, inhibiting microbial growth) that endanger human life and ecosystem of the river [2]. LAS is biodegradable (can be degraded by microorganisms), however, properties of biodegradable LAS began decreased at concentrations above 50 ppm [3].

In these decades, a growing interest has been attracted to using plasma electrolysis for removal of organic pollutants from water because it has high removal efficiency and environmental compatibility [4]. Plasma electrolysis is a one method that can be produce large amounts of hydroxyl radical (•OH) and used to degrade organic waste. Some researchers indicated that many organic pollutants can be destroyed by plasma electrolysis and the final products are either non-toxic or less toxic compounds, such as, carbon dioxide, water, inorganic salts, etc [5]. This is because plasma electrolysis method can production of reactive species such as hydroxyl radicals, hydrogen radicals (•H), oxygen radicals (•O) and molecules (H_2O_2, O_3, etc.) [6].

Plasma electrolysis process begins with the gas formation on Faraday electrolysis occur both at the anode and cathode. Oxygen gas is formed at the anode while hydrogen gas is formed at the cathode. Faraday electrolysis develops into plasma electrolysis on the power supply voltage is high enough. In addition, the plasma is obtained using asymmetric electrodes, glow discharge occurs at the electrodes with a smaller area. Plasma phenomenon is characterized by a drastic decrease in current as indicated by the appearance plasma on the anode [7].

In anode plasma electrolysis process produced reactive species such as H_2O_2, OH•, H•, and e-. Hickling theory states are two zones formation in the plasma electrolysis. In the primary reaction zone, water was ionized or activated, and the bombard with each other to break the bonds by electron transfer. This phenomenon produces compounds OH• and H•. Both of them are possible to interact to form H_2O_2 and H_2. Then, a mixture of •OH, •H, H_2O_2 and H_2 diffused out of the primary zone (that was called plasma later) and interacted each other with substrates or compounds dissolved in solution or liquid phase in the second reaction zone [8].

This research is also focused on the several parameters that can affect the process LAS degradation with plasma electrolysis, which include anode depth and amount of reactors. These two parameters can affect the production of hydroxyl radical [9], where hydroxyl radical is an oxidizing agent that can degrade LAS.

Experimental

The LAS waste (100%) used in the plasma electrolysis process was a synthetic LAS procured from Merck (Indonesia). Potassium hydroxide (KOH) was procured from Merck and used as an electrolyte solution. All aqueous solutions were prepare with distilled water from Bratachem, Indonesia. The plasma electrolysis process was operated in circulation mode and the experimental set-up is shown in Fig. 1. The reactor was made of a transparent acrylic tube with cylindrical solid cathode of stainless steel SS-314 which placed at the top reactor and anode made of tungsten is placed at the top reactor. A single anode and cathode used in each reactor. The reactor which operated in circulation system, using 6 L of LAS with initial concentration of 100 ppm.

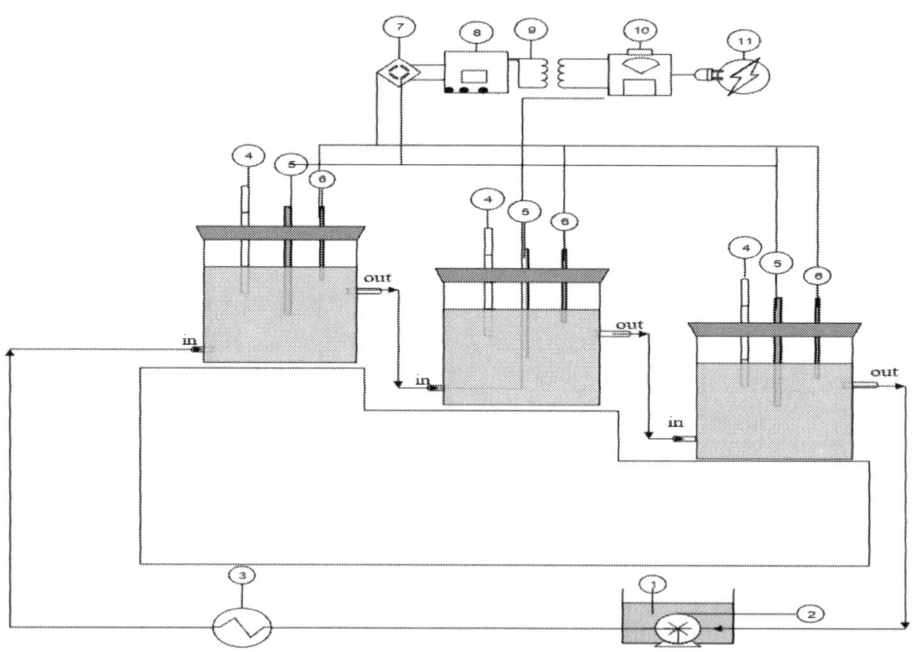

Figure 1 Schematic Diagram of Multi-Reactor Cascade: 1. LAS container, 2. Pump, 3. Cooling Coil, 4. Thermometer, 5. Cathode (Stainless steel), 6. Anode (Wolfram), 7. Diode bridge, 8. Power analyser, 9. Transformer, 10. Slide regulator, 11. Source voltage.

The anode used was a tungsten (wolfram) with 1.6 mm in diameter. The cathode was made of stainless steal that was 5 mm in diameter and placed along the 30 mm immersed in a solution of the system. The cathode separated from anoda with distance 40 mm. During the experiment, the temperature was maintained at 50 ± 1 °C by circulating LAS continous through a cooling coil. The high voltage applied in plasma electrolysis was set 600 Volts. The solution for treatment was prepared by dissolving a known amount of LAS in a KOH with the concentration solution at 0.03 M. In addition, the variety of the anode depths were 0.5 cm, 1 cm, and 1.5 cm from the surface of solution.

The discharge voltage and current in the plasma electrolysis process were measured using HELES UX-838TR multimeter. A small portion of LAS after the degradation was measured by Methylene Blue Active Substance (MBAS) method developed according to ASTM D 2330-02. This test method is to determine the LAS concentration and degraded products. The amount of Hydrogen peroxide (H_2O_2) as an indicator of the presence of hydroxyl radical was tested using the iodometric titration method. Iodometric titration was used as analytical method in order to represent the oxidizing agent formed through the release of gaseous iodine (I_2). Chemical Oxygen Demand (COD) was measured by the potassium dichromate standard method. Oxalic acid analysis was conducted to determine intermediate results LAS degradation. Oxalic acid analysis measured by titration permanganometry method.

Results and Discussion

Effect of Anode Depth on The LAS Degradation

Table 1 shows, the amount of LAS after degradation in the condition of the immersed anode 1.5 cm had the greatest value in comparison to the conditions of 0.5 cm and 1 cm. This is because the surface area of the anode that immersed deeper in the electrolyte solution is larger so that the sheath gas formation around the anode will be larger and more H_2O gas molecules ionized to form hydroxyl radical where is the location of the plasma formed [6]. The gas sheath around the anode is sufficiently stable at higher surface area so that more water vapor are produced when plasma is formed. The large gas sheath allows the amount of excited electrons increase, so that the current increases. This condition triggers the increasing of energy consumption. At anode depth of 0.5 cm, the energy used for gas formation with smaller volume due to position of anode that was near from solution surface, so the plasma formed and stabilized more quickly due to the lower influence of hydrostatic pressure.

Table 1 Effect of Anode Depth on LAS Degradation and Its Energy Consumption in 120 Minutes Process at 600 V with 0.03 M KOH Electrolyte Concentration Using One Reactor

Anoda depth (cm)	Time (minute)	LAS Degradation (%)	Energy Consumption for LAS degradation (kJ/mmol LAS)
0.5	5	37.05	87,58
	60	50.58	662,85
	120	56.21	1260,10
1	5	52.30	268,20
	60	54.42	1058,40
	120	57.70	2030,40
1.5	5	65.56	473,40
	60	71.91	1404,00
	120	74.17	1944,00

IOP Conf. Series: Materials Science and Engineering **316** (2018) 012008 doi:10.1088/1757-899X/316/1/012008

The stability of plasma can resist electron effectively. Based on these phenomena, the depth of 0.5 cm produce lower energy consumption rate per mmol degradated LAS than depth of 1 cm and 1.5 cm. However, the increasing of energy consumption is not proportional with the increasing of the LAS degradation. Figure 2 shows the increasing anode depth causes the plasma size and brightness increase.

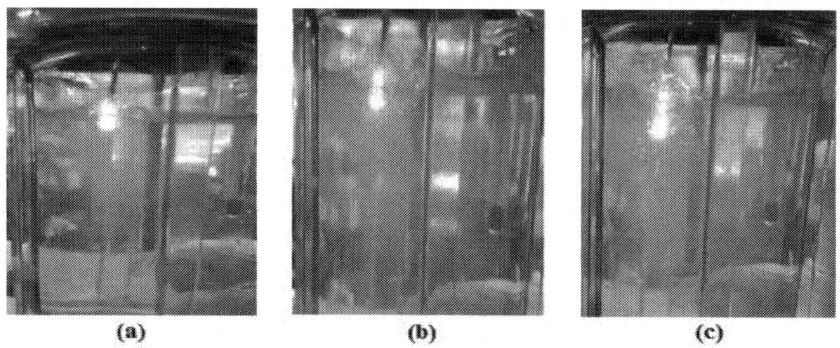

Figure 2 Visualization of Plasma at 600 V With 0.03 M KOH Electrolyte Concentration due to The Effect Anode Depth (a) 0.5 cm, (b) 1 cm, and (c) 1.5 cm.

LAS oxidation reaction with OH radicals is a complex reaction mechanism and produce the intermediates. The mechanism of LAS oxidation reaction mechanism consists of five stages [10]. Petterson proposed a simplified reaction mechanism, shown in Figure 3.

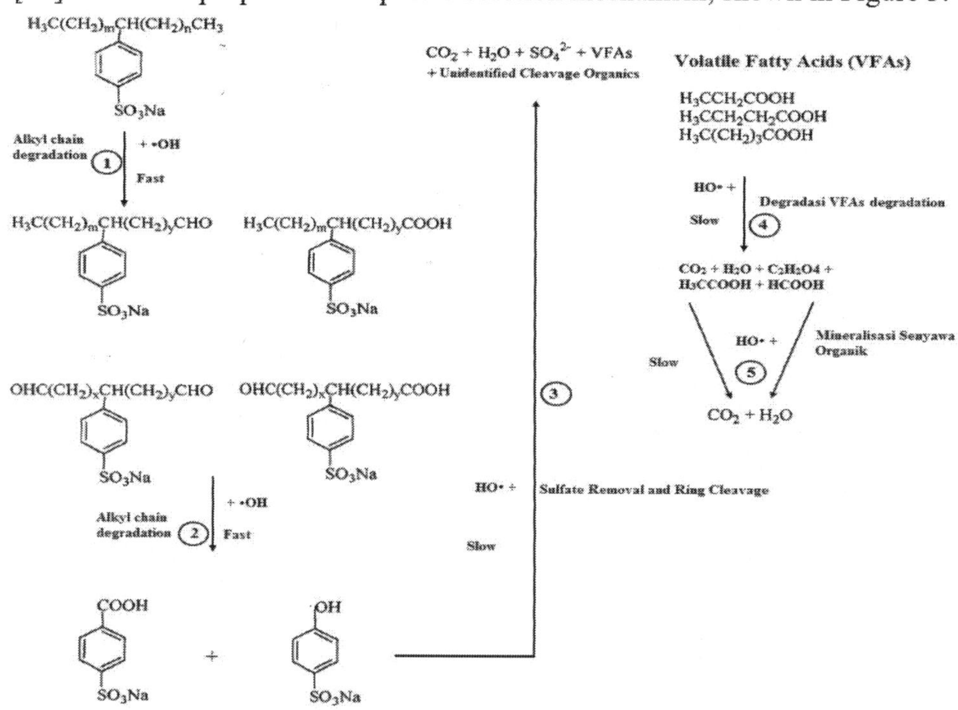

Figure 3 The mechanism of LAS degradation [10].

The first stages of the degradation produce alkyl chain sulfonated aromatic compounds such as sulfophenyl aldehyde (SPA). sulfophenyl carboxylic acid (SPC). sulfophenyl dialdehyde (SPDA). and carboxyl sulfophenyl aldehyde (CSPA). These compounds will be oxidized further in the second stage to produce aromatic compounds that have shorter alkyl chains such as carboxylic acid and sulfophenyl derivatives. These compounds will be oxidized further in the third stage to produce volatile fatty acids. In the fourth stage degradation of volatile fatty acids into carboxylic acid, oxalic acid and formic acid. The fifth stage is the mineralization of organic compounds that produce CO_2 and H_2O.

Effects of Multi-Reactor Cascade on The LAS Degradation

Table 2 shows the effects of multi-reactor cascade on LAS degradation and energy consumption. The result shows that LAS degradation increased with the number of reactor. After 120 minutes process, LAS degradation were 56.21%, 71.34% and 81.91%. This occurred because the increasing number of reactors will increase the number of points to form a plasma so that the interaction between •OH with increasing LAS [9]. Table 2 too shows that the energy consumption per mmol LAS lowest in five minutes. It occurs because the amount of LAS degraded very much in 5 minutes early ie 37.05% (one reactors), 48.43% (two reactors) and 66.65% (three reactor). Increasing the number of reactors will increase the amount of energy consumed per mmol LAS based on Table 2. During the 120 minute process, the amount of energy consumption three reactors higher than two reactors and one reactors because the increasing number of plasma formation point, so that joule heating energy required to form a sheath of gas is increasing. Based on Table 2, multi-reactor cascade effectively accelerated the LAS degradation process.

Table 2 Effects of Multi-Reactor Cascade on LAS Degradation and Its Energy Consumption in 120 Minutes Process at 600 V

Number of reactor	Time (minute)	LAS Degradation (%)	Energy consumption for LAS degradation (kJ/mmol LAS)
1	5	37.05	134.92
	60	50.58	757.90
	120	56.21	1279.37
2	5	29.19	313.19
	60	48.43	1476.30
	120	71.34	1555.02
3	5	66.65	499.31
	60	71.38	1588.79
	120	81.91	2227.34

3.3. Analysis of Chemical Oxygen Demand

COD are important indicators of waste degradation such as LAS. The COD changes during LAS degradation shown in Figure 4. COD degradation has increased achieve 33.88% during the 120 minutes process. But the percentage of COD degradation is not proportional to amount of LAS are degraded achieve 81.91% during 120 minutes process. This is caused by the intermediate products that are detected as COD and hydrogen peroxide formed in plasma electrolysis process that interfere the COD measurement shown in equation (1) [11] :

$$3H_2O_2 + Cr_2O_7^{2-} + 8H^+ \longrightarrow 3O_2 + 2Cr^{3+} + 7H_2O \quad (1)$$

IOP Conf. Series: Materials Science and Engineering **316** (2018) 012008 doi:10.1088/1757-899X/316/1/012008

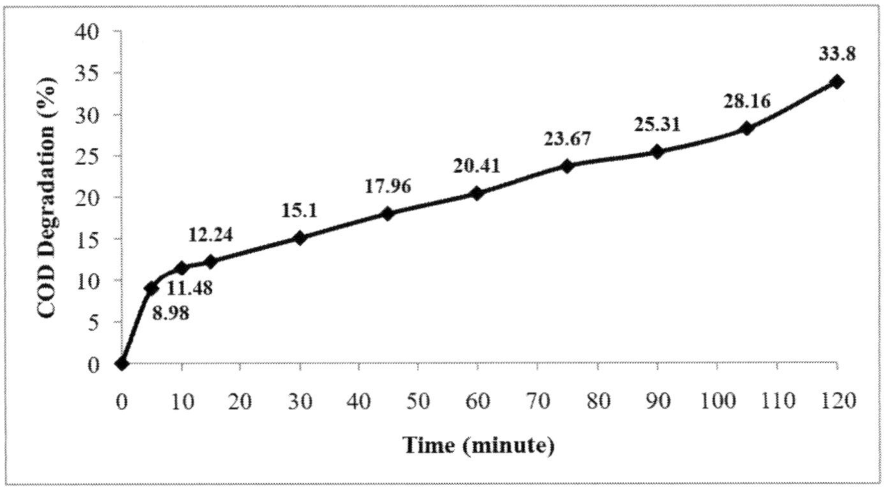

Figure 4 COD Degradation in 120 Minutes Process at 600 V with 0.03 M KOH with 0.5 cm Anode Depth Using Three Reactors.

Analysis of Oxalic Acid

Oxalic acid is one of the intermediate products of the most widely produced and the most stable of the LAS degradation process by •OH [12]. This occurred because the the reaction rate constants of oxalic acid with •OH smallest compared to other intermediate products which is $4.7 \times 10^7 M^{-1}S^{-1}$ [13].

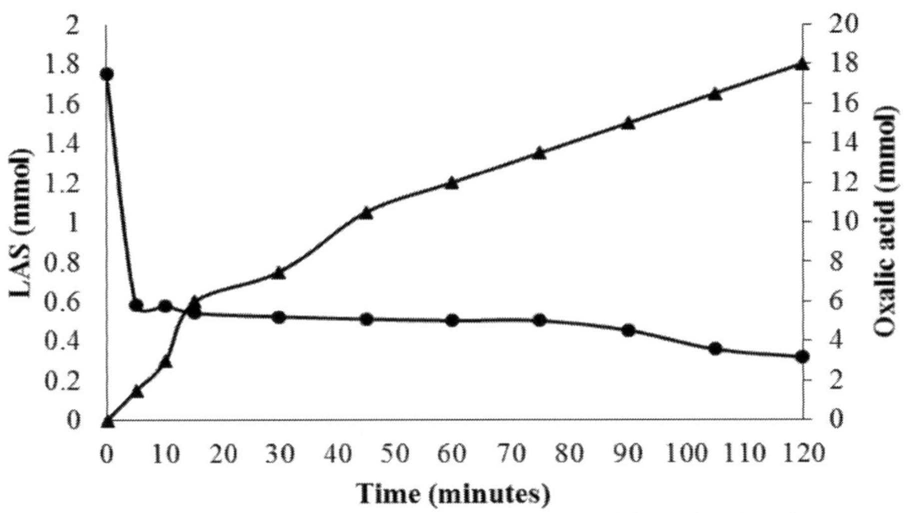

Figure 5 Comparison of LAS Degradation and Oxalic Acid Production for 0.03 M KOH Electrolyte with 0.5 cm Anode Depth Using Three Reactors.

Figure 5 shows that the LAS concentration decreased while the oxalic acid production increased along with increasing of operational time up to 120 minutes. It is caused by the decreasing of amount of LAS in the plasma electrolysis process and the interaction between hydroxyl radical with the organic compounds towards further degradation until produce final product, i.e. H_2O and CO_2.

Conclusions

This research has successfully degraded LAS using plasma electrolysis. The results show that the LAS degradation can be improved multi-reactor cascade plasma electrolysis. The greatest LAS degradation is achieved up to 81.91% with energy consumption of 2227.34 kJ/mmol of the energy consumption that is obtained during 120 minutes by using 600 Volt, 0.03 M of KOH, and 0.5 cm of the anode depth and initial concentration of LAS is 100 ppm.

Acknowledgement

This research was partially funded by "Hibah Publikasi Internasional Terindeks Tugas Akhir Mahasiswa Universitas Indonesia 2017" from Directorate Research and Community Service Universitas Indonesia. The authors declare no competing interests or any conflicts of financial interests.

This article's publication is supported by the United States Agency for International Development (USAID) through the Sustainable Higher Education Research Alliance (SHERA) Program for Universitas Indonesia's Scientific Modeling, Application, Research and Training for City-centered Innovation and Technology (SMART CITY) Project, Grant #AID-497-A-1600004, Sub Grant #IIE-00000078-UI-1.

References

[1] Okada, D, Y., Delforno, T, P., Esteves, A, S., Sakamoto, I, K., Duarte. I. C. S.. & Varesche. M. B. A. (2013). Bioresource Technology Optimization of linear alkylbenzene sulfonate (LAS) degradation in UASB reactors by varying bioavailability of LAS. hydraulic retention time and specific organic load rate. *Bioresource Technology* **128**. 125–133.

[2] Okada, D, Y., Delforno, T, P., Etchebehere, C,. & Varesche, M, B, A. (2014). International Biodeterioration & Biodegradation Evaluation of the microbial community of up flow anaerobic sludge blanket reactors used for the removal and degradation of linear alkylbenzene sulfonat. e by pyrosequencing. *International Biodeterioration & Biodegradation* **96**. 63–70.

[3] Mehrvar, M., Tabrizi, G. B., & Abdel-Jabbar, N. (2005). Effects of pilot-plant photochemical pre-treatment (UV/H_2O_2) on the biodegradability of aqueous linear alkylbenzene sulfonate (LAS). *International Journal of Photoenergy*, 7(4), 169–174.

[4] Jin, X, L., Wang, X, Y., Zhang, H, M., Xia, Q., Wei, D., Bin, & Yue, J. (2010). Influence of solution conductivity on contact glow discharge electrolysis. *Plasma Chemistry and Plasma Processing* 30. 429–436.

[5] Liu, Y., & Jiang, X. Plasma-induced degradation of chlorobenzene in aqueous solution. Plasma Chemistry and Plasma Processing 28. 15–24 (2008).

[6] Jiang, B., Zheng, J., Qiu, S., Wu, M., Zhang, Q., Yan, Z., & Xue, Q. (2014). Review on electrical discharge plasma technology for wastewater remediation. *Chemical Engineering Journal* 236. 348–368.

[7] Gao, J., Yu, J., Lu, Q., He, X., Yang, W., Li. Y., & Yang, Z. Decoloration of alizarin red S in aqueous solution by glow discharge electrolysis. Dyes and Pigments 76. 47–52 (2008).

[8] Yang, H., An, B., Wang, S., Li, L., Jin, W., & Li, L. (2013). Destruction of 4-phenolsulfonic acid in water by anodic contact glow discharge electrolysis. *Journal of Environmental Sciences* **25**. 1063–1070

[9] Saksono, Nelson., Nugraha, I., Ibrahim., Febiyanti, Irine. (2016). Hydroxyl Radical Production on Contact Glow Discharge Electrolysis for Degradation of Linear

Alkylbenzene Sulfonate. *Environmental Progress & Sustainable Energy*, 35(4), pp. 962-968

[10] Patterson, D, A., Metcalfe, I, S., Xiong, F., & Livingston, A, G. (2001). Wet Air Oxidation of Linear Alkylbenzene Sulfonate 1. Effect of Temperature and Pressure. Industrial & Engineering Chemistry Research 40. 5507–5516.

[11] Jin, X., Zhang, H., Wang, X., & Zhou, M. (2012), An improved multi-anode reactor for dye discoloration. Electrochimica Acta, 59, 474-478.

[12] Amano, R., & Tezuka, M. Mineralization of alkylbenzenesulfonates in water by means of contact glow discharge electrolysis. *Water Research*, 40, 1857–1863 (2006).

[13] Rivas, F, J., & Beltra, F, J. (2005). Iron type catalysts for the ozonation of oxalic acid in water. *Water Research*, 39, 3553–3564.

Carbon nanotubes shynthesis in fluidized bed reactor equipped with a cyclone

P Setyopratomo, M Sudibandriyo* and P P D K Wulan

Department of Chemical Engineering, Faculty of Engineering, Universitas Indonesia, Kampus Baru UI Depok, Depok 16424, Indonesia

*corresponding author's e-mail address: msudib@che.ui.ac.id

Abstract. This work aimed to observe the performance of a fluidized bed reactor which was equipped with a cyclone in the synthesis of carbon nanotubes (CNT) by chemical vapor deposition. Liquefied petroleum gas with a constant volumetric flow rate of 1940 cm^3/minutes was fed to the reactor as a carbon source, while a combination of metal components of Fe-Co-Mo supported on MgO was used as catalyst. The CNT synthesis was carried out at a reaction temperature which was maintained at around 800 – 850 °C for 1 hour. The CNT yield was decreased sharply when the catalyst feed was increased. The carbon efficiency is directly proportional to the mass of catalyst fed. It was found from the experiment that the mass of as-grown CNT increased in proportion to the increase of the catalyst mass fed. A sharp increase of the mass percentage of carbon nanotubes entrainment happened when the catalyst feed was raised from 3 to 7 grams. Agglomerates of carbon nanotubes have been formed. The agglomerates composed of mutually entangled carbon nanotubes which have an outer diameter range 8 – 14 nm and an inner diameter range 4 – 10 nm, which confirmed that the multi-walled carbon nanotubes were formed in this synthesis. It was found that the mesopores dominate the pore structure of the CNT product and contribute more than 90 % of the total pore volume.

1. Introduction

In the development of nanotechnology, carbon nanotubes (CNT) lately have become a particular concern of researchers and industrial communities. CNT have led to the development of intensive research in the field of science and technology due to their superior physical and chemical properties such as superior mechanical strength, excellent electron conductivity and other superior surface properties [1,2]. Based on the way how the energy source is introduced, three methods of synthesizing carbon nanotubes which have been widely developed are arc discharge, laser ablation and chemical vapor deposition (CVD) [3].

Among the three methods of synthesis, the CVD method is the most easily adapted for large-scale production in industrial applications since it requires lower reaction temperatures to produce at a low cost and provide high carbon nanotubes yield [4,5]. In the synthesis of carbon nanotubes by CVD method, the growth mechanism involves the decomposition of carbon source compounds into carbon atom [6].

In the presence of a catalyst, such carbon atom then deposited to form a structure of carbon nanotubes. The structure of the carbon nanotubes includes the wall number, diameter, length, orientation, and alignment, can be controlled by setting the reaction parameters during the growth of the carbon nanotubes [7]. Various carbon sources such as CH_4, CO, H_2/CO, C_2H_2, C_2H_4, and C_6H_6 have been widely used for the synthesis of carbon nanotubes by CVD method [8,9].

To date, fluidized bed reactor is considered as the most superior reactor type for large-scale production of carbon nanotubes due to the availability of sufficient space to stimulate the rapid growth of the carbon nanotubes, superior in facilitating excellent heat and mass transfer, easy to scale-up and suitable for continuous production [10,11]. Moreover, fluidized bed reactor provides a high space velocity which would trigger the achievement of a high carbon nanotubes yield [12]. Lately, the development of the synthesis of carbon nanotubes by CVD method using a fluidized bed reactor has focused on mass production [13]. Therefore, the capacity of a fluidized bed reactor during the synthesis of carbon nanotubes becomes a very important parameter.

In this work, we present the performance of a fluidized bed reactor which is equipped with a cyclone in synthesizing carbon nanotubes by CVD method. In this study it has been shown that with the use of cyclone the reactor is capable of synthesizing CNT with a high enough capacity which reach 23.39 g as-grown CNT per batch. The main observed parameters is the carbon nanotubes yield and the reactor capacity. The reactor capacity is represented by the mass of as-grown carbon nanotubes which are produced in a single batch operation. The carbon efficiency and some physical characteristic include the orientation, the morphology and the bulk density of the carbon nanotubes product are also investigated.

2. Experimental

2.1. Catalyst preparation

The combination of the transition metals Fe, Co and Mo which were supported on MgO was used as a catalyst in this work. Fe and Co act as the active components, while the role of the Mo component is as a promoter and to prevent rapid deactivation of the catalyst [14-17]. The catalyst was prepared using a wet impregnation method [8]. Ferric (III) nitrate nonahydrate ($Fe(NO_3)_3.9H_2O$), cobalt (II) nitrate ($Co(NO_3)_2.6H_2O$), ammonium heptamolybdate tetrahydrate ($(NH_4)6Mo_7O_{24}.4H_2O$) and magnesium oxide (MgO) were used as metal precursors. The catalyst was prepared to achieve a weight percent composition of Fe:Co:Mo:MgO (4:4:2:90).

2.2. The reactor

The fluidized bed reactor arrangement was schematically presented in figure 1. The reactor was made of a quartz tube 95 cm in height, 2.60 cm inner diameter and 0.20 cm thick. A tubular electric furnace which capable to supply 1.2 kW heat covered the reactor. The furnace was equipped with a temperature controller and able to raise the temperature of the furnace up to 900 °C. A K-type (chromel-alumel) thermocouple was embedded between the inner surface of the furnace and the outer surface of the reactor tube and serves to detect the furnace temperature.

A cyclone, which was made of stainless steel SS 304 having a diameter of about 7.8 cm and volume of about 1000 cm^3, was mounted on the top of the reactor. It is connected to the reactor exit gas line. The cyclone serves to separate the fine carbon nanotubes particles which were carried by the flowing gas out of the reactor. A thin plate, which is made of quartz porous material with a pore size of 400 mesh, was fitted at a distance 54 cm above the bottom of the reactor tube. It serves as a gas distributor and to retain the catalyst/product carbon nanotubes. The area under the quartz plate is a preheating zone, while the area above the quartz plate is a reaction zone. A certain amount of quartz wool was placed on a quartz plate that serves to avoid blockage of the quartz plate pores due to the growth of the carbon nanotubes.

2.3 Experimental procedure

The experiments were performed in a constant volumetric rate of feed gas mixture. The total feed gas flow rate was 1940 cm^3/minutes which consist of 77%vol argon, 13 %vol liquefied petroleum gas (LPG) and 10 %vol hydrogen. The furnace temperature was maintained at 900 ° C. While the amount of calcined catalyst fed was varied at 3, 5, 7 and 9 grams.

After reduced catalyst was fed to the reactor, reduction stage was started by flowing the hydrogen at a flow rate of 500 cm^3/minutes. The catalyst reduction process was held for 5 hours at furnace temperature 750 °C. Then the furnace temperature was raised to 900 °C. After the furnace temperature reach 900 °C, the synthesis of carbon nanotubes was started by introducing the feed gas mixture to the reactor while maintaining the furnace temperature at 900 °C. The synthesis reaction is terminated by stopping the flow of LPG gas. The furnace was then switched off and the reactor was cooled by flowing argon gas. As grown carbon nanotubes were removed from the reactor as well as from the cyclone and were stored for analysis.

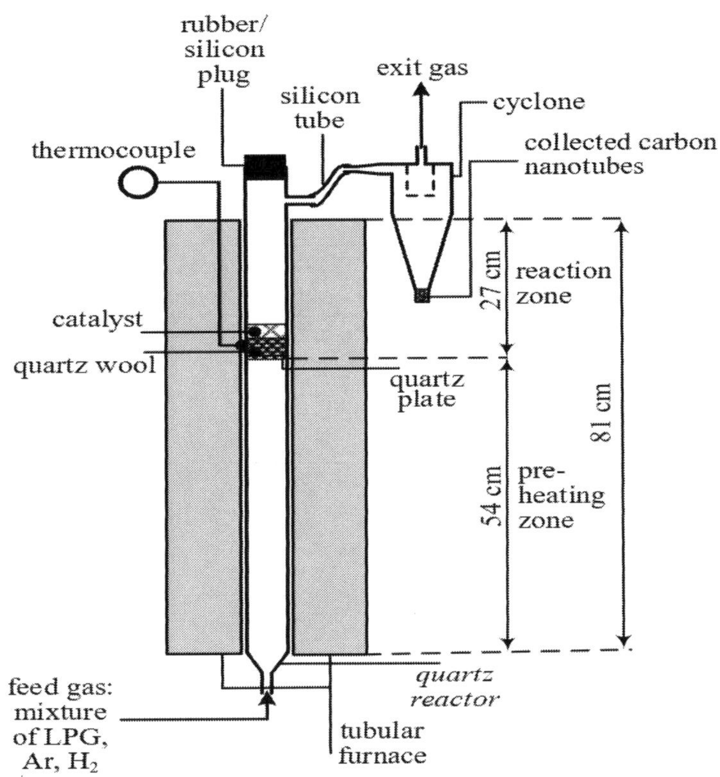

Figure 1. The schematic of fluidized bed reactor arrangement

2.4 Characterization techniques
The orientation of the CNT product was analyzed using scanning electron microscope - JSM-6510A/JSM-6510LA (Analytical/Analytical low vacuum SEM). While transmission electron microscope, TEM JEOL JEM 1400, were used to observe its diameter and morphology. The BET surface area and pore characteristic of the CNT were analyzed using Surface Area Analyser, Quanta chrome Nova 2000 Series - NovaWin - Instruments version 11.03.

3. Results and discussions

3.1 The CNT yield and the carbon efficiency
In this work, the CNT yield is defined as the mass of carbon nanotubes produced per unit mass of catalyst fed. The profile of the CNT yield at various of catalyst mass fed was presented in figure 2. It

can be seen that a sharp decrease in the CNT yield occurs when the catalyst feed was increased, that is from 3.61 g CNT/g catalyst at 3 grams catalyst feeding to 1.88 g CNT/g catalyst at 7 grams catalyst feeding. The availability of space for the growth of carbon nanotubes become a significant contributing factor in increasing the growth rate of the carbon nanotubes. The more the amount of catalyst fed will result in the less ratio of the available space for the carbon nanotubes growth to the mass of the catalyst. Thus it can be understood as the amount of catalyst feed is raised, then the CNT yield decreased. It should be noted that decreasing the CNT yield is associated with lowering the quality of CNT in regard to increasing the level of impurities content derived from the residual catalyst. Further addition of catalyst feed to 9 grams resulted in a slighter decrease in CNT yield to 1.60 g CNT/g catalyst. At high catalyst feeding ratio of the available space for the carbon nanotubes growth to the mass of the catalyst will be very low resulted in denser growing carbon nanotubes particles. This dense particle will create more intense collisions between particles and inhibit the occurrence of particles entrainment. Consequently, there will be more particles of carbon nanotubes have enough time to grow then avoid a sharp drop in CNT yield.

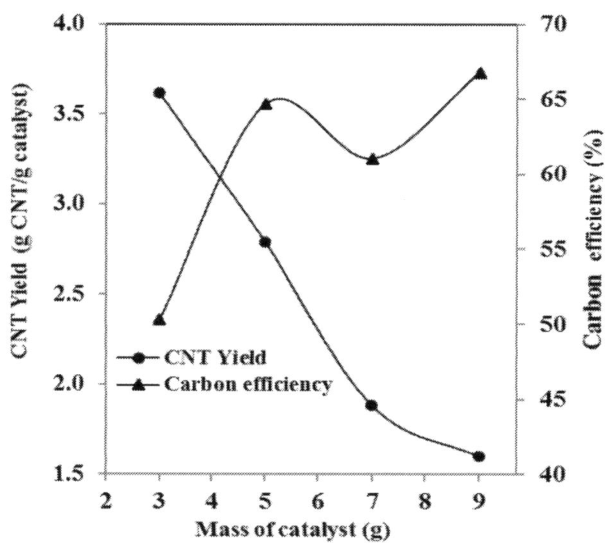

Figure 2. The CNT yield and carbon efficiency in various catalyst feed

The carbon nanotubes yield resulting from this work is comparable to the yields obtained by Zhang et al. [18] which is 3.8 g CNT/g catalyst and relatively greater than the yield achieved by some other researchers, i.e. Baddour et al. [19]: 0.96 g CNT/g catalyst, Huang et al. [20]: 1.48 g CNT/g catalyst, Danafar et al. [21]: 0.6-0.8 g CNT/g catalyst, Zhao, et al. [22]: 0.22 g CNT/g catalyst, and Hsieh et al. [23]: 0.7-3.0 g CNT/g catalyst. Figure 3 also expose the carbon efficiency in accordance with the amount of catalyst fed. The carbon efficiency is defined as the percentage of the amount of carbon available in the feed gas which is converted into carbon nanotubes. The liquefied petroleum gas which was used in this work has a chemical composition as follow: 1.7 %mol C_2H_6, 71.7 %mol C_3H_8, 11.4 %mol i-C_4H_{10} and 15.2 %mol C_4H_{10}. From these data, the available carbon which was supplied to the reactor will be 23.51 g carbon/hour. Base on these amount of the available carbon and the amount of CNT produced at the end of synthesis, the carbon efficiency was calculated.

In general, the carbon efficiency is directly proportional to the mass of catalyst fed. This can be explained that the more the mass of catalyst in the reactor, the contact time between the gas feed and the catalyst will be longer so that the conversion of hydrocarbons into carbon nanotubes will be higher. In this work, the highest carbon efficiency was 66.8% which was achieved when the mass of catalyst feed was 9 grams. From an economic viewpoint, a higher carbon efficiency is preferred because it reflects more efficient carbon source utilization. The carbon efficiency resulting in this work are comparable to those obtained in fluidized bed reactor studied by some other researchers, among others Morancais et al. [24] obtained carbon efficiency 48 - 75 % using Fe/Al_2O_3 catalyst at 450 -750 °C, while Venegonia et al. [25] got carbon efficiency 15-50 % applying $Fe\text{-}SiO_2$ catalyst at 750 -1350 °C.

3.2 The reactor performance and the CNT entrainment

Reactor capacity indicates how much the ability of the reactor for producing carbon nanotubes in one batch operation. The capacity of the reactor is a very important parameter for commercial-scale production applications since it will determine the economic level of the reactor operation. In this work, reactor capacity was expressed as mass of as-grown carbon nanotubes that can be produced in a single batch synthesis process. In this case, a mass of as-grown carbon nanotubes is a mass of carbon nanotubes obtained from the reactor at the end of the synthesis reaction, it includes the mass of the catalyst. Mass of as-grown CNT produced at various of catalyst mass feeding was presented in figure 3. It was shown that mass of as-grown CNT increase in proportion to the increase of the catalyst mass fed. It also can be seen that the reactor is capable of producing as much 23.4 grams as-grown in one batch synthesis when catalyst feed is 9 grams. It should be noted that the capacity of the produced as-grown carbon nanotubes was also counted the as-grown nanotubes products which were carried by exit gas stream (called as carbon nanotubes entrainment) and were separated by the cyclone.

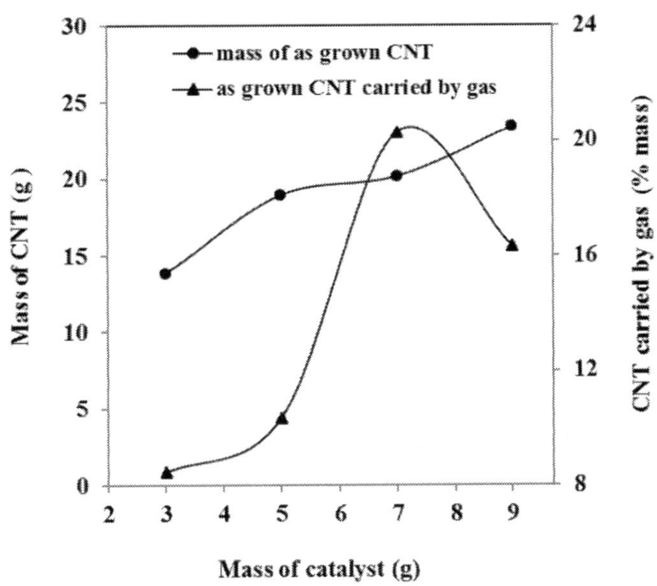

Figure 3. Mass of as-grown the CNT produced in various catalyst feed and the percentage which were carried by the flow of gas exit the reactor

Observations which was conducted during the experiment found that the entrainment of partly carbon nanotubes in the gas stream only occurs especially in a short time interval just after the synthesis reaction begins. It shows that the carbon nanotubes are in the early growth phase and are at the top section of the fluidized bed, such particles still have a small size so that it easily carried by the flow of gas.

The percentage of as-grown carbon nanotube entrainment which were separated in the cyclone was also shown in the figure. It can be seen in the figure that a sharp increase of the mass percentage of carbon nanotubes entrainment happened when the catalyst feed was raised from 3 to 7 grams. The lowest percentage of carbon nanotubes entrainment is 8.5 % mass when catalyst feed is 3 grams, and its highest value reaches 20.3 % mass which happens at 7 grams catalyst feeding. It indicates that the addition of an amount of the catalyst feed to a certain extent causes increasing the formation of carbon nanotubes with smaller particle size. The smaller the particle size of carbon nanotubes will be the greater the potential to be carried by the flow of gas. The experimental results also showed that further addition of catalyst feed from 7 to 9 grams decreased the percentage of carbon nanotubes entrainment. Adding a further amount of the catalyst feed will shorten the distance between the fluidized particles in the reactor chamber leading to the greater chance of collisions among the growing carbon nanotubes particles. The particles collision will become the barrier for the small particles of carbon nanotubes to be carried by the flow of gas out of the reactor.

3.3 The CNT orientation and morphology

The orientation of CNT product can be observed from SEM image with a magnification of 5,000 times which was presented in figure 4. From that figure, it can be observed that the agglomerates of carbon nanotubes have been formed in synthesis with the fluidized bed reactor. From the image can also be observed that each CNT agglomerate was composed of several smaller size sub-agglomerate. The orientation of carbon nanotubes is mutually entangled with each other to form a sub-agglomerate. Furthermore, during their growth carbon nanotubes in a single sub-agglomerate form a network with other sub-agglomerates then some sub-agglomerates are integrated into a bigger agglomerate.

It also can be seen that the size of an agglomerate was ranging from about 2 – 5 μm. The formation of a stable agglomerate size is certainly influenced by the interaction of the gas-solid during the occurrence of the fluidization phenomenon during the synthesis of carbon nanotubes. Clearer visualization of the orientation of the carbon nanotubes products was shown by SEM image with a magnification of 50,000 times which was presented in figure 5. From the image, it was seen more clearly that long carbon nanotubes were linked to one another to form a woven carbon nanotubes. The woven carbon nanotubes had lead to the formation of a stable agglomerate.

Figure 4. SEM image of the CNT at 5,000 times magnification produced with the catalyst feed 3g

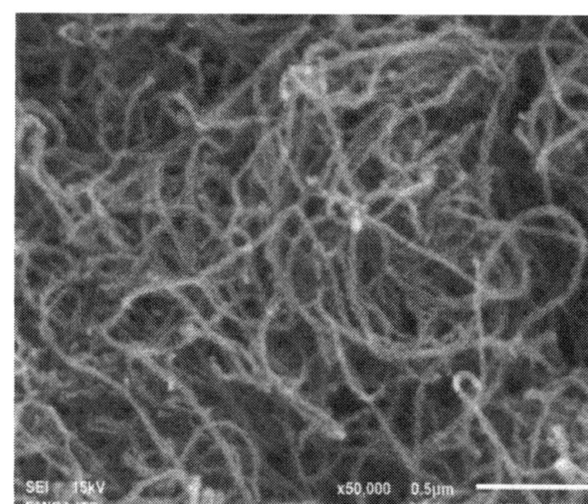

Figure 5. SEM image of the CNT product at 50,000 times magnification produced with the catalyst feed 3g

To investigate the morphology of the carbon nanotubes, TEM image capture was carried out and the results were displayed in figure 6 for synthesis with catalyst feeding 3 and 7 grams as representative

IOP Conf. Series: Materials Science and Engineering **316** (2018) 012009 doi:10.1088/1757-899X/316/1/012009

samples. By making manual measurements using a ruler on the basis on scaling shown in the images, it was found that the carbon nanotubes have an outer diameter range 8 – 14 nm and an inner diameter range 4 – 10 nm. It was confirmed that multi-walled carbon nanotubes have been formed in this synthesis. In the image was also found that a rest of catalyst particle stand at the end of carbon nanotubes. This indicated that the growth of carbon nanotubes follow the tip-growth mechanism.

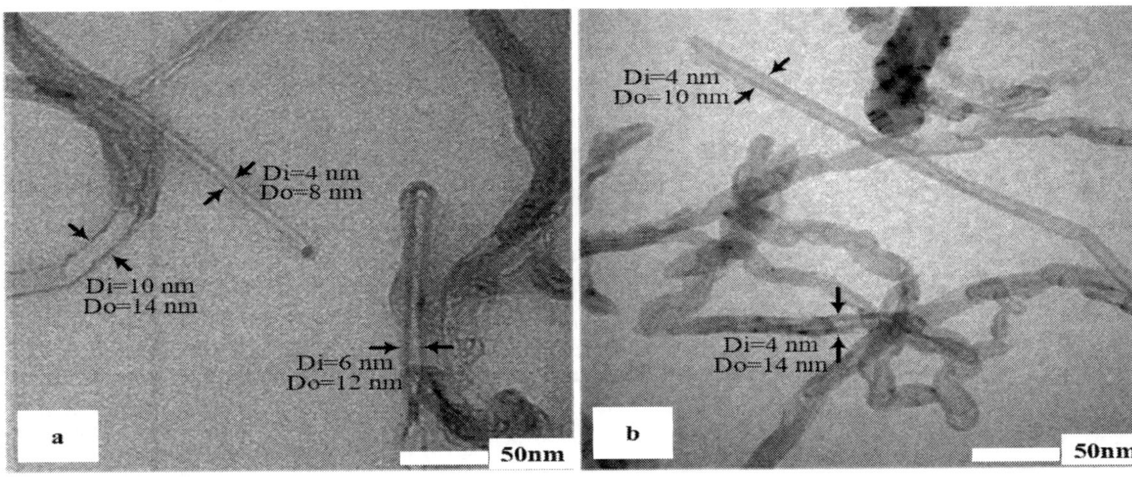

Figure 6. TEM image of the CNT produced with the catalyst feed: (a) 3g and (b) 7g

The morphology of the carbon nanotubes, resulting from the synthesis with catalyst feeding 3 and 7 grams, which were entrained by the flowing gas and collected in the cyclone were also investigated by taking its TEM image and the images were presented in figure 7. The diameter of the carbon nanotubes was measured using similar previous method applied in the determination of the diameter of carbon nanotubes which were collected from the reactor. The measurement results showed that carbon nanotubes which were carried by the gas stream have an outer diameter in the range 8 – 22 nm and inner diameters in the range 6 – 14 nm.

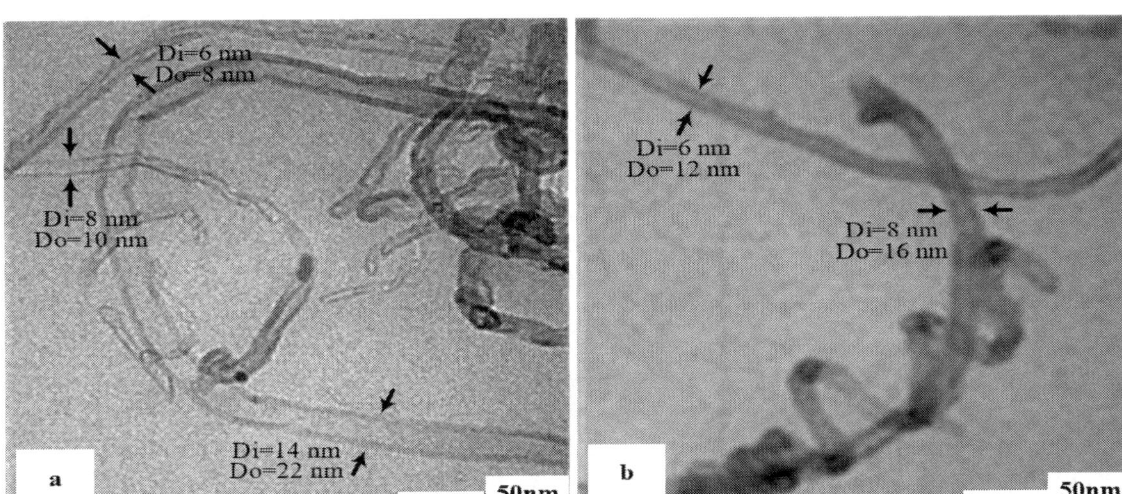

Figure 7. TEM images of the CNT carried by gas and separated in the cyclone with the catalyst feed: (a) 3g and (b) 7g

3.4 The surface area and pore volume of the CNT

Surface area and pore volume distribution were determined from the isotherm adsorption-desorption of N_2. The analysis was conducted to both CNT product which were collected in the reactor and which were carried by the gas stream and collected in the cyclone. The BET surface area was presented in figure 8, while the total pore volume was shown in figure 9. In the determination of the total pore volume, it was assumed that the pores were to be filled with liquid N_2. It was determined from the amount of nitrogen vapor adsorbed at a relative pressure $P/Po \approx 1$, where Po is the N_2 saturation pressure. From both figure can be seen that CNT which were collected in the reactor have BET surface area in the range 115 - 148 m^2/g and it was higher than BET surface area of the CNT carried by gas and separated in the cyclone which is in the range 71 - 79 m^2/g. The same thing happened to the total pore volume. The CNT which were collected in the reactor have a total pore volume in the range of 0.768 – 1.531 cm^3/g and it was higher than the total pore volume of the CNT carried by gas and separated in the cyclone which is in the range of 0.513 - 0.612 cm^3/g.

The micropores (pore size < 2nm) volume distribution of the CNT product collected in the reactor, which was determined using the Horvath-Kawazoe (HK) slit model, was expressed as cumulative micropores volume and presented in figure 10. It was shown that micropores with a pore size range 0.3 – 0.5 nm dominate the micro porosity. It can be seen from the figure that experiments which were conducted by varying the mass of the catalyst resulted in CNT with micropores volume in the range 0.0468 – 0.0598 cm^3/g. The Barret, Joyner and Halenda (BJH) method was used to determine the mesopores volume distribution of the CNT product collected in the reactor. Mesopores are pores with pore size range 2 – 50 nm. The result expressed as the cumulative BJH pore volume and presented in figure 11.

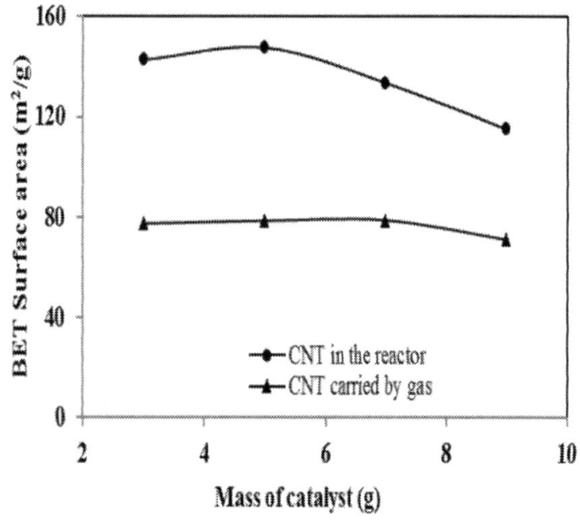

Figure 8. BET surface area of the CNT

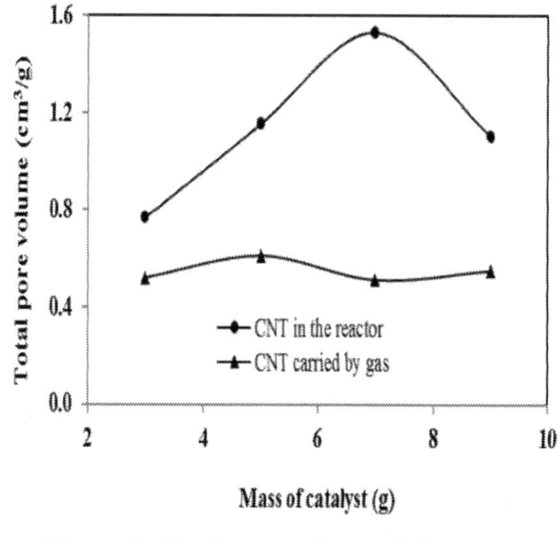

Figure 9. Total pore volume of the CNT

Even though the determination using BJH method include measurements on a limited range of macropores, but the cumulative pores volume curves tend to be flat in the macropores sizes range (pore size > 50 nm). It means that the determined pore size was dominated by the mesopores. At the level of the catalyst mass varied, the mesopores volume in the CNT product is in the range 0.7361 – 1.4933 cm^3/g. It means that the mesopores contribute more than 90 % of the total pore volume in the CNT product.

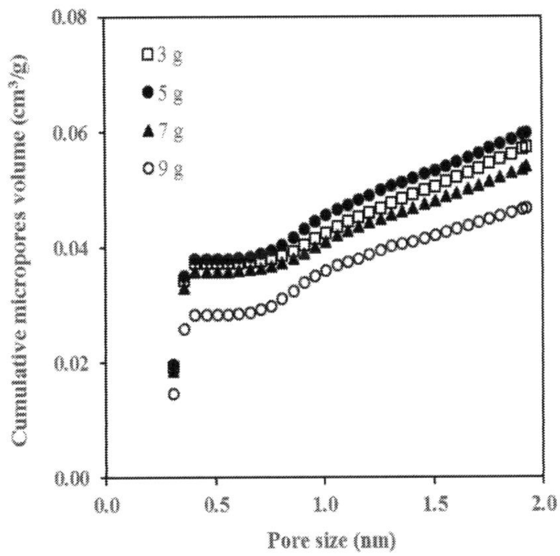

Figure 10. Cumulative micropore volume of the CNT collected in the reactor at all mass catalyst varied

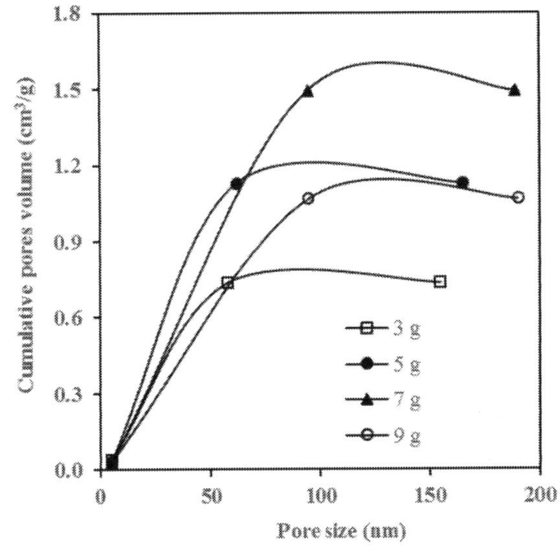

Figure 11. Cumulative BJH pore volume of the CNT collected in the reactor at all mass catalyst varied

4. Conclusions

The mass of as-grown CNT increase in proportion to the increase of the catalyst mass fed. A sharp increase of the mass percentage of carbon nanotubes entrainment happened when the catalyst feed was raised from 3 to 7 grams which indicate that the addition of an amount of the catalyst feed to a certain extent causes increasing the formation of carbon nanotubes with smaller particle size. The availability of space within the reactor for the growth of carbon nanotubes was considered become the factor that affecting the CNT yield. In general, the carbon efficiency is directly proportional to the mass of catalyst fed. Agglomerates of carbon nanotubes have been formed in synthesis using the fluidized bed reactor. The agglomerates composed of mutually entangled carbon nanotubes which have an outer diameter range 8 – 14 nm and an inner diameter range 4 – 10 nm, which confirmed that the multi-walled carbon nanotubes were formed in this synthesis. High mass catalyst feeding resulted in more dense fluidized catalyst particles and generate a more compact carbon nanotubes agglomerate. It was found that the mesopores dominate the pore structure of the CNT product.

Acknowledgement

The authors would like to thank those who have supported this work, especially The Directorate Research and Community Services, Universitas Indonesia, for providing the financial support through The Research Cluster Grant 2015.

References

[1] Chiangga S, Suttisiri N and Nilsaengrat P 2009 *Phys Procedia* **2** pp 107-11
[2] Tan L L, Ong W J, Chai S P and Mohamed A R 2013 *Catal. Today* **217** pp 1-12
[3] Wei F, Zhang Q. Qian W Z, Yu H, Wang Y, Luo L H, Xu G H and Wang D Z 2008 *Powder Technol.* **183** pp 10-20
[4] Popovska N, Danova K, Jipa I and Zenneck U 2011 *Powder Technol.* **207** pp 17-25
[5] Wang Y, Wei F, Luo G, Yu Y and Gu G 2002 *Chem. Phys. Lett.* **364** pp 568-72
[6] Son S Y, Lee Y, Lee D H, Kim S D, Sung S H, Hong S Y and Cho S J 2008 *J. Phys. Chem. Solids* **69** pp 1188-93

[7] Dasgupta D, Joshi J B and Banerjee S 2011 *Chem. Eng. J.* **171** pp 841-69
[8] Maccallini E, Tsoufis T, Policicchio P, Rosa S L, Caruso T, Chiarello C, Colavita E, Formoso V, Gournis D and Agostino R G 2010 *Carbon* **48** pp. 3434-45
[9] Shirazi Y, Tofighy M A, Mohammadi T and Pak A 2011 *Appl. Surf. Sci.* **257** pp 7359-67
[10] Tian G L, Huang J Q, Li J, Zhang Q and Wei F 2016 *Carbon* **108** pp 404-11
[11] Zhang Q, Zhao M Q, Huang J Q, Liu Y, Wang Y, Qian W Z and Wei F 2009 *Carbon* **47** pp. 2600-10
[12] Jeong S W, Kim J and Lee D H 2015 *Chem. Eng. Sci.* **134** pp 496-503
[13] Jeong S W, Lee J H, Kim J and Lee D H 2016 *J. Ind. Eng. Chem.* **35** pp 217-23
[14] Ago H, Uehara N, Yoshihara N, Tsuji M, Yumura M, Tomonaga N and Setoguchi T 2006 *Carbon* **44** pp 2912-18
[15] Dupuis A C 2005 *Prog. Mater Sci.* **50** pp 929-61
[16] Jeong S W, Son S Y and Lee D H 2010 *Adv. Powder Technol.* **21** pp. 93-99
[17] Pe´rez-Mendoza M, Valle´s C, Maser W K, Martı´nez M T, Langlois S, Sauvajol J L and Benito A M 2005 *Carbon* **43** pp 3002-39
[18] Zhang Q. Zhao M Q, Huang J Q and Wei F 2010 *Powder Technol.* **198** pp 285-91
[19] Baddour C E, Upham D C and Meunier J C 2010 *Carbon* **48** pp 2644-73
[20] Huang J Q, Zhao M Q, Zhang Q, Nie J Q, Yao L D, Su D S and Wei F 2012 *Catal. Today* **186** pp 83-92
[21] Danafar F, Fakhru'l-Razi A, Salleh M A M and Biak D R A 2011 *Chem. Eng. Res. Des.* **89** pp 214-23
[22] Zhao M Q, Zhang Q, Huang J Q and Wei F 2010 *J. Phys. Chem. Solids* **71** pp 624-26
[23] Hsieh C T, Lin Y T, Chen W Y and Wei J L 2009 *Powder Technol.* **192** pp 16-22
[24] Morancais A, Caussat B, Kihn Y, Kalck P, Plee D, Gaillard P, Bernard D and Serp P 2007 *Carbon* **45** pp 624-35
[25] Venegoni D, Serp P, Feurer R, Kihn Y, Vahlas C and Kalck P 2002 *Carbon* **40** pp 1799-1807

IOP Conf. Series: Materials Science and Engineering **316** (2018) 012010 doi:10.1088/1757-899X/316/1/012010

The effect of heating temperature on cytotoxicity and α-mangostin yield: Mangosteen pericarp juice and mangosteen extract

Kamarza Mulia, Fitria Hasanah and Elsa A Krisanti

Chemical Engineering Department, Faculty of Engineering
Universitas Indonesia, Depok, 16424, Indonesia

E-mail: kmulia@che.ui.ac.id

Abstract. The pericarp of mangosteen (*Garcinia mangostana* L.) contains bioactive xanthones, with α-mangostin being the major component, has been known to possess antitumor, antiviral, and other pharmacological activities. In this study, the effect of elevated temperature during the preparation step of fresh mangosteen pericarp juice and mangosteen extract, on their α-mangostin yield and cytotoxicities was investigated. The cytotoxicity activity of fresh juice and mangosteen extract was investigated using the brine shrimp test. Heating the fresh pericarp mangosteen in water at 65°C for 30 minutes prior to blending produced a juice with higher α-mangostin yield and cytotoxicity compared to the traditional way of blending the juice at room temperature. Increasing α-mangostin yield of 9%-w/w due to heating was also observed when mangosteen extract was heated at 65°C, consistent with the increased cytotoxicity in terms of LC_{50} value. It is concluded that the effect of temperature on α-mangostin yield was in line with the temperature effect on cytotoxicity activity in all samples of pericarp juice and mangosteen extract in ethyl acetate fraction.

1. Introduction

Mangosteen tree (*Garcinia mangostana Linn.*) is a tropical evergreen tree with green leaves and dark purple fruits. The pericarp of the mangosteen fruit is traditionally used for treatment of skin infection, diarrhea, abdominal pain, infected wounds, and chronic ulcer[1]. Previous studies had shown that the pericarp of the mangosteen fruit is rich in xanthone compounds. The first group of xanthones isolated was the mangostin group with α-mangostin as the most abundant bioactive compound present[2]. The extract of the pericarp of mangosteen has pharmacological activities including anti-inflammatory, cytotoxic, antioxidant, antitumoral, neuroprotective, immunomodulatory, anti-allergic, antibacterial, and antiviral properties[3,4]. Satong-aun et al. dried mangosteen pericarp at three temperatures and obtained the highest α-mangostin extraction yield of 40.3% (w/w) after drying at 65°C[5]. Suttirak and Manurakchinakorn reported that the amount of xanthones in the pericarp and their free radical scavenging activities markedly decreased after the drying step[6]. They recommended drying temperature of 75°C to inactivate degradative enzymes such as polyphenol oxidase and also minimize thermal degradation of xanthones.

Traditionally, juices of mangosteen pericarp are consumed to get the benefits of xanthones that reside within. Even though the efficacy of mangostins is widely known and the mangostin yield in the pericarp as a function of heating temperature has been reported, cytotoxicity data of these juices have been reported yet. It is anticipated that heating the pericarp of mangosteen will not only change the α-mangostin yield of the juices, but also their cytotoxicities. Also of interest is the effect of heating temperature on mangostin yield and cytotoxicity of the extracts obtained from the mangosteen pericarp since they contain a large amount of bioactive compounds, and therefore, have potential to be used for anti-cancer medication. Therefore, the objectives of this research are to determine: (1) the effect of preparation method and heating temperature on mangostin yield and cytotoxicity of the fresh mangosteen pericarp juice; (b) the effect of heating of the pericarp on mangostin yield and cytotoxicity of ethyl acetate fraction obtained from the ethanolic extract of mangosteen pericarp.

2. Material and methods

2.1. Materials

Mangosteen pericarp was obtained from Solo, Central Java, Indonesia, and was identified as *G. mangostana* by Herbarium Bogoriense, Research Center for Biotechnology-Indonesian Institute of Sciences (LIPI). Sodium tripolyphosphate (food grade) was obtained from Brataco Chemical, Indonesia, calcium chloride and sodium alginates were obtained from Merck. The standard compound α-mangostin (98%) was obtained from Aktin Chemicals, China. Chitosan (medical grade; deacetilation degree of 93.6%; viscosity of 23.3 cp) was obtained from Biotech Surindo, Indonesia. α-amylase enzyme was purchased from Claricem Indonesia and β-glucosidase enzyme (from almonds; lyophilized powder; ≥6 U/mg) was purchased from Sigma-Aldrich. The brine shrimp eggs (*Artemia Salina* Linn.) were purchased from Red Top.

2.2. Preparation of samples with mangostin

There were two types of samples that were assayed to determine their mangostin yield and cytotoxicities: (a) fresh mangosteen pericarp juice prepared using five methods; (b) extract of mangosteen pericarp obtained from the ethanol extraction followed by fractionation using ethyl acetate.

2.2.1. Preparation of mangostin pericarp juice.

Samples of the fresh mangosteen pericarp juice were obtained using five preparation methods listed in Table 1. The thick and soft part of mangosteen rind was separated from its hard skin and then blended or boiled/heated in various ways. All preparation methods used a same ratio of pericarp pulp to water of 1:6 (w/w), i.e. 50 gram pulp in 300 ml water.

2.2.2. Preparation of mangosteen extract in ethyl acetate fraction.

The extraction of dry mangosteen pericarp was carried out using a modified procedure reported by Jung et al.[7]. The maceration was performed using ethanol 96% for 7 days with mangosteen powder to ethanol ratio of 1:3 (w/v) and periodic stirring of the mixture. The macerated mixture were filtrated and evaporated under reduced pressure. The crude ethanolic extract was then fractionated using a mixture of water and ethyl acetate in 1:1 volume ratio. The evaporation of the ethyl acetate fraction under reduced pressure produced mangostin powder/paste.

2.3. Cytotoxicity assay using brine shrimp test (BST)

The cytotoxicity assay was performed using the brine shrimp lethality test (BST) reported by Meyer et al.[8]. In this method, *in-vitro* lethality assay of *A. Salina* was used to detect the cytotoxicity of mangosteen extract in each sample[9]. Brine shrimp eggs were placed in seawater and aerated at room temperature (24-28 ℃) for 72 h, giving a large number of larvae. Total larvae alive were counted after a sample is exposed to a lamp for 24 h and the lethal concentration (LC_{50}) was calculated.

2.4. Analysis of α-mangostin

The amount of α-mangostin in mangosteen pericarp juice and in the ethyl acetate fraction was determined quantitatively using UV spectrophotometry analysis (Spectroquant® Pharo 300, Merck). The juice was filtered prior to sampling at room temperature. The standard calibration curve was made based on the absorbance of α-mangostin standard solution (4-20 mg/L) obtained at the wavelength of 316 nm[10].

3. Result and discussion

The samples consisted of the mangosteen pericarp juice and ethyl acetate fraction of the mangosteen extract was prepared, and, the effect of the temperature imposed on the α-mangostin yield and the cytotoxicityies of these samples were determined.

3.1. α-Mangostin yield of the fresh mangosteen pericarp juice

Table 1 shows the α-mangostin concentration in samples of fresh mangosteen pericarp juice obtained using different treatment methods. The results show that the juice preparation method affects the extraction yield of active compounds from the pericarp pulp. The traditional way to obtain the juice of mangosteen pericarp is by mixing with water in a blender. Other traditional techniques are by drying the pericarp and then boil it in water, or heat the pericarp at 65°C and mix it with water in a blender. It was found that the lowest concentration of α-mangostin in juice was obtained using the traditional way, i.e. mixing the soft part of the pericarp with water in the blender (treatment 1). In contrast, heating up the pericarp at 65°C for 30 min and followed by blending with water (treatment 3), produced a solution with the highest α-mangostin yield of 179 ppm, significantly higher than the yields obtained using other treatment. It can be explained that even though boiling pericarp mixture at 100°C for 30 min (treatment 2, 4, and 5) might degrade α-mangostin in the plant matrix, heating up at 65°C might not degrade it significantly.

Heating temperature of 65°C might only delay the action of the polyphenol oxidase enzyme to degrade the phenolic compounds present in the pericarp of mangosteen[11]. This result, then indicates that heating the pericarp up to 65°C might enhance the solubility of α-mangostin in water, increase the diffusion of water in the mixture into the plant matrix, and at the same time delay the degradation action of the polyphenol oxidase enzyme. Al-Massarani et al. reported that during heating, plant extracts will degrade due to exposure to high temperature as well as to an enzyme called polyphenol oxidase that degrade the phenolic compounds in the extract[1]. However, the results of this study showed differently.

Table 1. α-Mangostin concentration in fresh mangosteen pericarp juice and the corresponding %-lethality of *A. salina*.

Treatment code	Treatment of the pericarp	α-mangostin yield (ppm)	α-mangostin yield (%-w/w)	*A. salina* lethality (%)
1	Blended (with water)	56	0.03	22
2	Blended and then boiled	88	0.05	56
3	Heated and then blended	179	0.11	86
4	Boiled only	66	0.04	42
5	Dried and then boiled	106	0.06	55

Note: Juice blend is 50 g pericarp in 300 ml water; boiling in water at 100°C for 30 min; heating in water at 65°C for 30 min.

3.2. Cytotoxicity of mangosteen pericarp juice

Figure 1 shows the relationship between the concentration of α-mangostin in various juices and the %-lethality of *A. salina* that represent the cytotoxicity of the sample in BST assay.

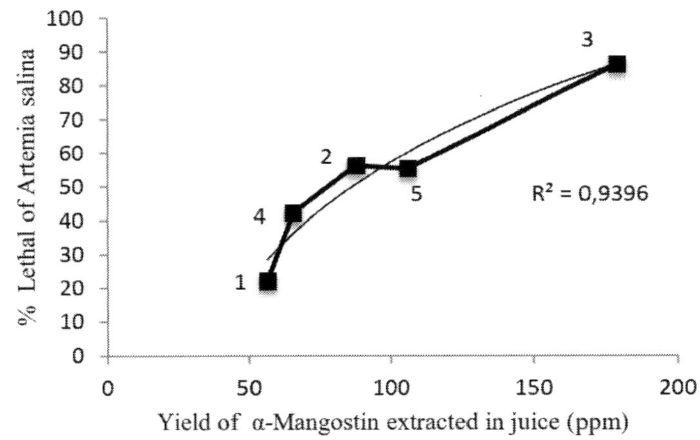

Figure 1. The lethal percentage of *A. salina* from fresh juice samples with various concentrations of α-mangostin.

The BST results of the fresh mangosteen pericarp juice given in Table 1 show that sample receiving treatment 3 is the most cytotoxic with 86% lethality of A. salina, where the pulp of pericarp was first heated in water and then mixed in a blender. Juice made by only mixing the pulp with water in a juice blender (treatment 1) had the lowest cytotoxicity of 22% lethality. The positive correlation between concentration of α-mangostin and the cytotoxicity is shown in Fig 1, with the R-square around 0.94. This means that α-mangostin yield in juice significantly affect the cytoxicity activity of pericarp extract juice. Sak reported the higher the concentration of phenolic compounds, such as α-mangostin, the higher the cytotoxicity[12].

3.3. Yield of α-mangostin in ethyl acetate fraction of mangosteen extract

The effect of heating temperature on the α-mangostin concentration in the extract samples is given in Table 2. Since the mangosteen extract was investigated in relation to its application for cancer medication via oral delivery system, the temperatures used were standard room temperature of 25°C, the body temperature of 37°C, and the heating temperature of 65°C to avoid degradation of α-mangostin[5]. The data show increased α-mangostin yields with increasing heating temperature of the mangosteen extract. The α-mangostin concentration in mangosteen extract heated at 65°C was around 50% (w/w), higher by 9% (w/w) compared to the concentration in the non-heated sample. The effect of temperature on the yield of α-mangostin in the mangosteen extract samples was similar with the result of that on pericarp juices as discussed previously.

Table 2. α-Mangostin yield and cytotoxicity assays of the mangosteen extract.

Temperature (°C)	α-mangostin yield (%-w/w)	*A. Salina* %-lethality* (%)	LC_{50} of extract (ppm)
25	40.8	57.0	0.71
37	45.1	74.5	0.51
65	49.8	97.1	< 0.1

* exposed to 1 ppm α -mangostin

The data obtained suggest that increased temperature of samples up to 65°C do not degrade the α-mangostin, which contradict with the result reported by Al-Massarani et al.[1]. It is known that the pericarp of mangosteen contained xanthones with structural molecules that are close to each other, such as α-mangostin, γ-mangostin and Gartinin[10]. The hypothesis explaining this phenomenon is that the increased temperature of mangosteen extract might cause the shifting of functional group in other types of mangostin compounds onto the structure of α-mangostin, hence, the increased yield of α-mangostin.

3.4. Cytotoxicity of ethyl acetate fraction of mangosteen pericarp extract

The %-lethality of *A. salina* exposed to an extract sample containing 1 ppm α-mangostin given in Table 2 is shown in figure 2. The highest percentage of lethality (97.1%), corresponding to the lowest LC_{50} value (< 0.1 ppm), was obtained for the extract sample heated at 65°C. The potent bioactive compound becomes toxic when it is given in high doses, but it becomes a drug in low doses[8]. An extract sample having an LC_{50} value of less than 1000 ppm, is considered to be active as an anti-cancer agent[9]. Based on its LC_{50} value, the extract of mangosteen pericarp in ethyl acetate fraction has the potential to be used for anti-cancer medication.

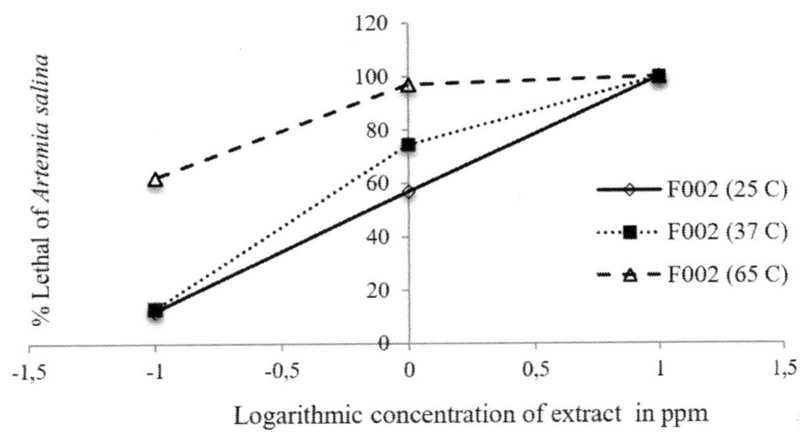

Figure 2. % Lethal of *A. salina* as function of log concentration of mangosteen extract at various temperatures.

Figure 3 shows the positive correlation between the cytotoxicity and the concentration of α-mangostin in the extract, based on the data of % lethal *A. salina* for 1 ppm extract concentration. High lethality of *A. salina* was in line with high yield of α-mangostin in the extract.

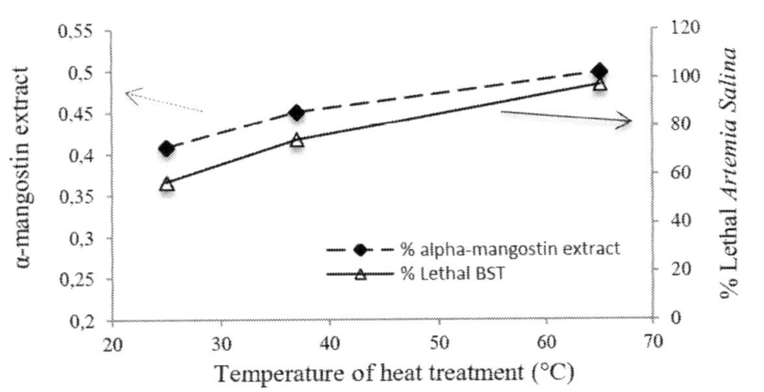

Figure 3. Effect of temperature on % lethal BST and α-Mangostin Yield in Extract

Increasing temperature imposed on the extract sample increase the yield of α-mangostin as well as the cytotoxicity activity, represented as the %-lethality of *A. salina* in BST. The BST assay result of mangosteen pericarp extract in this study is similar to the result reported by Wang, Sanderson and Zhang and Xu et al.[13,14], that higher the α-mangostin in the extract of pericarp higher the cytotoxicity activity.

4. Conclusion

Heating the fresh pericarp mangosteen in water at 65°C for 30 minutes prior to blending produced a juice with higher α-mangostin yield and cytotoxicity compared to the traditional way of blending the juice at room temperature. Increasing α-mangostin yield of 9%-w/w due to heating was also observed when mangosteen extract was heated at 65°C, consistent with the increased cytotoxicity in terms of LC_{50} values of 0.71 ppm at room temperature to < 0.1 ppm at 65°C. The results of this study indicate that the extract of mangosteen pericarp obtained from the ethyl acetate fraction of the ethanolic extract has the potential to be used for anti-cancer medication.

Acknowledgment

The authors are grateful for financial support from the Program Penelitian Unggulan Perguruan Tinggi (PUPT) 2016-2017 and the PITTA Universitas Indonesia 2017.

References

[1] Al-Massarani, S M., El Gamal, A., Al-Musayeib, N.M., Mothana, R.A., Basudan, O.A., Al-Rehaily, A.J., Farag, M., Assaf, M.H., El Tahir, K.H., and Maes, L. (2013), Phytochemical, Antimicrobial and Antiprotozoal Evaluation of Garcinia Mangostana Pericarp and α-Mangostin, Its Major Xanthone Derivative, *Molecules*, 19, pp.10599–10608.

[2] Pedraza-Chaverri, J.; Cárdenas-Rodríguez, N.; Orozco-Ibarra, M.; Pérez-Rojas, J.M., (2008) Medicinal properties of mangosteen (Garcinia mangostana), *Food Chem. Toxicol.*, 46, pp. 3227–3239.

[3] Moongkarndi, P., Kosem, N., Kaslungka, S., Luanratana, O., Pongpan, N., Neungton, N, (2004), Antiproliferation, antioxidation and induction of apoptosis by Garcinia mangostana (Mangosteen) on SKBR3 human breast cancer cell line, *J. Ethnopharmacol.*, 90, pp. 161–166.

[4] Yua, L.; Zhao, M.; Yang, B.; and Bai, W., (2009), Immunomodulatory and anticancer activities of phenolics from Garcinia mangostana fruit pericarp, *Food Chem.*, 116, pp. 969–973.

[5] Satong-aun, W., Assawarachan, R., and Noomhorm, A., (2011), The Influence of Drying Temperature and Extraction Methods on α-Mangostin in Mangosteen Pericarp, *Journal of Food Science and Engineering*, 1, pp. 85-89.

[6] Suttirak, W. and Manurakchinakorn, S., (2014), In-vitro antioxidant properties of mangosteen peel extract, *J Food Sci Technol.* 51 (12), pp. 3546-3558.

[7] Jung, H., Su, B., Keller, W., Mehta, R. and Kinghorn, A., (2006), Antioxidant Xanthones from the Pericarp of Garcinia mangostana (Mangosteen), *Journal of Agricultural and Food Chemistry*, 54, pp.2077–2082.

[8] Meyer, B.N., Ferrigni, N.R., Putnam, J.E., Jacobson, L.B., Nichols, D.E., and McLaughlin, J.L., (1982), Brine shrimp: a convenient general bioassay for active plant constituent, *Planta Medica*, 45, pp. 31-34.

[9] Mclaughlin, J. L., Rogers, L. L. and Anderson, J. E., (1998), The use of biological assays to evaluate botanicals, *Drug information journal*, 32, pp. 513-524.

[10] Aisha, A.F.A., Abu-Salah, K.M., Ismail, Z. and Majid, A.M.S, (2013), Determination of total xanthones in Garcinia mangostana fruit rind extracts by ultraviolet (UV) spectrophotometry. *Journal of Medicinal Plants Research*, 7(1), pp.29–35.

[11] Vámos-Vigyázó, L., (1981), Polyphenol oxidase and peroxidase in fruits and vegetables, *Critical Reviews in Food Science and Nutrition*, 15, pp. 381–388.

[12] Sak, K., (2014), Cytotoxicity of dietary flavonoids on different human cancer types, *Pharmacogn. Rev.*, 8(16), pp. 122-146.
[13] Wang, J.J., Sanderson, B.J.S., and Zhang, W., (2011), Cytotoxic effect of xanthones from pericarp of the tropical fruit mangosteen (*Garcinia mangostana Linn.*) on human melanoma cells, *Food and Chemical Toxicology*, 49, pp. 2385-2391.
[14] Xu, Z., Huang, L., Chen, X-H., Zhu, X-F., Qian, X-J., Feng, G-K., Lan, W-J., and Li, H-J., (2014), Cytotoxic Prenilated Xanthones from the Pericarps of Garcinia mangostana.

IOP Conf. Series: Materials Science and Engineering **316** (2018) 012011 doi:10.1088/1757-899X/316/1/012011

HYDROGEN GENERATION BY KOH-ETHANOL PLASMA ELECTROLYSIS USING DOUBLE COMPARTEMENT REACTOR

Nelson Saksono[1a,], Johannes Sasiang[1], Chandra Dewi Rosalina[1], Trisutanti Budikania[2]

[1]*Department of Chemical Engineering, Universitas Indonesia, Depok 16424, Indonesia*
[2]*Politeknik AKA Bogor, Indonesia*
[a)] *Corresponding Author: nelson@che.ui.ac.id*

Abstract

This study has successfully investigated the generation of hydrogen using double compartment reactor with plasma electrolysis process. Double compartment reactor is designed to achieve high discharged voltage, high concentration, and also reduce the energy consumption. The experimental results showed the use of double compartment reactor increased the productivity ratio 90 times higher compared to Faraday electrolysis process. The highest hydrogen production obtained is 26.50 mmol/min while the energy consumption can reach up 1.71 kJ/mmol H_2 at 0.01 M KOH solution. It was shown that KOH concentration, addition of ethanol, cathode depth, and temperature have important effects on hydrogen production, energy consumption, and process efficiency.

Keywords: Double compartment reactor, Ethanol, Hydrogen, KOH, Plasma electrolysis

Introduction

Hydrogen is widely manufactured from steam reforming and electrolysis process. The weaknesses of steam reforming are the high carbon emissions and not renewable feedstock (natural gas). Meanwhile, the main obstacles of electrolysis are the lack of production and high energy consumption [1].

Alternative way to produce hydrogen is plasma electrolysis. This method has a similar process like common electrolysis but it uses higher voltage that is able to create electrical sparks and form plasma. Plasma will produce large amounts of reactive species that can break H_2O binding. Mizuno used KOH solution at a temperature of 80°C and 230 V that can lower energy consumption eight times compared with Faraday electrolysis [2]. High discharged voltage causes the process efficiency more optimal [3].

Saksono, using 0.1 M KOH solution at 85°C and 300 V, obtained hydrogen production rate 13.4 times higher compared with Faraday electrolysis for the same amount of power electricity [4]. Higher discharged voltage increases excitation energy. It produces large amount of radicals so that the production of hydrogen increases.

Hydrogen atom and hydroxyl radical production are increased by addition of methanol. These active species are produced from the hemolytic dissociation of water molecules by plasma. The addition of 15 % (v/v) methanol in 0.01 M NaOH solution at 700 V increases hydrogen production nineteen times higher than without additives [5].

Therefore, high discharged voltage and additives are main criterias in designing a reactor to produce large amount hydrogens. In a single reactor, the distance between the anode and cathode is very close. At this condition, high discharged voltage will increase electrical current and power electricity that causes operation system less safety. Double compartment

IOP Conf. Series: Materials Science and Engineering **316** (2018) 012011 doi:10.1088/1757-899X/316/1/012011

reactor enlarges the distance between anode and cathode so that the electrical current which flow in the system will be much lower due to the increase of flow resistance. Utilization of double compartment reactor on the production of chlor gas showed reactor capability to increase production rate, minimize energy consumption, and reduce explosion possibility, especially at higher voltage and concentration [6].

The objective of using double compartment on this research is to achieve process condition at higher concentration with low electrical current flow. The addition of additives is needed to obtain higher hydrogen production rate. This study will examine independent variables such as discharged voltage, KOH concentration, ethanol concentration, cathode depth, and temperature effects on hydrogen production, energy consumption, ratio of energy consumption per mmol of product (Wr), and the ratio plasma electrolysis moles product per Faraday electrolysis moles product (G (H_2)).

Exsperimental

The design of the reactor is shown in Figure 1. The reactor has two compartments: one compartment is equipped with a tungsten cathode and the other is equipped with a stainless steel anode. The reactor is a semi batch system made from an acrylic housing filter connected to a globe valve. The electric source is connected to a 3 kVA slide regulator that is then connected to a transformer. A diode bridge is used to rectify the electric current. The current is measured from a multimeter Yuhua A830L. The hydrogen gas produced is then passed into gas chromatography 8A Shimadzu. Bubble soap flowmeter is used to measure hydrogen production rate. This study analyses the formation of plasma by varying KOH concentrations from 0.05 M to 0.1 M within 600–800 V. Furthermore, an examination of energy consumption is conducted by measuring the amount of electrical current and time of process.

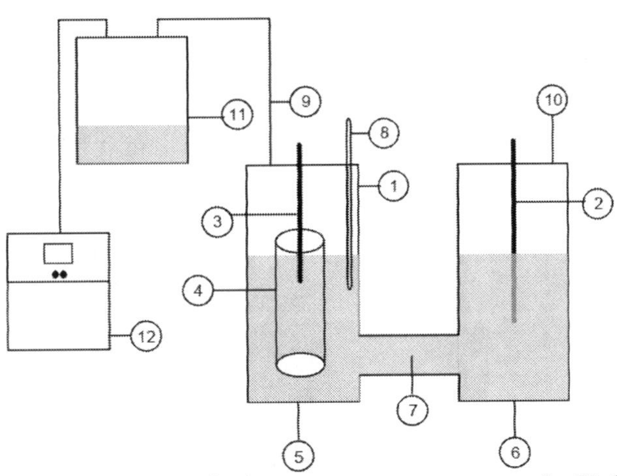

1. Reactor
2. Anode (*Stainless Steel*)
3. Cathode (Tungsten)
4. *Barrier*
5. Cathode compartment
6. Anode compartment
7. Electrolyte solution
8. Thermometer
9. Hydrogen Output
10. Oxygen Output
11. Condenser
12. Gas Chromatography

Figure 1 Experimental Set-up

Power electricity (P) is defined as the amount of electrical energy used in t second (eq. 1)

$$P = V \times I \quad (1)$$

V : Discharged voltage (V)
I : Electrical current average (A)

Hydrogen production mol rate (mmol/s) V_{H_2} is calculated by measuring product flow rate using bubble-soap flowmeter and then calculate it using equation 2:

$$V_{H_2} = \frac{V_{gH_2}}{22.4} \quad (2)$$

V_{gH_2} : Hydrogen gas product flow rate (mL/s)

Wr (kJ/mmol) is defined as ratio of energy consumption per mmol product (eq. 3) dan $G(H_2)$ is defined as ratio of plasma electrolysis moles product per Faraday electrolysis moles product (eq. 4).

$$Wr = \frac{V \times Q}{V_{H_2}} \quad (3)$$

$$G(H_2) = \frac{V_{H_2}/22.4}{Q/2F} \quad (4)$$

V : Discharged voltage (V)
Q : Electricity load (C)
F : Faraday Contant (96500 C/mole)

Results and Discussion

Effects of KOH Concentration

Figure 2 shows that the increasing of KOH concentration increases the hydrogen gas production rate. The highest production is 26.50 mmol/min at 0.1 M KOH solution with addition of ethanol 10% volume. Higher concentration more than 0.1 M will potentially cause an explosion.

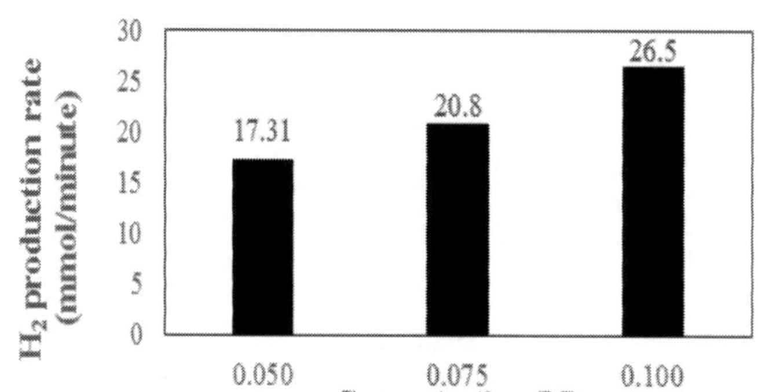

Figure 2 Hydrogen Production Rate in KOH + Ethanol 10%v at 800 V and 90°C

The increasing of KOH concentration in solution increases the number of OH^- and K^+ ion. Plasma transforms OH^- into OH•. OH radicals (OH•) are decomposed into hydrogen and oxygen as shown in the following reactions [3]:

IOP Conf. Series: Materials Science and Engineering **316** (2018) 012011 doi:10.1088/1757-899X/316/1/012011

$$4H^{+}_{(aq)} + 4e \rightarrow 4H\bullet \quad (5)$$
$$4OH^{-}_{(aq)} \rightarrow 4OH\bullet + 4e \quad (6)$$
$$4H+_{(aq)} + 4OH^{-}_{(aq)} \rightarrow 4H\bullet + 4OH\bullet \quad (7)$$
$$2H\bullet + 2H\bullet \rightarrow 2H_{2\,(g)} \quad (8)$$
$$4OH \rightarrow 2H_{2\,(aq)} + 2O_{2\,(aq)} \quad (9)$$

Large amount of ion in solution makes conductivity higher. Higher conductivity facilitates the mobilization of electrons so that electric current flowing in the system becomes higher. Therefore, energy consumption increases along with the increase of concentration.

At 0.1 M plasma spark brightestly and stable. The highest production rate is obtained at highest concentration. Wr decreases because the increase of production rate is higher than the increase of energy consumption. Figure 3 shows that at higher concentration, Wr which means energy consumption required per mole product becomes minimum. At this conditions $G(H_2)$ obtained is 90.

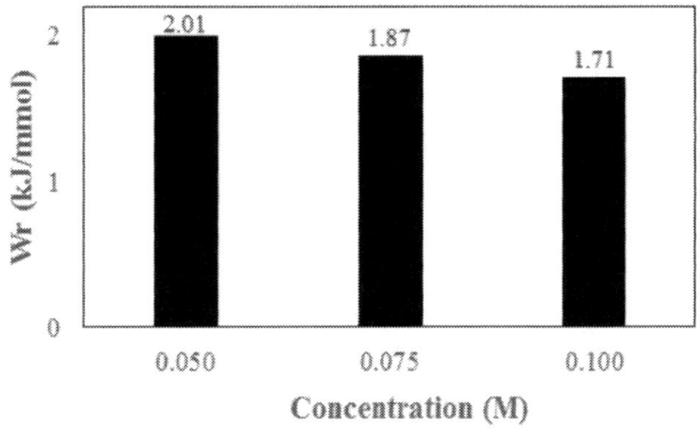

Figure 3 Wr in KOH + Ethanol 10%v at 800 V and 90°C

Effects of Ethanol Addition

The addition of ethanol increases the production rate significantly as shown in Figure 4. At 600 V, production rate reaches up 23 times higher compares without using ethanol. The ethanol additive function is to donate active species H• and OH• which in turn leads to the termination of the binding energy in the decomposition of water.

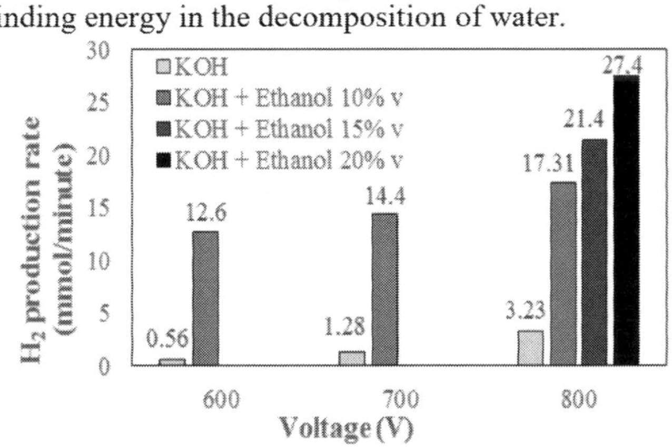

Figure 4 Hydrogen Production Rate in KOH 0.05 M at 90°C

Plasma dissociates ethanol in solution as shown in the following reactions [7] :

$$C_2H_5OH + e^* \rightarrow C_2H_5\bullet + OH\bullet + e \quad (10)$$
$$C_2H_5OH + e^* \rightarrow C_2H_4OH\bullet + H\bullet + e \quad (11)$$
$$C_2H_5OH + e^* \rightarrow C_2H_5O\bullet + H\bullet + e \quad (12)$$
$$C_2H_5OH + e^* \rightarrow \bullet C_2H_4 + H_2O\bullet + e \quad (13)$$
$$C_2H_5OH + e^* \rightarrow tran\_CH_3CHO + H_2 \quad (14)$$
$$C_2H_5OH + e^* \rightarrow cis\_CH_3CHO + H_2 \quad (15)$$
$$C_2H_5OH \rightarrow CH_3CHO + H_2 \quad (16)$$

Three reactions directly generate H_2 and two other reactions generate H•. The H• then recombines with each other to produce H_2. Dissociation reactions indicate that the presence of ethanol increases the production of H_2.

At higher voltage, energy that is used to excite electrons increases. This phenomenon is visualized by the presence of large and stable plasma in the reactor. Large and stable plasma accelerates the formation of radical species thus accelerating the reaction of hydrogen formation. At 800 V with addition of ethanol 15% v, G(H_2) obtained is 82.8 as shown in Figure 5.

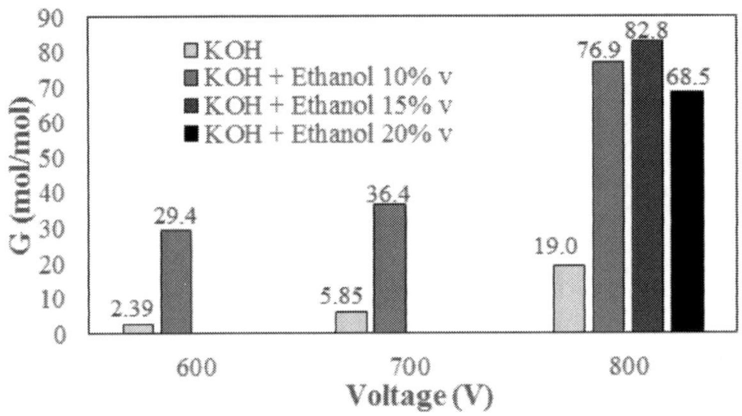

Figure 5 G(H_2) in KOH 0.05 M at 90 °C

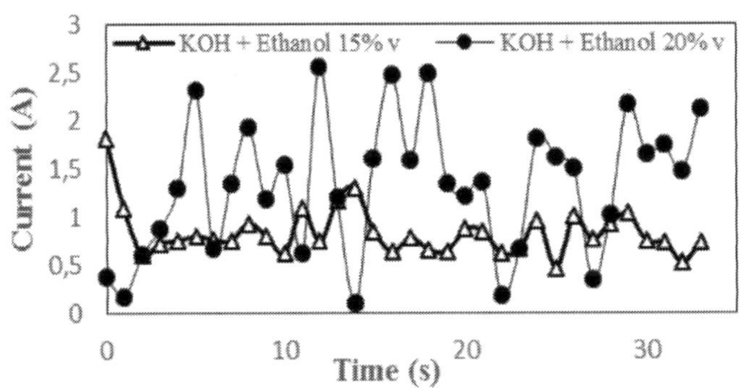

Figure 6 Electrical Current Characteristic Curve in KOH 0.05 M + Ethanol 20 %V at 800 V, Cathode Depth 1 cm

Addition of ethanol until 20%v causes G (H_2) decreased. This phenomenon can be described quantitatively by fluctuating current read on multimeter. The addition of ethanol 20%v makes the electrical current flowing in the system larger and more fluctuating which is visualized by high peaks as shown in Figure 6. At these high peaks, plasma inside reactor slightly dimmed accompanied the increase of current which affects to increase energy consumption.

Cathode Depth Effects

The brightest plasma spark happens when the cathode position is on the solution surface. Short electrode is utilized to achieve high current density to facilitate joule heating effect [8]. Joule heating effect around the cathode produces a gas sheath that surrounds the cathode. The presence of gas sheath triggers electron excitation to form plasma.

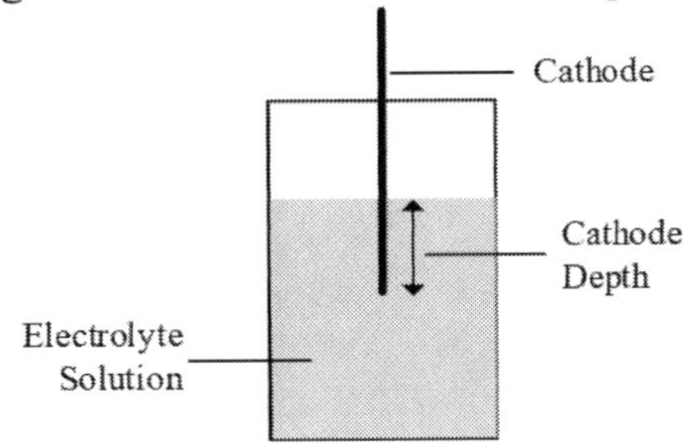

Figure 7 Cathode Depth

Cathode depth is the length of cathode submerged in solution as shown in Fig. 7. Fig. 8 shows that the highest hydrogen production rate is 17.3 mmol / min at 1 cm cathode depth. At solution surface, hydrogen production rate is lower than 1 and 2 cm although the brightest plasma spark was formed at solution surface. At solution surface, high energy electrons produced by plasma is not effective attacking OH^- and H^+. OH^- and H^+ ions are not only found on the surface but scattered in the solution. Therefore, at 1 cm position depth, plasma is more effective to form radicals that increase hydrogen production.

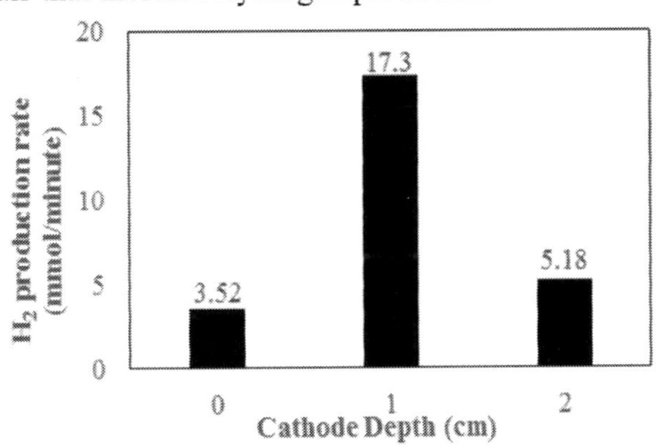

Figure 8 Hydrogen Production Rate in KOH 0.05 M + Ethanol 10%v at 800 V, 90 °C

At deeper cathode position, cathode will face higher hydrostatic pressure. At 2 cm below solution surface, the production of H_2 declines. At this position, the formation of bubbles on the submerged part of the cathode is more dominant than the formation of vapor around cathode surface solution so that the gas sheath formed was unstable. The presence of bubbles on the submerged part indicates that Faraday electrolysis was dominantly occured than plasma electrolysis. This phenomenon increased energy consumption as shown in Fig. 9.

IOP Conf. Series: Materials Science and Engineering **316** (2018) 012011 doi:10.1088/1757-899X/316/1/012011

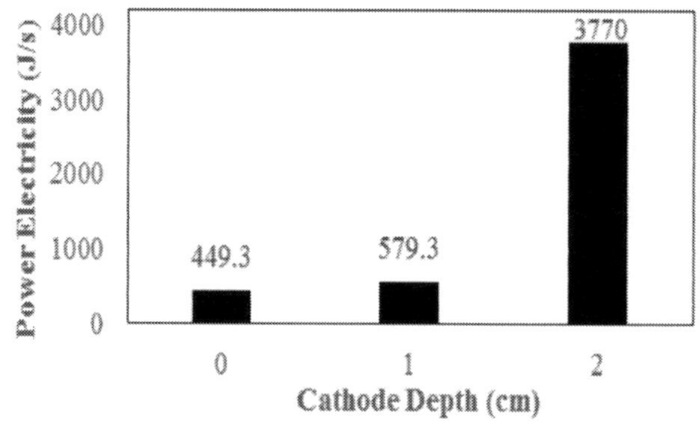

Figure 9 Power Electricity in KOH 0.05 M + Ethanol 10%v at 800 V, 90 °C

Temperature Effects

Fig. 10 shows that the highest hydrogen production rate is produced at 90°C. High temperature leads to the formation of bubbles (vapor bubbles on the surface of the solution). The bubbles will be ionized and form plasma when attacked by electron excitated [2].

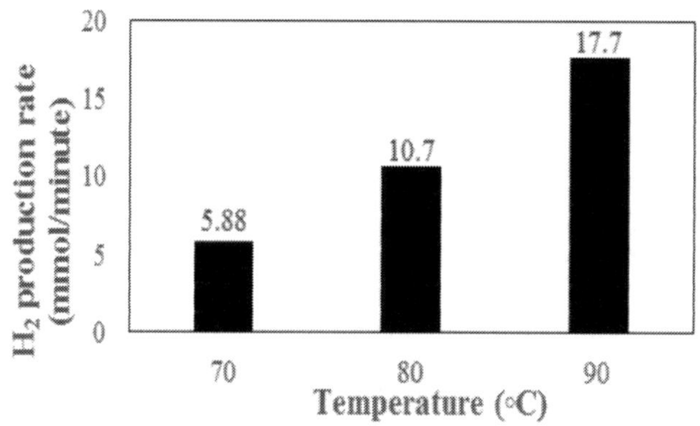

Figure 10 Hydrogen production rate in KOH 0.05 M + Ethanol 10%v Ethanol at 800 V, 90 °C

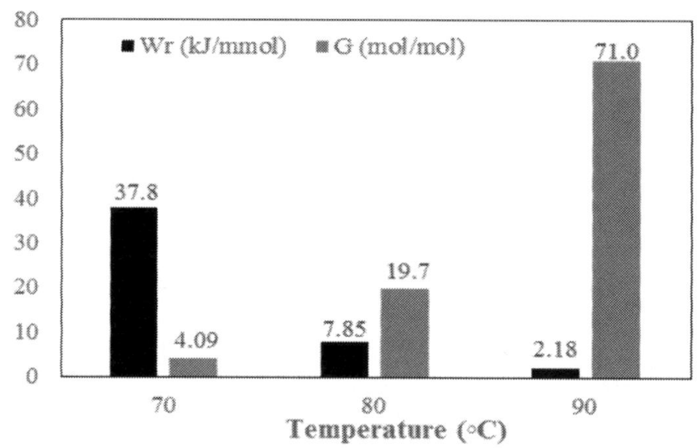

Figure 11 Wr and $G(H_2)$ in KOH 0.05 M + Ethanol 10% v at 800 V, 90 °C

IOP Conf. Series: Materials Science and Engineering **316** (2018) 012011 doi:10.1088/1757-899X/316/1/012011

At temperature of 90°C H_2O molecules are more easily broken down into its constituent atoms compared to the liquid phase with a lower temperature. In bubbler form, the space between water molecules becomes more tenuous so that collisions of high-energy electrons in the water molecules become easier. Figure 11 shows that the process becomes more optimal due to the increase of temperature.

Double Compartment Effects

The reactor consists of two double compartment where anode and cathode are separated in different room in order to extend the distance between electrode. Longer distance increases electrical current resistance and extend electron pathway so that electrical current flowing in system can be minimalized.

Utilization of double compartment reactor also aims to purify the products. In single compartment reactor, the hydrogen and oxygen are produced in one reactor. In double compartment reactor, hydrogen concentrates in cathode compartment while oxygen in anode compartment as shown in Figure 12.

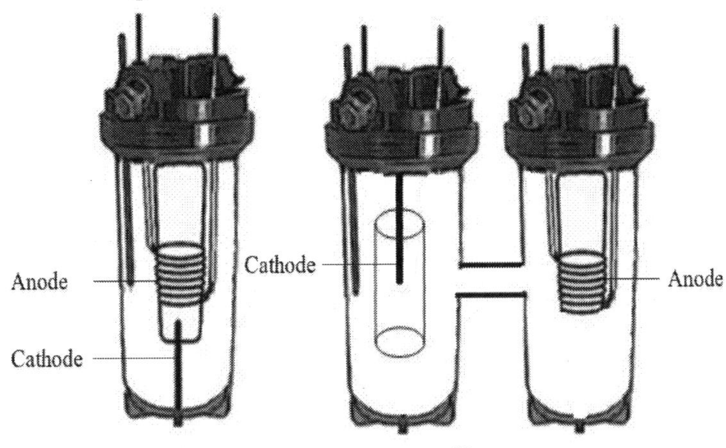

Figure 12 Plasma electrolysis reactor (a) single compartment (b) double compartment

Table 1 Comparison result between single and double compartment reactor in KOH 0.1 M, 1 cm cathode depth.

Process Variable	Single Compartment Reactor	Double Compartment Reactor
Max. discharged Voltage (V)	300	800
$I_{average}$ (A)	3.32	0.95
G (mol/mol)	21.8	90.05
Wr (kJ/mmol)	2.62	1.71

Double compartment reactor is equipped with cooling system. It aims to maintain the stability of operating temperature. Table 1 shows a comparison result between single and double compartment. Through double compartment reactor, high voltage (800 V) can be obtained with low electrical current flow and minimum Wr.

Conclusions

The application of plasma electrolysis in double compartment reactor can be implemented in the generation of hydrogen gas. The highest hydrogen production obtained is 26.50 mmol/min while the energy consumption can reach up 1.71 kJ/mmol H_2 at 0.01 M KOH solution. The experimental results showed that the use of double compartment reactor increased the productivity ratio 90 times higher compared to Faraday electrolysis process.

Acknowledgement

This research was partially funded by "Hibah Publikasi Internasional Terindeks Tugas Akhir Mahasiswa Universitas Indonesia 2017" from Directorate Research and Community Service Universitas Indonesia. The authors declare no competing interests or any conflicts of financial interests.

This article's publication is supported by the United States Agency for International Development (USAID) through the Sustainable Higher Education Research Alliance (SHERA) Program for Universitas Indonesia's Scientific Modeling, Application, Research and Training for City-centered Innovation and Technology (SMART CITY) Project, Grant #AID-497-A-1600004, Sub Grant #IIE-00000078-UI-1.

References

[1] Holladay, J. D., Hu, J., King, D., and Wang, Y. (2009). An Overview of Hydrogen Production Ttechnologies, *Catalysis Today,* pp. 244-260.
[2] Mizuno, T., Akimoto, T., Azumi, K., Ohmori, T., Aoki, Y. and Takahashi, A. (2005). Hydrogen Evolution by Plasma Electrolysis in Aqueous Solution. *Japanese Journal of Applied Physics*, 44(1A), pp. 396-401.
[3] Gao, Jinzhang, Aixiang, W., Yan, F., Jianlin, W., Dongping, M., Xiao, G., Yan, L., and Wu, Y. (2008). Analysis of Energetic Species Caused by Contact Glow Discharge Electrolysis in Aqueous Solution. *Plasma Science and Technology,* 10, (1).
[4] Saksono, Nelson., Ariawan, B., Bismo, Setijo. (2012). Hydrogen Production System Design by Using Non-Thermal Plasma Electrolysis in Glycerol-KOH Solution. International Journal of Technology, vol 3, pp. 8-15.
[5] Saksono. N, Kartohardjono, S., Yuniawati, T. (2016). High Peformance Plasma Electrolysis Reactor for Hydrogen Generation Using A NaOH-Methanol Solution. International Journal of Technology, vol 8, pp: 952-960.
[6] Saksono, N., Abqari, F., Bismo, S. and Kartohardjono, S. (2013). Effect of Process Condition in Plasma Electrolysis of Chloralkali Production. *International Journal of Chemical Engineering and Applications,* vol 4. No 5.
[7] Yan, Z.C., Li, C., and Wang, H.L. (2008). Hydrogen Generation by Glow Discharge Plasma Electrolysis of Ethanol Solutions. Guangzhou: College of Chemical and Energy Engineering, South China University of Technology
[8] Sengupta, Susanta K., Sfivastava, A., and Singh, R. (1997). Contact Glow Discharge Electrolysis: A Study on Its Origin in The Fight of The Theory of Hydrodynamic Instabilities in Local Solvent Vaporization by Joule Heating During Electrolysis. *Journal of Electroanalytical Chemistry*, 427, pp. 23-27.

IOP Conf. Series: Materials Science and Engineering **316** (2018) 012012 doi:10.1088/1757-899X/316/1/012012

Effect of Feed Gas Flow Rate on CO_2 Absorption through Super Hydrophobic Hollow Fiber membrane Contactor

Sutrasno Kartohardjono[1*], Kevin Alexander[1], Annisa Larasati[1] and Ivander Christian Sihombing[1]

[1]Process Intensification Laboratory, Department of Chemical Engineering, Faculty of Engineering, Universitas Indonesia, Kampus Baru UI, Depok 16424, Indonesia

[*]Corresponding Author: sutrasno@che.ui.ac.id

Abstract. Carbon dioxide is pollutant in natural gas that could reduce the heating value of the natural gas and cause problem in transportation due to corrosive to the pipeline. This study aims to evaluate the effects of feed gas flow rate on CO_2 absorption through super hydrophobic hollow fiber contactor. Polyethyleneglycol-300 (PEG-300) solution was used as absorbent in this study, whilst the feed gas used in the experiment was a mixture of 30% CO_2 and 70% CH_4. There are three super hydrophobic hollow fiber contactors sized 6 cm and 25 cm in diameter and length used in this study, which consists of 1000, 3000 and 5000 fibers, respectively. The super hydrophobic fiber membrane used is polypropylene-based with outer and inner diameter of about 525 and 235 μm, respectively. In the experiments, the feed gas was sent through the shell side of the membrane contactor, whilst the absorbent solution was pumped through the lumen fibers. The experimental results showed that the mass transfer coefficient, flux, absorption efficiency for CO_2-N_2 system and CO_2 loading increased with the feed gas flow rate, but the absorption efficiency for CO_2-N_2 system decreased. The mass transfer coefficient and the flux, at the same feed gas flow rate, decreased with the number of fibers in the membrane contactor, but the CO_2 absorption efficiency and the CO_2 loading increased.

Keywords: Absorption efficiency; Flux; Mass transfer coefficient; Membrane contactor; Super hydrophobic.

1. Introduction

Consumption of natural gas in the world reached 100 trillion cubic feet and is estimated to be 160 trillion cubic feet in 2035 [1]. Meanwhile, the consumption of natural gas as energy in Indonesia is the third largest after oil and coal. Increasing the amount of natural gas needs is directly proportional to the use of natural gas for the energy and industrial raw materials, as well as for domestic purposes. Therefore, there is a need to develop the existing natural gas processing to produce a better quality gas. Natural gas quality is usually determined by the composition of the substances contained in the gas. In general, the content of natural gas is methane (CH_4), ethane (C_2H_6), propane (C_3H_8), iso-butane (iCH_4H_{10}) and n-butane (nCH_4H_{10}), and non-hydrocarbons such as carbon dioxide (CO_2), nitrogen (N_2) and hydrogen sulfide (H_2S) [2]. The content of impurities such as CO_2 must be lowered as low as possible as it can cause corrosion in gas piping [3], reduce the calorific value of the gas [1], and clog the piping system [4]. The CO_2 content in natural gas can vary from place to place and in Indonesia ranging from 5 to 75% depending on the location. As for example, the content of 15% CO_2 in natural gas obtained in the Madura Strait, while the content of up to 75% in natural gas obtained in the Natuna Islands.

The technique has been used so far for the removal of CO_2 from natural gas is a chemical and physical absorption, solid adsorption, cryogenic distillation and membrane separation [5]. The most widely used technology is absorption. The absorption process is conventionally using packed towers and plate columns and it has high value for the mass transfer coefficient. However, this operating unit has disadvantages in its operation such as entrainment, flooding, foaming, small contact area per unit volume and high operating costs.

Therefore, there is a need of alternative technologies which more effective to absorb CO_2 from the natural gas. One of the alternative technologies for the removal of CO_2 from the natural gas is the membrane contactor. This technology is able to overcome the problems that arise in the previous separation technology and this technology also requires only a fairly low energy and the process is much simpler [6]. The separation process using a membrane contactor has several of important factors that affects the performance such as absorbent and feed gas flow rates and the concentration of absorbent [5]. This study will utilize super hydrophobic hollow fiber membrane module as a contactor to absorb CO_2 from its mixture with CH_4 or N_2 using physical absorbent polyethylene glycol (PEG). The physical absorbent has the advantages as it is not corrosive and energy required for the regeneration is lower compared to chemical absorption [7]. The objectives are to see the effect of feed gas flow rate and the number of fibers in the contactor on the mass transfer coefficient, the flux and the absorption efficiency.

2. Materials and Methods

The CO_2 absorption experiment is schematically shown in Figure 1 which has been presented elsewhere [8, 9]. There were three membrane contactors applied in this study consists of 1000, 3000 and 5000 fibers sized 6 and 25 cm in diameter and length, respectively. The contactors were super hydrophobic hollow fiber membrane modules which are provided by PT GDP Filter Bandung. The fibers are polypropylene based sized 525 and 235 μm in outer and inner diameter, respectively. The feed gas is a mixture or CO_2 and CH_4 (36:64) or CO_2 and N_2 (13:87), whilst the absorbent is solution of 5 vol. % PEG in water. The feed gas containing CO2 was flowed through the shell side of the contactor, whilst the absorbent was pumped through the lumen fiber in the contactor in a counter current arrangement. The inlet and outlet gases to and from the contactor were measured using mass flow meter Sierra Top Trak Instruments, whilst the inlet absorbent to the contactor were measured using liquid flow meter Krohne. Meanwhile, the inlet and outlet gases composition were analyzed using gas chromatography Bruker Scion 436-GC.

The overall mass transfer coefficient, K_{OVL}, and CO_2 absorption efficiency, $\%R$, were calculated using the following equations [10]:

$$K_{OVL} = \frac{Q_{Gin}}{A_m} ln\left(\frac{C_{in}}{C_{out}}\right) \tag{1}$$

$$\%R = \frac{C_{in} - C_{out}}{C_{in}} x100\% \tag{2}$$

Meanwhile, CO_2 flux through the membrane contactor, J, [9] and acid loading were calculated by:

$$J = (Q_{iGin} - Q_{iGout}) * RT/Am \tag{3}$$

$$Acid\ Loading = \frac{mol\ CO2\ Absorbed}{mol\ PEG} \tag{4}$$

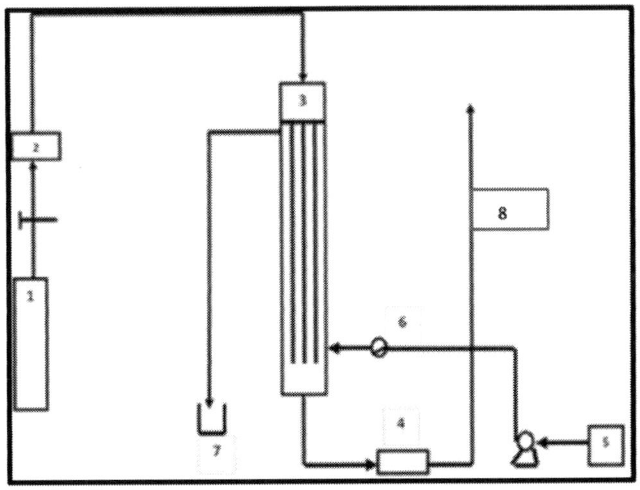

Figure 1 Schematic diagram of experimental set up: 1. Feed gas tank, 2. mass flow meter, 3. Hollow fiber membrane contactor, 4. Mass flow meter, 5. absorbent reservoir, 6. Liquid flow meter, 7. Outlet absorbent tank

3. Results and Discussion

Figure 2 shows the CO_2 overall mass transfer coefficients as a function of the feed gas flow rate at different number of fibers in the membrane contactor when absorbent flow rate in the lumen fibers was set at 300 $ml.min^{-1}$. Figure 2 reveals that the CO_2 overall mass transfer coefficient increases with increasing the feed gas flow rate in the shell side of the contactor due to more CO_2 can be absorbed by absorbent solution flowing in the lumen side of the contactor [11]. The similar trend also shows by the previous study using DEA vol.5% solution as absorbent for CO_2-N_2 (13:87) and CO_2-CH_4 (36:64) system [9] and by Wang et al., [11] using 2M DEA solution as absorbent for CO_2-N_2 system (20:80). In this study the overall mass transfer coefficient increases from 2.6 to 4.3 x 10^{-5} $cm.s^{-1}$ and from 1.3 to 5.8 x 10^{-5} $cm.s^{-1}$ for CO_2-CH_4 and CO_2-N_2 system, respectively, if the gas flow rate increase from 130 to 380 $ml.min^{-1}$. In the previous study [9] the overall mass transfer coefficient increased from 23 to 32 x 10^{-5} $cm.s^{-1}$ and from 1.39 to 37 x 10^{-5} $cm.s^{-1}$ for CO_2-CH_4 and CO_2-N_2 system, respectively, if the gas flow rate increased from 120 to 340 $ml.min^{-1}$ using DEA vol.5% solution as absorbent. Meanwhile, Wang et al., [11] reported that the overall mass transfer coefficient increased from 1.6 to 2.5 $m.s^{-1}$ if the gas velocity increased from 0.03 to 0.09 $m.s^{-1}$ for CO_2-N_2 system (20:80) using 2M DEA solution as absorbent.

The effect of the number of the fiber in the contactor can also be seen in Figure 2. Figure 2 shows that the mass transfer coefficient decreases with an increase in the number of fiber in the contactor at the same feed gas flow rate. The increase in the number of fiber will increase the surface area for gas-liquid contact, which will enhance the mass transfer coefficient, but will decrease the absorbent velocity in the lumen fiber, which will reduce the overall mass transfer coefficient. The decrease in the overall mass transfer coefficient with increasing the number of fibers in the contactor reveals that the effect of absorbent velocity more dominant than the effect of surface area for gas-liquid contact.

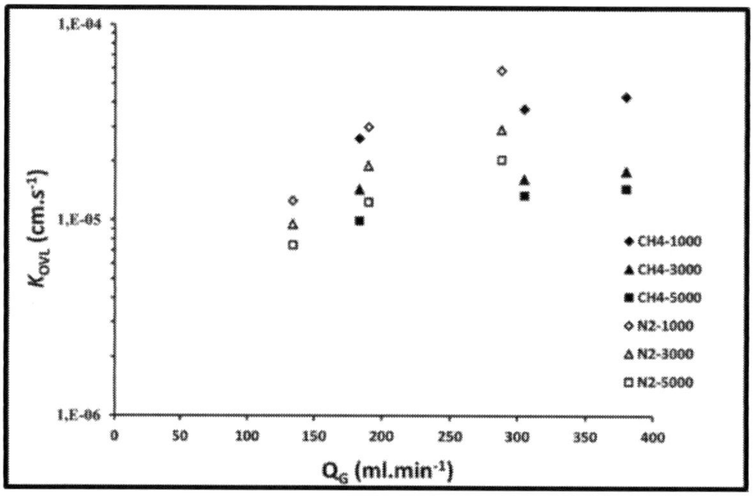

Figure 2 The CO_2 overall mass transfer coefficient, K_{OVL}, using 5 vol.% PEG solution as absorbent on the lumen side and a mixture of CO_2-CH_4 (CH_4-1000, CH_4-3000 and CH_4-5000) or a mixture of CO_2-N_2 (N_2-1000, N_2-3000 and N_2-5000) on the shell side of the contactors containing 1000, 3000 and 5000 fibers.

Figure 3 demonstrates the variation of the fluxes as a function of the feed gas flow rate at different number of fibers in the contactor when absorbent flow rate in the lumen fibers was set at 300 ml.min^{-1}. In addition to the mass transfer coefficient, the CO_2 flux increases with increasing the feed gas flow rate in the contactor due to more CO_2 can be absorbed by absorbent solution flowing in the lumen side of the contactor [11]. The similar trend was also shown by previous study [9] and Wang et al. [11]. In this study the CO_2 flux increases from 1.05 to 1.73 mol.m^{-2}.s^{-1} and from 0.49 to 2.18 mol.m^{-2}.s^{-1} for CO_2-CH_4 and CO_2-N_2 system, respectively, if the gas flow rate increases from 130 to 380 ml.min^{-1}. In the previous study [9] the CO_2 flux increased from 2.6 to 3.9 x 10^{-6} mmol.cm^{-2}.s^{-1} and from 2.7 to 8.8 x 10^{-6} mmol.cm^{-2}.s^{-1} for CO_2-CH_4 and CO_2-N_2 system, respectively, if the gas flow rate increased from 120 to 340 ml.min^{-1} using DEA vol.5% solution as absorbent. Meanwhile, Wang et al., [11] reported that the flux of CO_2 increased from 3 to 8 x 10^{-4} mol.m^{-2}.s^{-1} if the feed gas velocity increased from 0.03 to 0.09 m.s^{-1} for CO_2-N_2 system (20:80) using 2M DEA solution as absorbent. The CO_2 flux decreases with increasing the number of fibers in the contactor, as illustrates in Figure 3, indicating that the effect of absorbent flow rate more dominant than the effect of the surface area for gas-liquid contact.

The variation of CO_2 absorption efficiency, *R* (%), as a function of the feed gas flow rate at different number of fibers in the membrane contactor when absorbent flow rate in the lumen fibers was set at 300 ml.min^{-1} is presented in Figure 4. The different results were demonstrated by the two feed gas system. For the CO_2-N_2 system the CO_2 absorption efficiency increases with the feed gas flow rate. Meanwhile for the CO_2-CH_4 system the CO_2 absorption efficiency decreases with the feed gas flow rate. The reduction in the CO_2 absorption efficiency was also shown in the previous study [9] and by Yan et al. [5]. In this study, the CO_2 absorption efficiency for CO_2-CH_4 system decrease from 7.96 to 5.67% if the gas flow rate increases from 183 to 380 ml.min^{-1}. In the previous study [9], the CO_2 absorption efficiency for CO_2-CH_4 system decrease from 72.7 to 64.0% if the gas flow rate increase from 170 to 340 ml.min^{-1} using DEA vol.5% solution as absorbent. Meanwhile, Yan et al., [5] reported that the CO_2 absorption efficiency decrease from 94 to 65% and 89 to 54% if the gas velocity increase from 0.21 to 0.56 m.s^{-1} using 1M PEG and 1M MEA as absorbents, respectively. The number of fibers in the contactor has the same effect on the CO_2 absorption efficiency for both feed gas system. The CO_2 absorption efficiency increases with the number of fibers in the contactor.

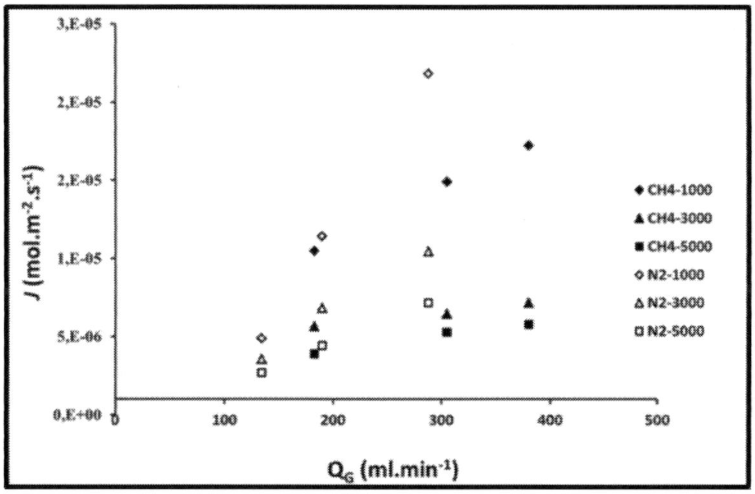

Figure 3 The CO_2 flux, *J*, using 5 vol.% PEG solution as absorbent on the lumen side and a mixture of CO_2-CH_4 (CH_4-1000, CH_4-3000 and CH_4-5000) or a mixture of CO_2-N_2 (N_2-1000, N_2-3000 and N_2-5000) on the shell side of the contactors containing 1000, 3000 and 5000 fibers.

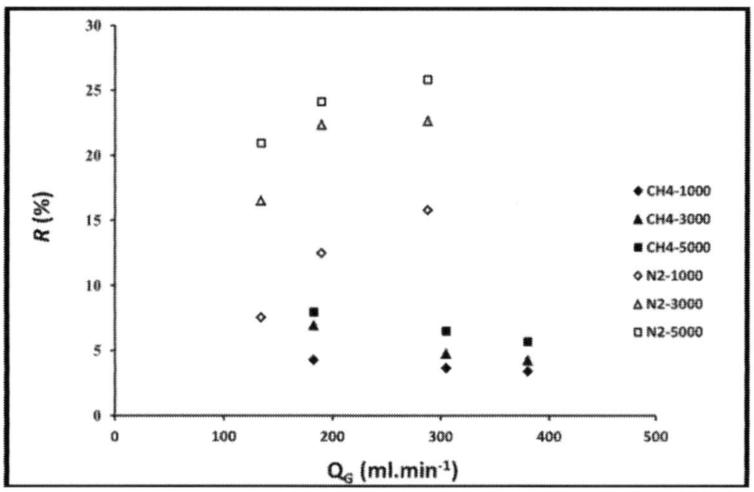

Figure 4 The CO_2 absorption efficiency, *R* (%), using 5 vol.% PEG solution as absorbent on the lumen side and a mixture of CO_2-CH_4 (CH_4-1000, CH_4-3000 and CH_4-5000) or a mixture of CO_2-N_2 (N_2-1000, N_2-3000 and N_2-5000) on the shell side of the contactors containing 1000, 3000 and 5000 fibers.

The variation of CO_2 loading as a function of the feed gas flow rate at different number of fibers in the membrane contactor when absorbent flow rate in the lumen fibers was set at 300 $ml.min^{-1}$ is presented in Figure 5. The CO_2 loading is the ratio between the amounts of CO_2 absorbed to the mole of absorbent flowing in the membrane contactor. As for example in the conventional column, the ideal CO_2 loading using amine solution as absorbent is 0.5 according to the equation below:

$$2RNH_2 + CO_2 + H_2O \xrightarrow{49\ oC} (RNH_2)_2H_2CO_3 \quad (5)$$

The CO_2 loading in this study increases with the feed gas flow rate and the number of fibers in the contactor due to more CO_2 absorbed by the absorbent as a result of the increase in the feed gas flow rate and the surface area for gas-liquid contact.

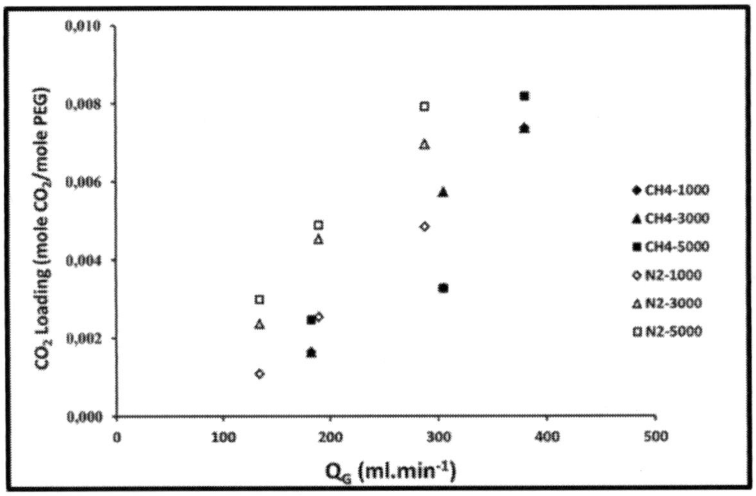

Figure 5 The CO_2 loading using 5 vol.% PEG solution as absorbent on the lumen side and a mixture of CO_2-CH_4 (CH_4-1000, CH_4-3000 and CH_4-5000) or a mixture of CO_2-N_2 (N_2-1000, N_2-3000 and N_2-5000) on the shell side of the contactors containing 1000, 3000 and 5000 fibers.

4. Conclusion

The process of gas-liquid contact through the membrane contactor can overcome the disadvantages of the conventional contactor using column or tower such as entrainment, flooding, unloading and foaming. The membrane gas-liquid contactor also requires low energy and the process is much simpler than conventional gas-liquid contactor. This study utilized the super hydrophobic hollow fiber modules as contactors to absorb CO_2 using 5 vol.% PEG as absorbent. Feed gas flow rate and the number of fibers in the contactor were the main observed variables to see their effects on the overall mass transfer coefficient, flux, absorption efficiency and the CO_2 loading. The experimental results showed that the mass transfer coefficient, flux, absorption efficiency for CO_2-N_2 system and CO_2 loading increased with increasing the feed gas flow rate, but the absorption efficiency for CO_2-N_2 system decreased. The mass transfer coefficient and the flux, at the same feed gas flow rate, decreased with increasing the number of fibers in the membrane contactor. Meanwhile, the absorption efficiency and the CO_2 loading, at the same feed gas flow rate, increased with increasing the number of fiber in the membrane contactor.

Acknowledgement

A part of this study was funded by Directorate General of Higher Education Republic of Indonesia through Universitas Indonesia with contract No. 1178/UN2.R12/HKP.05.00/2016.

References

[1] Zhang Y, Sunarso J, Liu S, and Wang R 2013 Current status and development of membranes for CO 2/CH 4 separation: a review *Int. J. Greenhouse Gas Control* **12** 84-107.

[2] Younger A and Eng P 2004 Natural Gas Processing Principles and Technology-part 2, *Gas Processors Association,* Tulsa Oklahoma.

IOP Conf. Series: Materials Science and Engineering **316** (2018) 012012 doi:10.1088/1757-899X/316/1/012012

[3] Jahn F, Cook M and Graham M 2008 *Hydrocarbon exploration and production* **55**: Elsevier.
[4] Othman M, Tan S, and Bhatia S 2009 Separability of carbon dioxide from methane using MFI zeolite–silica film deposited on gamma-alumina support *Microporous and Mesoporous Material* **121** 138-44.
[5] Yan S.-p, Fang M.-X, Zhang W.-F, Wang S.-Y, Xu Z.-K, Luo Z.-Y, *et al.* 2007 Experimental study on the separation of CO_2 from flue gas using hollow fiber membrane contactors without wetting *Fuel Proc. Tech.* **88** 501-11.
[6] Hoff K A, Juliussen O, Falk-Pedersen O, and Svendsen H F 2004 Modeling and experimental study of carbon dioxide absorption in aqueous alkanolamine solutions using a membrane contactor *Ind. & Eng. Chem. Res.* **43**, 4908-21.
[7] Songolzadeh M, Soleimani M, Takht Ravanchi M, and Songolzadeh R 2014 Carbon dioxide separation from flue gases: a technological review emphasizing reduction in greenhouse gas emissions *The Scientific World J.* **2014**.
[8] Kartohardjono S, Darmawan R, Karyadi M F and Saksono N 2016 CO_2 absorption through super-hydrophobic hollow fiber membrane contactors *J. Env. Sc. Tech.* **9** 214-9.
[9] Kartohardjono S, Sembiring K A, Rexy, Syandika R, Ghasani F and Saksono 2017 CO_2 absorption from its mixture through super-hydrophobic membrane contactor *J. Env. Sc. Tech* **10** 25-34.
[10] Wang D, Teo W and Li K 2004 Selective removal of trace H_2S from gas streams containing CO_2 using hollow fibre membrane modules/contractors *Sep. Pur. Tech.* **35** 125-31.
[11] Wang R, Zhang H, Feron P and Liang D 2005 Influence of membrane wetting on CO_2 capture in microporous hollow fiber membrane contactors *Sep. Pur. Tech.* **46** 33-40.

IOP Conf. Series: Materials Science and Engineering **316** (2018) 012013 doi:10.1088/1757-899X/316/1/012013

The effect of hydrochloric acid addition to increase carbon nanotubes dispersibility as drug delivery system by covalent functionalization

P P D K Wulan[1,3], S H Ulwani[2], H Wulandari[2], W W Purwanto[3], K Mulia[3]

[1]Research Center for Biomedical Engineering, Faculty of Engineering, University of Indonesia, Depok 16424, Indonesia

[2]Sustainable Energy Laboratory, Chemical Engineering Department, Faculty of Engineering, University of Indonesia, Depok 16424, Indonesia

[3]Chemical Engineering, Faculty of Engineering, University of Indonesia, Depok 16424, Indonesia

wulan@che.ui.ac.id

Abstract. This study is to obtain the effect of adding hydrochloric acid (HCl) to the mixture of sulfuric acid (H_2SO_4) and HNO_3 in CNT covalent functionalization. HCl expected to increase the dispersibility of functionalized CNT by improving the dispersion time period done with H_2SO_4 or HNO_3. Functionalization used mixture of H_2SO_4 and HNO_3 with volume ratio of 3:1. Covalent functionalization used 0.5 grams of MWCNT ultra sonicated in 50mL HNO_3 with and mixture of H_2SO_4 and HNO_3. Additions of 200 mL HCl used variation of molarity from 1M, 2M, 3M, 4M, 5M to 6M. CNT were oxidized to form carboxylic and hydroxyl bonds on the surface that increase dispersibility of CNT. FTIR spectrums showed the existences of carboxyl and hydroxyl group on spectra of 2600-3700 cm^{-1} and 900-1400 cm^{-1}. Dispersion tests, which showed functionalized CNT (f-CNT) dispersion capabilities, were done by dissolving f-CNT in water. The study resulted that 6M f-CNT (NSC6) gave the best dispersion with zeta potential value of -37.1mV. NSC6 gave the longest dispersion time which was 20 days until f-CNT settle again. SEM-EDS micrographs showed the surface structure of 6M f-CNT without significant damage and no longer contain impurities of Fe, Ni, and Cl.

1. Introduction

Cancer is one major cause of death in the world with approximately 8.2 million deaths in 2012 and grows up each year. Cancer treatments such as surgery, chemotherapy, radiotherapy and drugs still pose problems such as the content of toxic effects harming healthy cells surrounding the cancer cells called metastasis, hair loss and nausea, and lack of targeted specificity where there is an inability of drugs to reach the point of cell in particular cancers [7]. Instead of the primary cancer cells, most patient died because metastasis effect.

Therefore, we need a new technology in cancer drug which is effective and efficient. Drug delivery must be biocompatible, biodegradable, nontoxic, effective, high drug loading capacity, dispersed well, and have affordable price [11]. Carbon nanotubes (CNT) is being investigated as a drug delivery. CNT has unique physical and chemical properties to be a promising candidate for multifunction drug delivery systems [8]. Not only does CNT pave the way for the development of drug delivery agent, it also can be equipped with a targeting agent and a molecular disguises [9]. Targeting agents will help CNT to be well targeted and molecular disguises will avoid clearance by the immune system that detects foreign substances in the body.

Applying CNT as a drug delivery requires hydrophilic properties and a high dispersibility. Dispersibility is important since it is the key for CNT biocompatibility as a drug delivery [1]. Compatibility is a material ability to adapt to the environment where the material is placed, and cause no harm and no toxic. Biocompatibility of CNT will produce less side effects to the body. Functionalization is necessary to increase the solubility and dispersibility of CNTs in solution, forming CNT that biocompatible and low toxicity [3]. By functionalization, CNT will reach the cancer cells because it can circulate in the blood and does not lead to blockage of the arteries.

Functionalizing CNT can be done by covalent and non-covalent functionalization. Non-covalent functionalization of CNT is poorly suited to be used as a drug delivery as a result of weaker bond between the carbons and accentuate the electronic properties of CNT [7]. Covalent functionalization is preferable since it can improve the dispersibility of CNT, reduce toxicity, and form active sites on CNT so as to bring the drug to be delivered. Covalent modification of CNT ensures the emergence of the hydrophilic nature of the CNT [3]. Covalent functionalization commonly performed is the oxidation of CNT with strong acids. Strong acids will improve the properties of CNT by forming carboxylic groups on the surface of CNTs which will increase the dispersibility of CNTs in solution [6].

Strong acid used is nitric acid (HNO_3), or a mixture of sulfuric acid (H_2SO_4) and HNO_3. Each study of strong acids functionalization successfully produces carboxylic groups on the surface of the CNT. However, the results are not maximized. CNT resulting damage to the surface of CNT wall, and they form aggregates for their salt content makes CNT cannot be directly applied [12]. Other results showed that CNT produced yet stable and perfectly dispersed [10]. Subsequent research began to produce CNT which can be dispersed, but still suffered structural damage to the wall CNT [2].

Chloride acid (HCl) can be another strong acid used to functionalize CNT. HCl is a non-oxidative acid and cannot introduce oxygen-containing groups to CNT wall, but it can enhance the exposure of amorphous carbon. The exposure of amorphous carbon will lead to MWCNT with very low amount of graphitic nanoparticles, bearing higher oxygen existences. The presence of oxygen-containing groups facilitates the exfoliation of CNT bundles, and increases the solubility in polar media. HCl-treated MWCNT provided more available reaction sites, leading to enhanced sidewall functionalization.

This study is to obtain the effect of adding HCl to the mixture of H_2SO_4-HNO_3 which expected to increase the dispersibility of functionalized CNT. The previous literature studies of CNT functionalization have dispersion time less than 20 days. This study will improve the time required for CNT to settle again which demonstrates the CNT's dispersion capabilities. By the high dispersibility, CNT will be easier to circulate, not easily settle, and last longer in blood vessel without causing the blockage.

2. Methodology/Experimental

MWCNTs were functionalized with three different acids.

2.1. Preparation of materials and tools

Pristine MWCNTs were purchased from Chengdu Organic Chemicals Co. Ltd., Sichuan, China. MWCNT has purity >98% with outer diameter 10-20nm and length 10-30μm. MWCNT were placed in 8 different container for each 0.5 grams. Nitric, sulfuric, and chloride acid were purchased from Laboratory of Chemical Engineering, University of Indonesia. In this step, three kinds of solution were made in a room temperature. The solution were HNO_3 6M, 7 mixture solution of H_2SO_4/HNO_3 6M with volume ratio of 3:1 placed in 7 different beaker glasses, and HCl 1M, 2M, 3M, 4M, 5M, and 6M.

2.2. Functionalization of MWCNT

There are eight (8) samples made in the covalent functionalization of MWCNT to get the effect of the addition of different types of acid. First procedure is to immerse MWCNT into H_2SO_4/HNO_3 6M.

MWCNT in acid solution was treated in an ultrasonic bath for 2 hours in 40°C. After being left overnight, HCl 1M was added to the solution. Afterwards, the solution was filtered with 0.45 μm cellulose acetate membrane and neutralized with water until pH 5.5 was reached. MWCNT which have been separated was dried in an oven for 20 minutes in 140°C. The next trials had the same process, however, the addition of HCL were variated into 2M, 3M, 4M, 5M, and 6M. The second procedure had the same process, however, the addition of HCl was eliminated. The last procedure only use HNO_3 6M to functionalize MWCNT in spite of the mixture of acids.

Table 1. Sample number.

Sample Number	Acid Solution	Addition of HCl
NSC1	H_2SO_4/HNO_3	1M
NSC2	H_2SO_4/HNO_3	2M
NSC3	H_2SO_4/HNO_3	3M
NSC4	H_2SO_4/HNO_3	4M
NSC5	H_2SO_4/HNO_3	5M
NSC6	H_2SO_4/HNO_3	6M
NS	H_2SO_4/HNO_3	-
N	HNO_3	-

2.3. *Characterization of MWCNT*

2.3.1. FTIR spectroscopy. Fourier Transform Infrared spectroscopy (FTIR) data was obtained using Thermo Scientific iS5 Spectrometer. FTIR spectroscopy has been widely used in the determination of functional groups. There are 25 samples, including pristine MWCNT, 24 samples from 8 samples with each of 3 different stages that will be observed that the stages of ultrasonication, filtration and drying. KBr with 0.5% by weight of MWCNT of each samples were prepared for FTIR spectroscopy. MWCNTs were mechanically mixed to the KBr powder and pressed into discs shape. FTIR was used to analyse the changes of surface chemical bonding and structure in the frequency range of 4000–500 cm^{-1}.

2.3.2. Dispersion test. Dispersion test was performed to compare the stability of the suspension of MWCNT before and after functionalization. 4 milligrams of each sample MWCNT was dispersed in 10 ml of demineralized water with sonication for 5 minutes. These solutions were observed and recorded the time MWCNT takes to settle again.

2.3.3. Zeta potential. Zeta potential was done to prove the dispersion test results quantitatively. Zeta potential characterization was determining the accumulation of electric charge on CNT surface. Electric charge that shown was representation of dispersibility stabilization of CNT. 8 milligrams of each sample MWCNT was dispersed in 20ml of demineralized water with sonication for 5 minutes. These solution were put on the Malvern Zetasizer 4 to measure zeta value.

2.3.4. Scanning Electron Microscopy- Energy Dispersive Spectroscopy. Scanning Electron Microscope (SEM) with high resolution is a powerful tool for imaging the fine structure of materials and nanoparticles created by nanotechnology. Characterization of the nanotube composition was done using energy dispersive x-ray spectroscopy (EDS). MWCNT observations and their morphological analysis used SEM-EDS JEOL, JSM 6510 LA. Images were recorded by placing a drop of the samples dispersed in isopropanol on copper grids coated with carbon.

3. Results and discussion

3.1. FTIR spectroscopy

The presence of carboxyl and hydroxyl functional groups on the functionalized CNT (f-CNT) surface can be identified by different characterization. FTIR was conducted and analysed on the phase of ultrasonication in which influence the success of functionalization. FTIR spectra can complement the dispersion time data and identify chemical composition of pristine and f-CNT. The comparison of pristine and f-CNT spectra after treatment for each samples were showed on Figure 1 in the range of 500 to 4000 cm^{-1}. FTIR spectra from the pristine MWCNTs show a peak at 2068 and 2346 cm^{-1} which refers to the backbone of pristine MWCNT. Small peak at 1589 cm^{-1} on pristine MWCNT may refers to the C=O stretch of carboxyl group which can be created during purification process by the manufacturer.

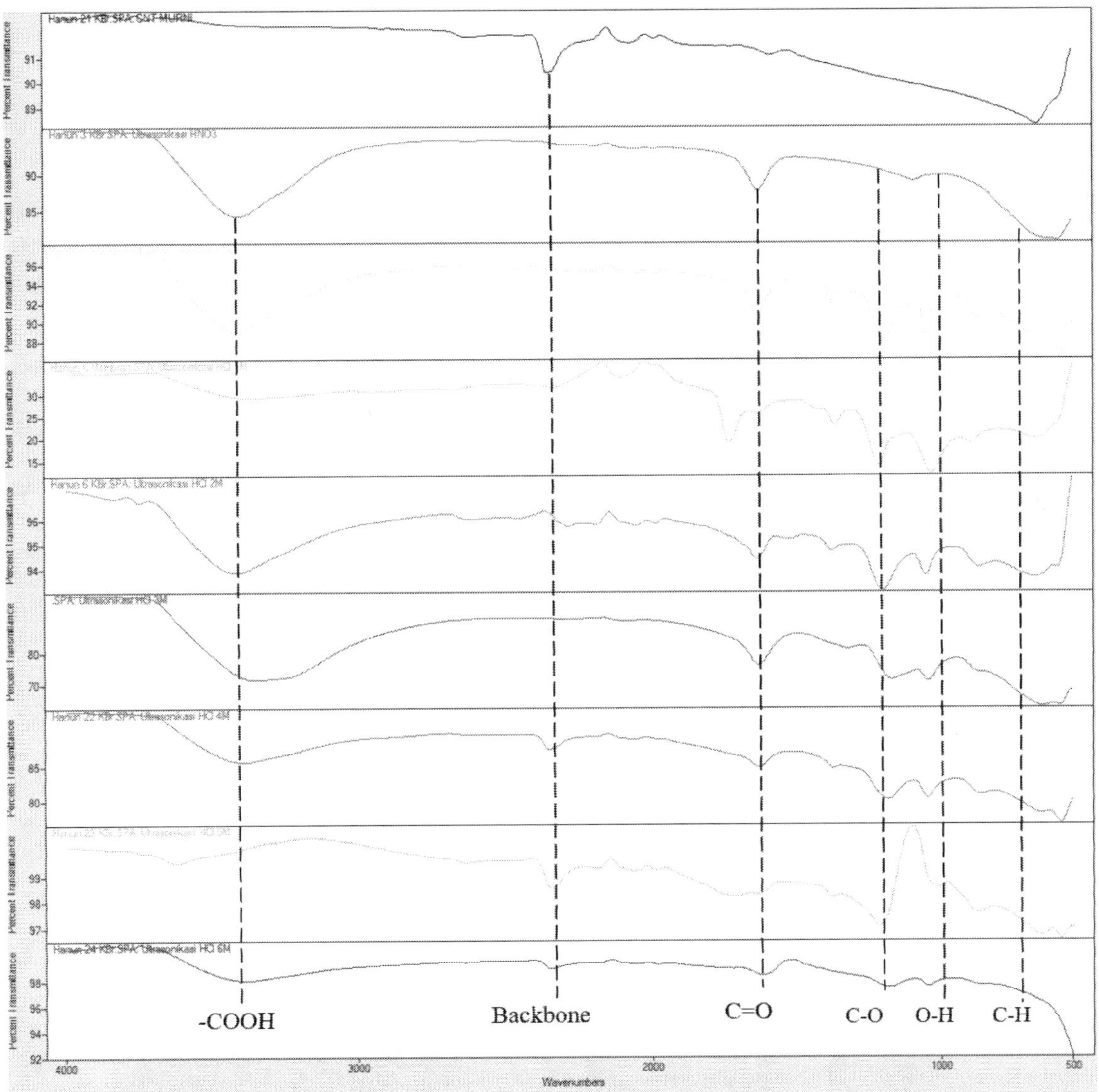

Figure 1. Comparison of FTIR spectra of pristine MWCNT and treated MWCNT.

Most of samples have been inserted carboxyl and hydroxyl groups successfully. Peaks between 2600 and 3700 cm-[1] are the characteristic of C-H vibrations and O–H bonds. These peaks can be related to carboxylic groups (-COOH). The strong peak of MWCNT backbone in 2346 cm^{-1} reduces in

IOP Conf. Series: Materials Science and Engineering **316** (2018) 012013 doi:10.1088/1757-899X/316/1/012013

functionalized MWCNTs spectra. It means, the structure of carbon bonds has been broken and added with new functional groups. However, backbone spectra should be existed to make sure that MWCNT has no fatal damage after functionalization.

Peaks between 1680-1800 cm-1 are the characteristics of C=O bond type from carbonyl and carboxyl groups. Peaks between 1000 and 1250 cm-1 are the characteristics of C-O bond type from alcohol, ether, and carboxylic acids functional groups. Peaks between 900 and 1400 cm^{-1} are the characteristics of O-H bond type from hydroxyl groups. Peaks between 680 and 880 cm^{-1} are the characteristics of C-H bending bond. From FTIR characterization, it can conclude that samples with addition of HCl of 2M, 5M, and 6M are three samples with the highest amount of functional groups.

3.2. Dispersion test

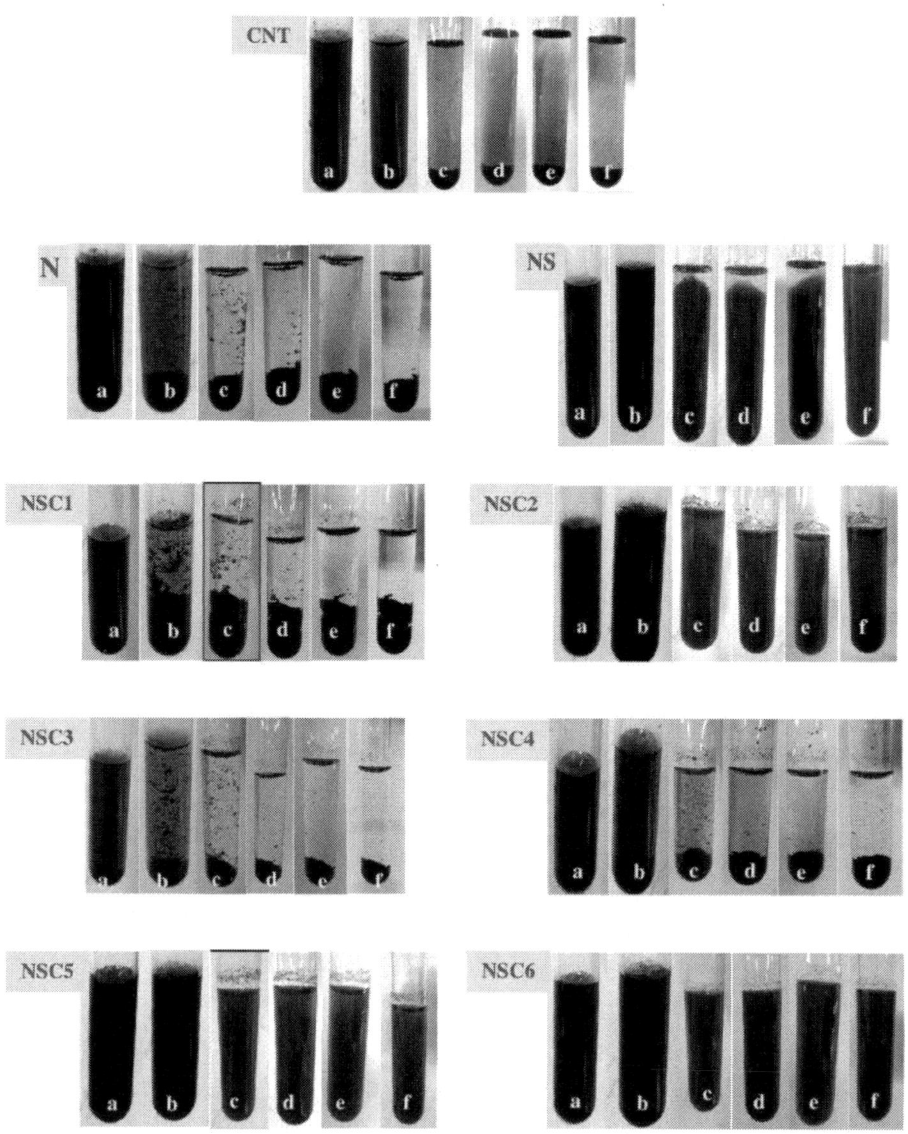

Figure 2. Dispersion test of samples: (a)after sonication, (b)day 1, (c)day 2, (d)day 3, (e)day 4, and (f)day 6

Dispersion test is occurred to identify the presence of hydrophilic functional groups on f-CNT surface and make sure that functionalization is succeed. The time required for CNT to settle in a polar solvent show how effective the functionalization works. This method is fast, cheap, and provide qualitative information. Water is chosen because it represents blood human well since blood contains 95% of water. F-CNT were upheld for several days after being sonicated in water for 5mins. Figure 2 shows photographs of pristine and f-CNT in water 144 hours (6days) after sonicated in water.

When f-CNT suffers oxidation, chemical elements are adsorbed to form functional groups. The functional groups will positive or negatively charged to keep the solution dispersed in water. From the observation, all samples can be dispersed in water. These show that functionalization are successfully occurs. However, the time f-CNT's need to settle are different for each sample. Some samples with addition of HCl still remained as stable suspension in water for more than 6 days compared to pristine CNT which only can settle for several hours. Acid solution without HCl addition only disperse less than 2 days. The settle of f-CNT may be caused by the agglomeration and lack of hydrogen bonding. In other way, stable dissolved f-CNT proof the presence more functional groups of carboxyl and hydroxyl on MWCNT surface. This fact leads to reduction of Van der Waals interactions among carbon bonds and increase the interactions of carboxylic group and water. The reduction of Van der Waals interactions can lead to increase the viscosity of the solution. Thus, the solution dispersity will last longer.

Variation of HCl molarity added are based on the amount of HCl moles contained on the solution. From dispersion test, best addition of HCl are NSC2 (2M), NSC5 (5M), and NSC6 (6M). Those sample support the results of FTIR spectra data with biggest amount of functional groups adsorbed will more stable in water. Furthermore, only NSC6 (6M) can disperse well for 20 days as shown in Figure 3. This result proof that the more amount of HCl moles possibly add the more hydrophilic functional groups into CNT surface and increase their dispersibility.

Figure 3. Dispersion test result of NSC2, NSC5, and NSC6 after 20 days

3.3. Dispersion test

High zeta potential values (negative or positive) prevent agglomeration of nanoparticles due to counteracting forces and electrical stabilization of nanoparticle dispersions. The potential zeta value and CNT dispersibility capability are considered as stable when the zeta value is above + 30mV or below -30 mV. If the zeta value is get near to zero, there will be a greater attraction than the repulsive force that causing agglomeration.

Table 2. Zeta potential result.

Sample Name	Zeta Potential (mV)
Pristine CNT	-16.8
NSC5	-21.9
NSC6	-37.1

Based on Table 2, NSC5 zeta value is -21.9mV and considered as unstable state since it is above -30mV. NSC6 shows zeta value -37.1mV. NSC6 value is considered as stable state. This results proofed that NSC6 has greater repulsive force, prevent agglomeration of particle, and gives the best dispersibility.

3.4. Scanning electron microscopy-energy dispersive spectroscopy

SEM was conducted to detect surface morphological changes shown on Figure 4a is pristine CNT and Figure 4b and 4c is chloride-sulphate-nitrate functionalized MWCNT. Figure 4a shows almost typical SEM images of the pristine CNT as well as those treated by HCl, HNO_3, and H_2SO_4. It means, the functionalization with 3 kinds of acid as oxidizing agents do not cause structural damage of the MWCNT. The etching of surface is due to the disordered sites which bonding of carbon and carbon have been replaced with others functional groups, carboxyl and hydroxyl groups (O=C-OH and C-OH). It is possibly because of the mild condition of acid which have been diluted before being immersed with CNT. Based on Figure 4, NSC6 shows clearer image than NSC5 that indicate few impurities (also showed with arrows). This result also supported by high zeta value of NSC6 has high dispersibility, less defects and impurities on surface.

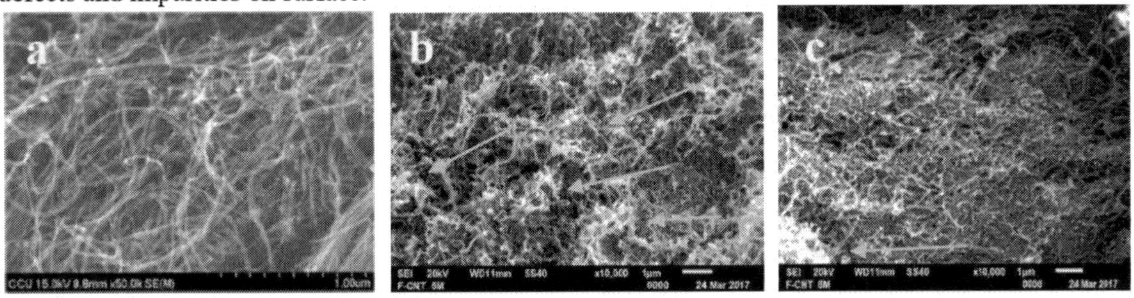

Figure 4. SEM images of CNTs: (a) pristine CNT, (b) NSC5 f-CNT, and (c) NSC6 f-CNT; with arrows show impurities

Table 3 supports that NSC5 have more impurities of Cu and Au than NSC6. Cu is impurity may come from catalyst used on manufacturing or impurity from tube used in EDS characterization. Au detected as the coating of CNT while doing EDS characterization. Both NSC5 and NSC6 have no Fe detected that can be said that both samples are considered as safe [4]. NSC6 gives more functional groups and few impurities because NSC6 have more moles in the solution of HCl. This fact in line with the result of FTIR and dispersion test that show NSC6 have the best dispersibility among others.

Table 3. EDS result of pristine CNT and f-CNT

Sample	Contents (%)						
	C	Cu	Au	Fe	Ni	Cl	O
Pristine	98.39	-	-	0.23	0.93	0.45	-
NSC5	95.06	1.83	3.11	-	-	-	-
NSC6	91.31	1.31	2.61	-	-	-	2.4

4. Conclusion

Additional of HCl to HNO_3 and H_2SO_4 solution successfully increase the dispersibility of CNT. HCl addition of 6M (NSC6) has the best dispersion in water with dispersion time improved from 1 day to 20 days. Sample NSC6 has the best dispersion with the least surface damage and no impurities. This proofed that the more amount of HCl moles clean more impurities and add more hydrophilic functional groups into CNT surface. Sample NSC6 can be implemented further to be drug delivery because it has qualified the requirement of drug delivery to have high dispersibility in water. The next requirement to be fulfil is the biocompatibility and toxicity of CNT in human body.

Acknowledgement
The author would like to thank The Directorate General of Higher Education, Ministry of National Education Indonesia by HIBAH Fundamental 2017 and Ministry of Research, Technology and Higher Education Republic of Indonesia by PKMP 2017. Thanks to Sustainable Energy Laboratory, Chemical Engineering Department, and Research Center for Biomedical Engineering University of Indonesia for the valuable cooperation.

References
[1] Albert Bianco, Kostas Kostarelos and Maurizio Prato 2005 *Applications of carbon nanotubes in drug delivery* (Italy: Elsevier Science Ltd) 9: 674-679
[2] Duha S Ahmed, Adawiya J H and M R Mohammad 2013 *Comparesion of functionalization of multi-walled carbon nanotubes treated by oil olive and nitric acid and their characterization* (Iraq: Energy Procedia) 36: 1111-1118
[3] Jacob M N, Sunny E I, Wilhelm E N and Clarence S Y 2011 *Covalent Functionalization for Multi-walled Carbon Nanotube (f-MWCNT)-Folic Acid Bound Bioconjugate* (South Africa: Asian Network for Scienticic Information) 15: 2700-2711
[4] Pardoyo, Sri Eny S, Siti N and Agus S 2015 *Synthesis of Carboxyl (CNT-COOH) and Ammine Carbon Nanotubes (CNT-NH_2)* (Semarang: Jurnal Sains dan Matematika Universitas Diponegoro)
[5] Praswasti PDKW and Hanifia W 2016 *Dispersion and Hydrophilic Characterization of Carbon Nanotube as Drug Delivery Using Olive Oil and Cetyltrimethyl Ammonium Bromide (CTAB)* (Indonesia: Submitted UI Proceedings on Science and Technology)
[6] S Shahnawaz, B Sohrabi and M Najafi 2013 *The investigation of functionalization role in multi-walled carbon nanotubes dispersion by surfactants* (Iran: Iran University of Science and Technology)
[7] Seyed Y M, Naghmeh N, Oshani D, Aaron T and Alexander M S 2011 *A new era of cancer treatment: carbon nanotubes as drug delivery tools* (United Kingdom: Dovepress International Journal of Nanomedicine)
[8] Sumio Iijima 2002 *Carbon nanotubes: past, present, and future* (Japan: Elsevier Science Ltd) 323: 1-5
[9] Swatantra K S K, Saurav G, Awani K R and Satyawan S 2013 *Carbon nanotubes as a novel drug delivery system for anticancer therapy: a review* (India: Brazilian Journal of Pharmaceutical Sciences)
[10] Ursula D-W, Jean-Michel B, Po-Wen C, Ralf G, Sergey L and Siegmar R 2002 *Chemical functionalization of single walled carbon nanotubes* (France: Elsevier Science Ltd) 2:497-501
[11] Zahra A M, Seyed F A, Ali Z and Mohammad R T 2015 *A comparative study on non-covalent functionalization of carbon nanotubes by chitosan and its derivatives for delivery of doxorubicin* (Iran: Elsevier Science Ltd)
[12] Zhuang Liu, Scott T, Kevin W and Hongjie D 2008 *Carbon Nanotubes in Biology and Medicine: In vitro and in vivo Detection, Imaging and Drug Delivery* (USA: Tsinghua University Press and Springer-Verlag) 2: 85-120

CFD study of mixing miscible liquid with high viscosity difference in a stirred tank

S Madhania[1,2], A B Cahyani[2], T Nurtono[2], Y Muharam[1], S Winardi[2] and W W Purwanto[1,*]

[1]Chemical Engineering Department, Faculty of Engineering, Universitas Indonesia, Kampus Baru UI Depok, Depok, 16424, Indonesia
[2]Chemical Engineering Department, Faculty of Industrial Technology, Institut Teknologi Sepuluh Nopember, Kampus ITS Sukolilo, Surabaya, 60111, Indonesia

*Corresponding author: widodo@che.ui.ac.id

Abstract. The mixing process of miscible liquids with high viscosity difference is crucial role even though the solution mutually dissolved. This paper describes the mixing behaviour of the water-molasses system in a conical-bottomed cylindrical stirred tank (D = 0.28 m and H = 0.395 m) equipped with a side-entry Marine propeller (d = 0.036 m) under the turbulence regime using a three-dimensional and transient CFD-simulation. The objective of this work is to compare the solution strategies was applied in the computational analysis to capture the detail phenomena of mixing two miscible liquid with high viscosity difference. Four solution strategies that have been used are the RANS Standards k-ε (SKE) model as the turbulence model coupled with the Multiple Reference Frame (MRF) method for impeller motion, the RANS Realizable k-ε (RKE) combine with the MRF, the Large Eddy Simulation (LES) coupled with the Sliding Mesh (SM) method and the LES-MRF combination. The transient calculations were conducted with Ansys Fluent 17.1 version. The mixing behaviour and the propeller characteristic are to be compared and discussed in this work. The simulation results show the differences of flow pattern and the molasses distribution profile for every solution strategy. The variation of the flow pattern which happened in each solution strategy showing an instability of the mixing process in stirred tank. The LES-SM strategy shows the realistic direction of flow than another solution strategies.

1. Introduction

Based on the type of fluid, the categories of the multiphase flow divided into gas-liquid, solid-liquid and liquid-liquid. The liquid-liquid mixing divided into immiscible liquids dispersion and miscible liquids blending, the latter of which is the primary focus of the present work. The mixing process of miscible liquids is crucial role even though the solution mutually dissolved. Homogenization of miscible liquids achieves slowly by molecular diffusion and natural convection, but agitated can obtain homogenous condition more quickly. The application of mixing of miscible liquids is in bioethanol industrial which uses molasses as the raw material.

Advances in computer technology and mathematical model have enabled researchers to conduct the fluid dynamic investigation computationally. The computationally study of mixing process more efficient than experimentally. The *Computational Fluid Dynamics* (CFD) has become a powerful the tools to conduct the fluid dynamic investigation, and it has been successfully used to predict the mixing behaviour inside stirred tank. The CFD study on mixing inside stirred tank mostly evaluating the top-entry impellers configuration. Little work has done on studying the side-entry impellers configuration by CFD method, i.e. the homogenization of crude oil [1], the effect of propeller layout on mixing time of crude oil [2], the effect of impeller layout on mixing intensity on the solid-liquid system [3] and the mixing of pseudoplastic solutions [4].

So far, the study on the computational fluid dynamic about the mixing of the miscible liquids inside stirred tank is limited, among others, the blending of ethanol and glycerol with an anchor impeller [5] and the homogenization of two miscible liquid with a pitch blade turbine [6]. The water-molasses process of mixing in a stirred tank under turbulence regime is subject of our research; the previous work has done to compare the mixing behaviour of top-entry impeller configuration and side-entry impeller configuration [7]. The specific concern on the appropriate solution strategy which was applied to determine the mixing behaviour inside the tank is the subject of this present paper. The main differences include the use of different modelling turbulence and impeller motion model.

Fluid flows inside stirred tank mostly under the turbulence regime, selection of turbulence models is crucial to obtain a velocity flow fields. The Direct Numerical Simulations (DNS), the Large Eddy Simulation (LES) and the Reynolds-Averages Navier-Stokes (RANS) were included in the turbulence approach. The DNS resolves all turbulent length and time scales by directly integrating the Navier-Stokes equation while in LES, large eddies are resolved directly, and small eddies are modelled [8]. On the other hand, the Reynolds-averaged Navier-Stokes (RANS) equations amount to averaging out the large eddies. The RANS model involved the Standard k-ε (SKE) model [9], the Renormalization k-ε (RNG) model [10] and the Realizable k-ε (RKE) model [11]. The RNG k-ε model more accurate and reliable for a wider class of flow than the SKE model, however, the RKE model provides the best performance [12].

The Sliding Mesh (SM) method and the Multiple Reference Frame (MRF) method was applied to determine the impeller motion. The SM method is a fully transient approach where the rotating mesh slides relatively to the stationary mesh, and no needs experimental boundary condition because the flow around the impeller blades is being calculated in detail [13]. The MFR is a steady flow field in which predicted the relative motion between the rotating frame and stationary reference frame. In MRF model, impeller region is modelled as stationary, and tank wall is modelled as rotating zones [14].

Another important factor that can not be ignored in stirred tank study is to characterise the impeller; the power number is one. Power number (N_p) is the amount of energy required per unit time in a mechanical agitation in stirred tank [15]. Power consumption (P) is the power delivered to the fluid (the product of the impeller speed, $2\pi N$, in rad/s) and measurement of torque (T_q) :

$$P = 2\pi N T_q \tag{1}$$

The power number (N_p) can be calculated by:

$$N_p = \frac{P}{\rho N^3 D^5} \tag{2}$$

The objective of this work is to compare the solution strategies which was applied in the computational analysis on mixing of two miscible liquid with high viscosity difference inside conical-bottomed cylindrical stirred tank equipped with a side-entry Marine propeller using the CFD method to capture the mixing behaviour and propeller characteristic.

2. Computational fluid dynamics methods

In this present work, the model equations numerically solved by adopting the finite volume. The three-dimensional and transient CFD-simulation was conducted with Ansys Fluent 17.1 version under the turbulence regime. The simulation is divided into four solution strategies, i.e. SKE-MRF, RKE-MRF, LES-SM and LES-MRF. The parallel calculation was performed using HP Z620 workstation with 12 cores and Windows 8 64 bit operating system.

IOP Conf. Series: Materials Science and Engineering **316** (2018) 012014 doi:10.1088/1757-899X/316/1/012014

2.1. Geometry system

The geometry system is a conical-bottomed cylindrical stirred tank with a diameter of 0.28 m and height 0.395 m. The geometry generated using Ansys Design Modeller software. The tank is equipped with a three-blade Marine propeller (d = 0.036 m) as can be seen in Figure 1, installed with side-entry configuration.

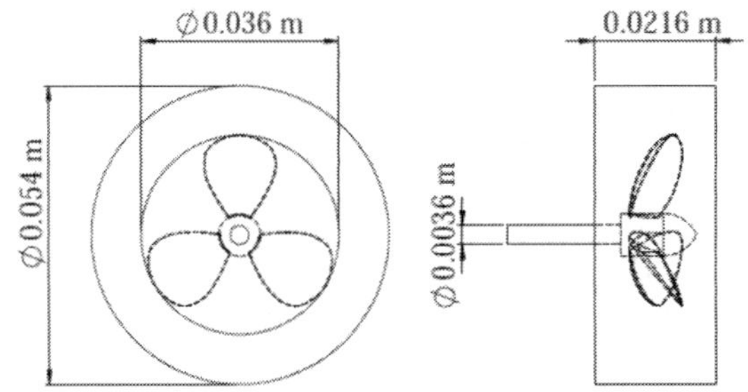

Figure 1. The geometry of Marine propeller and moving zone

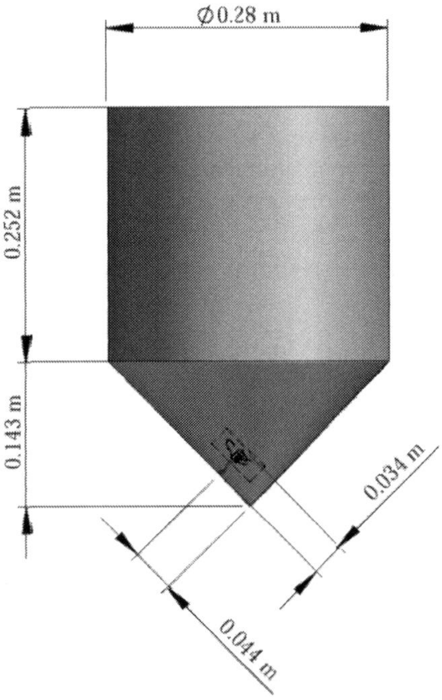

Figure 2. The stirred tank geometry

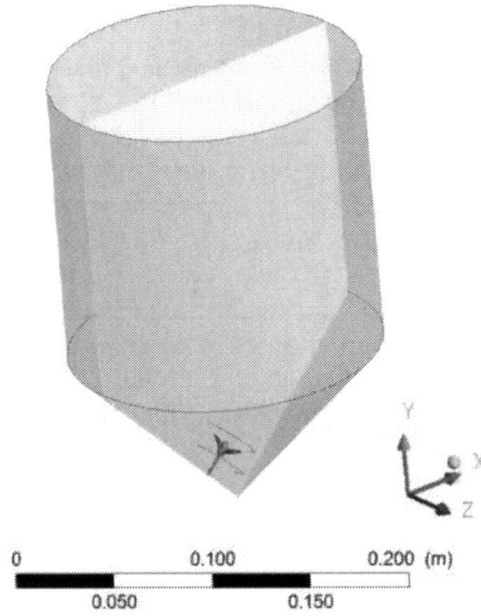

Figure 3. The observation plane

The propeller mounted on the conical side of the tank with and angle of 45° and located at a distance of 0.044 m from the bottom of the tank as shown in Figure 2. The flow domain is discretized through grids (122.468 nodes) by Ansys Meshing software and evaluate the quality of meshing using skewness methods. The result of simulation was analysed through the observation plane as shown in Figure 3.

2.2. Fluid flow approach and solution setup
The solver is pressure-based, and the gravitational acceleration is active at y-axis was set to -9.81 m/s^2. All the simulation was applied a Mixture model multiphase flow [12] under a turbulence regime with propeller rotational speed (N) set at 1000 rpm and time step 0,01 s. For the first solution strategy (SKE-MRF) is using the RANS (SKE) as the turbulence model [16] coupled with MRF as the rotational impeller method while the second solution strategy (RKE-MRF) is a combination of RANS (RKE) [11] with MRF. Moreover, the third solution strategy (LES-SM) was applied the Large Eddy Simulation (LES) approach with the WMLES S-Omega subgrid-scale turbulence model [17] coupled with the fully transient Sliding Mesh (SM) methods for simulating the propeller motion. The fourth solution strategy is using LES combined with MRF method. The solutions methods applied for every solution strategy as can be seen in Table 1.

2.3. Boundary condition
The fluid used are molasses (μ = 2.09 Pa.s) and water (μ = 0.001003 Pa.s). To define the initial condition, the height of molasses inside the tank is set at 0.158 m, and the height of water fills the tank is set at 0.235 above molasses. To define the boundary condition, the flow domain is divided into two parts, i.e. moving zone and static zone. The interfaces between moving zone and static zone treated as interface boundary condition. The shaft regarded as an absolute moving wall and the propeller is a moving wall in which relative to moving zone. The tank wall is a stationary wall and treated as the no-slip condition. The boundary condition at the liquid surface was set to symmetry because the shear stresses are zero.

Table 1. The solutions methods.

Solutions Methods	Solution Strategy			
	SKE-MRF	RKE-MRF	LES-SM	LES-MRF
Pressure – Velocity coupling	Semi-Implicit Method for Pressure-Linked Equation (SIMPLE)[18]		The Pressure-Implicit with Splitting of Operator (PISO)[19]	
Gradient	Least Squares Cell-Based*			
Pressure	Pressure Staggering Option (PRESTO)[20]			
Momentum			Bounded Central Differencing [21]	
Volume fraction	First Order Upwind*		First Order Upwind*	
Turbulent kinetic energy				
Turbulent dissipation rate				
Transient formulation	First Order Implicit*		Bounded Second Order Implicit*	

*[12]

3. Result and discussion
The mixing behaviour of two miscible liquid with high viscosity difference inside stirred tank is reviewed based on the flow pattern which is represented by the vector analysis of the flow velocity and the molasses distribution profile indicated by the density contour and prediction value of density mixture. Analysis of power number also carried out to show the propeller characteristic.

3.1. Flow pattern and molasses distribution profile

Based on Figure 4, it appears that the Marine propeller generates flow pattern with an axial direction and forms loop circulation of the fluid flow. However, there is a difference in the direction of the flow. Fluid flow with SKE-MRF solution strategy at 340 s which shown in Figure 4(a) indicates that the fluid is sucked and flow out towards the negative axial direction to the bottom of the tank towards the impeller, then it deviates changing its direction pushed upward towards the positive axial direction. Then the fluid flows radially to the negative axial direction towards the bottom of the tank and forms a double loop circulation inside the tank.

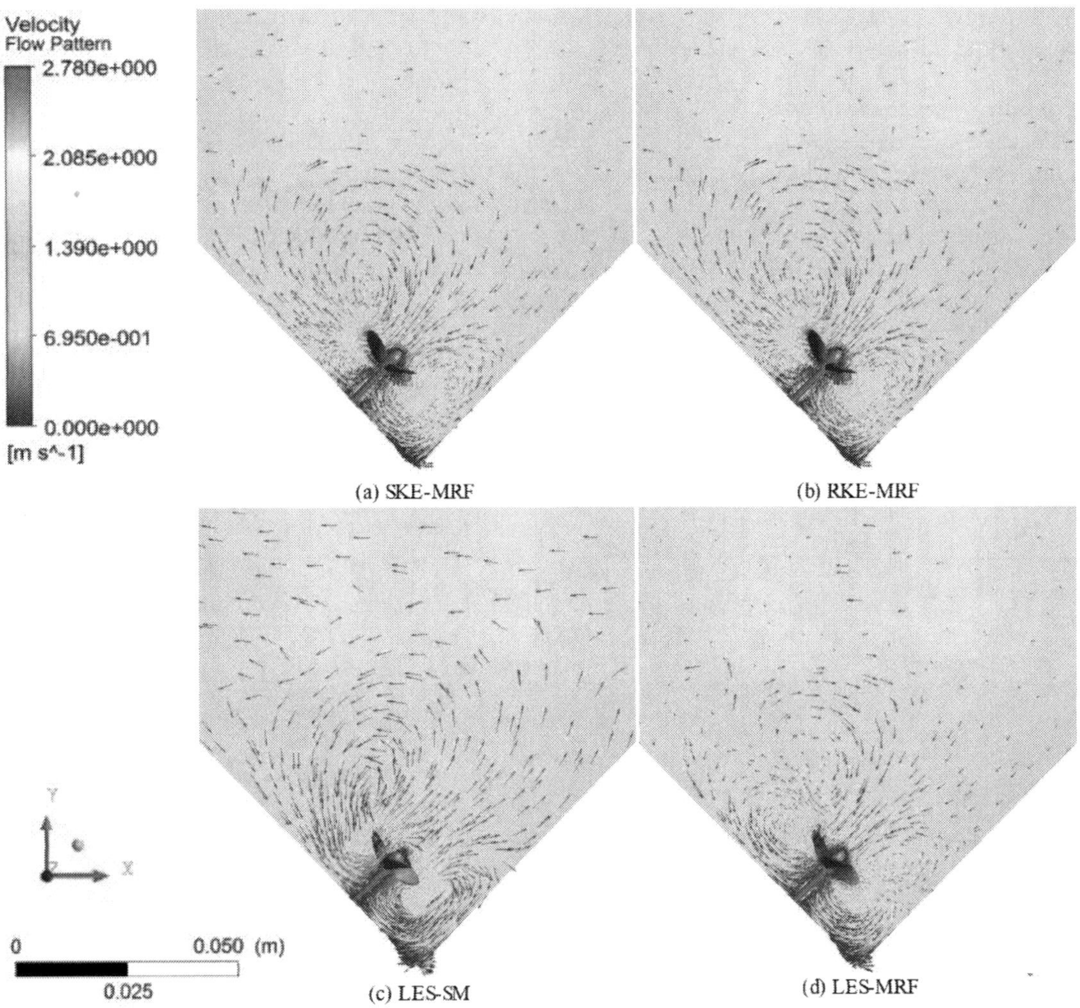

Figure 4. Flow pattern generated by Marine propeller for different solution strategy: (a) SKE-MRF; (b) RKE-MRF; (c) LES-SM; and (d) LES-MRF

The phenomenon of mixing generated by the SKE-MRF solution strategy is similar to the phenomenon of mixing produced by the RKE-MRF and the LES-MR solution strategy as shown Figure 4(b) and Figure 4(d), respectively. The loop circulation causes the mixing process of molasses and water which occurs in the tank as seen in Figure 5(a),(b) and (d). When the mixing process begins, water is sucked into the molasses-rich area at the bottom of the tank. It causes the molasses

IOP Conf. Series: Materials Science and Engineering **316** (2018) 012014 doi:10.1088/1757-899X/316/1/012014

concentration reduced. Then the water slowly diffused with molasses and formed some layer at the interface section. The contour analysis resulting from the LES-MRF solution strategy as seen in Figure 5(d) indicated a more quickly reach the homogenous condition than the two previous solution strategies (SKE-MRF and RKE-MRF).

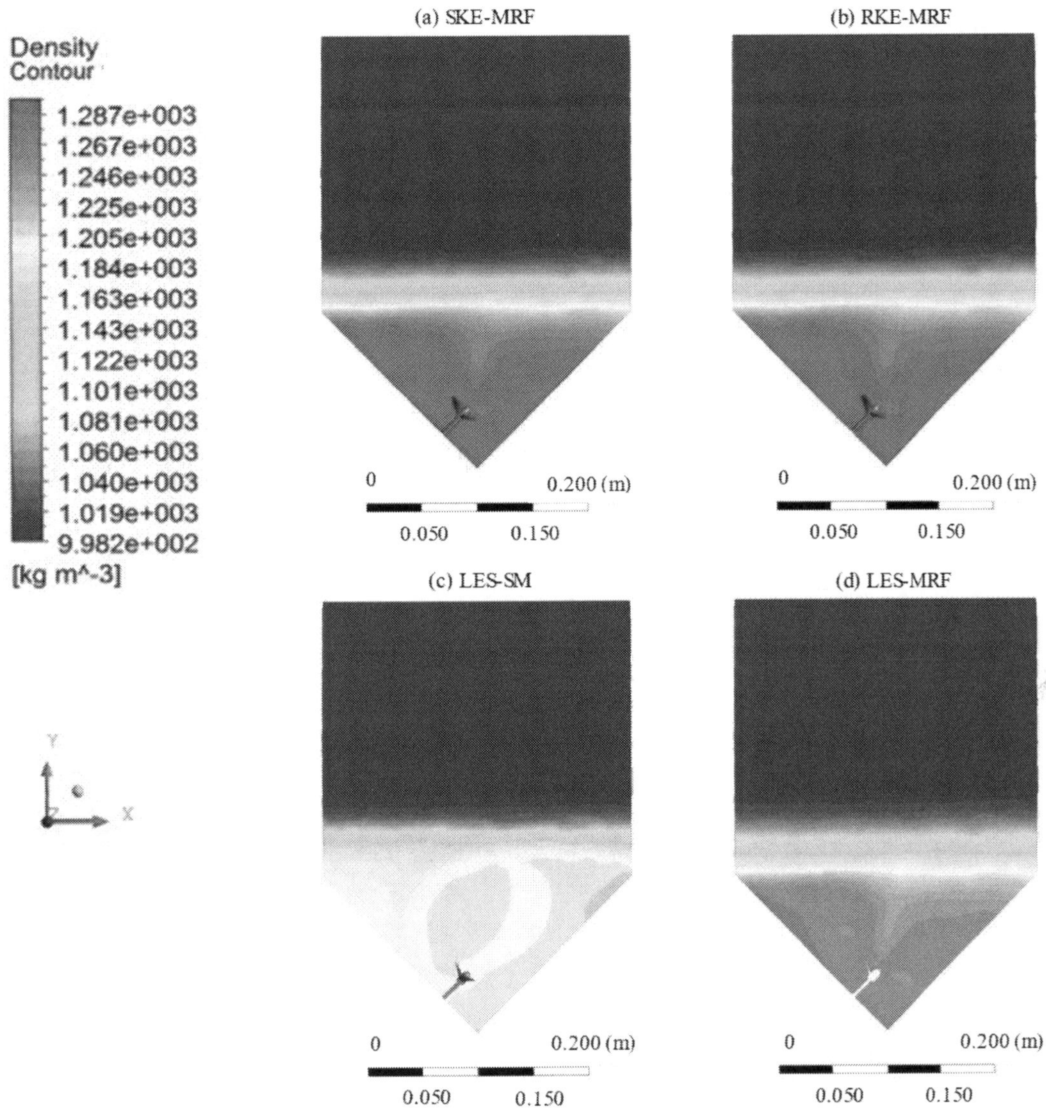

Figure 5. Density contour at 340 s for different solution strategy: (a) SKE-MRF; (b) RKE-MRF; (c) LES-SM; and (d) LES-MRF

Based on Figure 4(c), it can be seen that the direction of the flow is moving to the bottom of the tank towards the negative axial direction then deviates towards the impeller. The fluid is pumped out towards the negative axial direction then forms loop circulation around the tank There are some unstable flows and loops which are formed along the tank wall. The significant different in the direction of flow and flow pattern happened when using the Sliding Mesh impeller motion model. The motion of the impeller generated by the Sliding Mesh method shows the realistic direction of flow.

The Sliding Mesh method is suitable for the prediction of flow pattern in stirred tank [22]. Other than that, based on Figure 5(c) the density contour produced by LES-SM solution strategy shows the significant changes in the molasses distribution than another solution strategies.

The CFD simulation generates the prediction value of the density mixture as a function of time as shown in Figure 6. It indicates that the density changes yielded by LES-SM solution strategy show the significant decrease and it reveals that the homogeneous condition reached sooner than the other solution strategies.

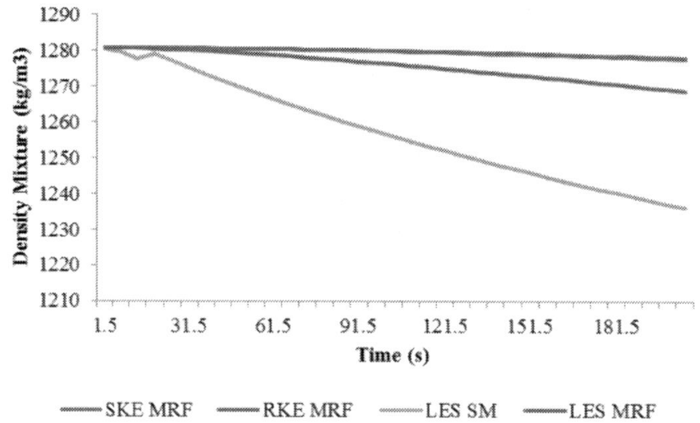

Figure 6. The density changes inside stirred tank as a function of time for different solution strategy

3.2. *Propeller characteristic*

The power number analysis represents the propeller characteristic in this work. The analysis of the power number generated by the propeller for all solution strategies is shown in Figure 7. Based on that graph, the power number prediction value generated by the LES-SM solution strategy is significantly different from three other solution strategies. The power number prediction value as a result of the LES-SM solution strategy is in accordance with the literature [23].

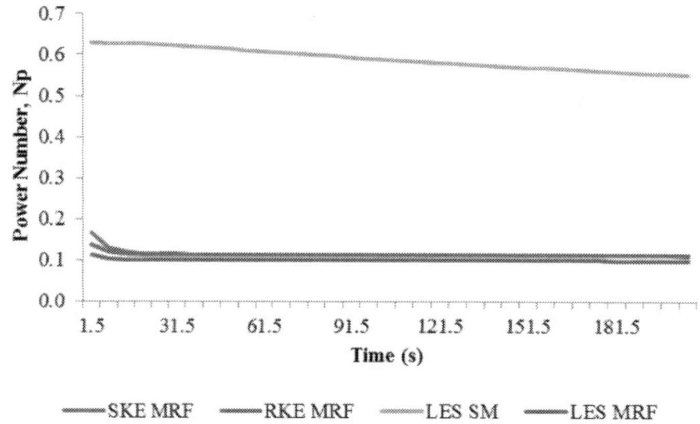

Figure 7. The power number produced by Marine Propeller for various solution strategies as a function of time

Furthermore, the graph trend on LES-SM solution strategy denotes that as the time goes by the value of power number decrease while the other solution strategies tend to indicate a constant trend. It represents that the power required for mixing process is reduced as the length of mixing time and the homogeneity of the mixture is increased.

4. Conclusion

In the presents work, the mixing of two miscible liquid with high viscosity difference inside conical-bottomed cylindrical stirred tank equipped with the side-entry Marine propeller in four different solution strategies (SKE-MRF, RKE-MRF, LES-SM and LES-MRF) at 1000 rpm was investigated via three-dimensional and transient CFD-simulation. The CFD-simulation result shows the mixing behaviour inside the stirred tank and the propeller characteristic for every solution strategy. The variation of the flow pattern which happened in each solution strategy showing an instability of the mixing process in stirred tank. The CFD analysis led to conclude that the flow direction generated by propeller as a result of the LES-SM solution strategy shows more realistic than another solution strategies. The prediction value of power number (N_p) generated by the LS-SM solution strategy is in accordance with the literature. Moreover, on the LES-SM solution strategy, signify that the density changes more quickly to reach the homogeneous condition than another solution strategies. The Sliding Mesh impeller motion approach generates the significant effect on the propeller characteristic and flows direction; this influences the molasses distribution inside the stirred tank. Further validation of these models by visual experimental and grid dependent test to determine the minimum grid is necessary will have to perform.

Acknowledgement

This work has been carried out by the financial support of the Universitas Indonesia under the project "Hibah Publikasi Internasional Terindeks Untuk Tugas Akhir Mahasiswa" (PITTA) Fiscal Year 2017 no. 849/UN2.R3.1/HKP.05.00/2017.

References

[1] Dakhel A A and Rahimi M 2004 *CFD simulation of homogenization in large-scale crude oil storage tanks* Journal of Petroleum Science and Engineering **43**(3–4) pp 151-161

[2] Rahimi M 2005 *The effect of impellers layout on mixing time in a large-scale crude oil storage tan.* Journal of Petroleum Science and Engineering **46**(3) pp 161-170

[3] Wu B 2011 *CFD investigation of turbulence models for mechanical agitation of non-newtonian fluids in anaerobic digesters* Water Research **45**(5) pp 2082-2094

[4] Sossa-Echeverria J and Taghipour F 2015 *Computational simulation of mixing flow of shear thinning non-newtonian fluids with various impellers in a stirred tank* Chemical Engineering and Processing: Process Intensification **93** pp 66-78

[5] Al-Qaessi F and Abu-Farah L 2009 *Prediction of mixing time for miscible liquids by CFD simulation in semi-batch and batch reactors* Engineering Applications of Computational Fluid Mechanics **3**(1) pp 135-146

[6] Derksen J J 2011 *Blending of miscible liquids with different densities starting from a stratified state* Computers & Fluids **50**(1) pp 35-45

[7] Madhania S et al 2017 *Mixing behaviour of miscible liquid-liquid multiphase flow in stirred tank with different marine propeller installment by computational fluid dynamics method* Chemical Engineering Transactions **56** pp 1057-1062

[8] Bakker A and Oshinowo L M 2004 *Modelling of turbulence in stirred vessels using large Eddy simulation* Chemical Engineering Research and Design **82**(9) pp 1169-1178

[9] Launder B E and Spalding D B 1974 *The numerical computation of turbulent flows* Computer Methods in Applied Mechanics and Engineering **3**(2) pp 269-289

[10] Yakhot V and Orszag S A 1986 *Renormalization group analysis of turbulence. I. Basic theory* Journal of Scientific Computing **1**(1) pp 3-51

IOP Conf. Series: Materials Science and Engineering **316** (2018) 012014 doi:10.1088/1757-899X/316/1/012014

[11] Shih T H et al 1995 *A new k-ϵ eddy viscosity model for high reynolds number turbulent flows* Computers & Fluids **24**(3) pp 227-238

[12] ANSYS-Fluent 2016 *Fluent 17.1 Documentation Fluent Theory Guide* Lebanon, N.H., USA: ANSYS Inc

[13] Marshall E M and Bakker A 2004 *Computational fluid mixing, in handbook of industrial mixing* John Wiley & Sons Inc. pp 257-343

[14] Luo J Y , Issa R I and Gosman A D 1994 *Prediction of impeller induced flows in mixing vessels using multiple frames of reference*, in *I. Chem. E. Symposium series* ***136***. 1994 Cambridge UK pp 549-556

[15] Zadghaffari R, Moghaddas J S and Revstedt J A 2009 *mixing study in a double-Rushton stirred tank* Computers & Chemical Engineering **33**(7) pp 1240-1246

[16] Yang C and Mao Z S 2014 *Chapter 3 - Multiphase stirred reactors, in Numerical Simulation of Multiphase Reactors with Continuous Liquid Phase* Academic Press: Oxford pp 75-151

[17] Shur M L et al 2008 *A hybrid RANS-LES approach with delayed-DES and wall-modelled LES capabilitie.* International Journal of Heat and Fluid Flow **29**(6) pp 1638-1649

[18] Van Doormaal JP and Raithby G D 1984 *Enhancements of the simple method for predicting incompressible fluid flows* Numerical Heat Transfer **7**(2) pp 147-163

[19] Issa R I 1986 *Solution of the implicitly discretised fluid flow equations by operator-splitting* J. Comput. Phys. **62**(1)pp 40-65

[20] Bender E 1980 *Numerical heat transfer and fluid flow*. Von S. V. Patankar. Hemisphere Publishing Corporation, Washington – New York – London. McGraw Hill Book Company, New York 1. Aufl., 197 S., 76 Abb., geb., DM 71,90. Chemie Ingenieur Technik **53**(3) pp 225-225

[21] Leonard B P 1991 *The Ultimate conservative difference scheme applied to unsteady one-dimensional advection* Computer Methods in Applied Mechanics and Engineering **88**(1) pp 17-74

[22] Bakker A et al 1997 *Sliding mesh simulation of laminar flow in stirred reactors* Chemical Engineering Research and Design **75**(1) pp 42-44

[23] Hemrajani R R and Tatterson G B 2004 *Mechanically stirred vessels, in handbook of industrial mixing* John Wiley & Sons Inc. pp 345-390

IOP Conf. Series: Materials Science and Engineering **316** (2018) 012015 doi:10.1088/1757-899X/316/1/012015

CFD Modelling of Adsorption Behaviour in AGN Tank with Polyethylene Terephthalate Plastic Waste Based Activated Carbon

Yuliusman[1], M K Afdhol[1], Alristo Sanal [1], Nasruddin[2]

1. *Department of Chemical Engineering, Faculty of Engineering, University of Indonesia, Kampus Baru UI, Depok 16424, Indonesia*
2. *Department of Mechanical Engineering, Faculty of Engineering, University of Indonesia, Kampus Baru UI, Depok 16424, Indonesia*

E-mail: usman@che.ui.ac.id, Muhammad.khairul.afdhol@gmail.com

Abstract. Indonesia imports fuel (fuel oil) in large quantities. Indonesia has reserves of methane gas in the form of natural gas in large numbers but has obstacles in the process of storage. To produce a storage tank to a safe condition then proclaimed to use ANG (Adsorbed Natural Gas) technology. Manufacture of activated PET based activated carbon for storage of natural gas where technology has been widely studied, but still has some shortcomings. Therefore to predict the performance of ANG technology, modeling of ANG tank with Fluent CFD program is done so the condition inside the ANG tank can be known and can be used to increased the performance of ANG technology. Therefore, in this experiment natural gas storage test is done at the ANG tank model using Fluent CFD program. This experiment is begin with preparation tools and material by characterize the natural gas and activated carbon followed by create the mesh and model of ANG tank. The next process is state the characteristic of activated carbon and fluid in this experiment. The last process is run the simulation using the condition that already been stated which is at 27°C and 35 bar during 15 minutes. The result is at adsorption contour we can see that adsorption is higher at the top of the tank because the input of the adsorbent is at the top of the ANG tank so the adsorbate distribution is uneven that cause the adsorbate concentration at the top of the ANG tank is higher than the bottom tank.
Keywords: Activated Carbon, PET, Methane Storage, ANG

1. Introduction

Indonesia imports BBM (Bahan Bakar Minyak) in a large scale. Badan Pusat Statistik (BPS) recorded that in February 2014, Indonesia imports BBM around 1,06 million tone or around US$ 1,13 billion. This is because BBM is the main fuel that been used by Indonesia people for transportation. We can see in year 2012, the number of vehicle already reach 94.373.324. The usage of BBM in Indonesia is increasing due to the fast growth of vehicle industries around 5-6% each year [1]. Natural gas have a big potential to substitute BBM because Indonesia have big gas reserves and Indonesia is at rank 14 worldwide with reserves around 103,3 trillion cubic feet, where natural gas is the main ingredient of methane gas production [2]. The main problem from this natural gas is in the storage and distribution process because in normal condition, natural gas is in gas phase so the storage capacity become small and not efficient to be used. The technology that generally used is CNG and LNG. However, CNG needs high pressure around >3500 psi or 25000 kpa (250 bar) and LNG needs low temperature (-161°C) that can harm the surrounding area. To solve this problem we need innovation to store natural gas such as storing in a cylinder tube with ANG (Adsorbed Natural Gas) and activated carbon as the adsorbent which make natural gas can be stored in low pressure around 500 psig (3447,37 kpa) and high storage capacity [3].

This activated carbon is made from PET plastic waste as raw material because the big availability in Indonesia which is the number is around 26500 ton each day based on Ministry of Environment of

Indonesia. Furthermore activated carbon that made from PET has highest surface area compared with PVC and tires [4].

ANG technology generally has not been applied in vehicle even this technology has more benefit compared to CNG. One of the factor is the low adsorbent performance to adsorbed natural gas compared with CNG technology in the same tube. Because of that to predict the performance of ANG technology modeling of ANG tank is done that can be used for increasing the performance of ANG technology. This kind of experiment is already done at adsorption of hydrogen gas at activated carbon experiment which look at temperature and adsorption contour. Because of that in this experiment natural gas storage test is done at CFD model of ANG tank that will produce natural gas storage capacity, temperature and pressure contour data inside ANG tank.

2. Methodology

2.1. Geometry Making

Simulation process is begin with geometry making at GAMBIT, this process is included in preprocessing process. At the geometry making there are 4 type of geometry which are vertex which is geometry making using dot, edge which is geometry making using line, face which is geometry making using area and the last is volume which is geometry making using volume or 3D. For area making, dot geometry making has to be done before, dot geometry making using coordinate, and than link the dot become line with edge and area making with face.

2.2. Mash Making

After geometry finish, followed with mash making. The purpose of mash making is to decide the iteration result accuracy at Fluent, The bigger the mesh than the iteration accuracy is better. At the meshing menu there are 3 type of mesh which are mesh edge which is mash making on line, mash face which is mash making on area and mash volume which is mash making on volume. Make mash on area isn't same with geometry making that have to start from dot first, on mash area making can use mash face directly. At mash face making there are mesh type option, type of mash and amount of mash.

2.3. Boundary Condition

After making a complete mesh subsequent determination of the boundary conditions. These boundary conditions will be read on the Fluent therefore better define the desired boundary conditions beforehand though the boundary conditions there are not clear because when the boundary conditions specified in GAMBIT it can be replaced at Fluent, but if the boundary conditions. In the specific boundary conditions there is a choice that serves to select the type of boundary conditions that we want on a particular line such as inlet pressure, discharge pressure, and other boundary conditions that have been provided by GAMBIT.

2.4. Running Simulation

Simulations performed with axis-symmetric 2D model so modeling tube only half. Used a triangular meshing with Pave type and the interval count 450 At Fluent, menu-read file can have read the case, read data and read case and data. When the initial preparation to do is read the case, read the case and the data can be done when an existing file data that has been made. Then read case on file that has been exported from the .msh GAMBIT. Make a check on the grid, grid-check of the menu, the menu grid can also do scale to the size of a dimension that has been made. Then specify the conditions to be desired on the menu define, in this menu there are conditions of a model that can be adapted to the conditions of research, there is a material that serves to define the desired material, there is a boundary condition that serves to define the boundary conditions are desirable and also user -defined which serves to insert the UDF. Once it can be done save data by file-write-case or file write-case and the data, these two types differ only in the data. After his return to the menu solve, workmanship solve next-animate that serves to show the contours or the other on the window so that after iteration is completed we can play back, then do iterations.

2.5. Flow Diagram

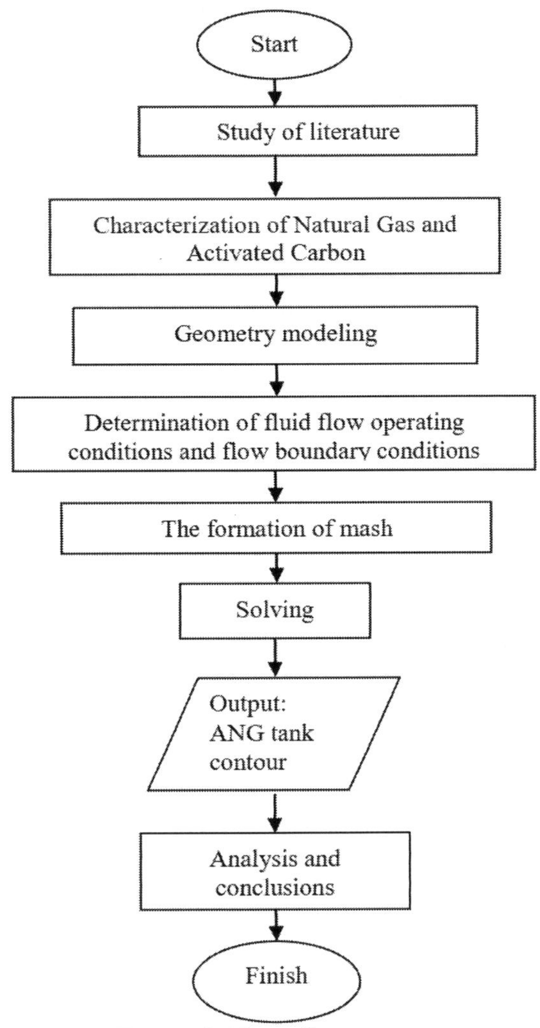

Figure 1. Flow diagram

3. Result And Discussion

3.1 Activated Carbon

In this experiment PET based activated carbon is used that produce by Alristo Sanal at year 2015 for thesis. Characteristic from activated carbon is used for analysis. The following is the result of activated carbon characteristic.

Table 1 Activated carbon characteristic data

Activated Carbon	***PET***
Surface area (Bil Iod) (m^2/g)	1591.72
Pore volume (cc/g)	0.176
Micropore volume (cc/g)	0.136
Pore diameter (A°)	15.2

Mesh size	10

3.2 Natural Gas Composition

In this experiment, adsorbate that is used for ANG test is natural gas from SPBG Pertamina JL.Raya Bogor. In this experiment gas composition test is done that become adsorbate in this experiment. This is done because adsorbate composition is needed in the adsorption calculation. Gas composition test is done in Lab Intensifikasi Proses Departemen Teknik Kimia Universitas Indonesia using Gas Cromatography (GC). The following is the result of gas composition from adsorbate that will be used :

Table 2 Natural gas analysis test data

No	*Name*	*Time*	*Quality (%Area)*	*Height (μV)*	*Area ($\mu V.Min$)*	*Area %*
1	CH_4	1.65	87.76	693,512.5	14,392.7	87.762
2	N_2	1.73	4.70	38,002.3	770.5	4.698
3	C_2H_6	1.94	3.78	29,539.4	620.0	3.781
4	CO_2	2.19	2.26	10,349.3	370.6	2.260
5	C_3H_8	2.92	1.28	5,394.5	209.6	1.278
6	C_4H_{10}	6.68	0.22	321.8	36.3	0.222
	Total		100	777,119.8	16,399.7	100

3.3 CFD Simulation

In this study conducted simulations using CFD program which aims to observe the adsorption process that occurs in the tube ANG. By knowing the adsorption process that occurs in the tube ANG we can predict the performance of the ANG technology. Results of the program are the CFD pressure contour, contour and contour absorption temperature on the tube ANG. Simulations use 2d field by using only half of the tube ANG, because on the other half of the tube ANG will have the same result. In this simulation gas will enter from the top of the tube and flows to the bottom of the tube.

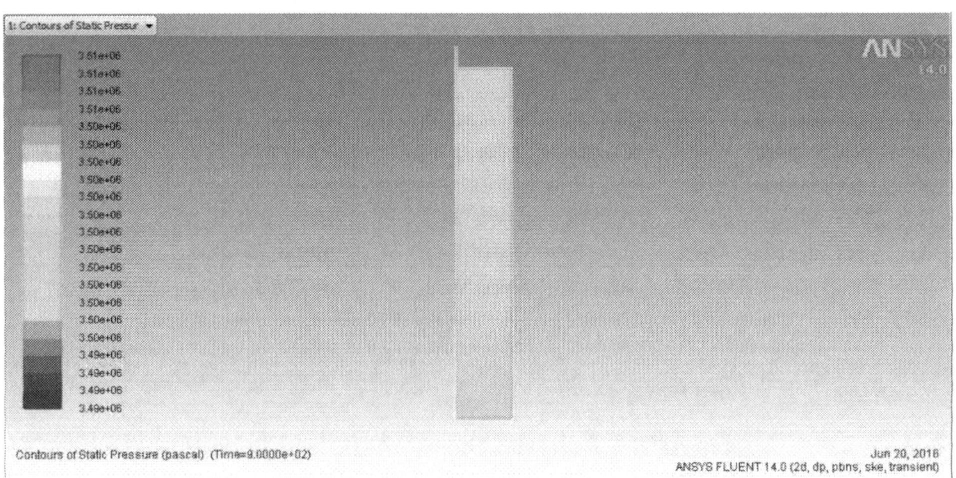

Figure 2. Pressrue contour of ANG tank

IOP Conf. Series: Materials Science and Engineering **316** (2018) 012015 doi:10.1088/1757-899X/316/1/012015

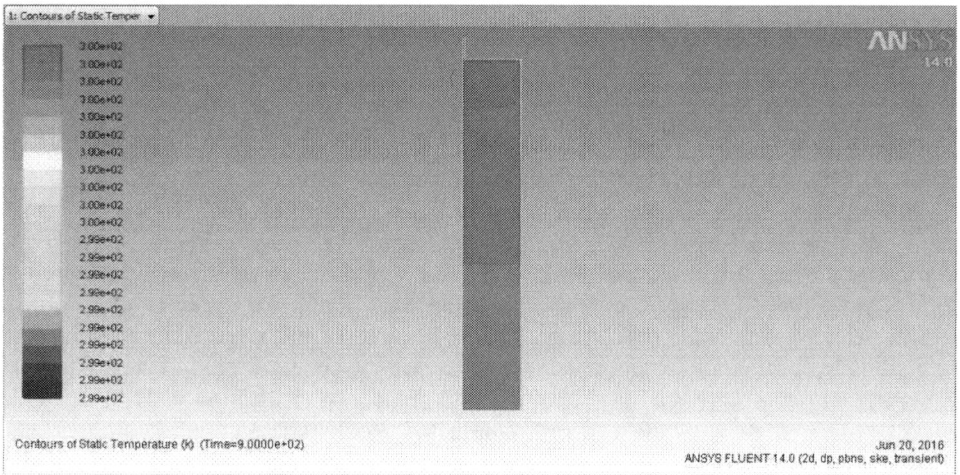

Figure 3. Temperature contour of ANG tank

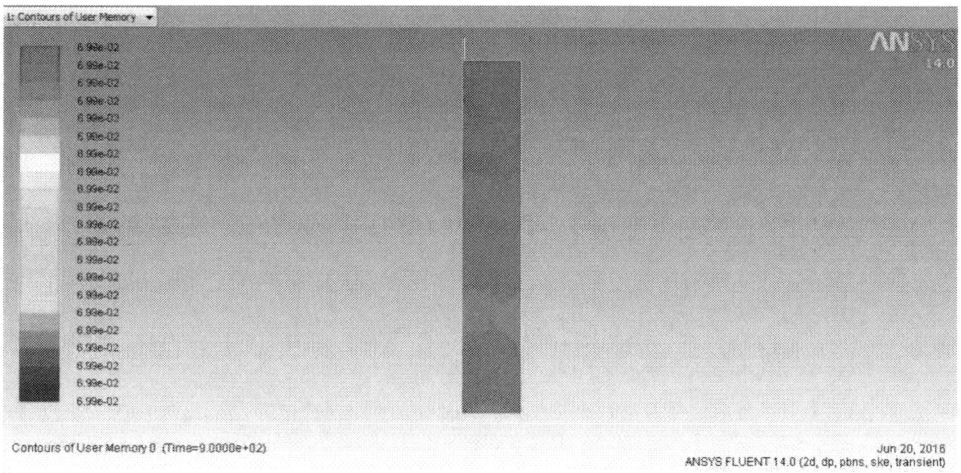

Figure 4. Storage capacity contour of ANG tank

Contours of pressure can be seen in Figure 2 where the pressure in the tube is at a pressure of 35 bar is uniformly distributed on the tube ANG. Contour temperature can be seen in Figure 3 where it can be seen the color difference where the color red is at the top of the tube. This indicates that conditions at the top of the tube has a higher temperature is due to the input of the adsorbent is at the top so that the distribution adsorbat uneven which causes the concentration of adsorbate the top of the tube is greater than the bottom of the tube in which the reaction of adsorption is exothermic so take off heat to the environment with greater concentration it will produce greater heat [4].

Contour absorption can be seen in Figure 4 where it can be seen the color difference where the color red is at the top of the tube. This indicates the absorption was more pronounced at the top of the tube marked with the colors redder than tube sections more because of the input from the adsorbent is at the top so that the distribution adsorbat uneven which causes the concentration of adsorbate the top of the tube is greater than the bottom of the tube , With a greater concentration it will produce greater absorption. When compared with the research conducted Nouh where research simulations using CFD program on ANG technology obtained similar results which were obtained by the distribution of storage capacity less prevalent. Incoming gas flow rate can be reduced so that natural gas can be absorbed and distributed better [4].

4. Conclusion

We can conclude from this experiment "CFD Modeling of Adsorption Behavior in ANG Tank with Polyethylene Terephthalate Plastic Waste Based Activated Carbon" that :

1. In CFD simulation we can get that the highest storage capacity is at the top of the tube which is the highest concentration of adsorbate is occur.
2. In CFD simulation we can get that the temperature is at the top of the tube which is the highest concentration of adsorbate is occur.
3. In CFD simulation we can get that the pressure is spread evenly at the ANG tank

5. Reference

[1] Statistic Center of Indonesia 2014 *Tabel jumlah kendaraan bermotor*. http://bps.go.id/ tab_sub/ view.php? Kat =2&tabel = 1&daftar =1&id_ subyek=17&notab=12.

[2] Yuliusman, Nasruddin, A Sanal, A Bernama, F Haris, I T Ramadhan. 2017. Preparation of activated carbon from waste plastics polyethylene terephthalate as adsorbent in natural gas storage. *IOP Conf. Series: Materials Science and Engineering, vol. 176*. 012055.

[3] Samid, Dedy Darmawan., 2011, *Optimasi Karakteristik Karbon Aktif Pada Sistem Adsorbed Natural Gas Dalam Kondisi Dinamis.*Universitas Indonesia.http://www.ui.ac.id.

[4] Yuliusman, Nasruddin, A Sanal, A Bernama, F Haris and M Hardhi. 2016. Improved of Natural Gas Storage with Adsorbed Natural Gas (ANG) Technology Using Activated Carbon from Plastic Waste Polyethylene Terepthalate. *IOP Conf. Series: Earth and Environmental Science, vol. 75*. 012018.

[5] Mat, Bin Hanapi., Zakaria, Zainal., Paou, Terry George. 2006. *The Development of Adsorbent Based Natural Gas Storage For Vehicle Application.* Universiti Teknologi Malaysia. eprints.utm.my.

[6] Mingjia Z, F Yang, F Meng, M Li, A Manivannan, and N Wu. 2014. Effects of Pore Structure on Performance of An Activated-Carbon Supercapacitor Electrode Recycled from Scrap Waste Tires. *ACS Sustainable Chemical Engineering*, 2 (7), pp 1592–1598.

IOP Conf. Series: Materials Science and Engineering **316** (2018) 012016 doi:10.1088/1757-899X/316/1/012016

Carbon monoxide and methane adsorption of crude oil refinery using activated carbon from palm shells as biosorbent

Yuliusman, M K Afdhol, Alristo Sanal
Department of Chemical Engineering, Faculty of Engineering, University of Indonesia, Kampus Baru UI, Depok 16424, Indonesia

E-mail: usman@che.ui.ac.id, Muhammad.khairul.afdhol@gmail.com

Abstract. Carbon monoxide and methane gas are widely present in oil refineries. Off-potential gas is used as raw material for the petrochemical industry. In order for this off-gas to be utilized, carbon monoxide and methane must be removed from off-gas. This study aims to adsorb carbon monoxide and methane using activated carbon of palm shells and commercial activated carbon simultaneously. This research was conducted in 2 stages: 1) Preparation and characterization of activated carbon, 2) Carbon monoxide and methane adsorption test. The activation experiments using carbon dioxide at a flow rate of 150 ml/min yielded a surface area of 978.29 m^2/g, Nitrogen at flow rate 150 ml/min yielded surface area 1241.48 m^2/g, and carbon dioxide and nitrogen at a flow rate 200 ml/min yielded a surface area 300.37 m^2/g. Adsorption of carbon monoxide and methane on activated carbon of palm shell systems yielded results in the amount of 0.5485 mg/g and 0.0649 mg/g and using commercial activated carbon yielded results in the amount of 0.5480 mg/g and 0.0650 mg/g

Keywords: Activated carbon, Adsorption, Carbon monoxide, Off gas, Methane, Palm shells

1. Introduction

Off-gas in refinery oil, specifically in the cracking unit which produces the products which include LPG, light naphtha, heavy naphtha, light and heavy kerosene, Automotive Diesel Oil (ADO), unconverted oil, and off-gas. Off-gas has not been widely used on the world. This is not economical because off-gas still contains many hydrocarbons such as ethane (C_2), propane (C_3) and butane (C_4), in which these hydrocarbons still has economic value because it can still be used as raw material in petrochemical industry. So far, carbon monoxide in refinery oil is generally reacted with steam to covert it into carbon dioxide and hydrogen. This process requires a higher cost, in terms of unit operating costs and steam production than adsorbing carbon monoxide, while methane is discharged directly through the flare. The adsorption process is an alternative to removing carbon monoxide and methane. Currently there have been several studies on adsorption by using activated carbon as biosorbent, such as adsorption of SO_2, NO and chlorobenzene [1], adsorption of CO_2 and CH_4 [2] and CO adsorption [3].

Activated carbon is a potent biosorbent used to adsorb carbon monoxide and methane gas present in off-gas [4]. Based on the structure pattern of the activated carbon, it is a material in the form of amorphous carbon consisting mostly of free carbon and has an inner surface that has a higher adsorption capacity than other adsorbents [5]. Organic and inorganic materials containing lignin, hemicellulose and cellulose can be utilized as feedstocks for activated carbon, as they are highly effective for adsorption [1].

One of the most potential materials and is being developed as a feedstock for the manufacture of activated carbon is the shell of palm [6]. When compared to the coconut shell, the palm shell has more mesoporous pore and the micropore is much larger [7]. The amount of palm shell that is being produced as one of the palm oil processing industry waste are quite large, reaching 12% of the weight of oil palm fruit. Indonesia is the largest palm oil producing country in the world. The average annual production of palm fruit is 5.6 million tons, so there are about 672 thousand tons of shells produced. This amount will continue to increase along with the increase of palm oil production. With the availability of such waste, a further process is needed to convert the waste of oil palm shells into a product of high economic value.

The quality of activated carbon of palm shells used as biosorbent in adsorption of carbon monoxide and methane is strongly influenced by the conditions of manufacture. Among other parameters that affect the process of making activated carbon such as activation solution, activation gas flow rate, activation temperature, and duration of the activation process. While the effectiveness of the adsorption process is strongly influenced by several parameters, including the mass of activated carbon, contact time, and surface area.

This research will be conducted by making activated carbon from palm shells and comparative performance test of commercial activated carbon activated carbon and activated carbon from palm kernel shells that have been activated with KOH as an activating agent or that have been activated through physical activation at 850°C temperature, where this temperature is the optimum temperature of activation [8]. This activation temperature is expected to produce activated carbon with a high surface area[9]. Commercial activated carbon and activated carbon from activated palm shells are then used to adsorb carbon monoxide and methane gas with batch and continuous systems. Tests of the adsorption parameters were carried out in the adsorption process so that the optimal conditions in adsorbing carbon monoxide and methane can be known.

2. Methodology/Experimental

2.1 Chemical and Physical Activation

The 177-micron palm shell is activated by mixing it with KOH solution. Impregnation is done on the palm shell with KOH (b/b). The mixture is stirred until it becomes a homogeneous structure. The stirring process is done in order to impregnate the activating agent on the raw material of palm shell. This mixing result will form a mixture of a slurry. After that, the mixture is heated in the oven for 1 hour at 110°C to remove the water content. Then the mixture of raw materials and activating agent in the form of a slurry is then put into the reactor. Nitrogen gas must be injected into the reactor before the reactor is used to ensure that there is no oxygen in it. The reactor operates at 850°C for 1 hour. Nitrogen, carbon dioxide, and carbon dioxide followed by nitrogen are emitted with flow rates of 100, 150 and 200 ml/min. After all the procedures are performed, the activated carbon is activated through chemical and physical means. This active carbon must go through the next treatment stage to get a truly pure result. Activated carbons are then washed with HCl 5 N to attain a neutral pH. Activated carbon is then washed with distilled water to remove the chloride. Characterization includes reactivated carbon-iodine and SEM-EDS test results. To change the amount of iodine in the surface area, a linear regression equation is used which refers to ASTM [6] D-4607-94. The equation is as follows:

$$\text{Iodine Number} = 0{,}6366 \times \text{surface area} + 174{,}34$$

2.2 Adsorption Test on Activated Carbon

The isothermal adsorption test on the batch system is performed by preparing 1 to 2.5 g of active carbon which is inserted rapidly into 10 mL glass bottles. The glass bottle is equipped with a silk-sealed septum and there is residual air that must be quickly removed as much as the amount of off-gas injected. This is done by using a syringe (1 mL), with variations in the concentration of CO: CH_4, the gases are imbued with BOC gas. The bottle is shaken at the equilibrium concentration at the batch reactor during the 15th, 30th, 45th, and 60th min. Subsequently, the samples (the unadsorbed concentration of CO and CH4) were taken and analyzed using Shimadzu-8A GC-TCD. Isothermal adsorption for CO and CH4 with activated carbon was determined using the equilibrium adsorption model of Langmuir model [10].

3. Result and discussion

3.1 Chemical Activation with KOH Activating Agent and Physical Activation

IOP Conf. Series: Materials Science and Engineering **316** (2018) 012016 doi:10.1088/1757-899X/316/1/012016

Chemical activation is the process of disconnecting carbon chains on active carbon feedstock by using KOH. The use of KOH as an activator because KOH is able to oxidize volatile compounds so as to increase the number of pores and surface area. Physical activation is the process of disconnecting the carbon chain on the raw material of activated carbon at high temperature by flowing steam, N_2 and CO_2. In this research, the physical activation process is done using N_2 and CO_2 with a variation of the flow rate of these gases and a combination between N_2 and CO_2 at flow rate 100, 150 and 200 ml/minute. Activators have a high temperature where the activator can react with carbon so as to develop pores and surface area[11]. Then we can determine the effect of activator gas and activator flow rate to the surface area of activated carbon as shown below:

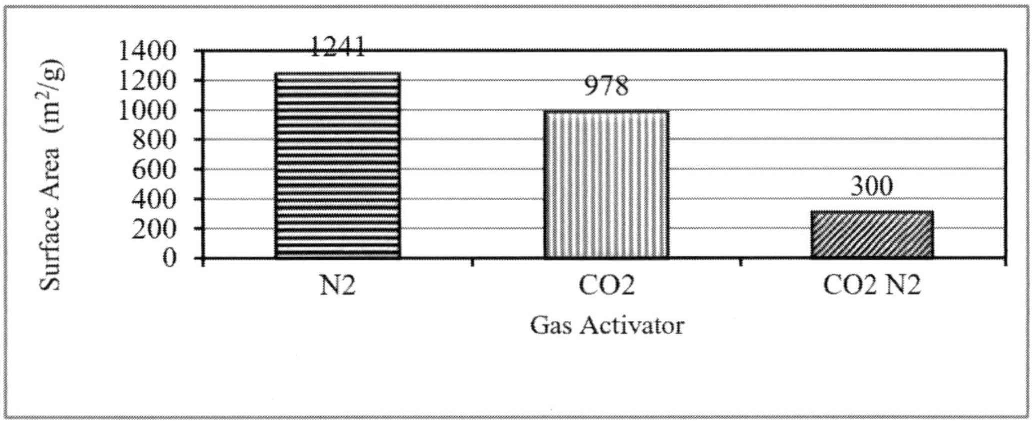

Figure 1 Effect of Gas And Flow Rate Activator Against The Area of Activated Carbon Surface

A high surface area is generated because the activation process removes volatile things and isolates the material from the air.

3.2 Test of iodine

Activated carbon with high iodine adsorption capability also has a larger surface area and has a larger micro and mesoporous structure.Figure 2 shows the differential ratio of the activation treatment to the iodine adsorption on the activated carbon.

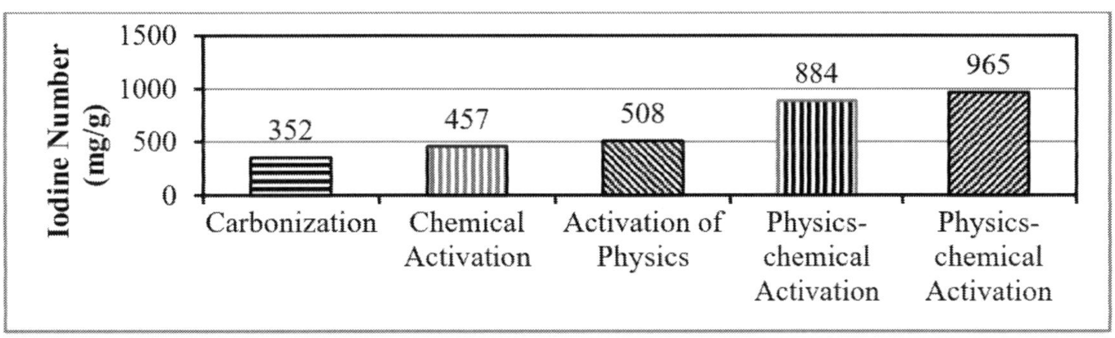

Figure 2 Effect of Activation Treatment on Iod Adsorption

From Figure 2, it can be seen that the activation treatment determines the adsorption capacity of iodine, in which activated carbon without activation has the smallest specific surface area. The principle of carbonization and activation is to form pores. A small number of pores causes the activated carbon to be unable to absorb iodine solution effectively; the amount of iodine absorbed in the carbonization process is 352 mg/g and in the physical-chemical activation process it was 884 mg/g [12]. The process of physical activation adsorbs 508 mg/g and the chemical activation process adsorbs

457 mg/g [13]. Whereas in this experiment iodine adsorption capacity with physical and chemical activation process is 964.67 mg/g.

Then as the optimum treatment, variations of the flow rate of the physical activator is done on the chemical-physical activation. Figure 3 shows the effect of physic activator on iodine adsorption.

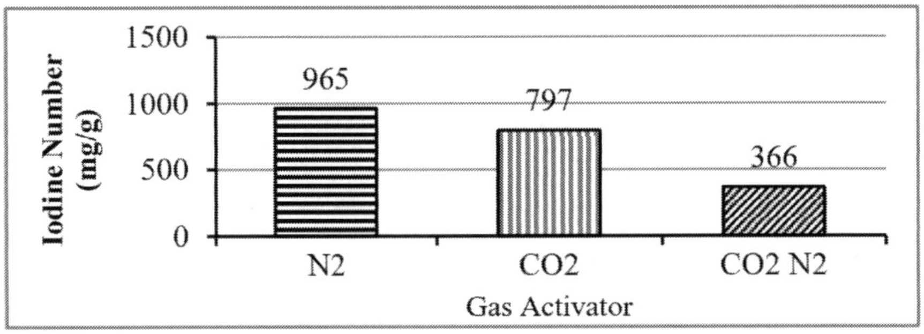

Figure 3 Effect of Gas Activator On Iodine Numbers

Figure 3 shows the effect of CO_2, N_2, and $CO_2 + N_2$ activators, where N_2 activator is able to increase iodine absorption capacity to 965 mg/g. Another study also mentioned that activation using KOH for chemical activation and $CO_2 + N_2$ for physical activation resulted in a surface area of 596 m^2/g, which also has a low iodine absorption due to the long period of physical activation process [8]. The difference of the activator gas and the activator flow rate on the activated carbon sample significantly affected the adsorption capacity of the activated carbons. The high adsorption of activated iods of carbon indicates the surface that has been formed by the activation process is also increasing - higher iodine adsorption indicates higher surface area. Additionally, it was also determined that increased activation time causes decreased iodine adsorption [14]. Increased activation time causes a decrease in microporous volume, thereby decreasing the iodine absorption capacity. In another study, the increase in iodine adsorption capacity also reflects an increase in the porosity of activated carbon by increasing the impregnation ratio. This may be due to the minimization of tars and other liquids, which can clog pores and inhibit the development of pore structures [8].

3.3. SEM (Scanning Electron Microscopy) Characterization

SEM characterization in the study was conducted to determine the pore conditions formed on commercial activated carbon and activated carbon from palm shells. Figure 4 shows the SEM characterization results.

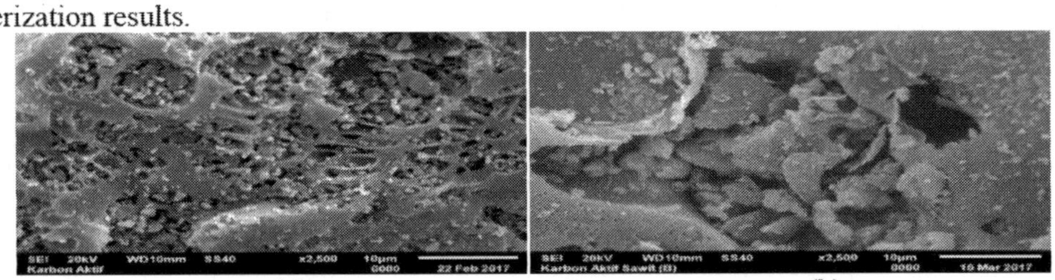

(a) (b)

Figure 4 SEM characterization results

Based on Figure 4 it can be seen that the pores have formed. This is due to the influence of KOH activators in activating activated carbon and also influenced the physical activation using nitrogen.During the activation process, KOH activates carbon based on dehydration function and eliminates hydrogen and oxygen atoms from carbon material in the form of water rather than as an oxygen organic component, thereby inducing porosity generation and increasing carbon content.

3.4. EDS (Energy-Dispersive X-Ray Spectroscopy)Characterization

EDS characterization aims to determine the composition of the commercial activated carbon and activated carbon created in the study.The composition under consideration is the composition of the weight of the commercial activated carbon content and the activated carbon of the oil palm shell. Table 1 shows the results of EDS characterization.

Table 1 EDS Characterization of Commercial and Palm Shells Activated Carbon

Activated Carbon	Mass Percent			
	C	O	Cu	Zn
Commercial	80.06	6.75	8.04	8.63
Palm Shells	60.44	13.73	8.040	8.63

According to the EDS characterization, the dominant element contained in activated carbon is a carbon element. Increasing the carbon element indicates that the activation of activated carbon is running well. Meanwhile, the difference in carbon content of both activated carbon is due to the production process of each activated carbon is different; on the activated carbon of the palm shell the influence of high temperature on the physical activation causes the release of other elements such as oxygen and impurities attached to the surface of activated carbon. Figure 5 shows the EDS characterization results.

Fig. 5 shows the EDS characterization results of the diffraction patterns of the content of O, Cu, and Zn on the surface of the activated carbon by 4.84%, 8.54%, and 4.57%, respectively. Figure 4.11 shows the EDS characterization of the diffraction patterns of the content of O, Cu and Zn on the surface of activated carbon, each with a mass percentage 13.73%, 8.040%, and 8.63% with KOH chemical activation and heating or physical activation at 850°C. This ensures the presence of ion-in on the surface of the activated carbon adsorbent. This can be attributed to the effect of the activation and screening washing step of the adsorbent used which may cause losses that may affect the adsorbent composition [15].

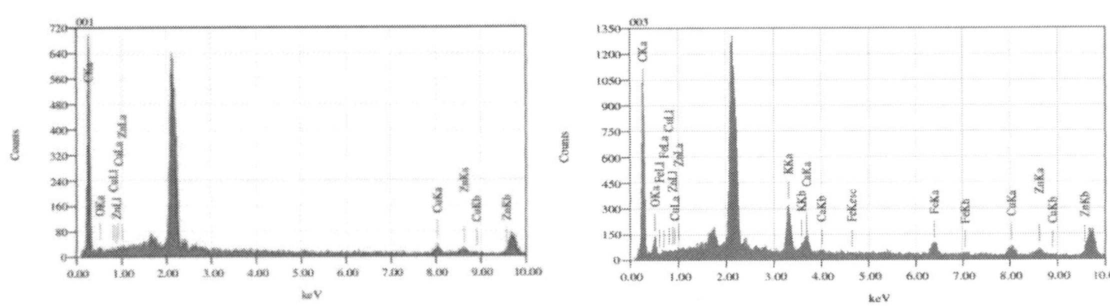

(a) (b)

Figure 5 EDS Characterization of commercial (a) and palm shell (b) activated carbon

3.5. Adsorption of CO and CH_4 On Commercial Active Carbon

Adsorption is carried out to determine the adsorption capacity of activated carbon in carbon monoxide and methane.Various amounts of carbon monoxide and methane are injected into glass bottles.The concentrations used in CO and CH_4 in this experiment were 75% and 25%, respectively.The effect of mass of activated carbon on CO adsorption simultaneously on commercial activated carbon is shown in Figure 6.

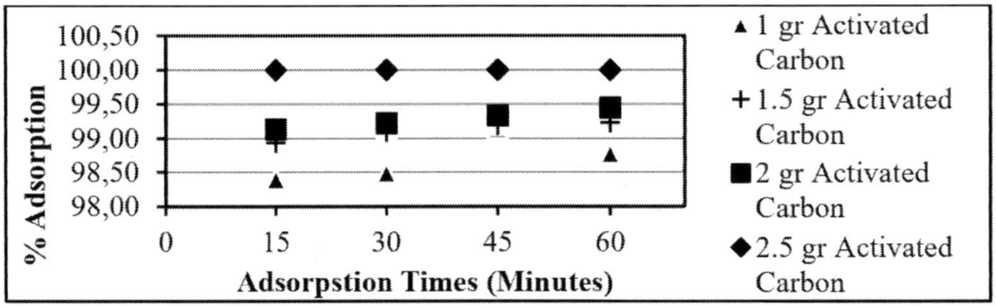

Figure 6 Adsorption Result of Carbon Monoxide

Based on the graph shown in Fig. 8 it is seen that the increase of carbon monoxide adsorbed on each of the activated carbon always increases as the adsorption time increases where 1 gram of activated carbon is able to adsorb at 98.39% in the 15-minute mark then increase until 98.77% in the 60-minute mark. Then, 1.5 gram of active carbon is able to adsorb 98.94% in the 15-minute mark then increase until 99.23% in the 60-minute mark. Lastly, 2 grams of activated carbon has the adsorption ability of 99.13% in the 15-minute mark then increase until 99.45% in the 60-minute mark and 2.5 grams of activated carbon is capable of adsorbing CO completely at 15 minutes to 60 minutes.

In other studies, the greater the contact area or the surface between the adsorbent, the gas the absorption rate will increase [4], so it can be said that 2.5 grams of active carbon has a better adsorption capacity. And, the effect of mass of activated carbon on CH_4 adsorption simultaneously on commercial activated carbon is shown in Figure 7.

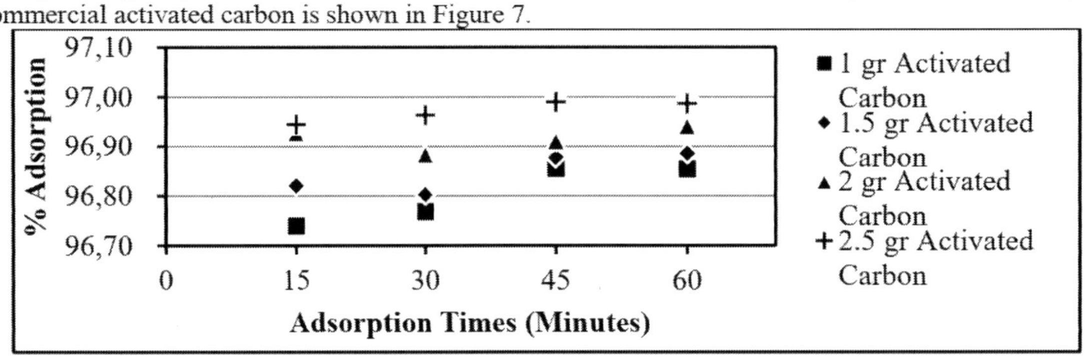

Figure 7 Adsorption Results of Methane

In Fig. 7 it is seen that the increase in the amount of CH_4 absorbed in each of the activated carbon always increases with the increase of adsorption time where on 1 gram activated carbon is able to adsorb 96.74% of methane in the 15-minute mark then increase until 96.85% in the 60-minute mark. Then, 1.5-gram active carbon able to adsorb 96.82% in the 15-minute mark then increases until 96.89% in the 60-minute mark. 2 grams of activated carbon is able to adsorb 96.93% in the 15-minute mark then increase until 96.94% in the 60-minute mark. Lastly, activated carbon in the amount of 2.5 grams has the ability to adsorb 96.94% in the 15-minute mark then increase until 96.99% in the 60-minute mark.

The surface of activated carbon has an important function in the adsorption process because it is an organic electrolyte, so it is important to consider the content of the surface of the activated carbon because it determines the adsorption capacity of the activated carbon by ion exchange. In a polar molecule having a strong dipole moment, adsorption can occur in the presence of an oxygen group on the surface. CO and CH_4 are adsorbed in the activated carbon as follows:

Table 2 Isothermal Equilibrium in CO and CH_4

Carbon Monoxide			Methane		
Co	**Ce**	**Qe (mg/g)**	**Co**	**Ce**	**Qe (mg/g)**
731	12	0.5476	260	0.91	0.0623
731	11	0.5477	260	0.83	0.0623
731	6	0.5480	260	0.60	0.0624
731	9	0.5478	260	0.60	0.0624

Table 2 shows the amount of how much CO and $CH_4$is absorbed in the activated carbon in which the adsorption process of the amount of CO and CH_4 keeps escalating over time [18], and indicates that the CO gas at the 15-minute mark is absorb as much as 0.5476 mg/g and increased to 0.5478 mg/g. Likewise, in CH_4 gas where increasing amount of CH_4 gas absorbed in activated carbon also occurs, that is 0.0623 mg/g to 0.0624 mg/g.

3.6. Isothermal Adsorption of CO and CH_4 On Activated Carbon from Palm Shells

Isothermal adsorption is carried out to determine the adsorption capacity of activated carbon in the gas CO and CH_4.The concentrations used in CO and CH_4 in this experiment were 75% and 25%, respectively. The effect of mass of activated carbon on CO adsorption simultaneously on activated carbon made from palm shells is shown in Figure 8.

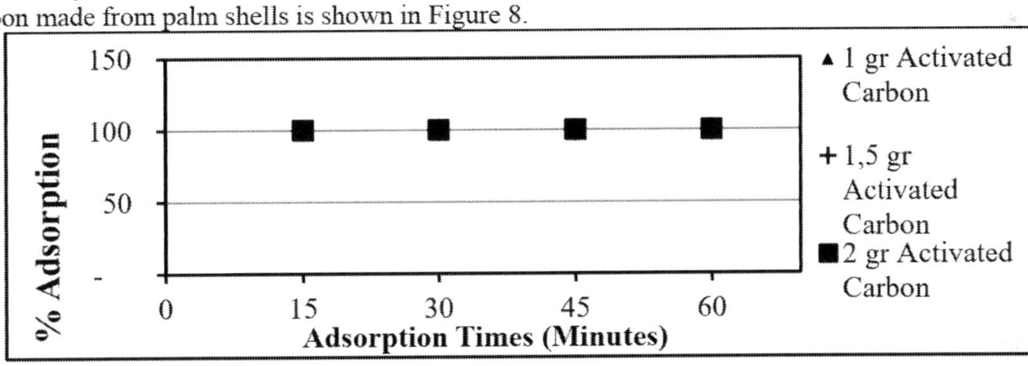

Figure 8. Adsorption Result of Carbon Monoxide

Based on the graph shown in Fig. 8 it is seen that carbon monoxide adsorbed by activated carbon occurs maximally in the active carbon mass of 1, 1.5, 2, and 2.5 gram with 100% adsorbed gas.From this phenomenon, it shows that the activated carbon produced by itself is able to adsorb the CO that is injected into the tube.This is due to the increasing number of adsorbent molecules capable of adsorbing adsorbates in units of time and the same volume of space.In other studies, the greater the contact area or the surface between the adsorbent and CO, the absorption rate will increase [4]. And then, the effect of mass of activated carbon on CH_4 adsorption simultaneously on conventional activated carbon is shown in Figure 9.

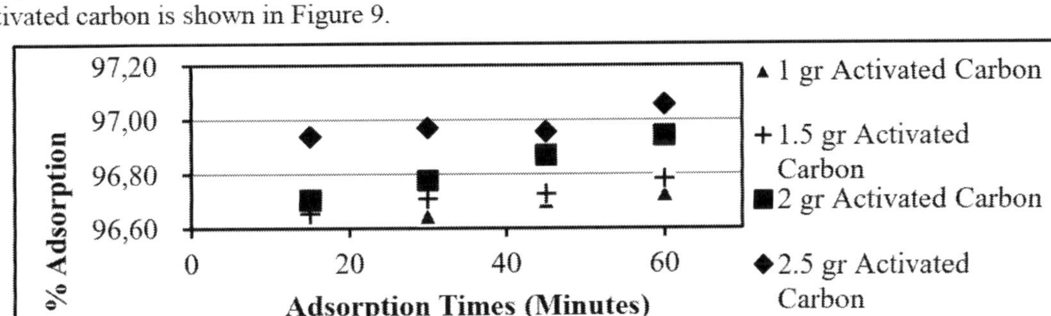

Figure 9. Methane Adsorption Results

Figure 9 shows that the increase of CH_4 adsorbed in each of the activated carbon always increases as the adsorption time increases where 1 gram of activated carbon is able to adsorb 96.67% in the 15-minute mark then increase 96.73%in the 60-minute mark.Then 1.5 gram of active carbon is able to adsorb 96.65% in the 15-minute mark then increase until 96.78%in the 60-minute mark. 2 grams of activated carbon is able to adsorb 96.70% in the 15-minute mark then increase until 96.94%in the 60-minute mark. Lastly, activated carbon of 2.5 grams has the ability to adsorb 96.94% in the 15-minute mark then increased until 97.05%in the 60-minute mark.

The surface of activated carbon has an important function in the adsorption process because it is an organic electrolyte, and determines the adsorption capacity of the activated carbon by ion exchange. The adsorption study on the activated carbon from the oil palm shell is done to obtain isothermal equilibrium.

Table 3 The isothermal equilibrium of the gas on the activated carbon of the palm shell

Carbon Monoxide			Methane		
Co	**Ce**	**Qe (mg/g)**	**Co**	**Ce**	**Qe (mg/g)**
731	-	0.5485	260	1	0.0648
731	-	0.5485	260	1	0.0648
731	-	0.5485	260	1	0.0648
731	-	0.5485	260	0.94	0.0649

Table 3 shows the amount of how much CO and CH_4 is absorbed in the activated carbon in which the adsorption process of the amount of CO and CH_4 keeps escalating over time [16], and shows that at 15 to 60 minutes, the amount of CO absorbed is 0.5485 mg/g.In CH_4 gas where increasing amount of CH_4 gas absorbed in activated carbon has increased, which is 0.0648 mg/g to 0.0649 mg/g.

3.7. Comparison of Adsorption Capability of Carbon Monoxide and Methane on Activated Carbon from Palm Shells with Commercial Active Carbon

Adsorption is carried out to determine the adsorption capacity of activated carbon in carbon monoxide compared to its adsorption capacity as shown in Figure 10 below.

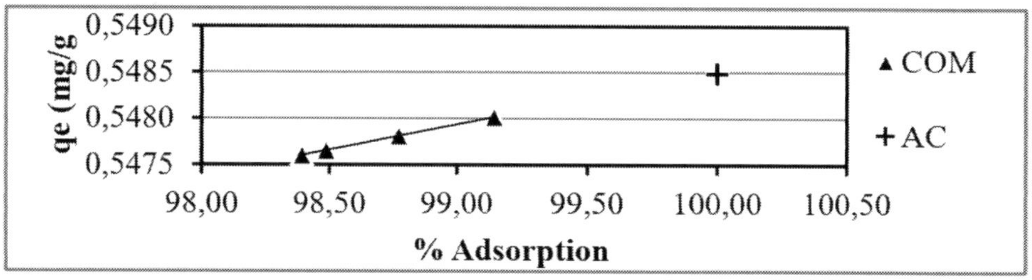

Figure 10 Representation of Langmuir Isotherm Carbon Monoxide Adsorption Model

Figure 10 shows the commercial activated carbon's maximum ability to absorb carbon monoxide. At minute 15 the amount of carbon monoxide absorbed is 0.5480 mg/g while the activated carbon from the palm oil shell which in its own production is able to absorb more carbon monoxide gas that is 0.5485 mg/g.This is due to the physical and chemical activation process carried out on the activated carbon of the palm shell that can increase the surface area and the pore area of the activated carbon obtained. Figure 11 shows the methane adsorption capacity ratio.

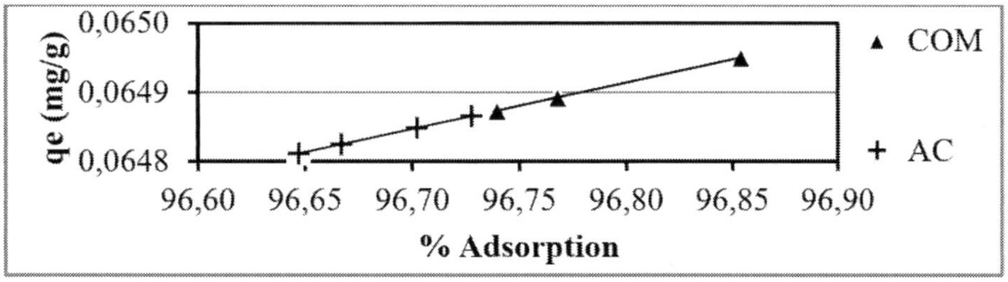

Figure 11 Representation of Langmuir Isotherm Methane Adsorption Model

Figure 11 shows the commercial activated carbon's maximum ability to absorb methane is 0.0649 mg/g while the activated carbon from the palm oil shell in production alone is able to absorb 0.0650 mg/g of methane.This is due to the difference in adsorption rates on carbon monoxide so that the activated carbon of palm shell carbon monoxide is faster adsorbed to cover the surface and pore of activated carbon so that the available surface area to accommodate less methane.While the commercial activated carbon absorbs less carbon monoxide then the methane is absorbed in the activated carbon will be more.

4. Conclusions

The activated carbons of the best palm shells produced with Nitrogen activators with a flow rate of 150 ml/min resulted in a surface area of 1241.48 m^2/g. While on carbon dioxide activators with a flow rate of 150 ml/min produced a surface area of 978.29 m^2/g and carbon dioxide activator mixed with nitrogen with a flow rate of 200 ml/min yielded a surface area of 300.37 m^2/g. From these results, it can be concluded that nitrogen with a flow rate of 150 ml/min produces a better surface area. Adsorption of activated carbon monoxide activated carbon from palm shell adsorbing carbon monoxide as much as 0.5485 mg/g and commercial activated carbon is 0.5480 mg/g. In methane gas commercial activated carbon is able to adsorb methane as much as 0.0625 mg/g, while activated carbon from palm shell adsorb 0.0649 mg/g.

5. Acknowledgement

The authors are grateful to the PITTA for their financial support. The authors would also like to thank all parties for support during the sample and manuscript preparation, respectively.

6. REFERENCES

[1] Li, Yuran, Guo, Yangyang, Zhu, Tingyu, & Ding, Song. 2016. Adsorption and desorption of SO2, NO and chlorobenzene on activated carbon. *Journal of Environmental Sciences, 43*, 128-135.

[2] Yadavalli, Gayatri, Lei, Hanwu, Wei, Yi, Zhu, Lei, Zhang, Xuesong, Liu, Yupeng, & Yan, Di. 2017. Carbon dioxide capture using ammonium sulfate surface modified activated biomass carbon. *Biomass and Bioenergy, 98*, 53-60.

[3] Yuliusman, W W Purwanto, Y S Nugroho. 2015. Smoke Clearing Method Using Activated Carbon and Natural Zeolite, *International Journal of Technology*, Vol. 6 (3), pp. 492-503.

[4] Yang, Ralph T. 2003. Sorbents for Applications *Adsorbents: Fundamentals and Applications* (pp. 280-381): John Wiley & Sons, Inc.

[5] Poulopoulos, S.G., & Inglezakis, V.J. 2006. *Adsorption, Ion Exchange and Catalysis: Design of Operations and Environmental Applications*: Elsevier Science.

[6] Zainal, Nahrul Hayawin, Aziz, Astimar Abdul, Idris, Juferi, Mamat, Ropandi, Hassan, Mohd Ali, Bahrin, Ezyana Kamal, & Abd-Aziz, Suraini. 2017. Microwave-assisted pre-carbonisation of palm kernel shell produced charcoal with high heating value and low gaseous emission. *Journal of Cleaner Production, 142, Part 4*, 2945-2949.

[7] Daud, Wan Mohd Ashri Wan, & Ali, Wan Shabuddin Wan. 2004. Comparison on pore

development of activated carbon produced from palm shell and coconut shell. *Bioresource Technology, 93*(1), 63-69.

[8] Yuliusman, Nasruddin, M K Afdhol, R A Amiliana, A Hanafi. 2017. Preparation of Activated Carbon from Palm Shells Using KOH and ZnCl2 as the Activating Agent. IOP Conf. Series: Earth and Environmental Science, vol. 75, 012009.

[9] Xiao, Yong, Long, Chao, Zheng, Ming-Tao, Dong, Han-Wu, Lei, Bing-Fu, Zhang, Hao-Ran, & Liu, Ying-Liang. 2014. High-capacity porous carbons prepared by KOH activation of activated carbon for supercapacitors. *Chinese Chemical Letters, 25*(6), 865-868.

[10] Park, Kyong-Mok, Nam, Hee-Geun, Lee, Ki Bong, & Mun, Sungyong. 2016. Adsorption behaviors of sugars and sulfuric acid on activated porous carbon. *Journal of Industrial and Engineering Chemistry, 34*, 21-26.

[11] Okafor Vincent, Solomon, Nwigbo, & Jude, and Dara. 2016. Study of Kinetics of Chemically Activated Carbon Produced from Palm Kernel Shell. *British Journal of Applied Science & Technology*. doi: 10.9734/BJAST/2016/27324

[12] Andriana. 2013. Preparasi Dan Karakterisasi Karbon Aktif Berbahan Dasar Tempurung Kelapa Sawit Dengan Bahan Pengaktif Koh Sebagai Adsorben Gas Co Dari Asap Kebakaran.

[13] Gloria, Beatrix. 2016. Adsorpsi Kontinyu CO2 Dalam Udara Menggunakan Karbon Aktif Dari Bambu Pentung.

[14] Suzuki. R. M, A. D. Andrade, J. C. Sousa, M. C. Rollemberg. 2007. Preparation and Characterization of Activated Carbon from Rice Bran. *Bioresource technology* 98(10): 1985-1991.

[15] Asuquo, Edidiong, Martin, Alastair, Nzerem, Petrus, Siperstein, Flor, & Fan, Xiaolei. 2017. Adsorption of Cd(II) and Pb(II) ions from aqueous solutions using mesoporous activated carbon adsorbent: Equilibrium, kinetics and characterisation studies. *Journal of Environmental Chemical Engineering, 5*(1), 679-698.

[16] Maneechakr, Panya, & Karnjanakom, Surachai. 2017. Adsorption behaviour of Fe(II) and Cr(VI) on activated carbon: Surface chemistry, isotherm, kinetic and thermodynamic studies. *The Journal of Chemical Thermodynamics, 106*, 104-112.

IOP Conf. Series: Materials Science and Engineering **316** (2018) 012017 doi:10.1088/1757-899X/316/1/012017

Synthesis and characterization of methyltrihydroxysilane water repellent

A Z Abidin[1], M N Harjandi, V Wirawan and S M Suharno[2]

Department of Chemical Engineering, Faculty of Industrial Technology Institut Teknologi Bandung, Jl. Ganesha 10, Bandung 40132, Indonesia

[1]mitraiqro@yahoo.com
[2]srimulyani.suharno@gmail.com

Abstract. Methyltrihydroxysilane ($CH_3Si\ (OH)_3$) as a water repellent has been synthesized from trichloromethylsilane and ethanol by varying their composition, reaction condition, and the addition of nanosilica. The properties of the material have been characterized using FTIR for identification of raw materials and water repellent product, SEM for identification of water repellent coating surface, and tensiometer for measurement of water repellent contact angle. The FTIR spectra confirm the reaction of the water-repellent formation. The water repellent product was applied by spraying or dip coating on the automotive window surface. This study shows that the best ethanol composition is 91% and the best contact angle of synthesized water repellent material is 149,46°. This contact angle is higher than that of a commercial product, which shows it as a property of the superhydrophobic material. Water repellency properties increase as the composition of trichloromethylsilane increases. It shows that the increasing of trichloromethylsilane composition can also increase methyltrihydroxysilane formation. However, glass surface becomes opaque as the composition of trichloromethylsilane increase because methyltrihydroxysilane will create the Si-O-Si layer that has a white color. The addition of nanomaterial also increases the surface roughness, but a binder is required to bind nanomaterial to the water-repellent layer. For an application, dip coating has better water repellency than spraying. This is because dip coating method creates more homogenous nanomaterial precipitation on the surface. On the other hand, the level of transparency is worse. Therefore, the water repellent of trichloromethylsilane is recommended for applications that do not need clarity such bathroom glass wall.

1. Introduction

Water repellent is a material having properties of rejecting water due to its high hydrophobicity. Hydrophobic material means that its surface tension less than 35 dyne/cm and its contact angle more than 90° (Arkles, 2006). Technically, water repellent can be made from both polymer and non-polymer to create low surface tension. In addition, it also uses nanotechnology to create large coating surface area. One of non-polymer material that can be made to be water repellent is silane, a chemical compound which has silicone as its structural center. Generally, the molecular structure of silane is $YSiX_3$ in which Y is organofunctional group and X is alkoxy or chloride group. Silane which consists of carbon-silicon bond (CH_3-Si) is well-known as organosilane (Materne et al, 2012). The carbon-silicon bond is really stable, nonpolar, gives low surface tension and hydrophobic effect (Gao and McCarthy, 2006). In addition, silane is also the most frequently used commercial adhesion promoter

(Harper and Petrie, 2003). Furthermore, silane will react easily with water and form silanol. Silanol can react with other silanol and form a very stable bond, siloxane (Si-O-Si). According to Gao and McCarthy (2006), trichloromethylsilane (TCMS) can be used to create a superhydrophobic surface.

Water repellent can also be made using lotus leaf effect that combines low surface tension and a large surface area in order to produce better hydrophobic properties, self-cleaning ability, and very rough surface. Hydrophobic properties can be generated by using a material, such as a polymer which creates low surface tension. Moreover, the rough surface can be generated by adding nanoparticle. Hydrophobicity can be determined by measuring contact angle between the water droplet and glass surface.

2. Experiment

2.1. Materials

Trichloromethylsilane (TCMS) was a commercial product from Alibaba (China), Nanosilica SiO_2 obtained from Center for Ceramics of Bandung (Indonesia), Sodium Silicate obtained from PT Ajidharmamas Tritunggal Sakti (Indonesia), demineralized water, acetone, ethanol, and toluene were obtained from Department of Chemical Engineering, Institut Teknologi Bandung.

2.2. Functional Group Identification of Trichloromethylsilane

Trichloromethylsilane functional group was identified using Fourier Transform Infrared (FTIR) Spectrometer Prestige 21 Shimadzu. TCMS was dropped between two plates of Potassium Bromide (KBr) or Sodium Chloride (NaCl).

2.3. Synthesis of Methyltrihydroxysilane Water Repellent

2.3.1. Method I. Trichloromethylsilane mixed with ethanol under the acid condition in the fume hood to make 91 to 98% TCMS solution, with or without nanosilica. The concentration of the solution varied from 2 to 9%-mole TCMS with 1%-mass of nanosilica. Having stirred for 20 minutes at 30°C using magnetic stirrer, the solution was then sprayed twice into a glass which would be heated in the oven at 120°C within 10 minutes.

2.3.2. Method II. Trichloromethylsilane mixed with ethanol in order to make 1 M TCMS solution. Glass dipped and enclosed in the solution for one day. Following this, the glass was taken out and washed with toluene, ethanol and ethanol-water solution (1:1). Having washed, the glass was then heated in an oven at 120°C for 10 minutes.

2.4. Characterization of Methyltrihydroxysilane

2.4.1. Functional Group Identification. The functional group of trichloromethylsilane that has been reacted with ethanol identified using Fourier Transform Infrared (FTIR) Spectrometer Prestige 21 Shimadzu. The sample dropped between two plates of Potassium Bromide (KBr) or Sodium Chloride (NaCl).

2.4.2. Contact Angle Measurement. The contact angle of methyltrihydroxysilane water repellent measured with Attension Theta Lite pendant drop tensiometer. The water repellent was applied to a glass surface using a syringe. After that, the glass surface was installed to pendant drop tensiometer and the contact angle was determined using OneAttention software. The contact angle was determined using Young equation on the relation of surface tension between solid and liquid (σ_{SL}), solid and vapor (σ_{SV}), and liquid and vapor (σ_{LV}) which is shown at equation 1.

$$\sigma_{LV} \cos\theta = \sigma_{SV} - \sigma_{SL} \quad (1)$$

The Young model for water droplet on a solid surface is also shown in Figure 1.

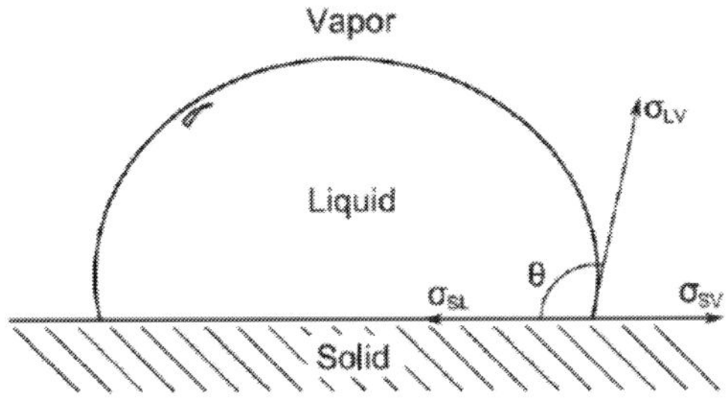

Figure 1. Young model for water drop on a solid surface (Stacy, 2009)

2.4.3. Coating Surface Identification. Methyltrihydroxysilane water repellent coating surface was identified using Scanning Electron Microscope (SEM) JEOL-JSM-6510LV. The sample was put at specimen support and then was set in order to make the upper part of specimen same with holder top part. Following this, the specimen was observed using software integrated to the SEM.

3. Results and Discussion

3.1. Synthesis of Methyltrihydroxysilane Water Repellent

Methyltrihydroxysilane water repellent was synthesized from the following reactions between trichloromethylsilane and ethanol:

$$CH_3SiCl_3 + 3\ C_2H_5OH \rightarrow CH_3Si(OC_2H_5)_3 + 3\ HCl \quad (2)$$

$$C_2H_5OH + HCl \rightarrow C_2H_5Cl + H_2O \quad (3)$$

$$CH_3Si(OC_2H_5)_3 + H_2O \rightarrow CH_3Si(OH)_3 + 3\ C_2H_5OH \quad (4)$$

In these reactions, ethanol acts as the reactant as well as TCMS solvent. Water repellent coating surface is very influenced by water presence. The more water composition, the more methyltrihydroxysilane [$CH_3Si(OC_2H_5)_3$] was formed. As a result, the coating of the methyltrihydroxysilane water repellent gave white color surface.

3.1.1. Characterization of Water Repellent Synthesis Method I. The absorption band of Si-O group that binds glass surface and forms the Si-O-Si bond is around 800-900 cm^{-1}. Based on Figure 2b, there is the Si-O group at trichloromethylsilane which is reacted with ethanol at 875 cm^{-1}. It shows that methyltrihydroxysilane layer has been formed. Besides, Si-O group is also detected at trichloromethylsilane at peak 898 cm^{-1} (Figure 2a). This is caused by a spontaneous reaction between trichloromethylsilaneand water in the air as follow:

$$CH_3SiCl_3 + 3\ H_2O \rightarrow CH_3Si(OH)_3 + 3\ HCl \quad (5)$$

In terms of composition variation, the bigger composition of trichloromethylsilane the more methyltrihydroxysilane generated. As a result, coating surface of the product will have more water repellent. However, the best contact angle of product generated based on the method I was still less than 90°. This means that the surface is not hydrophobic enough. Besides, increase of trichloromethylsilane concentration also increase the opacity of the glass. This is because methyltrihydroxysilane and glass generate Si-O-Si layer which has a white color. Figure 3 shows water droplet on the glass coated with the best product synthesized using the method I at 91% ethanol composition.

IOP Conf. Series: Materials Science and Engineering **316** (2018) 012017 doi:10.1088/1757-899X/316/1/012017

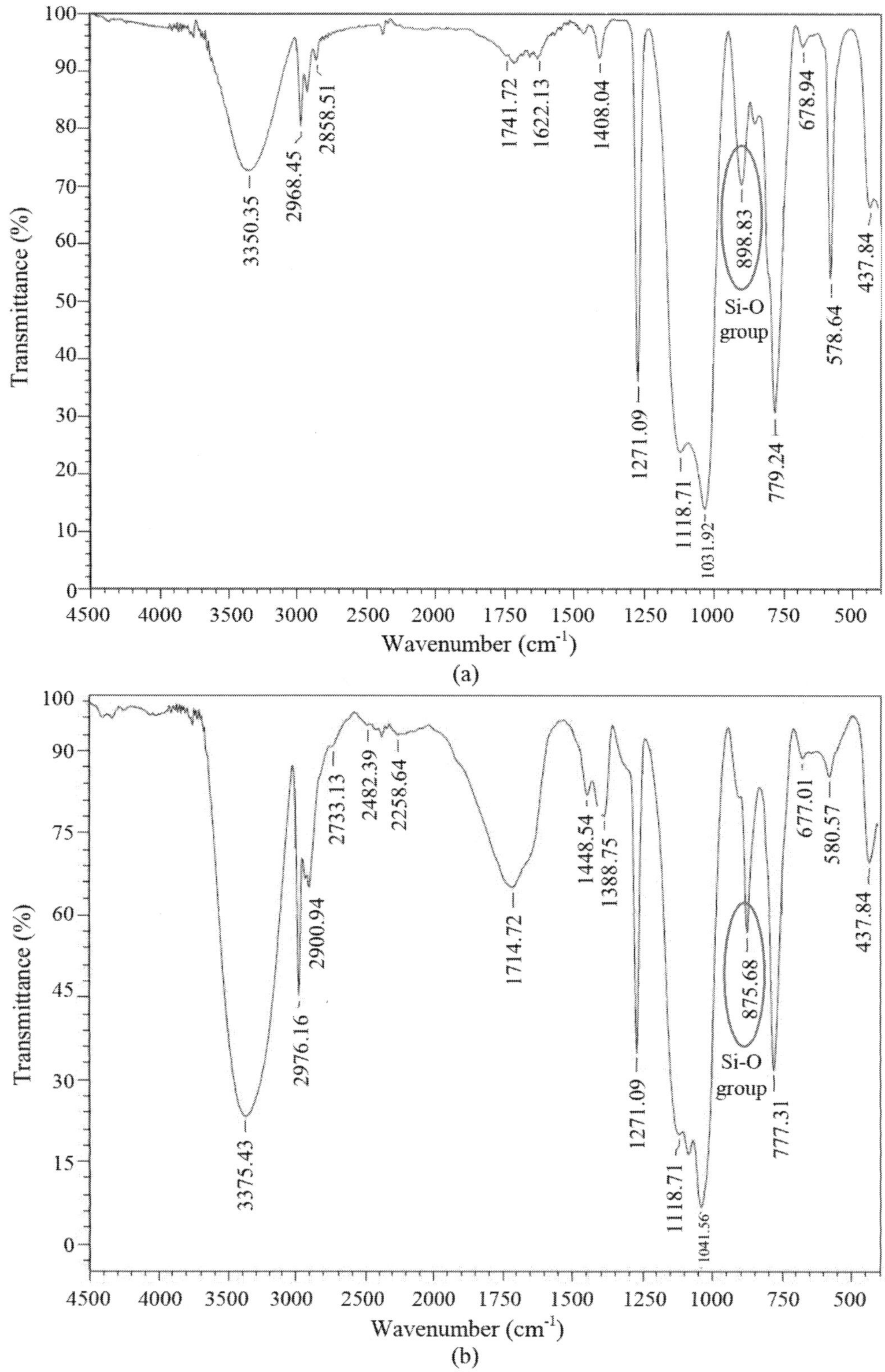

Figure 2. FTIR pattern of: (a) trichloromethylsilane; (b) trichloromethylsilane reacted with ethanol

Figure 3. Methyltrihydroxysilane water repellent with 91% ethanol of method I

There is also the addition of nanosilica at 1% of the total weight of ethanol and trichloromethylsilane. The generated surface color is also white. If it is wiped, the white color, as well as the water repellent, are disappeared. It shows that addition of nanosilica directly does not create a layer of a nanocomposite. Instead, nanosilica will settle on the glass surface, as well as inhibit the formation of the Si-O-Si bond. The best product of nanosilica addition is that with 91% ethanol compositions which can be seen in Figure 4.

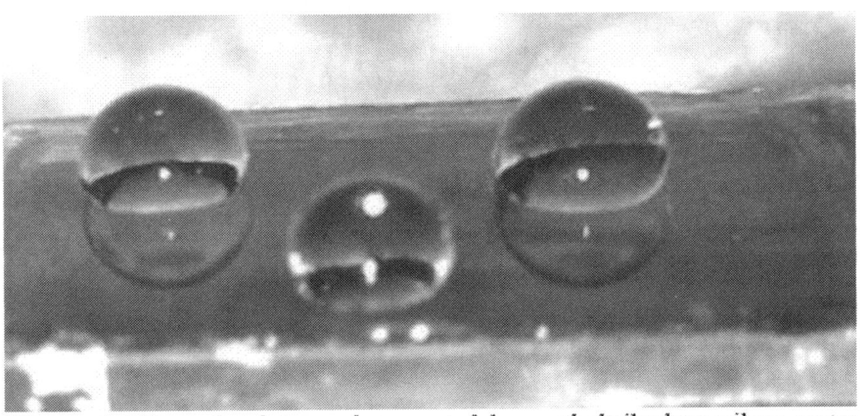

Figure 4. Water droplet on glass coated by methyltrihydroxysilane water repellent with 91% ethanol and 1%-mass of nanosilica

In terms of sodium silicate addition, it seemed that methyltrihydroxysilane water repellent was not generated. Meanwhile, the reaction between trichloromethylsilane, ethanol, and sodium silica produced cross-linked silicate polymer and did not generate the Si-O-Si bond between glass and trichloromethylsilane.

3.1.2. Characterization of Water Repellent Synthesis Method II. Water repellent synthesized by method II generates homogenous solid white layer. Glass that has been coated by methyltrihydroxysilane water repellent has contact angle more than 90°, which is 149,46° (Figure 5). This contact angle shows that the glass surface is superhydrophobic. The water repellent is a result of closed reaction which makes the water content in the reaction small. The surface is also homogenous as result of the reaction process occurs in the glass dipped for one day.

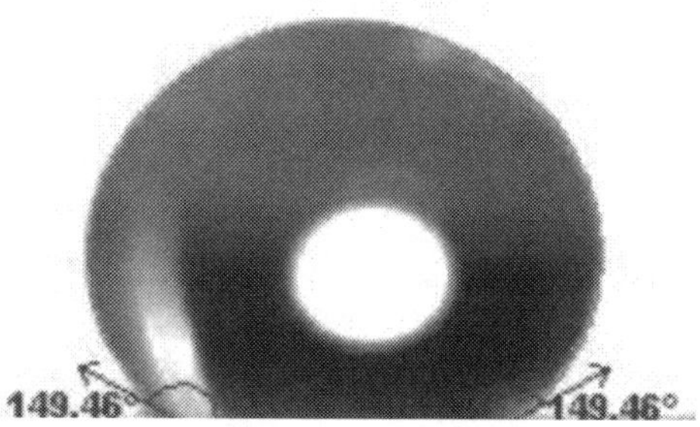

Figure 5. The contact angle of water droplet on glass coated by methyltrihydroxysilane water repellent method II

The glass coated with methyltrihydroxysilane water repellent method II was identified by Scanning Electron Microscope (SEM) at the enlargement of 500 and 2000 times. It can be seen in Figure 6, methyltrihydroxysilane water repellent is layered and gives whiter color in the glass.

(a) (b)

Figure 6. SEM surface morphology of glass coated by methyltrihydroxysilane water repellent method II: (a) enlarged 500 times; (b) enlarged 2000 times

3.2. Comparison of Synthetic and Commercial Product Performance

(a) (b)

Figure 7. Water droplet on glass coated with: (a) commercial water repellent, RainZ Dura; (b) synthesis water repellent, methyltrihydroxysilane

Figure 7 shows that both commercial and synthetic product has hydrophobicity. Besides, the contact angle for the commercial product is less than 90°, while for the synthetic product is approximately 149,46°. Therefore, the performance of synthetic water repellent is better than commercial in terms of contact angle and hydrophobicity. However, the transparency of the glass coated by both of them is different. The glass coated with commercial water repellent is transparent, while the glass coated with synthetic water repellent is opaque and white in color. By and large, synthesis water repellent is recommended for applications that do not need high clarity such glass wall of a bathroom.

4. Conclusion

Based on the results obtained, it can be concluded that the increase of trichloromethylsilane can increase hydrophobicity and opacity of glass. Besides, the addition of nanosilica also increases the hydrophobicity of the water repellent. Therefore, the methyltrihydroxysilane is really a potential material for water repellent regards to its hydrophobicity and contact angle, and especially it is best for applications that do not need transparency.

References

[1] Arkles B 2006 *Hydrophobicity, Hydrophilicity and Silane Surface Modification* (New York: Gelest)
[2] Gao L and McCarthy T J 2006 *J. Ame. Chem. Soc.* **128** 9052-9053
[3] Harper C A and Petrie E M 2003 *Plastic Materials and Processes* (New Jersey: John Wiley and Sons, Inc.)
[4] Materne T, Buyl F and Witucki G L 2012 *Organosilane Technology in Coating Applications: Review and Perspectives* (Midland: Dow Corning Corporation)
[5] Stacy R 2009 *Contact Angle Measurement Technique for Rough Surfaces* (Michigan: Michigan Technological University)

Acknowledgement

The authors are grateful to Center for Ceramics-BBK and Institut Teknologi Bandung which supported this research.

Preparation of hollow mesoporous carbon spheres and their performances for electrochemical applications

T Ariyanto[1,2*], G R Zhang[2,3], A Kern[2], B J M Etzold[2,3]

[1]Department of Chemical Engineering, Universitas Gadjah Mada, Jl Grafika 2, 55281 Yogyakarta, Indonesia
[2]Lehrstuhl für Chemische Reaktionstechnik, Friedrich-Alexander-Universität Erlangen-Nürnberg, Egerlandstrasse 3, 91058 Erlangen, Germany
[3]Ernst-Berl-Institut für Technische und Makromolekulare Chemie, Technische Universität Darmstadt, Alarich-Weiss-Strasse 8, 6428 Darmstadt, Germany

*Corresponding author: teguh.ariyanto@ugm.ac.id

Abstract. Hollow carbon materials have received intensive attention for energy storage/conversion applications due to their attractive properties of high conductivity, high surface area, large void and short diffusion pathway. In this work, a novel hollow mesoporous material based on carbide-derived carbon (CDC) is presented. CDC is a new class of carbon material synthesized by the selective extraction of metals from metal carbides. With a two-stage extraction procedure of carbides with chlorine, firstly hybrid core-shell carbon particles were synthesized, i.e. mesoporous/graphitic carbon shells covering microporous/amorphous carbon cores. The amorphous cores were then selectively removed from particles by a careful oxidative treatment utilizing its low thermal characters while the more stable carbon shells remained, thus resulting hollow particles. The characterization methods (e.g. N_2 sorption, Raman spectroscopy, temperature-programmed oxidation and SEM) proved the successful synthesis of the aspired material. In electric double-layer capacitor (EDLC) testing, this novel hollow core material showed a remarkable enhancement of EDLC's rate handling ability (75% at a high scan rate) with respect to an entirely solid-mesoporous material. Furthermore, as a fuel cell catalyst support the material showed higher Pt mass activity (a factor of 1.8) compared to a conventional carbon support for methanol oxidation without noticeably decreasing activity in a long-term testing. Therefore, this carbon nanostructure shows great promises as efficient electrode materials for energy storage and conversion systems.
Keywords: Carbide-derived carbon, Electric-double layer capacitors, Hollow carbon spheres, Methanol oxidation

1. Introduction

Porous carbons play an important role in electrochemical applications. They have been used as electrode material of electric double-layer capacitors (EDLC) and catalyst support for fuel cell electrodes [1]. Hollow carbon spheres are types of structured carbon materials which have received much attention. This is due to their attractive properties of high surface area, large void and short diffusion pathway [2]. A common method to synthesize hollow carbon sphere is templating routes using polymeric materials as carbon precursors [3,4]. Nevertheless, with this method, typically a low degree of carbon ordering results (amorphous structures). Therefore, the produced material features low electrical conductivity and poor electrochemical stability. These properties, however, are a major drawback in electrochemical devices [5,6]

In this work, hollow-core carbon material based on carbide-derived carbon (CDC) is presented. CDC is a new class of carbon material synthesized by the selective extraction of metals from metal carbides. The special advantage of CDC is its tuneable micro- and pore structures [7–9]. Therefore, it is possible to vary carbon micro- and pore structures within a particle in a core-shell arrangement which is the basis of the successful preparation of hollow core carbon spheres. The hollow core carbon produced was then tested for EDLC electrode and electrocatalysts support for methanol oxidation and the results are discussed.

2. Materials and methods

2.1. Materials

Commercial titanium carbide (TiC, Alfa Aesar) was employed as carbon precursor. Chlorine and hydrogen diluted by helium were used to perform selective extraction of titanium from the carbide matrix (called chlorination process) and subsequent surface annealing. All gases were purchased from Linde AG. In EDLC testing, tetraethylylammonium tetrafluoroborate (NEt_4-BF_4, Sigma Aldrich) in acetonitrile were used as electrolyte. For preparation of catalyst, hydrochloroplatinic acid ($H_2PtCl_6.H_2O$, ABCR GmbH) was used as platinum precursor.

2.2. Material synthesis

Hollow core CDC particles (named as hollow carbon spheres (HCS)) were produced from nonporous commercial TiC particles *via* two-step procedure as shown in **Figure 1** and described as follows.

i) A two-step chlorine etching to prepare an interim material of mesoporous/graphitic-microporous/amorphous carbon called as Meso-Micro Carbon.
ii) Amorphous carbon core removal by careful oxidative treatment at 430 °C.

Detail procedures of CDC synthesis are provided in [9,10]. As reference materials, CDC-800 and CDC-1200 were synthesized at fixed Cl_2 etching temperature of 800 °C and 1200 °C, respectively.

For preparation of fuel cell anode catalyst, a wet impregnation method was used to disperse platinum precursor ($H_2PtCl_6{\cdot}H_2O$) on HCS followed by a hydrogen reduction step at 260 °C

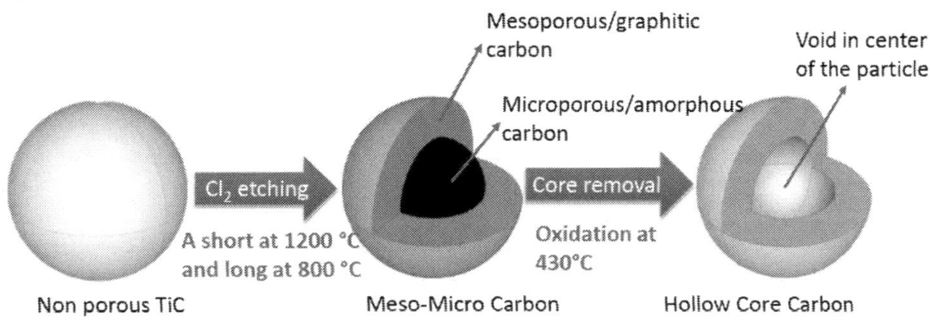

Figure 1. Schematic preparation of hollow core carbon particle from carbide.

2.3. Characterization methods

The properties of materials were characterized using N_2-sorption (Quantrachrome Quadrasorb Si-MP) , scanning electron microscopy (SEM, Philips XL 30) and temperature-programmed oxidation (TPO, Netzsch STA 409 PC Luxx).

2.4. Characterization methods

For supercapacitor testing, the working electrode was prepared by mixing carbon particles (95 wt.%) with 5% Nafion binders dispersed in isopropanol. After the mixing process, the ink was then applied onto glassy carbon electrode. Cyclic voltammograms and galvanostatic cycling data of the electrode were recorded with a PARSTAT 4000. For methanol oxidation, the electrode was prepared by combining catalyst (20% Pt/carbon) with 5% Nafion binders. The three-electrode system was employed to test the catalyst by recording cyclic voltammetry (CV) data. An Ag/AgCl electrode and a Pt wire were used as the reference and counter electrodes, respectively.

3. Results and discussion

3.1. Successful synthesis of hollow core carbon particles

A short reaction of chlorination at 1200 °C yielded the partially extracted carbide particles (carbon/carbide matrix) with 35% conversion. TEM characterization of this material indicated that the carbon shells enveloped the unreacted carbide cores (data not shown). The second conversion step of the unreacted core within chlorination at 800 °C yielded the carbon core. TPO characterization showed that Meso-Micro Carbon featured a bimodal microstructure showing a combination of material properties of CDC-1200 (graphitic) in the shell and CDC-800 (amorphous) in the core.

Selective removal of the amorphous core was carried out by mild oxidation process at 430 °C. The TPO profile of Meso-Micro Carbon was compared to the reference materials (see **Figure 2a**). The figure shows that CDC-800 reference material could be fully oxidized while CDC-1200 exhibits no mass change after 14 hours of the oxidation process. This shows that the shell material has different thermal stabilities compared to the core material which is likely due to different graphitization degree [11]. The oxidation profile of Meso-Micro Carbon shows a mass decrease till 70 wt.% indicating a complete removal of carbon in the core. The final oxidized particles were then mechanically crushed and SEM image was taken. **Figure 2b** shows clearly that carbons shell envelopes a hollow core. With all evidence, we can conclude that the hollow core material was successfully synthesized.

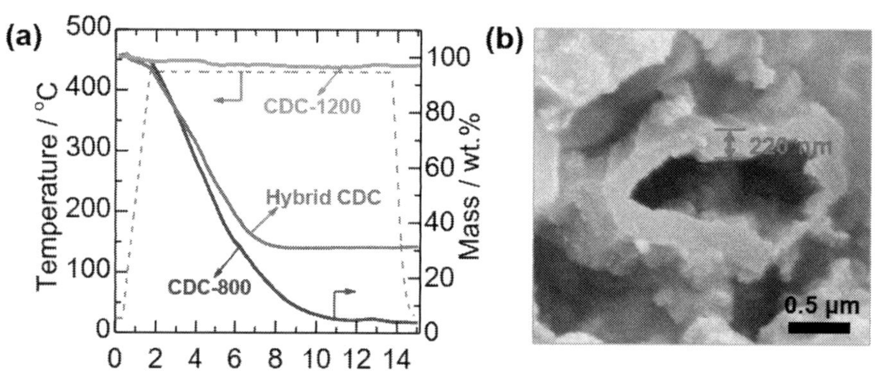

Figure 2. a) TPO profiles of materials. b) SEM image of a cracked carbon particle showing only carbon shell after selective removal of amorphous core. Figure 2B adapted with permission from [5]. Copyright 2015 Elsevier.

3.2. Electrochemical performances

Figure 3a compares the gravimetric capacitance of hollow core material and CDC-1200. At the low scan rate of 5 mV s^{-1} both materials feature a similar capacitance value (ca. 32 F g^{-1}). With increasing scan rate, the hollow core material exhibits better characteristics to maintain its initial capacitance than the CDC-1200. While only minor decrease in capacitance (30%) is seen for the hollow core material, the entirely solid mesoporous of CDC-1200 shows a huge drop of 62% at the highest scan rate of 5000 mV s^{-1}. Therefore, the hollow core material shows promising application for high rate EDLCs.

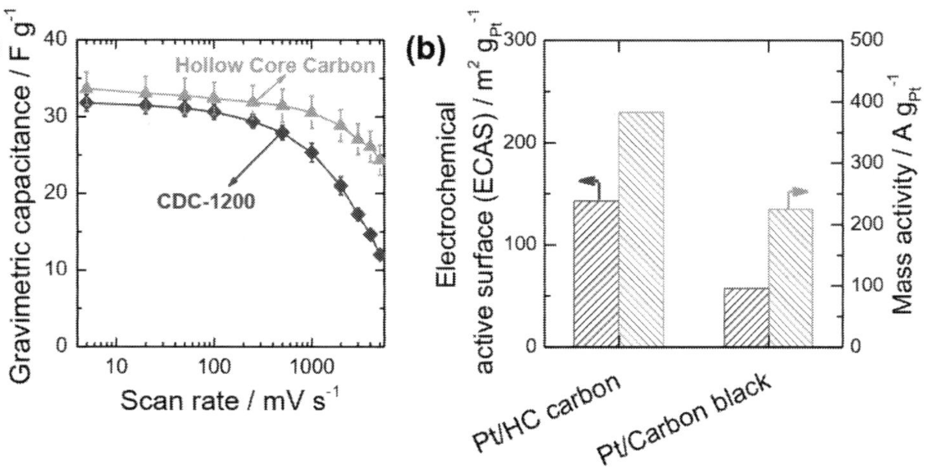

Figure 3. a) Specific capacitance of material evaluated by CV method. b) Active surface and performance of electrocatalysts of hollow core carbon support (Pt/HC carbon) and conventional support material of carbon black (Pt/CB) at 0.6 V.

The hollow core material was employed as catalyst supports. The catalyst characteristics and performances are summarized in **Figure 3b**. Pt/Hollow Carbon Spheres features high electrochemically active surface (ECAS) up to 140 m^2 g_{Pt}^{-1}. This is remarkably larger than that of the Pt immobilized on a conventional carbon black (60 m^2 g_{Pt}^{-1}). In the methanol electro-oxidation, Pt/HCS shows a remarkable activity of 390 A g_{Pt}^{-1}@ 0.6 V, which is 1.8 times that of Pt/CB.

4. Conclusions

Hollow core-mesoporous carbon spheres were successfully synthesized based on the CDC method. The novel material showed promising application for electric-double layer capacitor electrode and catalyst support for methanol electro-oxidation.

Acknowledgements

This work is supported by the funding from the German Federal Ministry of Education and Research (BMBF) under the project AktivCAPs and the German Research Council (DFG), which, within the framework of its “Excellence Initiative”, supports the Cluster of Excellence “Engineering of Advanced Materials” (www.eam.uni-erlangen.de) at the University of Erlangen-Nuremberg. TA acknowledges Directorate General of Higher Education (DIKTI) Indonesia for a PhD scholarship.

References

[1] Zhang J and Liu H 2009 *Electrocatalysis of direct methanol fuel cells : from fundamentals to applications* (Weinheim, Germany: Wiley-VCH)

[2] Li Y S and Shi J L 2014 Hollow-Structured Mesoporous Materials: Chemical Synthesis, Functionalization and Applications *Adv. Mater.* **26** 3176–205

[3] Prasetyo I, Rochmadi, Ariyanto T and Yunanto R 2013 Simple method to produce nanoporous carbon for various applications by pyrolysis of specially synthesized phenolic resin *Indones. J. Chem.* **13** 95–100

[4] Mezzavilla S, Baldizzone C, Mayrhofer K J J and Schuth F 2015 General Method for the Synthesis of Hollow Mesoporous Carbon Spheres with Tunable Textural Properties *ACS Appl. Mater. Interfaces* **7** 12914–22

[5] Ariyanto T, Dyatkin B, Zhang G-R, Kern A, Gogotsi Y and Etzold B J M 2015 Synthesis of carbon core–shell pore structures and their performance as supercapacitors *Microporous Mesoporous Mater.* **218** 130–6

[6] Sharma S and Pollet B G 2012 Support materials for PEMFC and DMFC electrocatalysts-A review *J. Power Sources* **208** 96–119

[7] Ariyanto T, Zhang G-R, Riyahi F, Gläsel J and Etzold B J M 2017 Controlled synthesis of core-shell carbide-derived carbons through in situ generated chlorine *Carbon N. Y.* **115** 422–9

[8] Ariyanto T, Kern A M, Etzold B J M, Zhang G-R, Ariyanto T, Ariyanto T, Etzold B J M, Zhang G-R, Ariyanto T, Kern A M, Etzold B J M and Zhang G-R 2017 Carbide-derived carbon with hollow core structure and its performance as catalyst support for methanol electro-oxidation *Electrochem. commun.* **82**

[9] Ariyanto T, Dyatkin B, Zhang G, Kern A, Gogotsi Y and Etzold B J M 2015 Synthesis of carbon core-shell pore structures and their performance as supercapacitors *Microporous Mesoporous Mater.* **218** 130–6

[10] Ariyanto T, Kern A, Etzold B J M and Zhang G-R 2017 Carbide-derived carbon with hollow core structure and its performance as catalyst support for methanol electro-oxidation *Electrochem. commun.* **82** 12–5

[11] Glasel J, Diao J Y, Feng Z B, Hilgart M, Wolker T, Su D S and Etzold B J M 2015 Mesoporous and Graphitic Carbide-Derived Carbons as Selective and Stable Catalysts for the Dehydrogenation Reaction *Chem. Mater.* **27** 5719–25

IOP Conf. Series: Materials Science and Engineering **316** (2018) 012019 doi:10.1088/1757-899X/316/1/012019

Reaction product of pyrogallol with methyl linoleate and its antioxidant potential for biodiesel

H Sutanto[1] , L Ainny[1], Lukman[1], B H Susanto[1], and M Nasikin[1]

[1]Chemical Engineering, Department of Chemical Engineering, Universitas Indonesia, Depok 16424, Indonesia

bambanghs@che.ui.ac.id

Abstract. The demand of biodiesel as an alternative fuel is increasing due to fossil fuel depletion. Biodiesel is a renewable diesel fuel in the form of fatty acid methyl ester or FAME as a result of an esterification of plant oils in a presence of catalyst. Compared to the conventional diesel fuel, biodiesel is more biodegradable, has higher lubricity, and lower toxic emissions. However, the high content of unsaturated fatty acid leads to a problem that biodiesel is prone to oxidation during storage period. This oxidation instability causes degradation of fuel quality and will affect engine performance. Pyrogallol and other phenolic derivatives have been used as the antioxidant additives to prevent biodiesel oxidation. As reported in many researches, pyrogallol is one of the best phenolic antioxidant. However, its low solubility in biodiesel needs an attention. Several reports indicate the increasing solubility of pyrogallol using molecule modification with the addition of alkyl groups to its benzene ring via electrophilic substitution. This paper discusses the idea about modification of pyrogallol molecule and methyl linoleate using 2,2-diphenyl-1-picrylhydrazyl (DPPH) radical in order to increase its solubility in biodiesel while keeping its antioxidant property. Three responses were analyzed to examine the antioxidant activity: iodine value, viscosity, and color intensity. The result shown that the addition of 0.1% reaction product exhibit antioxidant activity in biodiesel.

1. Introduction

Biodiesel or fatty acid methyl ester (FAME) is a renewable alternative fuel produced from plants to overcome fossil fuel depletion problem. It has been implemented and used globally for diesel vehicles. Biodiesel production and demand increases over time due to people's awareness regarding fossil fuel crisis and environmental issue. However, biodiesel has several weaknesses in which many researchers are interested in increasing its quality.

Other Biodiesel which contains mixture of FAME molecules is highly unstable; both thermal and oxidative instability. Molecule deterioration occurs when biodiesel is exposed to air, light, high storage temperature, as well as high humidity. The unsaturated fatty acid methyl ester contents are responsible for this weakness due to their high oxidation risk. The oxidized FAME will have higher viscosity, moreover it is highly possible to produce solid precipitates which would blockage the machines' filter or nozzle. This oxidative instability of biodiesel also causes its shorter shelf life compared to the conventional fuels. In the oxidized biodiesel, the acid value increases as the increasing rate of

[1] To whom any correspondence should be addressed.

decomposition of peroxide compound as the result of oil oxidation. This will cause disadvantage as well due to the corrosive effect to the engine materials which are made of metals.

Palm oils is the main source of fatty acid in the biodiesels production. This type of biodiesel has the unsaturated carbon chain length between 16 to 20 carbon number [1]. The longer the carbon chain, the more unstable the fatty acid molecules [2]. To increase the stability of palm oils' biodiesel, antioxidant as radical scavenging agent additives are required to prevent the oxidation to the oil components.

One of the commonly used antioxidant additive is pyrogallol (PY). Pyrogallol or 1,2,3-trihydroxybenzene is a three hydroxyl substituted benzene ring molecule [3], it is belong to phenolic group. Just like other phenolic compounds, pyrogallol has an antioxidant property. In the presence of radical species, the hydroxyl hydrogen of phenols will be abstracted by the free radical during the propagation phase [4]. The next propagation step will be stopped by radical phenol by delocalizing the radical electron into several available positions. This mechanism will protect the phenoxy radical to do another unwanted radical attack or further oxidations. Pyrogallol is the most effective antioxidant for biodiesel compared to the others [5]. However, the utilization of pyrogallol still have some weaknesses. It prevents the oxidation for only on the first six hours [6]. On the other hand, pyrogallol has a solubility problem when it is mixed with oil solution like biodiesel due to the polarity difference which causes the uneven additive distribution in the mixture [7]. Pyrogallol molecule modification is necessary to adjust its polarity thus the oil dispersion increased in order to increase the antioxidant activity.

Modification of phenolic compounds has been an interesting research topic. The objective is to increase solubility of phenol as antioxidant in oils especially in biodiesel. Based on previous works, pyrogallol is one of the best phenolic antioxidant, thus it is important to develop a new synthesized molecule with pyrogallol molecule basis by connecting it with a nonpolar molecule. The idea is to combine pyrogallol which is polar with a molecule of fatty acid methyl ester as the nonpolar part. Pyrogallol as a member of benzene group is a stable molecule but in the other hand it is a highly risking molecules to be attacked by radical species. Ester groups, amine, benzylic, and heteroatomic molecules are prone to oxidation by radical [8]. [9] did the analysis of radical reaction of pyrogallol using 2,2-diphenyl-1-picrylhydrazyl or DPPH. The result shows the products was the oxidative coupling product, dimer of pyrogallol. DPPH also can be used in oil solution, usually for antioxidant analysis of unsaturated fatty acid but using different solvent [10]. This new molecule can be synthesized by using several methods. One of the possible method is by using radical mechanism. The expected product is a result of oxidative coupling between two different molecules: pyrogallol and FAME. To represent FAME, methyl linoleate is suitable. It is a polyunsaturated fatty acid omega-6 with two double bonds at C-9 and C-12 which means it has a methylene hydrogen at interrupted position at C-11. This hydrogen is susceptible for radical attack. Methyl linoleate as a derivative of linoleic acid represents vegetable oil can be radicalized using DPPH [10]. The radical methyl linoleate and radical pyrogallol are expected to undergo oxidative coupling termination. This new molecule is expected to have a high solubility in oil without decreasing the antioxidant activity of pyrogallol.

2. Materials and methods

Methyl linoleate and pyrogallol were reacted based on stoichiometric calculation. Combination of [9] and [10] experiments using ethyl acetate solvent. Pyrogallol (p.a.) obtained from MERCK (CAS number: 87-66-1). DPPH or 2,2-diphenyl-1-picrylhydrazyl obtained from SIGMA-ALDRICH (CAS number: 1898-66-4). Methyl linoleate obtained from Santacruz. The solvents methanol and ethyl acetate as well as the silica TLC plate were obtained from MERCK. Wijs solution for iodine value obtained form MERCK. Viscometer canon feske pipe. Thin layer chromatography was used for the qualitative analysis of the desired product. The Rf values will be evaluated compared to the original reactants which are pyrogallol and methyl linoleate. The product mixture then mixed with biodiesel with different concentrations in order to examine its antioxidant property. Blank sample was biodiesel without any antioxidant addition. All samples were left open in the room atmosphere (light, air, pressure and temperature).

2.1. Synthesis reaction preparation
Synthesis reaction was prepared by mixing pyrogallol and methyl linoleate in DPPH solution. Pyrogallol solution was prepared by mixing 0.03mg solid pyrogallol with 2ml methanol. DPPH solution was a mixture of 0.07mg DPPH dissolved in 2 ml ethyl acetate producing a dark violet solution. While DPPH solution was constantly stirred, 2.5ml methyl linoleate was dropped slowly using burette into the DPPH solution to ensure all methyl linoleate were completely radicalized. The violet color of DPPH solution was gradually turned lighter as a mark of the radical production progress. After all methyl linoleate addition was finished, pyrogallol solution was added slowly. The solutions' color turned from light violet into yellowish color, and after 3 hours, it turned into darker yellow or brownish.

2.2. Application to biodiesel
The reaction mixture was added to biodiesel with different concentration (w/v): 0.05%, 0.07%, and 0.1%, while blank sample was without any addition of reaction mixture or 0.00%. The mixture as well as the blank were kept in a transparent glass bottles and left exposed to air and light with open cap to accelerate the oxidation reaction. As the responses, the measurement of iodine value, viscosity, and color intensity was determined weekly up to 4 weeks.

2.3. Iodine value determiation
Iodine value was determined using Wijs method.

2.4. Viscosity test
Viscosity test was performed by using viscometer at 40°C. Standard: ASTM D445-11a.

2.5. Color intensity test
Color intensity test was performed by using visible light region graduated color intensity of the electromagnetic spectrum.

3. Results and discussion

3.1. TLC analysis
The synthesis reaction mixture was spotted on TLC plate compared with the pure pyrogallol and methyl linoleate. The calculated Rf shown on the reaction mixture spot were 0.31 (a), 0.62 (c) and 0.84 (b) as can be seen on figure 1. Spot (a) 0.31 is equal to the Rf of pyrogallol, and (b) 0.84 is equal to the Rf of methyl linoleate. It shows that some numbers of unreacted reactants are still remain on the solution. From this result, it is qualitatively concluded that aside of the remaining unreacted pyrogallol and methyl linoleate, the new product was formed indicated by the new spot with Rf 0.62 (c). This new modification products is proposed to be the oxidative coupling product between pyrogallol and methyl linoleate. The new product is the result of termination step of both pyrogallol and methyl linoleate radical which were initially produced by using DPPH.

IOP Conf. Series: Materials Science and Engineering **316** (2018) 012019 doi:10.1088/1757-899X/316/1/012019

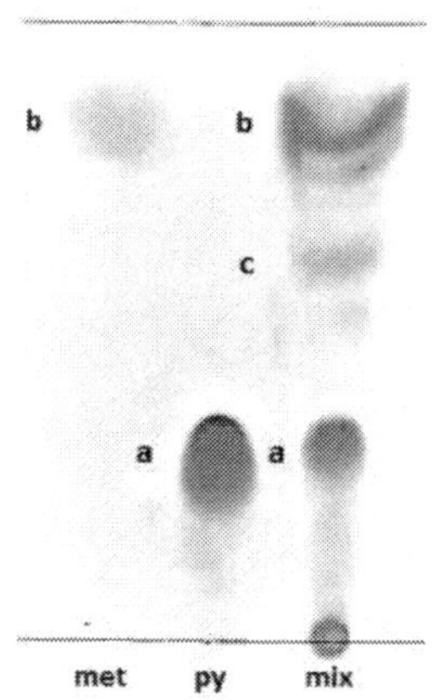

Figure 1. Thin layer chromatography result

3.2. Iodine value

Iodine value result can be seen in figure 2. Iodine value is a quantitative determination of oils' quality. The lower the iodine value means the higher the oxidation due to the decreasing of double bond number. When an unsaturated fatty acid is oxidized, one of the double bond will be transformed into single bond because of the formation of peroxide group. It is shown that the iodine value of the biodiesel sample were decreased over time for all concentrations. Compare to the 0.05 and 0.07, the lowest iodine value decreasing or ΔI was with 0.1% antioxidant, which was the most concentrated one. This result showing that the addition of antioxidant helps to prevent oxidation as also reported by [2] and [11].

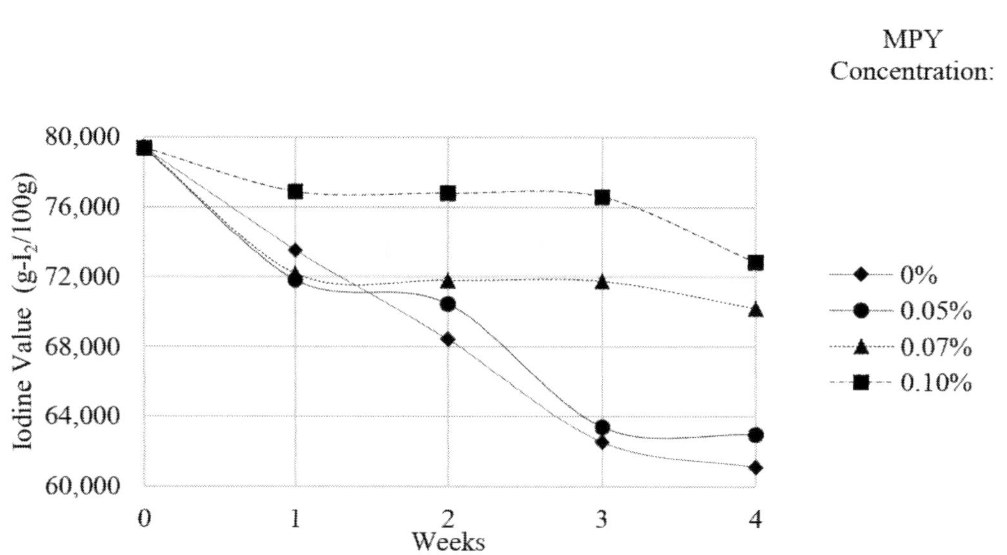

Figure 2. Iodine value

IOP Conf. Series: Materials Science and Engineering **316** (2018) 012019 doi:10.1088/1757-899X/316/1/012019

3.3. Viscosity value

Viscosity test result is shown in figure 3. Viscosity is commonly used as the marker of oils' oxidation process. The oxidized oil may transform into different types of molecule or the structure change occurrence. After the addition of 0.1% antioxidant, the viscosity value jump was lower than the blank. It is showing that the reaction product mixture kept the viscosity low, it is concluded that the antioxidant is effective in preventing oxidation. Formation of heavier yet bigger molecule such as oil polymers are responsible for the viscosity value increasing. The increasing viscosity as a marker of oils' oxidation also reported by [2], [11], and [6].

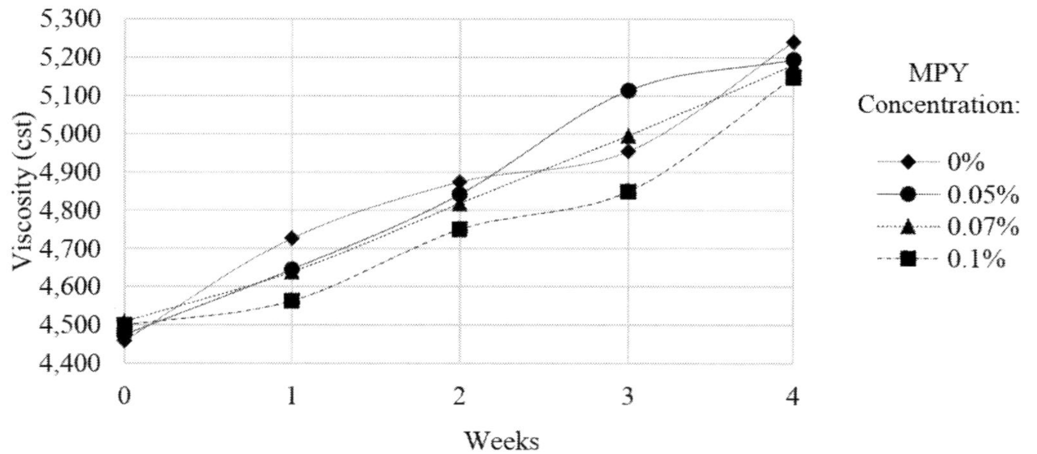

Figure 3. Viscosity value

3.4. Color intensity

Visually the color of all samples were different over time and also different on every sample. For the quantitative analysis, the color intensity was measured by measuring its optimum wavelength shifts. Within 4 weeks, the color intensity were decreased. The highest color intensity drop was the blank sample. The lowest color drop was the 0.1%. It shown on figure 4 that the higher the antioxidant concentration, the better the activity in retaining the biodiesel color. Color intensity decreasing indicates the quality degradation of the oil. The more oxidized molecule means the lighter the color due to the structural change to the FAME molecules. From this response, it is concluded that 0.1% addition of reaction mixture might act as antioxidant preventing structure deterioration from oxidation. It is proposed that glycerol molecule is responsible for the samples' color change. At high temperature or during oxidation, glycerol would cause the oils' color degradation as stated also by [12], [13], and [14].

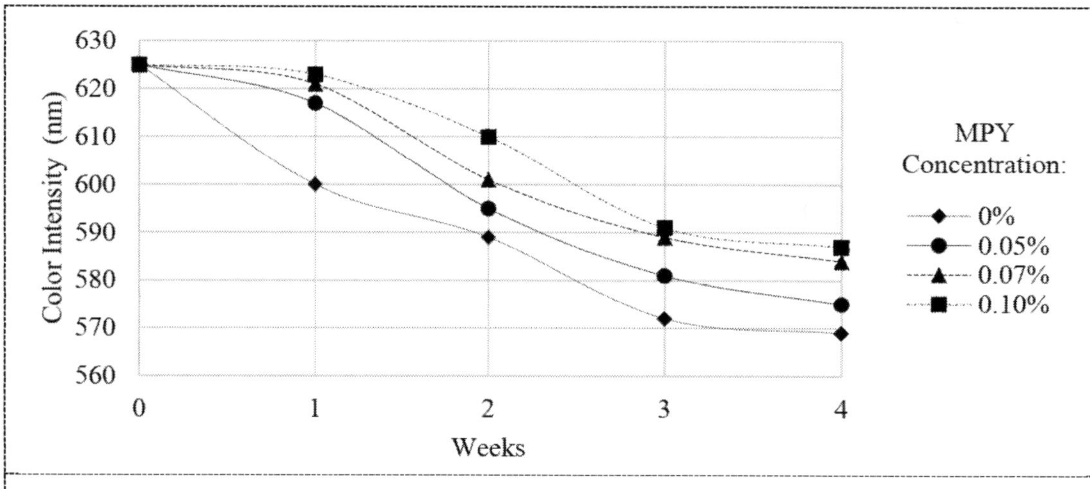

Figure 4. Color intensity.

4. Conclusion

The pyrogallol derivative molecule was synthesized using DPPH radical by adding methyl linoleate molecule via coupling oxidation mechanism. The product mixture showing antioxidant potential to be used to prevent oxidation in biodiesel. The antioxidant analysis using 3 different responses: iodine value, viscosity, and color intensity resulting the active radical scavenging effect by the addition of reaction product mixture. The best result shown by the 0.1% addition of product in biodiesel with the duration of 4 weeks.

5. References

[1] Darnoko, D, T, H, & P, G 2001 *Warta PKKS* p 17-27.
[2] Pullen, J, & Saeed, K 2012 *Renewable and Sustainable Energy* **16** p 5924-5950.
[3] XiCan Li 2012 *J. Agric. Food Chem* (america: American Chemical Society) **60** p 6418−6424.
[4] Rapopport. Zvi 2003 *Chemistry of Phenol, John Wiley & Sons*, Ltd ISBN: 0-471-49737-1.
[5] Supriyono, Sulistyo, H, Almeida, M F, & Dias, J M 2015 *Food Processing Technology* **132** p 133-138.
[6] Sorate, K A, & Bhale, P V 2015 *Renewable and Sustainable Energy Reviews* **41** p 777-798.
[7] Wilson, C P, & Houston, T 1995 United States Patent No. 2,052,858.
[8] Daley, R F, & Daley, S J 2005 *Organic Chemistry* p 530-567.
[9] Zhu, N, Wang, M, Wei, G J, Lin, J K, Yang, C S, & Ho, C T 2001 *Food Chemistry* **73** p 345-349.
[10] Prevc, T, Segatin, N, Ulrih, P N, & Cigic, B 2013 *Talanta* **109** p 13-19.
[11] Saluja, R K, Kumar, V, & Sham, R 2016 *Renewable and Sustainable Energy Reviews* **62** p 866-881.
[12] Banga, Sangita and Varshney, Pradeep K 2012 *Journal of Scientific & Industrial Research* **69** p 575-579.
[13] Keshvari, Mahtab, Asgary, Sedigheh, Jafarian-dehkordi, Abbas, Najafi, Somayeh, Ghoreyshi-Yazdi, Seyed Mojtaba 2013 *ARYA Atheroscler* **9** p 280-286.
[14] Kenreck, Gleen 2007 *Biodiesel Magazine*.

Acknowledgementt

This research was funded by Hibah Publikasi Internasional Terindeks untuk Tugas Akhir Mahasiswa (PITTA UI) Universitas Indonesia 2017).

IOP Conf. Series: Materials Science and Engineering **316** (2018) 012020 doi:10.1088/1757-899X/316/1/012020

Catalytic Pyrolysis of Waste Plastic Mixture

Ferdianta Sembiring[1], Chandra Wahyu Purnomo[1,3], Suryo Purwono[2]

[1]Advanced Materials and Mineral Processing Research Group,
[2]Energy and Environment Research Group, Chemical Engineering Dept. Universitas Gadjah Mada, Jalan Grafika no 2, 55281 Yogyakarta Indonesia.
[3]Renewable Energy and Waste Utilization Division PIAT UGM, Berbah Sleman Yogyakarta Indonesia.Type the author addresses here

chandra.purnomo@ugm.ac.id

Abstract. Inorganic waste especially plastics still become a major problem in many places. Low biodegradability of this materials causes the effort in recycling become very difficult. Most of the municipal solid waste (MSW) recycling facilities in developing country only use composting method to recover the organic fraction of the waste, while the inorganic fraction is still untreated. By pyrolysis, plastic waste can be treated to produce liquid fuels, flammable gas and chars. Reduction in volume and utilization of the liquid and gas as fuel are the major benefits of the process. By heat integration actually this process can become a self-sufficient system in terms of energy demand. However, the drawback of this process is usually due to the diverse type of plastic in the MSW creating low grade of liquid fuel and harmful gases. In this study, the mixture of plastics i.e. polypropylene (PP) and polyethylene terephthalate (PET) is treated using pyrolysis with catalyst in several operating temperature. PET is problematic to be treated using pyrolysis due to wax-like byproduct in liquid which may cause pipe clogging. The catalyst is the mixture of natural zeolite and bentonite which is able to handle PP and PET mixture feed to produce high grade liquid fuels in terms of calorific value and other fuel properties.

Keywords: Catalytic pyrolysis; Natural zeolite; Waste plastics.

1. Introduction

Plastic fraction inside the municipal solid waste (MSW) is difficult to be decomposed, and its quantity is increasing continuously by the time. The increasing demand of plastics materials in all sectors is due to its characteristics of relatively low manufacturing cost, durable, lighter and flexible. Therefore, plastic consumption around the world is increasing by 4 percent every year [1]. In general, waste plastic have the composition of 46% high and low density of polyethylene (HDPE and LDPE), 16% polypropylene (PP), 16% polystyrene (PS), 7% polyvinyl chloride (PVC), 5% polyethylene terephthalate (PET), 5% acrylonitrile-butadiene-styrene (ABS), and 5 percent other polymers [2]. This intensive usage of plastics in daily life will eventually end up in waste disposal problem. Plastic waste is a non-biodegradable material, so it has a negative impact on the environment. Based on data in 2011, the total number of plastic waste produced worldwide reached 280 million tons [3].

By the advancement of technology, plastic waste can be turned into alternative fuels using a heat treatment process called pyrolysis. Most of the research was conducted to characterize the fuel from pyrolysis using LDPE, HDPE, PP, and PS as the feedstock. Pyrolysis can be described as a chemical reaction of decomposition process by thermal and chemical of organic matter into carbon rich solid and vapor. Pyrolysis end products possess half of the carbon content from the feedstock organic matter, while, some part of the vapor phase can be condensate into liquid product and the rest is uncondensable [4].

The benefit of pyrolysis is producing high energy content products which are char, tar and gas from a low energy content of solid waste. In this process, the conversion ratio of the three products dependent on some operating conditions which are the heating rate, the maximum temperature, the processing time at the maximum temperature, the pressure, and the catalysts. One of the low cost catalyst that widely used for this application is natural zeolite.

The zeolite structure consists of tetrahedral unit of AlO_4 and SiO_4 which are interconnected through oxygen atoms. According to Smart and Moore [5], the role and activity of zeolite in catalyst is determined by the amount of acid sites located on the surface. The acid sites behave as Bronsted and Lewis acid. The Bronsted acid is the donor of H^+ while Lewis acid is able to bind with free electron pair. Prior to be used as a catalyst, natural zeolite should be activated in order to eliminate the impurities in the zeolite. Zeolite activation process also aims to modify the properties of the zeolites in terms of surface area and acidity. Increasing surface area and acidity will increase the catalytic activity of the natural zeolite. One of the advantages of natural zeolite is the ease of modification of the surface area and the acidity [6].

According to Xingzhong [7], there are three common methods for pyrolysis, namely thermal cracking, catalytic cracking, and cracking-catalytic reforming. The selected type of pyrolysis in this research was cracking-catalytic reforming using fixed bed reactor. On reforming cracking-catalytic or known as the two steps processes, plastics is decomposed at high temperatures first and then undergo catalytic reforming.

In terms of heating rate, there are two main types of pyrolysis i.e. fast pyrolysis and slow pyrolysis. Slow pyrolysis requires slower heating rate, and lower temperatures than the fast pyrolysis. In general, the slow pyrolysis is designed to the get more of solid products or char even the other product phases is still exist [4]. The pyrolysis reaction with Q as the heat required is represented by eq.1 [8].

$$C_xH_yO_z + Q \rightarrow \text{Char} + \text{Liquid} + \text{Gas} + H_2O \quad (1)$$

2. Materials and Methods

2.1. Material

The raw material in this study was using plastic wastes of polypropylene (PP) and polyethylene terephthalate (PET) from used water bottles obtained from waste disposal facility in Jogjakarta City Indonesia. One kilogram of the raw material used for every processing time, either single type or mixture of them. Before used, PET and PP bottles are crushed into 2-3 cm long plastic flakes.

The catalyst used in this research was a mixture of nature zeolite and bentonite obtained from Klaten, Central Java, Indonesia. The ratio between natural zeolite and bentonite was 70 %: 30 %. A catalyst formed into pellet with 5 mm in diameter and 10 mm in length. The catalyst/plastic waste ratio was 10% in each process. The maximum temperature was set at 400^0C, 450^0C and 500^0C.

2.2. Pyrolysis Methods

The pyrolysis process is done in the reactor chamber with the diameter of 23 cm and the height of 35 cm. At the top of the heating chamber, there is a smaller chamber to store the catalyst which is connected with a condenser and a phase separator. The reactor was made from stainless steel material equipped with a programmable electric heater. By using a constant heating rate, 1 kg of the plastic feedstock (PP only of PP and PET mixture) was heated to the maximum desired temperature from room temperature. The pyrolysis temperatures used for PP were 400^0C, 450^0C and 500^0C. And for PP-PET mixture was used 400^0C. During the process, the pressure inside the reactor was maintained constant at 1 atm. The plastic vapour produced from the heating chamber flew into the catalyst bed then to the condenser and finally to the separator. The maximum temperature was maintained until pyrolysis process finished that indicated by no liquid and gas product was collected from the separator. The amount of both products was collected and measured every 10 minutes consecutively up to 120 minutes. The pyrolysis liquid product (oil) was characterized by analytical Gas Chtomatography coupled with Mass Spectrophotometry QP2010S from SHIMADZU, while for analyzing the oil high heating value a bomb calorimeter was used (Galenkamp).

3. Results and Discussion

The pyrolysis product yields of PP with and without catalyst in three different temperatures is shown in Table. 1. It is shown by non-catalytic process that the temperature has a significant effect on the product proportion among liquid, char and gas. For the non-catalytic processes, the increase of the processing temperature from 400 to 500 °C will increase the liquid product yield while reducing the solid yield. Similar results obtained by Abadi et al. [9] and Jung et al. [10], who observed that the increase of pyrolysis temperatur will decrease char production in pyrolysis without catalyst. Higher temperature of the process can decompose better the volatile compound fraction of the feedstock to become condensable product.

However, the result become completely different for catalytic reactor, the higher pyrolysis temperature will reduce the liquid product. This effect is due to the cracking mechanism of the condensable gases entering the catalyst bed to become shorter carbon chain of non-condensable compounds, thus, the gas yield is increase. The effect is getting intense at the higher temperature since the catalytic effect will increase at higher temperature. Generally, a catalyst with high surface area can change the nature of the pyrolysis and affect on the pyrolysis results even in low catalyst contents [11-12].

Table 1. Polypropylene pyrolysis product yield at different processing temperature

Pyrolysis product	Non-catalyst product yield (g)			Catalystic product yield (g)		
	400^0C	450^0C	500^0C	400^0C	450^0C	500^0C
Liquid	782,67	796.09	810,21	784,19	755,83	532.06
Char	11,33	10,64	10,45	9,87	8,17	9.12
Gas	206	193,27	179,34	205,94	236	458,82

During the experiment using feedstock mixture of PP and PET, wax-like material was produced and accumulated inside the piping system and also in the bottom of liquid product. The wax was collected and weight separately. All the pyrolysis product weight using plastic mixture in various ratio is represented in Table 2. Higher PET fraction in raw material will produce lower liquid product but higher wax and solid content. Miandad et al. [1] has reported the similar finding that PET content in the pyrolysis feedstock will increase the char product and gas significantly and decrease the liquid oil product.

Table 2. Mixture of waste plastic pyrolysis yield at 400^0C using catalyst

Pyrolysis product	Ratio of raw material		
	PP:PET=9:1 product (g)	PP:PET=7.5:2.5 product (g)	PP:PET=6.5:3.5 product (g)
Liquid	723,82	515,39	390,3
Char	22,51	58,8	465,74
Gas	247,22	394,43	84,75
Wax	6,45	31,38	59,21

In terms of heating value (Table 3), the pyrolysis oil of PP without catalyst has slightly lower caloric content than the oil from catalytic cracking. The catalyst breakdown the long chain compound into shorter chain with higher heating value. Seo, et al [13] also reported the same results that the catalyst will decrease liquid oils product volume, but increase C3-C10 compound in liquid oils significantly while decrease C11-C18 compound and eliminate >C19 compound. For the feedstock mixture, PET addition tend to reduce the caloric value of the oil. This may be due to wax product existence.

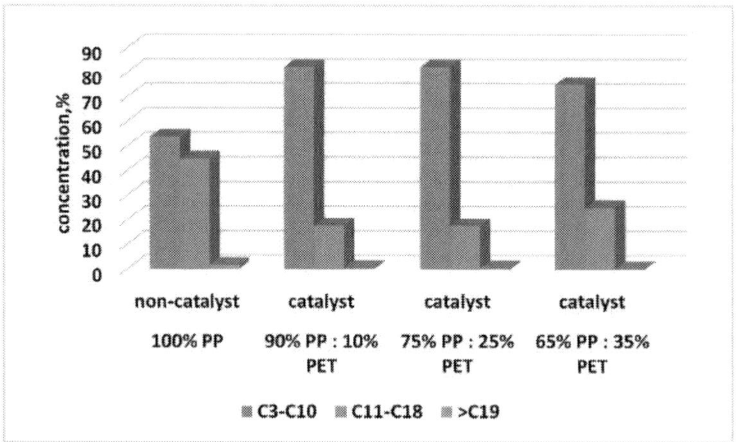

Figure 1. GC-MS result of selected liquid products at 400^{0}C

Table 3. Heating Value result of liquid oils product (high value)

Catalyst	Raw material	Temperature	Average HHV (cal/g)
Without catalyst	100% polypropilene	400^{0}C	10,711
Without catalyst	100% polypropilene	450^{0}C	10,829
Without catalyst	100% polypropilene	500^{0}C	10,991
Using catalyst	100% polypropilene	400^{0}C	10,942
Using catalyst	100% polypropilene	450^{0}C	11,067
Using catalyst	100% polypropilene	500^{0}C	11,190
Using catalyst	90% PP and 10% PET	400^{0}C	11,048
Using catalyst	75% PP and 25% PET	400^{0}C	10,805
Using catalyst	65% PP and 35% PET	400^{0}C	10,599

Similar with the heating value measurement data, GCMS results in Fig. 1 show that by adding catalyst, the lower carbon chain compound percentage is getting higher. The catalyst was successful in cracking the long chain of hydrocarbon vapor entering the catalyst bed before condensation. Meanwhile, PET tends to reduce the total heating value of the pyrolytic oil. It indicated that the catalyst require higher temperature to be able to treat feedstock with PET content.

4. Conclusion

The processing temperature, catalyst and raw material content affect significantly in the pyrolysis. In this study, the best temperature to produce the highest quantity liquid oils without catalyst is the highest temperature (500 °C), while using catalyst is the opposite. By using the catalyst, the amount of C3-C10 compound is increase which leads to the increasing of oil calorific value. PET is the type of plastic that is unsuitable to be treated with pyrolysis, but by mixing PET with PP as raw materials and using proper catalyst the problem can be minimized. Based on this study the lowest content of PET (10%) can give optimum results in terms of oil production and the heating value.

5. Acknowlegment
This work has been financially supported by PUPT 2017 research grant Kemenristekdikti, and also supported by MSW recycling facility in PIAT UGM.

References

[1] Miandad. R., Barakat, M.A., Asad, S., Aburiazaiza., Rehan, M., Ismail, I.M.I., Nizami, A.S., (2016), Effect of plastic waste types on pyrolysis liquid oil. International Biodeterioration & Biodegradation, Volume 30, pp. 1-14.
[2] Vasile, C., and Brebu., (2002), Solid waste treatment by pyrolysis methods, Journal of Environmental Protection and Ecology, Volume 1, pp. 230-235.
[3] Sriningsih, W., Saerodji, M.G., Trisunaryanti, W., Armunanto, R., Falah, I.I., (2014), Fuel production from LDPE plastic waste over natural zeolite supported Ni, Ni-Mo, Co and Co-Mo metals, Procedia Environmental Science, Volume 20, pp. 215-224.
[4] Brownsort, P.A., (2009), Biomass pyrolysis processes: review of scope, control and variability. Biomass, p.38.
[5] Smart, L. and Moore, E., (2001). Solid State Chemistry, Cheltenham: Nelson Thornes Ltd.
[6] Yuanita, D., (2009), Hidrogenasi Katalitik metil oleat menjadi stearil alkohol menggunakan katalis Ni/Zeolit alam, *In:* Prosiding Seminar Nasional Kimia, Yogyakarta, Universitas Negeri Yogyakarta.
[7] Xingzhong, Y. in: Scheirs, J., Kaminsky, W. (Eds.), (2006) Feedstock Recycling and Pyrolysis of Waste Plastics: Converting Waste Plastics into Diesel and Other Fuels, John Wiley & Sons, Chichester, UK, pp.729-756.9Abadi, M.S.A., Haghighi, M.N., Yeganeh, H., and McDonald, A.G., (2014), Evaluation of pyrolysis process parameters on polypropylene degradation products". Journal of Analytical and Applied Pyrolysis, Volume 109, pp. 272–277.
[8] Zadgaonkar, A., in: Scheirs, J., Kaminsky, W. (Eds.), (2006) Feedstock Recycling and Pyrolysis of Waste Plastics: Converting Waste Plastics into Diesel and Other Fuels, John Wiley & Sons, Chichester, UK, pp.705-728.
[9] Abadi, M.S.A., Haghighi, M.N., Yeganeh, H., and McDonald, A.G., (2014), Evaluation of pyrolysis process parameters on polypropylene degradation products". Journal of Analytical and Applied Pyrolysis, Volume 109, pp. 272–277.
[10] Jung, S.H., Cho, M.H., Kang, B.S., Kim, J.S., (2010), Pyrolysis of a fraction of waste polypropylene and polyethylene for the recovery of BTX aromatics using a fluidizedbed reactor, Fuel Process Technology, Volume 91, pp. 277–284.
[11] Marcilla, J.C., Quesada, G.S., Sánchez, R., Ruiz., (2005), Study of the catalytic pyrolysis behavior of polyethylene–polypropylene mixtures, Journal of Analytical and Applied Pyrolysis, Volume 74, pp. 387–392.
[12] Lin, H.T., Huang, M.S., Luo, J.W., Lin, L.H., Lee, C.M., Ou, K.L., (2010), Hydrocarbon fuels produced by catalytic pyrolysis of hospital plastic wastes in a fluidizing cracking process, Fuel Process Technology, Volume 91, pp. 1355–1363.
[13] Seo, Y.H., Lee, K. H., Shin, D.Y., (2003), Investigation of catalytic degradation of high-density polyethylene by hydrocarbon group type analysis, Journal of Analytical and Applied Pyrolysis, Volume 70, pp. 383-398.

STUDIES OF CARBONIZATION PROCESS ON THE PRODUCTION OF DURIAN PEEL BIOBRIQUETTES WITH MIXED BIOMASS COCONUT AND PALM SHELLS

Ellyta Sari[1], Pasymi[1], Umar Khatab[2], Reni Desmiarti[1], Rian Ariansyah[1], Hariadi[1], Sutra[1]

[1] Departement of Chemical Engineering, Universitas Bung Hatta, Padang West Sumatra, Indonesia
[2] Department of Civil Engineering, College of Technology Payakumbuh, West Sumatra, Indonesia
E-mail : sariellyta@yahoo.com

Abstract. Biobriquettes as alternative energy that can replace the role of kerosene. Biobriquettes made from agricultural waste biomass. Biobriquettes durian peel has been researched and developed continuously to obtain optimal quality in terms of calorific value, compressive strength and duration of ignition. In making durian peel biobriquettes needed other biomass mix to sustain duration of Ignition for biobriquettes durian skin quickly burned out. Stages of making biobriquettes durian skin are: material of drying, carbonization of biomass, grinding, mixing with adhesives, and printing. Carbonization process is a process that is important in obtaining the biomass charcoal. Carbonization is done by means of karbonisator pyrolysis. The purpose of this research is to study the process of carbonization to obtain biobriquettes durian skin that of quality in terms of value compressive strength, calorific value, and duration of ignition. Variations that done was kind mix of biomass,coconut shells and palm shells with the massa ratio 2 : 1, type of adhesive used tapioca powder and banana peels, carbonization of temperature 200°C, 300°C and 400 °C . The results showed that the highest compressive strength of the durian skin with a mixture of coconut shell and adhesive tapioca powder and carbonization temperature of 300 °C namely 12,7 g/cm^2. The calorific value of the highest on the mix of skin durian with coconut shells and adhesive banana skin with temperature of carbonization 400 °C ie 6040 cal/g, and duration of ignition highest on a mixture of skin durian with coconut shell and adhesive banana skin at a temperature of carbonization 300 °C is 73 minutes.

Keywords **:** Biobriquettes, Durian peel , Carbonization.

1. Introduction

Renewable alternative energy sources is being developed in Indonesia. One of these is biobriquettes from biomass. Biomass is composed of cellulose, hemicellulose and lignin that is commonly found in plant parts. Biomass is a renewable reseources that can be used sustainably.

Durian peel waste is one of biomass that could be used as a raw material source of renewable energy. Several studies of Biobriquettes of durian peel has been conducted. (Pramudya, et al.,2011) studied the variation of biomass and adhesive mixture. However, these studies without using the carbonization process of durian peel. In the biobriquettes manufacture, carbonization is a very important process because it is the main process in the biobriquettes manufacture which can affect the biobriquettes quality.

Carbonization process is done by burning biomass in open condition, the burning process then stopped when the charcoal formed, by spraying water on the charcoal that has been formed. This kind of process generated a lot of ash and smoke so that the charcoal making process is not optimal and produced only a little charcoal. Therefore it is necessary to look for solutions, so that the carbonization process authoring maximum results, which produces little ash, it also can take advantage of combustion smoke as liquid smoke which can be used as a natural food preservative,which is not harmful.

This is known as pyrolysis process (Mira, M, 2002). The characteristics of durian peel as an alternative fuel by the pyrolysis process has been investigated (Wahidin, 2014). In this study, the ash content obtained is still high (18.8%), caused not using a mixture of biomass. The process of making biobriquettes by pyrolysis carbonization been done by (Feri Fuji,2010), at a temperatures of 210, 250, 300, 350, and 390^0C using rice husk as raw materials, and the highest calorific value obtained at 5609.453 cal/g. In addition (Warapon R, *et.al,* 2011) also examined biobriquettes durian peel with a mixture of rice straw (9: 1) the calorific value obtain is 24,674MJ/kg (5902 cal/g). While (G. Siresh, *et al,*2016) examined the types of palm briquette powder and binder, providing a calorific value of 20 945 MJ /kg (5010 cal/g) at 80.79% efficiency.

(Paisal,2014) conducted a study ofbiobriquettes made from durian peel using an adhesive with banana peel waste without mixing other biomass, and the calorific values obtained 5074 cal/g. (Sari, E, *et.al* ,2015) also did research on biobriquettes durian peel with a mixture of biomass palm shells and rubber shell, using starch as an adhesive.From this study, the highest calorific value obtained of 5118 cal/g. (Merry, *et.al,*2015) also examined using an adhesive calcium hydroxide and obtained calorific value of 4968 cal/g. Accordingly it is need to do research with the pyrolysis carbonization process of durian peel, so that the higher ash levels can be reduced by using a mixture of other biomass.

Moreover carbonization pyrolysis also produces liquid smoke which can be used as a preservative. This research will be carried out at a pyrolysis carbonization temperature of 200-400°C with adhesive from banana peels waste and starch as a comparison, by add biomass from coconut shell and palm shell to increase the calorific value of biobriquettes.

The purpose of this study was to determine the effect of carbonization temperature to the biobriquettes result, the effect of biomass mixture with biomass durian peel to the calorific value, and the effect of adhesive type to quality of biobriquettes produced.

2. Methodology/ Experimental

This research was conducted at the Laboratory of Chemical Engineering, Universitas Bung Hatta Padang. The research parameters consist of Fixed Parameters, Variable Parameters, and Output Parameters. Fixed Parameters arebiomass of durian peel, adhesive: 10% from the total biomass mixture, biomass mixture with composition ratio 2: 1, with a total weight of biomass at 140 grams. Variable parameters is another biomass mixture,: palm shells, coconut shells; different types of adhesives: starch and banana peels; Carbonization temperature at 200, 300, and 400°C. Output parameters are compression strength, porosity, calorific value, ash content and burning time.

2.1. Tools

Pyrolysis carbonization reactor, condenser, biomass cutlery, sifter, Mixers, grinding, Biobriquettes Molds, basin, Zinc for drying, Stove, Pots, Spatula, scales, oven, desiccator and porcelain dish.

2.2. Materials

IOP Conf. Series: Materials Science and Engineering **316** (2018) 012021 doi:10.1088/1757-899X/316/1/012021

Durian peels from Padang, Biomass mixture consist of palm shells, Coconut shell, adhesive from starch flour, banana peel, and water.

2.3. Working Procedures

2.3.1. Carbonization Stage by means of Pyrolysis Carbonization

a. The durian peel was chopped up small, so that the drying process can be faster and more volume of material can be fed into the reactor
b. All material is dried under the sun for 4-5 days.
c. once the ingredients dry, biomass was entered in the carbonization reactor, then temperatur of heating was set and biomass was entered until almost full and then reactor was covered tightly.
d. The heater is switched on and the carbonization temperature is adjusted (according to variations at 200, 300 and 400°C) with a heating time of 2 hours each.
e. After two hours, pull out the carbonization results, then crush until smooth at size of 60 mesh.
f. The same thing is done on another biomass mixture; palm shell and coconut shell

2.3.2. Adhesives Manufacturing Phase

A. MakingAdhesives From Starch

Starch weighed as much as 10% of the total biomass feedstock, next insert the water at a ratio of 1: 10, then stirred. Starch solution was then heated on a stove until thickened.

B. Making Adhesives from Banana Peel

Banana peel retted ± 10 days, then was mashed. The smooth banana peel is mixed with 10% of the total weight of the biomass feedstock, then was add water gradually until homogeneous.

2.3.3. Briquetting Phase

The smooth durian peels charcoal is weighed according to the specified weight and the specified biomass variation. The refined durian peel is mixed with the adhesive in the container while stirring until evenly Further the mixture were molded using biobriquettes mold. Biobriquettes then dried under the sun until completely dry.

2.4. The tools image series can be seen in Figure 1.

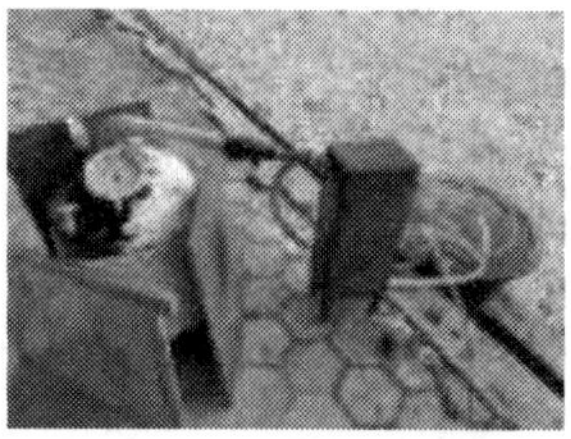

Figure 1. The series of pyrolysis carbonisator dual function tool

Information :

1. Reactor; 2. Furnace; 3.Thermometer; 4. Pressure Gauge; 5. Condenser; 6. The Hot Water flow; 7. The Cold Water flow in; 8. Condensate Flow; 9. Liquid smoke Container; 10. The cooling water hose

3. Results and Dicussion

Biobriquettes shape produced from the research of Manufacturing of biobriquettesdurian peel from the mixture of biomass and adhesive, can be seen in Figure 2.

Figure 2. Biobriquettes of Mixture of Biomass and Adhesives

3.1. Effect of Raw Materials to the Biobriquettes Porosity/ Density

Porosity/density is the ratio between the weight to the volume of briquettes. Density size is influenced by the size and homogeneity of the briquettes materials. Based on observations and calculations performed on the density value of each treatment,smaller particle size can expand the field of bonding between powders, thus increasing the density of briquettes (Masturin 2002). The effect of a biomass mixture to porosity can be seen in Figure 3 and Figure 4.

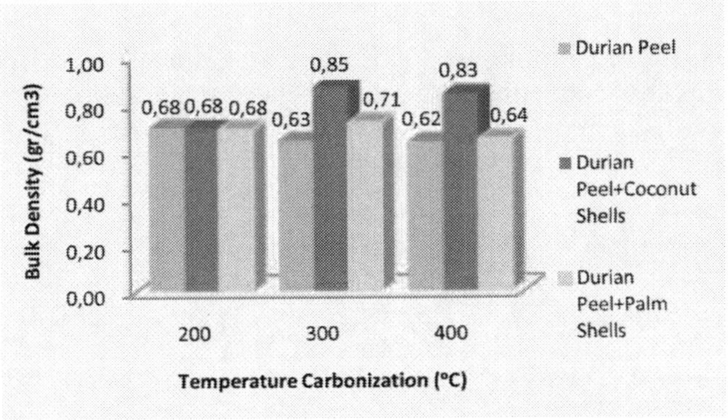

Figure 3. Effect of Biomass Mixture Using Adhesives Starch To The Porosity

From Figure 3 and Figure 4 can be seen that the mixing of biomass coconut shell and palm shells charcoal with durian peel (agricultural waste) at a smaller size (60 mesh) could expand the bonding between particles, thereby increasing the biobriquettesdensity, since the bonds between powder are stronger and more compact.

IOP Conf. Series: Materials Science and Engineering **316** (2018) 012021 doi:10.1088/1757-899X/316/1/012021

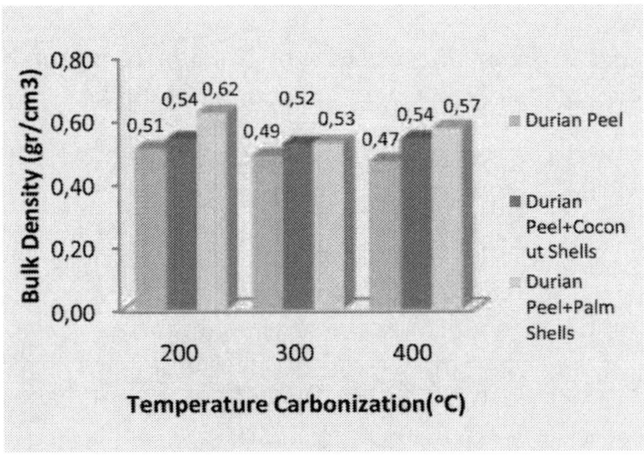

Figure 4. Effect of Biomass Mixture Using Adhesives Banana Peel to the Porosity

Figure 3 and Figure 4 show that biobriquettes porosity values ranged from 0.47 to 0.85 g/cm^3. Biobriquettes with adhesive starch obtained the best porosity of 0.85 g /cm^3 with a mix of biomass durian peel and coconut shell, while biobriquettes with adhesive banana peel earned best porosity of 0.62 g/cm^3 with a mixture of durian peel biomass and palm shells. Adhesive banana peel does not have a strong bond between the fibers, it is because the banana peel adhesion is not as strong of starch adhesion. This leads to a lower density value. Thus, the composition of the constituent material effect the biobriquettes density.

3.2 Effect of the Mixture of Biomass and Adhesive on CompressionStrength

Effect of biomass mixture and adhesive onbiobriquettescompression strength can be seen in Figure 5 and Figure 6.

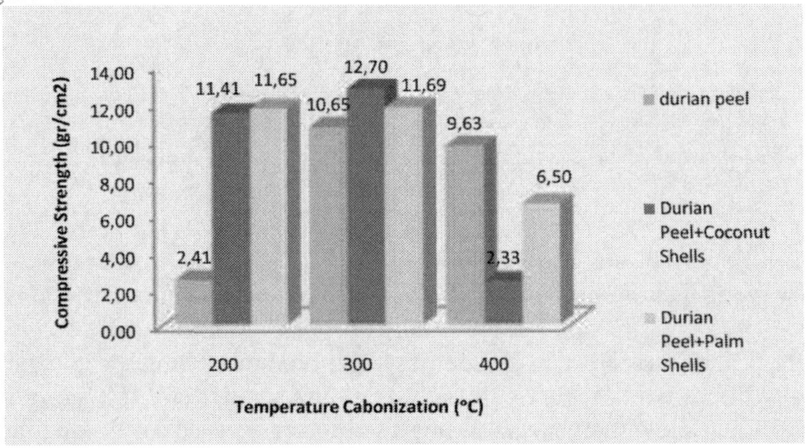

Figure 5. Effect of Biomass Mixture using Adhesives Starch To The Compression Strength

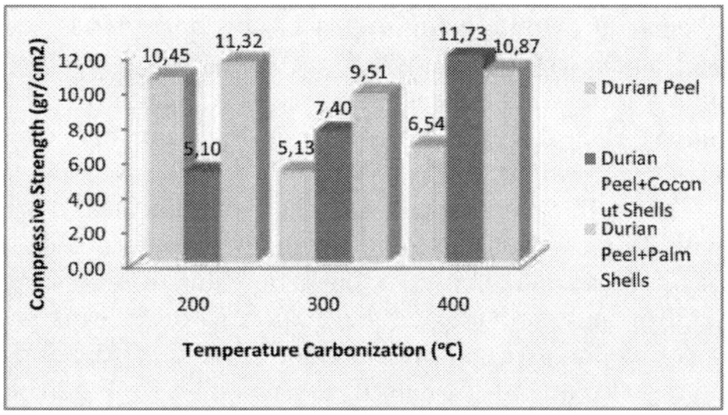

Figure 6. Effect of Biomass Mixture Using Adhesives Banana Peels To The Compression Strength

Figure 6 and 7 shows that the highest compression strength at 12.70 g/cm^2, using adhesive starch with a mixture of coconut shell biomass at temperatures of 300°C, while the highest compression strength using adhesive banana peel with a mixture of biomass palm shell at temperature of 400°C, namely 11.73 g/cm^2.

Adhesive starch has a better compression strength than adhesive banana peel, this is due to starch adhesion is higher than banana peel, moreover high density biomass would produce high compression strength biobriquettes as well, as proposed by (Sudrajat,1984), that charcoal biobriquettes from high density raw materials will also provide a high compression strength

3.3. The Effect of Biomas Mixture and Adhesive to the Calorific Value

The biomass mixture and adhesives used in biobriquettes manufacture also affect the calorific value. Each sample variation had been tested the calorific value.

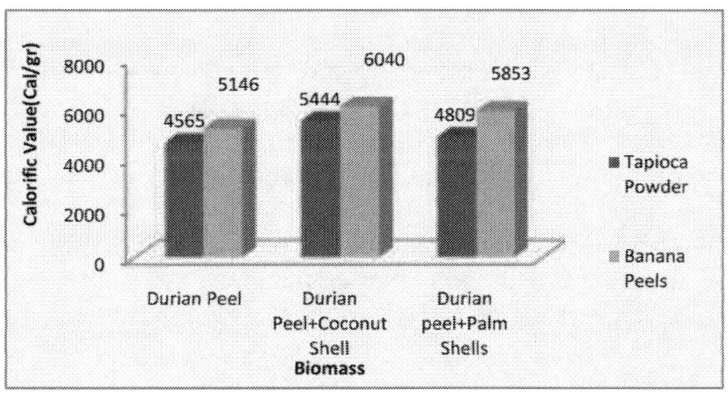

Figure 7. The Effectof Biomass Mixture and Adhesives To The Calorific Value

Figure 7 show that the increasing of calorific value is also influenced by a mixture of biomass used. Calorific value of Biobriquettes durian peel without other biomass mixture are 4564 and

5146 cal/g, and the calorific value were increased become 5444 and 6040 cal/g when biobriquettes durian peel mixed with coconut shell,.
Calorific value of biomass using a mixture of palm shells decreased to 4809 and 5853 cal/g. In the research by Pramudya, et.al (2011), the highest calorific value obtainedon durian peels and palm shells with jackfruit seed adhesive was 5495 cal / gr. Thereby it can be concluded that the used of adhesive type is can influenced the caloric value. Banana peel adhesive gives the higest calorific values than with the jackfruit seeds adhesive in research Pramudya ,et.al (2011).

The addition of biomass can increase the calorific value of biobriquettes durian peel. The addition of coconut shell biomass provides the best calorific value. This is due to coconut shell has a high carbon compound that gives additional calorific value to the durian peel biobriquettes produced. Adhesive type is also affects the calorific value obtained. Figure 7 shows that adhesive banana peel provides a higher calorific value than adhesive starch.

3.4. Effect of carbonization temperature and biomass mixture to the Burning Time
Effect of carbonisation temperature and biomass mixtures to the biobriquettes burning time can be seen in Figure 8 and Figure 9.

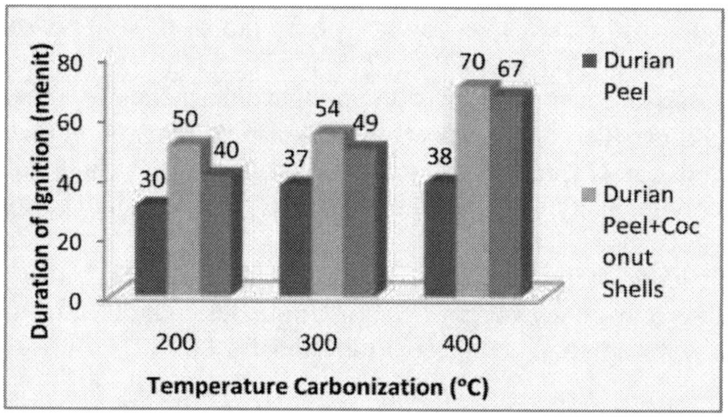

Figure 8. Effect of Carbonisation Temperature and Biomass Mixture Using Adhesives Starch To The Burning Time

Figure 8 and Figure 9 shows that the best burning time of biobriquettes obtained at carbonization temperature of 400°C ie 73 minutes on biobriquettes with adhesive banana peel and coconut shell mixture.This is because the burning temperature of 400 °C completely carbonized biomass as compared with 200°C and 300°C burning temperature, so that the carbon content tied will be higher at burning temperature of 400 °C, so the biobriquettes ignition time to ashes will be longer.

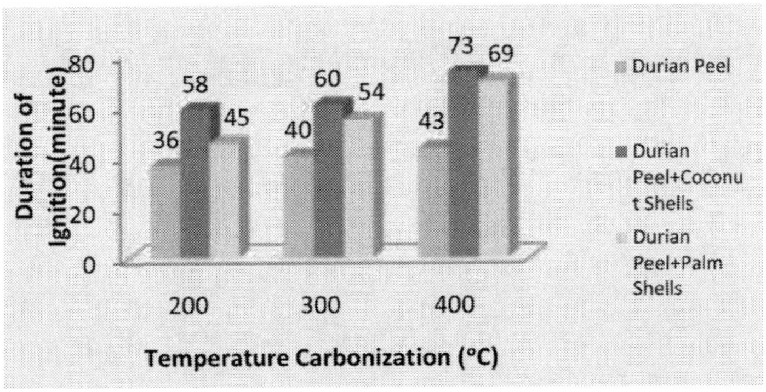

Figure 9. Effect of Carbonisation Temperature and Biomass Mixture Using Adhesive banana peels To The Burning Time

The effect of biomass mixture show that Biobriquettes mix with coconut shells provide the best ignition when compared with other. In research by Munas, et.all (2012) that the best burning time of biobriquettes is 40 minutes on biobriquettes with cocoa beans biomass without the addition of other biomass, but in this study the best burning time of biobriquettes is 73 minutes on biobriquettes with durian peels and coconut shells with banana peel s adhesive.

3.5. Effect of Carbonization Temperature and Biomass Mixture to Ash Content

Effect of carbonization temperature and the biomass mixture to ash content can be seen in Figure 10 and 11.

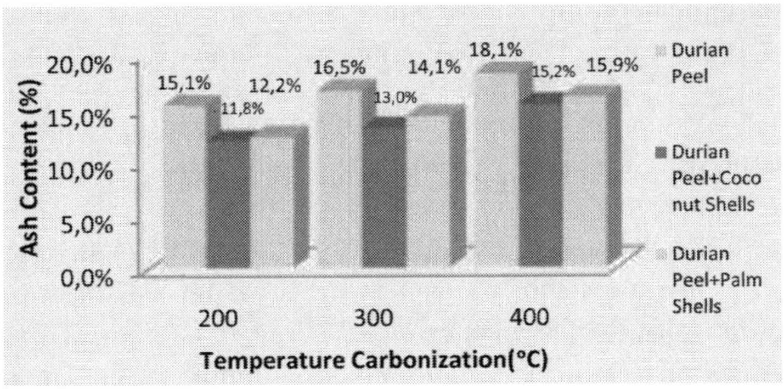

Figure 10. Effect of Carbonisation Temperature and Biomass Mixture using Adhesive starchs To Ash Content

Figure 10 and 11 shows that ash content obtained ranged between 10% - 24%. Ash content obtained is not yet meet the standards of <10%. Carbonization temperature of 400 °C give the highest ash content compared to the carbonization temperature of 200 °C and 300 °C.

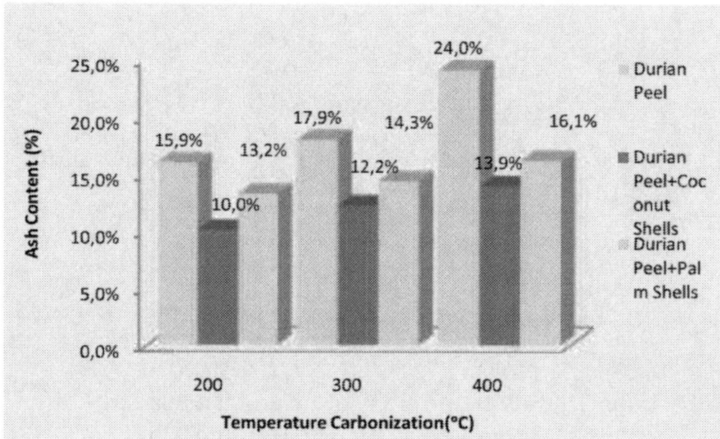

Figure 11. Effect of Carbonisation temperature and Biomass Mixture Using Adhesive banana peels To Ash Content

This is due to the ash content will increase at the higher carbonization temperature. Carbon will burn out at higher temperature and leave the ashes as a result from the combustion (Rosdiana, 2016), so it can be said that the carbonization temperature rise is directly proportional to the biobriquettesash content. On the influence of the biomass mixture, biobriquette mixture of durian peel with coconut shell biomass with adhesive banana peel provide the best ash content is 10%. In research by Pramudya, et. All (2011), biobriquettes mixture of durian peel with palm shell biomass with durian seeds adhesive provides the lowest ash content is 11.90%. This is due to mineral deposits that can not be burned in the coconut shell is lower compared to the others, so that the ash obtained is also low.

4. Conclusion

Results from the study of biobriquettes durian peel give some following conclusions:

a. Utilisation of durian peel waste, coconut shell, and palm shells on making biobriquettes may increase the economic value of raw materials.,
b. Banana peel waste can be use as an adhesive in the biobriquettes manufacture, but the adhesiveness given still low compared than adhesive starch, but adhesive banana peel can improve the calorific value of biobriquettes.
c. The addition of coconut shell and palm shell biomass improve the quality of biobriquettes durian peel.
d. Biobriquettes has the highest calorific value at 400°C carbonization temperature, with a mixture of coconut shell and adhesive banana peel of 6040 cal/g. Calorific values obtained reached the calorific standard based on ISO (5000 cal/g).

5. Acknowledgement

Thanks to all the parties concerned to this study: the Higher Education Research and

Technology which funded this research through grants applied, also to all students of guidance that help a lot of this research.

6. References

[1] Feri Fuji Hertanto 2010 *Optimasi Kondisi Operasi Pirolisis Sekam Padi untuk Menghasilkan Bahan Bakar Briket Bioarang Sebagai Bahan Bakar Alternatif* Universitas Diponegoro

[2] G Siresh Kumar, Dr C J Rao, Dr D Sreeramulu and SK Madhavi 2016 *Evaluation of Boiler efficiency of Bobriquettes by Indirect Method* International Journal of Mechanical Engineering and Technologi Vol. 7 Issue 6 pp 624-633

[3] John Taiwo Oladeji 2011 *Agricultural and Forestry Wastes and Opportunities for their use as an Energy source in Nigeria* .an overview World Rural Observations 2011;3 (4)

[4] M Faisal, M Arif Kurniawan and Deni K 2016 *Pengaruh komposisi biobriket dari TKKS, Ampas Tebu, dan serbuk gergaji dengan perekat kanji terhadap nilai pembakaran* Jurnal Teknik Kimia no 4 vol. 22

[5] Martynis M, Sundari E, Sari E 2012 *Pembuatan Biobriket dari Limbah Cangkang Kakao Briket dari kulit Kakao* Jurnal Litbang No 1 Vol. 2 Padang

[6] Nona Merry M.Mitan, Akmal H A, Nur Fathiah M N and Sian Meng Se 2015 *Binder Aplication in Durian Peels Briquettes a Solid Bofuel* Applied Mechanics and Materials vol. 761 pp 494-498 Trans Tech Publicatin Switzerland

[7] Paisal 2014 *Analisa Kualitas Briket Arang Kulit Durian Dengan Campuran Kulit Pisang pada Berbagai Komposisi Sebagai Bahan Bakar Alternatif* Politeknik Negeri Ambon

[8] Pasymi 2012 *Rancangan karbonisator tepat guna Untuk proses Karbonisasi temperatur rendah* Prosiding Seminar Resatek FTI ISSN 2087- 2526 Universitas Bung Hatta Padang

[9] Pound G S 1959 *Further Progress in Low Temperature Carbonisation of Coal and Production of Chemicals from Low Temperature Tar-1* Coke & Gas, 396-401

[10] Pramudya, Sari E, dkk. 2013 *Pembuatan briket kulit durian dengan campuran biomassa sawit dan perekat* E-jurnal Universitas Bung Hatta

[11] Rosdiana Moeksin, Nabila Zarwan and Muhammad Albusary 2016 *Pembuatan Biobriket dari Campuran Tempurung Kelapa dan Cangkang Sawit Biji Karet* Jurnal Teknik Kimia Universitas Sriwijaya Palembang

[12] Sari E, Praputri E, dkk. 2015 *Peningkatan Kualitas Biobriket Kulit Durian dari segi campuran Biomassa, bentuk fisik, Kuat Tekan dan Lama Penyalaan* Prosiding Simposium Nasional RAPI XIV-2015 ISSN 1412-9612 Fakultas Teknik UMS Solo

[13] Sudrajat 1984 *Pengaruh bahan baku, jenis perekat, dan tekanan kempa terhadap kualitas briket arang* Bogor

[14] Wahidin Nuriana, Nurfa nisa and Martana 2014 *Synthesis Preliminary Studies Durian Peel Bio Briquettes as an Alternative Fuels* Conference and Exhibition Indonesia Renewable Energy & Energy Conservations, Published by Elsevier, energi procedia 47 pp 295-302

[15] Waraporn Rattanongphisat and Sirinuch Chidaruksa 2011 *A bio-fuel briquette from durian peel and rice straw: Properties and Economic Feasibility* NU Science Jurnal 8 (2) pp1-11

IOP Conf. Series: Materials Science and Engineering **316** (2018) 012022 doi:10.1088/1757-899X/316/1/012022

DETERMINATION OF EXTRACTION PROCESS CONDITIONS OF GAMBIER CATECHIN (UNCARIA GAMBIER ROXB) FROM SOLOK BIO BIO LIMA PULUH KOTA DISTRICT – WEST SUMATERA

Elly Desni Rahman[1], Ellyta Sari[1], Burmawi[2], Frizka[1], Endah[1]

[1]Departement of Chemical Engineering, Universitas Bung Hatta, Padang West Sumatra, Indonesia
[2]Department of Mechanical Engineering, Universitas Bung Hatta, Padang West Sumatra, Indonesia
E-mail : ellydesnirahman@yahoo.com

Abstract. Catechin content is the determinant key of quality in gambier trade. The required Catechin content of gambier extracts as a herbal medicinal ingridient is greater than 90%. Mostly, Local gambier that produced by community is not uniform and low quality, thus lowering the price in the export markets. The quality improvement of gambier can be done by extraction and purification processes. This study aims to determine the best extraction process of catechin from Gambier (Uncaria Roxb) which derived from Solok Bio Bio Lima Puluh Kota, West Sumatra. The research methodology includes pre purification: raw materials preparation, washing, filtration, extraction, drying and testing. Washing was done on 100 gr gambier with a variation of water at 500, 600, 700, and 800 ml, heating for an hour at a temperature of 70^0C, screened, filtered, and allow to stand until a precipitate is formed, wash repeatedly, filtered, and dried. Further, extract with a solvent variation of : water, etyl acetate, heated at 70^0C temperature for 1 hour, then filtered. Filtrate then thickened by using a Rotary evaporator, dried at 50^0C temperature for 48 hours and analyzed. The results showed that the best conditions of the extraction process is by using a solvent etyl acetate, at a temperature of 70^0C, grading 97.40% catechins.

Keywords: Catechin; Gambier; Extraction

1. Introduction

Gambier contains important chemical substances, including tannins catechu acid (20-50%), catechin (7-33%), and Pyrocatechol (20-30%), while other chemical substances in a limited percentage [1]. Bachtiar [2] states that the chemical content of gambier most widely used are catechins and tannins. Gambier is not only used to eat the betel, but also used as a raw material for various industries, such as pharmaceuticals, cosmetics, batik, paint, tanners, bio-pesticides, growth hormones, pigments, and as a mixture of complementary foods [3]. Gambier can also function as antioxidants to inhibit the occurrence of Atherosclerosis [4]. Besides, etyl acetate from gambier extract acts as types of inhibitors and serve as an effective mild steel corrosion inhibitor in aqueous solutions [5].

Catechin content is the determinant key of quality in the trade of gambier. Price of gambier with catechin content of more than 99% (for analysis) is Rp. 888,000/ 10 mg, while catechin content of more than 90% is Rp. 984 000 per gram [6]. The catechins content required in pharmaceuticals industries is greater than 90% [7]. High-quality products will also provide a high price, so as to compete well in the global market, and vice versa.

But the instability price of gambier at this time clearly influences the community income and entrepreneurs economic conditions, especially gambier farmers and distribution chains that play an important role in the local gambier trades. Some factors influencing the problems in the business commodity of gambier include 1) Less qualified of product, and many gambier lost during processing, so requiring repair quality, 2) the bargaining position of farmers are low, and there is no guarantee that the stable prices can benefit farmers, 3) the farmer habit to mix gambier with other ingredients so lowering the selling price [8]. There is no knowledge and applied technology that can improve the productivity and quality of gambier in West Sumatra, so they can not access the market, both for domestic and export needs. However, the exporter usually does not directly export the local gambier to the destination country, but gambier extract should be processed first to be qualified gambier, then separated according to quality, after that the exporters export the qualified gambier extracts to the destination country [9]. Therefore, need to find the right extraction process in processing gambier to improve gambier quality, and high-quality of gambier can produce with a high catechin content. Thus, gambier prices could rise as expected, so as to improve the community economy

Gambier agribusiness has been going on for decades, but there has been no catechin production industry in West Sumatra until now. Actually gambier with catechin levels above 90%, with prepurification method using water and extraction with etyl acetate has been obtained on a small scale (laboratory) [10, 11]. Research on gambier extraction with water, followed by semi-polar solvents (ethyl acetate) resulted the highest catechin levels (97.82 ± 2.01%) [12]. While the research results of Youfa and Rahman [13] also demonstrated catechin levels above 90%, namely 96.73% and 99.33% in the absorption of charcoal and coconut shell after dissolved using acetic acid solvents, ethers and Pb acetate. Next, Ferdinal [14] got a simple catechin purification method of gambier, with 9.594 g catechin weighing result from 21.997g dirty catechin. The equipment used in laboratories include goblets, erlemenyer, funnel and big oven at material processing capacity approximately at 0.5kg. Process conditions have been carried out at a temperature of 70° C [15], handling materials in large quantities (industrial scale) would pose complex problems than the small amount material. Catechin purification method of gambier, usually start by utilizing the solubility difference properties. This process known as extraction that is taking solids from materials using liquid.

2. Methodology/Experimental

The samples used in this study are the local Gambier (Uncaria Gambier (Hunter) Roxb) from Nagari Solok Bio-Bio District Lima Puluh Kota, solvent of water and etyl acetate. Catechin was analyzed using a UV-Visible spectrophotometer to determine the catechin level.

2.1. Tools and materials used

2.1.1. Tool

Erlenmeyer, beaker, measuring cup, rod stirrer, vacuum pump, thermometer, hot plate, glass funnel and plastic funnel, filter cloth, Oven, mortar, analytical scale, centrifuges, Blender, Grail vaporizer, Pipette drops, Pipette peck of 10 ml, Rotary Vacuum Evaporator, distillation apparatus, cooling Leibig. UV-Visible spectrophotometer, porcelain cup, furnace, Aluminum foil.

2.1.2. Material

Material used are Gambier, Aquadest, water (H2O), etyl acetate, pure catechin, Ice cubes.

2.2. Parameters

2.2.1. Variables Parameter

1. Amount of water solvent prepurification :500, 600, 700, and 800 ml
2. Types of purification solvents (water, Etyl acetate)

2.2.2. Fixed Parameter

Total Gambier, prepurification temperature,Amount of leaching: 3 times and purification of 70 ° C.

2.2.3. Output Parameter

1. Acquisition Percentage, 2. Catechins content, 3. Catechin Weight

2.3. Working procedures.

Gambier powder weighed as much as 100 grams, then put into erlemeyer 1 L. Add water with a volume variation of 500, 600, 700, and 800 ml. Heat for 1 hour at temperature of 70^{0}C, then strain. Let stand filtrate until a precipitate is formed. Next wash sediment according predetermined 3times, using the washing solvent volume as the the initial amount. The precipitate was dried in the oven, then smoothed. Added a solvent (water/etyl acetate) with a solvent in accordance at the best ratio in the prepurification extraction. Then heated using reflux for 1 hour, at purification temperature of 70^{0}C, filtered in hot conditions. The filtrate was thickened using a rotary evaporator, then dried and analyzed. Repeat 3 times.

3. Results and Discussion

Catechin making process begins with a test sample of solok bio bio.Gambier measurable catechin level of 47.72%. A sample of 100 grams is done prepurification using water and heated at 70oC while stirring for 1 hour, then filtered and washed three times with water reatment at 16° C operating conditions. Once it is done filtration followed by purification using water or ethyl acetate solvent.

3.1. Relations with the amount of solvent prepurification weight gain after washing

Total acquisition catechin on prepurification process can be seen in Figure 1.

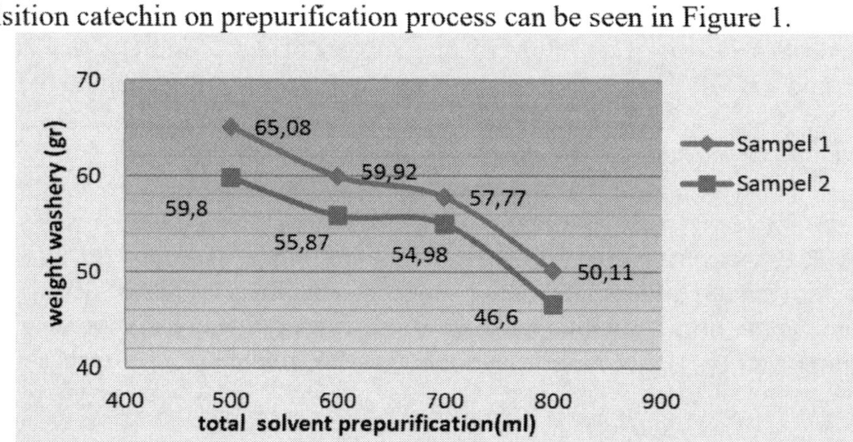

Figure 1. Relationship catechin number solvent for the acquisition of laundering

Figure 1 shows that the larger the volume of the solvent, the smaller the amount of acquisition catechin, can be seen in samples 1 and 2, from 65.08 into 50.11 and from 59.8 to 46.6, is due to the greater number of tannin who participated dissolved together with the solvent, but the solvent amount of 700-800 ml, a decrease in the acquisition of catechin very much when compared with a decrease of 500-600 ml and 600-700 ml, is caused by the amount of solvent is too large, estimated more catechin participating qualify along with tannin and solvent on filtration. From this research in getting the condition an effective amount of solvent volume is the volume of 700 ml with a tally of the sample 1 was 57.7 grams and for sample 2 of 54.98 grams.

3.2. Relationship between the amount of solvent pre-purification and the catechin levels

Catechin levels on the prepurification process can be seen in Figure 2.

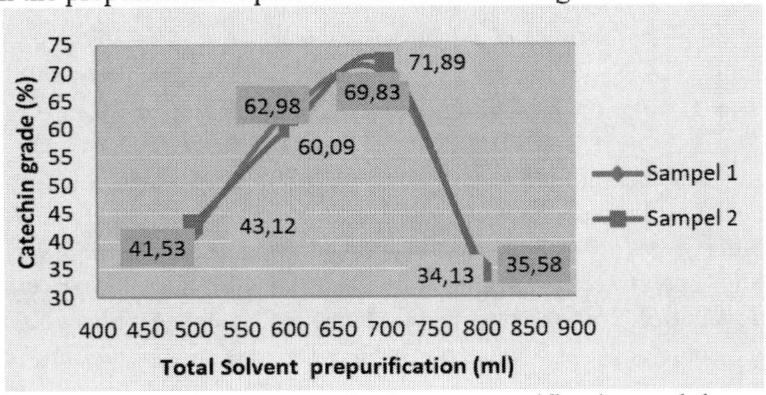

Figure 2. Relationship between the amount of solvent prepurification and the catechin levels

From Figure 2, it can be seen that catechin levels increased with the large amount of solvent in the process of the amount of solvent prepurification i.e. 500 ml to 700 ml for both samples, namely from 41.53% - 69.83% for sample 1, and 43.12% - 71.89% for sample 2, while the amount of solvent 800 ml decreased levels of catechin to 35, 58% for sample 1, and 34.13% for sample 2. Thus the study shows the number of 700 ml of solvent is the most good for prepurification process on levels of catechins, because the solvent volume of 800 ml, the amount of catechins that come together to qualify solvent, thus decreasing the percentage of catechin to material.

3.3. Relationship between the amount of solvent purification and the weight gain of

The amount of solvent is obtained from the amount of the acquisition of the volume of solvent prepurification to his best with a ratio 1; 7. Since the wash product from 65.08gr to 57.77gr for ethyl acetate solvent and from 59.8 gr to 54.98 gr for water solvent. So the amount of catechin recovery in the purification process can be seen in Figure 3.

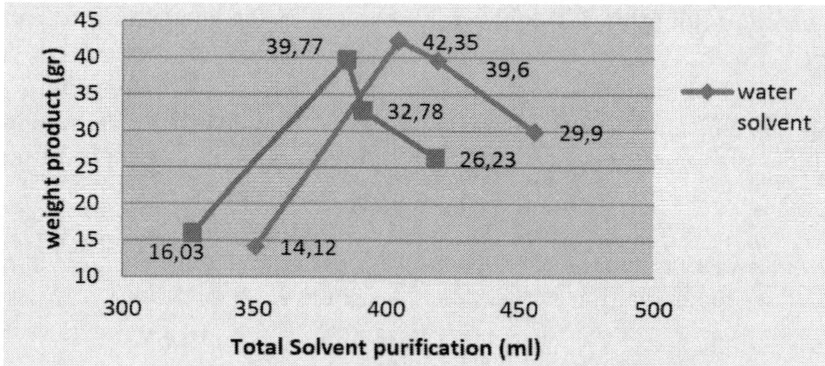

Figure 3. Relationship Number of solvents to the acquisition catechin

Figure 3 shows that the amount of catechins with solvent water acquisition larger than the solvent ethyl acetate is highest achievement in 42.35 grams of water, and the ethyl acetate tertinngi acquisition is 39, 77 grams. When viewed from the polar solvent, water is a polar solvent while ethyl acetate is semi-polar solvent, then allowing the solvent water produces more catechin acquisition compared with ethyl acetate, because polar solvents can dissolve polar compounds, semi-polar and even non-polar. Besides, the compound catechin its polar compounds vary greatly, depending on origin gambier. While the decline is due to the acquisition of catechinsolvent more will lead to the possibility of many catechins are shipped to a solvent.

3.4. Relationships amount of solvent purification on levels catechin
Catechin levels in the purification process can be seen in Figure 4.

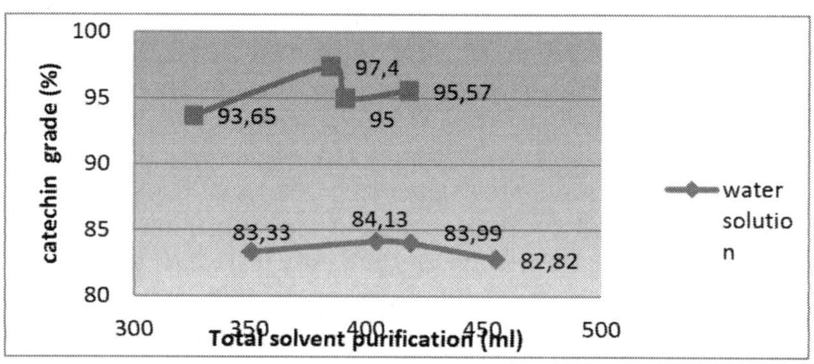

Figure 4. Relationship Number of Solvents on levels of solvent catechin with water and ethyl acetate

Figure 4 shows that the greater the amount of solvent volume, the higher the catechin content obtained in the 385 ml ethyl acetate solvent is 97.4% and the 405 ml of water solvent is 84.13%, but the higher solvent appears to decrease the catechin level. This is due to the higher volumes of solvents being inefficient, as many catechins escape with the solvent. When compared between water solvent and ethyl acetate, it appears that catechin levels are higher in ethyl acetate, in accordance with Sousa et al. [16], because the ethyl acetate solvent is capable of dissolving the compound to be extracted. Easily separated and purified again, Ethyl acetate is volatile compared to water solvents, resulting in a purer catechin.

3.5. Efficiency of pre-purification and purification process

Prepurification process is the initial process is done in Gambier to obtain initial catechin, performed by dissolution of the Gambier using solvent water and stirred at 70° C conditions of temperature conditions. Then filtered and washed 3 times in cold conditions 16°C. Fromthis process efiisiensi prepurification process work can be seen in Figure 5.

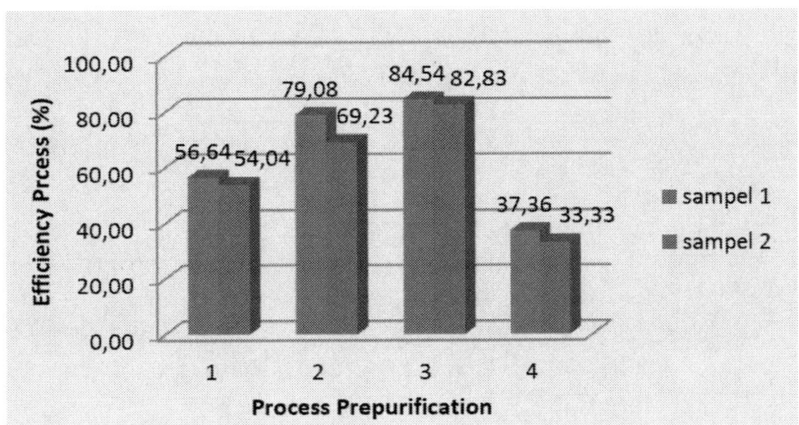

Figure 5. Efficiency process of prepurification

From Figure 5 it can be seen that the treatment to 3 provides the highest efficiency is 84.52% and 82.83%. This means that the extraction process beginning with material Gambier 100 g and 700 ml of solvent with 70oC and washing and filtering of 3 times can be used as an initial determination prepurification process operating conditions. Purification process is a process of continuation of materials or products from prepurification diluted with water and ethyl acetate solvent. The purpose of the purification process is to obtain process efficiency as shown in Figure 6.

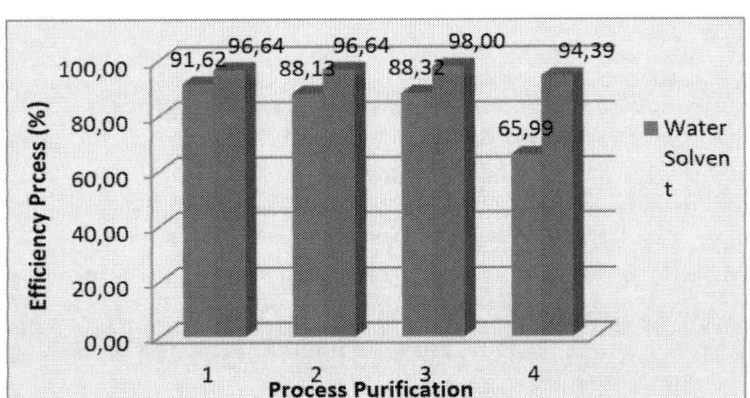

Figure 6. Efficiency process of purification

From Figure 6 it can be seen that the efficiency of work on the mean rat purification process are all high. But the highest point on the use of solvent acetic Ethy almost gives efficiency above 90% and the highest in treatment to 3 which provides process efficiency by 98%. And whatever the ultimate catechinnya quality is at a level of 97.4% catechin yangterlihat in Figure 4. This purification process is carried out at a temperature of 70 ° C and the ratio of raw material comparison of results of leaching and solvent is 1:7.

Overall, the extraction process catechin from Gambier starting from prepurification covering the initial extraction and leaching. Then followed by purification using ethyl acetate solvent or of the water at a temperature of 70° C. The efficiency of the extraction process catechin overall Gambier can be seen in Figure 7.

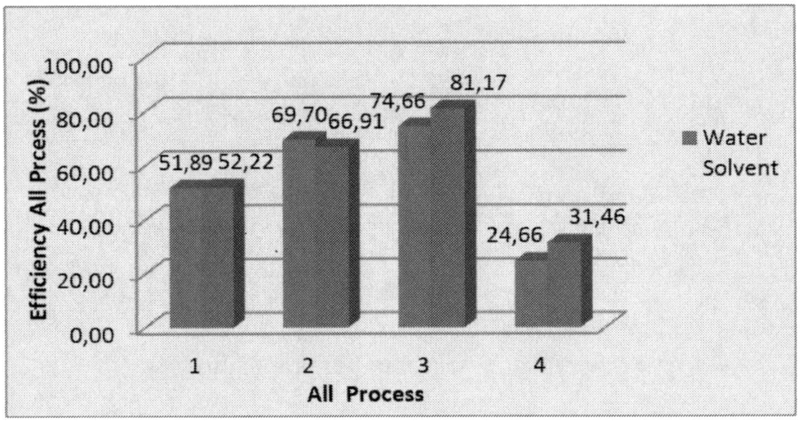

Figure 7. Efficiency of all process

From Figure 7 it can be seen that the working efficiency in the overall process highest in Ethyl Acetate solvent by 81.17%. It is seen from the many catechin acquisition of the initial content of the sample Gambier catechin 47.72% of 100 g which is 47.72 grams catechin obtained product with 97.4% of the catechin content of 39.77 g of product that is as much as 38.74 g catechinnya.

4. Conclusion

This study has been conducted to determine the best extraction process conditions and can be concluded that: For prepurification process, it was derived the best-laundering on the 700 ml volume, 70°C temperature, with the highest acquisition for sample 1 and 2 was 65.08 grams and 59,80 grams, with catechin levels were 71%, 89% and 69.83%. While the purification process it was obtained the best solvent is ethyl acetate, with the acquisition of catechin from water and ethyl acetate were 42.35 grams and 39.77 grams, respectively, with the acquisition of catechin levels were 84.13% and 97.4%. Thus, in the highest process work efficiency in the ethyl acetat solvent was 81,17%. This is from the number of acquisition of catechins from the initial content of catechin in gambier samples.

5. Acknowledgement

Thanks for all the parties concerned to this study: the Higher Education Research and Technology which funded this research through grants PUPT, also to all lecturer and students of guidance that help a lot of this research.

References

[1] Thorpe JF and Whiteley MA 1943 *ThorpeS Dictionary of Applied Chemistry Vol-VI* (Longmans, Green: London).

[2] Bakhtiar A 1991 *Manfaat Tanaman Gambier Makalah Penataran Petani dan Pedagang Pengumpul Gambier di Kecamatan Pangkalan Kabupaten 50 Kota 29-30 November 1991* (FMIPA Unand: Padang).

[3] Nazir N 2001 Gambir. *Budidaya, pengolahan dan prospek diversifikasinya. Diterbitkan atas kerjasama Yayasan Hasil Hutan Non Kayu (HUTANKU) (*Griya Andalas Ulu Gadut. Padang, 138).

[4] Yunarto N and Aini N 2015 *Health Science Journal of Indonesia* **6(2)** 105-110.

[5] Hussin MH, Kassim MJ 2010 *Journal of Physical* **21(1)** 1–13.

[6] Portier G 2010 *Extrasynthese Natural Product BP 62-69726* Genay Cedex France.

[7] Dirjen Pengawasan Obat dan Makanan 2000 *Parameter Standar Umum Ekstrak Tumbuhan Obat, Edisi 1* Departemen Kesehatan RI, Jakarta.

[8] Asben A 2008 *Agroindustri gambir di Sumatera Barat dari persepsi mutu*. Makalah Dep TIP Fateta SPS-IPB Bogor.

[9] Dinas Koperasi dan Perdagangan Kabupaten Lima Puluh Kota 2014 Sambutan Kepala Dinas Dalam Rangka Kegiatan Sistem Pembiayaan Perdagangan (Trade Financing) Payakumbuh: KOPERINDAG.

[10] Rahmawati N, Bakhtiar A and Putra DP 2013 Penelitian Farmasi Indonesia **1(01)** 6-10.

[11] Isnawati A, Raini M, Sampurno OD, Mutiatikum D, Widowati L, and Gitawati R 2012 *Buletin Penelitian Kesehatan* **40(4 Des)** 201-208.

[12] Yeni G, Syamsu K, Suparno O, Mardliyati E, and Muchtar H 2014 *International Journal of Applied Engineering Research* **9(24)** 24565-24578.

[13] Youfa R and Rahman ED 2006 *Majalah Ilmiah Teknologi Industri SAINTI* **3(2)** Desember 1829-7404.

[14] Ferdinal N 2014 *International Journal on Advanced Science, Engineering and Information Technology* **4(6)** 441-443.

[15] Rahman ED 2003 Bulletin Ilmiah Ekasakti **VIII(I)**.

[16] Sousa A, Ferreira IC, Barros L, Bento A, and Pereira JA 2008 *LWT-Food Science and Technology* **41(4)** 739-745.

IOP Conf. Series: Materials Science and Engineering **316** (2018) 012023 doi:10.1088/1757-899X/316/1/012023

Monoglyceride contents in biodiesel from various plants oil and the effect to low temperature properties

L Aisyah[1], C S Wibowo[1], S A Bethari[1], D Ufidian[1,2] and R Anggarani[1]

[1]Research and Development Center for Oil and Gas Technology "LEMIGAS", Indonesia
[2] Chemical Engineering Dept. of Universitas Indonesia, Depok, Indonesia
E-mail : riesta.anggarani@gmail.com

Abstract. Monoglyceride is a by-product component of biodiesel process that relates to sedimentation problem at low temperature environment. To prevent the problem in using biodiesel-diesel fuel blends, it is necessary to limit of the monoglyceride content. The factor affecting monoglyceride content in biodiesel is the transesterification reaction and also the plant that is used. In this study, we investigate the monoglyceride content in biodiesel made from 4 plant oils; kemiri sunan (*Reutealis trisperma*) oil, coconut oil, nyamplung (*Calophyllum inophyllum*) oil, and waste cooking oil. These oils are purified and checked for its critical properties then converted to biodiesel. The biodiesel tested refer to Standard National of Indonesia for biodiesel (SNI 7182:2015). The monoglyceride content of biodiesel from kemiri sunan (*Reutealis trisperma*) oil, coconut oil, nyamplung (*Calophyllum inophyllum*) oil, and waste cooking oil, are 8.86%, 0.69%, 4.0%, and 2.69% consecutively. The low temperature properties represented by viscosity (@40 0C) for the 4 samples in the same order as before are 6.1 cSt, 2.7 cSt, 4.71 cSt, and 4.90 cSt. The cloud point is measured with the result of 30 °C, -20 °C, -60 °C and 30 °C respectively. The conclusions indicate that monoglyceride content can affect the low temperature properties of biodiesel.

1. Introduction

The use of biodiesel blends with diesel fuel for transportation sector rise in many countries in the last 10 years, spread from European countries, United States, and mostly in Asian countries. From all countries implementing the biodiesel blends, Indonesia is the first country in the world to implement the biodiesel blends in the percentage of 20% on its regular diesel fuel nationally [1]. This policy was established in the Decree of Minister of Energy and Mineral Resources no.12/2015 [2], which set a target of 20% biodiesel blending for transportation sector that started effectively from January 2016. To support this program, the improvement on the biodiesel specification was established. One of the parameter being added on the specification is monoglyceride content.

Monoglyceride is one of the impurities in the biodiesel caused by the reversible reaction from triglycerides into glycerol which one of the step including the forming of diglyceride and monoglyceride [3]. Ramos et al studied that fatty acid composition in the vegetable oils influence the properties of biodiesel, and also mention that in Europe areas the olive, almond, corn, rapeseed, and sunflower oils are preferable to produce biodiesel. There are wide varieties all over the world on the feedstock of plant oils for biodiesel. Soybean oil that already known widely as biodiesel source in US and EU countries, is being enhanced on its production by using bentonite to promote methanolysis [4]. Another study performed on conventional feedstock, rapeseed oil, introducing their novelty on using 11 different linear or branched alcohols (C_1-C_6) rather than methanol [5]. A non-edible feedstock also being investigated for the alternative source of biodiesel, Karanj oil (*Pongamia pinnata*) which composed of 15.6% free fatty acid and produced biodiesel with 2.63%-wt monoglyceride [6].

In Indonesia, the most available feedstock for biodiesel production are palm oil and *Jatropha curcas* [7]. Some researchers investigated the potentiality of other plant oils for biodiesel feedstock. In a review article, Atabani [8] listed some non-edible plants producing oils that is potential to converted for biodiesel such as *Callophyllum inophyllum, Pongamia pinnata, Hevea brasiliensis*, *Ricinus communis*, *Azadirachta indica*, and many more plants. Even the edible oils has the same potentiality to be used in a single-cylinder diesel engine with palm oil for biodiesel, namely coconut oil [9]. Another promising source is *Callophyllum inophyllum*, which has seed oil content as high as 40-75% in a dry weight basis [10]. On this study we focusing on using 4 plant oils with high availability in Indonesia to be processed to biodiesel and investigating its monoglyceride content ; *Reutealis sperma* oil, Coconut oil, *Callophyllum inophyllum* oil, and waste cooking oil (WCO).

The physical and chemical characteristics of biodiesel depend largely on the composition profile of fatty acid methyl esters, while the fatty acid composition in the final products relate directly with the feedstock composition. Feedstock with highly saturated fatty acid structures (such as palm oil) produce biodiesel with poor cold flow properties [11]. Figure 1 shows us the average value of cold flow properties represented by Cloud Point and Pour Point of biodiesel from some feedstock. In countries having low temperature season, poor cold flow properties can affect the operability of diesel engine, starting from filter clogging to difficulty of fuel pumping from fuel tank to engine [12]. For blends of biodiesel and diesel fuel, cloud point (CP) had been considered as one of indicator parameter to evaluate the low temperature properties [13]. In a study by [14], they investigated the effect of saturated monoglyceride and other impurities on the precipitant formation on cold temperature of biodiesel, and concluded that saturated monoglyceride is the main component affecting cloud point and the final melting temperature (FMT). All the studies done then lead the biodiesel users to limit the monoglyceride content on the specification.

A study by Hoekman [11] revealed that biodiesel from Palm Oil produce highest cloud point and pour point among other 11 biodiesel samples from various feedstock (including; camelina, canola,coconut, corn, jatropha, rapeseed, safflower, soy, sunflower, tallow and grease). Considerable compositional variability is observed on this wide range of feedstock. Hoekman et al pointed that fatty acid composition is the main factor affecting the biodiesel parameters, most affected are the low temperature properties. Indonesia as a country that implement the biodiesel-diesel fuel blends policy to 20% volume is very strict to keep the biodiesel quality being blended to fulfill its specification. Biodiesel specification in Indonesia followed the SNI 7182:2015 [15]. The specification was prepared by the technical committee with consideration to domestic production of biodiesel quality. It already limiting the monoglyceride content by maximum 0.8% mass. However, in current time biodiesel used for blending mostly come from palm oil. Considering that Indonesia has a lot of plants which contain high oil content, it is necessary to study whether the derived biodiesel from various plants can fulfilled the specification especially cloud point and monoglyceride contents. These 2 parameters are considered to represent the low temperature properties along with viscosity that in practical perspective can detect the change of biodiesel properties causing by low temperature easily.

In this study we focusing on 4 (four) different plant oils to be converted through transesterification process to biodiesel and measuring the critical parameters of the produced biodiesel that relate to low temperature properties. The plant oils used in the study were: kemiri sunan (*Reutealis trisperma*) oil, coconut oil, nyamplung (*Calophyllum inophyllum*) oil, and waste cooking oil (WCO). We emphasize to measure the monoglyceride content on the biodiesels produced to know whether the 4 plant oils can fulfill the monoglyceride limitation on the biodiesel specification in SNI 7182:2015. The novelty of this research is based on the limitation of references that provide us the variation of monoglyceride content in biodiesel especially when produced from various feedstocks. By doing so, we hope that the result on this research which including only 4 feedstock can be explore widely for other potential feedstock available in Indonesia.

2. Methodology

2.1. Measurement of the properties of plant oils

The plant oils used in this study were collected from small scale oil milling. The oil were processed through pressing and mechanical purification from the remaining impurities. All of the plant oils were measured for the acid value, kinematic viscosity, saponification value, iodine value and density. This parameters are needed to determine the stage of transesterification and also the amount of alcohol and catalyst used for conversion.

2.2. Optimization of transesterification process

After getting the critical properties of the feedstock, we continue to optimize the transesterification process condition to get the highest yield of biodiesel. The fatty acid components in each feedstock is different, this lead to different process condition of each feedstock to get the best result of biodiesel yield.

2.3. Measurement of biodiesel properties

The biodiesel produced in the optimum process then continued to be measured of the critical parameters relate with the low temperature properties. In relation to the composition of the feedstock used to the low temperature properties we measured the monoglyceride content of all samples. The methods used in this study conform to SNI 7182:2015.

3. Results and discussion

Table 1 shows us the result of the measurements for the plant oils used in this study.

Table 1.Properties of plant oils.

No.	Test Parameters	Unit	Test Methods	Plant Oils			
				Coconut	*C.inophyllum*	*R.trisperma*	*WCO*
1	Acid Value	mg KOH/gr	ASTM D664	0.021	0.006	0.022	0.011
2	Viscosity @50 ^{0}C	cSt	ASTM D445	19.4	27.8	49.2	31.5
3	Saponification Value	mg KOH/gr	SNI 7182:2012	228	171	185	191
4	Iodine Value	gr I_2/100 gr	SNI 7182:2012	78.80	190.82	330.2	240.26
5	Density @40 ^{0}C	kg/m3	ASTM D4052	0.90893	0.92786	0.9192	0.90021

From table 1, we may observe that the acid value for the samples vary with the lowest belongs to *Calophyllum inophyllum* oil and the highest belongs to *Reutealis trisperma* oil. The acid value indicated the quantity of the fatty acid component of the plant oils, but cannot describe us the composition of fatty acid construct the oil. The information of the acid value is used to determine the transesterification steps needed for the biodiesel conversion. Because all samples possess low acid value then the transesterification can be performed in one stage.

The Saponification Value is used to calculate the average Molecular Weight (MW) of the oils, which then is used to calculate the methanol needed to convert the triglycerides of the oils into fatty acid methyl ester. The Iodine Value indicates the double bonds exist on the oils which also indicates the oxidation stability of oils. Among the samples, *Reutealis trisperma* (kemiri sunan) shows the highest Iodine Value that indicates this oil possess the highest double bonds among others. Kinematic viscosity is measured here as indicator the successfulness of the transesterification process, because the biodiesel produced should have required viscosity refer to the specification which is far lower than the origin plant oils.

All of the samples then being optimized for the transesterification process with the optimization parameter is the molar ratio of methanol to plant oil. Temperature of the reaction was fixed at 60 ^{0}C. Table 2 shows us the yield of biodiesel production for variations of molar ratio of reactants for each plant oils. For each plant oils we start the optimization with molar ratio 6:1 that is in general used for palm oil biodiesel.

Table 2. Optimization of molar ratio of methanol-plant oil.

No	Plant Oils	Reaction Temperature (°C)	Molar Ratio (CH_3OH: Oil)	Yield (%)
1	Coconut Oil	60	6:1	90
2	*Reutealis trisperma*	60	6:1	15
			10:1	20
			15:1	30
			20:1	60
3	*Calophyllum inophyllum*	60	6:1	10
			10:1	30
			15:1	85
4	Waste Cooking Oil	60	6:1	75

The optimum molar ratio for each plant oils are marked grey. We may observe that for coconut oil, first trial with molar ratio 6:1 resulting 90% yield of biodiesel which is high for biodiesel so we choose this as optimum. For *Reutealis trisperma*, the optimum molar ratio is higher, achieved at 20:1 means that more methanol is needed to convert the fatty acid in origin oil. For *Callophyllum inophyllum*, we have to mention that before optimizing the molar ratio on the main reaction, this plant oil require pre-treatment by extraction using methanol 1:1 up to 3 times extraction to eliminate the resin components on original oil. The waste cooking oil also use the optimum molar ratio of 6:1 to produce 75% yield to biodiesel.

As important part after transesterification process finish, the washing process should eliminate the remaining impurities from the final product. The impurities include unreacted methanol, glycerol, and also the catalyst. In this study, we performed washing process using warm water at temperature 60-70 ^{0}C in 3 stages of washing. After washing process we perform the last purification stage by heating the biodiesel to 80 ^{0}C for 30 minutes.

The biodiesel produced from the optimum process and passing the washing process then measured for the monoglyceride content, kinematic viscosity, and cloud point using the method listed in SNI 7182:2015.

In figure 1 is showed the result of monoglyceride content for each sample compare to the standard limitation established in SNI 7182:2015. Using the limit value of maximum 0.8 % mass, the biodiesel prepared in this study that can fulfill the limit is only from coconut oil. Other biodiesel fail to satisfy the limit. Monoglyceride content in biodiesel depends on some factors; reaction condition (temperature, molar ratio, type and quantity of the catalyst), and also other impurities that disturb the triglyceride conversion to fatty acid methyl ester). The failure of biodiesel from *Calophyllum inophyllum, Reutealis trisperma* and waste cooking oil may challenge us to perform variations over wider range to find optimum biodiesel yield not only from the quantity but also quality especially monoglyceride content.

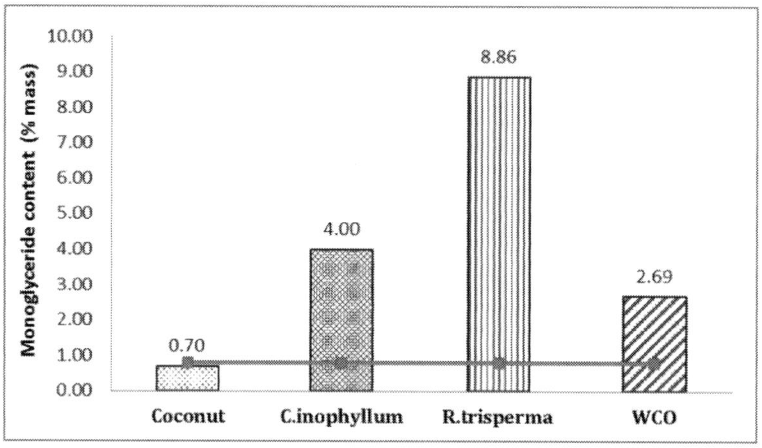

Figure 1. Monoglyceride content of the samples.

From figure 2 we can observe the kinematic viscosity of all biodiesel prepared from plant oils in this study. The red line is the limit value for kinematic viscosity established in SNI 7182:2015 with value of maximum 6.0 cSt. Only one sample biodiesel fail to fulfill the limit ; biodiesel of *Reutealis trisperma* which has kinematic viscosity 6.1 cSt. Investigation from the original plant oils in table 1 reveals that *Reutealis trisperma* oils has the highest kinematic viscosity among other samples with value of 49.2 cSt. Eventhough the biodiesel slightly exceed the limit, but the transesterification process in this study can reduce the kinematic viscosity of original oils.

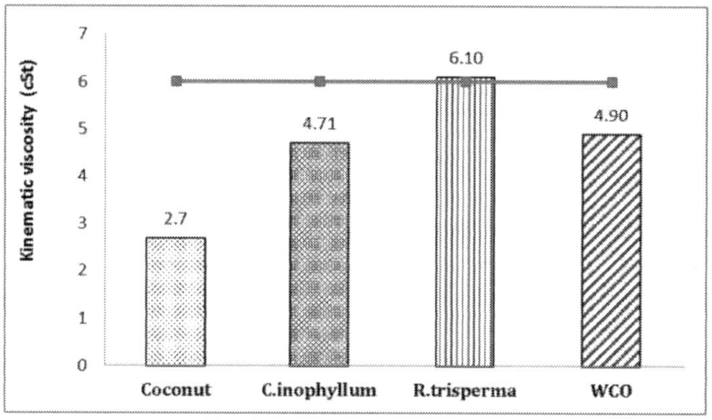

Figure 2. Kinematic viscosity of biodiesel from plant oils.

Cloud point of all biodiesel samples is showed in figure 3. Here we can observe that all samples fulfill the limit required in specification with maximum value 18 ^{0}C. Until here we can take a look to the monoglyceride contents of each samples. In figure 1 the monoglyceride content that fulfill the limit in Indonesian biodiesel specification is only biodiesel from coconut oil. We can observe the cloud point of coconut oil is -2 ^{0}C. The highest monoglyceride content which belongs to *Reutealis trisperma* related with the cloud point in 3 ^{0}C. So the relation of the monoglyceride contents with low temperature properties is not showed clearly using cloud point, we can try to show the realtionship using other method such as Cold Filter Plugging Point (CFPP). This fact may lead us that this can be explained from the fact that most of the study in the literature use plant oils from different climates with Indonesia. It becomes our benefit too that in tropical climate such in Indonesia won't be affected much with low temperature properties of the biodiesel. Only some areas in high altitude still need our concerns to provide good quality of biodiesel that is already covered in our specification.

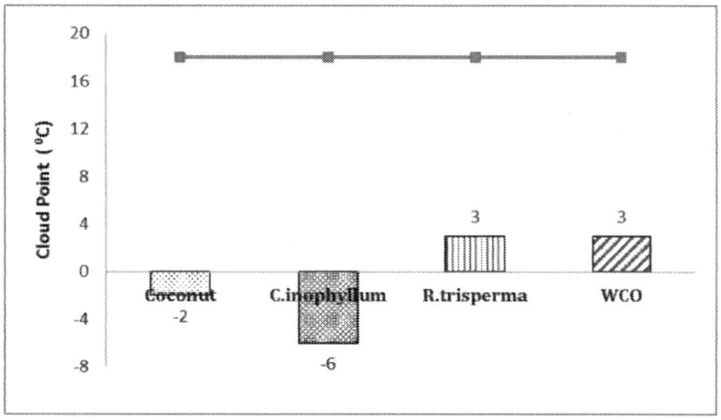

Figure 3. Cloud Point of biodiesel from plant oils.

4. Conclusion

The monoglyceride content from biodiesel prepared from 4 different plant oils (coconut oil, nyamplung/*Calophyllum inophyllum*, kemiri sunan/*Reutealis trisperma* and waste cooking oil) is measured with the result of only one biodiesel fulfill the maximum limit in Indonesian specification which belong to coconut oil (0.70 %mass). Kinematic viscosity of the samples can fulfill the limit in the specification except for biodiesel from *Reutealis trisperma* with value of 6.1 cSt. The cloud point of all samples can fulfill the limit in the specification.

References

[1]. Kharina, A., C. Malins, and S. Searle, *BIOFUELS POLICY IN INDONESIA: OVERVIEW AND STATUS REPORT.* 2016.

[2]. KESDM, *Peraturan Menteri Energi dan Sumber Daya Mineral Republik Indonesia No. 12/ 2015,* http://jdih.esdm.go.id/peraturan/Permen%20ESDM%2012%20Thn%202015.pdf: Jakarta.

[3]. Ramos, M.J., et al., *Influence of fatty acid composition of raw materials on biodiesel properties.* Bioresource Technology, 2009. **100**(1): p. 261-268.

[4]. Wu, L., et al., *Bentonite-enhanced biodiesel production by NaOH-catalyzed transesterification of soybean oil with methanol.* Fuel Processing Technology, 2016. **144**: p. 334-340.

[5]. Malins, K., et al., *Properties of rapeseed oil fatty acid alkyl esters derived from different alcohols.* Fuel, 2014. **137**: p. 28-35.

[6]. Khayoon, M.S., M.A. Olutoye, and B.H. Hameed, *Utilization of crude karanj (Pongamia pinnata) oil as a potential feedstock for the synthesis of fatty acid methyl esters.* Bioresource Technology, 2012. **111**: p. 175-179.

[7]. Silitonga, A.S., et al., *A review on prospect of Jatropha curcas for biodiesel in Indonesia.* Renewable and Sustainable Energy Reviews, 2011. **15**(8): p. 3733-3756.

[8]. Atabani, A.E., et al., *Non-edible vegetable oils: A critical evaluation of oil extraction, fatty acid compositions, biodiesel production, characteristics, engine performance and emissions production.* Renewable and Sustainable Energy Reviews, 2013. **18**: p. 211-245.

[9]. Habibullah, M., et al., *Biodiesel production and performance evaluation of coconut, palm and their combined blend with diesel in a single-cylinder diesel engine.* Energy Conversion and Management, 2014. **87**: p. 250-257.

[10]. Fadhlullah, M., S.N.B. Widiyanto, and E. Restiawaty, *The Potential of Nyamplung (Calophyllum inophyllum L.) Seed Oil as Biodiesel Feedstock: Effect of Seed Moisture Content and Particle Size on Oil Yield.* Energy Procedia, 2015. **68**: p. 177-185.

[11]. Hoekman, S.K., et al., *Review of biodiesel composition, properties, and specifications.* Renewable and Sustainable Energy Reviews, 2012. **16**(1): p. 143-169.
[12]. NREL, *Biodiesel Handling and Use Guide (Fifth Edition)*, 2016, US Department of Energy.
[13]. Chupka, G.M., et al., *Saturated monoglyceride effects on low-temperature performance of biodiesel blends.* Fuel Processing Technology, 2014. **118**: p. 302-309.
[14]. Chupka, G.M., et al., *Effect of Saturated Monoglyceride Polymorphism on Low-Temperature Performance of Biodiesel.* Energy & Fuels, 2011. **25**(1): p. 398-405.
[15]. BSN, *Standar Nasional Indonesia (SNI) 7182:2015 Biodiesel*, 2015.

Acknowledgement
We are so grateful for the work of our team in Physical and Chemical Properties of Fuel Laboratorium in LEMIGAS. We also thankful for the kindness of Mr. Tatang H.Surawidjaja and Mr. Tirto Prakoso of ITB in Chemical Engineering Dept to share their knowledge and vision of biodiesel application.

IOP Conf. Series: Materials Science and Engineering **316** (2018) 012024 doi:10.1088/1757-899X/316/1/012024

SYNTHESIS OF BIODIESEL FROM CRUDE PALM OIL BY USING CONTACT GLOW DISCHARGE ELECTROLYSIS

Nelson Saksono[1,a], Danar Aditya Siswosoebrotho[1], Jeremia J. C. Pranata[1], and Setijo Bismo[1]

[1]*Departemen Teknik Kimia, Universitas Indonesia, Depok 16424, Indonesia*
Telepon: +62217863516, Fax: +62217863515.
[a)] *Corresponding author*: nelson@che.ui.ac.id

Abstract

This research has evaluated the use of Contact Glow Discharge Electrolysis method in the synthesis of biodiesel. The purpose of this research is to get the synthesis process and biodiesel product. The solution used is the mix of Crude Palm Oil and methanol with molar ratio of 1:24, and catalyst of NaOH and KOH with variation of concentration 0.5% - 1.5%-wt. The result shows that the biodiesel can be made from transesterification reaction that may be initiated by radical methoxide. The use of electrolyte KOH is better than NaOH based on the yield of biodiesel and the energy consumption. The optimum yield reaches 97%, at the synthesis for 30 minutes with the use of KOH 1%-wt with the energy consumption of 1.32 kJ/mL.

Keywords: Biodiesel; Contact Glow Discharge Electrolysis; hydroxil radical.

Introduction

Conventional synthesis of biodiesel from Crude Palm Oil (CPO) commonly uses the transesterification method or alcoholysis of triglycerides by using a homogeneous catalyst NaOH or KOH. Base catalyst in the production process of biodiesel plants is the most commonly used because the process is fast and easy to do. However, this process has disadvantage which the reaction of saponification produces much soap and water. Moreover, this method has a slow reaction rate and the reaction stops before the 100% of feedstock is converted into biodiesel [1].

To fasten the reaction rate and avoid the formation of soap and water, this research uses a new method, which is Contact Glow Discharge Electrolysis (CGDE). CGDE is an electrochemical process in which plasma is generated by DC current between the electrodes and the electrolyte surface in the vicinity [2]. CGDE is one of plasma electrolysis methods that is very prolific producing hydroxyl radical (HO•) [3]. The hydroxyl radicals are supposed to react with methanol and form methoxyl radicals (CH_3O•). The methoxyl radicals are believed to react with triglycerides and form biodiesel. The basic principle is to maximize the production of hydroxyl radicals from electrolysis and then they will trigger the transesterification reaction without saponification reaction that forms soap [4]. Other main products of this process are the formation of hydrogen and oxygen gas which are much higher than the Faraday electrolysis process can do so that it is more environmentally friendly [5]. These two advantages can address the problems caused by the conventional method of biodiesel synthesis. In other words, this study can be an alternative method recommendation that is much better in the synthesis of biodiesel. At the end, it is expected to be a leading alternative method in the development of biodiesel industry in the future.

Experimental

The main tool in this study is a single anode Contact Glow Discharge Electrolysis (s-CGDE) batch reactor (adapted from Saksono [6]) with volume of 2 L. The series of reactor can be seen in Figure 1. The stirring uses a magnetic stirrer with the addition of baffles in the reactor. The function of baffles is to create a more perfect stirring. The reactor is equipped with cooling flow so the temperature inside the reactor can be maintained at 50-60°C.

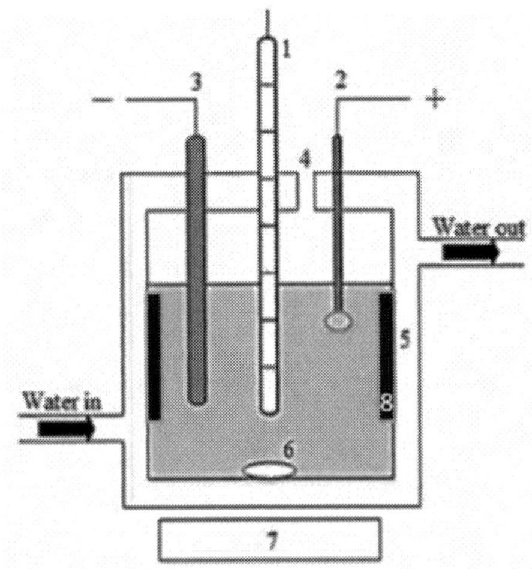

1. Thermometer
2. Cathode (SS316)
3. Anode (Tungsten)
4. Sample hole
5. Cooling jacket
6. Magnetic stirrer
7. Stirrer
8. Baffle

Figure 1 Single Anode Contact Glow Discharge Electrolysis Batch Reactor

Crude Palm Oil (CPO) that is used comes from a commercial product manufactured in Indonesia, while methanol (99.9% vol) is used as the transesterification agent. The molar ratio of CPO and methanol is 1:16 and 1:24. NaOH and KOH are used as transesterification reaction catalysts as much as 0.5% - 1.5% and 0.5% - 1% by weight of CPO, respectively. NaOH and KOH are also needed to increase the conductivity of the solution so that plasma can be generated.

Results and Discussion

Current-Voltage (I-V) Characteristics

Figure 2 shows current-voltage curves that are obtained from experiment of CPO-methanol-NaOH solution with CPO-methanol molar ratio of 1:16 and 1:24. Zone A (Point A1 and A2) is an electrolysis zone where the increase of voltage is proportional to the increase of current. Point B1 and B2 are the points where the plasma begins to appear, which are at 260 V, 0.89 A and 280 V, 0.65 A for molar ratio of 1:16 and 1:24, respectively. C1 and C2 are areas where the increase of plasma intensity is in line with the decrease of electric current; Point D1 and D2 are the points with the lowest current, which are at 700 V, 0.22 A and 740 V, 0.27 A for molar ratio of 1:16 and 1:24, respectively. Furthermore, the increase of voltage is accompanied by the increase of electric current and plasma is brighter. When the methanol content increases, the electrical current will decrease along the curve (Figure 2). This shows that methanol lowers the solution conductivity.

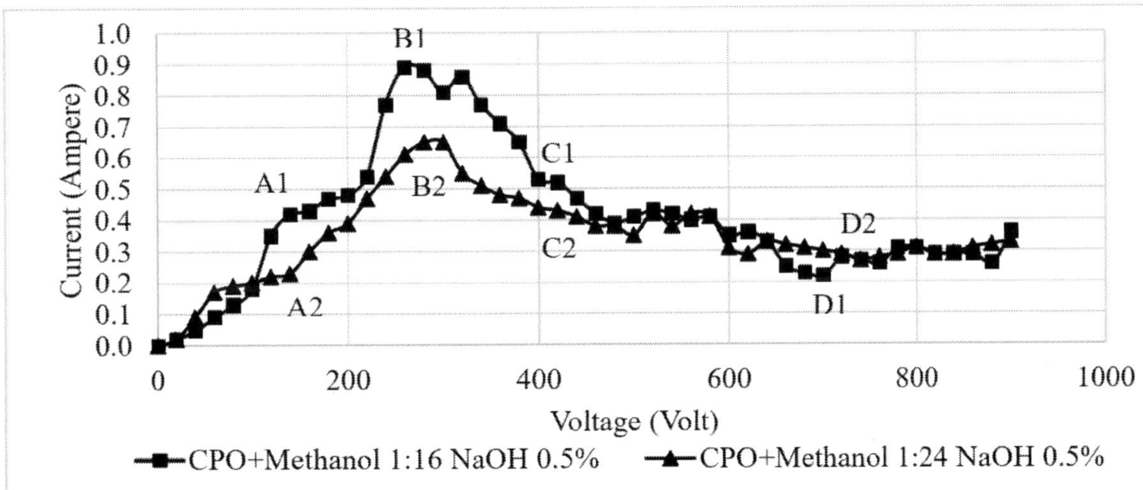

Figure 2 Profile of Electric Current-Voltage of CPO + Methanol + NaOH solution

Effect of NaOH and KOH concentration to Biodiesel Production

Table 1 shows that the solution with NaOH electrolyte concentration of 0.5% produces the highest yield compared to the concentration of 1% and 1.5%. This proves that the concentration of electrolytes is not proportional to the yield of biodiesel. The highest yield, which is 79%, is reached at the NaOH concentration of 0.5% with the synthesis duration of 30 minutes. The higher yield can be achieved when the duration is longer for the transesterification reaction will become more perfect.

Table 1 Yield of Biodiesel from Experiments with CPO:Methanol Ratio of 1:16 after 30 minutes of synthesis

Sample Solution	Yield (% v/v)
CPO-Metanol + 0.5% NaOH	79%
CPO-Metanol + 1.0% NaOH	73%
CPO-Metanol + 1.5% NaOH	73%
CPO-Metanol + 0.5% KOH	95%
CPO-Metanol + 1.0% KOH	97%

Table 2 also shows that the yield of synthesis using KOH electrolyte is much higher than the yield of synthesis using NaOH electrolyte because the reactivity and solubility of KOH in methanol is greater than NaOH in methanol. KOH is more reactive to form methoxyl which leads to the formation of biodiesel. Additionally, KOH is more active to form of hydroxyl radicals that take a role in breaking the chains of triglycerides.

Acid Number and Free Fatty Acid Content of Biodiesel Products

Acid Number is the mass of potassium hydroxide (KOH) in miligrams that is needed to neutralize a gram of chemical substance. Meanwhile, free fatty acid (FFA) content is the amount of free fatty acid in biodiesel as the result of the hydrolysis reaction. According to the Indonesian National Standard (SNI) [8], the allowable maximum limit of the acid number is 0.8 mg KOH/g with FFA content as much as 0.02% by weight.

The test results of acid number and FFA content in Table 2 indicate that the solution with 1.5% NaOH gives the lowest acid number and FFA content, which are 0.0019 and 0.0001, respectively. The tests show that all samples have values far below the maximum limit required in Indonesian National Standard (SNI).

Table 2 Chemical and Physical Characteristics of Biodiesel Products from Experiments with CPO:Methanol Ratio of 1:16 after 30 minutes of synthesis

Sample Solution	Acid Value (mg KOH/g)	FFA content (%-wt)	Density (g/mL)	Viscosity (cSt)
CPO-Metanol + 0.5% NaOH	0.00056	0.0003	0.872	5.407
CPO-Metanol + 1.0% NaOH	0.00037	0.0002	0.846	4.055
CPO-Metanol + 1.5% NaOH	0.00019	0.0001	0.876	3.898
CPO-Metanol + 0.5% KOH	0.00056	0.0003	0.864	4.589
CPO-Metanol + 1.0% KOH	0.00037	0.0002	0.865	5.375

Density dan Viscosity of Biodiesel Products

The test results indicate the lowest viscosity value, which is 3.898 cSt, is achieved on sample of CPO-methanol with 1.5% NaOH (Table 2). According to Indonesian National Standard (SNI), density of biodiesel should be in the range of 0.850 to 0.890 g/mL, while viscosity of biodiesel should be in the range of 2.3 to 6.0 cSt. The test results show that all samples meet the requirements of SNI.

Characterization of Methyl Ester in Biodiesel Products

Sample is analyzed using GCMS. The samples are biodiesel product from experiment using 1% NaOH and KOH electrolytes after 30 minutes of synthesis. The results are obtained in the form of chromatogram signal peaks as shown in Figures 3 and 4. The peaks represent as compound names and compositions in biodiesel sampel as shown in Table 3.

Table 3 shows that the total methyl ester contents in biodiesel products reach 99.89% and 99.8% when using NaOH and KOH electrolytes respectively. The products are dominated by Pentadecanoic acid methyl ester (Methyl pentadecanoate) and Octadecenoic acid methyl ester (Methyl octadecenoate). The results are in accordance with the feedstock content which is dominated by palmitic acid and oleic acid.

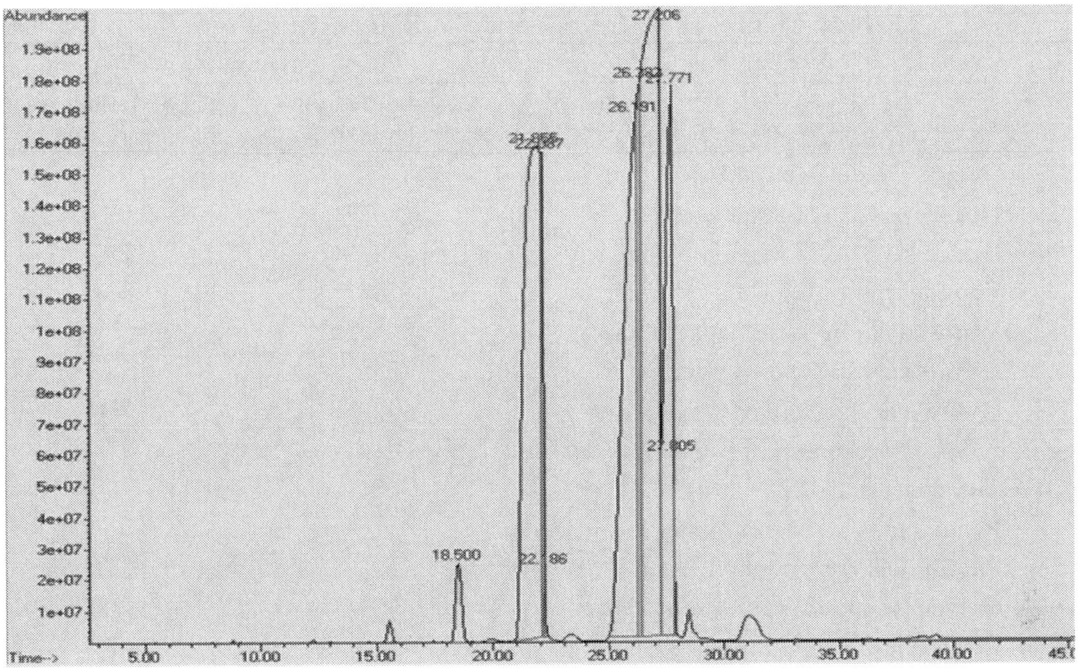

Figure 3 Characterization of Biodiesel from Experiments of CPO + Methanol + 1% wt NaOH by using GCMS

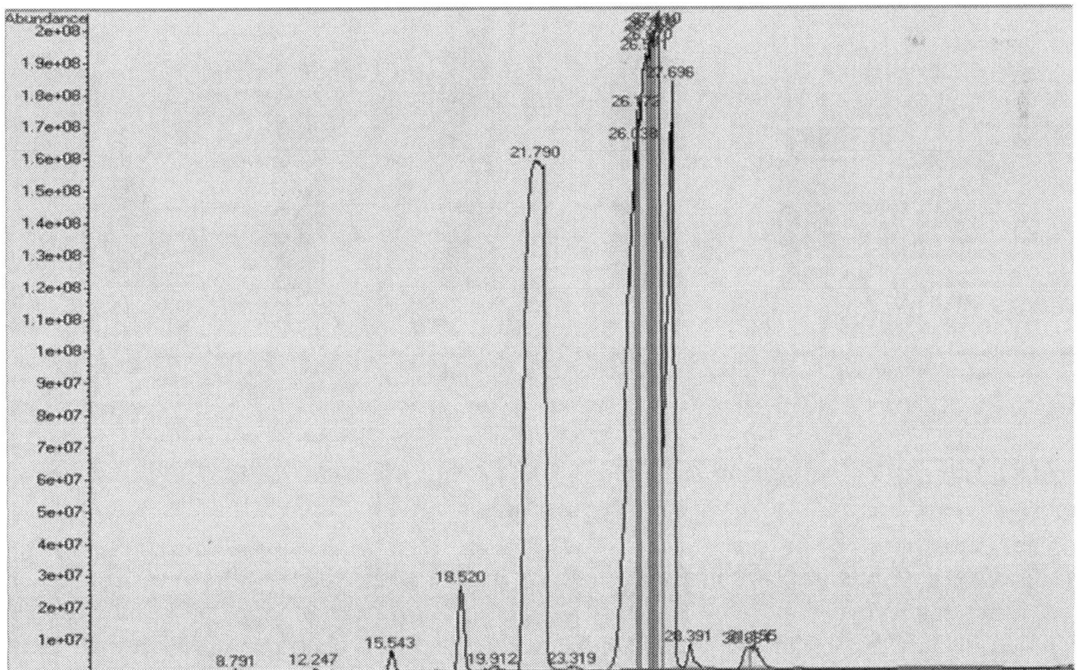

Figure 4 Characterization of Biodiesel from Experiments of CPO + Methanol + 1% wt KOH by using GCMS

Table 3 Methyl Ester Composition in Biodiesel Products

No.	Compound Name	Composition (%-wt)	
		CPO-Methanol-1% wt NaOH	CPO-Methanol-1% wt KOH
1	Octanoic Acid Methyl Ester	0.02	0.02
2	Decanoic Acid Methyl Ester	0.02	0.02
3	Dodecanoic Acid Methyl Ester	0.29	0.29
4	Tetradecanoic Acid Methyl Ester	1.47	1.43
5	**Pentadecanoic Acid Methyl Ester**	**26.42**	**26.27**
6	Hexadecanoic Acid Methyl Ester	0.55	0.23
7	**Octadecenoic Acid Methyl Ester**	**69.94**	**56.01**
8	Eicosanoic Acid Methyl Ester	1.16	1.14
10	Tetradecanoic Acid Methyl Ester	0.02	-
11	Pentanoic Acid Methyl Ester	-	0.05
12	Palmitoleic Acid Methyl Ester	-	0.27
13	Octadecadienoic Acid Methyl Ester	-	13.98
14	Docosanoic Acid Methyl Ester	-	0.08
15	Tetracosanoic Acid Methyl Ester	-	0.01
	Total	**99.89**	**99.80**

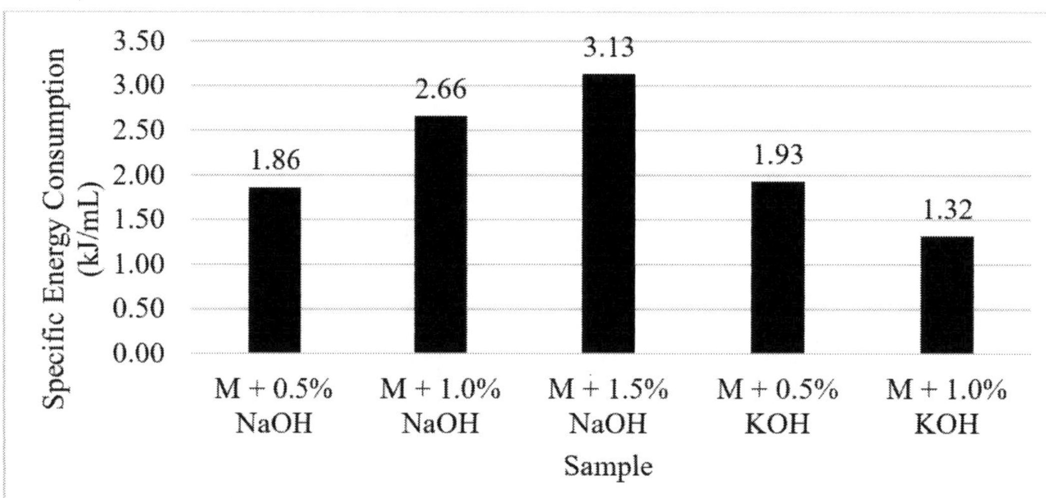

Figure 5 Energy Consumption at Various CDGE Samples with CPO:Methanol ratio of 1:16 in 30 minutes process

Specific Energy Consumption of Biodiesel Synthesis Using CGDE

The efficiency of the process can be measured from the amount of electrical energy consumed divided by volume of biodiesel produced. Figure 5 shows that the specific energy consumption of the solution using NaOH is a higher than using KOH. This is due to the yield of biodiesel produced using NaOH is lower than using KOH (Table 1). An increase of NaOH concentration in the solution will increase the specific energy consumption because it will increase the electric current due to the increased of solution conductivity. Besides, a low concentration of NaOH (0.5%) will produce the highest yield among the other higher concentrations of NaOH (Table 1). The highest specific energy consumption is achieved when using 1.5% wt NaOH, which is of 3.13 kJ per mL (Figure 5). The lowest specific energy consumption is achieved when using 1% KOH for 30 minutes, which is 1.32 kJ/mL. It is slightly higher than specific energy consumption of conventional methods, which is 1.17 kJ/mL [7].

Conclusions

This study has successfully produced biodiesel from CPO using CGDE method where the quality meets the requirement of Indonesian National Standard (SNI) in terms of acid number, free fatty acid content, density, and viscosity. The highest yield is 97% with methyl ester content reaches 99.8%, obtained from CPO-methanol solution with mole ratio of 1:16 and the addition of 1% KOH for 30 minutes process duration. In this condition, the specific energy consumption is 1.32 kJ/mL which is the lowest value in this experiment. It is slightly higher than the specific energy consumption of conventional methods, which is 1.17 kJ/mL.

This article's publication is supported by the United States Agency for International Development (USAID) through the Sustainable Higher Education Research Alliance (SHERA) Program for Universitas Indonesia's Scientific Modeling, Application, Research and Training for City-centered Innovation and Technology (SMART CITY) Project, Grant #AID-497-A-1600004, Sub Grant #IIE-00000078-UI-1.

Acknowledgement

This research was partially funded by "Hibah Publikasi Internasional Terindeks Tugas Akhir Mahasiswa Universitas Indonesia 2017" from Directorate Research and Community Service Universitas Indonesia. The authors declare no competing interests or any conflicts of financial interests.

References

[1] Boocock, D., Konar, S., Mao, V., Lee, C. and Buligan, Sonia. (1998). Fast Formation of High-Purity Methyl Esters From Vegetable Oils. *J.Am.Oil Chem.Soc.,* 75, pp. 1167-1172.

[2] Yan, Z., Li, C. and Lin, W. (2006). Experimental Study of Plasma Under-liquid Electrolysis in Hydrogen Generation. *The Chinese Journal of Process Engineering*, 6.

[3] Saksono, N., Nugraha, I., Gozan, M and Bismo, S. (2014). Plasma formation energy and hydroxyl production on contact glow discharge electrolysis. Int. J. Arts Sci., vol. 07, no. 03, pp. 71–77, 2014.

[4] Istadi, I., Buchori, L., Anggoro, D.D. (2014). Electro-Catalysis System for Biodiesel Synthesis from Palm Oil over Dielectric – Barrier Discharge Plasma Reactor. Chemical Reaction Engineering and Catalysis, 111-120.

[5] Saksono, N., Nugraha, I., Ibrahim., Febiyanti, Irine. (2016). Hydroxyl Radical Production on Contact Glow Discharge Electrolysis for Degradation of Linear Alkylbenzene Sulfonate. Environmental Progress & Sustainable Energy, 35(4), pp. 962-968.

[6] Saksono, N., Ariawan, B., Bismo, Setijo. (2012). Hydrogen Production System Design by Using Non-Thermal Plasma Electrolysis in Glycerol-KOH Solution. *International Journal of Technology,* 3, pp. 8-15.

[7] National Biodiesel Board. (2009). *Comprehensive Survey in Energy Use for Biodiesel Production*. [pdf] Jefferson City: National Biodiesel Board. Available at: http://biodiesel.org/docs/ffs-production/comprehensive-survey-on-energy-use-for-biodiesel-production.pdf?sfvrsn=4. [Accessed 26 Mar. 2017].
[8] Sisni.bsn.go.id, (2012). *SNI Biodiesel 7182:2012*. [online] Available at: http://sisni.bsn.go.id/index.php/sni_main/sni/detail_sni/15011. [Accessed 26 Feb. 2016].

IOP Conf. Series: Materials Science and Engineering **316** (2018) 012025 doi:10.1088/1757-899X/316/1/012025

Effect of Mechanochemical and Roasting Techniques for Extraction of Rare Earth Elements from Indonesian Low-Grade Bauxite

E Kusrini[1], S Harjanto[2], F Herdino[2], EA Prasetyanto[3], A Rahman[4]

[1]Department of Chemical Engineering, Faculty of Engineering, Universitas Indonesia, Kampus Baru UI Depok, 16424, Indonesia
[2]Department of Metallurgical and Materials Engineering, Faculty of Engineering, Universitas Indonesia, Depok, 16424, Indonesia
[3]Faculty of Medicine, Atma Jaya Catholic University of Indonesia, Jakarta, 14440 Indonesia
[4] Department of Chemistry, Faculty of Mathematics & Natural Sciences, Universitas Negeri Jakarta, Indonesia

E-mail: ekusrini@che.ui.ac.id

Abstract. In this research, the extraction of lanthanides from low-grade bauxite via mechanochemical and roasting methods has been studied. The addition of NaOH during mechanochemical process significantly increased the yield of collected rare earth elements. The effect of roasting process at temperatures in the ranges from 400°C to 1100°C was analyzed. The highest recovery values of lanthanide that extracted from low-grade bauxite at various temperatures and ratio low-grade bauxite and NaOH solid, with variations in the ratio of 1: 1 and 2: 1 were obtained for yttrium (~95.6%), lanthanum (~79.6%), cerium (~54.7%), neodymium (~81.8%), and samarium (~80.0%) from the theoretical value.

1. Introduction

Bauxite is important raw material for Indonesia economic development. However, seeing how much bauxite tailing produced, bauxite processing may give problem to surrounding environment such as plant grows abnormally, broken land structure, and may emit dangerous pollutants that endanger human life [1]. Therefore, there is a need to utilize residue bauxite for other applications rather than damaging the environment.

Utilization of low-grade bauxite can be categorized in three different field applications, namely (a) construction and chemical applications consist of civil and building construction, catalyst and adsorbent, ceramic, plastic, coating, and pigment; (b) environment and agronomic applications that consist of waste water and effluent treatment, waste gas treatment and agronomic applications; (c) metallurgical applications that consist of recovery of major metals, steel making and slag additive and recovery of minor metals (rare earth elements, REEs) [2]. Price of rare earth elements and rare earth elements oxide increased significantly, such as price of dysprosium was $300/kg in 2010 and increase in 2011 with $3600/kg [3], and lead many industries to investigate how to recover REEs in more effective way. Therefore, Indonesia, with a lot of natural resources, has to utilize this opportunity to increase economic value.

To increase REEs content that contains in low-grade bauxite, major elements such as aluminum, silica, and the other have to be downgraded. In this case, major elements considered as impurities in low grade bauxite. Removing impurities can be conducted by mechanochemical treatment through subjecting mechanic force and chemical activation which caused not only physically but also chemically changes on the low-grade bauxite. In other research had been conducted experiment that mechanochemical treatment on bauxite with soda ash (CaO and $Ca(OH)_2$) addition along with leaching process and then compared by bauxite without those treatments. This study aims to look at changes in the structure of bauxite and determine the recovery value of alumina. The results show that the particle size of bauxite decreased very rapidly in the first three minutes and the next minute not decrease significantly. In contrast to the particle size, the degree of amorphous increased as the length of time milling increase. In addition, comparison of bauxite with mechanochemical treatment, only mechanical force, and without treatment mechanochemical obtaining results that bauxite without mechanochemical treatment resulted in recovery of alumina values of the lowest 84.3%. For bauxite that given only mechanical force (milling) without giving reagent produced a slightly higher recovery compared to bauxite without mechanochemical treatment which was 86.65%. As for the treatment of bauxite performed mechanochemical obtain the value recovery of alumina which was 87.41% [4].

In another study also examined the effect of mechanochemical treatment of the bauxite which has a high silica content with the addition of limestone (CaO) with various concentrations. The results of these studies indicate that with the addition of limestone (CaO) can reduce levels of silica (quartz) from bauxite. The more the addition of CaO on mechanochemical process of bauxite, the more silica content decreased. In addition, when compared with the decrease of silica with and without the addition of CaO milling, result silica that decrease was greater if given the commission of milling [5].
In another research, there was also a study of mechanochemical on low grade bauxite as a result of Bayer process with adding NaOH and CaO as reagents. Based on the research, on the same ratio of reagent and red mud with constant milling time, was obtained that recovery of alumina got higher or more effective with NaOH rather than CaO as the reagent. When CaO added as reagent, value point of alumina recovery was 9%, while NaOH added as a reagent, value point of alumina recovery was 18% [6].

The decomposition of monazite and leaching of REEs using mechanochemical method has also reported by Kim and partners [7]. This process involves the mechanical milling of a mixture of monazite and NaOH [7].

To obtain high content of rare earth elements, it is necessary to conduct decomposition method because REEs don't exist as a single element. The decomposition method using roasting with alkali addition is a good method to decompose REEs. For instance, decomposition of monazite with roasting treatment with alkali addition at different temperatures. The roasting process was compared with and without alkali addition. Ratio of ore and alkali was 1:1. After roasting process, washing and filtration were conducted. It was obtained that decomposition of rare earth element with alkali addition was more effective than the one without alkali addition. The higher temperature process, the higher decomposition of rare earth element obtained with optimum temperature 400°C [8].

In another study, chemical beneficiation could reduce alumina and soda content from low-grade bauxite that contained Fe_2O_3-Al_2O_3-Na_2CO_3C. Chemical beneficiation step was roasting process of low-grade bauxite, soda, and carbon mixture that aims to reduce alumina content on Low-grade bauxite. It was obtained that roasting temperature <750°C had lower recovery of alumina (<50%). While roasting temperature >925°C obtained 70% of alumina recovery [9].

2. Experimental

2.1. Materials

Low-grade bauxite with alumina content is less than 44% was used. This sample was collected from Bintan Island, Indonesia. The sample was classified by its particle size distribution with sieving process with 25, 45, 100, 200 and 325 mesh.

2.2. Preparation
The next stage was the process of mechanochemical decomposition. In this process, low-grade bauxite and solid NaOH, with variations in the ratio of 1: 1 and 2: 1, mixed in a ball mill chamber. Ball mill used for this process was a variation of two balls tool steel 30 mm in diameter each weighing 111 g and 6 balls tool steel with a diameter of 20 mm weighing 32.5 g. Rotational speed applied in this process by 660 rpm or 11 Hz, and the milling time was 120 minutes. Then, the milling results sieved to determine the grain size of the low-grade bauxite which had milled with NaOH. Then, the results were roasted using a Carbolite Furnace with temperature variation 400°C, 500°C, 1000°C and 1100°C. The total sample was 12, which was comprised of three samples at each temperature variations. One sample was the result of milling with a ratio of low-grade bauxite and NaOH 1: 1, one sample was the result of milling with a ratio of low-grade bauxite and NaOH 2: 1, and a sample is a sample that does not do the milling process but only do the roasting process without performing the process of mechanochemical.

After those stages, samples of roasting were washed with hot water at 70°C with agitation for 30 minutes. Then the ore that had finished washing was filtered and dried in an oven with a temperature of 120°C for 2 hours until the water contained evaporates.

In this study, characterizing the process was conducted before and after the samples were undergo the process. The process of characterizing the samples to determine the effect of temperature differences in the addition of NaOH and roasting on mechanochemical decomposition process of the phase transformation occurring on low-grade bauxite. To determine the effect of variables on the process used some tools such as XRD (X-Ray Diffraction), XRF (X-Ray Fluorescence) and SEM (Scanning Electron Microscope).

3. Results and Discussion
3.1 Particle Distribution Before and After Sieving
In this study, low-grade bauxite was collected from Bintan Island, Indonesia which was by product of crushing and washing process from alumina mining. Distribution of sieving results before and after treatment was shown in Figure 1. Distribution particle size of low-grade bauxite was ranging from 0.044 to 0.707 mm. In Figure 1 above showed that the size distribution of the particle diameter of 0.074 mm or 200# had the most dominant distribution percentage that equal to 39%. As for the other particle diameter 0.044 mm, 0.149 mm, 0.354 mm, and 0.707 mm were 4.9%, 10.7%, and 15.4% respectively.

The results of the sieving process conducted a quartering process where homogenization process of distribution of particles occurred and then tested initial characterization using X-ray fluorescence (XRF) to determine content of the elements contained in the low-grade bauxite, X-ray Diffraction (XRD) to determine the compounds contained and Scanning Electron Microscope/Energy Dispersive Spectrometer (SEM/EDS) to determine the morphology of the low-grade bauxite.

After the initial characterization, the samples were subjected mechanochemical treatment in a planetary ball mill for two hours and then conducted a sieving process. Percentage of distribution particle size of low grade bauxite after mechanochemical process was shown in Figure 1.

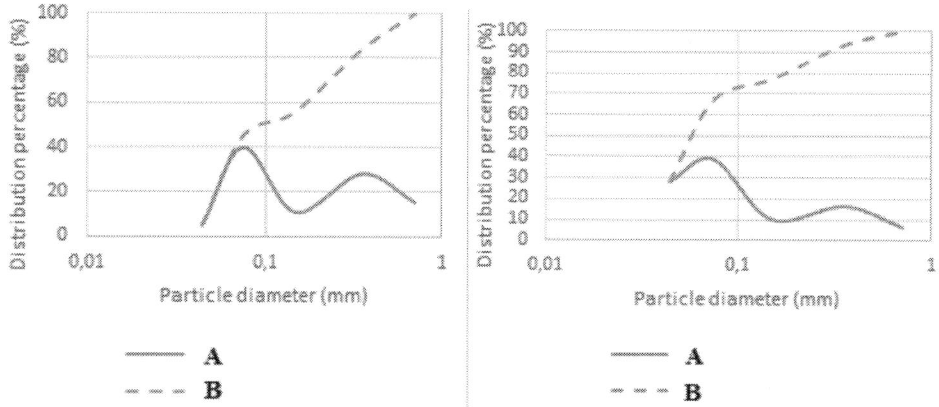

Figure 1 Particle Distribution before (left) and after sieving process (right), where A = particle distribution and B = distribution cumulative

In Figure 1, it showed that after being given the commission of mechanochemical, the particle size became smaller when compared to before by the commission of mechanochemical. When compared before and after given mechanochemical treatment, the percentage of 0.044 mm particle size distribution increased from 5% to 28% indicating that a reduction in size of the particle diameter of the low-grade bauxite. This was caused by the collision between the steel ball with ore due to the force of compression, shear, impact caused by the rotation of the planetary ball mill [10,11].

3.2. Morphology of low-grade bauxite before and after Mechanochemical and Roasting Treatment

From electron microscopy analysis, the low-grade bauxite showed variation of the particle diameter in the range of 0.074 to 0.354 mm. Overall Low grade bauxite samples were fine enough due to 0.074 mm particle size distribution was large enough for about 39%. However, there were still many coarse particles that present in low-grade bauxite.

Figure 2 was the result of the SEM characterization after mechanochemical process that indicating that the process of mechanochemical can refine and reduce the size of the grains. It can be seen in Figure 2 results low-grade bauxite particle after milling with a magnification of 100X shown that with the addition of NaOH mechanochemical bauxite became finer and smaller in size and had irregular shape. This was caused by collisions between particles of low-grade bauxite with steel balls in a planetary ball mill chamber that collided with one another, generating more impact and compression force capable of destroying the grain into smaller sizes and irregular shape. However, while looking in Figure 2, the sample residues shown the fine particles and clumping (agglomeration). The particles in the low-grade bauxite samples that have undergone a process of mechanochemical size will be smaller and then agglomerated thus reintegrated into more solid particles, then the particles low-grade bauxite will be very fine [12].

Figure 2 Morphology of low-grade bauxite (a) before and (b) after mechanochemical process with 100X Magnification

3.3. Elemental composition

Low-grade bauxite was characterized using X-ray Diffraction (XRD) which aims to determine the contained of compounds and transformation prior treatment and after the roasting process at a temperature of 400°C, 500°C, 1000°C, 1100°C. The XRD test result prior treatment was shown in Figure 3. It was shown that the most dominant compound contained in the low-grade bauxite was corundum (Al_2O_3) as much as 49%, 7% quartz (SiO_2) and 44% calcite ($CaCO_3$). Other minerals contained was very little quantity which was not detected by XRD. Fe_2O_3 which is usually contained in the low-grade bauxite, but in the XRD results, it did not appear. In addition to the initial characterization using XRD, in this study also use X-ray fluorescence to determine the chemical composition in quantitative or thoroughly into the low-grade bauxite sample which shown in Table 1. In addition to the initial characterization, XRF is also used to determine the chemical composition of the residues bauxite after mechanochemical treatment and the roasting process.

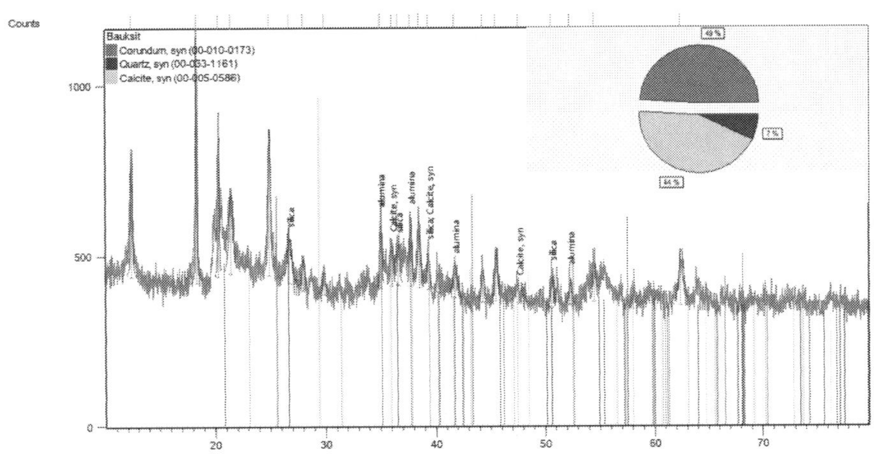

Figure 3 X-ray Diffraction pattern of low-grade bauxite before treatment

In the results of initial characterization XRF in Table 1 show that the dominant element contained in the low-grade bauxite was aluminum (Al) as much as 32.92%, 27.568% silicon (Si) and 36.383% iron (Fe) which indicated that the compositions usually contained in low-grade bauxite, a by-product processing of bauxite crushing and washing from waste Wacopek, Bintan. In this case, impurities such as aluminum, silica and other dominant element needed to be removed to increase the content of rare earth metals. As for the rare-earth metals contained in the low-grade bauxite, such as yttrium (Y) and cerium (Ce) in very low concentrations. Yttrium levels is 0.008% and for the concentration of cerium is 0.119%.

Table 1.Elemental analysis of low-grade bauxite

Element	Content (%)	Element	Content (%)
Na	0.055	Ti	1.188
Al	32.92	Fe	36.383
Si	27.568	Zr	0.086
S	0.392	Nb	0.007
Cl	0.377	Y	0.008
K	0.379	Ce	0.119
Ca	0.518		

3.4. Effect of Roasting Temperature on Total Content and Total Recovery of REEs

After the roasting process, washing and X-ray Fluorescence (XRF) characterization were conducted to determine the chemical composition of the low-grade bauxite quantitatively. Figure 4 shows graph of grade and recovery of roasting results. Figure 4 shows the effect of roasting temperature of total content and recovery of REEs for each temperature. It is shown that to increase the levels of rare earth metals optimal occurred at 1100°C by the addition of NaOH 33% and a grade for the lowest place at a temperature of 500°C with the addition of NaOH 0% or the sample residue bauxite without mechanochemical process. Overall, at a temperature of 1100°C with the addition of NaOH 33.3% or the ratio of low-grade bauxite and NaOH 2 : 1 had the most optimum grade value.

At 400°C, the addition of NaOH 50% (ratio of low-grade bauxite and NaOH 1: 1) gave higher value grade than at the temperature of 1100°C without addition of NaOH. Variation of temperature resulted highest grade at 1100°C. This is true also for the case of NaOH 0%, NaOH 33.3% and NaOH 50% addition. These results shown that in those temperature rare earth elements started to appear. While in the same NaOH addition with temperature of 500°C resulted in lowest grade value for NaOH 0% and NaOH 50% addition. This implied that roasting in higher temperature gave better effect on the recovery of rare earth elements.

As for the addition of NaOH same with different temperature values obtained the highest grade in either the addition of NaOH 1100°C temperature of 0%, 33.3% and NaOH 50%. This shows that at 1100C, the rare-earth metal elements began to appear. As for the addition of NaOH same with different temperatures lowest grade values obtained at a temperature of 500°C for the addition of NaOH 0% and 50% NaOH. This shows that roasting at a low temperature did not effectively enrich the rare earth metals content.

Figure 4 shown that the highest recovery value was obtained at a temperature of 500°C and 1100°C and with the addition of NaOH 33.33% while the lowest recovery values obtained at 500°C with the addition of NaOH 0%. In general, the greatest recovery value obtained on the addition of NaOH 33.33% for each temperature. In addition to the recovery value through mechanochemical process has a higher recovery value than the value of recovery is not through mechanochemical process. This was because the process of mechanochemical of alumina contained dissolved [13] and also reduce levels of CaO [6] so that the levels of rare earth metals contained can be increased. In this highest total recovery, from XRF we were obtained for yttrium (~95.6%), lanthanum (~79.6%), cerium (~54.7%), neodymium (~81.8%), and samarium (~80.0%) from the theoretical value.

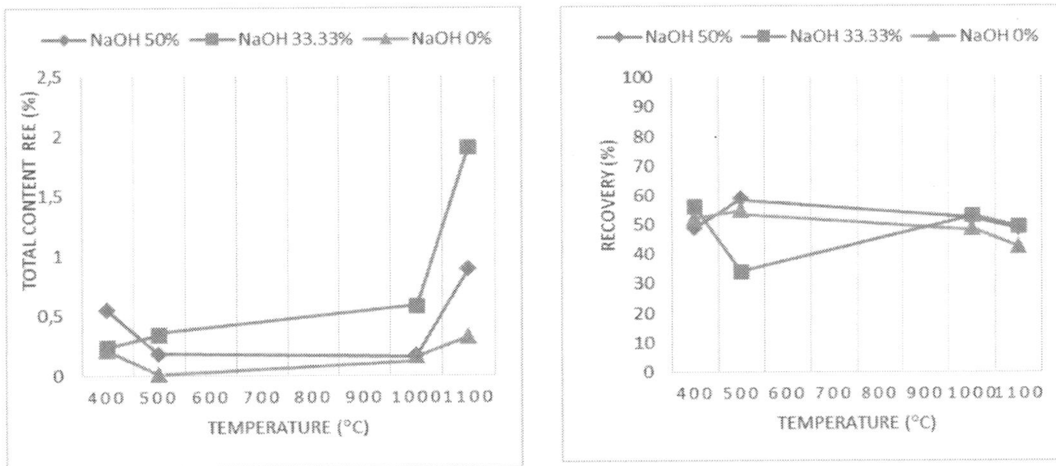

Figure 4 The effect of roasting temperature on total content (left) and recovery of REEs (right)

4. Conclusion

Mechanochemical process of low-grade bauxite with NaOH using a planetary ball mill could reduce and refine the grain size. On the low-grade bauxite by addition of NaOH 33% produced the highest overall grade value of rare earth metal element is obtained. In the roasting process at various temperatures, the recovery value residual got vary results. For the element yttrium (Y) has highest recovery value with the addition of NaOH 33.3% at a temperature of 400°C with the recovery value of 95.6%, for the element cerium (Ce) by addition of NaOH 33.3% at temperatures of roasting 1100°C had the recovery value of 54.7%, for the elements lanthanum (La) to NaOH 50% addition roasting temperature 1000°C had recovered value of 79.6%, for the element neodymium (Nd) with the addition of 50% NaOH at a temperature roasting 1100°C had recovery value of 81.8% and the element samarium (Sm) with the addition of NaOH 33.33% at the roasting temperature 1000°C had 80% recovery value.

5. Acknowledgments

The authors greatly acknowledge for financial support from Hibah Penelitian Unggulan Perguruan Tinggi (PUPT) No. 2740/UN2.R3.1/HKP05.00/2017 and Universitas Indonesia for the support.
This article's publication is supported by the United States Agency for International Development (USAID) through the Sustainable Higher Education Research Alliance (SHERA) Program for Universitas Indonesia's Scientific Modeling, Application, Research and Training for City-centered Innovation and Technology (SMART CITY) Project, Grant #AID-497-A-1600004, Sub Grant #IIE-00000078-UI-1.

6. References

[1] Sembiring S, 2008. *Chemical and Physical Properties of Land at Bauxite Mining Area in Bintan Island, Riau*. Forest Research Institute, Nort Sumatera, Indonesia.
[2] Klauber C, Grafe M, Power G. 2011 Low grade bauxite issues: II. options for residue utilization. *J. Hydrometallurgy* **108** 11-32
[3] EPA, U.S. Rare earth elements, 2012, a review of production, processing, recycling, and associated environmental issues, in: *Office of Research and Development* (ed.), Cincinnati
[4] Kumar R, Alex TC, Jha MK, Khan ZH, Mahaptra SP, Mishra CR, 2004, Mechnaochemistry and The Bayer Process of Alumina Production. *The Minerals, Metals & Materials Society*, 31-34

IOP Conf. Series: Materials Science and Engineering **316** (2018) 012025 doi:10.1088/1757-899X/316/1/012025

[5] McCormick PG, Picaro T, Smith PAI, 2002. *Minerals Engineering* **15**, 211-214.
[6] Tony P, 1997 *Red Mud Processing*. No. Paten: 1997029992
[7] Kim W, Bae I, Chae S, Shin H, 2009. *Journal of Alloys and Compounds* **486** 610-614.
[8] Yanhui X, Haijiao L, Zhijun M, Jianguo C, Wenyi Z, Liangcai L, 2012. *J Rare Earth* 2**30** 155-158.
[9] Kumar R., Alex TC, Mishra CR, Roy SK, Mehrotra SP, 2007. Envicon 45-51.
[10] Burmeister CF, Kwade A, 2013. *Chemical Society Review*, Issue **18** 7487-7740.
[11] Hea S, Qin Y, Walid E, Li L, Cuia J, Ma Y, 2014. Biotechnology Reports **3**, 54-59.
[12] Torabi O, Golabgir MH, Tajizadegan H, Torabi H, 2014. *Int. J. of Refractory Metals and Hard Materials*18-24.
[13] Gupta C K, Krishnamurthy N 2005, "*Extractive metallurgy of rare earths*", CRC Press, pp 20-250.

Performance Study of Fluidized Bed Dryer with Immersed Heater for Paddy Drying

S Suherman*, N F Azaria, and S Karami

Departement of Chemical Engineering, Faculty of Engineering, Diponegoro University Jl. Prof. Soedarto SH, Kampus Undip Tembalang, Semarang, 50275, Indonesia

suherman.mz@che.undip.ac.id

Abstract. This paper investigated the performance of fluidized bed dryer with immersed heater for paddy drying. The influence of drying temperature and the temperature of immersed heater on drying curve, thermal efficiency, and quality of paddy was investigated. The fixed operating conditions are drying time of 60 minutes, paddy weight of 200 grams and the air velocity of 0.4 m/s. The variables are drying temperature and the temperature immersed heater namely 50, 60, 70, 80, 90 (°C). The results show addition immersed heater will increase drying rates. No constant drying rate was found. Increasing the temperature will decrease the utilized energy. The thermal efficiency decreases with increasing temperature. The increasing temperature and use immersed heater will decrease the residual moisture content, increase damaged and yellow paddy grain, and increase red paddy grain.

1. The first section in your paper

Paddy is main food for Indonesian people. Normally, paddy is harvested with high moisture content 32% (db., dry base). Then paddy is predisposed to deteriorate rapidly. For tropical countries, it is an extremely serious problem due to high humid air condition can accelerate an excessive mould growth, and yellowing of grains [1]. Therefore, paddy has to be dried to maximum 16% for storage.

Many researches were conducted to get the best method drying for paddy. Mainly Fluid Bed Dryer (FBD) was chosen, in which other dryer methods are Spouted Bed Dryer (SBD), Pneumatic Dryer, Tray Dryer, Inclined Bed Dryer, Impinging Stream Dryer, and Heat Pump Dryer. Paddy drying consumes huge energy and is sensitive to product quality [2-3]. Mainly, drying paddy is performed by a fluidized bed dryer for due to (1) uniform product quality due to complete mixing, (2) high drying capacity due to high ratio of air mass to mass of product [4-5]. The drying rate of paddy in fluidized bed dryer was affected by drying air temperature and bed thickness [6]. The maximum drying temperature in fluidized bed was suggested as 115 ^{0}C to reduce moisture to 24-25% (db) for ensuring rice quality [4]. The drying temperature in pulsed fluidized bed should be less than 145 ^{0}C for initial paddy moisture content of 28% (db) to maintain rice quality [7]. A new model of rough rice drying in fluidized bed was successfully developed [8]. The energy consumption of commercial fluidized bed dryer decreased with increasing paddy moisture content and drying temperature [9]. Paddy drying is a highly energy-intensive process and sensitive to the quality of rice [10]. They recommend that the maximum drying temperature to reduce paddy moisture down become 22% (db) in a single pass is 150 ^{0}C to achieve acceptable quality of product. No impact of air velocity on drying characteristics, and employing the tempering stages substantially reduces the energy consumption [3]. Intermittent drying

IOP Conf. Series: Materials Science and Engineering **316** (2018) 012026 doi:10.1088/1757-899X/316/1/012026

of paddy was investigated to saving energy [11]. The moisture effective diffusivity and glass transition temperature changed during paddy drying [12]. Energy and quality aspects for fixed deep bed drying of paddy also were investigated [13].

However, the main feature of paddy drying is mainly paddy drying is only in the second-stage drying phase, namely the diffusion phase. So the effort for energy efficiency is more directed at the acceleration of drying time. This has been attempted by some previous researchers with recycleng air drying [4], intermittent drying method [11], and tempering method [3]. Meanwhile, other researchers tried to accelerate the drying rate on fluidized bed drying using the method of addition of immersed heater [14]. Therefore, this paper investigated drying paddy in fluid bed dryer with immersed heater. The drying kinetics, thermal efficiency and the product quality was measured. The variable is temperatures of drying.

2. Method

Paddy firstly is prepared by soaking in water and keeps in a refrigerator at a lower temperature to obtain the initial grain moisture content, i.e. 21.1% (wet base). The initial paddy moisture content was analyzed by oven for 24 hours at a temperature of 105°C. Figure 1 shows the apparatus set-up of lab scale batch fluid bed dryer with immersed heater in column of fluidization zone. The experiment starts by measuring the minimum fluidization velocity. Then, the intake air velocity is set 2 times above the minimum fluidization velocity. Inlet air temperature (T_{gi}) varied namely 50, 60, 70, 80 and 90°C. The process starts with an experiment that will turn on the blower air. Furthermore, air heater is positioned as desired, e.g. 50 °C. After drying air temperature reaches the temperature desired, 200 g paddy material is inserted into the bed. The temperature controller TIC (below distributor) maintains inlet air temperature at the desired level. During the experiment, inlet air temperature, inlet air humidity, outlet air temperature, and outlet air humidity is measured each 5 minutes. Finally, the drying process is stopped for 60 minutes. At the end of the experiment, samples were taken again for measuring the moisture content using oven. Furthermore, after the drying operation, the quality of dry paddy were analyzed with the manual method (physically election) by weighed the hollow grains, red and broken grains and percentage it to the sample weight so that the figures obtained percent grain of hollow, broken and red.

Figure 1. Experimental apparatus set up

3. Results and Discussion

3.1. Drying Curve

Figure 2 shows the profile of moisture content of paddy in fluid bed dryer in the variable with and without immersed heater. FBD with immersed heater give more drying rate, due to the additional heat from immersed heater. This additional heat from immersed heater will make the temperature in fluidizing column higher, then the diffusion of water vapor inside paddy higher. Furthermore, increase drying temperature will decrease drying time. This is due to increase drying temperature will increase diffusion rate of moisture inside particle, then the drying speed will be more faster. Similar results have been previously reported that increasing temperature will increase drying rates of paddy [11-13].

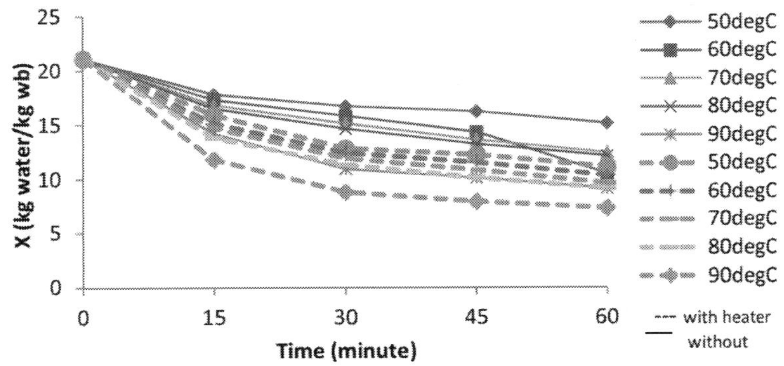

Figure 2. The influence of drying temperature on profile of moisture content of paddy in fluid bed dryer within variable with and without immersed heater

Figure 3 shows the profile the drying rate during the drying process. The absence of a constant drying rate indicates that in drying the paddy only has a second drying phase ie the rate of drying diffusion. The drying rate decrease by the time due to the moisture content inside of particle also decrease by the time. Therefore, the driving force from the center to the surface of particle decrease due to the moisture content in the center and the surface of particle became bigger. This results are similar that diffusivities of moisture inside particle paddy decreases during drying process (Midilli, 2002; Xing-jun, 2016).

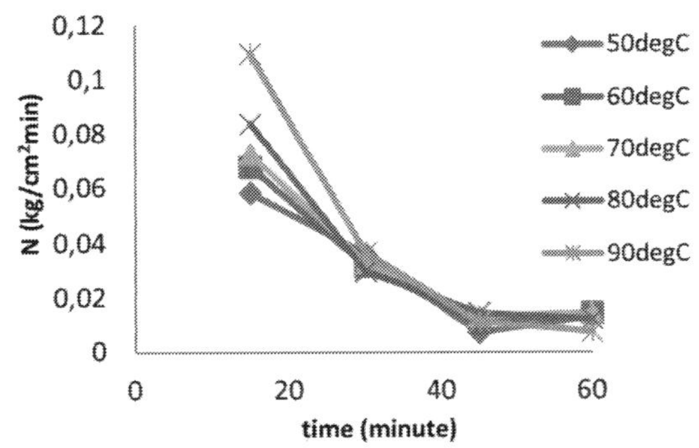

Figure 3. The influence of temperature on drying rate

3.2. Thermal Efficiency

Figure 4 shows the influence of temperature on utilized energy for evaporated moisture from paddy. Increasing the temperature will decrease the utilized energy. This is because, with the rise in dryer temperature, the supply rate of energy is getting bigger. Meanwhile, this increase is not proportional to the energy consumption for drying. As a result temperature rise decreases utilized energy. These results are similar that by increasing temperature, the total utilized energy is decrease [13]. Furthermore, by the time, the utilized energy decrease. It is caused, by the time, the sum of moisture inside particle decrease, then drying rate decrease. Therefore, the energy utilized for evaporation of moisture becomes lower.

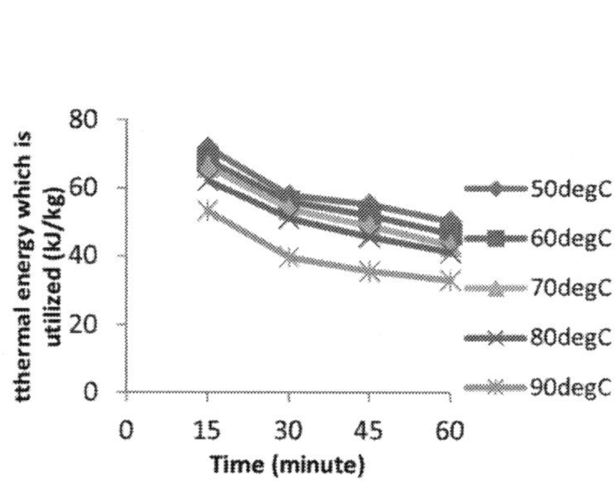

Figure 4. The influence of temperature on utilized thermal energy

Figures 5 show the thermal efficiency drying for different drying air temperature. The higher of drying temperature, the thermal efficiency decreases. It means, by increase drying temperature, the energy loss by carried air out of dryer become larger, then thermal efficiency decreases. Drying efficiency values in this study is less than 70%. The low value is due to the efficiency of the design of fluidized bed dryer is used less than optimal where the absence of insulation caused the loss of heat from the dryer. Furthermore, the thermal efficiency decreases with drying time. This is proportional with the profile of utilized energy, due to the supply energy for heating of drying air is constant. It can be concluded that the longer time of the drying, the energy utilized decrease, then the thermal efficiency decreases.

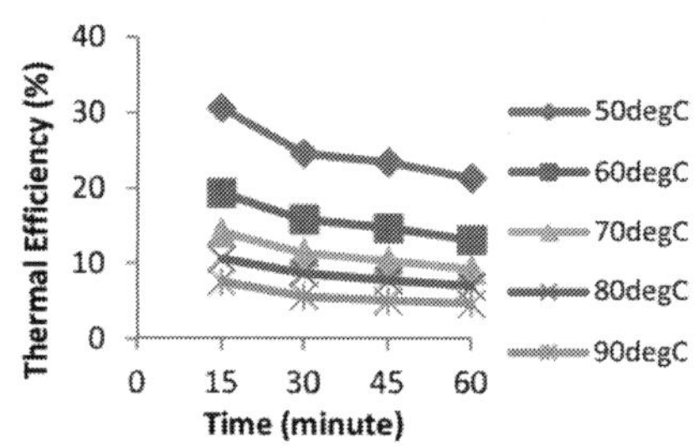

Figure 5. The influence of temperature on profile of thermal efficiency

3.3. *Product Quality*

Figure 6 shown the residual moisture content of each variable without or with immersed immersed heater has decreased along with increasing temperatures. The higher the drying air temperature, the lower relatie humidity of air, then capacity of drying air to capture water from the paddy is greater. Therefore, the residual moisture content becomes lower. Furthermore, with the immersed heater will increase the temperature again, so the relative humidity will also lower air, and residual moisture content is lower. These results are consistent with the results of [6].

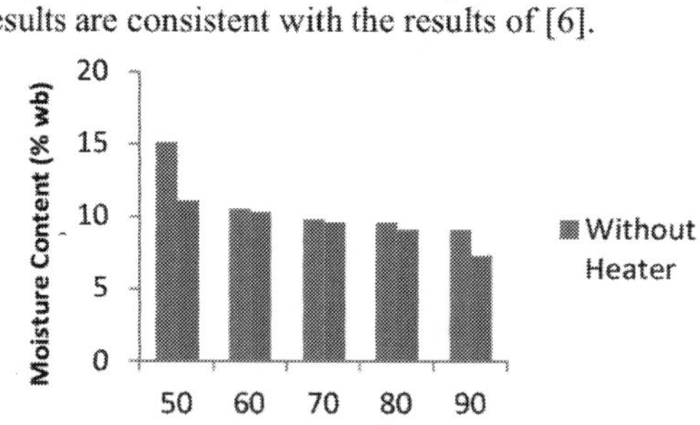

Figure 6. The influences of temperature on the residual moisture content

Figure 7 shows the influences of temperatures on the damaged grains and yellow grains. It can be seen by increasing temperature, the broken grains and yellow grains increased due to the skin of paddy will be burnt. These results are consistent with [2-3]. Furthermore, the use of immersed heater causes the damaged grain and yellowed grain more higher due to contact between the surfaces of grain with the coil of immersed heater.

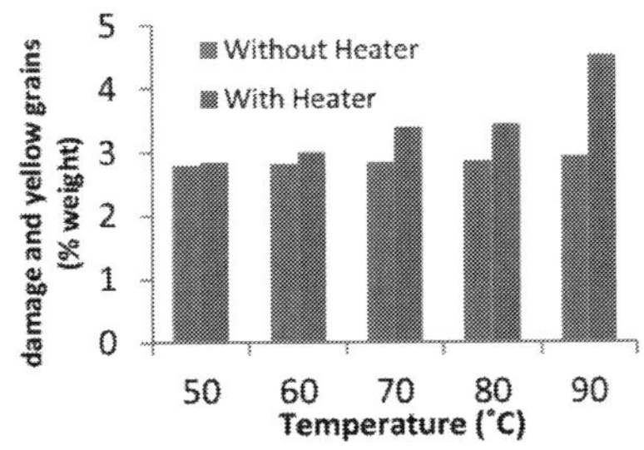

Figure 7. The influences of temperatures on the damaged grains and yellow grains

Figure 8 shows the influences of temperatures on the red grains. It can be seen increasing of temperature will increase the red grains. The use of immersed heater causes the damaged grain and yellowed grain higher. The reason is similar with the phenomenon of the damaged grains and yellow grains and consistent with results of [2-3].

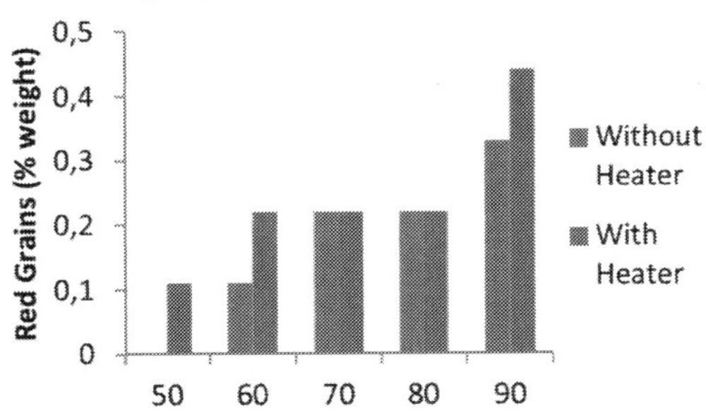

Figure 8. The influences of temperatures on the red grains

4. Conclusion

Fluid bed dryer with immersed heater was successfully drying of paddy. Addition immersed heater will increase drying rates. No constant drying rate was found. Increasing the temperature will decrease the utilized energy. The thermal efficiency decreases with increasing temperature. The increasing temperature and use immersed heater will decrease the residual moisture content, increase damaged and yellow paddy grain, and increase red paddy grain.

Acknowledgement

The authors wish to express their sincere thanks to the Diponegoro University for financial support within the Research for International Scientific Publication Year 2017.

References

[1] Sarker MSH, Ibrahim MN, Aziz NA, and Punan MS 2013 *Drying Tech.* **31** 286
[2] Sarker MSH, Ibrahim MN, Aziz NA, and Punan MS 2015 *Energy* **84** 131
[3] Morteza G, Moein A, Mehdi RH, and Hashemia SJ 2015 *Food Bioproducts Proc.* **94** 275
[4] Soponronnarit S and Somkiat P 1994 *Drying Tech.* **12** 1667

[5] Suherman S, Priyanto S and Ratnawati 2014 *Adv. J Food Sci. Tech.* **6** 403
[6] Tumambing JA, and Driscoll RH, 1991. *Proc. 14th ASEAN Sem. Grain Post-harvest Tech.* 193
[7] Prachayawarakorn S, Tia W, Poopaiboon K and Soponronnarit S 2005 *J. Stored Prod. Res.* **41** 479
[8] Midilli A, Kucuk H, and Yapar Z 2002 *Drying Tech.* **20** 1503
[9] Soponronnarit S, Prachayawarakorn S and Wangji M 1996 *Proc. 10th Int. Drying Sym.* 638
[10] Jittanit W, Saeteaw N, and Charoenchaisri A 2010 *J. Stored Prod. Res.* **46** 209
[11] Golmohammadi M, Assara M, Hamaneha MR, and Hashemia SJ 2015 *Food Bioprod. Proc.* **94** 275
[12] Xing L, Xin W, Yang L, Ping J, and Hui L 2016 *Comp. Elec. Agri.* **128** 112
[13] Tohidi M, Sadeghi M, and Harchegani MT 2017 *Renew. Sust. Energy Rev.* **70** 519
[14] Groenewold H, and Tsotsas E, *Chem. Eng. Sci.* **62** 481

IOP Conf. Series: Materials Science and Engineering **316** (2018) 012027 doi:10.1088/1757-899X/316/1/012027

Modeling of a Reaction-Distillation-Recycle System to Produce Dimethyl Ether through Methanol Dehydration

Y Muharam[1,2], **L M Zulkarnain** [1] **and A S Wirya**[1]

[1] Department of Chemical Engineering, University of Indonesia, Depok 16424, Indonesia

E-mail: muharam@che.ui.ac.id

Abstract. The increase in the dimethyl ether yield through methanol dehydration due to a recycle integration to a reaction-distillation system was studied in this research. A one-dimensional phenomenological model of a methanol dehydration reactor and a shortcut model of distillation columns were used to achieve the aim. Simulation results show that 10.7 moles/s of dimethyl ether is produced in a reaction-distillation system with the reactor length being 4 m, the reactor inlet pressure being 18 atm, the reactor inlet temperature being 533 K, the reactor inlet velocity being 0.408 m/s, and the distillation pressure being 8 atm. The methanol conversion is 90% and the dimethyl ether yield is 48%. The integration of the recycle stream to the system increases the dimethyl ether yield by 8%.

1. Introduction

Dimethyl ether (DME), previously known as a building block in the production of valuable chemicals such as methyl acetate and dimethyl sulfate, is finding its way to become a potential eco-friendly substitute for petroleum-derived diesel fuel and residential liquefied petroleum gas (LPG) replacing fuel [1, 2]. With its similar physical properties to LPG, it is possible to decrease the investment cost of its application as the material supply can be accomplished using existing LPG infrastructure.

Dimethyl ether is traditionally produced in a two-step process involving syngas conversion to methanol and subsequently methanol dehydration to DME on solid acid catalysts. When adequate supply of methanol is available, only a single step of methanol dehydration is necessary to produce dimethyl ether using solid acid catalysts, such as γ-alumina, zeolites, and ion-exchange resins in a fixed-bed reactor in the temperature range of 523 K – 673 K and at the pressure up to 10 bar [3]. The main reaction is considered as follows:

$$2CH_3OH \leftrightarrow CH_3OCH_3 + H_2O \qquad (1)$$

In the normal temperature range, no significant side reaction exists. The equilibrium conversion for pure methanol feed can also exceed 92%.

Modeling of packed-bed reactors for methanol dehydration to DME was usually accomplished for many industrial- and pilot-scale reactors. Bercic and Levec investigated the reaction kinetics of catalytic dehydration of methanol to DME [4], and then Shahrokhi & Baghmisheh developed a control system for methanol dehydration in a fixed-bed reactor [5]. In 2014, Ghavipour & Behbahani

[2] To whom any correspondence should be addressed.

IOP Conf. Series: Materials Science and Engineering **316** (2018) 012027 doi:10.1088/1757-899X/316/1/012027

developed a practical kinetic model that would be valid at 593 K [6]. Finally, Dadgar, et al. analyzed the influence of methanol dehydration on methanol synthesis in two-step DME synthesis [7].

Extensive researches on modeling of DME production reactors have been performed to obtain the maximum production rate of DME. None of them considers the integration of reaction, distillation and recycle stream. Through the use of a computer simulation, it is possible to determine the effects of the integration to the DME yield. In this study a model of reaction-distillation-recycle system was developed to predict the increase in the DME yield through methanol dehydration.

2. Methodology

The reaction-distillation-recycle system is shown in figure 1. A one-dimensional phenomenological model of a steady state, isothermal and heterogeneous reactor was developed. The mass transfer in the bed is assumed in axial direction and expressed by

$$v_z \frac{dC_i}{dz} - D_i\left(\frac{d^2C_i}{dz^2}\right) + ka(C_i - C_{i,s}) = 0 \tag{2}$$

The mass transfer in the catalyst particles is assumed in radial direction and described by

$$\frac{1}{r^2} D_{\text{eff}}\left[\frac{d}{dr}\left(r^2 \frac{d(C_{i,s})}{dr}\right)\right] + R_i = 0 \tag{3}$$

where C_i and $C_{i,s}$ are the concentrations of species i in the bed and the catalyst particles, respectively, D_i is the bulk diffusion of species i in the bed, D_{eff} is the effective diffusion of species i in the catalyst pore, k is the gas-solid interphase mass transfer coefficient, a is the specific area of gas-solid interphase, and R_i is the generation or consumption rate of species i.

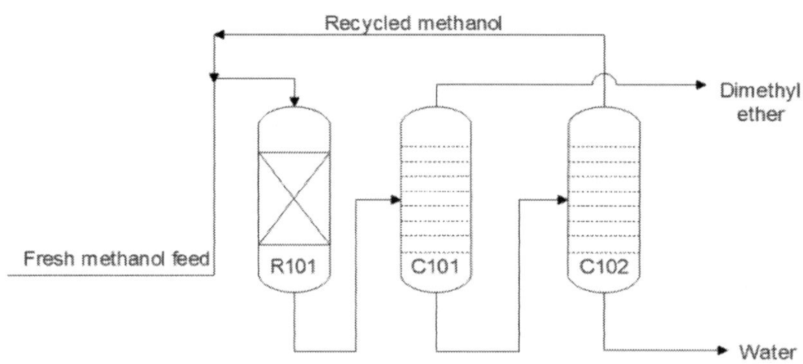

Figure 1. Process flow diagram.

The boundary conditions of those equations are as follows:

At $z = 0$, $C_i = C_{i,\text{inlet}}$

At $z = L$ (the reactor length), $dC_i/dz = 0$

At $r = 0$, $dC_{i,s}/dr = 0$

At $r = R_p$ (the catalyst particle radius), $D_{\text{eff}}(dC_{i,s}/dr) = k(C_i - C_{i,s})$

There have been extensive researches on the appropriate and accurate kinetic models of methanol dehydration. However, as several models sustain major deviation on operating temperature of below 593 K, Ghavipour and Behbahani proposed a new empirical kinetic model [6]. This model considers an equilibrium constant expressed by

$$K_{eq} = 863.2\exp[\frac{1}{8314}\{74{,}870\ln\frac{T}{298} - 761.555(T-298) + 1.127955(T^2-298^2) - 0.001176333(T^3-298^3) + 4.68550\text{x}\,10^{-7}(T^4-298^4) + 4478947026(\frac{1}{T}-\frac{1}{298})\}] \quad (4)$$

The methanol dehydration rate used in this study is

$$-r_M = k_0\exp(-\frac{E_a}{RT})\left(C_{MeOH} - \frac{C_W}{K_{eq}}\right) \quad (5)$$

where k_0 is the pre-exponential factor, E_a is the activation energy, C_{MeOH} denotes the methanol concentration, and C_W denotes the water concentration.

Since the reactor outlet goes to distillation columns to separate the mixture, it is necessary to incorporate a mathematical model for the distillation columns. In this instance, the multicomponent distillation was calculated using a shortcut method. A widely-used approximate method, the Fenske-Underwood-Gilliland Method, is used to estimate the minimum number of trays, the minimum reflux ratio, and the number of stages at a finite reflux.

The derivation of the Fenske's Equation is based on the assumption that the stages are at equilibrium and that no overhead product distillate is withdrawn from the column, resulting in equation (6):

$$N_{min} = \frac{\log\left[\left(\frac{LK_d}{HK_d}\right)\left(\frac{HK_b}{LK_b}\right)\right]}{\log\alpha_{average}} \quad (6)$$

The minimum reflux ratio for a multicomponent distillation means that at a specified reflux ratio, an infinite number of plates is required to achieve the desired separation. The Underwood's Equation works under this assumption so that there are constant molal overflows and constant relative volatilities at an average temperature of the column. This is followed by solving the Underwood's Equation for the variable φ:

$$q(1-F) = \sum_i \frac{\alpha_i n_{i,F}}{\alpha_i - \varphi} \quad (7)$$

When an appropriate value of variable φ is found, it is possible to calculate the minimum reflux ratio, as shown by equation (9):

$$R_m + 1 = \sum_i \frac{\alpha_i x_{i,D}}{\alpha_i - \varphi} \quad (8)$$

Gilliland uses an empirical correlation to calculate a final number of stage N from the values calculated through the Fenske's Equation and the Underwood's Equation. The procedure is quite simple as one simply enters a diagram with a known abscissa value and read the co-ordinate of the corresponding point on the Gilliland curve.

In order to determine the effect of a recycle stream, it is necessary to calculate two recovery fractions between two flowrates. The first fraction, X_1 is between the flowrate exiting the reactor and the that exiting the bottom product of the first distillation column. The second fraction, X_2 is between the flowrate exiting the bottom product of the first distillation column and the that exiting the distillate product of the second distillation column.

The iteration to calculate the recycle stream is as follows:

$$\begin{aligned} R_1 &= (Feed)(X_1)(X_2) \\ R_2 &= (Feed + R_1)(X_1)(X_2) \\ &\vdots \\ R_k &= (Feed + R_{k-1})(X_1)(X_2) \\ &\vdots \\ R_n &= (Feed + R_{n-1})(X_1)(X_2) \end{aligned} \tag{9}$$

The iteration lasts until there is no significant difference between the recycle stream calculated in the k iteration to that in the k-1 iteration.

3. Results and discussion

The simulation was carried out for the reactor length of 4 m, the inlet velocity of 0.408 m/s, the inlet temperature of 533 K and the inlet temperature of 18 atm.

Figure 2 shows the concentration profiles of methanol and dimethyl ether along the reactor without recycle. As exhibited in the figure, the methanol concentration decreases as it is dehydrated into dimethyl ether. The concentration curves that never reach the flat shape indicates that the reactor length is effective under these parameters. The methanol conversion reaches 93% as shown in figure 3. The conversion correlates to the Damkohler number of 11.18. As a rule of thumb, if the Damkohler number is higher than 10, a conversion of more than 90% may be expected. This makes the recycle stream of this process to have low flowrate, since more than 90% of methanol is converted into dimethyl ether. Nonetheless, the reaction occurs at optimum condition and thus can be tweaked for efficiency and sustainability in the future. As depicted in figure 4, the dimethyl ether yield increases along the reactor. At the position of 2 m from the reactor inlet, the dimethyl ether yield is 42% and continues to increase up to 48% at the reactor outlet.

In this study, the Fenske's Equation, the Underwood's Equation, and the Gilliland's Correlation were used to calculate the minimum number of theoretical stages at the reflux ratio of 1.5. Table 1 shows operating parameters of dimethyl ether–methanol separation. It can be observed from the table that the minimum number of the theoretical stage of dimethyl ether–methanol separation is relatively small, i.e. 4. This is due to the fact that dimethyl ether, which is the light-key component, has very high relative volatility to methanol, which is the heavy-key component, and water. The calculated relative volatility of dimethyl ether to methanol at average column temperature of 376 K is 9.1. Therefore, it is very easy for dimethyl ether to separate from methanol and water.

The minimum number of theoretical stages in the second column to separate methanol from water is more than three times larger than that in the first column. This is due to the fact that the boiling point of methanol and water is not that far apart from each other, and thus their relative volatility is not as high as dimethyl ether. The average relative volatility of methanol to water is 2.8, which is lower than that of dimethyl ether to methanol in the first column.

As water was stripped from the mixture in the second distillation column, the top product of the second column was recycled and fed into the reactor as additional feed. Although the addition of this recycle stream to the reactor brought water to the reaction, it should be noted that its fraction was much lower than that of methanol.

The recycle stream was calculated under the assumption that there was no major pressure or temperature losses during the whole process that would affect the conversion. It is possible to calculate the recycle stream in this method since the distillation calculation used is shortcut method, with the assumption that the compositions in the top and bottom products were constant. Therefore, the overall conversion of methanol in the process was also constant. The iterative calculation of the recycle stream could then be calculated and the results are tabulated in table 2.

From table 2 it is clear that the flowrates of all components out of reaction-distillation-recycle system decrease during the iteration course. However, if we observe the mole fractions in the final iteration, we can see that the mole fraction of dimethyl ether increases. This can be attributed to the overall increase of dimethyl ether yield due to methanol recycled to the reactor. This is also supported by the fact that there is an increase in smaller number of water fraction.

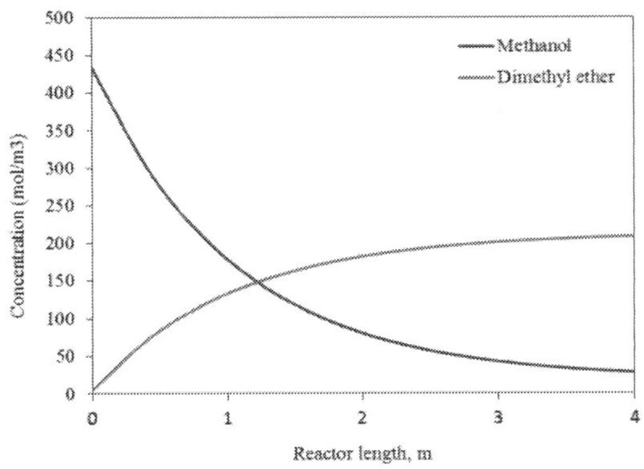

Figure 2. The methanol and dimethyl ether concentrations along the reactor without recycle.

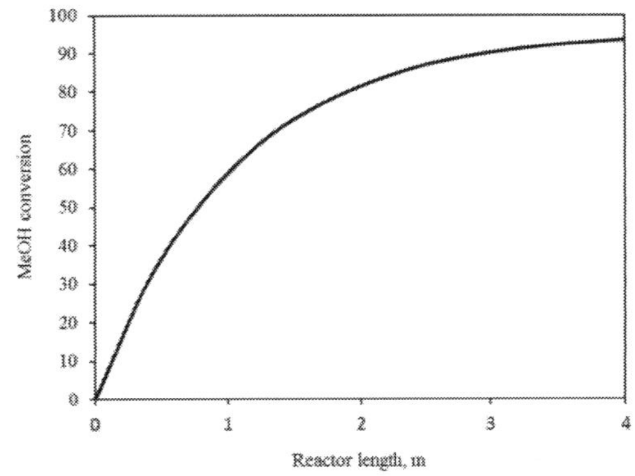

Figure 3. The methanol conversion along the reactor without recycle.

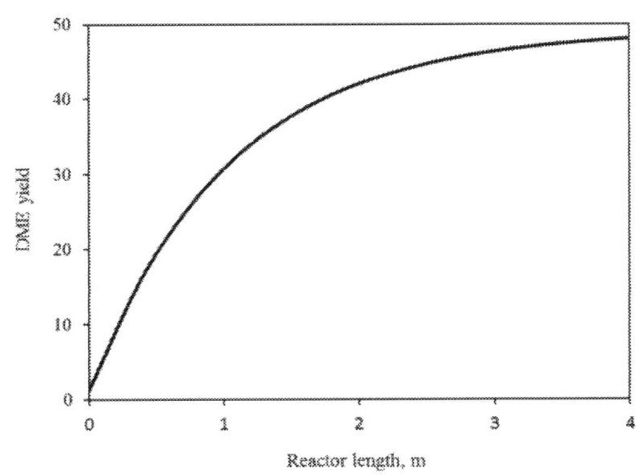

Figure 4. The dimethyl ether yield along the reactor without recycle.

Table 1. Operating parameters for dimethyl ether–methanol separation.

Operating parameters	Dimethyl ether–methanol	Methanol–water
Pressure	8 atm	8 atm
Inlet temperature	533 K	428 K
Top temperature	324 K	403 K
Bottom temperature	428 K	444 K
Reflux ratio	1.7 moles/s	10.5 moles/s
Minimum number of theoretical stages	4	15

The addition of recycle stream into the reactor has several effects on the whole process. At the first calculation, only fresh methanol was fed into the reactor. As the reactor outlet was processed further in the series of distillation columns, the recycle stream obtained as the top product of the second distillation column was then mixed with the fresh methanol. Therefore, the different composition of the feed to the reactor occurred during the process, as shown in Table 3. Although the feed of the reactor becomes less pure, thus decreasing the methanol conversion, it increases the overall yield of dimethyl ether. This is because moles of methanol recycled is higher than that unconverted. Therefore, the addition of small fraction of water in the feed does not affect the production rate to a large extent. Table 4 shows the change in the flowrates exiting the reactor after the iterative calculation.

As shown in Table 4, the integration of the recycle stream increases dimethyl ether yield from 192.58 moles/s to 208.26 moles/s or increases by 8%. This may be a small number but it should be noted that this study operates under the best condition for the reaction. If the operating condition of the reactor can be set such that less energy consumed and the same yield of dimethyl ether produced, then the capital investment of the recycle system would be feasible.

Table 2. Component flowrates and mole fractions in dimethyl ether–methanol separation.

Component	First iteration		Final iteration	
	Flowrate (moles/s)	Fraction	Flowrate (moles/s)	Fraction
Methanol	1.6390	0.7801	1.6170	0.7791
Dimethyl ether	0.3221	0.1534	0.3198	0.1541
Water	0.1397	0.0665	0.1387	0.0668

Table 3. The different composition of the feed to the reactor.

Component	Before recycle	After recycle
Methanol	1	0.9809
Dimethyl ether	0	0.0126
Water	0	0.0066

Table 4. Reactor outlet flowrates.

Component	Before recycle (moles/s)	After recycle (moles/s)
Dimethyl ether	192.58	208.26
Methanol	24.400	28.090
Water	192.58	205.61

4. Conclusion

Modelling of the reaction-distillation-recycle system for methanol dehydration to dimethyl ether has been performed successfully in this research. The addition of recycle stream from the second distillation column to the reactor increases the dimethyl ether yield by 8%.

5. Acknowledgement

We express our gratitude to the University of Indonesia, which funded this research through the scheme of Hibah Publikasi Internasional Terindeks untuk Tugas Akhir Mahasiswa No 863/UN2.R3.1/HKP.05.00/2017.

6. References

[1] Galvita V V, Semin G L, Belyaev V D, Yurieva T M, Sobyanin V A 2001 *Appl. Catal. A* **216** 85
[2] Lee T S, Sung J Y and Park D J 2012 *Fire Safety J.* **49** 62
[3] Liu D, Yao C, Zhang J, Fang D and Chen D 2011 *Fuel* **90** 1738
[4] Bercic G and Levec J 1993 *Ind. Eng. Chem. Res.* **32** 2478
[5] Shahrokhi M and Baghmisheh G R 2005 *Comput. Chem. Eng.* **60** 4275
[6] Ghavipour M and Behbahani R M 2014 J. *Ind. Eng. Chem.* **20** 1942
[7] Dadgar F, Myrstad R, Pfeifer P, Holmen A and Venvik H J 2016 *Catal. Today* **270** 76

IOP Conf. Series: Materials Science and Engineering **316** (2018) 012028 doi:10.1088/1757-899X/316/1/012028

Effect of sodium tripolyphosphate concentration and simulated gastrointestinal fluids on release profile of paracetamol from chitosan microsphere

Kamarza Mulia, Andrie, Elsa A Krisanti

Chemical Engineering Department, Faculty of Engineering
Universitas Indonesia, Depok, 16424, Indonesia

E-mail: kmulia@che.ui.ac.id

Abstract. The problem to overcome in oral drug administration is the significant pH changes present in the human digestive system. In this study, ionotropic gelation method employing 2-8% (w/v) tripolyphosphate solutions were used to crosslink chitosan microspheres for a controlled release of paracetamol as a model drug. The release profiles of paracetamol from chitosan microspheres were determined using simulated gastrointestinal fluids having pH values of 1.2, 6.8, and 7.4. The results showed that the paracetamol loading and the encapsulation efficiency values increased with increasing concentration of tripolyphosphate solutions used in the preparation step. Paracetamol released at pH 1.2 and 6.8 buffer solutions was significantly higher than that at pH 7.4; also, more paracetamol was released in the presence of α-amylase and β-glucosidase enzymes. The release profiles showed zero-order release behaviour up to 8 hours where the highest drug release was 39% of the paracetamol loaded in the chitosan microspheres, indicating a strong crosslinking between chitosan and TPP anions. The relatively low accumulated drug release could be compensated by employing suitable enzymes, lower TPP solution concentration, and addition of other biodegradable polymer to reduce the TPP crosslink.

1. Introduction

Most of the drugs in the clinical phase fail to achieve clinical outcomes due the absence of ability to reach the target site. An effective approach to overcome this failure is by the improvement of controlled drug release systems that release drugs or bioactive compounds at the targeted sites. One of the issues to be solved in is the extreme changes of pH in the human digestive system. Drugs that do not survive the changing pH along its path to reach the targeted site would not produce effective medications [1]. A controlled drug release system consists of three components: a therapeutic agent, a targeting moiety, and, a carrier system. A wide range of materials such as natural or synthetic polymers, lipids, and surfactants have been employed as drug carriers [2]. Among the natural polymers used as drug carriers, chitosan has received significant attention due to its abundant availability, unique mucoadhesivity, biodegradability, non-toxicity, low-immunogenicity, and biocompatibility [3-4]. Chitosan, a linear amino polysaccharide, is obtained by deacetylation of chitin, a natural polysaccharide found in the exoskeleton of crustaceans such as crab and shrimp.

Chitosan beads or microspheres prepared using the complexation reaction between anions and chitosan has been developed by many researchers due to its relatively easy process [5]. This anion complexation, usually called ionotropic gelation method, used a wide variety of anions to crosslink chitosan to produce chitosan-drug matrix in the form of microspheres or nanospheres. Among the anions that can be used in this method, tripolyphosphate (TPP) produces chitosan-drug microspheres with the strongest mechanical resistance, as well as release profile curves approaching the zero-order curve [5]. The non-toxic multivalent tripolyphosphate forms a gel with chitosan due to ionic interactions between the positively charged amino groups on chitosan and a negatively charged TPP anion. This ionic interaction is dependent on the pH of the solution [6].

The aims of this study are to obtain chitosan-drug microspheres with high drug loadings and encapsulation efficiencies, as well as having release profiles suitable for drug release targeted in the digestive system, specifically in the colon area. Digestive system condition is represented by using synthetic gastrointestinal fluids which are buffer solutions with pH values of 1.2, 6.8 and 7.4 to simulate conditions in the stomach, small intestine, and large intestine or colon, respectively. The effect of enzymes existing in the digestive system on drug release profile was also observed by the addition of two enzymes, α-amylase and β-glucosidase, into the buffer solutions. The cumulative release of paracetamol from chitosan microspheres in 24 h duration was determined using UV-Vis spectrophotometry.

2. Experimental

2.1. Materials
Chitosan was obtained from Biotech Surindo (Cirebon, Indonesia) as a medical grade powder with degree of N-deacetylation of 93.6%. Paracetamol with purity >99% was obtained from Mallinckrodt Pharmaceuticals. Sodium tripolyphosphate (STTP) with a purity of 90% was obtained from Bratachem Chemical. KCl-HCl buffer solution (pH 1.2), KH2PO4-NaOH buffer solution (pH 6.8), and KH2PO4-NaOH buffer solution (pH 7.4) were prepared using the standard USP. The enzymes α-amylase and β-glucosidase were obtained from Clarichem Indonesia and Sigma Aldrich, respectively.

2.2. Preparation of chitosan –paracetamol microspheres
Paracetamol-loaded chitosan microspheres were prepared following the procedure used by Yu et al. [7]. A solution of 1% (w/v) acetic acid-chitosan mixture was prepared by dissolving chitosan in 2.5% acetic acid solution (w/v). Then, paracetamol was dissolved in the aqueous chitosan solution with drug to chitosan weight ratio of 1:5. The mixture was homogenized at 400 rpm for 15 min. various amounts of STPP were dissolved in 100 ml of distilled water to form 2, 4, 6 and 8% (w/v) solutions. The drug-chitosan mixture was injected and slowly dripped into the STPP solution as it was homogenized at 400 rpm. After all of chitosan mixture has been added into STPP solution, it was left for 30 min to let TPP anions crosslink the chitosan beads. The produced solid particles were filtered, washed repeatedly, and freeze dried. The filtered solution was analysed for the amount of residual drug to determine the total drug loading and the encapsulation efficiency. The dried particles were ground to form TPP-chitosan-drug microspheres and the resulting microspheres were sieved through a wire mesh screen.

2.3. Preparation of gastrointestinal synthetic fluids
The solutions obtained from microsphere filtration and washing was diluted 100 times and the absorbance of paracetamol was measured using a UV spectrophotometer at 243 nm. Using the standard curve, the amount of paracetamol encapsulated can be calculated based on the drug concentration before and after microsphere formation. The following equations are used to calculate percentage of drug loading and encapsulation efficiency:

$$\%\text{ drug loading}=\frac{\text{mass of encapsulated drug}}{\text{mass of microspheres}}\times 100\%$$

$$\%\text{ encapsulation efficiency}=\frac{\text{mass of encapsulated drug}}{\text{mass of initially used drug}}\text{ x } 100\%$$

2.4. Release profile determination
Batches of accurately weighed microspheres, approximately 25 mg each, were placed into test tubes. Synthetic gastrointestinal fluids prepared earlier were added into each test tube. Samples were taken from the release solution every hour until 8 h and the last sample was taken after 24 h. After each sampling, the same volume of fresh solution was added to replace the sample taken. Drug presence in each sample taken from the synthetic fluids was analysed using UV/VIS Spectrophotometer at 243 nm.

3. Results and discussion

3.1. Particle size
Micrographs of the TPP-chitosan-drug particles confirmed that microspheres having diameter less than 100 μm were obtained.

3.2. Drug loading and encapsulation efficiency
Microsphere of chitosan-paracetamol is formed due to crosslinking that takes place during its preparation. The more crosslinking occurred, the stronger the microstructure of chitosan-drug formed. It is noted that not all paracetamol, which was previously interacting with chitosan by hydrogen bonding, was entrapped into the chitosan-tripolyphosphate matrix. Varying TPP concentration used during microsphere preparation resulted in drug loadings and encapsulation efficiencies given in Table 1.

Table 1. Paracetamol loading and encapsulation efficiency in chitosan microspheres prepared using various TPP concentrations.

TPP concentration (%, w/v)	Encapsulation efficiency (%)	Paracetamol loading (%)
2	30.3	4.5
4	43.5	3.0
6	42.4	6.2
8	50.1	7.8

The data given in Table 1 indicated that increased TPP concentrations correspond to the increased value of drug loading and encapsulation efficiency. The ionic crosslinking caused by TPP anions increased the strength of the microsphere networks and more paracetamol is entrapped within the microspheres. The large amounts of microspheres formed using the 4%-w/v TPP solution caused a decline in the drug loading.

3.3. In-vitro drug release
The release profile of paracetamol was obtained in simulated gastrointestinal fluids which were buffered solutions of pH 1.2 (SGF), 6.8 (SIF), and, 7.4 (SCF). The release experiments were repeated using buffer pH 6.8 containing α-amylase and buffer pH 7.4 containing β-glucosidase to see the effect of enzymes on chitosan degradation. Figure 1 shows that the paracetamol release profile in simulated gastric fluid is increasing with time, up to 8 h. The highest amount of drug released decreases with increasing TPP solution concentration and up to 30% paracetamol was released from microspheres prepared using the 2% TPP solution. In ionically crosslinked TPP-chitosan microspheres, the pH of the release fluids affects the crosslinking density, degree of swelling, and release of paracetamol. If the pH decreases, the protonation of the amino groups of chitosan leads to chain repulsion and polymer network swelling. This explains the relatively high drug release in acidic media of pH 1.2.

IOP Conf. Series: Materials Science and Engineering **316** (2018) 012028 doi:10.1088/1757-899X/316/1/012028

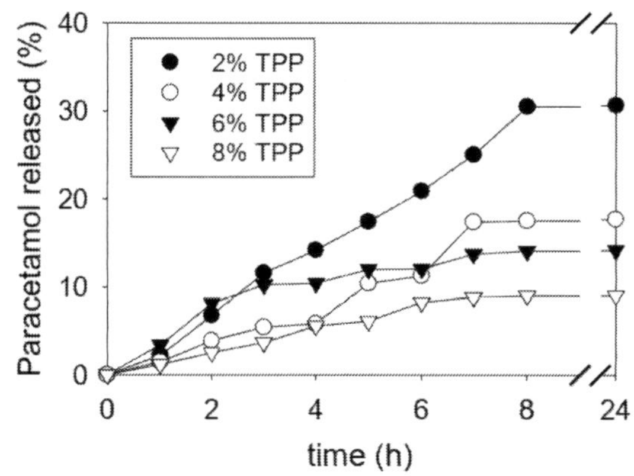

Figure 1. Paracetamol released in the simulated gastric fluid of pH 1.2.

Paracetamol profile release from chitosan microspheres in simulated intestinal fluid is shown Figure 2a for release without enzyme and Figure 2b for containing α-amylase enzyme (b). The release profiles in SIF show similar trends as those in SGF, except that the amount of paracetamol released is much lower. This result indicated that in buffer pH 7.4 chitosan microspheres were only slightly swelled due to electrostatic repulsion between the ionized groups in chitosan chain.

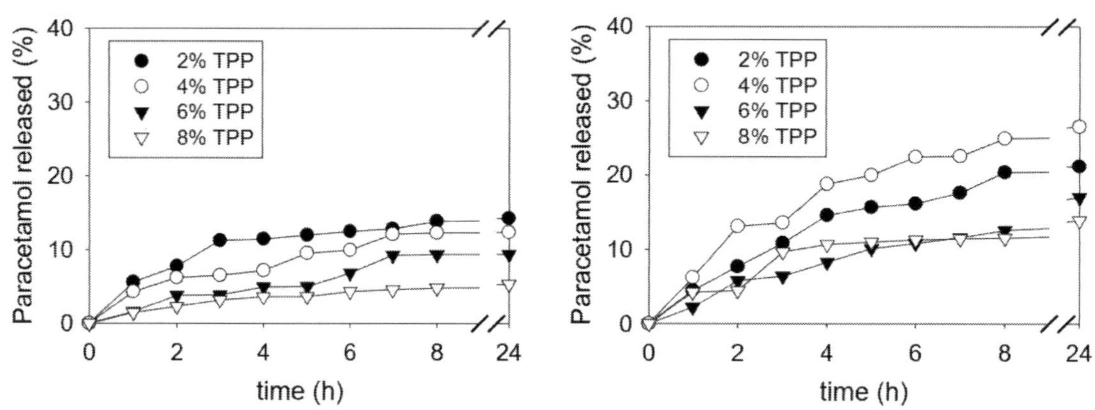

Figure 2. Paracetamol released in the simulated intestinal fluid of pH 7.4 without enzyme (a) and with α-amylase enzyme (b).

The addition of α-amylase enhanced the release of paracetamol up to twice the amount released without added enzyme. The enzyme was found to be effective to hydrolyse β-1,4 linkages in chitosan in pH of 7.4 although it was reported to be most effective in pH of 5.0 [8-9]. It is noted that the presence of α-amylase changed the position of the release profiles in Figure 2 (b) relative to those in Figure 2 (a).

Paracetamol release profiles obtained in simulated colonic fluid shown in Figure 3 (a) closely matched those shown in Figure 1 for release in SGF where up to 30% of the paracetamol in the microspheres prepared with the 2% TPP solution were released.

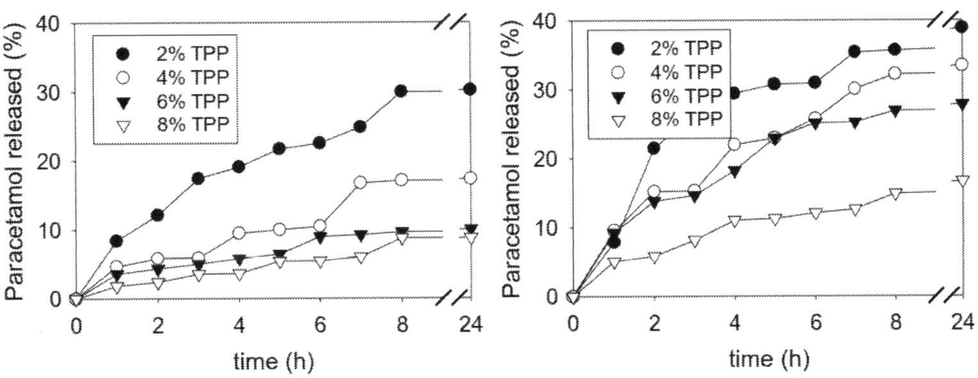

Figure 3. Paracetamol released in the simulated colonic fluid of pH 6.8 without enzyme (a) and with β-glucosidase enzyme.

Figure 3 (b) shows paracetamol release from microspheres immersed in SCF containing β-glucosidase enzyme. In the presence of this enzyme, the maximum drug release of 39% was observed in this study. This observation indicates that β-glucosidase, which pH optimum is 5.0 [10], is still active accelerating the scission of β-glucosidic link in chitosan at pH 6.8. The paracetamol release profiles in SCF without β-glucosidase is similar to those in SGF, and they are significantly higher than those in SIF without α-amylase. The difference is due to the pH of the release media where an acidic media facilitates swelling and more extensive drug release. It is noted that while pH of the SCF and the SIF were 6.8 and 7.4, respectively; the difference of only 0.6 units was sufficient to double the amount of drug released in the lower pH media.

In general, the release profiles obtained in this study show zero-order release behavior up to about 8 h. Afterwards, the release profiles become flat up to 24 h, indicating that no more drug is released during this period. The maximum paracetamol release observed in this study was only 39% of the amount loaded in the microspheres and the rest of the drug was entrapped in the partly degraded chitosan micro particles. The incomplete drug release indicated that strong crosslinking between chitosan and TPP anions were formed. To increase the percentage drug release, an additional polymer whose hydrophilicity is different from chitosan could be incorporated into the microspheres to perturb the ionic crosslinking between chitosan chains, decreasing the crosslinking density and making available more protonable amino groups [11].

4. Conclusions

The results indicated that increasing concentration tripolyphosphate solutions increased the loading of paracetamol in chitosan microspheres, while the least extensive crosslinks formed at lower concentrations of tripolyphosphate facilitated more drugs to be released. The release profiles showed zero-order release behaviour up to 8 hours where the highest drug release was 39% of the paracetamol loaded in the chitosan microspheres, indicating a strong crosslinking between chitosan and TPP anions. The relatively low accumulated drug release could be compensated by employing suitable enzymes, lower TPP solution concentration, and others biodegradable polymer to reduce the crosslink due to TPP. In conclusion, this study indicated that the chitosan microspheres prepared by this method can be developed for preparing controlled drug release formulation.

Acknowledgment

The authors are grateful for financial support from the DRPM Universitas Indonesia through the PITTA Grant 2017.

This article's publication is supported by the United States Agency for International Development (USAID) through the Sustainable Higher Education Research Alliance

(SHERA) Program for Universitas Indonesia's Scientific Modeling, Application, Research and Training for City-centered Innovation and Technology (SMART CITY) Project, Grant #AID-497-A-1600004, Sub Grant #IIE-00000078-UI-1.

References

[1] Wang X-Q and Zhang Q 2012 pH-sensitive polymeric nanoparticles to improve oral bioavailability of peptide/protein drugs and poorly water-soluble drugs *Eur. J. Pharm. Biopharm.* **82** 219–29
[2] Patel M P, Patel R R and Patel J K 2010 Chitosan mediated targeted drug delivery system: a review. *J. Pharm. Pharm. Sci.* **13** 536–57
[3] Sinha V ., Singla A ., Wadhawan S, Kaushik R, Kumria R, Bansal K and Dhawan S 2004 Chitosan microspheres as a potential carrier for drugs *Int. J. Pharm.* **274** 1–33
[4] Kumar M N V R, Muzzarelli R A A, Muzzarelli C, Sashiwa H and Domb A J 2004 Chitosan Chemistry and Pharmaceutical Perspectives *Chem. Rev.* **104** 6017–84
[5] Shu X Z and Zhu K J 2002 Controlled drug release properties of ionically cross-linked chitosan beads: the influence of anion structure. *Int. J. Pharm.* **233** 217–25
[6] Ko J A, Park H J, Hwang S J, Park J B and Lee J S 2002 Preparation and characterization of chitosan microparticles intended for controlled drug delivery. *Int. J. Pharm.* **249** 165–74
[7] Yu C Y, Yin B C, Zhang W, Cheng S X, Zhang X Z and Zhuo R X 2009 Composite microparticle drug delivery systems based on chitosan, alginate and pectin with improved pH-sensitive drug release property *Colloids Surfaces B Biointerfaces* **68** 245–9
[8] Pan S, Wu S and Kim J 2011 Preparation of glucosamine by hydrolysis of chitosan with commercial α-amylase and glucoamylase *J. Zhejiang Univ. Sci. B* **12** 931–4
[9] Wu S 2011 Preparation of water soluble chitosan by hydrolysis with commercial α-amylase containing chitosanase activity *Food Chem.* **128** 769–72
[10] Zhang H and Neau S H 2001 In vitro degradation of chitosan by a commercial enzyme preparation: effect of molecular weight and degree of deacetylation. *Biomaterials* **22** 1653–8
[11] Yao K De, Peng T, Feng H B and He Y Y 1994 Swelling kinetics and release characteristic of crosslinked chitosan: Polyether polymer network (semi-IPN) hydrogels *J. Polym. Sci. Part A Polym. Chem.* **32** 1213–23

IOP Conf. Series: Materials Science and Engineering **316** (2018) 012029 doi:10.1088/1757-899X/316/1/012029

Textile Dye Removal from Aqueous Solution using Modified Graphite Waste/Lanthanum/Chitosan Composite

E Kusrini[1], B Wicaksono[1], Y Yulizar[2], EA Prasetyanto[3], C Gunawan[4]

[1]Department of Chemical Engineering, Faculty of Engineering, Universitas Indonesia, Kampus Baru UI Depok, 16424, Indonesia
[2]Department of Chemistry, Faculty of Mathematics and Natural Sciences, Universitas Indonesia, Kampus Baru UI Depok, 16424, Indonesia
[3]Faculty of Medicine, Atma Jaya Catholic University of Indonesia, Jakarta, 14440 Indonesia
[4]ithree Institute, University of Technology Sydney, Sydney, NSW 2007, Australia

E-mail: ekusrini@che.ui.ac.id

Abstract. We investigated various pre-treatment processes of graphite waste using thermal, mechanical and chemical methods. The aim of this work is to study the performance of modified graphite waste/lanthanum/chitosan composite (MG) as adsorbent for textile dye removal from aqueous solution. Effect of graphite waste resources, adsorbent size and lanthanum concentration on the dye removal were studied in batch experiments. Selectivity of MG was also investigated. Pre-heated graphite waste (NMG) was conducted at 80°C for 1 h, followed by mechanical crushing of the resultant graphite to 75 μm particle size, giving adsorption performance of ~58%, ~67%, ~93% and ~98% of the model dye rhodamine B (concentration determined by UV-vis spectroscopy at 554 nm), methyl orange (464 nm), methylene blue (664 nm) and methyl violet (580 nm), respectively from aqueous solution. For this process, the system required less than ~5 min for adsorbent material to be completely saturated with the adsorbate. Further chemical modification of the pre-treated graphite waste (MG) with lanthanum (0.01 – 0.03 M) and chitosan (0.5% w/w) did not improve the performance of dye adsorption. Under comparable experimental conditions, as those of the 'thermal-mechanical-pre-treated-only' (NMG), modification of graphite waste (MG) with 0.03 M lanthanum and 0.5% w/w chitosan resulted in ~14%, ~47%, ~72% and ~85% adsorption of rhodamine B, methyl orange, methylene blue and methyl violet, respectively. Selective adsorption of methylene blue at most to ~79%, followed by methyl orange, methyl violet and rhodamine B with adsorption efficiency ~67, ~38, and ~9% sequentially using MG with 0.03 M lanthanum and 0.5% w/w chitosan.

1. Introduction

Availability of graphite waste from ex-electrolysis process is very abundant with availability of 1,000 ton per month and having very low price [1]. So far, the graphite electrode waste is only used as filler in steel manufacturing. It is widely known that the main compound in graphite electrode waste is carbon and has the ability to adsorb some dye, heavy metal ions and CO_2 [2]. From our previous research, these graphite waste contained the carbon is about 80% that can be utilized as sorbent [2].

In textile industry, the synthetic dyes that commonly used are rhodamine B, methyl orange, methylene blue (MB) and methyl violet (MV) and so on. These synthetic dyes are organic compounds that are not easily biodegraded by nature and could be easily deposited in the body. Both methylene blue and methyl violet are cationic dye and toxic [3]. Textile dyes deposited *in vivo* will accumulate in the food chain and may harm to human body [3].

IOP Conf. Series: Materials Science and Engineering **316** (2018) 012029 doi:10.1088/1757-899X/316/1/012029

In this study, we investigated various pre-treatment processes – thermal, mechanical and chemical treatments, for the utilization of graphite waste as sorbent for the selected of dye models in textile wastewater. Chitosan and lanthanum were selected as modifier to increase the adsorption capacity of graphite waste. Chitosan is used as adsorbent because of chitosan has amine group (NH_2) [4] which is able to bind chemically with the dye textiles. Besides, chitosan is a biodegradable material [5,6]. Lanthanum is used to help the adsorption of anion because lanthanum is cationic and contains empty orbitals that can bind the dye molecules [7]. Herein, we study on the preparation of modified graphite waste/lanthanum/chitosan composite as adsorbent for textile dye removal from aqueous solution. The aim of this work is to evaluate the adsorption efficiency of the selected dye models, rhodamine B, methyl orange, methyl violet, and methylene blue, from aqueous solution onto the prepared adsorbents. Effect of graphite waste resources, adsorbent size and lanthanum concentration on the dye removal were studied in batch experiments. Selectivity of modified graphite waste/lanthanum/chitosan composite (MG) was also investigated.

2. Method

2.1. Modification of adsorbent. Graphite waste was oven-dried at a temperature of 80°C for 1 h. Graphite waste that was dried then mashed with a grinder and filtered with a filter size of 150 microns. These steps are repeated with variations in the size of 100 and 75 μm. Graphite (5 g) was immersed and soaked in 50 mL of 0.01 M La^{3+} for 1 h. Results filtered with filter paper soaked cake then oven-dried at a temperature of 80°C for 2 hours. The dried cake weighed and then placed in an oven at 80°C for 1 hour on a regular basis to make sure the cake does not contain any water at all. The solids graphite re-soaked in a solution of 0.5% chitosan in 1% acetic acid solution for 3 h. After filtration, soaked cake then oven-dried at a temperature of 80°C for 2 h. The dried cake weighed and then dried again in an oven at 80°C for 1 hour on a regular basis to make sure the cake does not contain any water at all. All method was repeated using the La^{3+} ions concentration from 0.02 and 0.03 M.

2.2. Characterization. Nitrogen adsorption measurement was conducted at 77.2 K (-195.8°C), using an ASAP 2020 V4.02 unit gas adsorption analyzer, at an equilibration interval of 5 sec, with a sample mass from 0.28 to 0.30 g. The surface area of the adsorbent was determined using the Brunauer, Emmet, and Teller (BET) equation, and the microspore volume of the adsorbent was measured using the t-plot method. The pore size distribution was calculated according to the Barrett, Joyner, and Halenda (BJH) model, and the surface morphology was determined by field emission scanning electron microscopy and energy dispersive x-ray (FESEM-EDX) analysis.

2.3. Batch adsorption. 0.1 g of sorbent was soaked in 20 mL of dye with a concentration of 10 ppm for 5 minutes. The resulting mixture is filtered and the filtrate was taken to measure its concentration. The amount adsorbed of dye in filtrate was measured by using UV-Vis spectroscopy at wavelengths of 464, 554, 580, and 664 nm. The process is repeated using a textile dye concentrations of 20, 30, 40, and 50 ppm. Mixture of textile dye solution (rhodamine B, methyl orange, methyl violet, and methylene blue) is made with a concentration of 10 mg/L by the way each one mg dye dissolved into 100 mL of distilled water. 20 mL of the selected dye model is taken and mixed with 0.1 g of sorbent for 5 minutes. The resulting mixture was filtered and the filtrate was taken to measure its concentration. The absorption of filtrate was analyzed using UV-Vis spectroscopy with wavelength (λ) at 464, 554, 580, and 664 nm.

$$Q_e = \frac{V(C_0 - C_E)}{W} \quad (1)$$

The value of C_0 is the initial concentration taken from the initial concentration before the sorbent was soaked in textile dyes solution. C_E obtained from UV-Vis spectroscopy measurements were compared with a calibration curve. V is the volume of solution used. W is the mass of adsorbent used.

3. Results and Discussion

3.1. Characteristics adsorbent. The morphology of non-modified graphite (NMG), modified graphite (MG), and modified graphite after adsorbs dyes (MGA) were observed with FESEM-EDX (see Figure 1A-C).

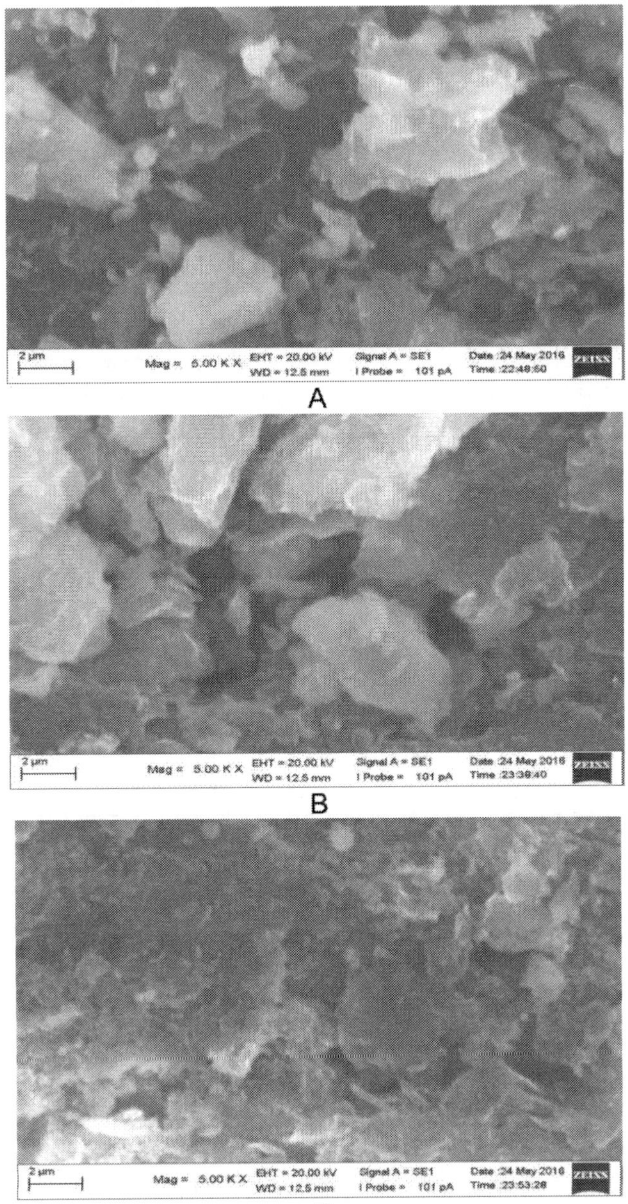

Figure 1. FESEM images of (A) non modified-graphite (NMG) with 200 mesh (75µm), (B) Modified graphite (MG) with lanthanum concentration of 0.03 M, and (C) Modified graphite after adsorption (MGA) with magnification of 5000x

Figure 1(a) shows that the NMG has uneven surface with small crytalline impurities. The NMG has pores ranging between 1 - 4 μm, with impurities crystal size ranges between 2 - 5 μm. After modification, the graphite surface shows larger size crystals with a lighter colour (Figure 1(b)) compared to Figure 1(a). This large crystals are lanthanum nitrate crystals with a size of 3 - 6 μm, while the bright side shows chitosan that blanketed graphite and lanthanum in which the composition can be seen in MG has a pore diameter of 1.5 - 5 μm. Figure 1(c) shows the surface of MGA becomes dark and some possible dyes aggregates appear. After carefully examining the pore diameter, pore diameter graphite after adsorbs reduced to range from 0.5 to 2.5 μm, which indicates pore filling by bye molecules.

The elemental composition of sorbents was sumarized in Table 1. The carbon (C) content of NMG and MG slightly decreased because the addition of lanthanum and chitosan modification, and shows the lanthanum ions and chitosan were covering the graphite waste. Lanthanum was detected in MG and MGA indicating lanthanum is attached on the surface of graphite waste. After adsorbtion of the dyes, the composition of carbon in MGA increase due to molecularly adsorbed dyes. Fe content was not observed in both MG and MGA. Very little content of Si was observed in both NMG and MG. It is similar observed for non-modified graphite waste as previously reported [2].

Table 1. EDX composition of graphite waste before and after adsorption process

Component (wt %)	NMG	MG	MGA
C	81.25	75.79	83.09
O	6.43	3.97	4.41
F	8.05	12.78	7.19
Na	1.91	1.52	0.48
Al	0.67	0.99	0.55
Si	0.12	0.06	-
S	0.76	0.08	0.23
Ca	0.62	0.65	0.86
Fe	0.20	-	-
La	0.0	0.3	0.3

In accordance with Table 2, physical properties of MG such as surface area, pore volume and pore size are larger than the NMG. This properties enhancement increases possible contact of dye molecules with graphite. However, there is also a possibility that the concentration of lanthanum is used is still less optimum for improving dye adsorbs chemical substances. The results of the BET surface area calculations suggests that the graphite waste modified by lanthanum and chitosan has larger surface area (up to 84%).

Table 2. The physicochemical properties of graphite waste before and after modification

Parameter	NMG	MG
BET Surface area (m^2/g)	5.9	10.8
BJH adsorption cumulative surface area of pores (m^2/g)	3.5	9.0
t-Plot microspore area (m^2/g)	2.1	0.5
Microspore volume (cm^3/g)	1.1×10^{-3}	1.3×10^{-4}

BJH adsorption cumulative volume of pores (m^2/g)	1.6×10^{-2}	4.2×10^{-2}
Adsorption average pore (nm)	11.9	16.1
BJH adsorption average pores diameter (nm)	18.5	18.7
Average nanoparticle size (nm)	1025.1	556.7

3.2. Effect of Conditions in Adsorption of Textile Dyes

3.2.1. Effect of Graphite Waste Resources. At the same condition treatment of graphite waste (cathode, anode and a mixture of sequential), adsorption ability of textile dyes of sorbents are shown in Figure 2. The types of graphite waste sources are used in this research that influenced the ability to adsorbs textile dyes. This is caused by the type of the structure possessed by each graphite with different percentage amount of impurities. In general, non-modified graphite (NMG) waste from cathode has better performance compared to other sources of graphite waste.

3.2.2. Effect of Adsorbent size. At the same condition treatment of graphite waste non modification (NMG) from three source (cathode, anode and mixed) with different particle sizes from 75 to 150 μm in sequence, adsorption of textiles dyes of adsorbent are shown in Figure 2. It is caused by the surface area increases as the particle size gets smaller. However, graphite waste from anode shows some anomaly where 100 μm particle size gave better adsorption than 75 μm.

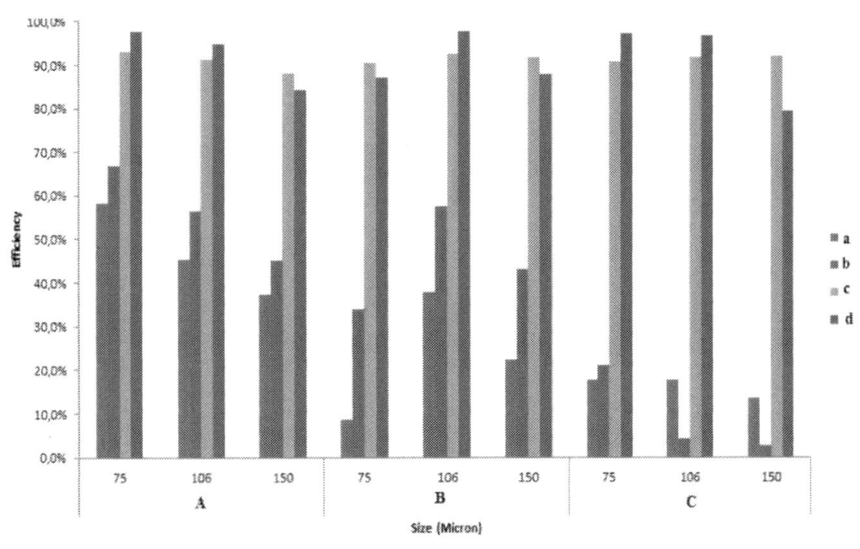

Figure 2. Adsorption efficiency of NMG with three graphite waste sources for (A) cathode, (B) anode and (C) mixture, where a = rhodamine B, b = methyl orange, c = methylene blue, and d = methyl violet,

3.2.3. Effect of lanthanum concentration on dye removal. At the same condition treatment of graphite anode, cathode, and a mixture of sequential, adsorption of textiles dyes of adsorbent dyes are shown in Figure 3. Figure 3 illustrates the adsorption of MG with varying concentrations of lanthanum and 0.5% chitosan and its comparison to a prior modified. Researcher using modified graphite (MG) with chitosan 0.5% as control (lanthanum concentrations of 0 M). Figure 3 also shows that the concentration of lanthanum is used to modify the graphite waste into the adsorbent effect on adsorption ability of the adsorbent. If the concentration of lanthanum increases the adsorption ability, this is in accordance with the initial hypothesis which lanthanum provides an empty orbital space to be chemically bound to the textile dye that acts as a ligand complexes on lanthanum.

The best results in this study were at concentrations of lanthanum of 0.03 M. However, as we can see in Figure 3, the performance decreases when the concentration of 0.02 M and then its adsorption ability climbed back in the concentrations of 0.03 M. This anomaly occurs because of the concentration of 0.02 M is the minimum concentration to indicate chemical adsorption adsorbent works well and balanced with physical adsorption adsorbent.

Adsorption capability of the entire MG when compared to NMG decreased. It can be caused by the presence of competition space textile dye adsorption by chitosan used in the modification. Both types of these compounds are organic substances that having empty electron, which found to be capable on binding to the ligand and lanthanum as the van der Waals with graphite waste, but it has possibility that the concentration of lanthanum was too small, thus, there is no positive change. It can be concluded chitosan does not have any positive impact on the ability of the textile dye adsorbs.

MG sorbents have different capabilities for different textile dyes. This is caused by the structure of the textile dyes used in this study. MG sorbents are more binding on methyl violet and methylene blue when compared with rhodamine B and methyl orange. This is because methyl violet and methyl orange contains the same properties, which are a cationic dye, while methyl orange dye which is anionic. Rhodamine B is also a cationic dye such as methyl violet and methylene blue, rhodamine B, but the particle size is greater than the methyl violet and methylene blue because it has a molecular mass that is much larger. Both of these causes leads to adsorption on methyl violet and methylene blue amounted to 84.9% and ~72%, respectively, compared to rhodamine B and methyl orange by ~14% and ~47%.

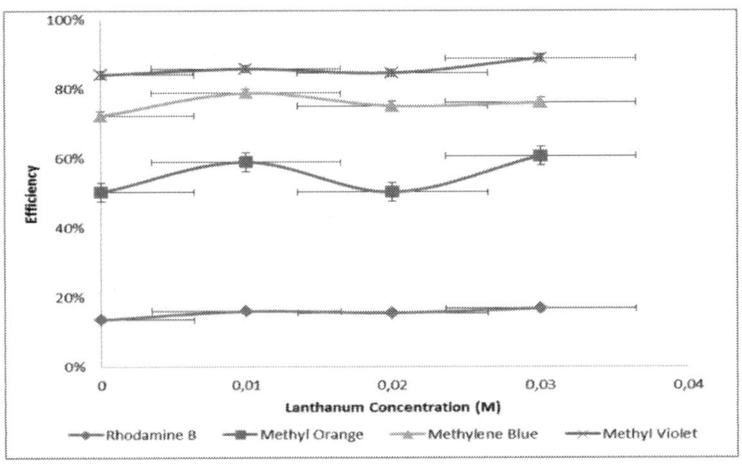

Figure 3. Effect of lanthanum concentration in modified graphite (MG) on dye removal in textile wastewater

3.3. Selectivity. The final concentration of each dye in Figure 4 illustrates the adsorbent selectivity against multiple types of dyes. The adsorbent has a selective adsorption of methylene blue at most to ~79%, followed by methyl orange, methyl violet and rhodamine B with adsorption ~67, ~38, and ~9% sequentially using MG with 0.03 M lanthanum and 0.5% w/w chitosan. The adsorption efficiency of methylene blue (MB) and methyl violet (MV) using modifed graphite (MG) are lower that found using nanocomposite of hydrolyzed polyacrylamide grafted onto xanthan gum and incorporated-nanosilica had efficiency of 99.4 – 81.6% and 99.1 - 87.7%, respectively [3]. In the explanation for the modified graphite (MG) adsorption test and according to the literature review, the type of content and particle size affects the ability of the dye adsorbs an adsorbent. In this test of selectivity, adsorbent found to be more compatible with dye, which has a mass of dye molecules that are smaller or smallest particle size of the dye. Sorbent prefer this type of cationic textile dyes because the surface of the adsorbent is negative due to the influence of the hydroxyl (OH) groups of chitosan is more dominant in neutral or alkaline pH. As we know that, the adsorption efficiency for removal dye from aqueous solution depend on the adsorbent dose, initial concentration of dye, pH, particle size of sorbent, type of sorbent, the chemical modifier of sorbent as well as temperature of the solution. It is similar was observed for removal of congo red dye from aqueous solution using modified xanthan gum/silica hybrid nanocomposite [8] and activated carbon/cobalt ferrite/alginate composite beads [9].

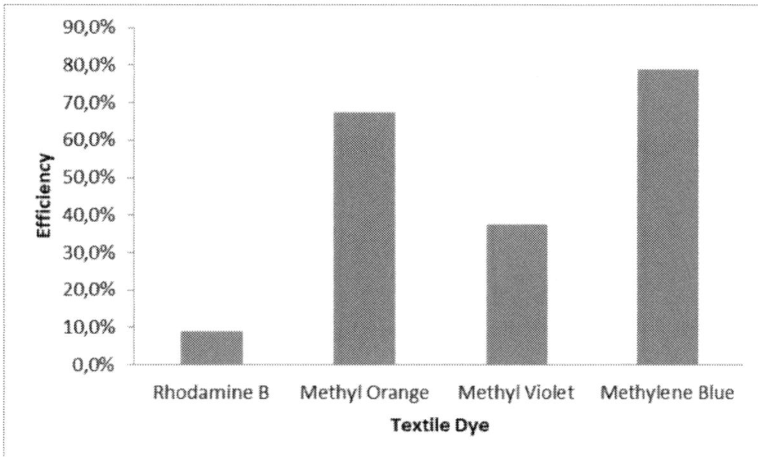

Figure 4. Selectivity of modified graphite (MG) with 0.03 M lanthanum and 0.5% w/w chitosan on textile dye removal

4. Conclusion

In conclusion, the thermal-mechanical pre-treatment of graphite waste (NMG) features the potential as the much-sought effective yet low cost technology for removal of dyes in textile wastewater. Further chemical modification of the pre-treated graphite waste with La (0.01 – 0.03 M) and chitosan (0.5% w/w) did not improve the extent of dye adsorption.

5. Acknowledgments

The authors greatly acknowledge the Universitas Indonesia for providing financial support through a grant, Hibah PITTA No. 774/UN2.R3.1/HKP.05.00/2017.
This article's publication is supported by the United States Agency for International Development (USAID) through the Sustainable Higher Education Research Alliance (SHERA) Program for Universitas Indonesia's Scientific Modeling, Application, Research and Training for City-centered Innovation and Technology (SMART CITY) Project, Grant #AID-497-A-1600004, Sub Grant #IIE-00000078-UI-1.

6. References

[1] Kusrini E, Utami CH, Nasruddin, Prasetyanto EA, Bawono AA, 2017. *IOP Conference Series: Materials Science and Engineering*, Accepted.
[2] Kusrini E, Sasongko AK, Nasruddin, Usman A, 2017. International Journal of Technology 8, 1436-1444.
[3] Ghorai S, Sarkar A, Raoufi M, Panda AB, Schönherr H, Pal S, 2014. *Appl Mater & Interfaces*, *6* (7), 4766–4777.
[4] Kusrini E, Sofyan N, Suwartha N, Yesya G, Priadi CR, 2015. J Rare Earths 33, 1104 – 1113.
[5] Pillai C, Paul W, Sharma C, 2009. *Progress In Polymer Science*, 641-678.

[6] Rinaudo M, 2006. Chitin and Chitosan: Properties and Applications. *Progress In Polymer Science*, 603-632.
[7] Alvarez E, Jimenez L, Ruiz L, Flores P, Mendez J, 2015. *J Colloid Interface Sci*, 194-202.
[8] Ghorai S, Sarkar AK, Panda AB, Pal S, 2013. Bioresour Technol 144, 485-491.
[9] Ai L, Li M, Li L, 2011. *J Chem Eng Data 56* (8), 3475–3483.

Preparation and Characterization of Graphite Waste/CeO_2 Composites

E Kusrini[1*], C S Utami[1], Nasruddin[2], E A Prasetyanto[3], Aji A Bawono[1]

[1]Department of Chemical Engineering, Faculty of Engineering, Universitas Indonesia, Depok, 16424, Indonesia
[2]Department of Mechanical Engineering, Faculty of Engineering, Universitas Indonesia, Depok, 16424, Indonesia
[3]Faculty of Medicine, Atma Jaya Catholic University of Indonesia, Jakarta, 14440 Indonesia

*E-mail: ekusrini@che.ui.ac.id

Abstract. In this research, the chemical modification of graphite waste with CeO_2 was developed and characterized. Graphite waste was pretreated with mechanical to obtain the size 200 mesh (75 μm), and thermal methods at 110°C oven for 6 hours. Here, we demonstrate final properties of graphite before modification (GBM), activated graphite (GA) and graphite/CeO_2 composite with variation of 0.5, 1 and 2 g of CeO_2 (G0.5; G1; G2). The effect of CeO_2 concentration was observed. The presence of cerium in modified graphite samples (G0.5; G1; G2) were analyzed using SEM-EDX. The results show that the best surface area was found in G2 is 26.82 m^2/g. The presence of CeO_2 onto graphite surface does not significantly increase the surface area of composites.

Keywords: Adsorption; Cerium; Graphite waste; Modification

1. Introduction

Adsorption is one of the most widely used processes in industry [1]. The adsorption process is widely used for gas separation processes, removal of pollutants in water, purification of solvents, and as catalysts. Activated carbon is known as the most widely used adsorbent because of its large surface area so that the adsorption ability is good [2,3]. High adsorption capacity of activated carbon is usually associated with high surface area, pore volume, and porosity [1,4].

Although commercial activated carbon is widely used, its use is limited due to high costs [1]. The high initial cost and the need for an expensive regeneration system make it less economical as an adsorbent. The actual cost is an important parameter for comparing the adsorbent material. Adsorbents can be considered as inexpensive if it requires less processing and is abundant in nature. Taking into account these criteria, research on the non-conventional alternatives of adsorbents have been widely conducted. Natural materials, industrial waste materials, agricultural waste and biosorbents can be used as cheap adsorbents for many applications.

The graphite waste is very abundant with availability of 1,000 ton per month and having very low price. From our previous research, these graphite waste samples contained the percentage of carbon is about 75 - 83% that can be utilized as adsorbent with modification with lanthanum for removal of dye [5]. The pinned graphite structure makes it applicable as an adsorbent. However, graphite still has low performance as an adsorbent due to its limited surface area. Thus, the recent research has focused on improving sorbent and expanding efficient system for cost-effective for CO_2 capture [1,3]. Therefore, more efficient sorbent for CO_2 capture is very important to be produced.

Cerium oxide has been the subject of thorough investigations, and has attracted much attention especially in the field of catalysts and catalytic supporters [6]. CeO_2 has attracted a lot of attention because of its excellent nature and has been widely applied in recent years. CeO_2 can provide favorable catalytic activity and provide a large area of chemical surface absorption and improve reducibility [7]. By adding CeO_2, the graphite waste contains more Ce^{3+} ions that provide redox-active sites. CeO_2 can provide favorable catalytic activity and provide a large area of chemical surface absorption [8]. Therefore, in this research, we modified the graphite waste using $Ce(NO_3)_2.6H_2O$ with weight variations of 0.5, 1 and 2 g to increase the BET surface area. The effect of $Ce(NO_3)_3.6H_2O$ concentrations on BET surface area of prepared graphite waste/CeO_2 composites were investigated.

2. Materials and Methods

2.1. Materials

Graphite was obtained from waste of ex-electrolysis cathode graphite from Medan (North Sumatra, Indonesia) which was received in powder form (GBM) and was treated by HNO_3 (65%). Cerium oxide was synthesized over graphite via co-precipitation method using $Ce(NO_3)_2.6H_2O$ and NH_3 aqueous solution (25%).

2.2. *Pretreated and Preparation of Graphite/CeO_2*

Graphite/CeO_2 obtained from waste graphite pulverized into 200 mash particle size and heated at 110°C for 6 h. The sample was then activated with 1 M HNO_3 solution. The mixture was stirred continuously for 8 hours at 80°C. The sample filtered and washed with distilled water until the pH neutral filtrate (pH ± 7). The obtained solid was heated at 110° C for 24 hours, and then calcined at 500° C for 3 hours. The resulted sample (GA) was then modified using CeO_2. CeO_2 was precipitated from a solution of cerium nitrate. 0.5, 1 and 2 g of cerium nitrate and NH_3 solution were drop-wise into the sample solution for 30 min and stirred slowly with a magnetic stirrer and reached pH at 9. After one hour of mixing, the sample was dried for 12 hours at 110°C, and then it was calcined at 500°C for 5 hours.

2.3. Characterization

The morphology and structure of the samples were investigated by Scanning Electron Microscopy (SEM) analyzer (JSM-651OLA). For elemental analysis Energy Dispersive X-ray analysis (EDX) was performed by JED-2300, JEOL detector. Also, the specific surface areas of the samples were studied in a Quantachrome CHEMBET 3000 unit applying the standard Brunauer, Emmett and Teller (BET) method. The specific surface area (BET) of nanocomposites was characterized by N_2 adsorption and desorption isotherms obtained at -196°C using a Quantachrome ChemBET-3000. To investigate diagnosing surface functional groups, Fourier Transform Infrared Spectroscopy (FTIR) was conducted in the range of 400–4000 cm^{-1} wave numbers.

3. Results and Discussion

In this study, graphite before modification (GBM), activated graphite (GA) and modified graphite with variation of 0.5, 1 and 2 g of CeO_2 (G0,5; G1; G2) were characterized by FTIR, SEM-EDX and BET test

3.1. Morphology and composition study

SEM micrograph images of (a) GBM, (b) GA, (c) G0.5 (d) G1, and (e) G2 are shown in Fig. 1 (a-e). In general, Fig. 1(a) shows the smoother surface, almost flat and compact, where the porosity is not observed. SEM micrograph image of GA showed fragile surface (see Figure 1(b)). The activation causes the replacement of K^+, Na^+ and Ca^{2+} ions originally inside the graphite waste with H^+ ions. Also, this substitution results in the release of Al, Mg, Fe and other impurities on the lattice of the layered carbon structure. This makes graphite active. The HNO_3 also oxidized the surface of the graphite resulted hydrophilic surface so that the polar molecules (organic compounds) will have better interaction than the nonpolar molecules. SEM micrograph images of all the graphite/CeO_2 composites shows amorphous, fragile and irregular rounded granules (see Figs. 1(c), (d) and (e)). The formed structure has an irregular shape of particles with certain degree of agglomeration.

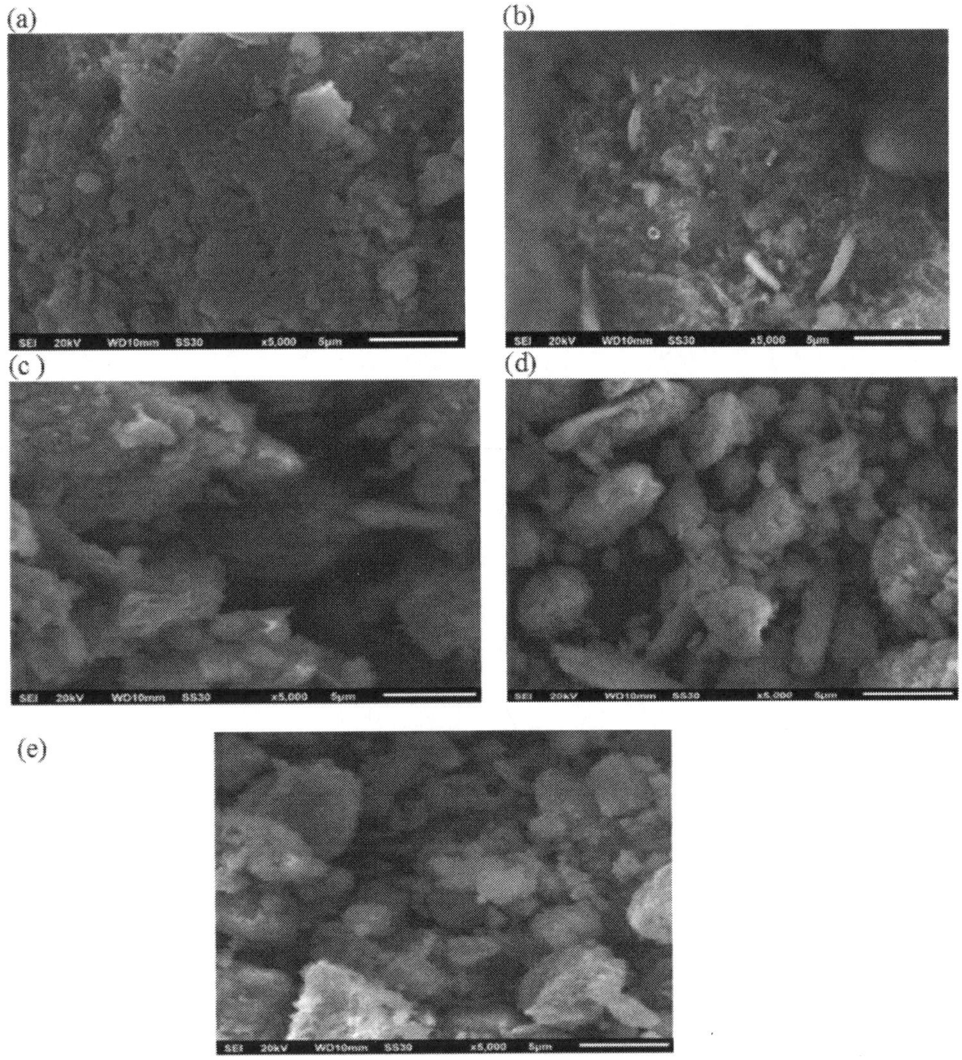

Figure 1. SEM micrograph images of (a) GNM, (b) GA, (c) G0.5 (d) G1, and (e) G2

The EDX composition of all samples is shown in Table 1. The carbon content of GBM is only 21.29% and other impurities elements in graphite such as sodium, oxygen, aluminum, fluorine, calcium, iron, and silicon. The high Al impurity content is due to the remaining aluminum from aluminum smelting process since the graphite waste used in this research was electrode for that process. The other impurity elements may be derived from the mineral additives of the graphite-making process. After activation with the acid treatment by HNO_3 1 M, the carbon content has increased to be 78.85%, then the impurity content becomes reduced (see Table 1). Most of the impurities are soluble in the HNO_3 and washed out after the treatment. Activated graphite was then impregnated with CeO_2, it shows that there is an increase of CeO_2 content from G0.5 to G1 and to G2. This corresponds to the amount of mass of cerium nitrate added at the time of modification of the graphite.

Table 1. EDX composition of graphite and after modification

	GBM	GA	G 0.5	G 1	G 2
C	21.29	78.85	7.1	6.99	4.03
N	-	-	24.26	22.09	21.62
O	19.15	16.02	42.18	32.07	37.38
Ce	-	-	19.57	25.44	26.49
Al	15.92	0.92	-	-	-
Si	8.27	3.41	-	-	-
F	18.18	-	-	-	-
Na	9.67	-	-	-	-
Ca	3.88	-	-		-
Fe	3.64	-	-	-	-

3.2. FTIR Analysis

Figure 2 showed the FTIR spectra of the synthesized *G0.5, G1 and G2* composites. The FTIR spectra with a broad band below 600 cm^{-1} shows the Ce-O stretch mode to confirm the presence of cerium in modified graphite composites [9]. FTIR results show that the Ce-O stretching modes on modified graphite composites at wavenumber 462, 464 and 458 cm^{-1}. While at a wavenumber at 1000 cm^{-1} indicated the stretching vibration of C-O.

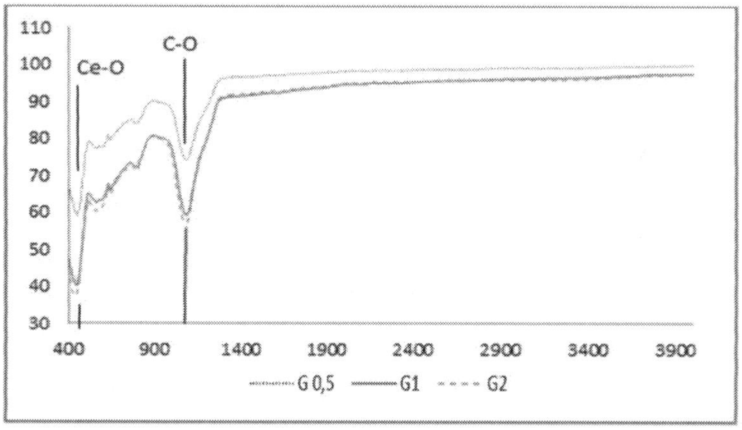

Figure 2. FTIR spectra of G0.5 G1, G2 composites

3.3. Surface area Analysis

Surface area is a key parameter in nanostructured catalyst and has a great influence on catalytic performance. The larger the surface area, the more the reaction site is available and hence higher conversion is obtained. Table 2 shows the result of BET surface area, it can be seen that non-modified graphite (GBM) has surface area 26.35 m^2/g, and after activation using HNO_3 the surface area of the GA is 16.63 m^2/g. The decrease in the activation process is because HNO_3 treatment extracted most of the metallic impurities which may form pillar structure in between graphitic sheet. After treatment the graphitic sheets collapsed and decrease the surface area of the GA.

After the activation process, the GA sample was chemically modified with $Ce(NO_3)_3.2H_2O$ *via* precipitation method. The largest BET surface area was observed for G2 composite (26.82 m^2/g). This is due to ceria (CeO_2) was attached to the graphite surface and probably intercalates the graphitic layer. From the analysis of pore distribution, it was found that the pores of all samples of the tested materials were distributed in the mesoporous range. Modification surface of graphite waste using $Ce(NO_3)_3.2H_2O$ have changed the structure of graphite waste, it gives more Ce^{3+} ions on surface, thus they provide more redox-active sites for adsorption. On the other hand, we also observed by modified with $Ce(NO_3)_3.2H_2O$, the BET surface area of composites does not significantly increase.

Table 2. BET Surface Area of Sorbents

	BET (m^2/g)	Pore Volume (cm^3/g)	Pore Size (nm)
GBM	26.35	0.09	14.05
GA	16.64	0.05	11.53
G0.5	20.03	0.06	11.67
G1	18.57	0.06	12.04
G2	26.82	0.08	12.25

4. Conclusion

The graphite/CeO_2 composites were synthesized and characterized. Further studies should be carried out to examine the performance of this graphite waste/CeO_2 Composites for further commercial application.

5. Acknowledgments

The authors greatly acknowledge the Universitas Indonesia for providing financial support through a grant, Hibah PITTA No. 774/UN2.R3.1/HKP.05.00/2017

6. References

[1] Lee S-Y, Park S-J, 2015. *Journal of Industrial and Engineering Chemistry* 23, 1–11.

[2] Vences-Alvarez E, Velazquez-Jimenez L, Chazaro-Ruiz, L, Diaz-Flores P, Rangel-Mendez, J, 2015. *Journal of colloid and interface science*, 194-202.

[3] Vilarrasa-García E, Cecilia JA, Azevedo DCS, Cavalcante Jr, CL, Rodríguez-Castellon E, 2017, *Microporous and Mesoporous Materials* **249**, 25-33.

[4] Md. Ahmaruzzaman, 2008. *Advances in Colloid and Interface Science* **143** 48–67

[5] E. Kusrini, B. Wicaksono, Y. Yulizar, C. Gunawan, C.H. Utami, 2015. Modification of Graphite Waste via Thermal-Mechanical and Chemcial Processes for Adsorbents, *Makara Journal Technology*, Under Review.

[6] Vivier L, Duprez D, 2010. ChemSusChem **3** 654-678.

[7] Jamalzadeh Z, Haghighi M, Asgari N, 2013. *Frontiers of Environmental Science & Engineering* **7**, 365–381.

[8] Abbasi Z, Haghighi M, Fatehifar E, Saedy S, 2011. *Journal of Hazardous Materials.* **186** 1445–1454.

[9] Khan SB, Faisal M, Rahman MM, Jamal A, 2011. *Science of the Total Environment* **409** 2987-2992.

IOP Conf. Series: Materials Science and Engineering **316** (2018) 012031 doi:10.1088/1757-899X/316/1/012031

Improvement of cement plant dust emission by bag filter system

Chandra Wahyu Purnomo[1], Wiratni Budhijanto[1], Muziibu Alfisyah[2], Triyono[2]

[1]Chemical Engineering Department. Universitas Gadjah Mada, Jalan Grafika no 2, 55281 Yogyakarta Indonesia.
[2]PT Indocement Tunggal Prakarsa, Palimanan, Cirebon, West Java Indonesia

chandra.purnomo@ugm.ac.id

Abstract. The limestone quarry in PT Indocement Tunggal Prakarsa (ITP) in Cirebon is considered as a complex quarry in terms of chemical composition and material hardness. From the beginning of the plant operation up to the end of 2015, the dust removal was rely on electrostatic precipitator (EP) system. Whenever limestone from specific quarry zones were incorporated into Raw Mill (RM) feed or there was an upset condition, the dust emission increased significantly. Beside higher demand of electricity, an EP system requires lower gas inlet temperature in order to remove the dust effectively which requires larger cooling water in the previous gas conditioning tower to cool down gas from 400 °C to about 100 °C. By considering the drawbacks, the EP system was replaced by a bag filter (BF) system. The BF allows higher temperature of gas inlet and it has higher dust removal efficiency. In this study, the efficiency of the two different dust removal systems is compared. The effect of process variables i.e. RM feed, kiln feed, inlet temperature and pressure, and small size particle fraction to the dust emission are studied by multivariate linier regression analysis. It is observed that the BF system can reduce significantly the dust emission from 30 to 6 mg/m^3 and in the same time reducing CO_2 emission by 0.24 ton/year from the electricity consumption saving.

Keywords: Dust emission; Bag Filter; Electrostatic precipitation.

1. Introduction

The air emission in cement industry is considered as the major environmental issue particularly for carbon dioxide, sulfur dioxide, nitric oxide and dust emission [1,2]. For instance in China, cement industry contributes at about 14% of the total PM2.5 emission throughout the countries [3]. Many efforts have been made to control this type of emission such as raw material modification, energy efficiency, and modification of the dust removal system in this industry [1,3]. On the other hand, the regulation of the dust emission is getting stringent for this business from time to time which drives the former constructed plants to improve their dust removing facility. The limit of dust emission for cement industry in the world is varied for instance in EU the limit is 0.15 kg/ton clinker [4,5], while in India the limit is depend on the production capacity from 250 mg^2/Nm^3 (plant capacity $\leq$ 200 tons/day) to 150 mg^2/Nm^3 [6].

In terms of dust emission reduction, there are two main gas cleaning systems implemented in cement industry i.e. electrostatic precipitator (EP) and bag filter (BF). It has been recorded that these two efficient gas cleaning system have a significant contribution to particulates emission reduction of cement plants from time to time [7]. Between the two dust removal system, EP technology has been widely applied due to ease of operation and maintenance and also lower pressure drop requirement. Most of early constructed cement plants were equipped with this dust cleaning system due to its established technology. In fact, an EP system requires a high electric power requirement and the performance during operation is strongly affected by operating condition (gas temperature, flowrate and dust content) and particulate properties such as particle size and dust resistivity [6]. Meanwhile, particulate resistivity is affected by chemical composition as well as gas temperature. Higher operating temperature will increase the resistivity thus the dust is getting harder to be separated by negatively charged electrodes.

Table 1. Design Specification of Electrostatic Precipitator and Bag Filter

Description	Unit	EP	BF
Size (LxWxH)	m x m x m	21.5 x 12.5 x 15.8	
inlet gas temperature	°C	120	180
Inlet dust content	g/Nm^3	95	200
Outlet dust content	mg/Nm^3	50	10
Gas volume	m^3/h	688900	547000
Removal efficiency	%	99.950	99.995
Total collecting area	m^2	15300	17603

Nowadays, BF system is getting famous to be used in cement plants due to stringent regulation and environmental standards [8,9]. By using special fabric for special separation purpose, the high efficiency of this system can be guaranteed. Even tough, BF is susceptible from leakages and higher maintenance cost due to bag replacement after certain years [10]. Most of the newly constructed cement plants are equipped with this air pollution prevention system, while older plants have to replace or improve their existing de-dusting system by BF system. PT Indocement Tunggal Prakarsa (ITP) is no exception, ITP has replaced EP by BF system in their 9th clinker production line. The specification of the two system is summarized in Table 1.

Table 2. XRF composition and compressive strength of selected quarry samples

Quarry zone ID	XRF result in weight %								Compressive strength (kg/m^2)
	SiO_2	Al_2O_3	Fe_2O_3	CaO	MgO	SO_3	Na_2O	K_2O	
A 1.1	12.88	2.66	0.37	46.39	0.94	0.15	0	0.03	664
A 1.4	0.62	0.89	0.28	54.84	0.29	0.03	0.02	0.01	128
BC150 3	11.54	3.12	1.11	43.78	2.15	0.6	0.13	0.18	691
BC120K2	0.1	0.5	0.18	51.63	3.36	0.16	0.06	1.26	96
C90SM 2	10.91	1.93	0.77	44.79	1.99	0.39	0.02	0.1	768
C90SM 3	8.98	1.69	0.46	48.14	0.89	0.18	0.02	0.11	432
SLS 1	5.87	2.23	0.77	47.92	3.1	0.44	0.08	0.1	248

Meanwhile, the quarry of ITP is considered as a complex quarry due to the variation of the chemical compound and also the material hardness in terms of compressive strength among the zones as listed in Table 2. It is common that there is a large raw material properties variation even in the same zone (A, B, C or D). The limestone content in the Raw Mill feed basically is a mixture from different zones to produce desired feed composition for clinker production. The limestone properties variation will affect the dust properties such as dust resistivity and particle size distribution. In an EP system, dust resistivity is a critical factor in determining the effectiveness of this dust removal system [6].

In this study, the two system of dust removal (EP and BF) are compared using real plant data. Some operating condition variables correlated with the separation performance are observed by a statistical method to determine the main variables for each separation equipment system.

2. Materials and Methods

Multivariate linier regression analysis is used to statistically interpret the correlation between dust emission data as the single dependent variable (Y) and selected plant data as the independent variables. The plat data are RM feed in ton/day (X1), kiln feed in ton/day (X2), inlet gas temperature (Tin) in °C (X3), inlet gas pressure or Pin in mmHg (X4), and residue in wt. % (X5). Residue in here is the mass fraction in percent of RM product with particle size less than 200 mesh. The both dust removal system (EP and BF) are analyzed using daily real plant data before (July to September 2015) and after (November 2015 to January 2016) system replacement that can be obtained in the supplement data (Table A.1 and A.2) rsepectively. The complete equation of the regression result is represented in Eq. 1. Then, P-value will be used to remove insignificant variables with the limit of 15%. Thus, the variable which has P-value lower than 0.15 will be rejected and the model is re-evaluated using only the remaining variables. The regression was done using Microsoft Excel 2013.

$$\text{Dust emission (Y)} = C0 + C1(\text{RM feed}) + C2(\text{kiln feed}) + C3(\text{Tin}) + C4(\text{Pin}) + C5(\text{residue}) \quad (1)$$

Beside the dust emission, the power consumption between the two systems is also compared using data of electrical capacity and average usage.

3. Results and Discussion

The comparison performance of the former EP versus the new BF system in terms of outlet dust emission (Y) is shown in Figure 1. In general the BF performance is far above EP in minimizing the dust emission. Indeed, EP design specification stated that outlet dust is 50 mg/m^3 in Table 2, thus most of the real dust emission is still below the targeted value. However, the fluctuation of the emission from the EP system is quite frequent during 80 days of observation while in the new BF system the dust emission is much stable at below 10 mg/m^3.

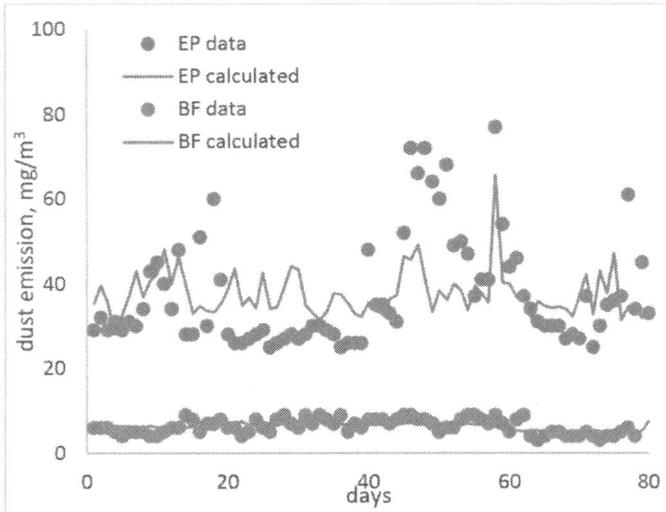

Figure 1. Dust emission data of EP and BF system, real plant data and calculated from regression

The five process variables (X_1-X_5) were used to model the dust emission (Y) for each system using simple multivariate regression method. The variables were chosen because they are easy to be collected from the plant data record and mostly available in a real time basis. The method is used to determine the important variables which has significant effect on the dust emission rather than to predict the emission. Because the emission is actually an intertwined parameter, many factors get involved in this air quality indicator.

Table 3. The regression result of EP system

	5 variables		3 variables	
Multiple R	0.4245		0.4219	
R Squared	0.1802		0.1780	
Significance F	0.0096		0.0016	
Regression	Coefficient	P value	Coefficient	P value
C_0	242.2133	0.050245	282.8821	0.000378
C_1	-0.36878	0.095111	-0.34985	0.102161
C_2	-0.56704	0.013869	-0.60454	0.001136
C_3	0.364608	0.65621	0	-
C_4	0.012802	0.889859	0	-
C_5	7.161029	0.035398	7.19346	0.030258

The regression result of EP system is listed in Table 3. By using complete set of variables, the calculated R^2 value is 0.18 with a quite good significance F test value which indicate that some of the selected dependent variables (X_1 to X_5) have correlation with the fluctuation of the independent variables. Meanwhile, the P value of some variables i.e. X_1, X_3 and X_5 are higher than 0.15 which represent a low correlation to the related independent variables. If these variables are omitted, the significance is increased with negligible reduction of R^2 value. It is confirmed that the RM feed (X_1), kiln feed (X_2) and residue (X_5) are the most important variables to determine the dust emission of EP system. Among the three variables, the small size particle fraction is the most significant variable due to the highest value of the corresponding coefficient. It can be said that the increase of small fraction particles will significantly increase the dust emission or lowering the efficiency of EP dust removal.

$$\text{Dust emission } (Y_{EP}) = 282.8821 - 0.3498.X_1 - 0.6045.X_2 + 7.1935.X_5 \quad (2)$$

Table 4. The regression result of BF system

	5 variables		2 variables	
Multiple R	0.5457		0.4900	
R Squared	0.2978		0.2401	
Significance F	$9.0559.10^{-5}$		$2.5565.10^{-5}$	
Regression	Coefficient	P value	Coefficient	P value
C_0	14.2749	0.0057	7.1962	0.0103
C_1	-0.0072	0.5609	0	
C_2	-0.0116	0.4516	0	
C_3	0.0272	0.1414	0.0306	0.0917
C_4	0.0374	0.0007	0.0387	0.0001
C_5	-0.3910	0.2036	0	

For a new installed dust removal system of bag filter, the regression results are listed in Table 4. Quite high R squared value indicated that the variation of dust emission can be explained by the selected independent variables. The very low of significance F value confirm that there is a strong correlation between dependent and independent variables. From the P test value, only two variables (X_3 and X_4) which contribute significantly to the dust emission. Equation 3 shows the final equation of the dust emission correlation in the BF system.

$$\text{Dust emission } (Y_{BF}) = 7.1962 + 0.0306.X_3 + 0.0387.X_4 \quad (3)$$

In terms of electric energy, the BF has lower energy demand compared to EP. The energy saving mainly from the omitted electrode side while the energy surplus is from the motor fan since BF requires higher pressure drop than EP. The other excess energy of BF system are from additional conveyor and air dryer. The energy reduction in one side with new energy demand on the other sides caused the energy saving is not so high that can be observed in Table 5.

Table 5. Electrical energy real demand and installed capacity

	EP		BF	
Equipment	Capacity, kW	Usage, kW	Capacity, kW	Usage, kW
motor fan	890	830	1800	1200
electrode	680	630	-	-
compressor	-	-	396	130
conveyor	30	27	67	60
air dyer	-	-	22	20
total		1487		1410

The reduction of power consumption here is approximately 70 kW, by having the correlation of CO_2 specific emission of each KWh generated is equal to 726 gCO_2 according to International Energy Agency (IEA), the CO_2e reduction can be estimated. Thus by modification EP to BF system, the CO_2 reduction is equal to 240 kg per year.

4. Conclusion

The EP replacement to BF system is successfully reduce and solve the problem of emission fluctuation in ITP plant. The dust reduction is significant from around 35 to 6 mg/m^3. By multivariate regression analysis it is observed that the raw material properties and quantity were statistically significant to the dust emission during electrostatic precipitation usage. However after replacement to fabric filter, only the process variables of inlet temperature and pressure are significant to the particulate emission. In terms of power consumption, BF system has a reasonably lower power demand than EP system by approximately 70 kW which is equal to reduction of CO_2 emission by 0.240 ton/year.

References

[1] Lu YL, Geng J, and He GZ 2015 *Adv. Climate Change Res.* **6** 202-209.
[2] Wang W, Jiang D, Chen D, Chen Z, Zhou W, and Zhu B 2016 *J. Clean Prod.* **112** 787-796.
[3] Zhang S, Worrell E, Crijns-Graus W, Krol M, Bruine M, Geng G, Wagner F, Cofala J, 2015 *Energy Procedia* **75** 584 – 589.
[4] Salas DA, Ramirez AD, Rodríguez CR, Petroche DM, Boero AJ, Duque-Rivera J, 2016 *J Clean Prod.* **113** 114-122
[5] Chen W, Hong J, Xu C, 2015 *J. Clean Prod.* 103 61-69.
[6] Bapat JD 2001 *J. Hazard Mater.* **B81** 285–308.
[7] Hua S, Tian H, Wang K, Zhu C, Gao J, Ma Y, Xue Y, Wang Y, Duan S, Zhou J 2016 *Atmos. Environ.* **128** 1-9.
[8] Anonim 2011 *Filtration Industry Analyst* **6** 6.
[9] Anonim 2013 *Filtration + Separation* **50** 13
[10] Kurtz O, Meyer J, Kasper G 2016 *Particuology* **30** 40 – 52

IOP Conf. Series: Materials Science and Engineering **316** (2018) 012032 doi:10.1088/1757-899X/316/1/012032

Numerical and analytical investigation of steel beam subjected to four-point bending

F. M. Farida[1,2], A Surahman[1], and A Sofwan[3]

[1] Civil Engineering Department, Civil Engineering and Environmental Faculty Institut Teknologi Bandung, Bandung, 40132, Indonesia
[2] Civil Engineering Department, Engineering Faculty, Universitas Negeri Jakarta, Jakarta 13220, Indonesia
[3] Engineering Division, P.T. LAPI, Institut Teknologi Bandung, Bandung, 40132, Indonesia
Email: fransisca_farida@unj.ac.id

Abstract. A One type of bending tests is four-point bending test. The aim of this test is to investigate the properties and behavior of materials with structural applications. This study uses numerical and analytical studies. Results from both of these studies help to improve in experimental works. The purpose of this study is to predict steel beam behavior subjected to four-point bending test. This study intension is to analyze flexural beam subjected to four-point bending prior to experimental work. Main results of this research are location of strain gauge and LVDT on steel beam based on numerical study, manual calculation, and analytical study. Analytical study uses linear elasticity theory of solid objects. This study results is position of strain gauge and LVDT. Strain gauge is located between two concentrated loads at the top beam and bottom beam. LVDT is located between two concentrated loads.

1. Introduction

Four-point bending test is one of the most common material testing procedures for investigating the properties and behaviour of materials with structural applications. It has also been realized that the real behaviour of steel beam subjected to four-point bending not only the specific performance of four-point bending but also the overall structural response. In order to get structural response, the application of measurement tools such as strain gauge and Linear Variable Displacement Transducer (LVDT) are practicaly important.

In this paper, a suitable location of strain gauge and LVDT measurement tools for a steel beam subjected to four-point bending were investigated by using numerical and analytical studies. Results of numerical and analytical studies help to improve in experimental works. Numerical and analytical studies prior to experimental studies have been conducted by [1], [2], [3], [4]and [5]. They have conducted experimental investigations, distribution of stresses and displacements of cold formed thin-walled beams. Based on these studies, study by [6] investigates the behaviour of web crippling of cold-formed steel lipped channel beams subjected to end-one-flange (EOF), interior-one-flange (IOF), end-two-flange (ETF), and interior-two-flange (ITF). Study by [6] tested 48 cold-formed steel lipped channel beams considering different boundary and loading conditions, bearing lengths and section heights.

The beam, which consider to be used in this presented paper, is simply supported and subjected to pure bending. [7] is used as a reference for four-point bending test. The distance of two concentrated loads is one third of support span. The main outcome of this study is the placement of strain gauge and LVDT on steel beam based on the numerical results and analytical analysis.

2. Methodology

This study used numerical and analytical study. Figure 1 presents the methodology in this study. It is assume that each concentrated load, which is applied on the steel beam, is 100 tons. Profile of steel section is IWF. Figure 2 shows a cross-sectional IWF profile steel beams. Height of IWF profile is 200 mm. Width of IWF profile is 150 mm. Thick of wings and web are 9 mm and 6 mm, respectively. These measurement are based on table of Gunung Garuda, condition of the furnace in PusLitBangKim Cileunyi Bandung, and [7]. According to [7], the proportion of the height of beam to the length of beam is 16: 1. The length of the beam, that can be burned in furnaces located in Bandung Cileunyi PUSKIM, is 3800 mm.

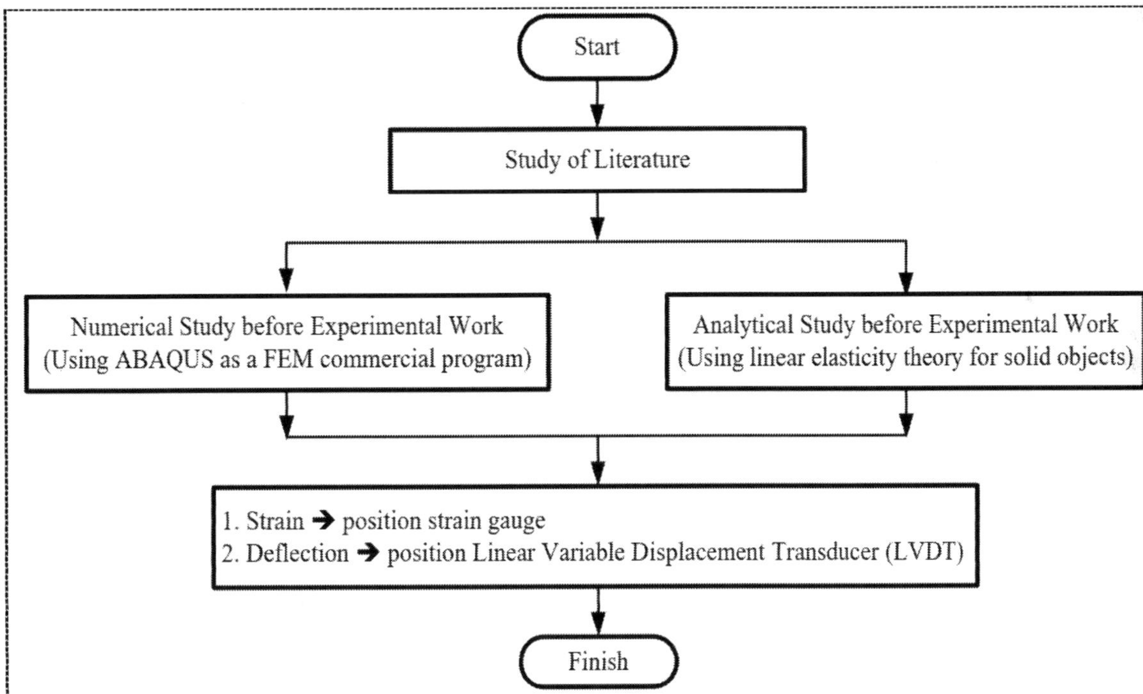

Figure 1. Study method

IOP Conf. Series: Materials Science and Engineering **316** (2018) 012032 doi:10.1088/1757-899X/316/1/012032

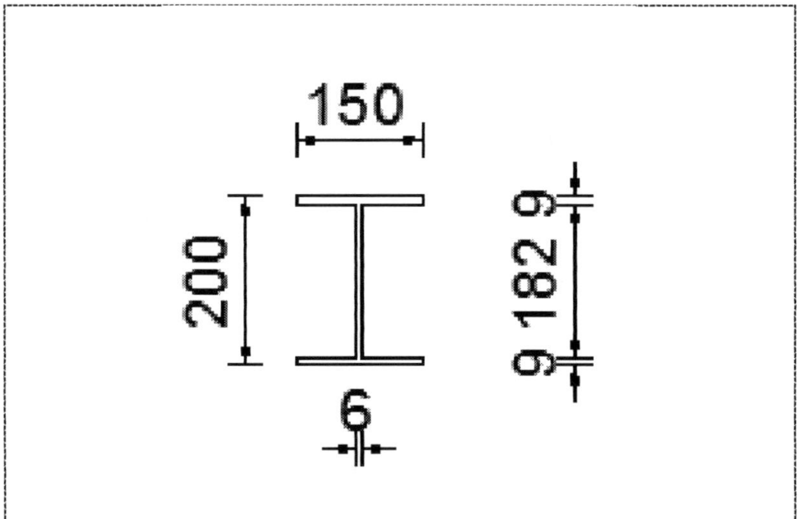

Figure 2. F IWF profile steel beam section.

3. Results and discussion

Isotropic and solid material, that confirms to the Von Mises yielding criterion, Young's modulus E, and poisson ratio are used in modeling the materials. Assumptions used in this study are E = 200.000 MPa, and Poisson's ratio $\upsilon = 0.3$. The commercial FEM analysis used in this study is ABAQUS 6.14.5

The beam uses linear elasticity as the simplest form of elasticity available in Abaqus [8]. Isotropic material behavior is defined in this linear elastic model. Isotropic material behavior is valid for small elastic strains [8]. To ensure smooth deformation in the modelling, the model was discretized using linear hexahedral. Steps in this model are initial and final step. Boundary condition is in initial step. Meanwhile, loads are in final step. Two concentrated loads with the same magnitude are placed on the one third and two third of the beam length. Figure 3 to Figure 5 presents the stress occurs in the beam.

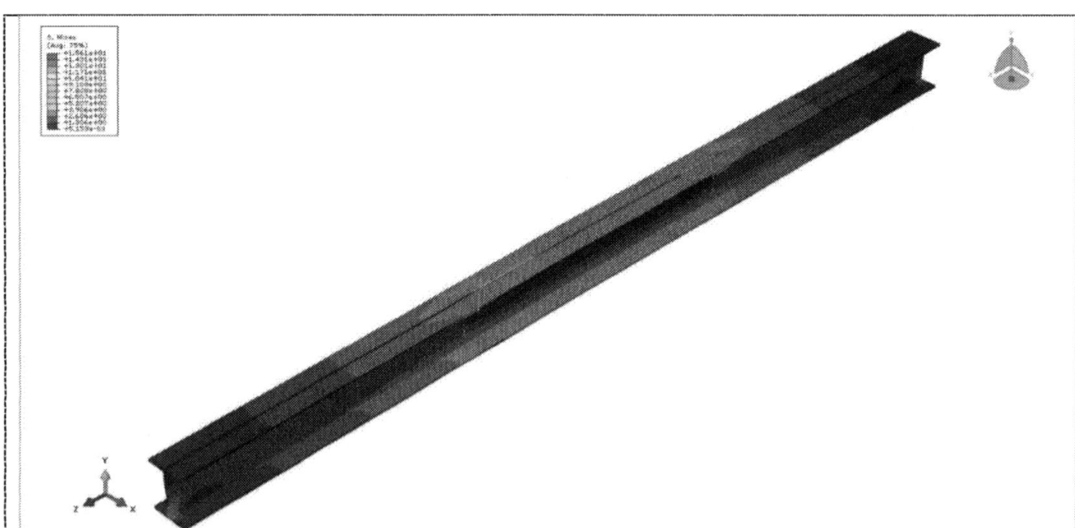

Figure 3. Stress occurs at the beam before two concentrated loads applied.

Figure 4. Stress occurs at the beam after two concentrated loads applied.

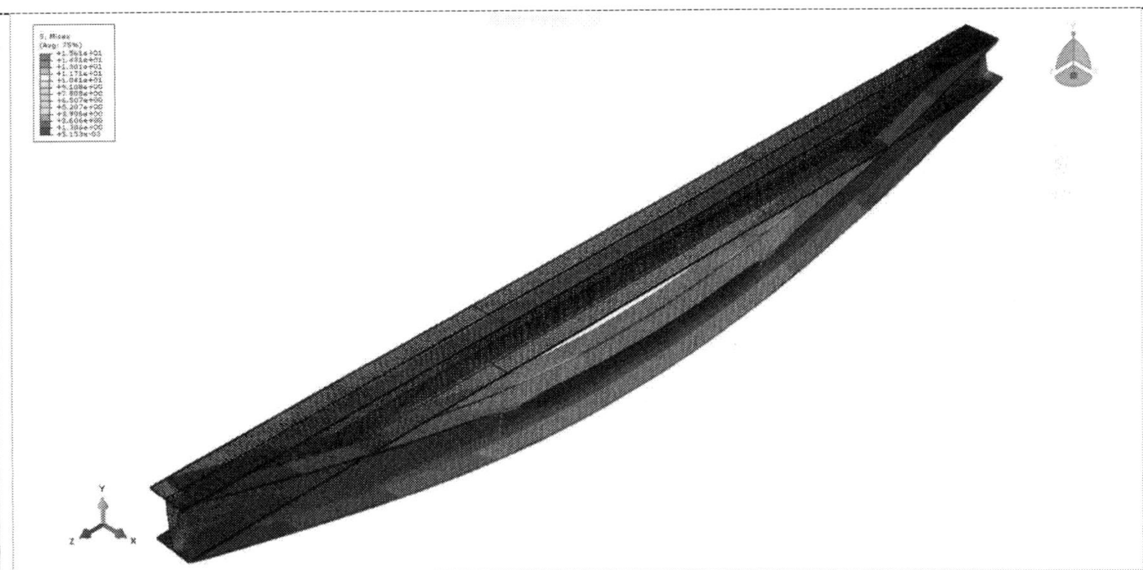

Figure 5. Stresss occurs at the beam before and after two concentrated loads applied.

IOP Conf. Series: Materials Science and Engineering **316** (2018) 012032 doi:10.1088/1757-899X/316/1/012032

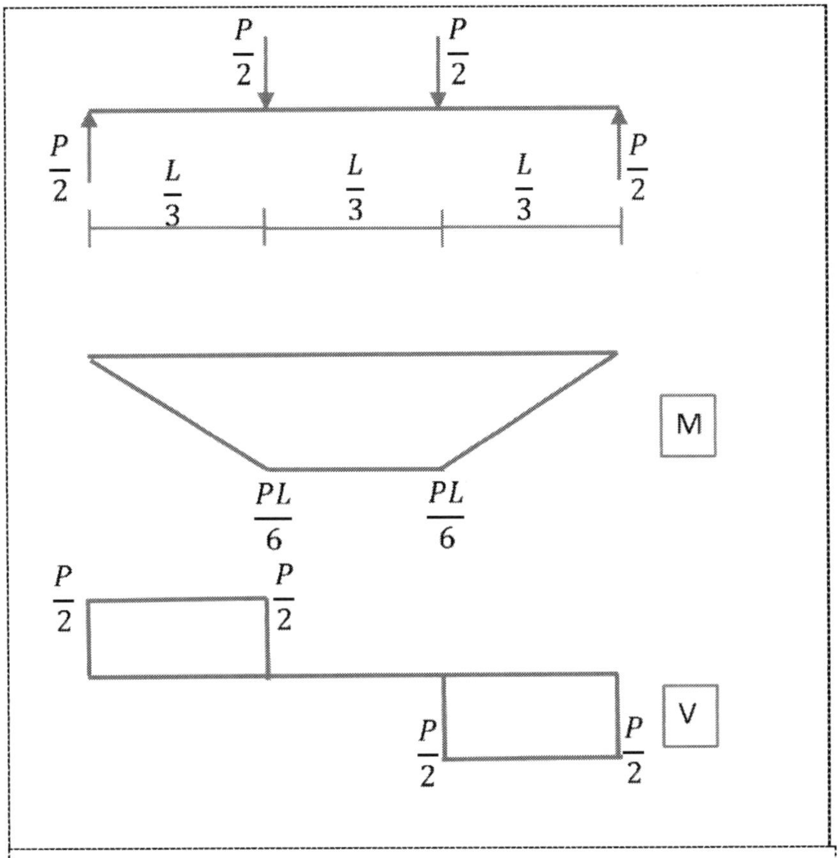

Figure 6. Moment and shear from manual calculation at beam with two concentrated loads.

Figure 6 shows that positive moment occur on the beam with two concentrated loads. The bottom side of the beam section has tension moment. On the other side, the upper side of the beam section has compressive moment. Strain on the bottom side of the beam section is tension strain. Meanwhile, strain on the upper side of the beam section is compressive strain. The same result appears by using finite element method. In this study, Abaqus version 16.4 is used as finite element method program. Strain gauge position, which will be placed between two concentrated loads at upper side and down side of beam section, is based on manual and numeric calculation. Figure 7 presents strain gauge location based on manual and numerical calculation.

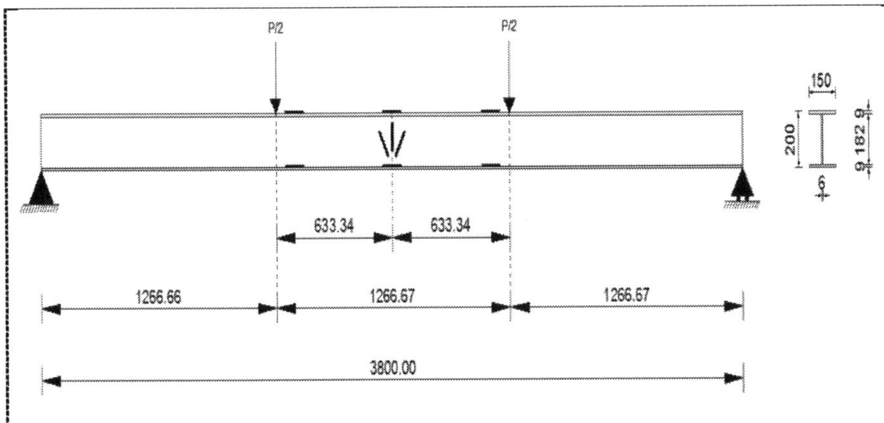

Figure 7. Strain gauge position based on manual and numeric calculation.

Flexural beam element analysis, that uses linear elasticity theory for solid objects, gives exact solution. In case of primary beam axis as shown in Figure 8, it gets Equation (1).

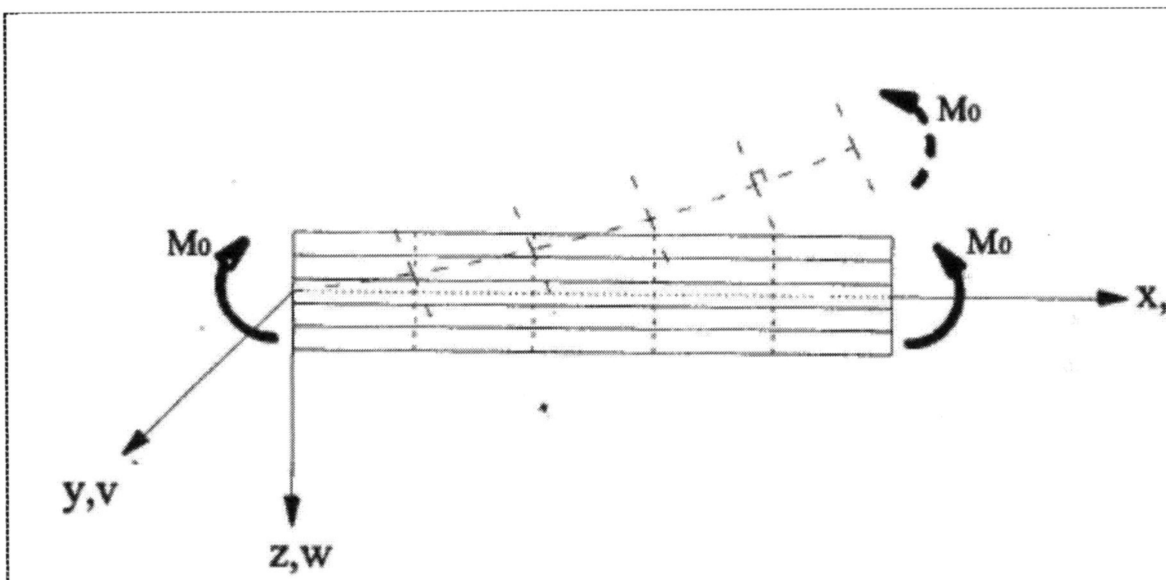

Figure 8. Flexural beam.

$$M_o = \int_A z\sigma_{xx} dA \tag{1}$$

In Equation (1), A is beam section area. Stress component ≠ 0 is in Equation (2):

$$\sigma_{xx} = \frac{M_0 z}{I} \tag{2}$$

Deflection component at Equation (3) comes from Hooke law.

IOP Conf. Series: Materials Science and Engineering **316** (2018) 012032 doi:10.1088/1757-899X/316/1/012032

$$w = \frac{M_0}{2EI}[-vz^2 - x^2 + vy^2] \tag{3}$$

In Equation (3), Figure 4, Figure 5, and Figure 6, the biggest value of moment places is located between two concentrated loads on the steel beam. Based on this result, position of LVDT is between two concentrated loads. Figure 9 shows position of LVDT.

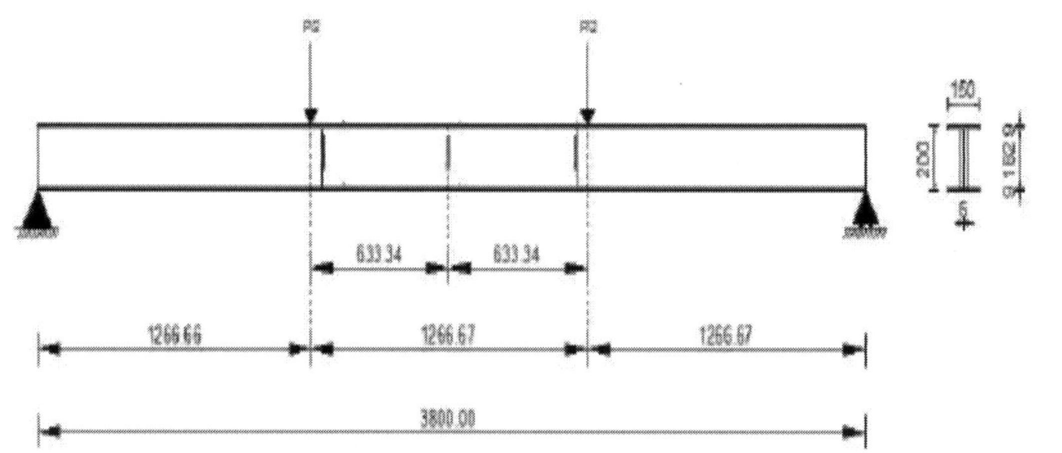

Figure 9. Position of LVDT based on linear elasticity theory of solid objects, numerical study, and manual calculation.igure with short caption (caption centred).

4. Conclusion

This study analyzes flexural beams by using numerical and analytical method. The intention of this study is to predict strain and deflection of steel beams subjected to four-point bending. Earlier prediction regarding strain will be used to place strain gauge on the beam. In addition, previous prediction regarding deflection is used to place Linear Variable Displacement Transducer (LVDT). Strain gauge and LVDT location in this studi :

1. Strain gauge is located between two concentrated loads at the top beam and bottom beam.
2. LVDT is located between two concentrated loads.

5. References

[1] Biegus, A., and Czepizak, D. 2008 *Thin-Walled Struct.* **46**, 303-309

[2] Mahendran, M., and Jeyaragan, M. 2008 *Proceedings of the 5th international conference on thin-walled structures* (1) 433-441.

[3] Paczos, P., and Wasilewicz, P. 2009 *Thin-Walled Struct* **47** 1354-1362.

[4] Magnucki, K., and Paczos, P. 2009 *J. Const. Steel Res.* **65**, 1731-1737.

[5] Belingardi, G., and Scattina, A. 2013 *Int. J. Adhes. Adhes,* **40** 31-37.

[6] Chen, Y., Chen, X., and Wang, C. 2015 **Thin-Walled Struct**. **87**(4), 1-52.

[7] ASTM D6272-02, (2002), *Standard test method for flexural properties of unreinforced and reinforced plastics and electrical insulating materials by four-point bending PA 19428-2959*, West Conshohocken, United States.

[8] Abacus. It is obtained from internet site: http://perso.mines-albi.fr/~penazzi/emboutissage/getting-start-cae.pdf. Downloaded on 24th Febuari 2017.

STUDY GRADATION AND MOISTURE CONTENT OF SAND EMBANKMENT ON PEAT SUBJECTED VIBRATION POTENTIAL LIQUEFACTION

Soewignjo Agus Nugroho
Civil Engineering Undergraduate Program University of Riau, Kampus Bina Widya, Pekanbaru 28293, Indonesia

Agus Ika Putra
Soil and Rock Mechanics Laboratory, Engineering Faculty,Riau University, Subrantas KM 12.5, Pekanbaru 28293, Indonesia

Muhamad Yusa
Civil Engineering Department University of Riau Kampus Bina Widya Subrantas KM 12.5 Faculty Riau University, Pekanbaru, Indonesia

ABSTRACT

In recent years large earthquakes often occur on the island of Sumatra. There is a phenomenon of the damage occurred during the earthquake, one of the effects is a phenomenon of loss of soil strength due to vibration called liquefaction. Some cases of liquefaction occur in some areas in Aceh, Nias Island, Padang and Pariaman. Pekanbaru is located close to the fault area that causes the occurrence of earthquake wave propagation. Pekanbaru are also at risk for geotechnical problems because of earthquake such as liquefaction. Evaluation of liquefaction potential could using by in-situ test and by laboratory test. The laboratory test to evaluation liquefaction potential among which method of experiment shaking table. In this study, liquefaction phenomenon was conducted by creating a physical model of a laboratory scale using a one-way vibration machine, with a review of how big the influence of sand gradation, sand shaped and grain-size, and surface water level in the sand against liquefaction potential. Evaluate of liquefaction potential based on the surface reading of the soil movement, elapsed time for final settlement and an excess pore water dissipation (EPD) during testing. Based on the results of performed test, indicated that fine sand on fully saturated conditions have the potential of maximum settlement for 20.67% and maximum ascend of pore water for 46.67%. This result mean that poorly graded fine sand on fully saturated conditions has more liquefaction potential than medium sand, coarse sand, and well graded sand

Keywords: liquefaction; settlement; excess pore water pressure; gradation; grain size.

1. INTRODUCTION

Indonesia lies between three tectonic plates, Eurasian plate, Pacific plate and Indian-Australia plate. Geologically, the islands of Indonesia are in a meeting two main earthquake belt, namely Sircum Pacific earthquake belt and earthquake belt Alpide Trans Asiatic. This led to Indonesia, including areas that have relatively high earthquake activity.

Pekanbaru is in the fault area that caused the earthquake wave propagation and also at risk of earthquake and liquefaction. Liquefaction is a phenomenon of loss of power ground layer due to vibrations emanating from earthquakes and explosions. When subjected to vibration properties of the soil layer becomes like a liquid and is unable to sustain the burden of buildings in or on it.

Intensive Study on liquefaction occupied after the earthquake that occurred in Alaska (April, 1964) and the earthquake in Niigata, Japan (June, 1964). Knowledge and liquefaction study is very important in order to understand the potential and the characteristic of liquefaction in an earthquake-prone region. Establish liquefaction concepts and theories about this, various studies were conducted in order to find a method that can explain this phenomenon of liquefaction, including the laboratory model test.

This study was conducted simulating the phenomenon of liquefaction by creating a physical model laboratory scale using the machine vibrating one-way analysis, the study of how much influence the variation of gradation, grain size and condition of the water table in the sand against liquefaction potential and how the relationship between these variations to the time required that can lead to liquefaction.

2. METHODOLOGY/ EXPERIMENTAL

Liquefaction is a process of loss of soil strength due increasing in pore water pressure caused by the cyclic load, so that the total overburden pressure is almost entirely replaced by pore water pressure. Liquefaction can cause ground movement direction of the horizontal (lateral spreading) or sand boiling.

Seed et al. [1], based on laboratory and field tests, and stated that the majority of cohesive soils are not liquefaction after earthquake. Soil at risk of liquefaction during an earthquake found in soil layers comprising a small grained sand to medium grained sand and silt with low plasticity [2]. Kramer [3] concluded that most of the incident occurred in the soil liquefaction were graded poorly. Poulos et al. [4] states that if a layer of soil is dense, it is not necessary in the evaluation of liquefaction, liquefaction occurs in saturated sand and silty sand with relative density less than 50% (loose sand) to limit of 70% (medium sand). Limit of earthquake acceleration value can result in liquefaction at a location of the liquefaction potential and relationship between these variations to the time required so as to cause liquefaction is 0,10g, and local magnitude is 5.0 [2]. The liquefaction potential is increasing when intensity of earthquakes and periods of vibration also increased. Saturated sand at a depth of more than 15-18 m is not easy to liquefaction. Saturated sand is not liquefaction if effective vertical overburden pressure value is 190 kN/m2 or more.

Liquefaction usually occurs in saturated soil [5]. Resistance to liquefaction decreases with increasing degree of saturation, and soil with a small degree of saturation can only liquefaction with a great earthquake tremor and a long duration [6]. A small portion of liquefaction occurs in unsaturated soil. Liquefaction does not occur on dry land, but only deflection in soil.

2.1. Method of Shaking Table

Shaking table during the test to move in the direction of the tool length alternating with the steps that remain. Steps that remains is a model of acceleration of the shear forces that occur in the soil layer in an earthquake. With certain acceleration, it can be evaluated whether the soil samples the water-saturated sand, occurs whether or liquefaction, based on a reading of pore water pressure in the test sample.

Test do to evaluate the liquefaction based dynamic cyclic loading by shaking table made by [7].

Some of the advantages obtained in the shaking table test [8], among others:

a. Box where soil samples placed, were made of thick glass or flexi glass so that all processes and phase voltage during the test can be seen and read.

b. Allows the presence of a water-saturated sand samples were homogenized in large quantities, so it is closer to the real situation on the ground.

c. Effect of stress concentration can be limited and occurs only in a relatively small location, so that its effects can be ignored.

d. Allows to be equipped with recording equipment (transducer) is adequate, so that the distribution of pore water pressure that occurs during the test and after the test can be observed and recorded carefully

2.2. Test procedure sequences

Sand taken from district Kampar, Riau Province. Research conducted at the Soil and Rock Mechanics Laboratory University of Riau. Soil classified by varied of grained size according to gradation. Some Data testing of physical characteristic, soil properties, and particle-size distribution, specific gravity (Gs), minimum dry unit weight (γmin) and maximum dry unit weight (γmax).

Test vibrating box 1-direction can be seen in Figure 1.

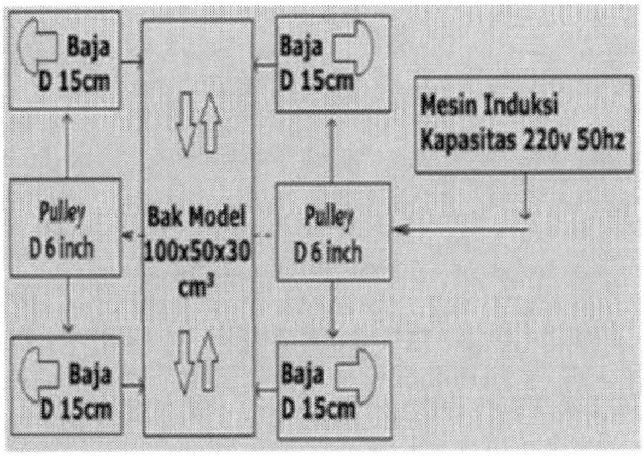

Figure 1 Vibration Table 1-direction

3. RESULTS AND DISCUSSION

Grain size distribution curve based on gradation analysis results for samples of coarse sand, medium sand, fine sand, and well graded-sand shown in Figure 2

IOP Conf. Series: Materials Science and Engineering **316** (2018) 012033 doi:10.1088/1757-899X/316/1/012033

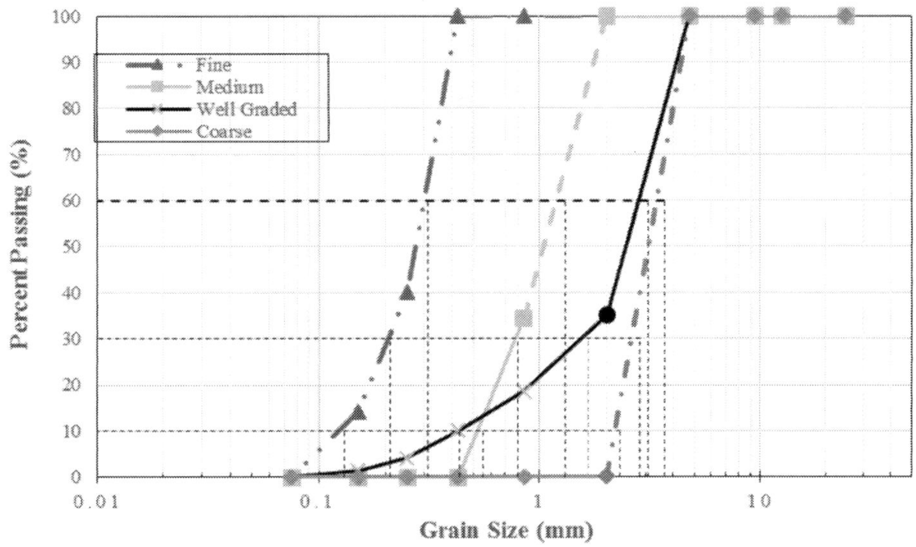

Figure 2 Grain Size Distribution Curve of samples

The test results of samples quality control of coarse sand, medium, fine, and well graded sand shown in Table 1

Table 1 Quality Control test Result

G.W.L	Well Graded			Poorly Graded								
				Coarse			Medium			fine		
	w	γ	E	w	γ	e	W	γ	e	w	γ	e
	(%)	(gr/cm^3)		(%)	(gr/cm^3)		(%)	(gr/cm^3)		(%)	(gr/cm^3)	
Dry	0.00	1.762	0.51	0.00	1.599	0.65	0.00	1.669	0.59	0.00	1.506	0.77
	0.00	1.741	0.53	0.00	1.611	0.64	0.00	1.660	0.60	0.00	1.483	0.80
	000	1.732	0.54	0.00	1.604	0.64	0.00	1.674	0.59	0.00	1.465	0.82
Unsaturated	5.15	1.722	0.54	4.25	1.552	0.70	5.17	1.623	0.64	7.39	1.462	0.82
	5.12	1.730	0.54	4.25	1.562	0.69	5.14	1.610	0.65	7.19	1.448	0.84
	5.15	1.723	0.54	4.25	1.569	0.68	5.16	1.613	0.65	7.33	1.475	0.81
Saturated	19.00	1.719	0.55	26.9	1.509	0.75	28.40	1.478	0.80	34.17	1.369	0.94
	19.50	1.710	0.55	26.9	1.507	0.75	28.46	1.481	0.79	34.07	1.369	0.95
	18.80	1.727	0.54	26.9	1.505	0.75	28.50	1.479	0.80	34.02	1.368	0.95

Particle size effect

Based on the test results in Table 1 and a density calculation results, obtained that level of density well graded sand samples were in range of 0.298% to 61.919%. This indicates that sample be in loose condition (very loose) to medium dense which range to sand in conditions of very loose to loose is 0% to 70% .

The influence of grain size reduction, time reduction, and excess pore water pressure dissipation shown in Figure 2 for the condition of dry sample, Figure 3 on the condition of the sample is not saturated (unsaturated) and Figure 4 for samples submerged conditions (saturated).

Uniform gradation condition of Fine graded sand has larger potential for decline than uniform graded of medium graded sand and uniform graded of coarse graded sand. Figure 1 also demonstrated the magnitude of decline in the sample maximum occurs in fine graded sand, on dry condition is 1.4 cm, 1.5 cm conditions are not saturated and saturated conditions is 3.1 cm. This is due to the fine graded sand void ratio greater than medium graded sand and coarse graded sand

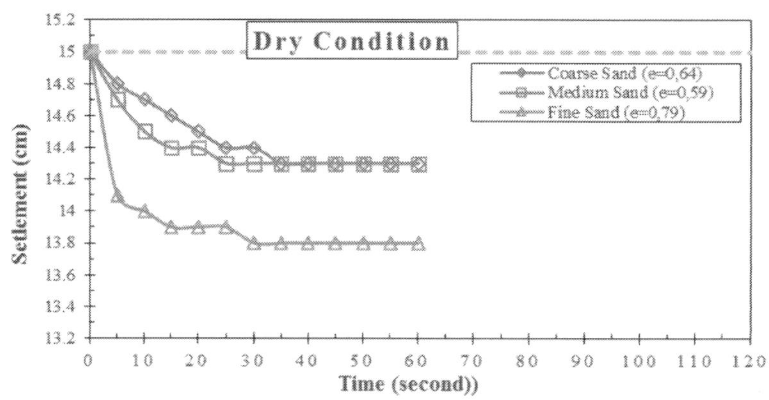

Figure 3 Particle Size Dry Condition

Total settlement soil on a fine sand tend to take longer than medium sand and coarse sand. Proved by length of time the maximum decrease occurred in the fine sand, the dry conditions in 35 seconds, unsaturated condition is 20 seconds, and saturated condition is 90 seconds. Void ratio also affects the total reduction of time. Large void ratio on fine sand will cause the time required is greater

Measurement of excess pore water pressure dissipation shows that in unsaturated and saturated condition, fine graded sand has a potential increasing pore water greatest. Proved by increasing maximum pore water pressure occurs in fine sand, on the condition of unsaturated 1 cm, and saturated 7 cm. When vibrated, settlement on soft sand is large where sand was compacted and volume was reduced. So, water in pores of the soil was receives stress, pore water pressure increasing result. While the abrasive of coarse graded sand pore water flow relatively quickly without causing significant pressure.

Excess pore water pressure dissipation on fine graded sand tend to take longer than the medium and coarse graded sand, proved the length of time the maximum pore water rise

occurs in fine sand. Coarse sand so smaller total reduction of excess pore water pressure quicker exit than fine sand and sand being shorter for water passage.
The test results in terms of the influence of grain size is consistent with the results of the liquefaction study [2], that the risk of soil liquefaction during earthquakes are found in soil layers comprising a small grain of sand to medium sand and silt that contain low plasticity

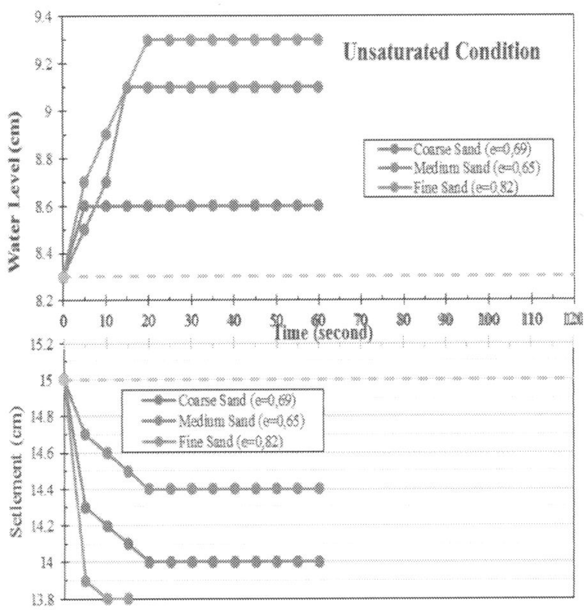

Figure 4 Unsaturated Condition

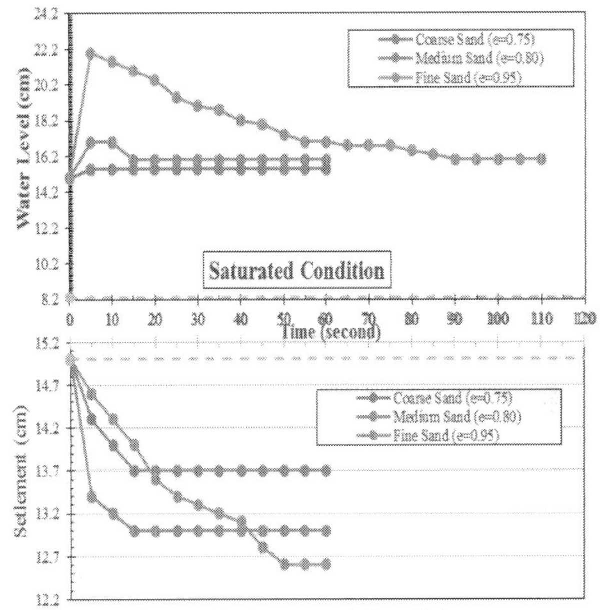

Figure 5 Saturated Condition

Particle size distribution effect
(Influence of gradation on the settlement, a decline, and the excess pore water pressure dissipation shown in Figure 6 to Figure 8
Samples in dry conditions (Figure 6), the condition is not saturated (unsaturated), and saturated, poorly graded sand (fine sand) has the potential for a larger decline than the well graded sand, demonstrated the magnitude of the decline in either poorly graded smaller than fine graded sand, which is dry condition 0.2 cm, 0.2 cm unsaturated and saturated 1.9 cm. This is due to well graded sand, void ratio smaller than coarse sand, medium, and fine.
At saturated conditions, when viewed from the void ratio of coarse sand which is larger than the sand gradation is good, but the sand grading both the decline is greater than the coarse sand, resulting samples graded either a mixture of coarse sand, medium, and fine, granular lesser charge cavity that existed at the time the sample is vibrated, so good sand grading denser than the coarse sand.

IOP Conf. Series: Materials Science and Engineering **316** (2018) 012033 doi:10.1088/1757-899X/316/1/012033

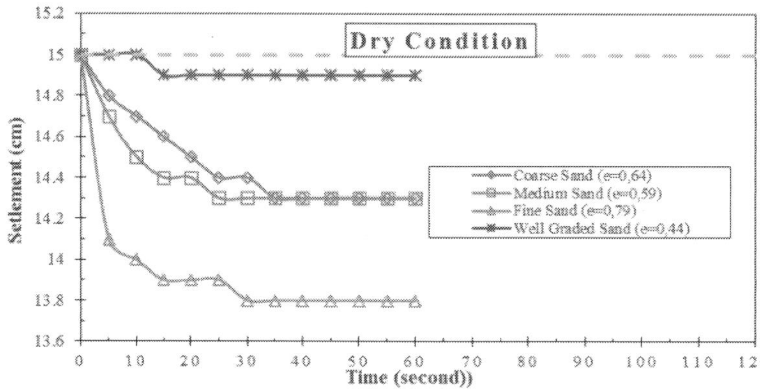

Figure 6 Gradation in Dry Condition

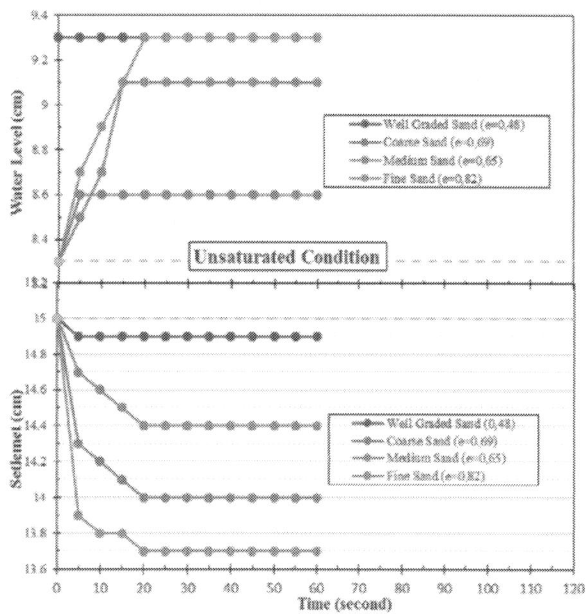

Figure 7 Gradation in Unsaturated Condition

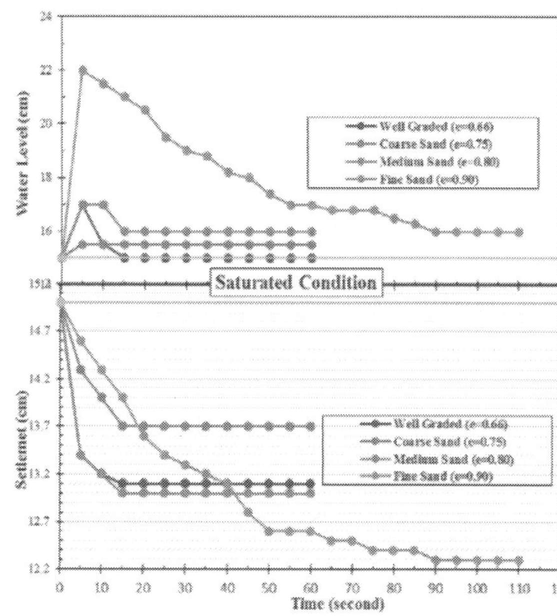

Figure 8 Gradation in Saturated Condition

The decline in total on the sand grading land both also tend to be faster stabilized namely in dry conditions 15 second, 5 second unsaturated and saturated 15 seconds. Figures pore either small sand grading will cause the time required is vanishingly small.

Measurement of excess pore water pressure which dissipation in Figure 6 and Figure 8 shows that the potential increase in pore water on sand grading either smaller than fine sand, proved in conditions not saturated, sand grading either did not experience significant pressure, while at the saturation magnitude increase in pore water sand grading better than fine sand 2 cm to 7 cm.

Excess pore water pressure sand grading both dissipation also tend to be faster than its stable of fine sand, which is the second to 15, compared to the fine sand on the second to

90. Well graded smaller so that the total reduction of excess pore water pressure quicker exit than fine sand because the water path is shorter.
The test results are reviewed gradations consistent with research on liquefaction [3], that most of incident liquefaction occurs in poorly graded soil).

4. CONCLUSION
The discussion can draw Final Conclusions as follows:
Test results of different quality control samples before vibrated, showing the density value are on Range 0.298% to 61.919%. Soil were in Limit Very loose to Medium on range less than 70% means that potentially liquefaction.
Seen From the diameter of the sand, fine graded sand Potential Decrease in air samples and the increase in pore big, proven by fine sand samples saturated condition occurred Decline Maximum 3.1 cm and maximum pore air rise 7 cm, and the length of time the Maximum Decrease Also Occurs on fine sand saturated condition i.e. 90 seconds.
Seen From gradation, poorly graded sand potential settlement and increasing pore water pressure is large, proven poorly graded sand (fine graded sand) on saturated condition shows the maximum decline of 3.1 cm, gradation both saturated 1.9 cm. While well graded sand show a minimum decrease in dry conditions and saturated 0.2 cm. well graded sand Also showed the smallest decrease in time compared poorly graded sand, i.e. at dry conditions 15 seconds, unsaturated 5 seconds, and 15 seconds is saturated.
Sand with emaks and emin wide difference, more Potential decline. Fine graded sand saturated condition, difference of emaks and emin is 0.374, shows Maximum decrease of 3.1 cm. For saturated coarse graded sand, Difference of emaks and emin is 0.272 showed decline 1.3 cm.

5. ACKNOWLEDGEMENT
Thanks for the assistance of the Fund Kemenristekdikti research on competitive grants scheme 2016 through LPPM of University of Riau. Thank you for the good cooperation to Suratman, Andy Hendri, Murialdi, and Resty Vesariany.

6. REFERENCES
[1] Seed, H. B., Idriss, I. M., and Arango, I., (1983), Evaluation of Liquefaction Potential Using Field Performance Data, Journal of the Geotechnical Engineering Division, ASCE, Vol. 109, No. GT3 [online], USA.
[2] Ishihara, K., (1985),Stability of Natural Deposits During Earthquakes, Proceedings of the Eleventh International Conference on Soil Mechanics and Foundation Engineering [online], San Francisco, USA.
[3] Kramer, S. L., (1994), Geotechnical Earthquake Engineering [online], Upper saddle River, New Jersey, USA.
[4] Poulos, S. J., Castro, G., and France, J. W., (1985), Liquefaction Evaluation Procedure, Journal of the Geotechnical Engineering Division, ASCE, Vol. 111, No. GT6 [online], USA.

[5] Seed, H. B., and Idriss, I. M., (1971), Simplified Procedure for Evaluating Soil Liquefaction Potential, Journal of Soil Mechanics and Foundation Engineering, Vol. 97 (9) [online], 1249-1273.
[6] Sherif, M. A., Ishibashi, I., and Tsuchiya, C., (1977), Saturation Effects on Initial Soil Liquefaction, Journal of the Geotechnical Engineering Division, ASCE, Vol. 103, No. GT8 [online]
[7] Tokimatsu, K, (1979), Generation and Dissipation of Pore Water Pressure in Sand Deposits during Earthquakes. Thesis [online],. Tokyo Institute of Technology, Oh-okayasa, Tokyo, p. 81, 88-109, Tokyo
[8] Hasmar, Halim, (2007), Evaluasi Potensial Likuifaksi Akibat Gempa Bumi Tektonik Lapisan Pasir Jenuh Air dengan Metode Shaking Table. Jurnal LOGIKA [online]. Vol. 4, No. 1 [online], 2007

Paper for Publication in IOP: Conference Series

Leachate Treatment using three Years Aged Lysimetric Bioreactor Models

Djoko M. Hartono [a]*, Gabriel Andari Kristanto [a], Irma Gusniani Sofian [a], Ahmad Fauzan [a], Ghanis Mahdiana[a]

[a], *Environmental Engineering Study Programme, Civil Engineering Department, Faculty of Engineering, Universitas Indonesia, Kampus Universitas Indonesia, 16424, INDONESIA*

E-mail: djokomh@eng.ui.ac.id

Abstract

This study was conducted as a response to address the problem of land availability for Cipayung landfill that no longer able to accommodate waste generation Depok City and to protect water pollution in receiving water body. Law No. 8/2008 explained that local governments and cities are required to create a sanitary landfill as a final waste processing system to replace open dumping that had been done by almost all the final processing of waste in cities in Indonesia. Sanitary landfill is the final waste processing system that works best and is environmentally friendly. The sanitary landfill will generate leachate. Leachate is the result of precipitation, evaporation, surface runoff, water infiltration into the waste, and also including the water contained in the waste. The purpose of this study was to determine the utilization of leachate generated by three years aged reactor.This study use a modeling tools as bioreactor landfill tank or so called lysimetric, that made of the polymer material that susceptible to high heat and pressure. This bioreactor landfill tank has a diameter of 0.83 m, with a surface area of 0.54 m^2 and a height of 2.02 m, with the examination duration of 115 days. This tank consists of several layer, such as sand layer, solid waste layer, water layer and piping system. These layer has 3 year aged. The In this research, leachate recirculation in bioreactor landfills was conducted with waste layered loading systems with percolation system. This research has been conducted since the beginning of 2016, sampling, field measurement and analysis of leachate and waste quality carried out for approximately 115 days of field measurements.Several parameter were measured such as pH, BOD, COD, nitrate, nitrite and TSS. From the analysis of the leachate quality parameters of pH, BOD, COD, nitrite, TSS, showed a reduction in the concentration of the three reactors. The concentration of parameters measured at the initial stage until the final stage, showed a reduction in the concentration of the parameters, even reaching 90% reduction for BOD (biological oxygen demand), COD, (chemical oxygen demand) nitrite, and TSS (total solid suspended) parameters. So it can be concluded that the recirculation of leachate of the sanitary landfill can reduce the concentration of pollutants in the leachate that will be discharged into water bodies, thereby reducing the pollution of the receiving water body.
This research is funding by PUPT Kemristekdikti and DRPM UI

Keywords: bioreactor, lysimeter, leachate, sanitary landfill, parameter

1. INTRODUCTION

This study was undertaken in response to the Law of the Republic of Indonesia No. 18 of 2008 which states that the method of waste disposal in landfills must meet environmental protection by changing the method of final disposal from open dumping into the sanitary landfill. The sanitary landfill will generate leachate. Leachate is the result of precipitation, evaporation, surface runoff, water infiltration into the waste, and also including the water contained in the waste. Leachate potentially could increase pollution of the receiving water body. The increasing volume of municipal solid waste (MSW) generated would cause problems such as land availability, management and treatment of waste in the landfill land, which is limited and costly has been a concern of the government so that the expected capacity of the landfill can be optimized by increasing the percentage reduction in the volume of waste. In fact, local governments had trouble getting vacant land for the landfill. The existing landfill could no longer accommodate the waste generated. Besides leachate generated in the landfill could potentially pollute the receiving water bodies that have an impact on humans and other living things. Therefore, landfill issues not only about the availability of land, but also the environmental pollution caused by leachate.

The purpose of this study was to determine the utilization of leachate generated by three years aged reactor. The reactor is as much as possible copy the sanitary landfill condition such as the structure, materials as well as environmental condition.

2. METHODOLOGY

This study was conducted by modeling of landfill using lysimeter bioreactor filled with a layer of waste with percolation system, the system of drainage of a fluid passing through a layer to the underlying layer due to the force of gravity. Containers used for lysimeter bioreactors is a water reservoir. The limited variations in the size of the water tank on the market were one of the problems in representing the height of the landfill. Water reservoir used has a diameter of 0.83 m with a surface area of 0.54 m^2 and a height of 2.02 m. With these measures, the expected mass of waste can reach more than 50 kg (Sidauruk, 2013).

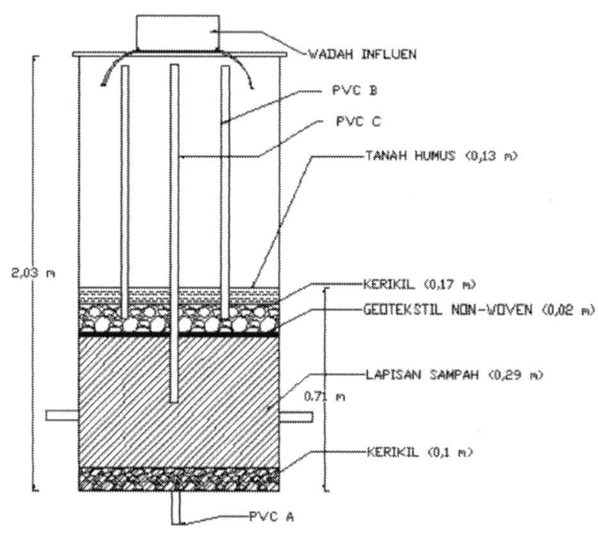

Figure 1. Bioreactor Model

The function of drainage layer is to filter leachate from waste and to prevent clogging on the pipe in the layer of gravel and sand. HDPE Geotextile used to avoid direct contact with the outside air so that waste layers maintained in anaerobic conditions. Geotextile is placed on the first layer of waste. The waste layer then covered by humus soil that serves as a daily cover with consideration of the cell has been completed the operation.

The feedstock used in bioreactor landfills is organic waste that originated from the waste processing unit of Kemiri Muka Market, Depok. The composition of waste had previously sorted and separated from metal or glass material. Organic waste is not shredded before entered into the lysimeter bioreactor to replicate the operating system of waste in landfills. The organic fraction of waste which is not shredded has high water content so as to produce a relatively large air space. Therefore, in the initial period, lysimeter can undergo aerobic conditions that can support anaerobic microorganisms. Recirculation of leachate conducted based on the frequency of leachate feeding. Leachate recirculation performed by collecting leachate coming out and flowed back through four recirculation pipe so the leachate flows back to waste layers. The total volume of leachate to be recirculated is 1.5 liters.

The feedstock used aged 3 years commencing from 2013. There has been a compaction by gravity that changes the volume, height, and density of the waste composition. In addition to the force of gravity, decomposition of waste also resulted in decreased of waste elevation significantly.
Decomposition of the waste is not occurring naturally in whole. However, to accelerate the decomposition of the addition of leachate regularly. Data that will be obtained in this study include the leachate quality including pH, Biological Oxygen Demand (BOD), Chemical Oxygen Demand (COD), Total Suspended Solid (TSS) and the content of Carbon (C), Hydrogen (H), Oxygen (O), Nitrogen (N), Ash, Carbon per Nitrogen (C/N) ratio, and heat of waste that passed by leachate. The first test conducted was the characteristics of waste as the main composition of the bioreactor. Further test conducted was leachate quality improvement that has been in contact with the waste in the bioreactor. The tests were conducted to determine the changes that occur before and after the use of bioreactors. Based on the data of initial and final characteristics, the effectiveness of using three years aged waste as a feedstock for the bioreactor, improvement of leachate quality and reaction that occurs between leachate and waste contacted can be analyzed.

3. RESULT AND DISCUSSION

Initial characteristic of waste as the feedstock of bioreactor are listed below:

Table 1. Initial Characteristic of Waste

Parameter	Unit	Result	Method
Water Content	%, ar	1.2	ASTM D.3173-11
Carbon	%, ar	2.24	ASTM D 5373-14
Hydrogen	%, ar	2.08	ASTM D 5373-14
Nitrogen	%, ar	0.23	ASTM D 5373-14
Oxygen	%, ar	11.59	ASTM D 3176-15

Source: Sucofindo, 2016

Initial characterization of waste in the bioreactor will be compared with the characterization of waste after leachate drainage. Chemistry content and heat changes of waste is a parameter that will be observed. Organic waste consisting of vegetable waste, food waste, and garden waste containing carbon (C) in the form of simple or complex compounds. Cellulose is one of the complex compounds that require a relatively long decomposition process yet can be broken down by the enzyme cellulose produced by bacteria.

Table.1 Final characteristics of Waste

Parameter	Unit	Result	Method
Water Content	%, ar	3.4	ASTM D.3173-11
Carbon	%, ar	1.51	ASTM D 5373-14
Hydrogen	%, ar	1.66	ASTM D 5373-14
Nitrogen	%, ar	0.19	ASTM D 5373-14
Oxygen	%, ar	12.32	ASTM D 3176-15

Source: Sucofindo, 2016

Final characterization of waste was measured after leachate passed trough waste in the bioreactor. The water content of waste increased from 1.2% to 3.4% after drainage of leachate. Increasing in water content is consistent with the literature, that waste drainage by leachate will have higher water content compared with decomposed waste without leachate drainage.

Meanwhile, carbon, hydrogen, and nitrogen content are decreased. Carbon content decreased from 2.24% to 1.51%. Reduction in carbon and hydrogen content occurs due to the reaction of methanogenesis when waste drained by leachate. Methanogenesis is a formation reaction of methane gas. There are two types of methanogens. First, methane is formed through the conversion CH_3COOH to CH_4. Second, the formation of methane is carried out by methanogenic bacteria to consuming CH_2 and H_2 compound thereby reducing CO_2 and produce CH_4.

3.1. COD Removal Capability

COD Removal Capability of Bioreactor is reactor ability to improve leachate quality by decreasing COD concentration that contained in leachate. COD removal capability is expressed in a percentage of the difference in COD concentration in the influent and effluent leachate. Mathematically, it can be shown as follows:

$$\text{COD Removal Capability} = \frac{(COD\ influent - COD\ effluent)}{COD\ influent}\ X\ 100\%$$

Based on the results of the study, COD removal capability of the bioreactor is passable. The following graph shows the rate of removal of COD by the bioreactor. The average of COD removal rate is 95.83%. The removal rate of COD tends to decrease during 115 days examination. In the beginning of examination (June 2016) the COD removal rate is 94.01 % and reached the highest COD removal rate 99.07% on July 2016.The COD removal rate decreased up to 95.31 % by the end of examination on October 2016.

IOP Conf. Series: Materials Science and Engineering **316** (2018) 012034 doi:10.1088/1757-899X/316/1/012034

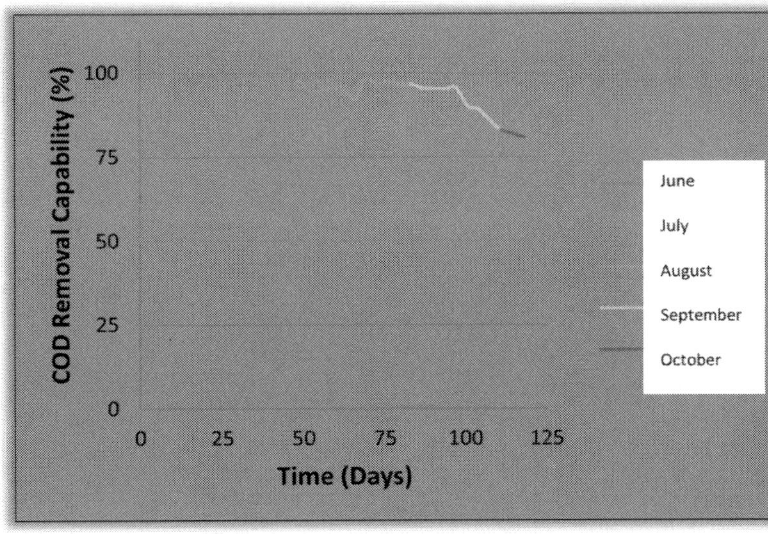

Figure 2. COD Removal Rate based on Time

In the beginning of examination the microbe activity is relatively high since the contact space of microbe and leachate is enough. The decrease on the COD removal rate is caused by the clogging of waste layer. Clog come from the organic material which is not decomposed by microbe. The chance of microbe to contact with leachate nutrition is decrease since the waste layer is clogging.

3.2. Nitrite Removal Capability

Nitrite removal capabilities trend is not much different with COD removal capability. In the initial phase, the concentration of nitrite tend to be unstable, it can be observed from the high fluctuations on Figure 3. Figure 3 shows the rate of Nitrite removal by the bioreactor. Mathematically, it can be shown as follows:

$$\text{Nitrite Removal Capability} = \frac{(Nitrite\ influent - Nitrite\ effluent)}{Nitrite\ influent}\ X\ 100\%$$

IOP Conf. Series: Materials Science and Engineering **316** (2018) 012034 doi:10.1088/1757-899X/316/1/012034

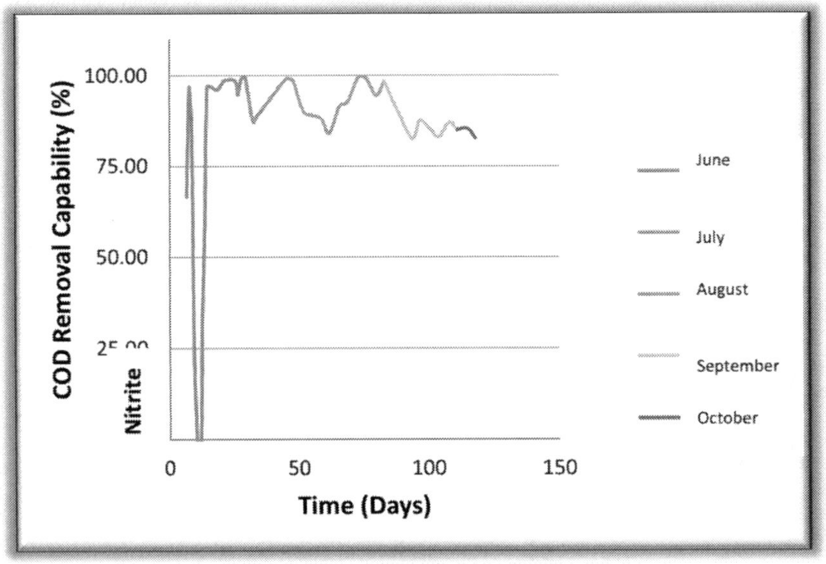

Figure.3 Nitrite Removal based on Time
Source: Research, 2016

Decreasing of nitrite concentration can occur due to the denitrification reaction which converts nitrite to nitrate. At the initial stage at June, fluctuations are very high, where the average is 77.68 % and standard deviation reaches 34.24. The high standard deviation expressing very unstable removal capability. At the last stage at October the Nitrite removal rate decrease up to 84%and standard deviation reaches 2.12 %.However, in the subsequent months, nitrite removal rate is stable with low value. Judging from the average of each month, the removal rate decreased gradually which indicates the ability of the bioreactor in improving the quality of leachate also reduced.

3.3. Nitrate Formation Capability

Unlike other parameters tested in this study such as COD, BOD and Nitrite, nitrate concentration increases with a decrease in the concentration of nitrite. This occurs due to the decrease in nitrite concentration caused by denitrification process that converted nitrite to nitrate. This phenomenon cannot be described with a graph showing nitrate removal capability but by a graph of increasing nitrate concentrations. Although the nitrate concentration always increases in concomitant with decreasing in nitrite concentration, but the rate of increase in the nitrate concentration will gradually decline as seen in Figure 4.

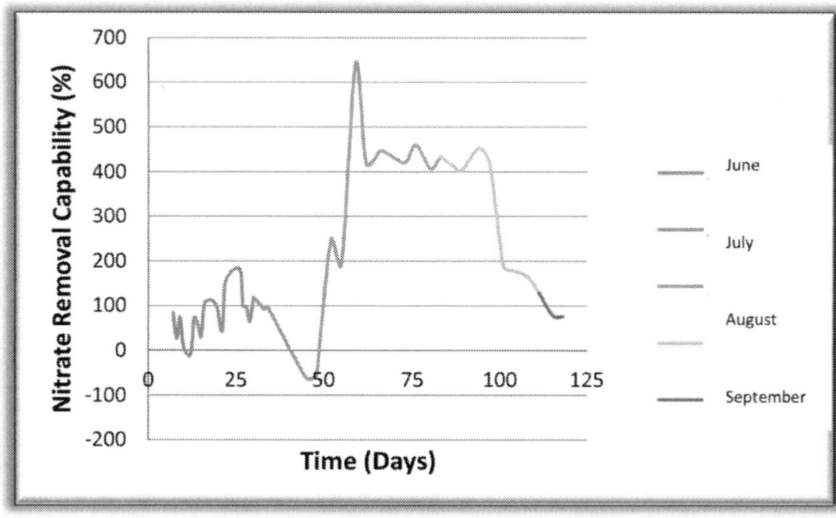

Figure 4. Nitrate Formation Rate based on Time
Source: Research, 2016

Based on Figure 4, the average nitrate formation rate at June is 77.95% and a standard deviation of 50.07. The rate decrease at July up to 35.11 % and a standard deviation of 85.68. The rate decreased gradually until October. This can occur due to high denitrification reaction and unstable conditions of the bioreactor in June until August. However, in August to October, it can be observed that bioreactor can reduce the concentration of nitrate produced gradually.

3.4. TSS Removal Capability

TSS removal capability of the bioreactor is quite good. In Figure 5, shown that fluctuations in TSS removal capability only occurred in the early stage of the process. In August to October TSS concentration decreased gradually. The following chart show TSS removal capability of the bioreactor.

IOP Conf. Series: Materials Science and Engineering **316** (2018) 012034 doi:10.1088/1757-899X/316/1/012034

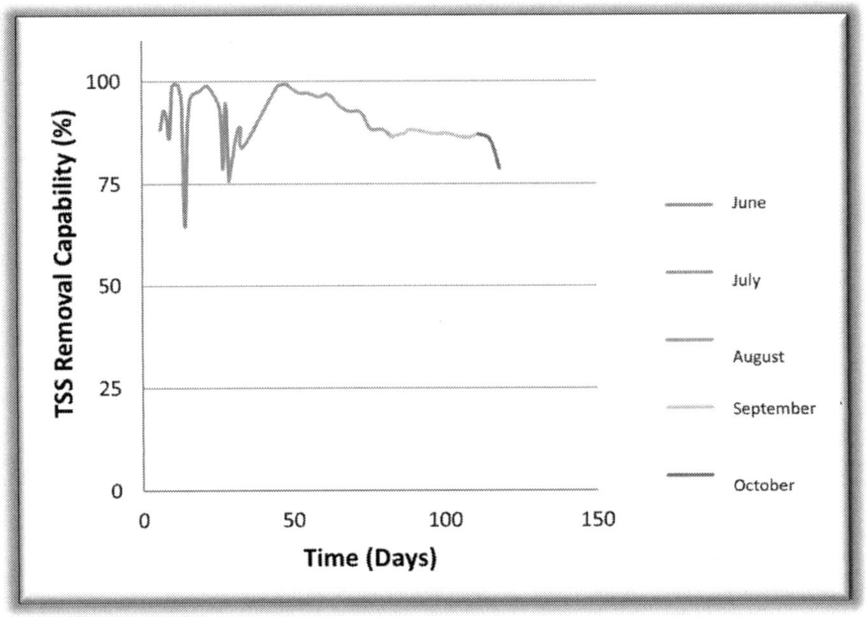

Figure 5. Percentage of TSS Removal based on Time

Source: Research, 2016

In the initial phase at June the TSS removal rate is 90.59 % and deviation standard is 9.87.The rate decrease at July up to 89.07 % and deviation standard is 6.78.Increasing rate occur d at August up to 92.92 % and at the end of phase the TSS removal is 82.23 %

High deviation standard can be observed in June and July, which indicates unstable removal process. However, in August to October, there is a very significant decrease of average removal value. Therefore, stability in TSS removal process has been achieved since August.

TSS is suspended material which contributes to viscosity and turbidity in leachate. The higher TSS concentration contained, the higher density levels and turbidity observed visually. TSS in the leachate is usually derived from decomposition process of organic material after physical, chemical, or biological treatment process. The high value of BOD and COD is one of the aspects that contribute to the high value of TSS. The higher BOD and COD contained, the higher TSS that contained.

3.5 BOD Removal Capability

BOD is used to calculate the concentration of organic material from a water sample which is described by oxidation process by microorganisms through a biological process. Theoretically, BOD content of leachate ranged from 40% -60% of COD concentration. Therefore, the ability of BOD removal is not much different than the COD removal capability. The high percentage removal of BOD concentration caused by the activity of microorganisms in the bioreactor which is growing due to the use of waste as media. Beside influent leachate that was added to bioreactor contains many microorganisms that decompose organic matter aerobically and

IOP Conf. Series: Materials Science and Engineering **316** (2018) 012034 doi:10.1088/1757-899X/316/1/012034

serves minerals needed by microorganisms. Mathematically calculation of BOD removal can be shown as follows:

$$\text{BOD Removal Capability} = \frac{(BOD\ influent - BOD\ effluent)}{BOD\ influent}\ X\ 100\%$$

Figure 6 shows the rate of BOD removal by the bioreactor based on research time.

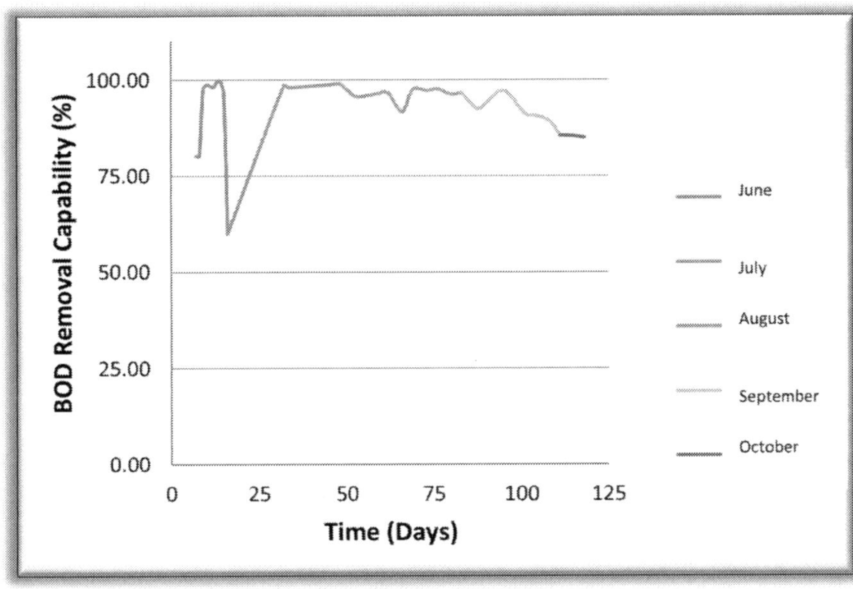

Figure 6. Percentage of BOD Removal based on Time
Source: Research, 2016

4. CONCLUSION

1. Bioreactor lysimeter can improve the quality of leachate from five of the six tested parameters, namely COD, BOD, TSS, Nitrite, and pH. The concentration of COD, BOD, TSS, and Nitrite has decreased significantly and reach more than 90% removal. While the acidity (pH), effluent leachate quality getting closer to neutral. However, this does not apply to the Nitrate. Nitrate concentration increased up to many times. This occurs because Nitrite had a reaction in the bioreactor and converted into Nitrate. Therefore, Nitrate concentration is not decreasing, on the contrary, has increased dramatically.

2. Characteristics of waste before and after drainage by leachate, performing decreasing in carbon, hydrogen and nitrogen content, but increased in oxygen content. Based on decreasing of those contents in waste, indicates that the bioreactor has a particularly active period for increasing effluent leachate quality. These bioreactors so that it can no longer be a solution of leachate generated in a landfill.

5. ACKNOWLEDGEMENT

Thanks to the Directorate of Research and Community Services University of Indonesia for awarded research grant through the funding of Higher Education Prime Program of Ministry of Research Technology and Higher Education.

6. References

[1] Adam, Gary A.M. 2015. Analisis Pengaruh Sifat Fisik-Kimia Sampah Terhadap Reduksi Volume Sampah Dan Karakteristik Air Lindi Pada Bioreaktor Landfill Aerobik Dan Anaerobik. Depok: Universitas Indonesia.
[2] ASTM D 5373-14,
[3] ASTM D3175-09
[4] ASTM D 6349-13
[5] Chung et al. 2015. *Acceleration of Aged-Landfill Stabilization by Combining Partial Nitrification and Leachate Recirculation: A Field-Scale Study*. Journal of Hazardous Materials 285 (2015) 436–444
[6] Dini. 2014. Analisis Pengaruh Resirkulasi Air Lindi pada Bioreaktor Landfill Metode Pengisian Sampah Berkala Terhadap Dekomposiis Sampah dan Konsentrasi Ammonia, Nitrat, Nitrit. Depok: Universitas Indonesia.
[7] Sidauruk, Ingen A. 2013. Analisis Pengaruh Resirkulasi Air Lindi terhadap Umur Landfill Ditinjau dari Parameter Fisik (Penurunan Sampah, pH dan Temperatur) dan Parameter Kimia (*Biochemical Oxygen Demand* dan *Chemical Oxygen Demand*) pada Bioreaktor Landfill. Depok: Universitas Indonesia.
[8] Sidauruk, Ingen A. 2014. Pengaruh Resirkulasi Air Lindi terhadap Kualitas Air Lindi dan Fisik Sampah pada Bioreaktor *Landfill*. Depok: Universitas Indonesia..
[9] SNI 06-6989.11 -2004
[10] SNI 06-6989.23 -2005
[11] SNI 698.72:2009
[12] Tchobanoglous, G., Theisen, H., & Vigil, S. 1993. *Integrated Solid Waste Management : Engineering Principles and Management Issues*. Singapura: McGraw Hill International Edition.
[13] Tchobanoglous, George. dan Frank Kreith. 2002. *Handbook of Solid Waste Management*. New York: McGraw-Hill.
[14] Undang-Undang Republik Indonesia Nomer 18 Tahun 2008, tentang Pengelolaan Sampah

IOP Conf. Series: Materials Science and Engineering **316** (2018) 012035 doi:10.1088/1757-899X/316/1/012035

Bearing capacity of helical pile foundation in peat soil from different, diameter and spacing of helical plates

F Fatnanta[1], S Satibi[1] and Muhardi[1]

[1)] Civil Engineering Dept, Faculty Engineering, Riau University, Campus Binawidya Jl. HR Subrantas, Pekanbaru 28294 Indonesia
e-mail: fatnanto1964@gmail.com

Abstract. In an area dominated by thick peat soil layers, driven piles foundation is often used. These piles are generally skin friction piles where the pile tips do not reach hard stratum. Since the bearing capacity of the piles rely on the resistance of their smooth skin, the bearing capacity of the piles are generally low. One way to increase the bearing capacity of the piles is by installing helical plates around the pile tips. Many research has been performed on helical pile foundation. However, literature on the use of helical pile foundation on peat soil is still hardly found. This research focus on the study of axial bearing capacity of helical pile foundation in peat soil, especially in Riau Province. These full-scale tests on helical pile foundation were performed in a rectangular box partially embedded into the ground. The box is filled with peat soil, which was taken from Rimbo Panjang area in the district of Kampar, Riau Province. Several helical piles with different number, diameter and spacing of the helical plates have been tested and analysed. The tests result show that helical pile with three helical plates of uniform diameter has better bearing capacity compared to other helical piles with varying diameter and different number of helical plates. The bearing capacity of helical pile foundation is affected by the spacing between helical plates. It is found that the effective helical plates spacing for helical pile foundation with diameter of 15cm to 35cm is between 20cm to 30cm. This behaviour may be considered to apply to other type of helical pile foundations in peat soil.

Keywords : *bearing capacity helical piles, helical pile foundation, peat soil, various number and spacing of helical plates.*

1. INTRODUCTION

Peat soil is a type of soil with high organic content [1]. Peat soil mainly comprises of water-saturated pores, which turns it to be highly compressible. Due to the high water content, peat soil can be categorized as very soft soil with very low bearing capacity.

To overcome these problems, light construction on peat soil is generally built using cerocok pile or conventional pile foundation. To save costs, the piles are generally installed far above the hard soil layer. Consequently, the axial load carrying capacities of the piles are determined solely by their skin frictions, which are generally low. To increase the pile bearing capacity, the area of the pile tips is increased by installing helical plates at around the tip of the pile foundation.

Many research has been performed on helical pile foundation. However, literature on the use of helical pile foundation on peat soil is still hardly found. Hence, there are still some gaps on the information about the application of helical pile foundation on the peat soil. Method and parameters considerations related to helical pile foundation design on inorganic soil cannot be adopted directly in

IOP Conf. Series: Materials Science and Engineering **316** (2018) 012035 doi:10.1088/1757-899X/316/1/012035

the design of helical pile foundation on peat soils. Therefore, this study was conducted to investigate the behaviour of helical pile foundation in peat soil.

2. METHODOLOGY

2.1 Materials and Equipment

The full scale helical pile loading tests were conducted in a site near the Soil Mechanics Laboratory of Civil Engineering Department, University of Riau. The tests were performed in a rectangular pit filled with peat soil taken from Rimbo Panjang area in the district of Kampar. Before each helical pile loading test was performed, vane shear test on the peat soil close to the test pile was done to obtain the shear strength parameter [2]. The equipment used for the tests is listed in **Table 1** and the schematic configuration of the tests is depicted in **Figure 1.**

Table 1. Equipment used for helical pile loading tests

Mechanical jack	Capacity: 5 tons
Proving ring	Capacity: 28 kN and 50 kN
Dial gauge	Maximum displacement: 50 mm
A rectangular box filled with peat soil	Dimension: 3 x 6 m, height 2.8 m

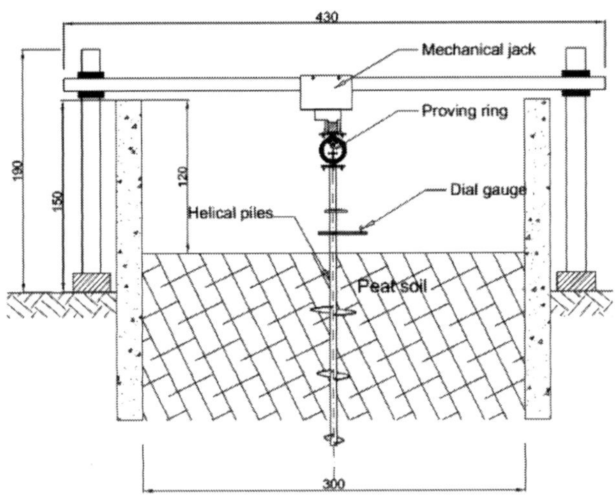

Figure 1. Schematic configuration of helical pile loading test

2.2 Helical Pile Test Nomenclature

The Helical piles were screwed down to the depth of 2 m in the peat soil. The piles diameter was 6 cm. The installed helical plates were 15, 25, and 35 cm in diameter. The 15cm diameter helical plate is referred as small plate (S), the 25cm diameter helical plate is referred as medium plate (M) and the 35cm diameter helical plate as large helical plate (L), as shown in **Figure 2.** The number of helical plates and their spacings were varied. The numbers of installed helical plates were 1, 2, or 3 plates. The spacings between plates were 20 cm, 30 cm or 50 cm.

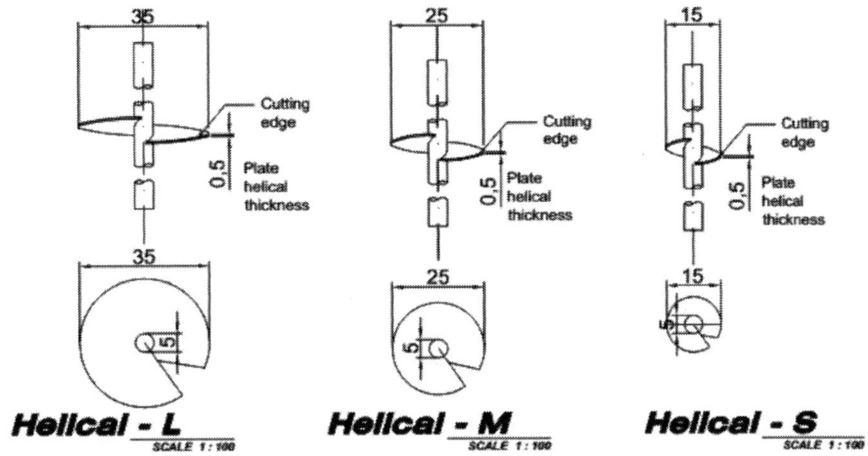

Figure 2. Helical Plate Diameters

If a helical pile is referred as LMS 30, then the helical pile is installed with three plates and a spacing of 30 cm. The upper plate L (large) is 35 cm diameter, the middle plate M (medium) is 25 cm diameter, and the lower plate S (small) is 15 cm diameter (see **Figure 3**).

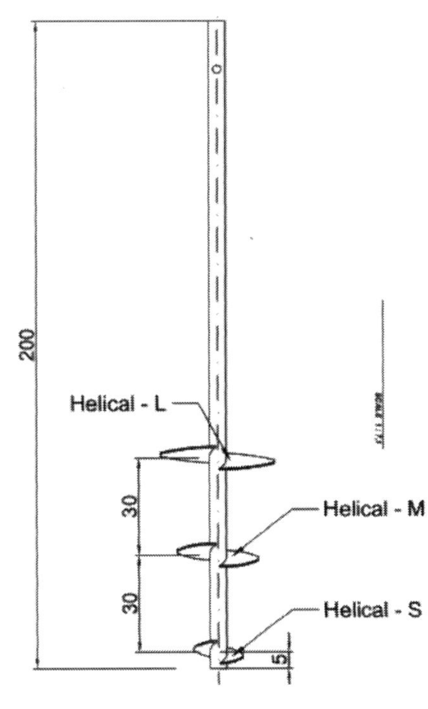

Figure 3. Helical pile type LMS 30

Each helical pile is named based on the number, diameter, and spacing of the helical plates. Several types of helical pile used in this study are tabulated in **Table 2.**

IOP Conf. Series: Materials Science and Engineering **316** (2018) 012035 doi:10.1088/1757-899X/316/1/012035

Table 2. Types of tested Helical Piles

Helical Plate Diameter, cm			Spacing, cm	Nomenclature
Upper	**Middle**	**Lower**		
-	35	35	20	LL 20
-	35	35	30	LL 30
-	35	35	50	LL 50
-	35	25	20	LM 20
-	35	25	30	LM 30
-	35	25	50	LM 50
35	25	15	20	LMS 20
35	25	15	30	LMS 30
35	25	15	50	LMS 50
35	35	35	50	LLL 50
-	35	35	50	LL 50
-	-	35	-	L
-	-	-	-	Pile without helical plate
-	-	-	-	Cerocok pile

2.3 Test Procedure and Interpretation of Ultimate Bearing Capacity of Helical Pile

The helical piles were screwed down to the depth of 1,5 m in the peat soil. Before the loading was conducted, field vane shear tests were performed in the peat soil at 50, 100, and 150 cm depth. The axial loading tests for the helical pile foundation was performed using constant rate of penetration method. Mechanical jack, proving ring, and dial gauge were used for the pile test [3]. The rate of penetration was 1 mm per 30 seconds (2 mm/minute). The load was recorded at every 1 mm of settlement. The test was performed until the maximum settlement of 45 mm, which was based on the capacity of the dial gauge. **Figure 4** shows typical load-settlement plot based on the data acquired from load test.

The bearing capacity of a foundation is limited by two limit states. First, the ultimate shear failure must be adequate and second, the settlement of the foundation shall not exceed tolerable values [4]. In the case of peat soil, foundation design is most likely governed by settlement, rather than shear strength. Therefore, the bearing capacity is calculated based on 25 mm of [5]. This method is simpler and has the same interpretation for load versus settlement data. For example, in a load-settlement plot shown in **Figure 4**, the bearing capacity of the helical pile is 8.3 kN which is selected at 25 mm of settlement.

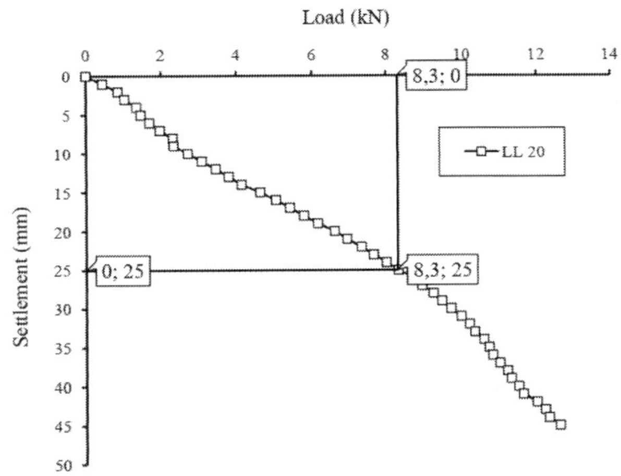

Figure 4. Typical load vs settlement plot for helical pile

3. RESULT AND ANALYSIS

3.1 Soil Properties

Soil samples were taken at depths of 50 – 100 and 100 – 150 cm. The peat soil properties are tabulated in **Table 3.**

Table 3. Peat Soil Properties

Characteristics	Unit	Depth (cm)	
		50 - 100	100 - 150
Water content	%	190.949	236.837
Dry density	gr/cm^3	0.273	0.236
Specific gravity	-	1.355	1.480
Ash content	%	35.544	54.463
Fibre content	%	5.281	6.114

Figure 5 shows the peat soil shear strength test results for each foundation type. In **Figure 5** it can be seen that the shear strength of the peat soil is between 3.83 to 5.25 kPa. Vane shear test is often used for measuring the shear strength of an organic soil. In engineering practices, some reduction factors are introduced to produce safe design. The reduction factors were developed in response to local experience and practice [6-9].

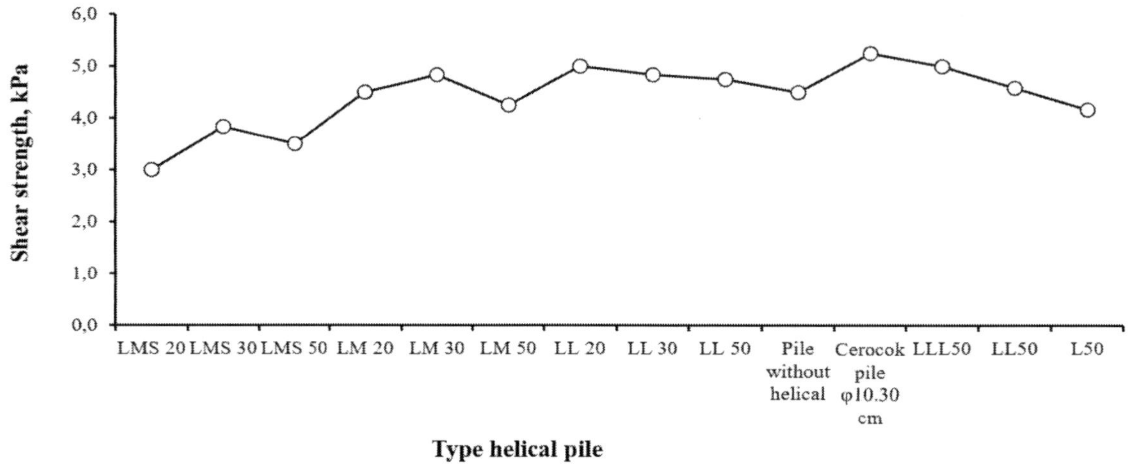

Figure 5. Peat Soils Shear Strength

3.2 Number of Plates

Sprince and Pakrasinsh [10] stated that the bearing capacity of a helical pile is affected by the number of helical plates. This research is conducted to investigate the impact of the number of helical plates to the bearing capacity of helical pile in peat soil. The results show that as the number of helical plates is increased, the bearing capacity is also improved. The improvement of the bearing capacity is caused by the larger surface area of the foundation as the number of helical plates increase. LLL50 helical pile has larger bearing capacity than L50 and LL50, as shown in **Figure 6.** This indicates that the helical pile bearing capacity is dominated by helical plate resistance. This result agrees with other research on helical pile foundation such as by Rao et. Al. [11] and L.A and Jong [12].

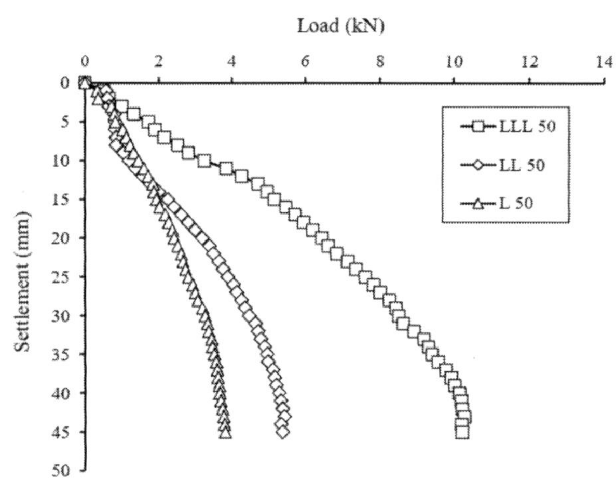

Figure 6. Load vs settlement with various number of helical plates [13]

3.3 Plate Spacing

The bearing capacity of helical plate foundation is affected by plate spacing. The use of effective plate spacing will give maximum bearing capacity [14]. In this study, tests were conducted on LL20, LL30, LL50, LM 20, LM 30, and LM 50 helical pile foundations. Test results show that LM 30 and LL20 helical foundation provide maximum bearing capacities, as shown in **Figure 7.** The result also indicates that effective plate spacing is also influenced by helical plate diameter. The bottom helical plate plays significant role to increase the bearing capacity of helical foundation. LM50 foundation bearing capacity is smaller than LL50. It occurs because the bottom helical plate in LM 50 foundation is of small diameter, thus the end resistance is small.

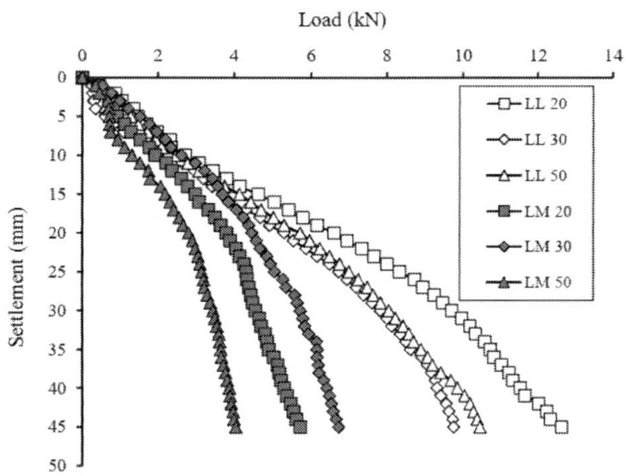

Figure 7. Load vs settlement with various spacing of helical plate [15]

3.4 Helical Pile Compared with Cerocok Pile (Wooden pile) And Pile Without Helical Plate

The addition of helical plates is aimed to improve the pile sectional area such that the bearing capacity of the foundation will increase. However, the magnitude of the improvement in bearing capacity needs to be checked by comparing helical pile bearing capacity with conventional pile without helical plate. In addition to that, since one of the backgrounds of this study is to find alternative to cerocok pile foundation, it is important to show how much the bearing capacity of a helical pile foundation is improved relative to cerocok pile foundation.

Figure 8 shows that the installation of helical plates significantly increases the bearing capacity of foundation. The bearing capacities of LL20 and LMS 50 helical pile foundations is significantly larger than cerocok pile foundation and pile without helical plate of similar size. The load carrying capacity of cerocok pile is mainly provided by the frictional resistance between the relatively rough pile skin and the soil layer. The pile foundation without helical plate exhibits the smallest bearing capacity. This is due to its smallest diameter of 6 cm and the smoother pile skin surface compared to cerocok pile.

3.5 Prediction vs Measurement

According to Perko [12] there are two methods to approximate the bearing capacity of helical foundation. The first one is individual bearing method, and the second one is cylindrical shear method. Individual bearing method assumes that each helical plate contributes to bearing capacity, whereas cylindrical shear method assumes that the bearing capacity is provided by the friction between the soil and helical plates.

The comparison analysis of pile bearing capacities in this study is conducted using both methods, the individual bearing method and the cylindrical shear method. The peat soil shear strength reduction factor is taken as 0,65, which is slightly higher than 0,5 – 0,57 as suggested by [6] and [7]. The change in reduction factor was done to obtain the smallest error between the predicted and measured bearing capacity.

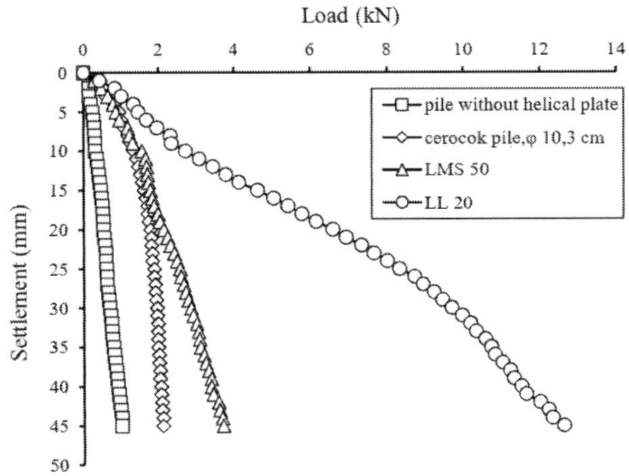

Figure 8 Load vs settlement plate with and without helical plate and cerocok pile [13]

In **Figure 9,** it can be observed that the helical pile bearing capacities predicted using individual bearing method is close to the test results. This is indicated by the smallest error value of 18,80%. Therefore, it can be concluded that the shear strength reduction factor of 0,65 of the peat soil shear strength test results and the individual bearing method provides best fit between the measured and predicted values. If the value of shear strength reduction factor is changed, then the error value increases. However, individual bearing method still having lowest error value.

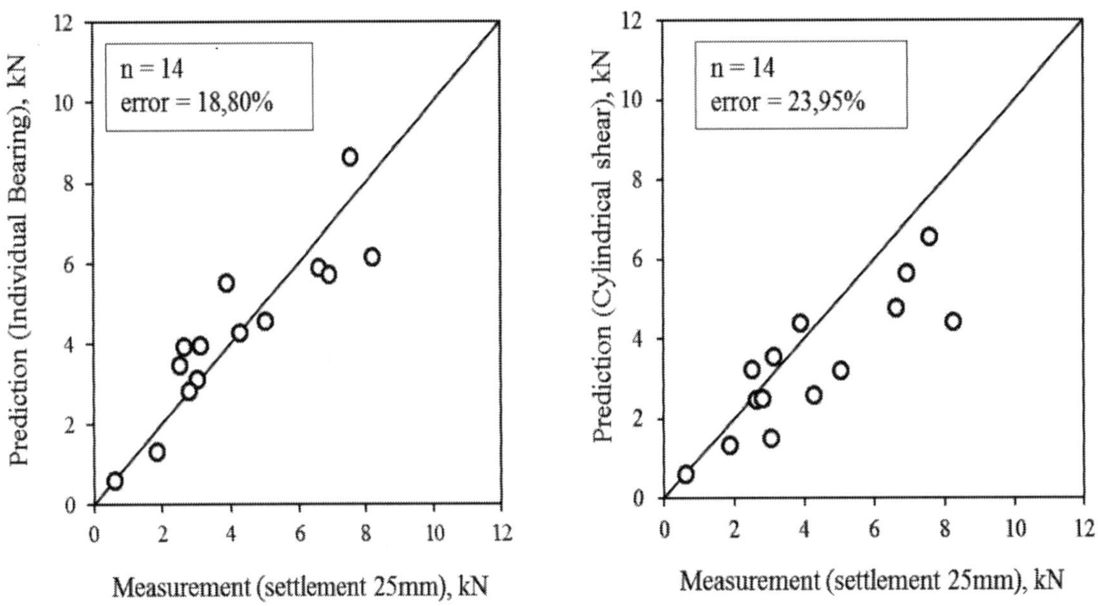

Figure 9 Measured vs. Predicted Bearing Capacity of Helical Pile.

4. CONCLUSION

The conclusion of this research are as follows:

The pile foundation with three uniform helical plates produces the largest bearing capacity compared to helical pile foundations with one and two plates. The bearing capacity of a helical pile is affected by helical plate spacing. The effective spacing for L (35 cm), M (25 cm), and S (15 cm) helical pile diameters as used in this study is between 20 to 30 cm.

The bottom helical plate plays significant role to increase the bearing capacity of a helical pile foundation. The bearing capacity of LM 50 foundation is smaller than LL 50, which shows that smaller bottom plate gives smaller end point resistance.

The peat soil shear strength reduction factor s_u used in this research is 0,65, which is determined based on the the smallest error of 0.188.

Helical pile foundation exhibits significantly greater bearing capacity than 10,30 cm diameter cerocok pile. Therefore, the application of helical pile foundation in peat soil can be more profitable than cerocok pile foundation.

The individual bearing method provides better prediction on the bearing capacity of helical pile compared to cylindrical shear method.

5. REFERENCES

[1] Mochtar, N.E. & Imananto, E.I., 1999. Aplikasi Model Gibson dan Lo untuk Tanah Gambut Berserat di Indonesia. *Jurnal Teknik Sipil*, 6(1), pp.1 – 8.
[2] ASTM D2573-01., 2001, *Standard Test Method For Field Vane Shear Test in Cohesive Soils*. United States of America: Annual Book of ASTM Standards sec 4.
[3] ASTM D1143-81., 1994, *Standard Test Method For Piles Under Static Axial Compressive Load*, United States of America: Annual Book of ASTM Standards sec 4.
[4] NAFVAC DM 7.02, 1986, *Foundations and Earth Structures*, Naval Facilities Engineering Command 200 Stovall Street Alexandria Virginia
[5] Terzaghi, K and Peck, R. B, 1967. *Soil Mechanics in Engineering Practice*, John Wiley & Sons, New York.
[6] Golebiewska A., 1983, *Vane Testing in peat*, Proceeding 7th Danube ECSMFE, Kishinev, Vol 1, pp. 113-117
[7] Landva A.O., and La Rochelle P., 1983, Compressibility and shear strength characteristics of Radforth Peats, Testing of Peats and Organics Soils, ASTM STP 820, P.M. Jarret (ed), pp. 157-191
[8] Hanrahan E.T., 1994, *Discussion on Major canal reconstruction in peat by Pigott et al.*, Proceeding Instn. Civil Engineering Water Marit & Energy, Dec 1994., pp. 399-403
[9] Hanzawa H., Kishida T., Fukusawa T., Asada H., 1994, *A case studey of the application of direct shear and cone penetration test to soil investigation, design and quality control for peaty soils*, Soil and Foundations, Vol. 34 No. 4 pp. 13-22.
[10] Sprince, A dan Pakrastinsh, L.,. 2010. Helical Pile Behaviour And Load Transfer Mechanism In Different Soils. *Journal of Riga Technical University*, vol. 10, No. 10: 121 – 130.
[11] Narasimha Rao. S, Prasad, Y. V. S. N, Shetty, M.D. 1991. *The Behaviour of Model Screw Piles in Cohessive Soils*. Soils and Foundation, vol. 31, No. 2: 35 – 50.
[12] L., A. & Jong, O.E., 1995. Pengaruh Posisi "Helical Plate" pada Dinding Tiang Pondasi Terhadap Peningkatan Daya Dukungnya. *Jurnal Universitas Kristen Petra*, Surabaya.
[13] Perko, Howard, 2009, *Helical Piles (A Pratical Guide to Design and Installation)*, New Jersey: John Wiley & Sons, Inc.
[14] Parlan, 2016, *Pengaruh Jumlah Pelat Helical Terhadap Daya Dukung Tekan Pondasi Tiang Helical pada tanah Gambut*, Tugas Akhir Program Studi S1 Teknik Sipil Fakultas Teknik Universitas Riau.
[15] Ditra, Ralan, 2016, *Analisis Pengaruh Variasi Jarak Pelat Helical terhadap daya Dukung Tekan Helical Pile pada tanah Gambut*, Tugas Akhir Program Studi S1 Teknik Sipil Fakultas Teknik Universitas Riau.

IOP Conf. Series: Materials Science and Engineering **316** (2018) 012036 doi:10.1088/1757-899X/316/1/012036

Deformation of high performance concrete plate under humid tropical weather

C Niken[1], T Elly[2], FX Supartono[3], I Laksmi[4]
[1,4]Senior Lecturer, Faculty of Engineering, University of Lampung, Indonesia
[2]Head of Structure and Material, Faculty of Engineering, Universitas Indonesia, Indonesia
[3]Senior Lecturer and Head of Partono Fondas Consultant, Indonesia

E-mail: chatarinaniken@yahoo.com

Abstract. This paper presents the relationship between surrounding relative humidity and temperature on deformation behavior of one sample concrete plate with compressive strength of 60MPa. This research was done in Indonesia that is in humid tropical weather. A specimens measuring 3000 mm×1600 mm × 150 mm were used. The behavior was obtained by using four embedded vibrating wire strain gauges (VWESG). As a result there is a very strong relationship between humidity and deformation at the age range of 7 until 21 days. The largest deformation occurs in the corner and the fluctuation of deformation in side position is larger than in the corner and in the middle. The peaks of surrounding relative humidity were fully followed by the deepest valley of deformation on time in the corner, while in another position the range delay time was 8 - 11 hours. There is a strong relationship between surrounding temperature and deformation at the range of 7 until 14 days. The influenced of surrounding relative humidity to concrete behavior is faster and longer than surrounding temperature. The influence of surrounding temperature in humid tropical weather was shorter than in non-humid tropical weather.

1. Introduction

Surrounding relative humidity (SRH) and temperature are never constant. The humidity and temperature penetrates to inside concrete through the concrete pores through the connected to the air surface. Element structure that has largest surface contacts to surrounding humidity and temperature is plate. Sarman, et al., 2015 observed concrete flat roof defects in equatorial climates [1]. Kim, et al., 2014 obtained finite element modeling of environmental effect on rigid pavement deformation [2].

To accommodate the need of increasing durability and load, many designers used high performance concrete (HPC). HPC is defined commonly as concrete with a minimum compressive strength of 60 MPa [3]. To obtain high strength concrete with good performance, more cement but less water is used in HPC than in normal concrete.

The response of materials depends on environmental conditions. The air temperature and relative humidity are changing time by time during 24 hours; therefore, its influence is not stable. The influence of the environmental to concrete depend is on concrete pores. The growth of product hydration changes the amount and size of pores.

Humid area is an area with high humidity. Humidity more than 40%, was mentioned less creep and shrinkage by ACI 209R, 1997 [4]. Although, high humidity is revealed less deformation, but high fluctuation of it, absolutely give a serious impact to concrete performance. Based on the observation of Niken et al, 2013, starting November 2009 until December 2011 in Jakarta as the representation humid tropical weather, the average humidity and temperature in Jakarta is 72% and 27°C respectively as shown in Fig. 1 and 2 [5]. Large deviation between maximum and minimum humidity occurs over the year (Fig.1).

A. Jakarta B. New York

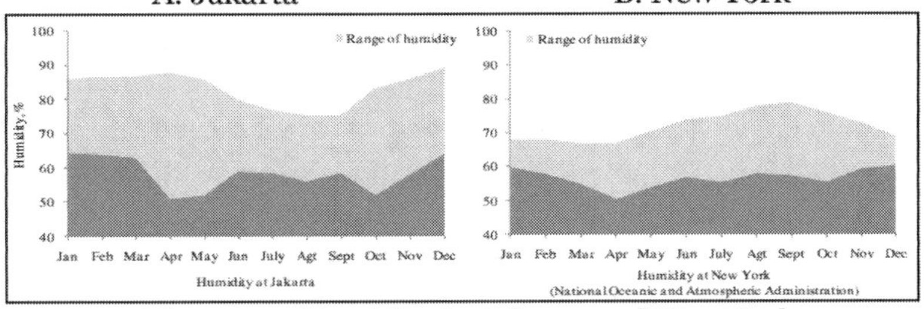

Figure 1. Humidity in Jakarta and New York

A. Jakarta B. New York

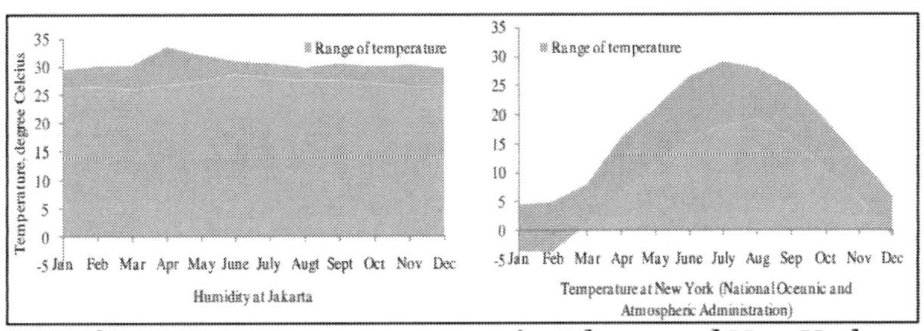

Figure 2. Temperature range in Jakarta and New York

Temperature effect is usually considered to be less important than relative humidity. Although, it is less important, at the temperature of 50°C, creep strain is approximately two to three times than creep strain in the temperature less than 19 - 24 °C [4].

Surrounding humidity and temperature influence internal humidity and concrete deformation. The influence of internal humidity to concrete has already learned [6]. Various studies of the influence of temperature on concrete behavior have been published among others [7, 2, 8, and 9]. As a guide, ACI 209 R, 1997 has accommodated the influence of humidity and temperature on concrete shrinkage and creep by giving a factor [4].

Researchers as mentioned above, consider the influence of humidity on concrete is separated with the influence of temperature, even though there is a relationship between humidity and temperature by Dew point temperature [10]. Researchers who had studied about the influence of humidity and temperature together on concrete are Xu, et al., 2016 and Saha & Eckelman, 2014 [11,12]. They obtained degradation of material due to moisture and temperature.

2. Methodology/Experimental

The research was conducted in Jakarta, Indonesia, with humid tropical weather. This research was performed experimentally using one specimen of 3000 mm×1600 mm × 150 mm, with four embedded vibrating wire strain gauges (EVWSG) as shown in Fig. 3.

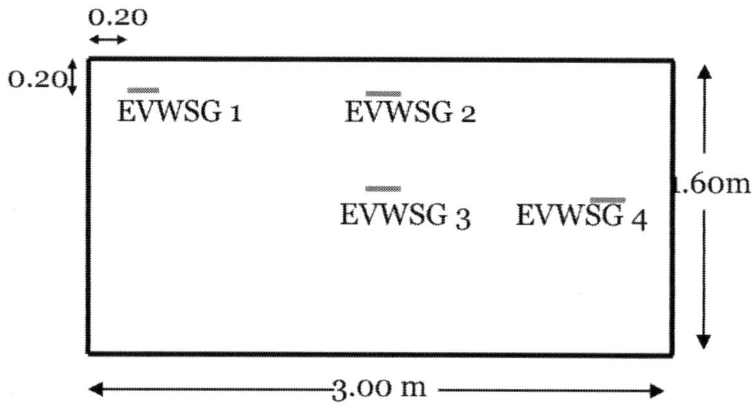

Figure 3. The scheme of plate and the placement of vibrating wire strain gauges

HPC with target compressive strength of 60 MPa and slump flow diameter of 35 ± 2 cm was applied. The plate was cast at 22.10 pm with surrounding temperature of 25.4°C and relative humidity of 94%.

2.1. *Materials*

The mix design was conducted in compliance with ACI 211.4R, 1993 [13], with a limit of 500 kg/m^3 cement content to meet the shrinkage factor closest to 1 [4]. Ordinary Portland Cement (OPC) produced by Indocement Ltd was used. Condition of the aggregate was saturated surface dry (SSD). Fine aggregate in the form of river sand was brought from Sungai Liat (Bangka, Sumatra, Indonesia); specific gravity (SSD) was 2.605; and absorption was 0.4%. The sand had been filtered and cleaned using a mixture of standard graphs obtained from the mid-gradation, according to ASTM C.33/C.33M-08 [14]. Coarse aggregate in the form of volcanic rock fragments was obtained from Banten, West Java, Indonesia. Composition of the coarse aggregate used was 70% size of 13 - 19mm; specific gravity (SSD) was 2.563; absorption was 1.543%; and 30% size of 6 - 12mm; specific gravity was 2.636; and absorption was 2.26%.

Added material used was silicafume of 8% cement weight produced by Sika Indonesia Ltd. To achieve the desired high strength with low water to cementitious material ratio and good workability, polycarboxylic superplasticizer under the commercial name Visco Crete 10 from Sika Indonesia Ltd was added to the concrete mix as the high range water reducer (HRWR). The dose of HRWR of 1.4% cement weight was added according to that generally used in Indonesia. Local water was supplied by the Structure and Material Laboratory of Universitas Indonesia. An electrical scale was used especially for cementitious materials, HRWR and water to obtain the accurate water to cementitious material ratio. The mix composition is OPC 500 kg/m^3, silica fume 40 kg/m^3, water 142.6 kg/m^3, sand 800 kg/m^3, coarse aggregate 935 kg/m^3, and HRWR 7.6 kg/m^3.

During the concrete mix design stage, all of the aggregate were assumed to be under saturated surface-dry condition. A tilting drum mixture of 0.3m^3 capacity was used. The mixing started with all cementitious material in dry condition, followed with 50% fine aggregate. Subsequently, 50% water was added to the revolving mixture. Then these materials were mixed for approximately 1.5 minutes. Next, 50% water was slowly poured in, which was mixed with HRWR homogenously. Thereafter, 100% coarse and 50% fine aggregate were added. With all the materials placed according to its order into the mixer, the concrete was mixed for approximately 3 minutes. The slump flow of the mixture was measured before pouring by using Abram's cone upside down.

2.2. *Methods*

Shrinkage was measured as strain change against time by installing four embedded vibrating wire strain gauge (EVWSG) in the specimen (Fig. 3). The EVWSG has abilities to detect the strain up to 3000 με with accuracy of about .025% and concrete temperature between -80°C and 60°C with about .5% accuracy. Surrounding relative humidity and temperature were obtained by using room humidity and thermometer.

Right after casting specimens were covered with plastic membrane to eliminate water evaporation. The specimens were cured after de-molding (one day after casting) by covering wet sacks over the specimens to the age of 7 days. After this treatment, specimen was allowed to intersect with surrounding relative humidity and temperature but with protection again raindrops. Observation was performed right after pouring as follows: 0 - 24 hours, every 15 minutes; 24 - 48 hours, every 60 minutes; days 3 - 10, every 2 hours; days 10 - 14, one time a day using a read out. Two beams of 3000 mm×600 mm × 200 mm were cast on the plate at the plate age of 14 days; therefore, on the plate occurs creep. Fifteen days is selected because load generally applied on the rigid pavement or slab. Observation was done right after applied load as follows: days (specimen age) 14 - 16, every 15 minutes; days 16 - 17, every 60 minutes and days 17 - 21, every 2 hours.

Data were analysis in the age range of 7 until 21 days. The deformation of every position is sought for the cause of its behavior difference by literature study and the properties of water and heat movement.

3. Result and Discussion

3.1. *Result*

The observation results of all deformation were shown in Fig. 4.

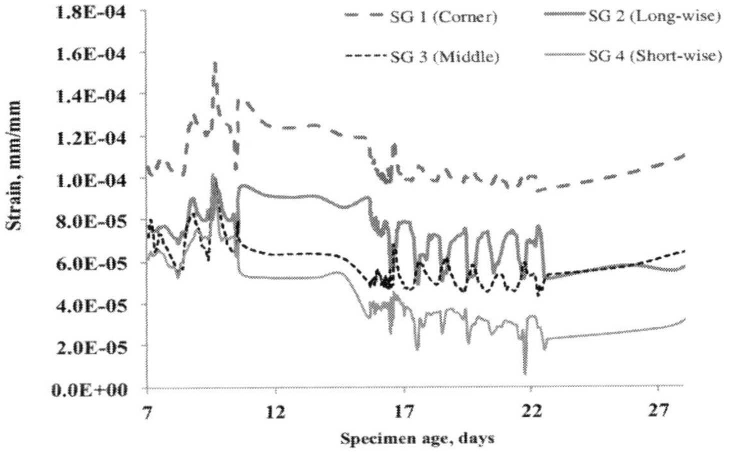

Figure 4. Plate deformation of all vibrating wire strain gauges

Observation results of surrounding humidity, temperature and concrete temperature were shown in Figure 5. Concrete temperature is lower than surrounding temperature since 17 days age as shown in Fig.5.

The lowest SRH for every fluctuation happen at the time range of 10.00 am to 04.00 pm. Only 30% the lowest of SRH occur at the time range of 10 am until 11.00 pm.

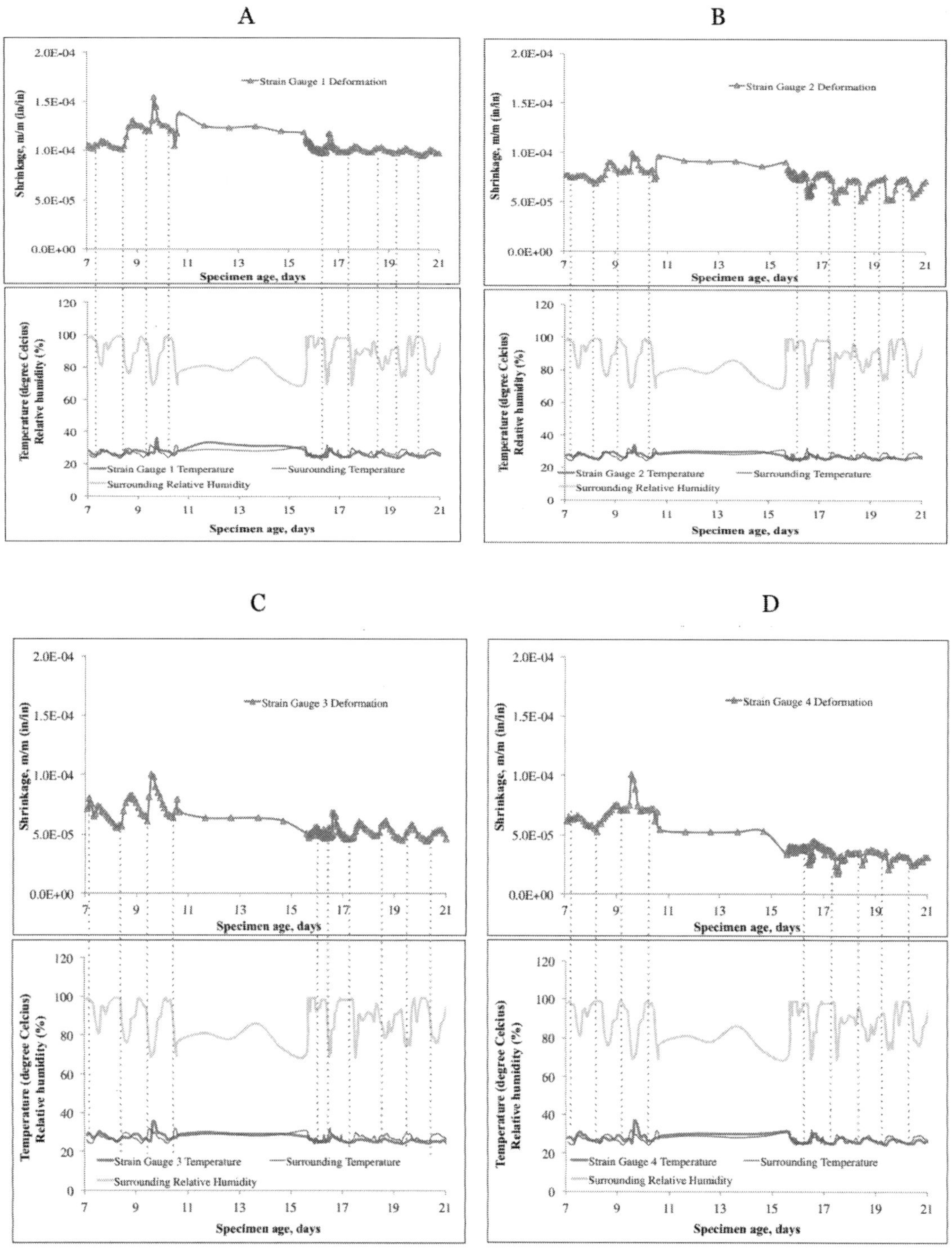

Figure 5. Relationship between concrete deformation, concrete temperature, surrounding relative humidity and surrounding temperature (7 - 21 days), (A) Strain gauge 1 (corner), (B) Strain gauge 2 (longwise), (C) Strain gauge 3 (middle), (D) Strain gauge 4 (short wise).

3.2 Discussion

Deformation in all position shows similar type (Fig.4). The largest deformation occurs in the corner position and the smallest is in the short-wise position. Although, the plate is placed on the soil, corner deformation is larger than deformation in another position (Fig.4). The difference was mentioned by Huang, 2003 [15].

At this age of 7 – 21 days, hydration is a major process in concrete. Cement becomes a bond material in concrete formation through the hydration process. Deformation caused by hydration process, and the process was determined especially by water. Cement it-self, is very sensitive to the water even water content in the air. By ensnaring the water in air, cement grains forms a bond itself. At humid tropical weather there are a lot of water vapor available in the air together with high temperature over the year (Fig. 1A and 2A); therefore, the hydration process occurs under this situation. High humidity condition makes the growth of silica bridges as a bond becomes optimum [16]. Because in humid weather area the high humidity is occur over the year, hydration rate or the form of bond to be faster than in non-humid area.

There are two types deformation: shrinking and swelling as displayed in Fig. 5 and 6. The growth of product hydration causes volume shrinks as a consequence of the smaller concrete paste ingredients volume at mature time than at mixing time. Swelling is caused the available of Ca^{2+} as a product of the reaction [17] and capillary pore tension. Pores were formed because during concrete manufacture, some air was trapped. When the concrete formed at the beginning, it is many and large size. Hard cement paste, has enormous areas of internal pores that is about $500m^2/cm^3$ [18]. The amount and size of pores are decreases according to concrete age [19]. The pores become entrance and exit of air water vapor and air temperature. In plate, environment moisture and heat enters the concrete from the top surface, and other than that, at longwise and short wise position the moisture and heat enter also from one side of the plate, and for the corner position enters through two plate sides that are perpendicular to each other.

Moisture that enters from one side will accelerate moisture movement from top surface; therefore, the deformation becomes more fluctuate according to SRH fluctuation (Fig. 5B and 5D) at the age of 15 – 21 days. While in the corner, moisture also enters through two sides (perpendicular to each other) leads the moisture to more spread sideways and make the mode of fluctuation smoother (Fig. 5A and 5C). This matter will change internal moisture of the concrete. The first change of internal moisture occurs in the position directly contact with surrounding environment; and then distribute to the deeper part of concrete. If the water moves in and out from the micro-pores, macro-diffusion will occur. The infiltration and distribution of air water vapor depend on the number and size of pores, and will be accelerated by the surrounding air pressure. This infiltration water will increase internal moisture and then lead the increase of hydration process rate.

The several of water that infiltrates, will stay in pores. The characters of this water depend on the pores size and humidity [20, 21]. The pores size was distributed over a wide range. Pores with diameters of 10 - 20 nm are related to pores space of C-S-H hydrat cluster, and pores with diameters of 1 - 2 nm is related to internal porosity of hydrat [22]. Pores with a diameter of 10 nm are often formed in the normal concrete [20].

The use of low water to cement ratio or silica fume (nano particles materials) in high performance concrete and high strength concrete (HSC) makes the pores condition as mention above becomes change, big capillary pores decrease, medium capillary pores (diameter of 50 -10nm) and little gel pores (diameter of 10 - 2.5nm) increase. If low cement to water ratio was used together with silica fume, the size of pores getting smaller again, because silica fume is a nano material [23]. Big capillary pores have no menisci; thus, no capillary tension force; while medium capillary pores and little gel pores have strong menisci [20]. This condition makes capillary tension force in HPC and HSC higher than in normal concrete. More over, water in medium capillary pores and small gel pores can evaporate even in high relative humidity situation; on the other hand, the concrete also can shrink in the similar condition [20].

The peaks of surrounding relative humidity were fully followed by the deepest valley of deformation (Fig. 5). Delay times between peak of surrounding relative humidity and deepest valley of deformation means infiltration of moisture and temperature meet some barriers. In the longwise deformation, largest fluctuation occurs (Fig. 5B). The fluctuation type was similar to deformation in the short-wise position (Fig.5D).
The peaks of surrounding temperature were also followed by the peaks of deformation (Fig.5). Delay times between the both were also occurs. The delay time can be seen in Fig. 6.

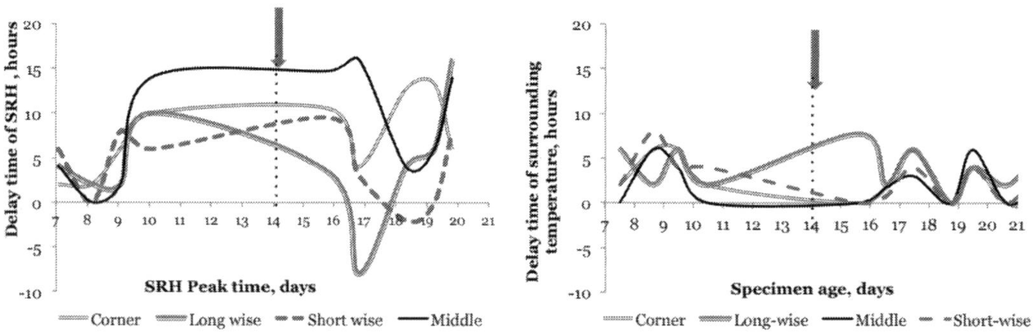

Figure 6. Delay time of (A). Peak of surrounding relative humidity and deepest deformation; (B). The peak of surrounding temperature and the peak of deformation

Delay time between peak of deformation and peak of surrounding temperature is about half delay-time correspondent to surrounding relative humidity (Fig.6A and Fig.6B); thus, surrounding temperature is easier to penetrate the concrete than surrounding relative humidity. This result shows that surrounding temperature meets significant barriers that are the available of pores water and product hydration. Because the barrier has a fluctuation in the form of sinusoidal type according to surrounding relative humidity; therefore, the delay time is also has a similar fluctuation too (Fig. 6B). The long delay time means that surrounding relative humidity meet strong barrier. The strong barrier caused by same cause of surrounding temperature and the movement of particles as a consequence of the ease of surrounding temperature infiltration.

Temperature in Jakarta is more than 24°C and occurs over the year (Figure 2); thus, the heats energy level is also high. The highest temperature was reached at 9.7 days with 36.7°C (Fig. 5). This temperature is 12.7°C exceeded normal range temperature of ACI 209R, 1997 [4]. It can be meant 1.5 – 2 times than creep strain in the temperature less than 19 - 24 °C, if linear interpolation was used. High temperature infiltration makes atoms and molecules move rapidly. The move was depends on its strength barrier. The barriers are pores water and silica bridges which growth rapidly. If the barriers become strong and increase in number, surrounding temperature more difficult to penetrate and its influence to concrete deformation decrease.

Relative humidity in Jakarta is more than 40%, therefore shrinkage factor by relative humidity is more than 1. Water vapor that enter in concrete is more difficult to exit than inner temperature, because the barriers growth; therefore, several water still stay in the pores. Generally, pores water temperature is lower than surrounding temperature because no direct connection. This condition together with decreasing hydration heat, and water vapor that still stay in the concrete will make concrete temperature to be lower than surrounding temperature. This condition makes concrete temperature become lower than surrounding temperature. This phenomenon occurs at the age more than 16 days (Fig. 5).

High heat energy occurs together with high surrounding relative humidity. High heat energy makes the particles move rapidly, and high surrounding relative humidity leads product hydration growth rapidly. Product hydration would be a barrier for particles to move; thus, the influence of temperature to concrete deformation can shorter than in non-humid weather. The particles movement meets strong barrier at the age of 14 days because inner C-S-H were formed [20].

We also note that there is different phenomenon of the environment during night and day. At noon, temperature and relative humidity fluctuations are influenced by direct sunrays radiation. While during the night, the temperature and relative humidity are mainly influenced by output energy of the earth's surface [24]. From this observation, the lowest surrounding relative humidity occurs in day time, in between 10 am to 04.00 pm. This condition makes high fluctuation of surrounding relative humidity at noon and smooth fluctuation at night (Fig. 5 and 6). To avoid, the influence of high surrounding temperature at the beginning hydration process, concrete was casting in night.

4. Conclusion

The conclusions of this research are:

1. Deformation in all position shows similar type.
2. The largest deformation occurs in the corner position and the smallest is in the short-wise position.
3. Moisture that enters from one side will accelerate moisture movement from top surface and the deformation becomes more fluctuate according to surrounding relative humidity fluctuation
4. Moisture enters through two sides (perpendicular to each other) leads the moisture to more spread sideways and make the mode of fluctuation smoother
5. The peaks of surrounding relative humidity were followed by the deepest valley of deformation
6. The peaks of surrounding temperature were followed by the peaks of deformation
7. Delay times between peak of surrounding relative humidity and deepest valley of deformation means infiltration of moisture and temperature meet some barriers
8. Delay time between peak of deformation and peak of surrounding temperature is about half delay-time correspondent to surrounding relative humidity
9. Delay time correspond to temperature has a sinusoidal fluctuation
10. Concrete temperature is lower than surrounding temperature at the age more than 16 days
11. The influence of temperature to concrete deformation is shorter than in non-humid weather

References

[1] Sarman AM, Nawi MNM and Che-Ani AI 2015, *Int. J. Eng. Res.* **10** 3 7319-7324.
[2] Kim S, Ceylan H, and Gopalakrishnan K 2014, *J. Front. Struc. Civ. Eng.* **8** 2 101-114.
[3] Nishiyama M 2009, *J. Adv. Con. Tech.* **7** 2 157-182.
[4] ACI 209R-92 1992 *Prediction of Creep, Shrinkage, and Temperature Effects in Concrete Structure* (American Concrete Institute) p 1 - 47.
[5] Niken C, Elly T and Supartono FX 2013, *J. Civ. Envir. Res.* **3** 2 35-46.
[6] Bažant ZP and Chern JC 1985, *J. Matériaux et Construction.* **18** 103 1-20.
[7] Klovanych and Sergei 2015, *J. Comput. Con.* **16** 6 897-908.
[8] Ceylan H, Kim S, Gopalakhrisnan K and Wang K 2014, *J. Trans. Res. B.* **2037** 03 30-39.
[9] Ladaoui W, Vidal T, Seller A and Bourbon X 2011, *J. Mater. Struc.* **44** 9 1629-1639.
[10] Linden J and Trewin S 2013, *Understanding the Role of Temperature, Relative Humidity and Dew Point in Creating a Sustainable Preservation Environment* (Image Permanence Institute: RIT presentation).
[11] Xu C, Mudunuru MK, and Nakshatrala KB 2016, *J. Continuum Mech. Therm.* **28** 6 1847-1885.
[12] Saha M and Eckelman MJ 2014, *J. Urb.Clim.* **9** 101-114.
[13] ACI 211.4R 1993 *Guide for Selecting Proportion for High Strength Concrete with Portland Cement and Fly Ash* (American Concrete Institute) p 272 - 283.

[14] ASTM C33/C33M-08 2009 *Construction, Concrete and Aggregates: Standard Specification for Concrete Aggregates* (West Conshohocken) p12 - 18.
[15] Huang CS 2003, *Int. J. Mech. Sci.* **45** 1-20.
[16] Gartner E, Kurtis KE, Monteiro PJM 2000, *J.Cem.Con. Res.* **30** 5 817-822.
[17] Paulini P 1990, *J. Cem. Con. Res,* **20** 910-918.
[18] Bažant ZP, Fellow, ASCE, Haugaard AB, Baweja S and Ulm FJ 1997, J.Eng.Mech. **Nov** 1188-1194.
[19] Kurtis K 2009 *Portland Cement Hydration* (Georgia Institute of Technology, Atlanta, Georgia).
[20] Taylor HFW 1997 *Cement Chemistry* (Thomas Telford, London).
[21] Bentz DP and Aïtcin PC 2008, *Concrete Int.* 51-54.
[22] Morin V, Cohen-Tenoudji F, Feylessoufi A, and Richard P 2002, *J. Cem.Con. Res.* **32** 1907-1914.
[23] Zhang M-H and Li H 2011, *J. Constr. Build. Mater.* **25** 2 606-615.
[24] Bahtiar ET, Arinana, Nugroho N, and Nandika, D 2014, *Asian J. Sci. Res.* **7** 501-512.

Performance of Asphalt Concrete Wearing Course (AC-WC) Utilizing Reclaimed Asphalt Pavement from Cold Milling Bound with 80/100 Pen Asphalt

I N.A. Thanya[a], I W. Suweda[b], and G.K. Putra[c]
[a,b,c] *Civil Eng. Dept. Udayana University, Bukit Jimbaran-Bali, Denpasar, 80361, Indonesia.*
Email corresponding author: aryathanaya@unud.ac.id

Abstract. Demands on natural aggregate materials for road pavement can be reduced by utilizing reclaimed asphalt pavement (RAP). This research was aimed at evaluating the performance of AC-WC mixture using RAP materials from cold milling, bound with 80/100 pen asphalt. The RAP aggregate gradation was adjusted by adding the required amount of natural aggregates to meet the specification in Indonesia. The RAP and added aggregates were hotmixed and compacted with Marshall hummer at 2x75 blows. The asphalt content were varied. It was found that the optimum asphalt content was 6.05 % with the following Marshall characteristics: stability 1237.08 kg; flow 3.36 mm; Marshall quotient 324,73kg /mm; void in mix (VIM) 3,360% ; void in mineral aggregate (VMA) 15.103; and void filled with bitumen (VFB) 77.759% and residual stability 91.04; all met the Indonesian specification. The cantabro abration loss (CAL) at 30°C was 9,02%. The indirect tensile stiffness modulus (ITSM) at 20 °C was 7961.4 MPa; dynamic creep with 100 kPa pressure at 40°C gave slope 0.0112 microsstrain/pulse which is suitable for heavy load traffic. The fatigue test results was obtained at increased stress level, i.e. at 900, 1100, and 1300 kPa. Based on the equation derived from the fatigue strain and repeated loading relationship, at 100 microstrain (με) the repeated load was 434,661.58 times, and at one million (10^6) repeated loading, the samples could withstand strain of 92,38 microstrain. The performance of the samples were overall better than AC-WC mixture using virgin aggregates bound with 60/70 pen asphalt.

Keywords: AC-WC; cantabro; creep; fatigue; ITSM; Marshall; 80-100 pen asphalt; RAP

1. Introduction

Effort in finding out more alternative materials resources for road materials had been done for some years. A potential material that can be used is the reclaimed asphalt pavement (RAP). The use of RAP for road pavement, requires sufficient amount of RAP materials. In developed countries recycling of road asphalt pavement had been done in regular bases so sufficient RAP materials would be available. However, in developing countries, recycling of road pavement has not yet been carried out systematically and regularly. Only heavily damage road section is usually milled in limited amount then overlaid using asphalt mixture with materials from natural aggregates. This practices results in increasing elevation of road way surfaces compare to the road shoulder and surrounding areas which can cause driving and drainage problems.

Researchers and road authorities had done some studies utilizing RAP. Performance hot mix recycled asphalt using RAP greatly affected by binder content. High RAP content can worsen fatigue resistance of the mixture [1]. Meanwhile, studied the use of glass fibre modified warm mix asphalt that contain RAP up to 50% had also been done [2]. The mixture gave a good resistance to rutting and moisture susceptibility. RAP also effect the fatigue performance of the mixture. Higher RAP content up to 40% worsen the fatigue performance [3].

In Spanish General Technical Specifications for Highway Rehabilitation allows max 50 RAP content. [4] G. Valdés et al. (2011) used up to 60 % RAP (10 % higher than the specification). The new added binder was using 250 pen grade new bitumen. Overall the mixture gave satisfactory results.

I is suggest that the preheating temperature of RAP is 100°C; the preheating temperatures of aggregate recommended to be 190°C, 205°C and 215°C at RAP contents of 30%, 40% and 50% respectively, and adding the rejuvenator to the RAP first before mixing with the aggregates and the newly added asphalt [5].

Meanwhile some studies had been done in using higher RAP content. Common methods in using a high percentage of RAP and the principles are use of softer binder, addition of softening additives or rejuvenators [6]. It was also found that it is possible to use RAP up to 90%, through the use of rejuvenators and development of careful blend design [7]. The performance of a seven year old road section using 100% recycled RAP, gave satisfactory performances [8]. The performance hot asphalt mixture with 100 % RAP in a road section compared to the control section, showed little difference. This results encourages the utilization of RAP [9].

This research was aimed at evaluating the performance of Asphalt Concrete Wearing Course (AC-WC) utilizing RAP from Cold Milling bound with 80/100 pen asphalt. Softer grade asphalt was used to rejuvenate the binder in the final mixture.

2. Methodology

2.1. The materials used, aggregate gradation and properties
The materials used was RAP from cold milling stored at a stockpile in central of Bali. Added material used was natural aggregates taken from a quarry site in eastern part of Bali. The RAP and aggregate gradation of Asphalt Concrete-Wearing Course (AC-WC) mixture is shown in Figure 1. The RAP aggregate gradation was modified by adding the correct amount of materials required to meet the middle gradation of AC-WC mixture. The RAP was 42% and the added aggregates were 58% of the total weight of RAP and added aggregates.

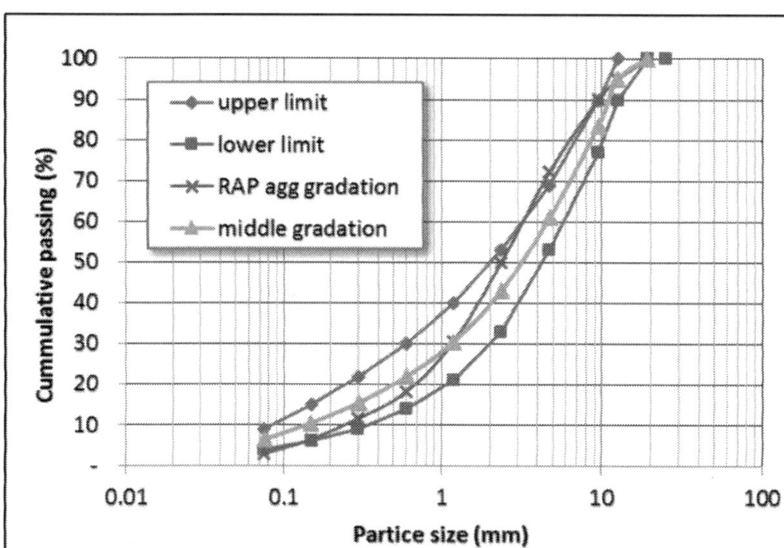

Figure 1. Aggregate gradation of AC-WC mixture.

The asphalt content of the RAP was extracted, and gave 6.01% asphalt content by weight of the RAP or about 2.55% by weight of total mixture. The added asphalt was 3.5%. It had been calculated that the The properties of the added aggregate materials is given in Table 1. The added asphalt used was a 80/100 pen grade asphalt penetration that had met all specification.

Table 1. Properties of the added aggregates.

Properties	Result			Spec
Secific gravity	Bulk	SSD	Apparent	
	2,53	2,59	2,70	
Water absorption		2,46%		Max. 3%
Angularity		50,94%		Min. 40%
Mud content		0,40%		Max.1.0%
Sand Equivalent		92,42%		Min. 60%

2.2. Sample production and testing
The RAP and added aggregate materials were proportioned and hotmixed with the 80/100 pen asphalt. The added asphalt was varied and the samples were tested by Marshall test procedures and compacted 2x75 Marshall blows. The properties of the samples were analysed and the optimum added asphalt was determined. Samples with optimum asphalt content were tested for its Cantabro Abrasion Loss (CAL), stiffness, dynamic creep, and fatigue. The results were compared with AC-WC mixture using natural aggregates and 60/70 pen asphalt. For the CAL, 3 samples were rotated for 300 rotation without steel ball which is usually used for testing aggregate abrasion values. The Loss Angeles abrasion test was carried out at room temperature 30°C The percentage loss of weight due to abrasion then determined.

For testing indirect tensile stiffness modulus-ITSM (stiffness), dynamic creep, and fatigue test were done using a Universal Testing Machine 30 (UTM 30) Dynapave. The ITSM test was done at temperature of 20°C; loading pulse width 250ms (millisecond); pulse repetition period 3000ms; conditioning pulse count 5; estimated poisson's ratio 0.35; estimated modulus 3000 MPa; target strain 50 micronstrains; seating force 20N; dan havershine loading shape [10]. The dynamic creep test was done at temperature of 40°C; havershine loading shape; 100 kPa loading stress; 5 kPa seating stress; 1000ms cycle duration; 1000ms cycle repetition time; 20 kPa preload stress; 600s preload time; cycle count 3600. The samples was cylindrical with diameter of 4 inches (101.6 mm) , height 60-70 mm, loaded trough a top steel plate [11]. The fatigue test was done with 10 N seating force; 100ms cycle width; 500ms cycle repeat time; dan temperature of 20°C. The test was done at three level of stresses. Initially 700 kPa stress was applied, but the sample did not fail until after 40.000 load cycles. Then the stresses were increased at 900, 1100, dan 1300 kPa [12].

3. Results and discussion

The properties of mixture at optimum asphalt content are given in Table 2.

IOP Conf. Series: Materials Science and Engineering **316** (2018) 012037 doi:10.1088/1757-899X/316/1/012037

Table 2. The properties of the AC-WC RAP mixture at optimum asphalt content.

Properties of AC – WC RAP mix (Soaked at 60°C in a water bath)		Optimum asphalt content = 6,05%		Specification
		30-40 minutes	24 hours	
Stabilitas	(Kg)	1358.76	1237.08	Min. 800
Flow	(mm)	3.92	3.360	Min.3
Marshall quotient	(kg/mm)	351.82	324.73	Min. 250
VIM	(%)	3.919	3.360	3,0-5,0
VMA	(%)	15.594	15.103	Min. 15
VFB	(%)	74.904	77.759	Min.65
Retained stability	(%)	= (1273.08/1358.76) x100 % = 91.04%		Min 90%

The Cantabro Abrasion Loss (CAL) at 30°C gave 9.02 % which is below the allowed 16% [13]. This indicates that the samples have a good bond. The stiffness of the AC-WC with RAP and 80/100 pen asphalt was 7961,4 MPa was higher than the AC-WC 60/70 pen using natural aggregates with stiffness of 5022 MPa. This is because the samples were stiffer due to the RAP content with aged asphalt and rejuvenated by the 80/100 pen asphalt. Although the mix contain RAP, it appear to be still flexible enough.

The dynamic creep test gave results as shown in Figures 2 and 3. The AC-WC, RAP, 80/100pen, gave lower strain hence higher stiffness, i.e. stress over strain, compared to AC-WC mix using natural aggregate with 60/70 pen asphalt [14].

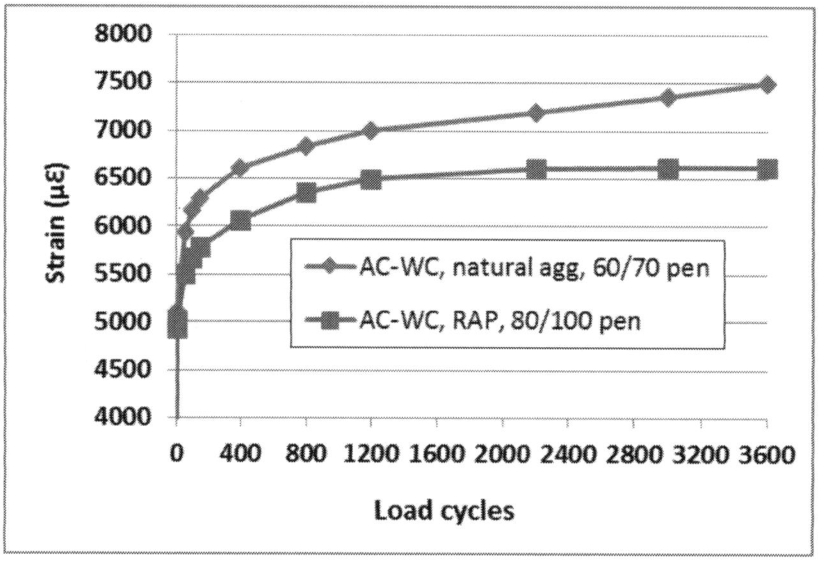

Figure 2. The strain of the sample under dynamic creep test.

The linear part of the strains as shown in Figure 2 are plotted in Figure 4 where linear trend line equation was produced. The x coefficient of the equation of mix AC-WC, RAP, 80/100pen, i.e. 0.0112, was far lower then the AC-WC mix using natural aggregate with 60/70 pen asphalt (i.e. 0.2156). This coefficient indicate the dynamic creep slope of the mixture. In line with data in Table 3, the AC-WC, RAP, 80/100pen can accommodate higher traffic load and suitable for pavement for heavy traffics.

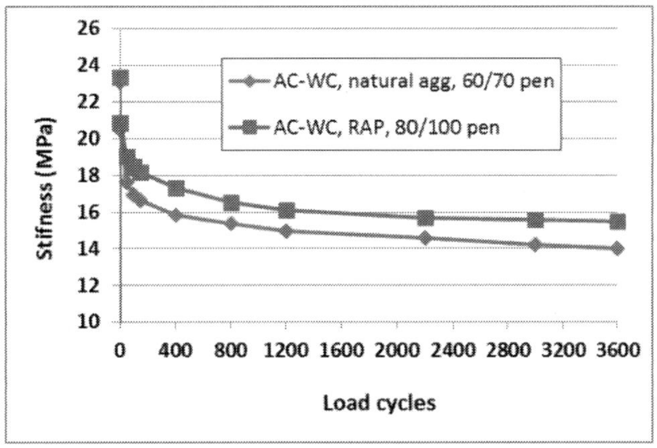

Figure 3. The stiffness of the sample under dynamic creep test.

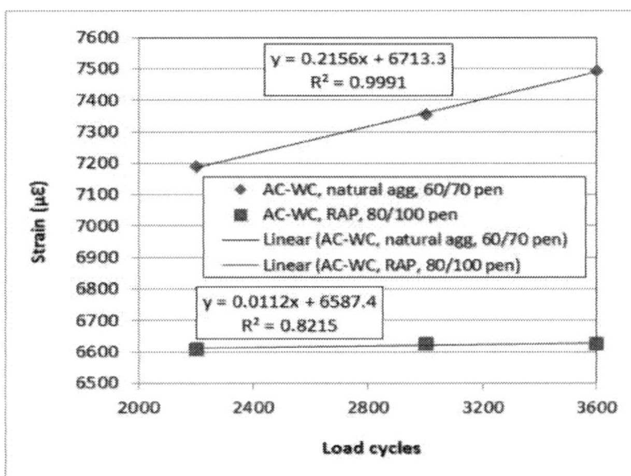

Figure 4. The slope of the mixture from dynamic test.

Table 3. Typical of minimum slope value of dynamic creep test [15].

Annual avg pavement temp (°C)	Heavy traffic load $> 10^6$ ESA	Medium traffic load 5×10^5 to 10^6 ESA	Light traffic load $< 5\times10^5$ ESA
> 30	< 0.5	0.5 - 3	> 3 - 6
20 - 30	< 1	1 - 6	> 6 - 10

10 - 20	< 2	2 - 10	Not Applicable

The fatigue test gave results as shown in Figures 5 and 6. If the lines in Figure 5 extrapolated, at certain stress the AC-WC, RAP, 80/100pen mix fails at far higher load cycles compared to the AC-WC mix using natural aggregate with 60/70 pen asphalt [15], which indicates that the mix is a lot stronger.

Based on results in Figure 6 using the related equation, at 100 micro strain commonly used for pavement design the load cycle required is shown in Table 4. Table 4 also shows, the strain can be accommodated by the mixtures at 10^6 load cycles, where the trend line equation was obtain by swapping the x and y axis in Figure 6, to produced graph as shown in Figure 7. It is shown that mixture of AC-WC, RAP, 80/110pen can withstand 92.28 microstrains which is higher than 70.31 microstrains of mixture AC-WC, natural agg, with 60/70 pen asphalt.

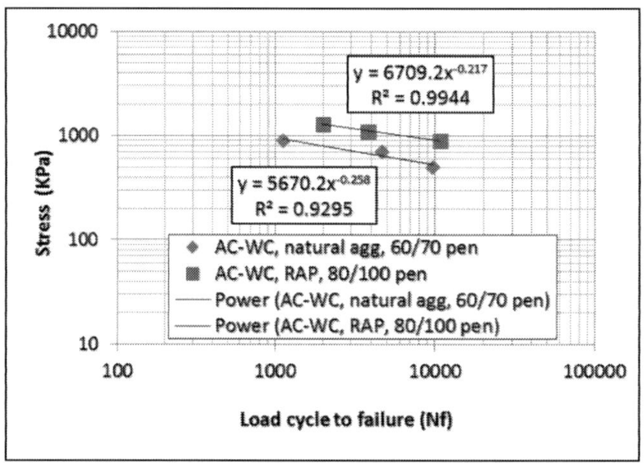

Figure 5. Load cycle vs. stress of fatigue test.

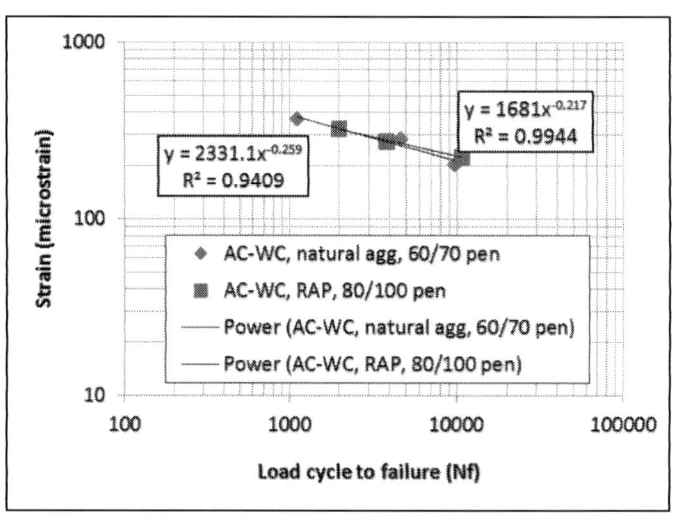

Figure 6. Load cycle vs. strain of fatigue test.

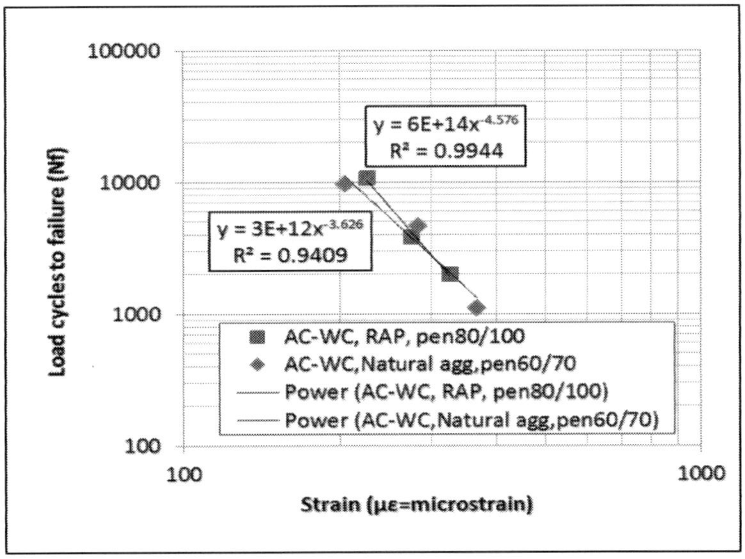

Figure 7. Strain vs. load cycle of fatigue test.

Table 4. Load cycles 100 $\mu\varepsilon$ and strain of the mixture at 10^6 load cycles.

Mixture	Equation based on strain (Figure 6)	Equation based on load cycles (Figure 7)	Load cycles at 100 $\mu\varepsilon$	Strain ($\mu\varepsilon$), at 10^6 load cycles
AC-WC, RAP, pen80/100	$\varepsilon(\mu\varepsilon) = 1681 \times Nf^{-0.217}$	$Nf = 6.0 \times 10^{14} \varepsilon^{-4.576}$	434,661.58	92.38
AC-WC, natural agg, pen 60/70	$\varepsilon(\mu\varepsilon) = 2331 \times Nf^{-0.259}$	$Nf = 3.0 \times 10^{12} \varepsilon^{-3.626}$	172,631.98	73.71

4. Conclusion

Considering the results and analysis, it can be concluded as follow:

1. The AC-WC, RAP, 80/100pen properties can meet Marshall Specification, and gave Cantabro abbrassion loss of 9.01 %, far lower than maks 16% at 30°C.
2. Compared to AC-WC, Natural agg, with 60/70pen asphalt mixture, the AC-WC, RAP, 80/100pen mixture gave:
   - higher stiffness (ITSM), i.e 7961,4 MPa compared to 5022 MPa
   - lower deformation and lower creep slope value under dynamic creep loading. The slope was 0.0112 compared to 0.2156. This indicate that the mixture can withstand higher load repetition.
   - the studied mixture can accomodate higher load cycles and can withstand higher strain of fatigue test. At 100 microstrain (με) the repeated load required was 434,661.58 times, and at one million (10^6) repeated loading, the samples could withstand strain of 92,38 microstrain.

5. References

[1] Dong, F., Yu, X., Bo Xu, B., Wang, T. 2017 *Comparison of high temperature performance and microstructure for foamed WMA and HMA with RAP binder*, Construction and Building Materials, Volume 134, 1 March 2017, Pages 594-601.

[2] Fakhri, M., Hosseini, S.A. 2017 *Laboratory evaluation of rutting and moisture damage resistance of glass fiber modified warm mix asphalt incorporating high RAP proportion*, Construction and Building Materials, Volume 134, 1 March 2017, Pages 626-640 http://www.sciencedirect.com/science/article/pii/S0950061816320864, visited 12-01-2017.

[3] Daniel, J.S., Gibson, N., Tarbox, S. , Sean Tarbox , Copeland, S.A., and Andriescu, A. 2013 Effect of long-term ageing on RAP mixtures: laboratory evaluation of plant-produced mixtures, Journal Road Materials and Pavement Design Volume 14, 2013 - Issue sup2*: Papers from the 88th Association of Asphalt Paving Technologists' Annual Meeting*, April 7–10, 2013. http://www.tandfonline.com/doi/full/10.1080/14680629.2013.812840

[4] Valdés, G., Jiménez, F.P., Miró, R., Martínez, A., Botella, R. 2011 *Experimental study of recycled asphalt mixtures with high percentages of reclaimed asphalt pavement (RAP)*, Construction and Building Materials 25 (2011) 1289–1297, Elsevier.

[5] Yu, B., Gu, X., Wu, M., and Ni, F. 2016 *Application of a high percentage of reclaimed asphalt pavement in an asphalt mixture: blending process and performance investigation*, Road Materials and Pavement Design, May 2016, Southeast University (China), © 2016 Informa UK Limited, trading as Taylor & Francis Group

[6] Al-Qadi, I. L., Elseifi, M. A., and Carpenter, S. H. 2007 *Reclaimed asphalt pavement: A literature review* (Report Number FHWA-ICT-07-001). Springfield: Illinois Center for Transportation.

[7] Carrión, J.B., Presti, D.L., and Airey, G.D. 2015 *Binder design of high RAP content hot and warm asphalt mixture wearing courses*, Road Materials and Pavement Design, University of Nottingham-UK. DOI: 10.1080/14680629.2015.1029707. http://dx.doi.org/10.1080/14680629.2015.1029707, accessed on 25-5-2017.

[8] Mallick, R.B., O'Sullivan, Anne, K,. Mingjiang, T., Robert, F. 2010 *Why Not Use Rejuvenator for 100% RAP Recycling?*, Transportation Research Board, TRB 89th Annual Meeting Compendium of Papers DVD. https://trid.trb.org/view.aspx?id=910199

[9] Rowe, G.M., Barry, J., Crawford, K. 2015 Evaluation of a 100 % Rap Recycling Project in Fort Wayne, Indiana, *8th RILEM International Symposium on Testing and Characterization of Sustainable and Innovative Bituminous Materials*, Volume 11 of the series RILEM Book series pp 941-951. http://link.springer.com/chapter/10.1007/978-94-017-7342-3_75

[10] British Standard 2012 Bituminous mixtures – Test methods for hot mix asphalt, Part 26: *Stiffness*, BS EN 12697-26:2012. http://www.sciencedirect.com/science/article/pii/S0950061816320268, visited 10-01-2017

[11] British Standard 2005 Bituminous mixtures - Test methods for hot mix asphalt, Part 25*: Cyclic compression test*, BS EN 12697-25:2005.

[12] British Standard 2012 Bituminous mixtures - Test methods for hot mix asphalt, Part 24: *Resistance to fatigue*, BS EN 12697-24:2012.

[13] Hamzah, M. O., Hasan, M. R. M., Che Wan, C. N., and Abdullah, N.H. 2010 Comparative Study on Performance of Malaysian Porous Asphalt Mixes Incorporating Conventional and Modified Binders, *Journal of Applied Sciences 10(20)*, pp 2403-2410, 2010. http://www.h-a-d.hr/pubfile.php?id=619, visited 26/02/2017

[14] Thanaya, I N.A., Purbanto, I G.R., Nugraha, I N.S. 2016 Study on the Characteristics of Asphalt Concrete Wearimg Course (AC-WC) Mix Using Asphalt 60/70 Penetration with

Addition of Latex. *Civil Engineering Study Program Research Report*, Udayana University-Bali, Indonesia.
[15] Alderson, A. 1995 *Gyropac and Matta Training Course, Australian Road Research Board Ltd.*

IOP Conf. Series: Materials Science and Engineering **316** (2018) 012038 doi:10.1088/1757-899X/316/1/012038

THE INFLUENCE OF SAND's GRADATION AND CLAY CONTENT OF DIRECT SHEART TEST ON CLAYEY SAND

Gunawan Wibisono
Civil Eng. Department, Engineering Faculty, University of Riau, Pekanbaru Indonesia
Soewignjo Agus Nugroho
Civil Eng. Department, Engineering Faculty, University of Riau, Pekanbaru Indonesia
Khairul Umam
Civil Eng. Department, Engineering Faculty, University of Riau, Pekanbaru Indonesia

ABSTRACT
The shear strength of clayey-sand can be affected by several factors, e.g. gradation, density, moisture content, and the percentage of clay and sand fraction. The same percentage of clay and sand fraction in clayey-sand mixtures may have different shear strengths due to those factors. This research aims to study the effect of clay content on sand that cause the change of its shear strength. Samples consisted of different clay and sand fractions were reconstituted at a certain moisture content. Sand fractions varied from well-graded to poorly-graded sand. Shear strength was measured in terms of the direct shear test. Prior to the test, surcharge loads were applied to represent overburden pressures. Shear strength results and their components (i.e. Cohesion and internal angle of friction) were correlated with physical properties of samples (i.e. grading coefficient of curvature, coefficient of uniformity, and density). Results showed that samples classified as well-graded and dense sand had higher shear strength. In the other hand, the shear strengths decreased when the mixtures became poorly-graded and less dense. The inclusion of the clay fraction increased cohesion component and decreased internal angle of friction.

Keywords : Clay, Sand, Density, Gradation, Shear Strength

1. INTRODUCTION
Several places of Indonesia's coastal areas, that are not adjacent to oceans, are dominated by alluvial deposits. The deposits often comprise of soft cohesive soil, which is known to have a low bearing capacity. Road embankments sometimes have to be laid upon this kind of difficult soil. Embankment fill materials that meet standard specifications, are also scarcely available in such areas. The low quality subgrade and fill materials can lead into embankment failures.

In order to ensure fill materials are adequate to withstand loads imposed and become workable, Bina Marga [1] specified properties of materials that can be used as fill. The strength of embankment soil is determined by its mechanical properties, i.e. internal angle of friction angle (ϕ) and cohesion (c). The soil friction is affected by size, shape, and texture of coarse particles, contributed mainly by sand type of soil. Cohesion (c), which is provided by cohesive or fine particle soil, is the force that holds together molecules or like particles within a soil. Soil cohesion is affected by the distance between molecules and particle size distribution, water content, and pore water pressure. The percentage of cohesive (clay) and cohesion-less (sand) soil within soil mass, therefore, governs the strength of the soil. This research attempts to observe the effect of clay content on shear strength of clayey sand mixtures.

2. REVIEW OF LITERATURES

Prihatin & Wahyudi [2] studied a correlation of gradation, density and shear strength of fill materials. The result indicated that the higher ratio of curvature coefficient and uniformity coefficient (Cc/Cu) or the more homogenous the soil particles, the lower γ_{dmax}, cohesion, and internal angle of friction, and vice versa. Kara et al [4] carried out a research related to the effect of sand particle sizes on its shear strength changes using a direct shear test. It showed that the larger the size of particles, the higher the internal angle of friction. The sand, however, has no cohesion and hardly become denser, which results in lower bearing capacity. To solve this problem, some binder materials such as cohesive soil (clay) should be added into the sand. Hakam et al [5] observed the effect of clay additions to the shear strength of beach sand. The result showed that to some extend if the mixture become denser, the internal angle of friction increased, and vice versa. The further addition of clay would increase the mixture cohesion, but decreased the internal angel of friction.

3. RESEARCH METHODOLOGY

This research aims to observe the effect of clayey-sand mixtures gradation and density on their shear strength changes. The procedure of the research is as follows.

3.1. Samples Preparation

The samples were prepared based on the planned variations of sand gradation according to ASHTOO and USCS specifications, and also the variety of clay content in sand mixture. The planned sand gradation can be seen in Figure 1.

Sand can be categorized as well-graded sand if Cc > 6 and Cu between 1 and 3 ($1<Cu<3$). If only one of them is complied, either Cu or Cc, then the sand can be categorized as poorly-graded sand. Gap grading is a condition where the particle sizes are lacking on one or two numbers of sieves. Gap-graded sand from pure sand soil is a gradation achieved by dispelling the particle size of #20 and #40. Gradation A1-b and A3 are soil classifications based on AASHTO system that are categorized as poorly-graded sand. Uniform fine and coarse-graded sand are the uniform sand with particle size from #40 to #60 and #4 to #40, consecutively. From each gradation of sand above, clay fraction with percentages of 0%, 10%, 20%, and 40%, were then added to the mixtures.

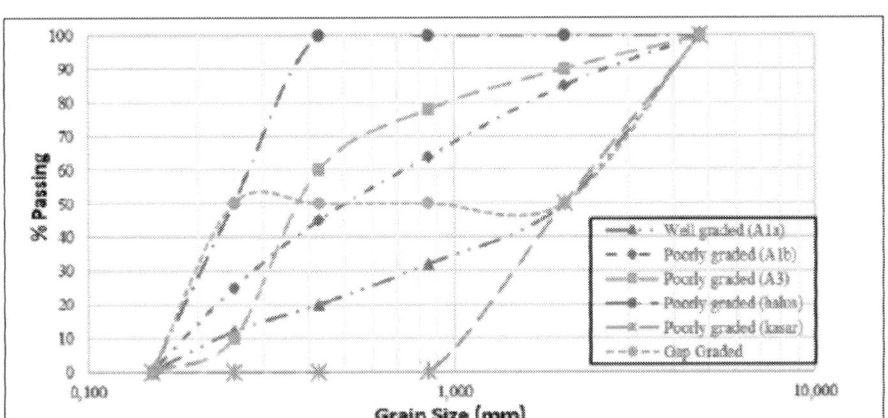

Figure 1. Planned Sand Gradations

IOP Conf. Series: Materials Science and Engineering **316** (2018) 012038 doi:10.1088/1757-899X/316/1/012038

3.2. Samples Densities

The samples were varied into two kinds of densities. The first was 15,24 kN/m^3, which was applied for all sample variances. Secondly, a density that was formed by giving preloading surcharge of 0,16 kN/m^2 for 24 hours on molds. This load resulted in different densities on each variant. A consolidation test was done based on ASTM D-2435. After the desired densities were achieved, shearing test was performed.

3.3. Direct Shear Test

Shearing test was performed using direct shear test according to ASTM D-3080. The purpose of this test was to obtain mixture shear strength parameters, i.e. cohesion (c) and the internal angle of friction (ϕ). The direct shear tests were done on each mixture.

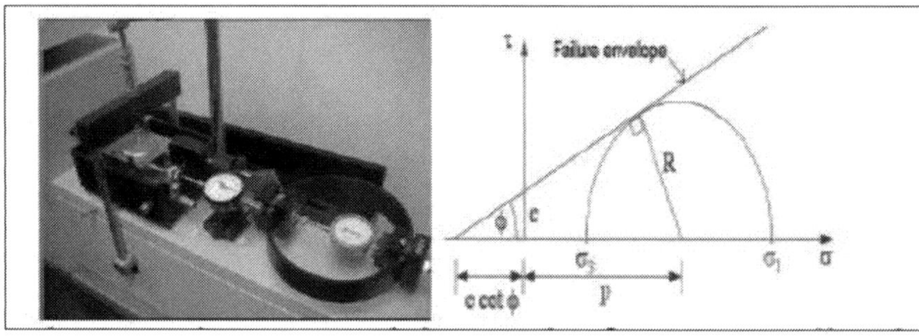

Figure 2. Direct Shear Test Apparatus and Mohr-Coulomb Normal-Shear Stress Chart

4. RESULT AND DISCUSSION

Prior to mixing, the index properties (gradation, specific gravity, and Atterberg limits) of sand and clay used in this research were obtained. The main results of direct shear tests were separated into two parts, with and without preloading.

4.1. Soil Index Properties

The result of soil properties test can be observed in Table 1.

Table 1. The Result of Soil Properties Test

Properties	Unit	Clay	Sand	90%	80%	70%	60%
Specific Gravity (Gs)	-	2.675	2.636	2.637	2.638	2.647	2.651
Atterberg Limits							
-Liquid Limit	%	30.80	-	-	-		
-Plastic Limit	%	20.50	-	-	-		
-Plasticity Index	%	10.30	-	-	-		

It can be seen that specific gravity increased with the increasing of clay content on mixtures. Based on USCS, clay was categorized as low plasticity clay.

4.2 Direct Shear Test Results

Shear strength components, i.e. internal angle of friction and cohesion, were obtained from the tests. In conjunction with other properties tested, i.e. densities and grading, these parameters were then analyzed to discover their relationships.

4.2.1 Correlation between Clay or Sand Content and the Internal Angle of Friction

4.2.1.1 The Internal Angle of Friction without Preloading (Without Consolidation)

The relation between clay content and internal friction angle of samples with the same density of 15,24 kN/m^3 can be seen in Figure 3. Most of internal friction angles showed a trend of improvement as clay fraction increased, except for uniformly fine-graded sand (UF) mixture. This is due to the replenishment of clay content. The internal friction angle of UF mixtures decreased for clay content more than 10%. The further addition of clay fraction caused a reduction in interlocking fine particles. The same trend was also discovered by Arifin and Saputra (2015), where the addition of sand in clay increased the internal angle of friction, peaked at about 50% of sand, and decreased at further inclusion.

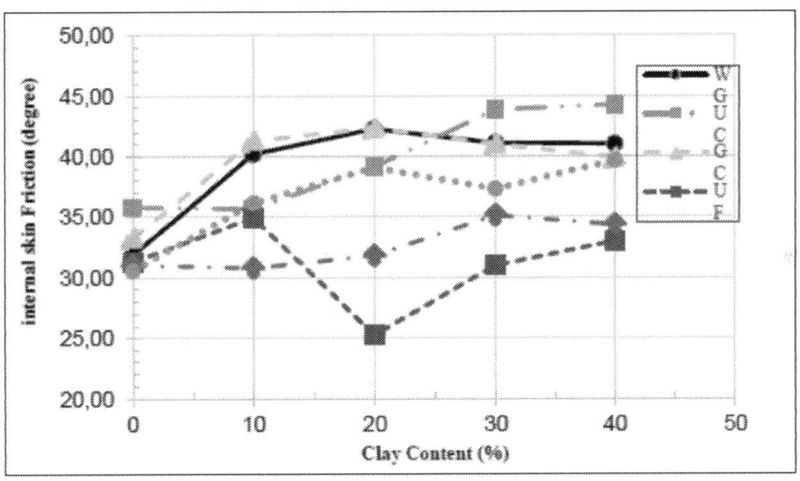

Figure 3. The Correlation between Internal Angle of Friction and Clay Content without Preloading

4.2.1.2 The Internal Angle of Friction with Preloading (Consolidation)

Figure 4 shows fluctuated trends of the internal angle of friction as clay content increased on mixtures with preloading. On mixtures with fine particles less than 12% (i.e. SW, Gap, and UF), the higher the percentage of clay fraction added, the lower the internal angle of friction achieved. On mixtures of A1-b, A3 and UC, however, showed the other way around. Perhaps the addition of clay fraction increase shear plane area due to void fill by fine clay particles. Further addition of clay (more than 12%) made the internal friction angle become higher, except for SG and UF. The clay fraction improved mixtures gradation to become more compact and denser. The addition of clay on SG and UF mixtures, however, decreased the internal angle of friction. This is because the increasing of fine particles, not bigger particles that are needed. Figure 3 and 4 shows that high values of internal friction angle are dominated by coarse sand materials (especially the size of > 0,850 mm) like WG, UC, and SG. For A1-b, A3, and UF, the sand particle sizes were smaller, as a result the internal angle of friction become lower. This corresponds to a previous research conducted by Kara et al. [4], which concluded that the increasing of granular particle size resulted in higher internal friction angle, or vice versa.

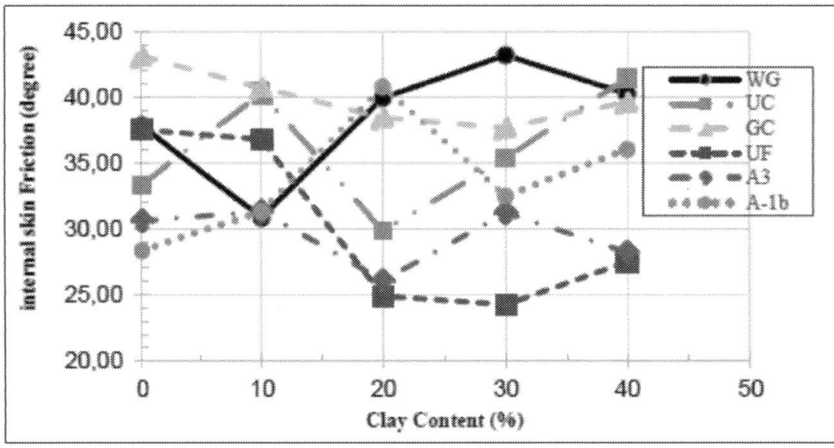

Figure 4. Correlation between Internal Angle of Friction and Clay Content with Preloading

4.2.1.3. The Difference of Shear Strength between Mixtures with and without Preloading

The difference is shown as ratio of the internal angle of friction values between mixtures with and without preloading, as shown in Figure 5. The ratio of higher than 1 means that compression/compaction due to surcharge load (preloading) increases the shear strength component. The ratio of lower than 1 shows the opposite.

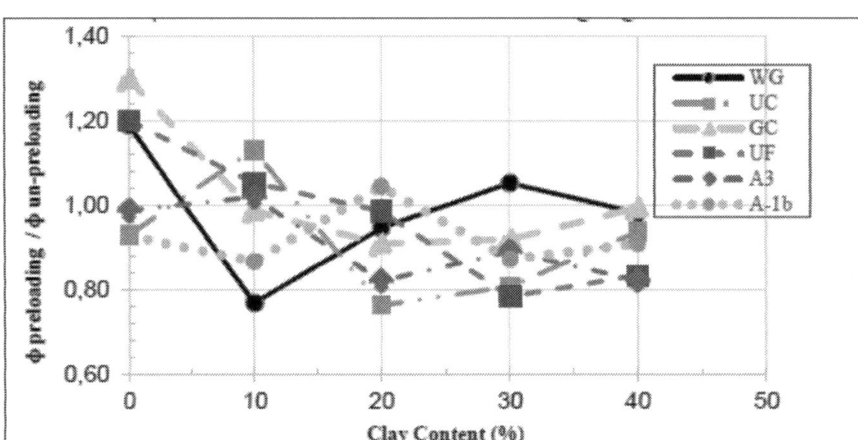

Figure 5. Ratio of the Internal Angle of Friction of Mixtures with and without Preloading

As can be seen from Figure 5, the majority of ratio values is situated below the value of 1, especially at clay content more than 12%. The more compact or denser mixture due to preloading reduced the internal angle of friction. The clay particles reduced the interlocking particles of sand particles.

4.2.2. Correlation between Sand or Clay Content and Cohesion (c)

4.2.2.1. Cohesion of Mixtures without Preloading

The effect of clay fraction inclusion in the mixtures is shown in Figure 6. The cohesions of mixtures with less than 5% clay content are the apparent (untrue) cohesions. As can be seen from the graph, cohesion values were fluctuated as clay content increased. In this test, samples were in relatively less dense condition, therefore the contact between clay particles were still remained low. The shear strength of mixtures were still dominated by friction of sand particles.

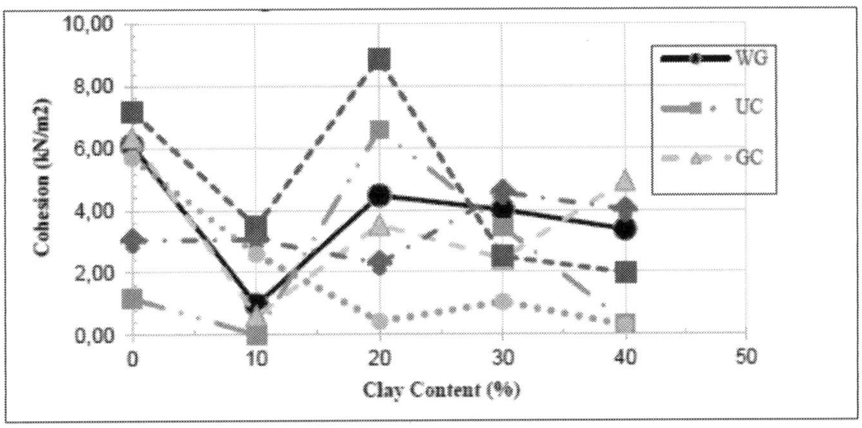

Figure 6. Correlation between Clay Content and Cohesion of Mixtures without Preloading

4.2.2.2. Cohesion of Mixtures with Preloading

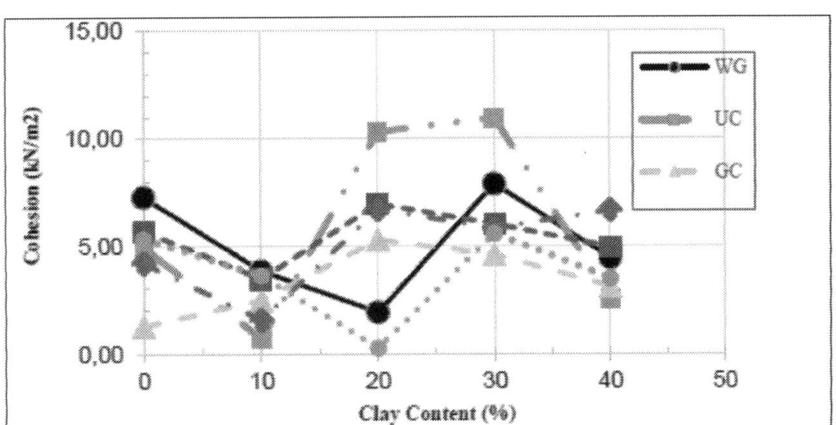

Figure 7. Correlation between Clay Content and Cohesion of Mixtures with Preloading

The effect of clay fraction addition on the change of cohesion values within mixtures with preloading can be seen in Figure 7. In general, the inclusion of clay up to 30% increased the mixtures cohesion component. This indicates that in denser mixtures, fine particles of clay also play an important role in affecting mixture shear strengths. Some coarse particles of sand were covered by fine particles, resulting in lower interlocking particles/friction. In this test, the maximum cohesion occurred when 30% of clay fraction was added to the mixtures.

4.2.2.3. The Difference of Cohesion of Mixture with and without Preloading

Similar to 4.2.1.3, the comparison of cohesion is shown as ratio of cohesion values between mixtures with and without preloading.

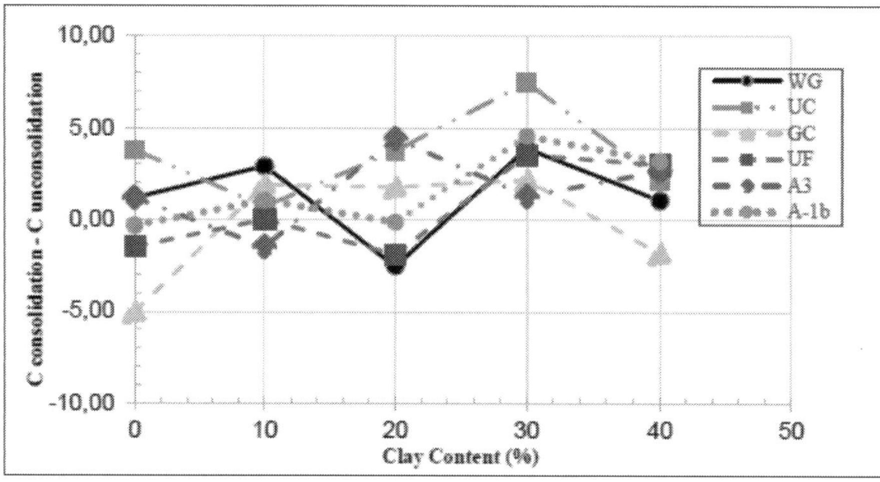

Figure 8. The Cohesion Comparison between Mixtures with and without Preloading

As can be seen in Figure 8, the additional compaction (preloading) on samples enhanced the cohesion component. This improvement occurred significantly for the compressed samples with more than 12% clay content. This also indicated that on that percentage, transition frm sand to clay influence on mixture shear plane was taken place. It also confirmed the previous explanation related to the effect of internal friction angle, where 12% clay fraction addition on compressed mixtures increased the friction component.

4.2.3. Correlation between Gradation Coefficient and Shear Strength

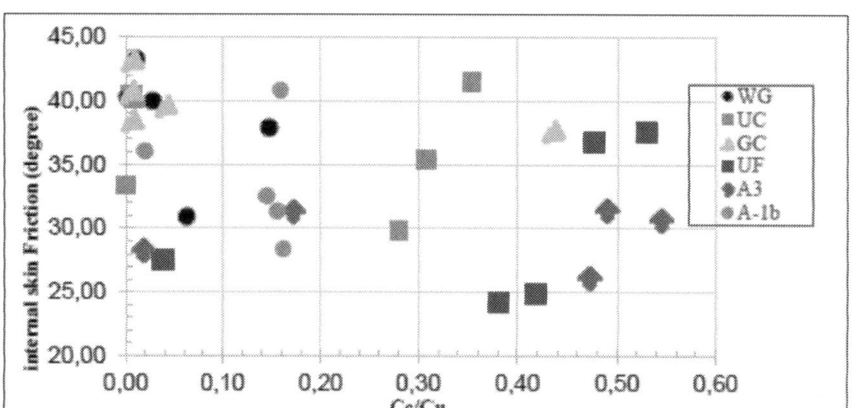

Figure 9. The Effect of Mixture Gradation (Cc/Cu) on the Internal Angle of Friction

IOP Conf. Series: Materials Science and Engineering **316** (2018) 012038 doi:10.1088/1757-899X/316/1/012038

According to Prihatin and Wahyudi (2009), if the Cc/Cu become lower, the internal friction angle tends to increase. Figure 9 shows scattered values of gradation coefficient (Cc/Cu) and shear strength of mixtures. The gradation (Cc/Cu) influence on mixtures shear strength cannot be represented only by particles size distribution, but also by other factors i.e. the coherence of fine particles, the void filling of finer materials, and the percentage of coarse sized particles.

Figure 9 shows that mixtures contained coarser sand (i.e. WG and UC) had higher shear strengths, whereas UF and A3, which were dominated by fine sand particles, seemed to have lower shear strength. It demonstrates the influence of particles size on shear strength. The larger the coarse particles, the higher the shear strength.

4.2.4. Correlation between Clay Content and Mixture Dry Density (γ_d)

The mixture densities as seen in Figure 10, are the results of 0,16 kg/m^2 surcharge load.

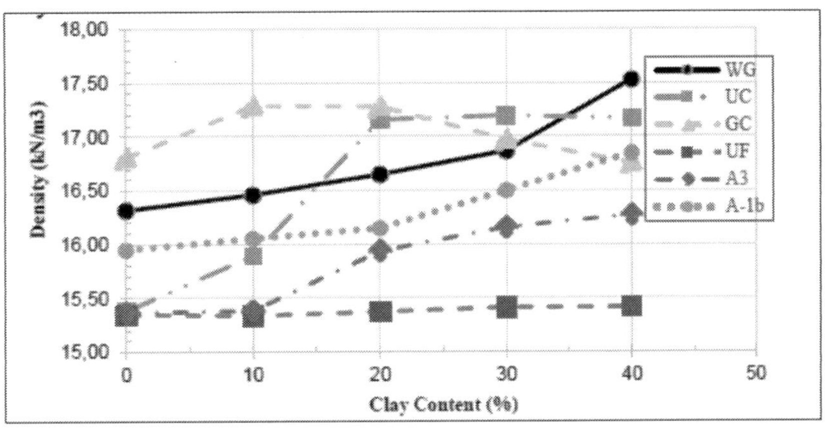

Figure 10. The Correlation between Clay Content and Mixture Dry Density

As can be seen in Figure 10, the addition of clay fraction increased the mixture densities, except for GC. The most significant change occurred on UC mixture, i.e. from 15,38 kN/m^3 to 17,19 kN/m^3 on 30% of clay addition. UC mixture is a uniform-graded sand with coarse sized particles, thus the clay particles could occupy the mixture voids.

On UF mixtures, the addition of clay almost had no impact on the change of mixture density. There was only slightly change from 15,35 kN/m^3 to 15,42 kN/m^3. This was due to lacking of coarse sized particles in the mixtures that could improve the particle gradation.

The clay fraction addition of more than 20% resulted in the decreasing of SG mixture density. This is because SG mixture lacked of medium-sized sand particles, so the addition of clay of more than 20% did not improve the mixture density.

It is found that significant improvement on mixture densities, was achieved at the addition 20% clay fraction. Further addition could no longer effective in increasing densities.

4.2.5. Correlation between Mixture Dry Density (γ_{dry}) and Shear Strength (ϕ and c)

Hakam et al (2010) stated, that on direct shear test, the denser the samples, the higher the internal angle of friction, or vice versa. The higher the clay content, the higher the cohesion (c), and the lower the internal angle of friction (ϕ).

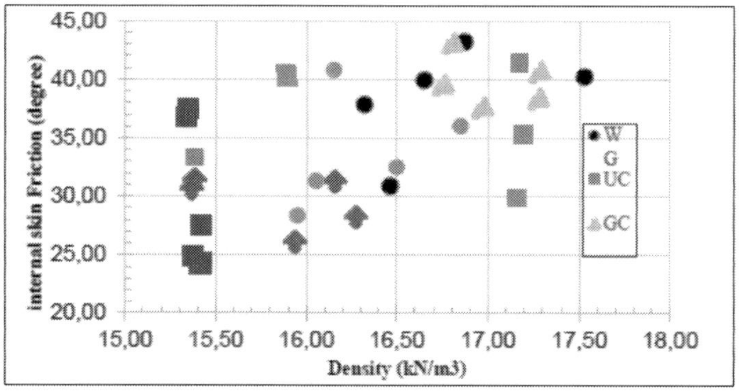

Figure 11. Density (γ_d) and the Internal Angle of Friction (ϕ)

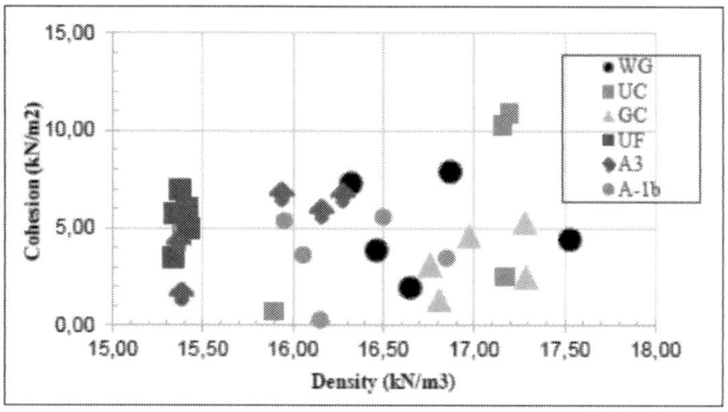

Figure 12. Density (γ_d) and Cohesion(c)

Figure 11 shows that there is an influence of density on mixture internal friction angle (ϕ). The relation of them, however, were fluctuated and inconsistent. The same thing also happen in the correlation of density and cohesion component.

5. CONCLUSION

The test results and discussions illustrate the following conclusions.

1. The addition of clay fraction changes clayey sand mixture gradation.

2. Shear strength components (cohesion and the internal angel of friction) are strongly affected by the particle size distribution of soil mixtures. The larger the sand particles, the higher the friction, and vice versa.

3. Gradation coefficient (Cc and Cu) does not affect the shear strength significantly.

4. The addition of clay fraction in sand mixtures increases the internal angle of friction, and decreases cohesion, at a certain extent where coarse particles still dominates the shear plane. The further of clay addition make coarse particles no longer dominant on the shear plane, resulting in the opposite effects (lower internal friction angle and higher cohesion).

5. Cohesion is more affected by fine materials content rather than coarse particles.

6. The addition of clay up to 40% tends to increase mixture density.

7. The addition of clay in coarse sand mixtures, increases density more than in fine sand mixtures.

REFERENCES

[1] Bina Marga, (2012), Panduan Geoteknik 4, Panduan Geoteknik Indonesia Timbunan Jalan pada Tanah Lunak, Pedoman Kimpraswil No. Pt. T-10-2002-B, Jakarta
[2] Prihatin, K. & Wahyudi K.., (2009), Korelasi Antara Gradasi Material Timbunan Reklamasi dengan Kepadatan dan Kuat Geser Tanah, ISBN No. 978-979-18342-0-9. E19-E28, Jakarta
[3] Arifin, Y., F., and Saputra, M., I., (2015), Kuat Geser Sisa Campuran Lempung dan Pasir Yang di Padatkan, Prosiding Seminar Nasional Teknik Sipil 1 (SeNaTS 1), Jakarta
[4] Kara, E., M. et. Al, (2013), Contribution of Particles Size Ranges to Sand friction, Engineering Technology & Applied Science research (ETASR), 3 (4), pp.497-501
[5] Hakam, A., Yuliet, R. & Donal R., (2010), Studi Pengaruh Penambahan Tanah Lempung Pada Tanah Pasir Pantai Terhadap kekuatan Geser Tanah, Jurnal Rekayasa Sipil, 3 (4), pp. 11-22

Using Cementitious Materials Such as Fly Ash to Replace a Part of Cement in Producing High Strength Concrete in Hot Weather

Gidion Turuallo

Harun Mallisa

Civil Engineering Department, Tadulako University, Palu, Indonesia

turuallo@yahoo.co.uk

Abstract. The use of waste materials in concrete gave many advantages to prove the properties of concrete such as its workability, strength and durability; as well to support sustaianable development programs. Fly ash was a waste material produced from coal combustion. This research was conducted to find out the effect of fly ash as a part replacement of cement to produce high strength concrete. The fly ash, which was used in this research, was taken from PLTU Mpanau Palu, Central Sulawesi. The water-binder ratio used in this research was 0.3 selected from trial mixes done before. The results of this research showed that the strength of fly ash concretes were higher than concrete with PCC only. The replacement of cement with fly ash concrete could be up to 20% to produce high strength concrete .

Keywords: Cementitious Materials; Sustainable Development; Fly Ash; High Strength Concrete and Workability

...

1. Introduction

Concrete is the most widely used construction material because of its versatility, economy, and availability of raw materials, strength, and durability. Researchers are continuously working for improving the properties of concrete with the use of innovative chemical admixtures and supplementary cementing materials (SCMs). Furthermore, the use of SCMs in concrete can conserve energy and give environmental benefits as reduction in carbon dioxide emission as the result of decrease in Portland cement production.

Using supplementary cementing materials in concrete mixes will reduce cement contents, improve workability, increase strength and enhance durability of concrete [1-5]. The use of supplementary cementing materials (SCMs), which are by products such as fly ash, ground granulated blast furnace slag, silica fume, micro silica, and others will not only prevents them from being land-filled but also improve the properties of concrete in both fresh and hardened states [2, 6].

This research was conducted to find out the use of cementitious materials such as fly ash as a part replacement of Portland Composite Cement (PCC) to produce high strength concrete cured in hot temperature. Fly ash is a by-product from burning pulverized coal in electric power generating plants. The size of fly ash particles is commonly finer than Portland cement, which it ranges from 10 to 100 micron [7, 8].

Some researchers [9-12] found that the strength development of fly ash concretes at early ages, which were cured under standard curing temperature (20^0C); were slower when compared to the concrete with Portland cement only. In addition, Metha found that the pozzolanic hydration of fly ash cured at standard curing temperature started at 11-days after mixing and its contribution to the strength development of concrete appeared after 28-days [13]. However, the strength development of fly ash concrete at early ages especially for concrete with a lower water-binder, which were cured at elevated curing temperature (higher than standard curing temperature); were similar to that of concrete with Portland cement only [14, 15]. It is believed that this is due to the rate of pozzolanic reactions were increased as the increase of curing temperature. Large quantities of fly ash contributed a significant improvement of strength development of concrete.

2. Methodology/experimental work

The reaction of fly ash in hydration process is a secondary raection. Fly ash reacts after the hydration process of cement. The silica in fly ash reacted with the calcium hydroxide (Ca(OH)2), which is produced in the hydration of cement formed calcium silicate hydrate (C-S-H), which contributed to the strength development of concrete as the following reaction [2, 6]:

Cement reaction : $$C_3S + H \rightarrow C - S - H + Ca(OH)_2 \quad (1)$$

Pozzolanic reaction : $$Ca(OH)_2 + S \rightarrow C - S - H \quad (2)$$

One of the benefits of using fly ash in concrete is that it can increase the workability of concrete. The replacement of cement with fly ash will increase the volume of cementitious materials in mixture. The increase of paste volume in concrete results in a reduction in aggregate particle interference and it will also increase the workability of concrete [9, 16]. The mix proportion of of concrete was calculated based on the Modified Maximum Density Theory [17].

2.1. Materials

A single batch of sand, coarse aggregate, PCC cement, and fly ash were used throughout the experimental work. The sand was taken from Palu River, while the coarse aggregate was a crushed aggregate, which was taken from Loli; with a nominal size ranging from 5 to 20 mm. The superplasticizer was Type Naptha 511P, while fly ash was from PLTU Mpanau Palu, which is categorized as Class F fly ash with chemical composition as shown in Table 1 below:

Table 1 Chemical composition of fly ash taken from PLTU Mpanau Palu

Chemical compounds	Percentage by weight
S_iO_2	51,55
Fe_2O_3	25,54
Al_2O_3	17,26
CaO	2,09
K_2O	1,23
TiO_2	0,95
MnO	0,48
Others	0,90

The absorption, density, sieve analysis, and the maximum void content of both coarse and fine aggregates were measured. The sand is in air dry conditions; therefore, some water was added for absorption to meet the saturated service dry (SSD) condition of the sand; similar for the coarse aggregate. The PCC cement was produced by PT Semen Tonasa, which is Type I cement with the specific gravity of 3.1.

2.2. Mix proportion
The mix proportion of concretes was calculated using the Modified Maximum Density Theory (MMDT), where the measurement of void content and proportion of sand to the total of aggregates are 23.25 % and 35% respectively. These gave the volume of paste and total aggregates were 0.273 and 0.727 respectively. The calculation of mixes with water-binder ratio of 0.30 are presented in Table 2 below:

Table 2. Calculated mix proportion of concrete

Material	*Weight of materials per m^3 concrete (kg)*				
	Percentage of fly ash, w/b = 0.30				
	0	10	15	20	25
Cement	437.713	393.942	372.056	350.170	328.285
Fly ash	-	43.771	65.657	87.543	109.428
Sand	688.900	688.900	688.900	688.900	688.900
Coarse aggregate	1256.00	1256.00	1256.00	1256.00	1256.00
Water	126.698	126.698	126.698	126.698	126.698
Superplasticizer	0.119	0.119	0.119	0.119	0.119

2.3. Specimen
The water-binder ratio that was taken in calculating mix proportions was 0.30. There were five mixes prepared, which were referred to the percentage of fly ash replacing cement in concretes, i.e. 0, 10, 15, 20, and 25%. The specimens for compressive strength test were cube with the size of 150mm x 150mm x 150mm. Three cubes were prepared for each testing age. The testing ages were 1, 3, 7, 21, and 28-days to find out the strength development of concretes.

3. Results and discussions
As described above, the percentage of fly ash to replace a part of cement were 0, 0.10, 0.15, 0.20 and 0.25 with the symbols of PCC, 10FA, 15FA, 20FA and 25FA respectively. The strength developments of concrete were investigated up to 28-days, which can be seen in Table 3 below:

Table 3. Strength development of concretes Concrete ages

Mix	3-days		7-days		21-days		28-days	
	Weight (kg)	Strength (MPa)	Weight (kg)	Strength (MPa)	Weight (kg)	Strength (MPa)	Weight (kg)	Strength (MPa)
PCC	8.823	36.366	8.814	57.689	8.824	60.644	8.820	62.478
10FA	8.803	36.267	8.768	57.244	8.846	58.178	8.868	60.267
15FA	8.601	42.044	8.841	56.800	8.858	59.956	8.862	66.756
20FA	8.768	43.733	8.650	52.444	8.862	57.644	8.872	61.556
25FA	8.781	39.289	8.739	45.111	8.860	54.844	8.874	56.622

Table 3 above shows that the strength of fly ash concretes at three days are similar to that of PCC concrete, even the strength of concretes with replacement fly ash above 15% have strengths higher than that of PCC concrete. This proves that there is no delay in strength gain of fly ash concrete at early ages (three days) with replacement cement levels up to 25%. At 28-days, the strength of concrete with 15% fly ash is higher 6.8% above the strength of concrete with PCC only, and it is like to continuously increase. It is believed as the results of the curing temperature above of the standard curing temperature (20^0C) applied in Europe and America. It means that the use fly ash in concrete is suitable in counties with hot weather such as Indonesia. The strength of fly ash concrete obtained, however, are lower than that of fly ash concrete obtained in Europe and America with the same water-binder ratio. This is caused the used PCC has already mixed with pozzolanic materials in clinker.

4. Conclusions

Results of this research can be concluded as follow:

1. The slump of concrete mixes increased as the percentage of fly ash in concrete increased, even when using fly ash of 25%; the slump is collapse as the concrete flows.
2. The maksimum levels of fly ash for cement replacement to produce high strength concrete is 20%.

5. Acknowledgement

The authors would like to thanks to the ministry of Technology Research and Higher Education of Indonesia for funding through Tadulako University Palu. The authors also thanks to PT Beton Jaya Palu for all facilities provided.

6. References

[1]. Bilodeau, A., V.K. Sivasundaram, K.E. Painter, and V.M. Malhotra, *Durability of Concrete Incorporating High Volumes of Fly Ash From Sources in the USA.* ACI materials journal, 1994. **91**(1): p. 3-12.

[2]. Siddique, R. and M.I. Khan, *Supplementary cementing materials.* 2011: Springer Verlag.

[3]. Bilodeau, A. and V.M. Malhotra, *High-volume fly ash system: the concrete solution for sustainable development.* ACI Materials Journal, 2000. **97**(1): p. 41-48.

[4]. Turuallo, G., *Early age strength development of ggbs concrete cured under different temperatures*. 2013, PhD Thesis, Liverpool University: Liverpool, UK.

[5]. Bilodeau, A., V.K. Sivasundaram, K.E. Painter, and V.M. Malhotra, *Durability of Concrete Incorporating High Volumes of Fly Ash From Sources in the USA.* 1994, 1994. **91**: p. 3-12.

[6]. Siddique, R., *Waste materials and by-products in concrete.* 2007: Springer Verlag.

[7]. ASTM-C618-03, *Standard Specification for Coal Fly Ash and Raw or Calcined Natural Pozzolan for Use in Concrete.* 2004, ASTM International: West Conshohocken, PA 19428-2959, United States.

[8]. 2002, A.-R.-. *Use of Fly Ash in Concrete.* 2002, ACI Committee: Washington, USA.

[9]. Ravina, D. and P.K. Mehta, *Properties of fresh concrete containing large amounts of fly ash.* Cement and Concrete Research, 1986. **16**(2): p. 227-238.

[10]. Sivasundaram, V., G.G. Carette, and V.M. Malhotra, *Properties of concrete incorporating low quantity of cement and high volumes of low calcium fly ash.* ACI Special Publication SP-114, 1989. **1**: p. 45-71.

[11]. Joshi, R.C. and R.P. Lohtia, *Fly ash in concrete: production, properties and uses.* Taylor & Francis. 1997: Taylor & Francis.

[12]. Naik, T.R. and S.S. Singh, *Influence of fly ash on setting and hardening characteristics of concrete systems.* ACI materials journal, 1997. **94**(5): p. 355-360.
[13]. Mehta, P.K., *Concrete technology at the crossroads--problems and opportunities.* International Concrete Special Publication, 1994. **144**: p. 1-30.
[14]. Ravina, D., *Efficient concrete utilization of coarse and fine fly ash in precast concrete by incorporating thermal curing.* International Concrete, 1981. **78**(3): p. 194-200.
[15]. Turuallo, G. and H. Mallisa, *Sustainable development: Early age strength of HSC using fly ash to replace part of cemen*, in *Sriwijaya International Conference on Engineering, Science and Technology (SICEST).* 2016, Civil Engineering Department, Riau University: Bangka Belitung, West Sumatera.
[16]. Dhir, R.K., F.H. Hubbard, J.G.L. Munday, M.R. Jones, and S.L. Duerden, *Contribution of PFA to concrete workability and strength development.* Cement and Concrete Research, 1988. **18**(2): p. 277-289.
[17]. Domone, P.L. and M.N. Soutsos, *An approach to the proportioning of high-strength concrete mixes.* Concrete International-Design and Construction, 1994. **16**(10): p. 26-31.

IOP Conf. Series: Materials Science and Engineering **316** (2018) 012040 doi:10.1088/1757-899X/316/1/012040

Ductility of polystyrene waste panel

Dewi Sulistyorini and Iskandar Yasin

Fakultas Teknik Universitas Sarjanawiyata Tamansiswa, Jl.Kusumanegara 157 Yogyakarta, Indonesia

E-mail: dewi.sulistyorini@ustjogja.ac.id

Abstract. Polystyrene waste panel is one of alternative materials that uses polystyrene waste. This experiment is to utilize the polystyrene waste as a non structural panel to be evaluated the ductility. The specimen consisted of cement 250 kg/m^3, polystyrene waste was as aggregate, water cement ratio was 0.4 and wire mesh diameter was 0.6 mm with the grid 6 mm x 6 mm placed on the top and bottom of the panels. The polystyrene panels were compressed at 2 MPa. Six specimens had dimension 80 cm length, 30 cm width and the thickness planned were two varieties, they were 0.5 cm and 1 cm. Flexural testing is used to examine load and deflection to measure the ductility. The load and the deflection showed that the maximum load for the specimen with 0.5 cm thickness is 0.4, 0.56 and 0.37. And for 1 cm thickness is 0.4, 0.36, 0.64. It shows that the thickness variation does not give effect on the maximum load. Result showed the average of Displacement Ductility Index of polystyrene waste panels with 0.5 cm thickness was 1.692 and for 1 cm thickness, the average was 4.043. So the average of the panel with 0.5 cm thickness planned is under 1.99 and the panel with 1 cm thickness planned is upper 3, therefore, it is considered imperative for adequate ductility.

1. Introduction

Polystyrene waste panel as one of alternative materials that use polystyrene waste [1]. It is estimated that EPS foam (or polystyrene) products accounts for less than 1% of the total weight of landfill materials, the fraction of landfill space it takes up is much higher considering that it is very lightweight. Furthermore, it is essentially non biodegradable materials, taking hundreds perhaps thousands of years to decompose. Even when already disposed of in landfills, EPS can easily be carried by the win and litter the streets or end up polluting water bodies [2].

Expanded polystyrene waste in a granular form is used as lightweight aggregate to produce lightweight structural concrete with the unit weight varying from 1600 to 2000 kg/m^3[3].

According to[4] that EPS dosage has the most significant effect on compressive strength of EPS lightweight aggregate concrete, than water and cement ratio, while the content of cement and sand ratio play a comparatively less important part. This experiment is to utilize the polystyrene waste as the material of non structural panel to be evaluated the ductility.

2. Experimental study

This research was done by trying out the mixture in order to get the appropriate composition of the panel. All panels are tested under one point loading scheme based on SNI 03-4154-1996 to get relationship load – deflection.

IOP Conf. Series: Materials Science and Engineering **316** (2018) 012040 doi:10.1088/1757-899X/316/1/012040

2.1 Material

The basic materials used in this research are cement, water, polystyrene waste in the form of granules and 0.6 mm diameter wire mesh with grid 6 mm x 6 mm is cut to size of 80 cm x 30 cm. The basic materials as shown Figure 1 other than water.

Cement	Polystyrene waste granules	Wire mesh

Figure 1. Material for polystyrene waste panel

2.2 Specimen

There are six polystyrene waste panels with two various thicknesses as shown in Table 1 and specimen detail shown in Figure 2.

Table 1. Specimen dimension

Specimen	Length [cm]	Width [cm]	Thickness planned [cm]
T0.5-1	80	30	0.5
T0.5-2	80	30	0.5
T0.5-3	80	30	0.5
T1-1	80	30	1
T1-2	80	30	1
T1-3	80	30	1

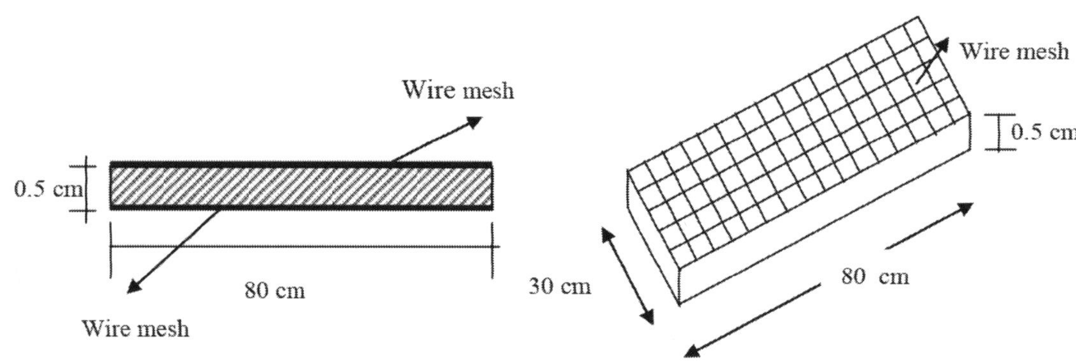

Figure 2. Specimen detail

2.3 Composition

Polystyrene waste panels are made of a mixture of cement, polystyrene waste, and water. After trying out the mixture, it is obtained the appropriate composition as it is shown in Table 2. The cement contented 250 kg/m3, water cement ratio is 0.4 and the unit weight of polystyrene waste is 15.9 kg/m3. Then the panel is compressed 2 MPa. Table 2 shown the composition for polystyrene waste panel after it was tried out the mixture.

Table 2. Composition for polystyrene waste panel

Specimens thick [cm]	Cement content [kg]]	Water [liter]	Polystyrene [kg]	Compressing [MPa]
0.5	1.5	0.6	0.1	2
1	3	1.2	0.2	2

2.4 Methodology

First put the wire mesh on the bottom of the mold. Second pour the mixture. Third put the wire mesh on the top and cover it. Then compress it by hydraulic jack at 2 MPa and let it stand for 15 minutes until the specimen is incompressible. The last take it out from the mold and save it for 28 days in a dry state.

2.5 Test procedure

To monitor the behavior of the tested panel under the applied loading, two LVDT were placed on the mid-span to measure the deflection of the test specimen and connected with data logger. Figure 3 show the test of panel (panel T0.5-2). All panels were tested under a one-point loading scheme.

Figure 3. Test Set Up

3. Result and discussion

Table 3. The test result for load and deflection of polystyrene waste concrete

Specimen	Dimension			Maximum Load [KN]	Deflection [cm]
	Length [cm]	Width [cm]	Thickness [cm]		
T0.5-1	80	30	0.61	0.4	7.854
T0.5-2	80	30	0.58	0.56	7.026
T0.5-3	80	30	0.57	0.37	6.970
T1-1	80	30	1.25	0.4	1.304
T1-2	80	30	1.57	0.36	1.874
T1-3	80	30	2.17	0.64	1.425

IOP Conf. Series: Materials Science and Engineering **316** (2018) 012040 doi:10.1088/1757-899X/316/1/012040

The test results showed in Table 3, after making tree specimens with 0.5 cm thickness planed, the result obtained of thickness was variety, they were 0.57, 0.58, 0.61 whereas for 1 cm thickness planned, the results were much more different from the plan. They are 1.25,1.57, and 2.17 cm. It is shown that the thinner specimen the more constant. The maximum load for all specimens was almost the same. The deflection for 0.5 cm thickness was bigger than 1 cm thickness. Therefore, maximum load is not as comparable as the deflection.
Displacement ductility index is defined as the ratio of deflection at ultimate load to deflection at yield point. The value of ultimate load, according to [5], is 0.8 from maximum load and [6] mentioned that the deflection at yield point can get from the load intersection where the load reached 75% ultimate load. As shown in Figure 4 below.

$$\mu = \frac{\Delta u}{\Delta y} \tag{1}$$

Where μ is the displacement ductility index, Δu *is* the lateral deflection at ultimate load and Δy isthe lateral deflection when yield is first reached.

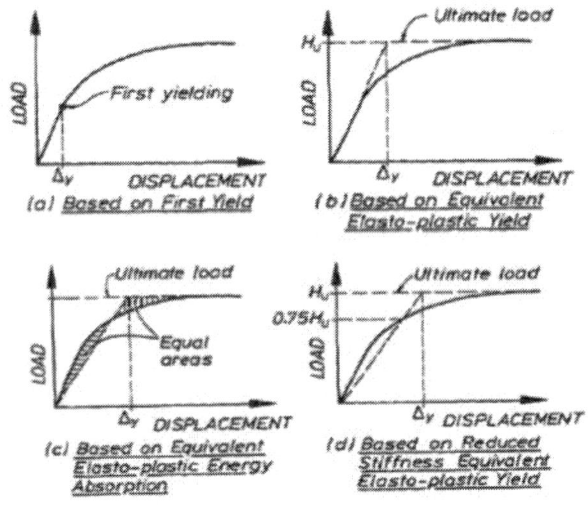

Figure 4. Deflection on the yield point

According to [7] mentioned that the displacement ductility index, μΔ, in the range of 3 to 5 is considered imperative for adequate ductility especially for seismic design and redistribution of moments. Generally, high ductility ratios indicate that a structural member is capable of undergoing large deformations prior to failure.

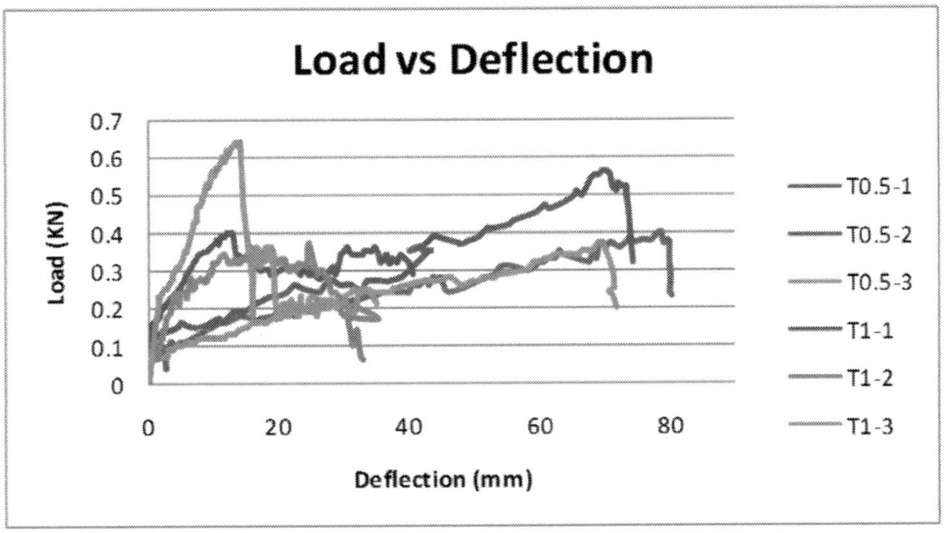

Figure 5. Load vs Deflection

From the figure 5, the load and the deflection show that the maximum load for the specimen with 0.5 cm thickness is 0.4, 0.56 and 0.37. And for 1 cm thickness is 0.4, 0.36, 0.64. It shows that the thickness variation does not give effect on the maximum load.

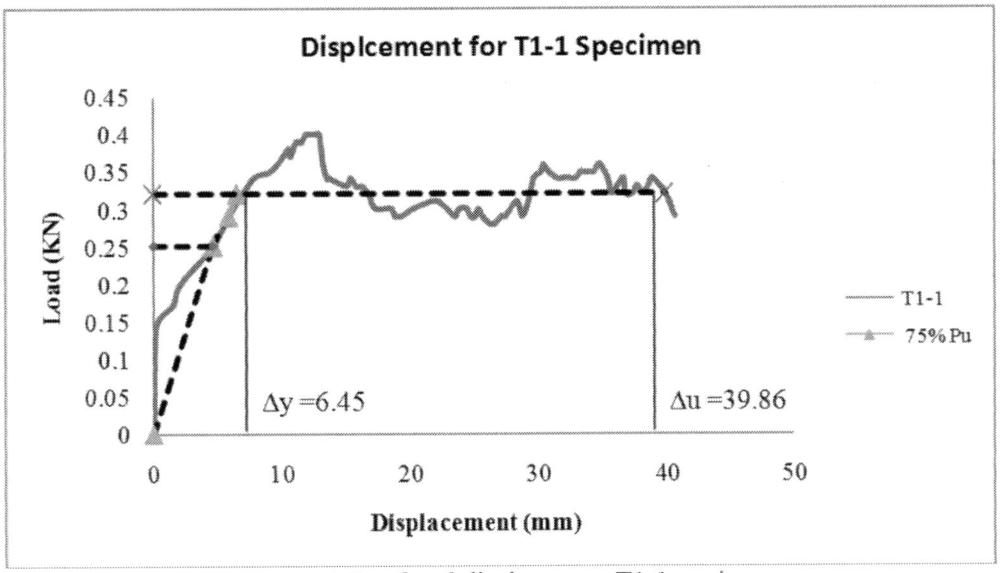

Figure 6. Load and displacement T1-1specimen

Displacement ductility index for T1-1 specimen as shown in Figure 6, according to [5] the ultimate load is 0.8 maximum load, it is the same as 0.32 KN. Therefore from the figure 6, the result of the displacement (Δu) is 39.86 mm. The yield load is 75% ultimate load, it is the same as 0.24 KN. The result of the displacement (Δy) is 6.45 mm. Therefore, the displacement ductility index (μ) for T1-1 is 6.179. The deflection ductility index, (μ), for all panels tested in this study experimentally is presented in Table 4 and Figure 7.

Table 4. The result of deflection ductility index polystyrene waste panels

Specimen	Δu [mm]	Δy [mm]	Ductility Index
T0.5-1	80.04	48.33	1.656
T0.5-2	73.86	59.00	1.252
T0.5-3	71.28	32.89	2.167
	Average		**1.692**
T1-1	39.86	6.45	6.179
T1-2	27.24	6.44	4.239
T1-3	14.58	8.24	1.711
	Average		**4.043**

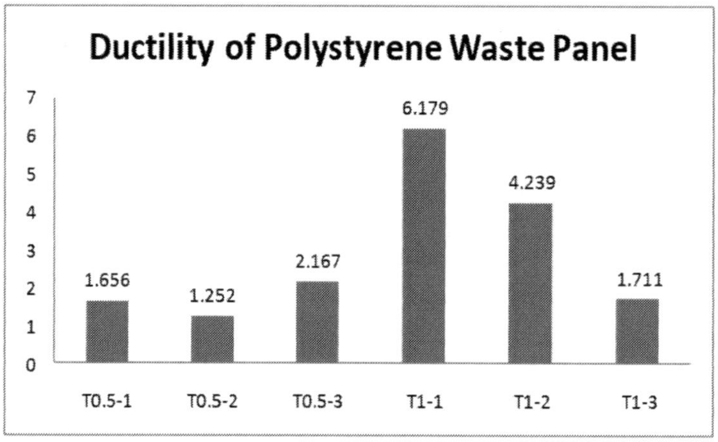

Figure 7. Ductility of Polystyrene waste Panel

A high ductility index indicates that a structural member is capable of undergoing large deformations prior to failure. Panel with ductility index in the range of 3 to 5 is considered imperative for adequate ductility especially in the areas of seismic design and redistribution of moments [7]. Panels with ductility index only up 1.99 lacked adequate ductility and cannot redistribute moment.

Based on the test result, the average of displacement ductility index of panel with 0.5 cm thickness planned is under 1.99 and the panel with 1 cm thickness planned is upper 3, therefore, it is consideredimperative for adequate ductility.

4. Conclusion

Based on the test, the following conclusions can be drawn :
(1). The polystyrene waste panel with 0.5 cm thickness planned, the results are more stable than 1 cm thickness planned but the variation of the thickness is not comparable to the variation of the maximum load. So in this experiment, the wire mesh is the main role in restraining the load.
(2). Displacement Ductility Index of the polystyrene waste panel with 1 cm thickness planned upper 3 indicates that the panel is capable of undergoing large deformation.
(3). In general polystyrene waste has the potential to be used as non structural panels and it is expected to reduce polystyrene waste.

Acknowledgments
We would like to thank S.Suprapto for his support and suggestion regarding to the polystyrene waste panel research. We also would like to thank A.O.Irlan for her support and cooperation during the laboratory.

References
[1] Sulistyorini,D 2016 Concrete Panel From Polystyrene Waste., In : Proceedings of International Conference Civil Offshore Environmental Engineering 2016, 15 Agustus, Malaysia
[2] Micelle Rose Rubio 2016 Dealing With Polystyrene Waste Online Available from : http://www.EcoMENA.org/polystyrene-wastes
[3] Ben Sabaa and Ravindrarajah 1997 Engineering Properties of Lightweight Concrete Containing Crushed Expanded Polystyrene Waste., In Symposium MM: Advances in Materials for Cementitious Composites ,Boston, USA
[4] Yi Xu, Linhua Jiang, Jinxia Xu, and Yang Li 2012 Mecanical properties of expanded polystyrene lightweight aggregate concrete and brick Online available from : http://www.sciencedirect.com/science/article/pii/S0950061811004636
[5] Park and Pauly 1988 Reinforced concrete structure in Amir, M. Yusuf 2011, Flexural Behavior on The state of Serviceability and Limit hollow Core Reinforced Concrete Beam., Jurnal Ilmiah Semesta Teknika Vol. 14 (1): 41-51.
[6] Park 1988 Ductility Evaluation from Laboratory and Analytical Testing State ofthe Art Report, 9th World Conference on Earthquake Engineering, Tokyo-Kyoto,Japan, pp605-616.
[7] Ashour, S.A 2000: Effect of compressive strength and tensile reinforcement ratio on flexural behavior of high- strength concrete beams. Engineering structures, 22 (5), 413 423 in Apeh and Okoli 2016., Evaluation of Ductility Index of Concrete Beams Reinforced with Rebars Milled from Scrap Metals, Concrete Research Letter, Vol. 7 (2)

IOP Conf. Series: Materials Science and Engineering **316** (2018) 012041 doi:10.1088/1757-899X/316/1/012041

The drag forces exerted by lahar flows on a cylindrical pier: case study of post Mount Merapi eruptions

Zainul Faizien Haza

Fakultas Teknik Universitas Sarjanawiyata Tamansiswa, Jl.Kusumanegara 157 Yogyakarta, Indonesia

E-mail: zainul@ustjogja.ac.id

Abstract. Debris flows of lahar flows occurred in post mount eruption is a phenomenon in which large quantities of water, mud, and gravel flow down a stream at a high velocity. It is a second stage of danger after the first danger of lava flows, pyroclastic, and toxic gases. The debris flow of lahar flows has a high density and also high velocity; therefore it has potential detrimental consequences against homes, bridges, and infrastructures, as well as loss of life along its pathway. The collision event between lahar flows and pier of a bridge is observed. The condition is numerically simulated using commercial software of computational fluid dynamic (CFD). The work is also conducted in order to investigate drag force generated during collision. Rheological data of lahar is observed through laboratory test of lahar model as density and viscosity. These data were used as the input data of the CFD simulation. The numerical model is involving two types of fluid: mud and water, therefore multiphase model is adopted in the current CFD simulation. The problem formulation is referring to the constitutive equations of mass and momentum conservation for incompressible and viscous fluid, which in perspective of two dimension (2D). The simulation models describe the situation of the collision event between lahar flows and pier of a bridge. It provides sequential view images of lahar flow impaction and the propagation trend line of the drag force coefficient values. Lahar flow analysis used non-dimensional parameter of Reynolds number. According to the results of numerical simulations, the drag force coefficients are in range 1.23 to 1.48 those are generated by value of flow velocity in range 11.11 m/s to 16.67 m/s.

1. Introduction

Mount Merapi located in the middle of Java Island is a very active volcano which has erupted regularly every two to five years. One of danger emerged by its eruption is a lava flows in the form of debris flow, which is in Indonesian term; it is called 'lahar'. Term of 'lahar' describes a hot or cold mixture of water and rock fragments flowing down the slopes of a volcano and (or) river valleys. When moving, a lahar looks like a mass of wet concrete that carries rock debris ranging in size from clay to boulders more than 10 m in diameter. Lahars vary in size and speed. Small lahar flow less than a few meters wide and several centimeters deep may flow a few meters per second. Large lahars hundreds of meters wide and tens of meters deep can flow several tens of meters per second--much too fast for people to outrun [1].

The debris flows is not a danger which followed an eruptions immediately. The debris flows happened when volcanic materials deposited on top of mountain post eruptions was washed away by heavy rainfall. A debris flow consists of thick sedimentary rocks, gravel, sand, mud, and volcanic ash mixed with water. The debris flow has a high density and velocity; therefore it has potential

IOP Conf. Series: Materials Science and Engineering **316** (2018) 012041 doi:10.1088/1757-899X/316/1/012041

detrimental consequences against of homes, bridges, and infrastructures, as well as loss of life along its pathway.

After Mount Merapi eruptions, debris flows are frequent and numerous disasters occurred. A protection system to reduce losses due to debris flow was created as sediment retaining that called check dam or sabo dam. The purpose of the sabo structures was to reduce the excess sediment discharge to prevent river degradation further downstream. Recently, sabo works shifted to the control of debris flows. The system of dam can be used to reduce the impact of debris torrents [1]. According to Mount Merapi protection system, sabo dams are built on rivers which are the pathway of debris flows; those are Kali Krasak, Kali Batang, Kali Putih, Kali Gendol and other rivers as seen in Figure 1.

Sabo dam built at some locations in mountainside of Mount Merapi are designed based on hydrological data (rainfall data and watershed maps), location maps, situation maps (topographic maps and geometry of the river), river sedimentation and soil mechanics data. The data are used as basis of the design of the main dam, sub dam, and dam stability control [2].

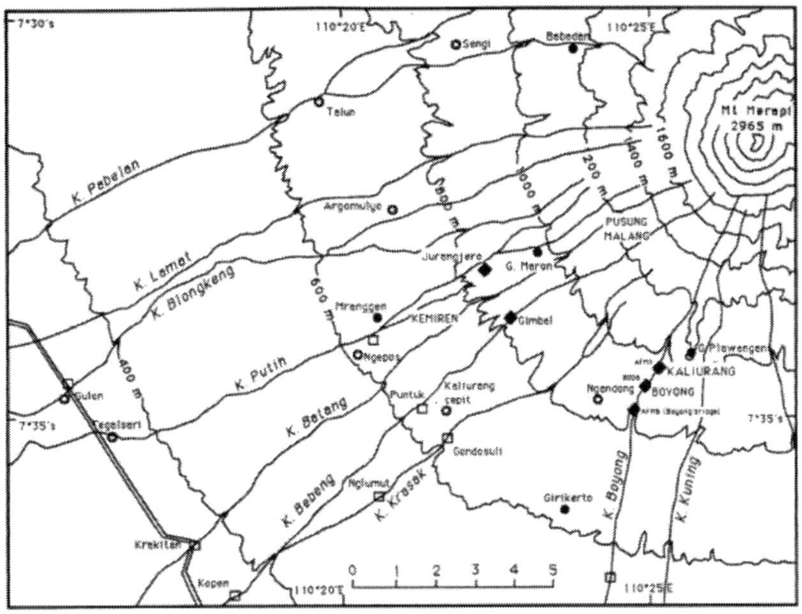

Figure 1. Pathway Map of Mount Merapi lahar flows, Middle Java, Indonesia [3]

Moreover, one of the structures affected by the flow of lahar is the bridge pier, which it is exposed to be directly hit by lahar flow. The situation is compounded by the conditions of the bottom of the piers that already experiencing scour, so that the piers are in a state of hanging. Obviously this situation is very dangerous for the stability of the bridge as shown in Figure 2.

According to [4], there are at least five bridges, the dams, and four houses in Magelang damaged by Merapi lahar flows. It describes the destructive force of lahar flows. Since Mt. Merapi is a strato volcano type, which has a very steep slope, the characteristics of lahar flow is fast-moving with great force. The combination of the flow of volcanic material such as volcanic ash, gravel, gravel, pieces of rock with steep slopes make the lahar flows are also controlled by the acceleration of the gravity force.

Figure 2. Bridge pier damages on Pabelan River, Pabelan, Jawa Tengah [5]

According to [4], there are at least five bridges, the dams, and four houses in Magelang damaged by Merapi lahar flows. It describes the destructive force of lahar flows. Since Mt. Merapi is a strato volcano type, which has a very steep slope, the characteristics of lahar flow is fast-moving with great force. The combination of the flow of volcanic material such as volcanic ash, gravel, gravel, pieces of rock with steep slopes make the lahar flows are also controlled by the acceleration of the gravity force.

In the process of flowing lahar, energy generated due to surface run-off is changed according to the square of the velocity. Grain transport capacity of volcanic material will change with the rank of 5 in time in one-dimensional unit. It means that the surface flow rate increases to two times, the number of transported grains material increases 32 times more [4]. These events are normally (in geological time scale) occurred in many areas around the Mount Merapi lahar flows pathways. It has potential detrimental consequences against facilities such as bridge, dam, and river structure.

IOP Conf. Series: Materials Science and Engineering **316** (2018) 012041 doi:10.1088/1757-899X/316/1/012041

Figure 3. Collapsed bridge due to the destructive of lahar flow [6]

The aim of this research is to investigate the drag force exerted by lahar flows in the collision against the cylindrical bridge pier by employing numerical simulation of ANSYS Fluent 14.0. The determination of drag force and drag coefficient is referring to rheological properties of lahar. The current numerical modeling is carried out in view of providing the basis for developing methods for prediction of the impact forces of lahar flow on bridge pier.

2. Numerical simulation

In order to numerically model the collision between lahar flow and bridge piers, fluid dynamics approach is implemented using gravity flow concept, which has been widely used in research on fluid flow. In the current study, this method is applied to accommodate the additional effect of flow plasticity on the drag force related to the dynamic pressure which is proportional to lahar density and the squared flow velocity. Therefore, drag force generated by lahar flow on bridge piers can be expressed by traditional fluid dynamic force and rheology properties of non-Newtonian fluid flow as the following equation [7].

$$F_d = \frac{1}{2} \cdot \rho_f \cdot C_d \cdot A \cdot u^2 \quad (1)$$

where F_d is the drag force components perpendicular to bridge piers axis, ρ_f is the lahar density, which is according to [8], in range 1230 kg/m^3 to 1840 kg/m^3. Whereas C_d is the drag coefficient, A is area of bridge piers, which is facing opposite to lahar flow direction, and u is flow front velocity of lahar flow. As seen in Eq. (1), drag force analysis of this interaction is determined by properties of lahar flow as density and velocity together with bridge piers physical attribute of outside diameter (OD) and surface area.

The current numerical simulation is utilizing the ANSYS Fluent 14.0, which provided the broad physical modelling capabilities needed to model the flow. By using the advanced post-processor, examining the results and settings, analysis can be performed in further. The problem formulation is referring to the constitutive equations of mass and momentum conservation for incompressible and viscous fluid, which in perspective of 2D; it is expressed as Eq. (2) and Eq. (3), respectively. Equations of mass conservation is based on principles that rate of inflow is equal to outflow rate [9].

$$\frac{\partial \rho}{\partial t}+\nabla \cdot (\rho \vec{u}) = S_m \tag{2}$$

where ρ is lahar density, u is velocity, and S_m is the mass added to continuous phase from the dispersed second phase (specified mass sources).

While, the momentum conservation is formulated as follow.

$$\frac{\partial}{\partial t}(\rho \vec{u}) + \nabla \cdot (\rho \vec{u}\vec{u}) = -\nabla p + \nabla \cdot (\bar{\tau}) + \rho \vec{g} + \vec{F} \tag{3}$$

where p is the static pressure, $\rho \vec{g}$ and $\vec{F}$ are the gravitational body force and external body force.

Whereas, $\bar{\tau}$ is the stress tensor, which is formulated as follow.

$$\bar{\tau} = \mu\left[(\nabla \vec{u} + \nabla \vec{u}^T) - \frac{2}{3}\nabla \cdot \vec{u} I \right] \tag{4}$$

where μ is lahar viscosity and I is the unit tensor.

Lahar flow analysis used non-dimensional parameter of Reynolds number (Re), which is the ratio between inertia force and viscous force. Inertia force is expressed by using density parameter (ρ_f), velocity (u), and slope of lahar flow velocity (du/dx). While viscous force is expressed by using dynamic viscosity parameter (μ). Reynolds number is expressed as the following.

$$R_e = \frac{(\rho_f \cdot u)(du/dx)}{\mu \cdot (d^2u/dx^2))} \tag{5}$$

The geometry model was created by using ANSYS DesignModeler (DM). The meshing process was implemented by using the ANSYS Meshing (ANSYS ICEM CFD) with 2D behaviour of plane stress and Lagrangian reference frame. The 18,871 nodes and 18,425 elements were created by using size function of on proximity and curvature with proximity minimum size of 1.2512e-003 m and maximum face size of 2e-002 m. However, the refinement of mesh was definitely depending on the performance of the hardware; higher capability provided better meshing process.

Figure 3 illustrates of the preparation of pre-processing stages. As seen in this figure, lahar flow is modeled as a fluid flow with density of ρf enters into the empty zone as a velocity inlet model. Whereas three sides of model are stated as free flow boundary condition. In addition, applying such boundary condition will allow the lahar flow will be having wave behavior.

According to [3], for several lahars in the years following the 1984 Merapi eruption, a maximum peak flow velocity of 11 m/s was recorded at Jurangjero, Putih River (865 m, 4.5% slope), for the 28 November 1985 event. In the Boyong River, the maximum recorded peak flow velocity was 15 m/s at Kaliurang (850 m, 4.6% slope) on 20 May 1995. Therefore, in the current research, the velocity inlet used varies into 40 km/h to 60 km/h.

IOP Conf. Series: Materials Science and Engineering **316** (2018) 012041 doi:10.1088/1757-899X/316/1/012041

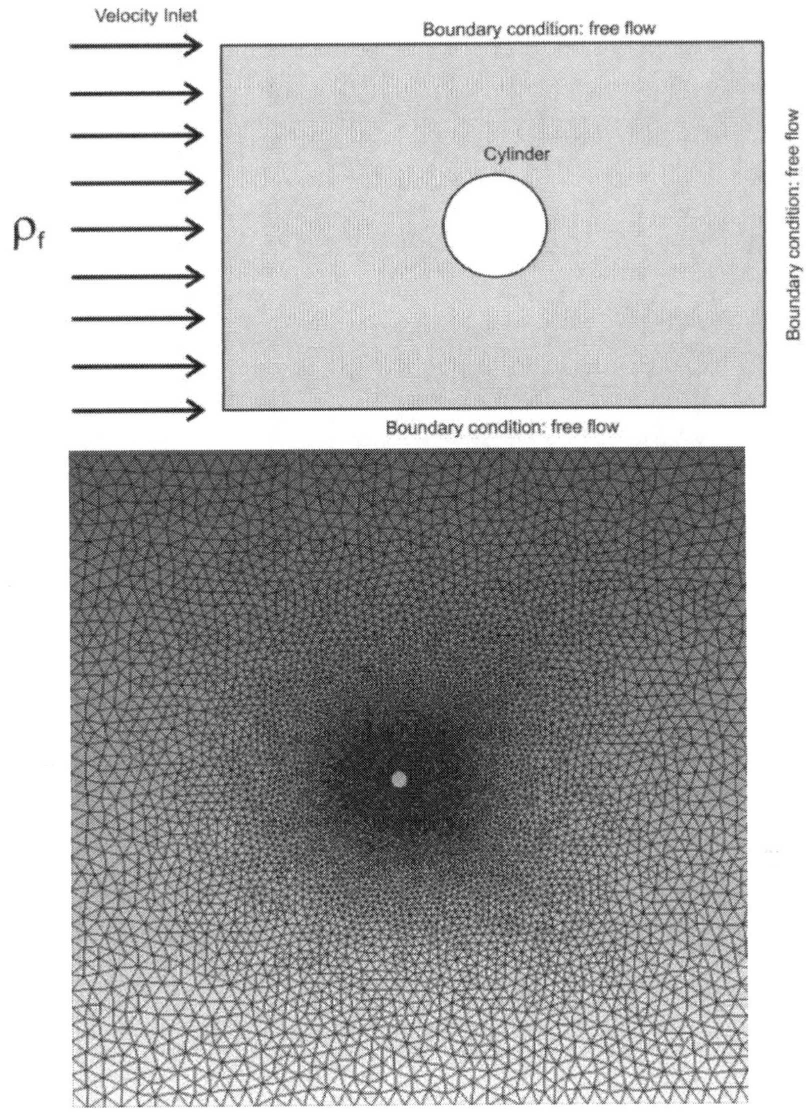

Figure 4. Geometry model and meshed geometry

3. Result and discussion

Non-dimensional parameter of Reynolds numbers (Re) were calculated by using Eq. (5). Values of Re related to lahar flow velocities are as shown in Table 1 below.

Table 1. Reynolds number (Re).

Dia. of Pier (m)	Linear Characteristics [m]	Flow Velocity (u) [m/s]	Density of Lahar [kg/m^3]	Viscosity of Lahar [Pa.s]	Re [x10^7]
2,5	0,942857	11,11	1840	0,000861	0,923
2,5	0,942857	13,89	1840	0,000861	1,54
2,5	0,942857	16,67	1840	0,000861	2,46

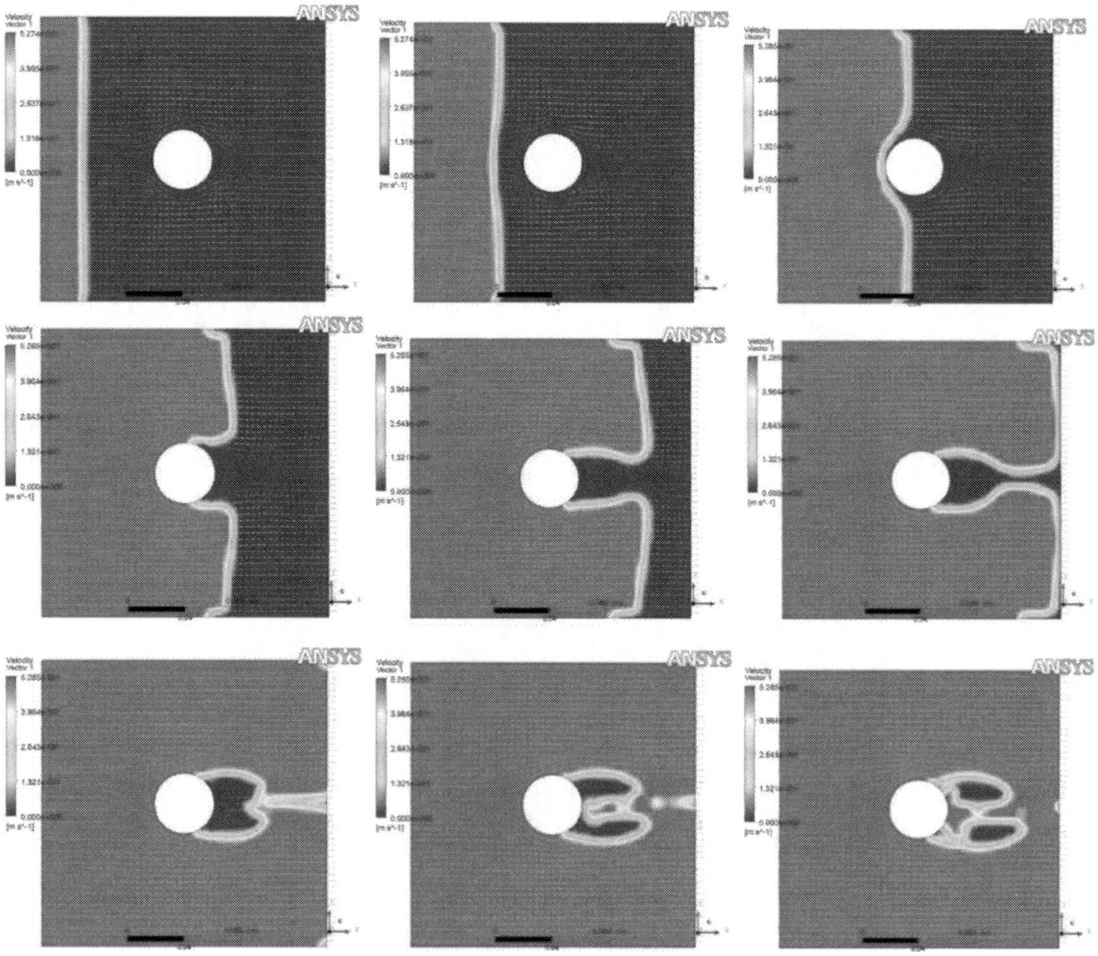

Figure 5. Image capturing in sequence views of the collision between lahar flow and pier

The coefficient of drag force is one of the parameters of drag force equation (Eq. (1)). In using CFD simulations, the used C_d value is the maximum C_d value obtained from each model. The C_d value obtained from simulation is a combination of the parameters of the velocity waves (u) and the size of the diameter of the pillar. The distribution of C_d value of numerical simulation results are shown in Table 2 below.

Table 2. Drag force coefficient (C_d)

Diameter of pier (m)	Flow Velocity (u) [m/s]	Reynolds number (Re) [x10^7]	Drag Force Coefficient (Cd)
2,5	11,11	0.923	1.23
2,5	13,89	1.54	1.3
2,5	16,67	2.46	1.48

Normally, the collision between non-Newtonian fluid flows with cylinder occurs within a relatively short time [10]. One of the outcomes of simulations using CFD on the collision event between lahar flows with bridge pier is a graphical form of distribution values of C_d when the collision takes place in the modeling process. Figure 5 below shows the distribution of C_d values based on the magnitude of the lahar flows velocity used in the numerical simulations.

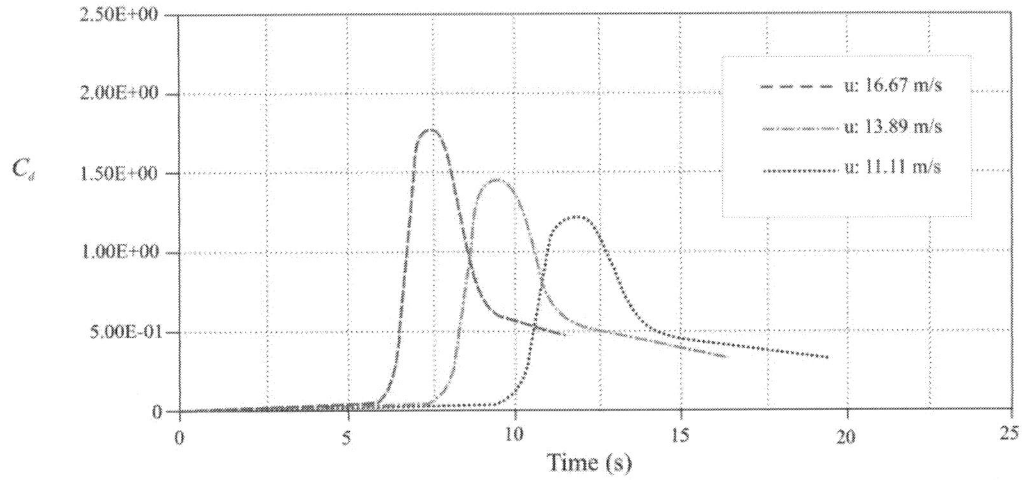

Figure 6. Drag force coefficient (C_d)

4. Concluding remark

The application of commercial software of ANSYS Fluent to simulate the collision event between lahar flows and pier was presented. Simulation setup was implemented as the pressure-based Navier-Stokes (*pbns*) with absolute velocity formulation and mixture-model for multiphase model with two phases of Eulerian phase. It produced the simulation model in the form of sequential views images of lahar flow head impaction and the propagation trend line of the drag force coefficient values. Furthermore, he values of the coefficient of drag force are depending on the diameter of bridge pillar used as well as the velocity of the lahar flows. The larger the diameter used and the greater the velocity of lahar flows will produce the higher drag coefficient. According to the results of numerical simulations, the drag force coefficients are in range 1.23 to 1.48 those are generated by value of flow velocity in range 11.11 m/s to 16.67 m/s.

Acknowledgments

This research is possible thanks to all Staff and Facilities of Laboratory of Program Studi Teknik Sipil Universitas Sarjanawiyata Tamansiswa Yogyakarta.

References

[1] United Stated Geological Survey n d Lahars and Their Effects *Online* Available from: http://volcanoes.usgs.gov/hazards/lahar/index.php. *Accessed*: 11August 2015

[2] Chanson H 2004 Sabo check dams – mountain protection system in Japan *Intl. J. River Basin Management* **2** 4 301-07

[3] Lavigne F, Thouret J C, Voight B, Suwad H, Sumaryono A 2000 Lahars at Merapi volcano, Central Java: an overview *Journal of Volcanology and Geothermal Research* **100** 423–56

[4] Pusat Studi Bencana IPB n d Bahaya Banjir Lahar *Online* Available from: http://psb.ipb.ac.id/index.php/news/92-bahaya-banjir-lahar *Accessed*: 9 November 2016

[5] Istiarto 2011 Jembatan Pabelan, Magelang, Jawa Tengah *Istiarto Blog* blog post: 1 May 2011 Available from: http://istiarto.staff.ugm.ac.id/index.php/2011/05/jembatan-pabelan-magelang-jawa-tengah/.

[6] Fathani T F and Legono D 2013 The Application of Monitoring and Early Warning System of Rainfall-Triggered Debris Flow at Merapi Volcano, Central Java, Indonesia In: Wang F,

Miyajima M, Li T, Shan W, Fathani T F (eds) *Progress of Geo-Disaster Mitigation Technology in Asia* Springer Berlin Heidelberg 263-75
[7] Pazwash H and Robertson M 1975 Forces on Bodies in Bingham fluids *Journal of Hydrulic Research* **13** 35-55
[8] Japan International Cooperation Agency 2003 *Density Test: The study on Sabo and Flood Control for Western River Basins* JICA
[9] Sleigh P A and Goodwill I M 2000 *The St. Venant Equations* Leeds School of Civil Engineering University of Leeds England
[10] Haza Z F, Harahap I S H, Dakssa L M 2013 Experimental studies of the flow-front and drag forces exerted by subaqueous mudflow on inclined base *Journal of Natural Hazards* **68** 2 587-611

IOP Conf. Series: Materials Science and Engineering **316** (2018) 012042 doi:10.1088/1757-899X/316/1/012042

Estimating design flood and HEC-RAS modelling approach for flood analysis in Bojonegoro city

R M S Prastica, C Maitri, A Hermawan, P C Nugroho, D Sutjiningsih, E Anggraheni

Civil Engineering Department of Universitas Indonesia, Kampus UI Depok, Postcode 16424, West Java, Indonesia

rianmantasasp@gmail.com, rian.mantasa@ui.ac.id

Abstract. Bojonegoro faces flood every year with less advanced prevention development. Bojonegoro city development could not peak because the flood results material losses. It affects every sectors in Bojonegoro: education, politics, economy, social, and infrastructure development. This research aims to analyse and to ensure that river capacity has high probability to be the main factor of flood in Bojonegoro. Flood discharge analysis uses Nakayasu synthetic unit hydrograph for period of 5 years, 10 years, 25 years, 50 years, and 100 years. They would be compared to the water maximum capacity that could be loaded by downstream part of Bengawan Solo River in Bojonegoro. According to analysis result, Bengawan Solo River in Bojonegoro could not able to load flood discharges. Another method used is HEC-RAS analysis. The conclusion that shown by HEC-RAS analysis has the same view. It could be observed that flood water loading is more than full bank capacity elevation in the river. To conclude, the main factor that should be noticed by government to solve flood problem is river capacity.

1. Introduction

Flooding is one of the most happening hazards worldwide, including Indonesia. Flood disaster happens frequently when wet season comes to Indonesia. Flood problems make material losses and cause discomfort social activities. They affect to the low city development in every sectors. Flood could be derived from natural factors and due to human activity itself. Each year, national media reported that flood disasters' result was dramatically increased from previous flooding.

When the wet season approaches to Indonesia, several prevention acts are well planned. But, nature has more destructive power. Schanze (2006), Nyarko (2000), and Merz et al. (2007) concluded that flooding is the undesired rise of water level either by dam failure or long rainfall duration with high intensity. The most popular region with both drought and flood disasters is Bojonegoro. Bojonegoro is located in the downstream of Bengawan Solo River, East Java. It could not be denied that in the past several years, Bojonegoro suffered drought in dry season with minimum availability of water. However, abundant water is found in wet season in the form of flood hazard. In the beginning year of 2017, flooding in Bojonegoro destructed 203 houses and affected over 1100 people. Some experts argued that the factors due to overflowing river, poor drainage system, ecological destruction, forest clearing both legally and illegally, or river capacity are several answers for 'what is the main reason of flood in Bojonegoro?' question. Furthermore, this research aims to prove one of those well-known factors: river capacity.

Hydrologic models are used to simulate peak runoff and design storm hydrographs. There are several methods to compute runoff in a watershed. Nigussie (2015) proposed Snyder's, Soil Conservation Service, Mockus, Nakayasu, Rodriguez-Valdez, and Gupta-Waymire synthetic unit hydrograph (SUH) to seek peak runoff. Models such as the Rational method, USDA Soil Conservation Service (SCS) method, and Nakayasu method have been widely used in practice to estimate peak runoff in Korean watersheds (Jung and Moon, 2001). Furthermore, Kang (2009) advised five different models, including Rational, WFRP, Nakayasu, SCS, and Clark methods to estimate peak discharges.

Rivera (2007) conducted and performed a study of hydrological modelling to determine potential regions to be flooded during high rainfall events in the Aguán river basin, Central Honduras. This procedure was made by using HEC-RAS model (Hydrological Engineering Center-River Analysis System). Final result demonstrated the exact location of areas with high, moderate, and low risk to be flooded in specific high flood events. Paringit (2015) used HEC-RAS to reach periodic outflow while considering surface roughness to generate flood extents.

2. Methodology

To prepare the study of flood analysis, methods that could help this research steps are systematically needed. Firstly, literature study is conducted from various journals, feasibility study reports, research reports, government regulations, and other publications. This research uses secondary data. Hydrological data are obtained from Balai Besar Wilayah Sungai Bengawan Solo. This research assumes each river cross sections have same characteristics as trapezoidal channel. Lack of primary and secondary data availability about river cross sections makes HEC-RAS analysis is difficult to obtain the best result.

According to the availability data, two stages of methods analysis would be conducted in this research, namely the analysis of design peak runoff using Nakayasu synthetic unit hydrograph and the simulation using HEC-RAS software. The research aims to compare flood loading with river capacity. Manning's equation helps this research to analyze river capacity. The analysis using HEC-RAS modeling goals to determine hazard area, to obtain runoff height information, and to show the cross sections characteristics of Bengawan Solo River downstream.

3. Results and Discussion

3.1. Analysis of flood

Bengawan Solo River flows into Bojonegoro district, in East Java province. It is classified in the Sub-basin Bengawan Solo downstream area. Bengawan Solo River divides 2307.06 hectares area of Bojonegoro into two areas with 0-8% slope interval. It is located in low land with elevation of +12.00 m until +15.00 m . Based on the topography, Bojonegoro district in northern part is lower than the southern part. Therefore, all of the streams flow to the estuary in the northern part of Bengawan Solo River. Based on hazard data from Regional Disaster Management Agency of Bojonegoro, 146 villages in 16 districts were frequently hit by flood due to overflowing of Bengawan Solo River.

3.1.1. Thiessen polygon analysis. Generally, hydrological analysis is one part of the preliminary analysis to design hydraulic structures. Hydrological analysis is required to determine the characteristics of hydrology from Bengawan Solo river basin to Bojonegoro City. Hydrological analysis is used to determine the volume of designed flood discharges. Point of origin derives from the flood-impacted region from Bengawan Solo Basin Main Office data. Geographic Information System (GIS) analysis results that the location of this research watershed total area has 11,056.71 km^2. Downstream part of Bengawan Solo River length is approximately 315,099 m. Figure 1 shows the flood-impacted region in 2016. This data is managed by Balai Besar Wilayah Sungai Bengawan Solo. Figure 2 depicts GIS user interface result for Bengawan Solo watershed that used in this research.

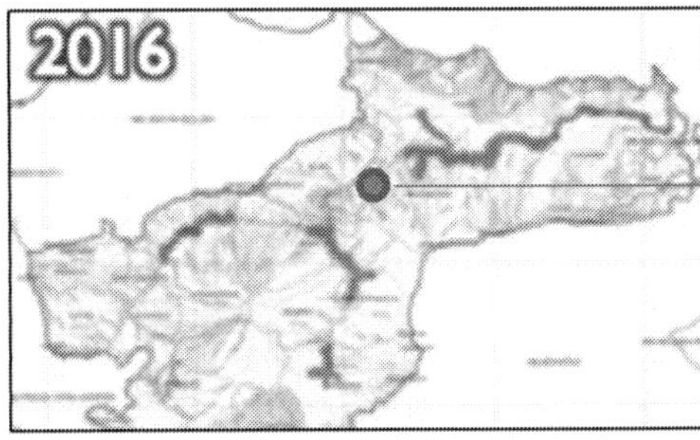

Figure 1. Flood-impacted Area in Bojonegoro

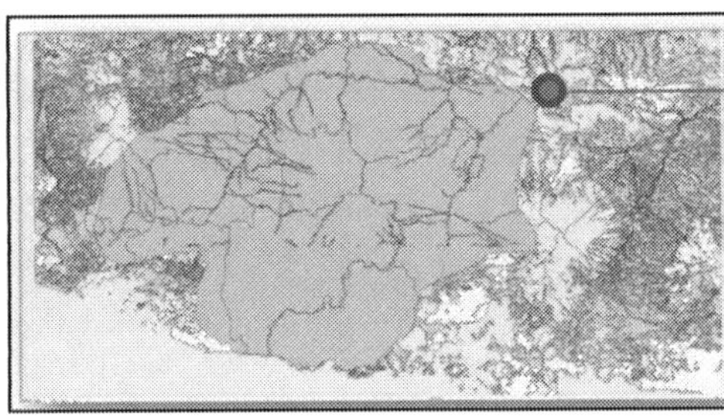

Figure 2. GIS analysis of Bengawan Solo watershed

3.1.2. Distribution compatibility test. This analysis is conducted in Normal, Log-normal, Gumbel, and Log Pearson type-III. Both Squared-chi and Smirnov-kolmogorov test result that all of those distributions meet satisfying result. The result of squared-chi test for those distributions are 1.33, 1.33, 3.00, and 0.5, respectively. They are less than the critical-chi: 5.99. Furthermore, Smirnov-Kolmogorov results 0.091, 0.116, 0.161, and 0.099, respectively. They are less than the critical value: 0.380. The analysis of GIS is not only for calculating areas, but also finding the run off coefficient. In this research, the weighed coefficient is computed for return period of 5-year, 10-year, 25-year, 50-year, and 100-year. The coefficients are 0.3306, 0.3541, 0.3925, 0.4259, and 0.4667, respectively. Furthermore, these coefficients could be brought to acquire flood discharges.

Table 1. Maximum daily rainfall data (mm)

Year	Rain gauge station		
	Yogya	Semarang	Sawahan
2004	106	85	102
2005	98	98	165
2006	131.5	152	77
2007	117.8	108	150
2008	102	119	215.5
2009	85	235	180
2010	131	139	156
2011	81	106	182
2012	118	110	123
2013	131.3	146.5	128

Year	Rain gauge station		
	Yogya	Semarang	Sawahan
2014	85.3	141.7	147.5
2015	73.9	71	130

3.1.3. Nakayasu synthetic unit hydrograph. This synthetic unit hydrograph method formula as written in the equation 1, aims to find the return period discharges. Table 2 and figure 3 show the amount of peak discharges estimation that chanced in each return period.

$$Qp = \frac{C.A.Ro}{3{,}6(0{,}3Tp + T_{0,3})} \quad (1)$$

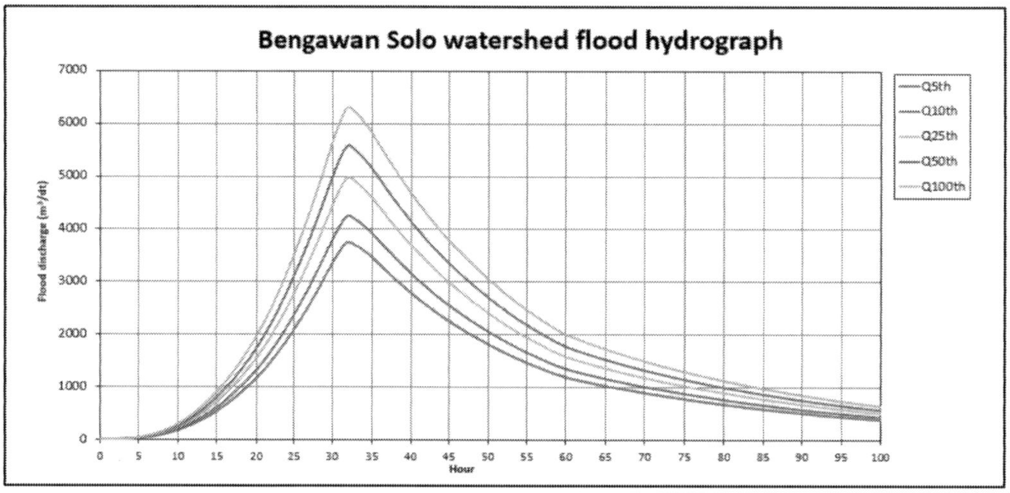

Figure 3. Bengawan Solo watershed flood hydrograph

Table 2. Runoff Estimation Analysis

Return Period (years)	Peak runoff (m^3/s)
5	3750.98
10	4242.70
25	4969.26
50	5579.20
100	6297.57

3.2. River capacity analysis

According to the simulated discharge data in table 2, the comparison analysis with the maximum capacity of river to load up water discharge is conducted. The river capacity is analysed with Manning equation (Chow, 1959). Equation 2 figures the formula to calculate Manning coefficient. In addition, the simulated river design has 180 m base width, slope of 0.00017, z = 1, and n = 0.03. The maximum daily height in the downstream part of Bengawan Solo River is 10 m (Kurniawati, 2016) and the result of maximum projected water loading of the river is approximately 3605.1352 m^3/s.

$$Q = A \times R^{2/3} \times S^{1/2} \times n \tag{2}$$

Based on the calculation and the comparison analyses, Solo River downstream area could not accommodate the existing overflow. So, when rainfall experiences return period discharge of 5-year, 10-year, 25-year, 50-year, and 100-year, flood would happen. The result could be observed in the table 3.

Table 3. River capacity analysis result

H (m)	A (m^2)	P (m)	R (m)	V(m/s)	Q storage (m^3/s)
10	1900.00	208.2843	9.1221	1.8974	3605.1352
9.5	1800.25	206.8701	8.7023	1.8388	3310.2400
9	1701.00	205.4558	8.2792	1.7787	3025.5063
8.5	1602.25	204.0416	7.8526	1.7170	2751.1091
8	1504.00	202.6274	7.4225	1.6537	2487.2387
7.5	1406.25	201.2132	6.9889	1.5887	2234.1026
7	1309.00	199.7990	6.5516	1.5217	1991.9286
6.5	1212.25	198.3848	6.1106	1.4526	1760.9673
6	1116.00	196.9706	5.6658	1.3813	1541.4969
5.5	1020.25	195.5563	5.2172	1.3074	1333.8273
5	925.00	194.1421	4.7646	1.2306	1138.3074
4.5	830.25	192.7279	4.3079	1.1507	955.3333
4	736.00	191.3137	3.8471	1.0671	785.3601
3.5	642.25	189.8995	3.3821	0.9792	628.9182

3.3. HEC-RAS analysis

Another proposed method to analyze the factor of flood is performed by using HEC-RAS software. The procedure in using this program are (a) entering the river geometric data, (b) entering steady flow data, and (c) entering return period discharge. The lack of cross sections data in this river makes the result could not create excellent performance. This research assumes that all of the cross sections in this river have the same patterns. The river base level starts at 0 meter and full bank elevation plus surveillance height reaches 13.33 meters. However, this analysis result makes the previous discussed method stronger. According to HEC-RAS analysis, it could be noticed that point in the downstream (sta.0), flood discharge is still under the maximum capacity. But, in the central point (sta.49), design peak runoff with return period of 25, 50, and 100 years begin to overspill. Furthermore, the upstream point (sta.98), all of return periods' flood discharge experience spilling over. The expected location of simulated points are reserved in figure 5. The samples of river's cross sections test result by using HEC-RAS could be comprehended in figure 4 and table 4.

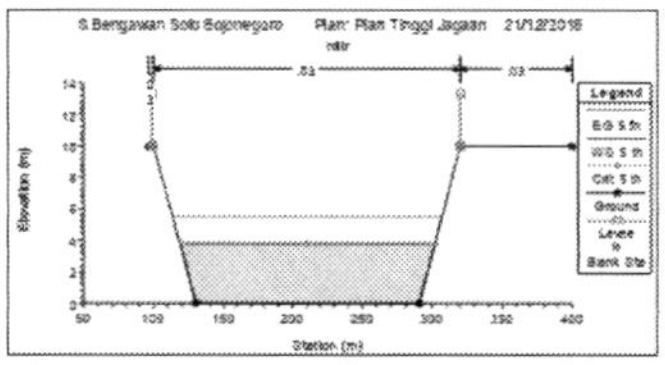
River cross section (HEC-RAS result) Sta.0 return period of 5-year

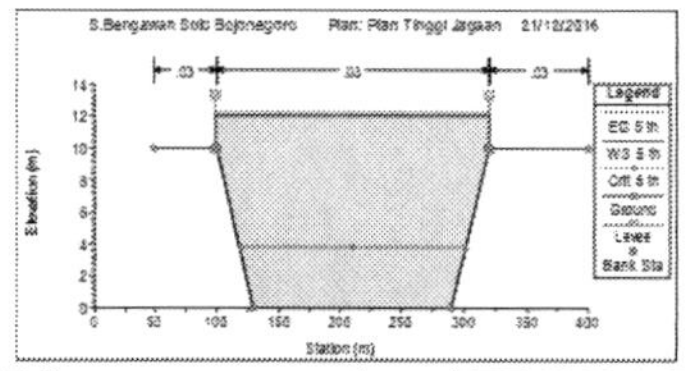
River cross section (HEC-RAS result) Sta.49 return period of 5-year

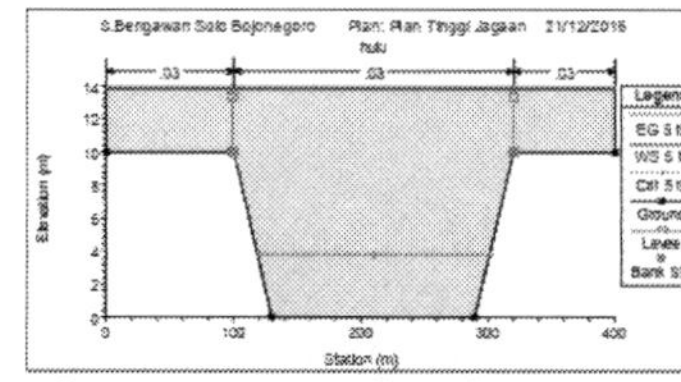
River cross section (HEC-RAS result) Sta.98 return period of 5-year

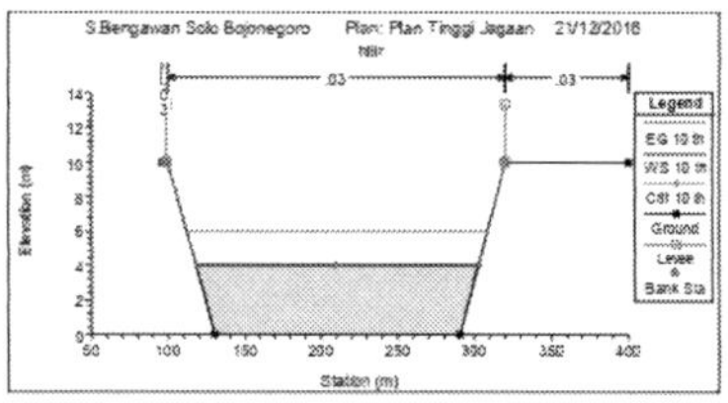
River cross section (HEC-RAS result) Sta.0 return period of 10-year

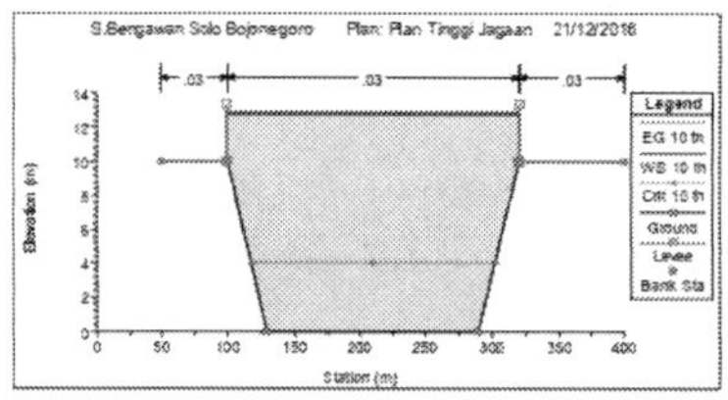
River cross section (HEC-RAS result) Sta.49 return period of 10-year

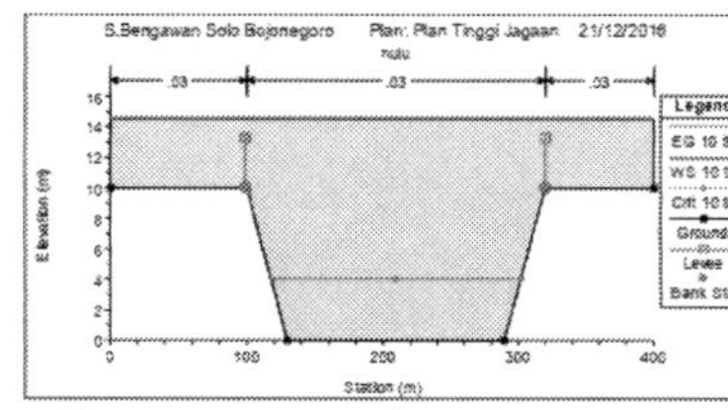
River cross section (HEC-RAS result) Sta.98 return period of 10-year

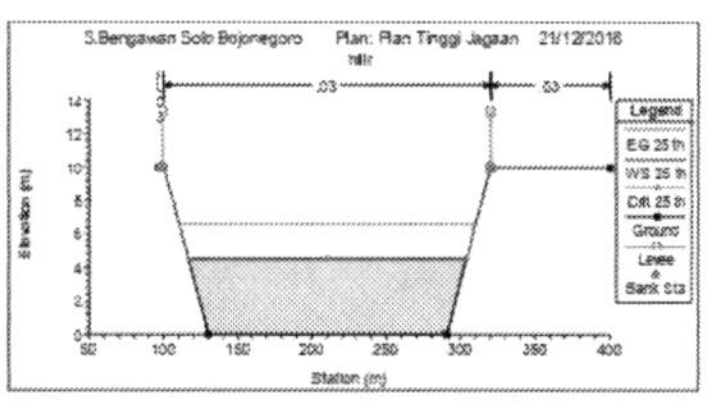
River cross section (HEC-RAS result) Sta.0 return period of 25-year

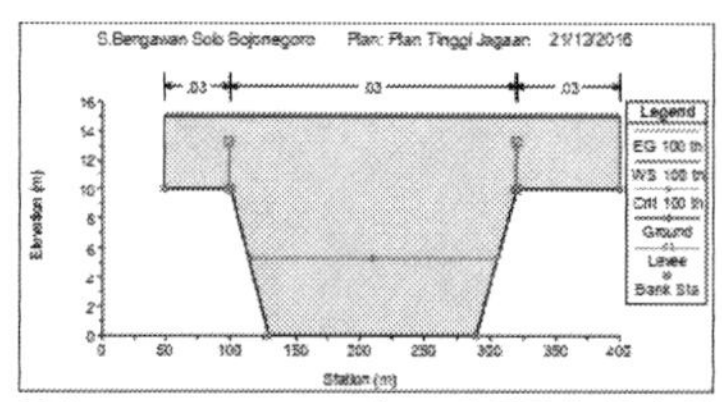
River cross section (HEC-RAS result) Sta.49 return period of 25-year

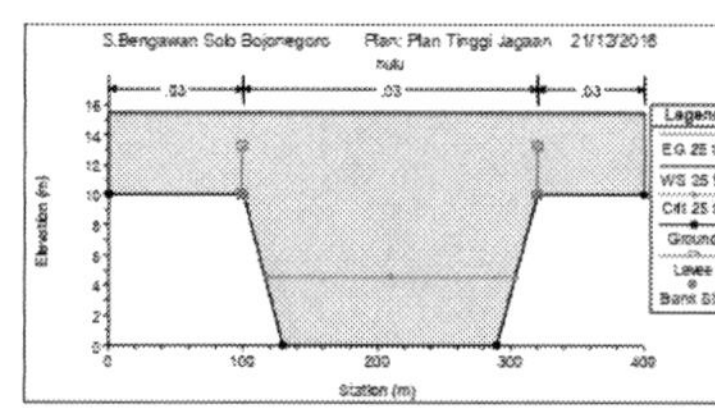
River cross section (HEC-RAS result) Sta.98 return period of 25-year

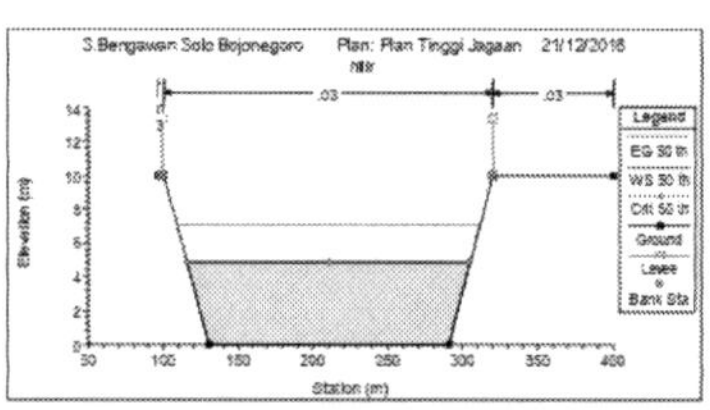
River cross section (HEC-RAS result) Sta.0 return period of 50-year

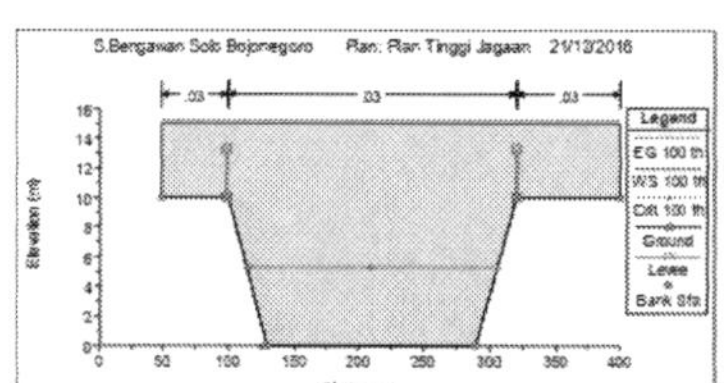
River cross section (HEC-RAS result) Sta.49 return period of 50-year

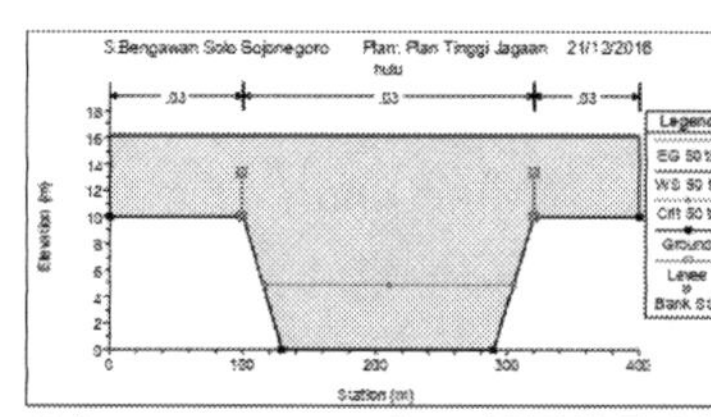
River cross section (HEC-RAS result) Sta.98 return period of 50-year

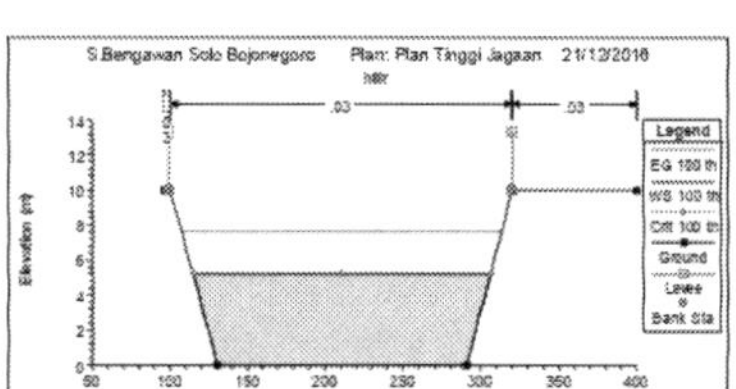
River cross section (HEC-RAS result) Sta.0 return period of 100-year

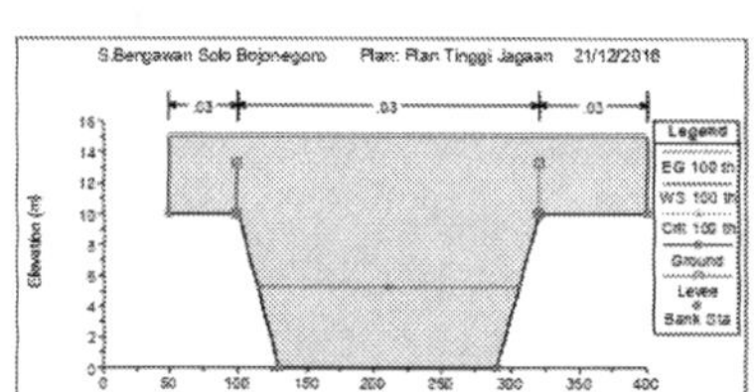
River cross section (HEC-RAS result) Sta.49 return period of 100-year

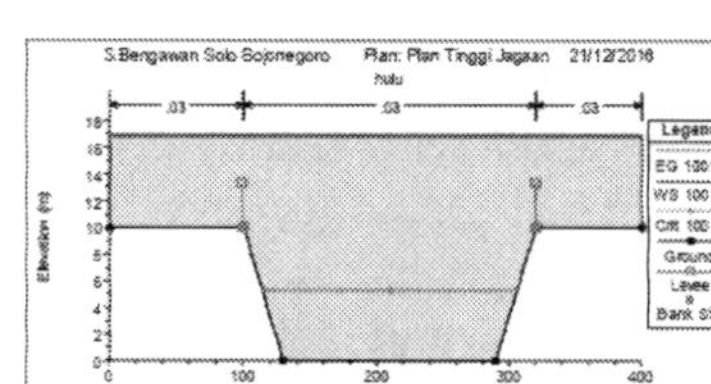
River cross section (HEC-RAS result) Sta.98 return period of 100-year

Figure 4. Cross sections of downstream part of Bengawan Solo River from HEC-RAS analysis

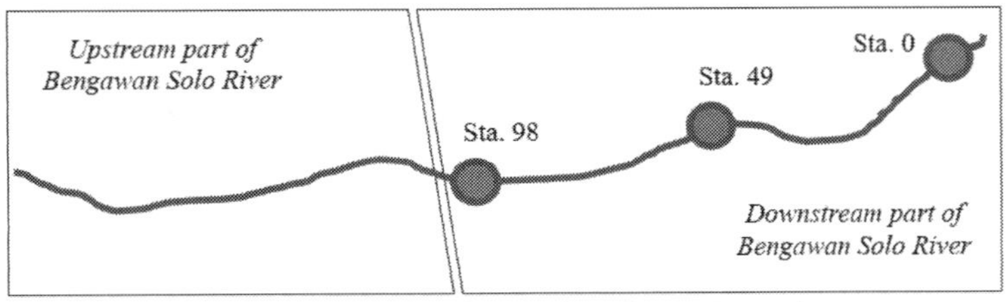

Figure 5. Sketch of Simulated Sta. Location

Table 4. HEC-RAS cross section result in Sta.98, Sta.49 and Sta.0

Sta. of River	Return period (years)	Peak runoff (m^3/s)	Runoff level (m)	Velocity (m/s)	Discharge level (m)
98	5	3750.98	13.85	1.23	-0.52
98	10	4242.70	14.50	1.30	-1.17
98	25	4969.26	15.41	1.39	-2.08
98	50	5579.20	16.11	1.46	-2.78
98	100	6297.57	16.89	1.54	-3.56
49	5	3750.98	12.07	1.59	1.26
49	10	4242.70	12.74	1.69	0.59
49	25	4969.26	13.65	1.71	-0.32
49	50	5579.20	14.29	1.80	-0.96
49	100	6297.57	14.99	1.90	-1.66
0	5	3750.98	3.74	5.86	9.59
0	10	4242.70	4.05	6.09	9.28
0	25	4969.26	4.48	6.39	8.85
0	50	5579.20	4.83	6.62	8.50
0	100	6297.57	5.23	6.86	8.10

4. Conclusion

To conclude, the flood in Bojonegoro city is caused by the water loading capacity of the river could not accommodate all of the simulated design peak runoff. The HEC-RAS software analysis agrees that the flood water level from all of simulated design peak runoff is more than full bank elevation of Bengawan Solo River in Bojonegoro City. So, Bojonegoro government should concern to this factor to reduce flood problems' impacts in Bojonegoro City. This research suggests deeper analysis about HEC-RAS river cross sections with detailed primary data to get more comprehensive and better result.

References

[1] Chow V T 1959 *Open-channel hydraulics* (New York: McGraw-Hill)
[2] Jung S W and Moon J W 2001 The comparison of existing synthetic unit hydrograph method in Korea *Journal of Korea Water Resources Association* **34** (6) pages 659-672.
[3] Kang M S 2009 Design of drainage culverts considering critical storm duration *Biosystems Engineering* Volume **104** Issue 3 pages 425-434
[4] Kurniawati O T 2016 *Thesis: The analysis of water level of downstream part of Bengawan Solo*

due to floodway (Malang: Engineering Faculty of Brawijaya University)
[5] Merz B, Thieken A H and Gocht M 2007 Flood risk mapping at the local scale: concepts and challenges In: S Begum, M J F Stive and J W Hall (eds) Flood risk management in Europe *Advance in Natural and Technological Hazard Research* **25** page 231-251
[6] Nigussie T A 2015 Performance Evaluation of Synthetic Unit Hydrograph Methods in Mediterranean Climate: A Case Study at Guvenc Micro-watershed, Turkey *Landscape Dynamics, Soils and Hydrological Processes in Varied Climates* part III pages 293-315.
[7] Nyarko B K 2000 Flood Risk Zoning of Ghana: Accra Experience *International Archives of Photogrammetry and Remote Sensing* Vol **33** Part B7
[8] Paringit E C 2015 Automation of Cagayan de Oro River flood watch system using HEC-HMS and HECRAS *Proceedings of ACRS 2015 - 36th Asian Conference on Remote Sensing: Fostering Resilient Growth in Asia in Quezon City, Metro Manila, Philippines on 24th – 28th October 2015*
[9] Rivera S et al 2007 Predicting flood hazard areas: A SWAT and HEC-RAS simulations conducted in Aguan river basin of Honduras, central America *American Society for Photogrammetry and Remote Sensing - ASPRS Annual Conference 2007: Identifying Geospatial Solutions* Volume **2** pp 594-603
[10] Schanze J 2006 Flood risk management a basic framework In: Schanze J, Zeman E and Marsalek J (eds) Flood Risk Management: hazards, vulnerability and mitigation measures *NATO Science Series* **67** pp 1-20

Acknowledgements
The present work was performed as a part of Engineering Hydrology lecture project in Civil Engineering Department of University of Indonesia, Indonesia. The authors wish to thank to Hydraulic Laboratory of Engineering Faculty, University of Indonesia and *Balai Besar Wilayah Sungai Bengawan Solo*. The first author's special gratitude goes to LPDP as scholarship and financial assistance to pursue Master Degree in Universitas Indonesia.

IOP Conf. Series: Materials Science and Engineering **316** (2018) 012043 doi:10.1088/1757-899X/316/1/012043

Comparative analysis layers method of t-beam reinforcement

Dimas Langga Chandra Galuh

Fakultas Teknik Universitas Sarjanawiyata Tamansiswa, Jl.Kusumanegara 157 Yogyakarta, Indonesia

E-mail: galuh1985@gmail.com

Abstract. Wire rope commonly used by heavy equipments in order to operate lift weights. wire ropes have a high tensile strength, that in this study tried to use these materials as reinforcement material in the T-beam negative moment area. In this analysis the T beam reinforcement using method layers of this by comparing the layers1 method (ignoring the wing reinforcement) and layers2 method (considering the wing reinforcement). Results of the test showed that the maximum loading for BK, BP1, and BP2 specimens were 88.5 kN, 180 kN, and 259 kN, respectively. The comparative flexural capacity by layers1 method (ignoring the wing reinforcement) for BK, BP1, and BP2 specimens were 1.28, 0.76, and 0.78. The comparative flexural capacity by layers2 method (considering the wing reinforcement) for BK, BP1, and BP2 specimens were 1.28, 0.79, and 0.80, respectively. Comparative ratios of layers2 method for BK, BP1, and BP2 to layers1 method were 1, 0,96 and 0,975, respectively. This result means of comparative analysis layers methode similar and methods layers2 most closely with test results.

1. Introduction

Wire rope usually used to lift heavy materials on a construction or industrial. The form of wire rope in the form of steel wire strands, has great tensile strength and the flexible nature of the wire rope, so easily rolled up and stored. steel wire rope is 6 or more strands having a layer of zinc or without coated with zinc [1]. Strengthen structures on reinforced concrete T-beam negative moment by using wire rope and composite mortar based on the advantages possessed by these materials.

2. Methodology/experimental

Analysis using layers method can be used to determine the capacity of reinforced concrete beam section is based on the stress-strain diagram of the constituent materials [2]. Formulas in beam material used in analyzing the layers method include [3]:

- The concrete stress-strain model based on Popovics model (high strength), as shown in the following equations :

$$f_{ci} = -\left(\frac{\varepsilon_{ci}}{\varepsilon_p}\right) f_p \frac{n}{n-1+\left(\varepsilon_{ci}/\varepsilon_p\right)^{nk}} \text{ for } \varepsilon_{ci} < 0 \quad (1)$$

$$n = 0{,}8 + \frac{f_p}{17} \quad (2)$$

IOP Conf. Series: Materials Science and Engineering **316** (2018) 012043 doi:10.1088/1757-899X/316/1/012043

$$k = \begin{cases} 1{,}0 & \text{for } \varepsilon_p < \varepsilon_{ci} < 0 \\ 0{,}67 + \dfrac{f_p}{62} & \text{for } \varepsilon_{ci} < \varepsilon_p < 0 \end{cases} \tag{3}$$

$$E_c = \frac{f_p}{|\varepsilon_p|} \cdot \frac{n}{(n-1)} \tag{4}$$

Where:
f_p : peak concrete compressive stress (MPa); ε_p : concrete compressive strain corresponding to f_p; f_{ci} : the stress occurs in the concrete (MPa); ε_{ci} : strain occurs at peak; n : curve fitting parameter for stress-stain response of concrete in compression; k : post-peak decay parameter for stress-strain response of concrete in compression; E_c : concrete elasticity modulus (MPa).

- The reinforced concrete stress-strain model was based on the trilinieri curve models, as shown in the following Equation :

$$f_s = \begin{cases} E_s \varepsilon_s & \text{for } |\varepsilon_s| \le \varepsilon_y \\ f_y & \text{for } \varepsilon_y < |\varepsilon_s| \le \varepsilon_{sh} \\ f_y + E_{sh}(\varepsilon_s - \varepsilon_{sh}) & \text{for } \varepsilon_{sh} < |\varepsilon_s| \le \varepsilon_y \\ 0 & \text{for } \varepsilon_u < |\varepsilon_s| \end{cases} \tag{5}$$

$$\varepsilon_u = \varepsilon_{sh} + \frac{(f_u - f_y)}{E_{sh}} \tag{6}$$

where:
ε_s : reinforcement strain; ε_y : yield strain; ε_{sh} : strain at the onset of strain hardening; ε_u : ultimate strain; E_{sh} : strain hardening modulus (MPa); f_y : yield strength (MPa); f_u : ultimate strength (MPa); E_s : elastic modulus (MPa).

- Wire rope stress strain model based on the Ramsberg-Osgood, as shown in the following Equation :

$$f_s = E_s \varepsilon_s \left\{ A + \frac{1 - A}{\left[1 + (B\varepsilon_s)^C\right]^{1/C}} \right\} \le f_u \tag{7}$$

$$A = \frac{E_{sh}}{E_s} \tag{8}$$

$$B = \frac{E_s(1 - A)}{f_s^*} \tag{9}$$

$$C = \text{transition coefficient} \tag{10}$$

where:
ε_s : reinforcement strain; E_s : elastic modulus (MPa); E_{sh} : strain hardening modulus (MPa); f_u : ultimate strength (MPa); f_s* : value at which the second linear branch intercepts the stress axis at zero strain; C : 10 [4].

2.1. Beam Specimens

There were three specimens of reinforced concrete beam: the controlling beam (BK), the reinforcing beam type 1 (BP1) and type 2 (BP2). Specifications of reinforced concrete beam specimens are presented on Table 1, Figure 1 and Figure 2 [5].

Table 1. Specifications of specimens off reinforce concrete beam

Speci-men	L (mm)	b_f (mm)	t_f (mm)	b_w (mm)	Main Reinforcement		Cross Bar Reinfor-cement	Reinforcement	
					Tensile	Com-presive		Tensile	Compressive
BK	2400	400	75	150	3D13	2P8	P8-40	-	-
BP1	2400	400	115	150	3D13	2P8	P8-40	4Ø10	-
BP2	2400	400	115	150	3D13	2P8	P8-40	4Ø10	2P8

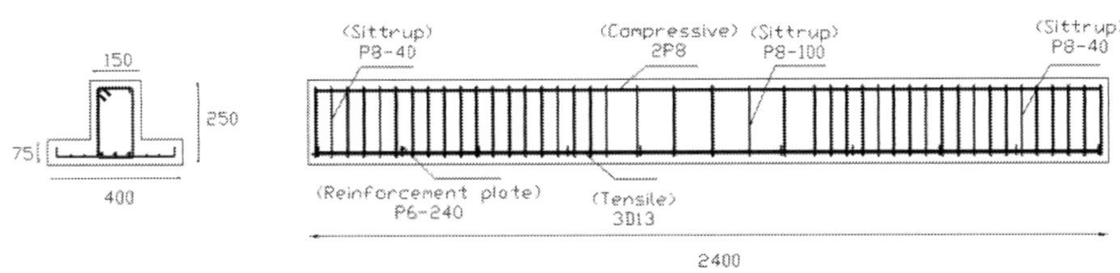

Figure 1. Controlling Beam (BK)

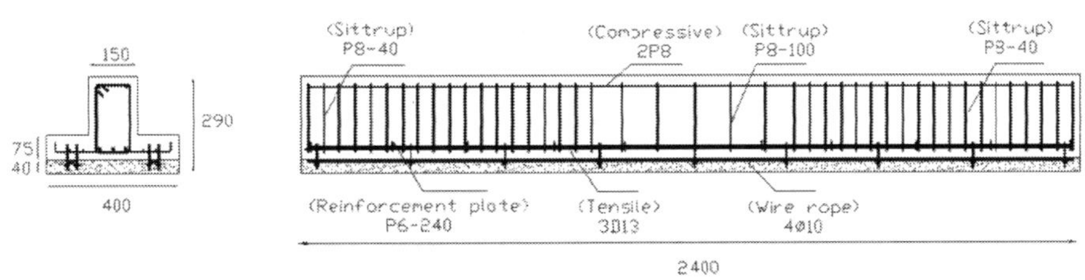

Figure 2. Reinforcing Beam (BP1)

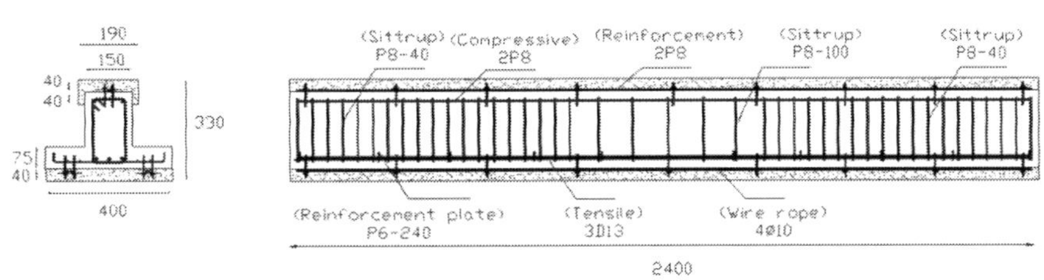

Figure 3. Reinforcing Beam (BP2)

3. Result and discussion

Load capacity calculation results received in the state crack, yield, and ultimited. The results given in of the calculation method of these layers only compare the condition of crack and ultimited with steel cables used parameter is the modulus of elasticity, rupture strain, and ultimate strength taken in accordance with the experimental results that 35725,02 MPa, 25,6 mm/m and 740,48 MPa respectively. Load - deflection occurs can be seen in Table 1 and Table 2 calculated based on the load provided by the load cell (load one point). Maximum load ratio is shown in Table 3 and Figure 1.

Table 2. The capacity of beam load and deflection by ignoring the wing reinforcement (layers method 1)

No.	Specimen	Load Capacity (kN)		Deflection (mm)		Ratio
		crack	max	crack	max.	P_{max}.
1.	BK	15,17	113,37	0.90	14,68	1
2.	BP1	22,87	137,50	0.15	13,32	1,21
3.	BP2	32,17	200,91	0.12	18,50	1,77

Table 3. The capacity of beam load and deflection with calculation on wing reinforcement (layers method 2)

No.	Specimen	Load Capacity (kN)		Deflection (mm)		Ratio
		crack	max	crack	max.	P_{max}.
1.	BK	15,17	113,37	0,90	14,68	1
2.	BP1	22,87	143,07	0,15	12,32	1,26
3.	BP2	32,17	208,28	0,12	17,59	1,84

Tabel 4. Comparison beam load capacity on two conditions

No.	Specimen	P_{max}. (kN)		Ratio
		layers method 1	layers method 2	P_{max}.
1.	BK	113,37	113,37	1
2.	BP1	137,50	143,07	1,04
3.	BP2	200,91	208,28	1,04

Based on Table 4 shows that the results obtained are close to each other. The maximum load value analysis ignores reinforcement layers with wings on BK, BP1 and BP2 113,37 kN, 137,5 kN, and 200,91 kN respectively. While the value of the maximum load analyzes take into account the reinforcement layers with wings on BK, BP1 and BP2 113,37 kN, 143,07 kN, and 208,28 kN respectively. Improvement occurring in the analysis layers 1, 1,04, and 1,04 respectively. This increase was influenced by the presence of the reinforcing wings be included in the calculation. Relations load deflections under both conditions can be seen in Figure 4 and Figure 5.

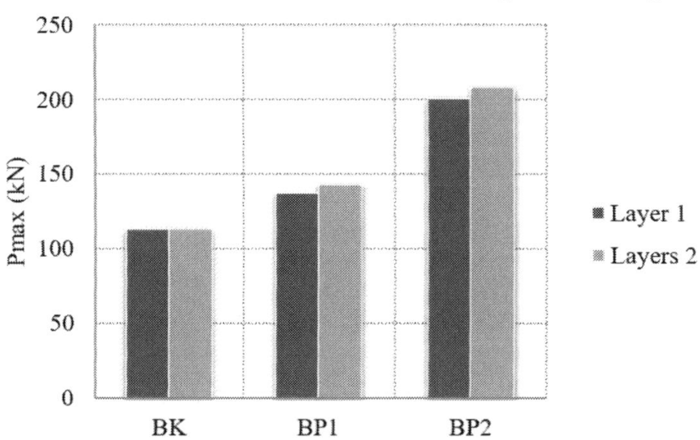

Figure 4. The ratio of beam load capacity on two conditions

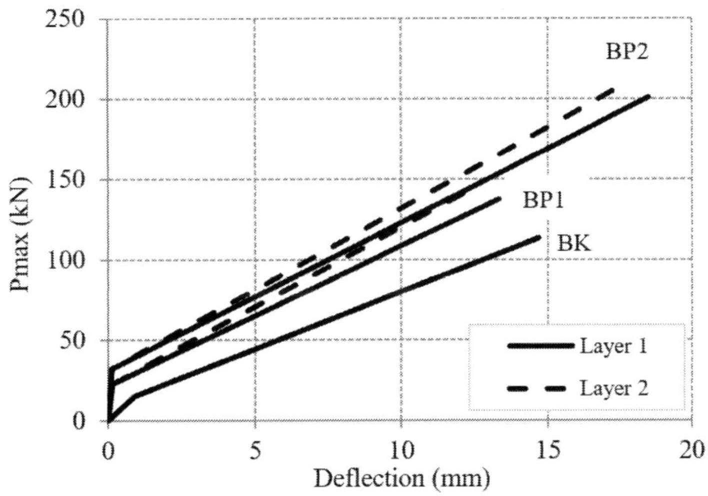

Figure 5. Comparison of the load and deflection relationship beam

4. Concluding

The comparative flexural capacity by layer1 method (ignoring the wing reinforcement) for BK, BP1, and BP2 specimens were 1.28, 0.76, and 0.78. The comparative flexural capacity by layer2 method (considering the wing reinforcement) for BK, BP1, and BP2 specimens were 1.28, 0.79, and 0.80, respectively. Comparative ratios of layer2 method for BK, BP1, and BP2 to layers1 method were 1, 0.96 and 0.975, respectively. The means of comparative analysis of the layer methods are similar. The method of layers2 is the closest with the test results.

Acknowledgments

This research is possible thanks to all Staff and Facilities of Laboratory of Program Studi Teknik Sipil Universitas Gadjah Mada Yogyakarta.

References

[1] Badan Standar Nasional, 2008, *SNI 0076-2008 Tali kawat baja*, Jakarta
[2] Park, R. dan Paulay, T., 1975, *Reinforced Concrete Structure*, John Wiley & Sons Inc, Canada
[3] Wong, P.S. dan Vecchio, F.J, 2002, *Vector2 & Form Works User's Manual*, http://www.ecf.utoronto.ca/~bentz/manual2/final.pdf accessed on February 9, 2011.
[4] Bentz, E. dan Collins M. P., 2001, *Response-2000 User Manual*, http://www.ecf.utoronto.ca/~bentz/manual2/final.pdf accessed on February 9, 2011.
[5] Y. Haryanto, *Efektifitas Wire Rope Sebagai Perkuatan Pada Daerah Momen Negatif Balok Beton Bertulang Tampang T*, Dinamika Rekayasa journal, Vol. **8**, No. 1, February 2012.

IOP Conf. Series: Materials Science and Engineering **316** (2018) 012044 doi:10.1088/1757-899X/316/1/012044

Performance Analysis of Trans-Jakarta Bus Suburban Service Move-Across Greater Jakarta

ESW Tangkudung[1] and C Widyadayinta[2]

[1]Universitas Indonesia, Kampus UI Depok 16424, Indonesia
[2]Universitas Indonesia, Kampus UI Depok 16424, Indonesia

E-mail: ellen@eng.ui.ac.id

Abstract. Trans-Jakarta have developed their services scope as Suburban Service or Feeder move-across service that operate from greater Jakarta into Jakarta central vice versa. One of the route is Ciputat – Bundaran Hotel Indonesia (Tosari) and integrated with corridor 1 (one) and 8 (eight). This service is not travel on the exclusive lane or bus-way. Objective of Government Jakarta to provide this service is to decrease private car to enter the central of Jakarta. The objective of this study is to find the performance of the service. Survey have conducted static and dynamic on work day to get variable of travel time and delay, waiting time of passenger at the bus stop, headway and ridership of the bus. Service Standard Minimum of Trans-Jakarta have compared with the result of variable headway, travel speed, and waiting time at bus stop as concern of all the passengers. Analysis use correlation test method and linear regression model have done. The performance of Trans-Jakarta bus suburban service, based on travel speed indicator is fairly bad, only 8.1% of trip could comply with Minimum Service Standard. Bus performance based on the indicator of density in the bus is good, where all points are below the maximum limit i.e. 8 people/m^2 at peak hour and 5 people/m^2 at off-peak hour.

1. Introduction

Bus is a public transportation that can accommodate large number of passengers that also have high accessibility compared to rail-based mass transportation. Therefore, it developed to a Bus Rapid Transit (BRT) as a public transport that offers fast, comfortable, and secure transport service, either from its infrastructure, vehicle and schedule. Trans-Jakarta is an implementation of BRT system in Jakarta, which has been operating officially since February 2004.

Based on data of BPTJ (Badan Pengelola Transportasi Jabodetabek) in 2015, the number of trips in Jabodetabek region reached 47.5 million trips per day. Of the many trips, 50 percent of the trip is done by people outside DKI Jakarta. In addition, based on statistic data of Dinas Perhubungan, there are only one-third of 25.7 million vehicles per day in Jakarta which are public transportation users and the rest are private vehicle users.

After 12 years, Trans-Jakarta develop its service coverage by creating Suburban Service. Since June 2016, Trans-Jakarta opening of new routes, serving not only in Jakarta, but rather provide the services from Greater Jakarta to Jakarta central and vice versa. One of the route is Ciputat – Bundaran Hotel Indonesia (Tosari) which integrated with corridor 1 (one) and 8 (eight). Objective of Government Jakarta to provide this service is to decrease private car to enter the central of Jakarta. These services are not travel on the exclusive lane or bus-way and there are no shelters, only stops, except in the corridor. So, performance analysis based on Minimum Service Standard (MSS) of Trans-Jakarta needs to be done to develop services for the people.

Minimum Service Standard of Trans-Jakarta given by the service provider should be accomplished as the minimum quality which will be received by the passengers. There are four substances of MSS, namely service reliability, safety and security, convenience, and comfort. Reliability and comfort partially affect passengers' satisfaction (Silaningsih, 2015). For each substance, there are several indicators. Headway, travel speed, density in the bus, and waiting time at bus stop as concern of all passengers, are the analyzed indicators in this paper.

2. Methodology

2.1. Preparation phase

This phase begins by conducting a study on the existing condition of the bus service. From the existing conditions, identification of problem formulation and goal setting, which is to analyze the performance of the bus. The literature study was conducted. At this phase also studied performance indicators and minimum service standards of Trans-Jakarta as a reference in conducting analysis.

2.2. Data collection phase

The required data are: 1) time and duration of survey, 2) departure and arrival time, 3) delay, 4) mileage, 5) number of passengers, 6) passenger waiting time at the bus stop, 7) bus arrival and departure time at the bus stop. Observations were made for approximately two months, in January to March. Observations are made on weekdays (Monday to Friday) at peak hours and off-peak hours.
Primary data collection:

- Dynamic survey: surveys conducted in a moving bus vehicle with the method of recording the travel time and the number of passengers boarding and alighting through a route. In this survey also recorded where and how long the vehicle stopped due to delay.
- Static survey: surveys conducted from outside the vehicle (at the bus stop). The information recorded is the passenger waiting time and arrival/departure time of the bus (headway).

2.3. Data processing phase

Data processing is done to compare the existing data with Trans-Jakarta Minimum Service Standard (MSS), with the help if Microsoft Excel software. Indicators of MSS used are travel speed, number of passenger in bus, passenger waiting time, and headway. Therefore, the data obtained is processed first to be able to compare its value with existing indicators.

- Travel speed is obtained by using the travel time obtained by founding the difference between the time of arrival and the departure time. Travel time is the time required by the vehicle to travel a certain distance or the total time used to serve a route in one way, including delay and dwelling time. The amount of travel speed then can be calculated by the formula

$$V = \frac{L}{T}$$

  where V= travel speed (km/hour), L= distance (km), T= travel time (hour). Delay is the time the vehicle is in a state of stalled due to traffic conditions or others that causing the vehicle to stop.
- Density in the bus is obtained by using the ridership data (number of passengers per square meter). The number of passengers on the bus is obtained by adding and reducing the initial passenger with the boarding-alighting passengers. So to compare the field performance with the MSS, the number of passengers in the bus divided by the floor area of the bus.
- Waiting time at bus stop is the time required by the passenger to wait until he/she can get into the bus. The average waiting time for public transport can be measured from half of the headway, assuming random passenger arrival rate and public transport headways have normal distribution. If the waiting time is too long, it will reduce the interest of public transport users. The passenger waiting time at the bus stop can be directly compared with existing MSS.

- Headway is interval between two successive vehicles. Headway is obtained by viewing the bus arrival time at the bus stop.

Correlation test, using the help of SPSS software, will do to find the variable of performance as the base of priority to develop services for the people. Correlation is statistical method used to determine whether there is a relationship between variables. The correlation coefficient is a number indicating the direction and strength of the linear relationship between the dependent variable (Y) and the independent variable (X). Interpretation of correlation coefficient (Hasan, 2001) is as follows:

- If ρ (correlation coefficient spearman's rho) is positive, then the variables are positively correlated. The closer the value of this ρ to +1 the stronger the correlation, and vice versa.
- If ρ is negative, then the variables are negatively correlated. The closer the value of this ρ to -1 the stronger the correlation, and vice versa.
- If ρ is 0 (zero), then the variables do not show correlation.
- If ρ is either +1 or -1, then the variable shows a perfect positive or negative correlation.

2.4. Analysis and decision making phase

After doing data processing, field data can be compared with MSS. If the value obtained is still at the range (maximum or minimum) then the performance of Trans-Jakarta bus can be said good and vice versa. From the data, if there is 50 percent or more data that said the performance is good, then we can say the overall performance can be said good. Analysis with correlation test will result the relationship between variables and we can determine what variables or indicators that most influential on the performance, which is the parameter of travel time.

3. Results and Discussion

3.1. Transjakarta suburban service performance

- Travel speed

Figure 1 and Figure 2 show the travel time, where the red line indicates the maximum speed limit of 30 km/h while the green line shows the minimum speed limit of 18 km/h, according to MSS Transjakarta. From the Figures can be seen that only a few points are between the two lines. The percentage indicates good performance is only 7/62 or 11.3% at peak hour and 3/62 or 4.8% at off-peak hours, or overall only 10/124 or 8.1%. So it can be said that the performance of the bus based on travel speed indicator is fairly bad because it has not met the standard.

The performance based on travel time depends on the situation on the road, where this service is not travel on the exclusive lane busway. The bus must compete with other road users, ranging from private cars, motorcycles, urban transportation (angkot) to intercity buses. This route passes through markets and campuses that at certain hours will affect traffic flow. The study was conducted from January to March, at which time construction work was underway, one of which is MRT Jakarta. Other construction work at Pondok Indah area also take part of reducing the travel speed.

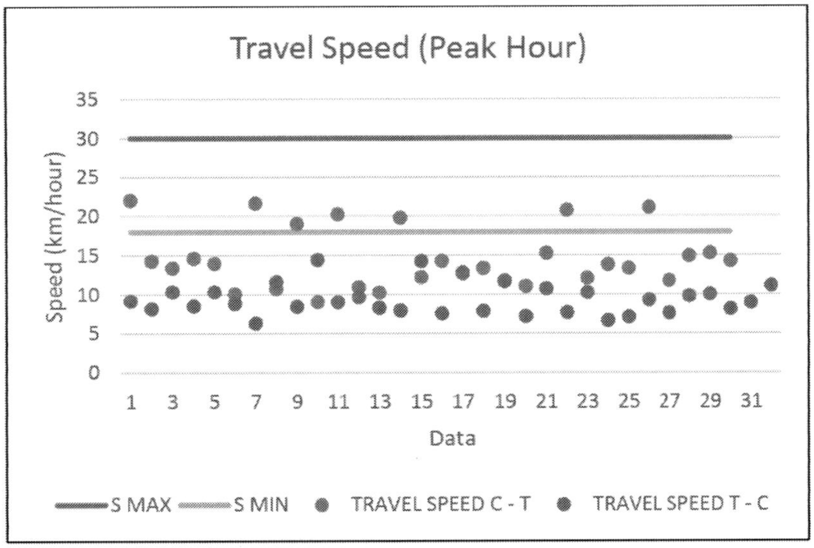

Figure 1 Travel Speed Data on Peak Hour

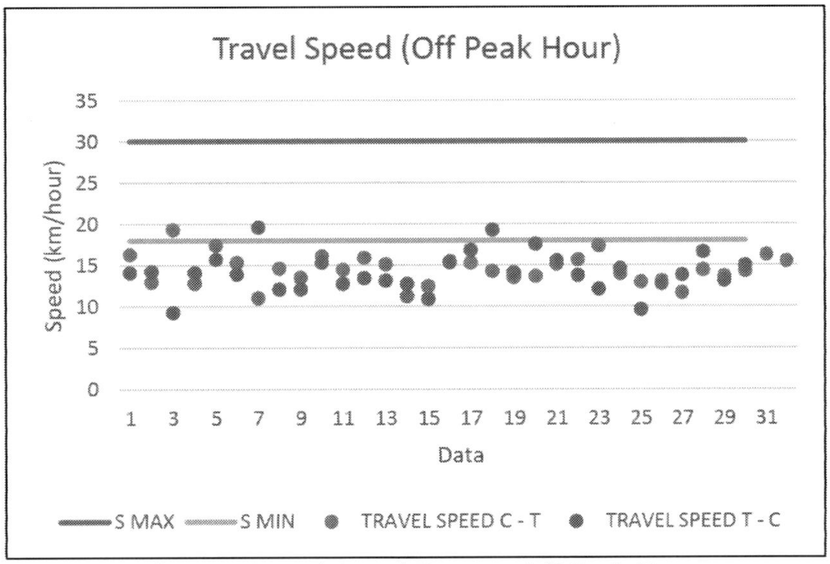

Figure 2 Travel Speed Data on Off-Peak Hour

- Density in the bus

From Figure 3 and Figure 4 can be seen that bus performance based on the indicator of density in the bus has met the standard, where all points are below the maximum limit of 8 people/m^2 at peak hour and 5 people/m^2 at off-peak hour. Although it still met the Minimum Service Standard of Trans-Jakarta, the density during peak hour can reach up to 5 people/m^2 or load factor of 121%, which is not acceptable for most passenger.

IOP Conf. Series: Materials Science and Engineering **316** (2018) 012044 doi:10.1088/1757-899X/316/1/012044

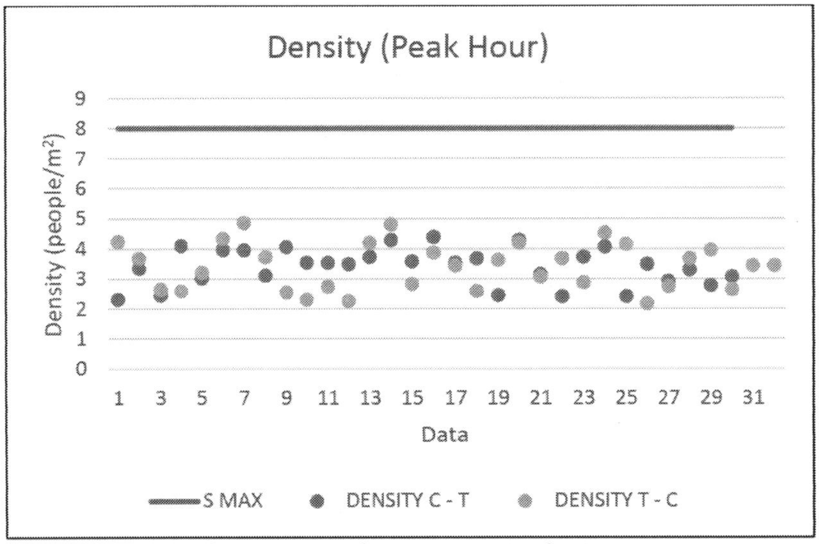

Figure 3 Density Data on Peak Hour

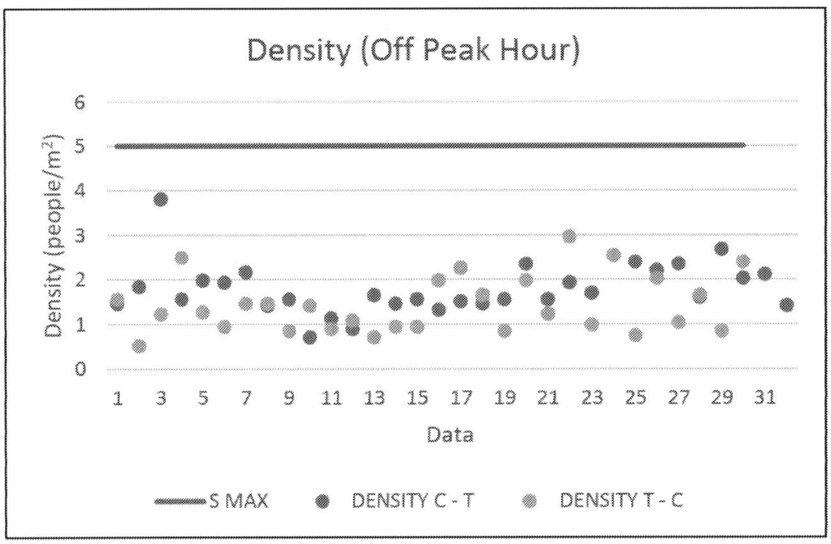

Figure 4 Density Data on Off-Peak Hour

- Waiting time at bus stop and Headway

The waiting time is calculated starting the moment the passenger arrives at the bus stop until he/she boards the bus. According to field observations, the waiting time for passengers varies, depending on the headway of the bus at that time, the density in the bus, where if it is too full, the passenger will be required to wait for the next bus, or the passenger personally, whether in a hurry, waiting for the next bus to get a seat (especially at the Tosari and Pool Ciputat). Headways at each stop will vary, especially when on a regular path, not in the main busway corridor. Maximum waiting time limit is 5 minutes during peak and 15 minutes during off-peak hour. While the maximum headway limit is 5 minutes during peak and 10 minutes during off-peak hour. From 20 bus stops that we observed, note that the performance is bad for headway and waiting time because they have not met the standards as seen from Table 1. Where, N=amount of all data, n=the amount of data that meets the standard, %=percentage of data that meets the standard.

Table 1 Performance Based on Headway and Waiting Time

		Standard (Peak Hour)	Standard (Off-Peak Hour)	N	n	%	Performance
Headway	P OP T	≤ 5 minutes	≤ 10 minutes	21 9 30	0 1 1	0 11,1 3,33	X
Waiting time	P OP T	≤ 5 minutes	≤ 15 minutes	21 9 30	2 9 11	9,52 100 36,67	X

The performance of the Trans-Jakarta Ciputat-Tosari based on the four MSS indicators above, is overall quite bad. Variable of speed, headway and waiting time show poor performance while passenger density shows good performance. Off-peak hour waiting time indicate good performance, especially at corridor stops that are also assisted by other buses that can serve the passengers..

3.2. Influence of variables on performance

Performance of a public transport service, at the most could viewed by the travel time indicator. Therefore, to find the most influential variable on performance, look for relationships between variables, where travel time becomes dependent variable and other variables become independent variables. Non-parametric tests are an option where normality and homogeneity assumptions are not met. So in this study, rank spearman correlation is used to find the relationship and test the significance of one dependent variable and one independent variable. Based on IBM SPSS output, we can see Correlation Coefficient and Sig value, at α=5%. In this correlation test, the variables tested include travel time, delay time, speed, density, headway and waiting time.

Based on the result, it can be seen that the travel time variables at most have a negative relationship with the variable speed and positive relationship with variable delay. For variable speed, spearman rho correlation value is all at interval 0.8 to 1 so it can be said a very strong negative relationship level. So, if the travel time is getting bigger then the speed will decrease, and vice versa. For variable delay, spearman rho correlation value varies on peak and off-peak hour. The rho value of 0.653 at peak hour indicates a strong level of relationship. The rho value of 0.450 in the peak hour of Ciputat direction indicates the moderate level of relationship. While the value of rho of 0.377 in off-peak hour direction of Ciputat shows a low level of relationship. Positive value indicates a positive-correlated variable, i.e. if the delay is shorter, the travel time will also shorter.

4. Conclusion

The Performance of Trans-Jakarta bus suburban service which move-across Greater Jakarta, on route Ciputat – Bundaran Hotel Indonesia (Tosari), based on travel speed indicator is fairly bad, only 8.1% of trip could comply with Minimum Service Standard. Bus performance based on the indicator of density in the bus is good, where all points are below the maximum limit of 8 people/m^2 at peak hour and 5 people/m^2 at off-peak hour.

For variable of speed, headway and waiting time show poor performance while passenger density shows good performance. Off-peak hour waiting time indicate good performance, especially at corridor stops that are also assisted by other buses that can serve the passengers.

References

[1] Arifin AM 2015 *Analisis Tingkat Kepuasan Penumpang pada Fasilitas Pelayanan Bus Transjakarta Berbasis Standar Pelayanan Minimal (SPM)*

[2] Bluman AG 2012 *Elementary Statics A Step By Step Approach Eight Edition*. New York: McGraw-Hill

IOP Conf. Series: Materials Science and Engineering **316** (2018) 012044 doi:10.1088/1757-899X/316/1/012044

[3] BPS Provinsi DKI Jakarta 2016 *Statistik Transportasi DKI Jakarta 2015-2016*
[4] Ceder A 2016 *Public Transit Planning and Operation, Modeling, Practice and Behavior, 2nd Ed.* CRC Press, Taylor and Francis Group, FL, pp 23-55
[5] Giannopoulos GA 1990 *Bus Planning and Operation in Urban Areas: A Practical Guide*
[6] Harinaldi 2005 *Prinsip-Prinsip Statistik untuk Teknik dan Sains*. Jakarta: Erlangga
[7] Hasan I 2001 *Pokok-Pokok Materi Statistik 2 (Statistik Inferensif) Edisi kedua*. Jakarta: PT Bumi Aksara
[8] Luthfiani A 2007 *Kenyamanan Pengguna Bus Rapid Transit (Studi Kasus: Koridor 1 Transjakarta)*
[9] Nurvia I 2007 *Analisis Persepsi Penumpang Terhadap Tingkat Pelayanan Busway (Studi Kasus Koridor 1)*
[10] *Pedoman Standar Pelayanan Minimal Transjakarta*. ITDP Indonesia, PT. Inresh Indonesia
[11] Satyadi R 2016 *Evaluasi Standar Pelayanan Minimal Operasional Transjakarta Koridor 9 dan Koridor 12*
[12] Silaningsih E 2015 *Transjakarta Company's Srategy and Minimum Service Standard to Raise Passengers' Satisfaction*

IOP Conf. Series: Materials Science and Engineering **316** (2018) 012045 doi:10.1088/1757-899X/316/1/012045

Incorporating Servqual-QFD with Taguchi Design for optimizing service quality design

M. Arbi Hadiyat

Industrial Engineering, University of Surabaya, Indonesia

Email: arbi@staff.ubaya.ac.id

Abstract. Deploying good service design in service companies has been updated issue in improving customer satisfaction, especially based on the level of service quality measured by Parasuraman's SERVQUAL. Many researchers have been proposing methods in designing the service, and some of them are based on engineering viewpoint, especially by implementing the QFD method or even using robust Taguchi method. The QFD method would found the qualitative solution by generating the "how's", while Taguchi method gives more quantitative calculation in optimizing best solution. However, incorporating both QFD and Taguchi has been done in this paper and yields better design process. The purposes of this research is to evaluate the incorporated methods by implemented it to a case study, then analyze the result and see the robustness of those methods to customer perception of service quality. Started by measuring service attributes using SERVQUAL and find the improvement with QFD, the deployment of QFD solution then generated by defining Taguchi factors levels and calculating the Signal-to-noise ratio in its orthogonal array, and optimized Taguchi response then found. A case study was given for designing service in local bank. Afterward, the service design obtained from previous analysis was then evaluated and shows that it was still meet the customer satisfaction. Incorporating QFD and Taguchi has performed well and can be adopted and developed for another research for evaluating the robustness of result.

1. Introduction

Since firstly proposed by [1], Service Quality (SERVQUAL) has been applied for evaluating the quality level of services, and as reference for improving the service design. Based on the 5 dimensional customer attributes, the SERVQUAL measures the 5 gaps, including the customer gap i.e. the difference between customer expectation and perception. Attributes with large negative gaps represents some dissatisfactions and lead to the opportunity for improving them. However, many researches evaluate the effectiveness of SERVQUAL related to large amount of variables in questionnaire, and also their likert scale that is considered difficult to interpret [2]. Some papers proposed modifications of SERVQUAL by developing similar procedures or dimensions for particular service, such as DINESERV [3] for evaluating restaurant service, SERVPERF [4] for omitting the customer expectation rather than calculate its different with the perception, and HEDPERF [5] for evaluating education service.

Considering the customer gaps in SERVQUAL, the improvement should accommodate all those negative gaps, especially the larger ones. The researchers often qualitatively generate the improvement plans according to the gaps and then implement them, see [6], and [7]. A well-known tool for generating those plans is Quality Function Deployment (QFD), as used by [8] and [9]. QFD would produce the "hows", i.e. solutions related to the negative gaps in SERVQUAL called "whats", then by subjectively weighting and associating between hows and whats, priority in implementing the improvement would be found.

Once the prioritized improvement plans implemented, they should fulfill and satisfy the customers, and reduce the negative SERVQUAL gaps. However, there's no guarantee that the prioritized plans produced by QFD would robustly satisfying the customers because the customer can only accept what the QFD hows determined, which has qualitatively generated and subjectively weighted. Customers don't have a chance to select which plans they prefer to implement, so those plans can't be reliably satisfying over time and fall to un-optimized deployment. [10] and [11] proposed the Taguchi method in selecting best combination in the QFD hows for designing the deployed service for customer by determining additional plans levels, similar to factors in experimental design method

Incoporating Taguchi with SERVQUAL and QFD has not been discussed before, this paper proposed new approach in increasing robustness of service design through Taguchi method. The framework is started from identifying the variables, calculating the gaps, generating the improvement plans through QFD, and then applying Taguchi method for optimizing the plans into service design. A case study has been conducted and the deployed service design is obtained, and related analysis is performed for evaluating this framework.

2. Literature review

2.1. SERVQUAL-QFD

The Parasuramans's SERVQUAL [1] has been popular method in evaluating the quality of service. There are 5 gaps as the main idea in this analysis, represent unsatisfactory of customer and inconvenience managed service, see figure 1. One of SERVQUAL advantage adopted in this research is the gap analysis for measuring the customer satisfaction, i.e. the difference between expectation and perception of customers. The negative gap shows the unsatisfactory of customers, and the positive ones represent fulfillment of their expectation. All negative gaps considered to be improved, with the scope of 5 dimensions, i.e. tangible, empathy, reliability, responsiveness, and assurance.

The incorporated QFD takes place to generate improvement plans due to negative gaps which called what, see [12] The improvement plan (hows) then associated with the whats followed by calculating weight of prioritized plan. Solutions for customer gaps refer to the highest priority represented by highest relative weight. All this analysis could be done in a tools called House of Quality, see figure 2.

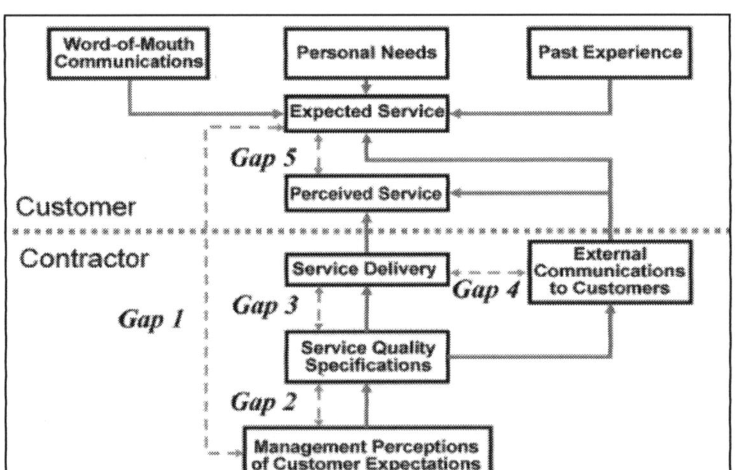

Figure 1. SERVQUAL gaps (taken from [1])

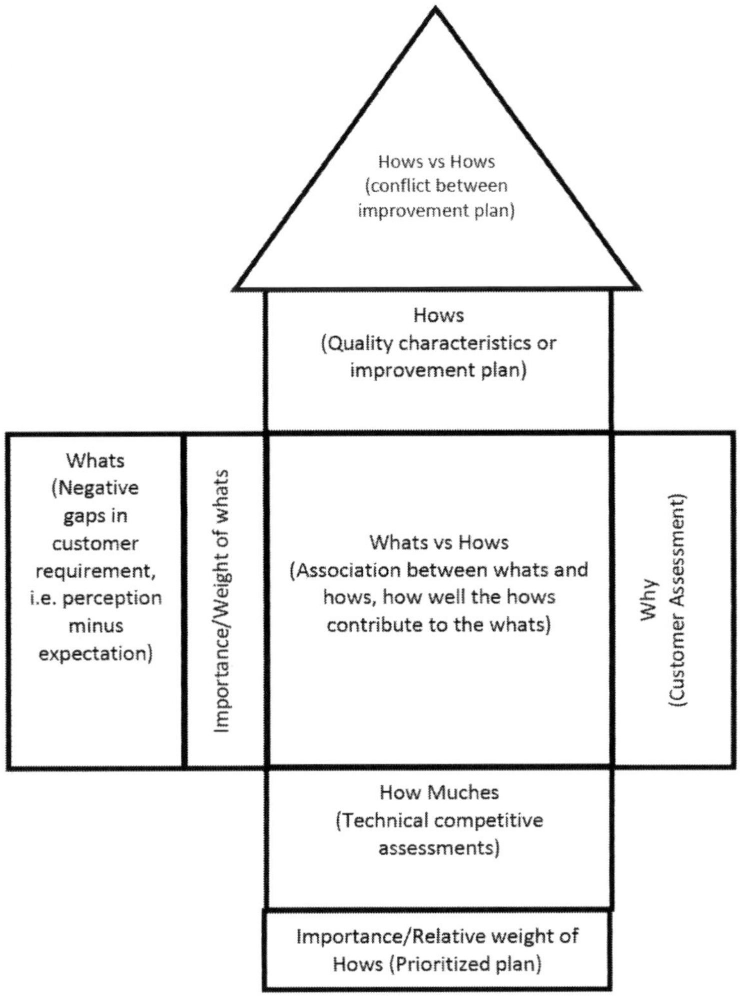

Figure 2. House of Quality in QFD

2.2. Taguchi Method

The Taguchi method usually applied in hard engineering for finding the best experiment factor combinations that optimize the response, see [13] and [14], and its development as in [15], [16] and [17]. First step, researchers determine the factors to be optimized including their levels. Following by conducting experiment refer to orthogonal array design. By calculating averages for each factor level and rank the difference between each of them, Taguchi could find the optimal level combination that optimize the interested response transformed in term of signal-to-noise ratio (SNR), considering the objective of optimization as follows:

- Signal-to-noise ratio for larger the better
- Signal-to-noise ratio for nominal the best
- Signal-to-noise ratio for smaller the better

Only larger the better signal-to-noise ratio would be used in this paper, since the objective in Taguchi was maximized customer response. Formula of signal-to-noise ratio for larger the better shown in (1)

$$SNR = -10 \log \left[\frac{1}{n}\sum_{i=1}^{n}\frac{1}{y^2}\right] \tag{1}$$

IOP Conf. Series: Materials Science and Engineering **316** (2018) 012045 doi:10.1088/1757-899X/316/1/012045

In this research, the Taguchi method would be used for designing robust service determined before by SERVQUAL-QFD. [11] has successfully applied Taguchi in designing retail service, but the factors and also their levels was determined qualitatively by brainstorming the customers, so the subjective consideration still dominates them. One needs to conduct the Taguchi pre-determined process in selecting factors and levels especially for service analysis. In this research, the pre-process will be incorporated with the SERVQUAL-QFD analysis, so the factors should be more subjectively determined.

3. Framework and methodology

As mentioned above, the SERVQUAL-QFD and Taguchi method need to be incorporated and then complemented by additional steps. This research framework is shown in figure 3. The SERVQUAL and Taguchi part of the framework needs customer feedback, while the QFD part still needs subjective weighting determination of prioritized plans. However, these subjective processes can be neutralized by customer response in Taguchi part. Steps in implementing the framework briefly explained below.

1. Generating questionnaire variables refer to SERVQUAL dimensions, then conducting survey
2. Once the feedbacks got, calculate the customer gaps for each variables
3. Selecting the variables with negative gaps to be whats component in QFD, and weighting them refer to those gaps value.
4. Subjectively generate QFD hows (the improvement plans), and associate them with whats by using relationship weighting, so the prioritized plans should be resulted from.
5. Prioritized plans as factors in Taguchi design. Determine the two levels of each factor, and then assign them in the orthogonal array.
6. Each run in orthogonal array represents the combination of prioritized plan, which should gain confirmation responses from customer.
7. Optimizing the response using Taguchi technique, then the optimal combination of plans should be obtained. This results the robust service design.

Robust service design has accommodated almost all parts of customer responses. So, it should decrease the number of complaints and has longer timeframe implementation. One should maintain this design and periodically evaluate it. Someday, there's a chance where customer responses for this design will be decreased and needs to be re-improved.

Table 1. Variables with negative gaps

no	Variables	Expectation (E)	Perception (P)	customer gaps (P-E)
1	comfort waiting room	4.85	4.03	-0.82
2	variations of bank service	4.83	4.25	-0.58
3	responsiveness of customer service	4.71	4.22	-0.49
4	responsiveness of teller	4.74	4.22	-0.52
5	competency of customer service and teller	4.63	3.91	-0.72
6	willingness to answer customer questions	4.79	4.14	-0.65
7	friendly and polite service	4.69	4.23	-0.46
8	simplicity in conveying complaints	4.9	4.31	-0.59
9	full attention from bank officer to customer	4.9	4.35	-0.55
10	no discrimination in servicing customer	4.68	4.23	-0.45

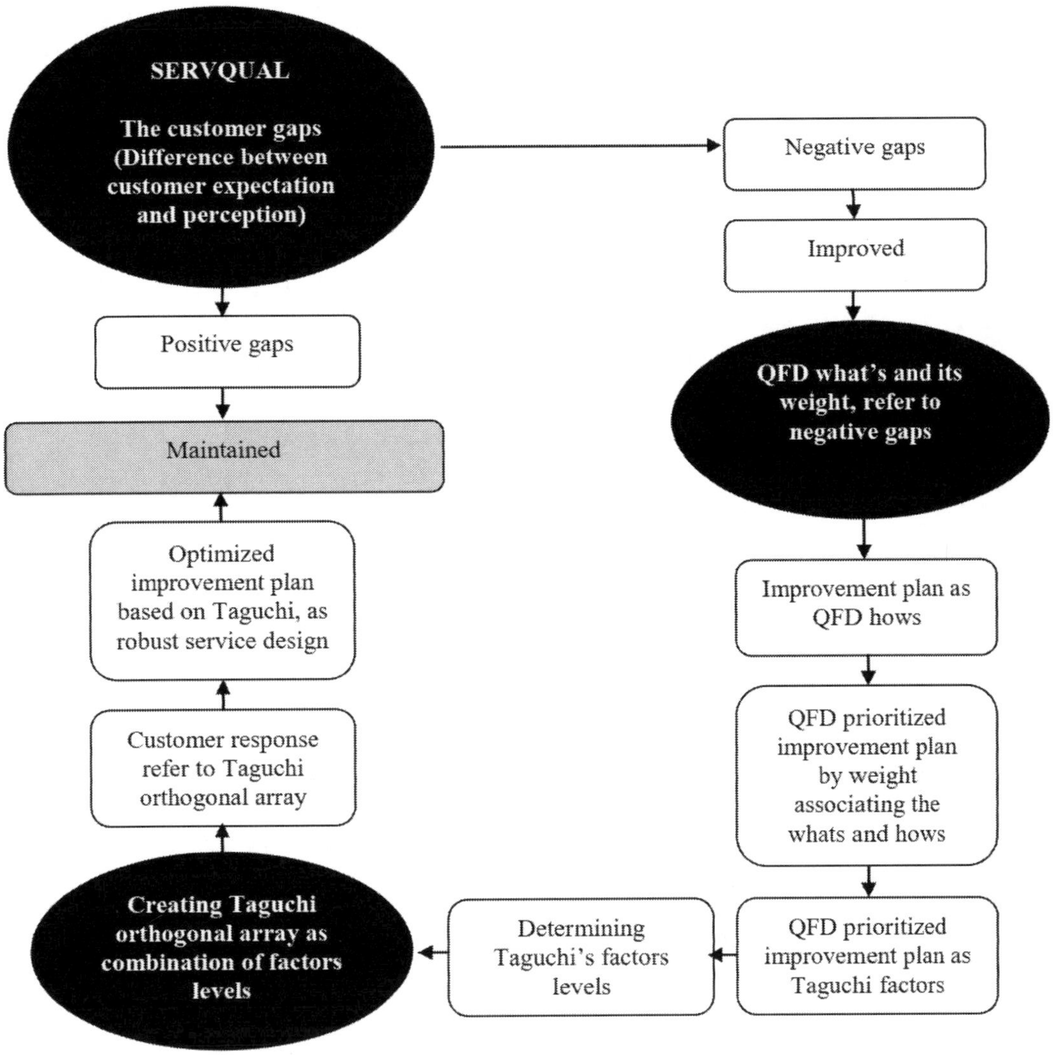

Figure 3. Incorporating SERVQUAL-QFD-Taguchi framework

4. Results and discussion

A case study was taken from [18], which has evaluated the service quality of local bank in Indonesia, with 22 variables generated considering 5 dimensions in SERVQUAL. The customer gaps had calculated, and 10 variables with negative gaps (unsatisfactory) were found, list of those variables shown in table 1. All negative gaps would be the whats components in QFD, and their gaps value as weights. The improvement plan (hows component) then determined subjectively related to the whats. There were 6 hows generated, and the associate weight with whats component also put in QFD, this result could be seen in figure 4. The prioritized plans were then obtained by calculating relative weight for each improvement plan, see table 2. Without incorporating Taguchi, these prioritized plans were the final solution for improving bank service design.

However, there were no such confirmation process form customers so those solutions had no guarantee to survive over a long periodic evaluation. One should make sure that the solution would robust and reliable at longer time. The incorporated Taguchi then take place for increasing the service design robustness. The prioritized plans in table 2 then converted as factor in Taguchi experimental design. Levels from each factor then determined considering the high level and low level as in Taguchi. Table 3 shows these levels.

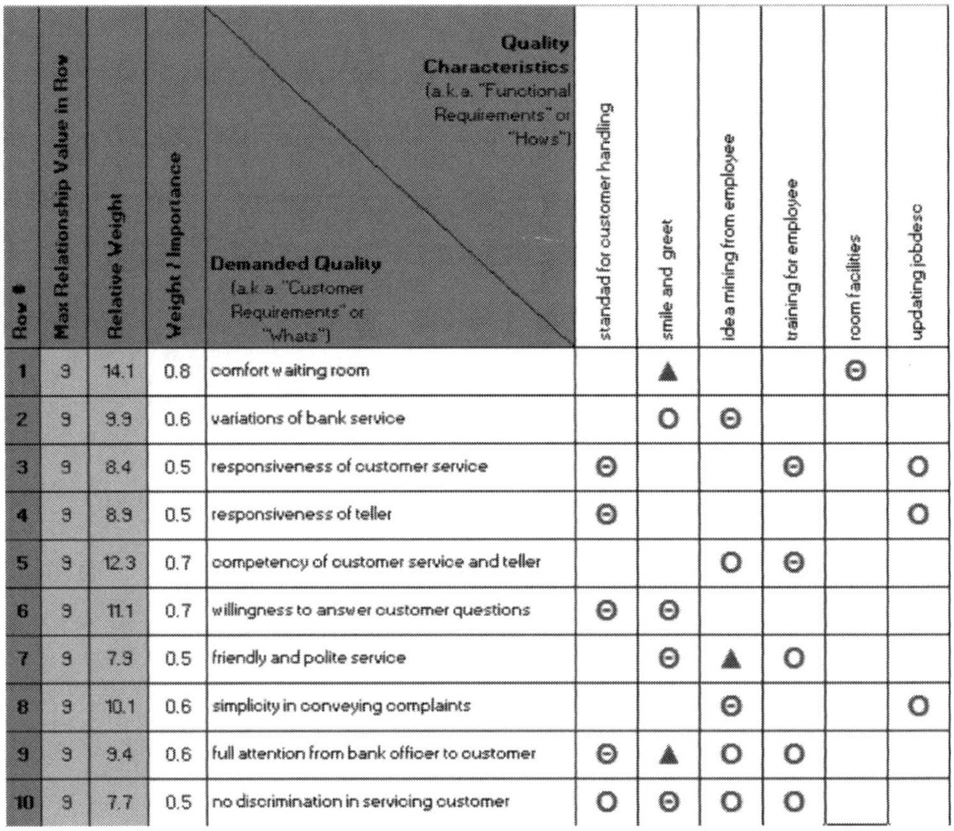

Figure 4. Association between whats (negative gaps) and hows (improvement plan) in QFD. Symbols represent association level.

Table 2. prioritized plan

the hows in QFD	QFD relative weight	prioritized plan
standard for customer handling	25.9	YES
smile and greet	20.9	YES
idea mining from employee	19.7	YES
training for employee	18.6	YES
room facilities	9.0	NO
updating jobdesc	5.9	NO

Only 4 factors selected into Taguchi analysis, omitting prioritized plans with low QFD relative weight as they didn't gives significant influence to customer. All these 4 factors and their levels then assigned into L_8 orthogonal array experiment design, selected design refer to the number of factors [13] with no assumption of interaction between factors.

IOP Conf. Series: Materials Science and Engineering **316** (2018) 012045 doi:10.1088/1757-899X/316/1/012045

Table 3. Taguchi factors and level

Taguchi factors	Prioritized plan (the hows in QFD)	level 1	level 2
A	standard for customer handling	create SOP then publish it to customer	create SOP without publish it, as it's confidential
B	smile and greet	mandatory for any level of employee, including the securities	mandatory only for customer service and teller, as they are directly communicate with customer
C	idea mining from employee	employee periodic meeting for idea mining	directly idea conveying with reward
D	training for employee	employee periodic training	employee training as needed

Combinations of levels in L_8 had assumed to be experiment runs in Taguchi, responses taken from customer were their perceptions for each level combination. For example in first run, customers were asked to gives their perception if the bank deploying this service designs:

a. create SOP then publish it to customer
b. mandatory for any level of employee, including the securities
c. employee periodic meeting for idea mining
d. employee periodic training

Customers fill a likert scale questionnaire represent how high they perception about this first run service design. This technique was similar to [10] and [11].

Table 4. Customer responses in Taguchi L_8

run	factors				unused columns			average customer responses	S/N ratio
	A	B	C	D					
1	1	1	1	1	1	1	1	2.2	6.848454
2	1	1	1	2	2	2	2	4	12.0412
3	1	2	2	1	1	2	2	3.3	10.37028
4	1	2	2	2	2	1	1	2.1	6.444386
5	2	1	2	1	2	1	2	2	6.0206
6	2	1	2	2	1	2	1	1.2	1.583625
7	2	2	1	1	2	2	1	2.1	6.444386
8	2	2	1	2	1	1	2	3.1	9.827234

Perceptions of customers for this first L_8 run would be the experiment responses, another experiment runs then treated in same way as well. Final responses from customers forming a completed Taguchi L_8 orthogonal array are shown in Table 4. Transformation from responses to signal-to-noise also calculated as the objective of this experiment was maximizing customer responses
As standard Taguchi analysis, response table and graphic then created to find optimal responses. Table 5 and figure 5 shows optimal levels combination that had optimized customer responses. Final optimized service design was:

a. factor D level 1 : employee periodic meeting for idea mining

IOP Conf. Series: Materials Science and Engineering **316** (2018) 012045 doi:10.1088/1757-899X/316/1/012045

b. factor A level 1 : create SOP then publish it to customer
c. factor B level 2 : mandatory only for customer service and teller, as they are directly communicate with customer
d. factor C level 2 : directly idea conveying with reward

This final service design then could be deployed to customer and still need to evaluate periodically. It should be survives in a longer time until customer response decreased, if it be then the banks should re-improve the service design and find new one.

Table 5. S/N Taguchi response table

Level	factors			
	A	B	C	D
1	8.926	6.623	7.421	5.33
2	5.969	8.272	7.474	9.565
delta	2.957	1.648	0.053	4.235
rank	2	3	4	1

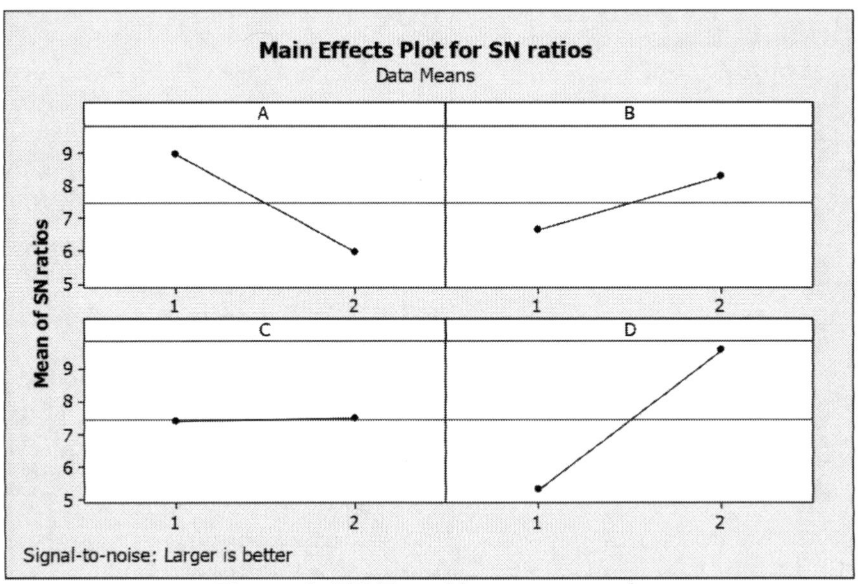

Figure 5. S/N Taguchi response graph

5. Conclusion
The framework for incorporating SERVQUAL-QFD-Taguchi has successfully implemented in this research, optimal service design has founded. It should have more reliable positive responses from customer, since it was obtained by optimization process. Further, the development of this research should include the reliability estimation of service design that has deployed, so the service life cycle could be predicted.

6. References

[1] Parasuraman A, Zeithaml V A, Berry L L 1988 SERVQUAL: a multi-item scale for measuring consumer perceptions of the service quality *J. of Retailing* **64** 12

[2] Buttle F 1995 SERVQUAL: review, critique, research agenda, *European J. of Marketing* **30** 8

[3] Stevens P, Knutson B, Patton M 1995 DINESERV: A tool for measuring service quality in restaurants *The Cornell Hotel and Restaurant Administration Quarterly* **36** 56

[4] Cronin J J and Taylor S A 1992 Measuring service quality: A re-examination and extension *J. of Marketing* **56** 55

[5] Abdullah F 2006 The development of HEdPERF: a new measuring instrument of service quality for the higher education sector *Int. J. of Consumer Studies* **30** 69

[6] Anderson E A 1995 Measuring service quality in a university health clinic *Int. J. of Health Care Quality Assurance* **8** 32

[7] Magi A and Julander C R 1996 Perceived service quality and customer satisfaction in a store performance framework: An empirical study of Swedish grocery retailers *J. of Retailing and Consumer Services* **3** 33

[8] Sahney S, Banwet D K, Karunes S 2004 A SERVQUAL and QFD approach to total quality education: A student perspective *Int. J. of Productivity and Performance Management* **53** 143

[9] Apriani D and Juliani Y D 2015 Integrating SERVQUAL and QFD framework for product and Service design case study: beehive café Bandung *J. of Business and Management* **4** 990

[10] Kumar A, Motwani J, Otero L 1996 An application of Taguchi's robust experimental design technique to improve service performance *Int. J. of Quality & Reliability Management* **13** 85

[11] Raajpoot N, Javed R, Koh K 2008 Application of Taguchi design to retail service *Int. J. of Commerce and Management* **18** 184

[12] Green P 2014 Measuring service quality in higher education: A South African case study *J. of Int. Education Research* **10** 131

[13] Belavendram N 1995 *Quality by Design: Taguchi Techniques for Industrial Experimentation* (Hertfordshire UK: Prentice-Hall) p 68

[14] Roy R 2001 *Design of experiments using the Taguchi approach* (New York: Wiley-Interscience) p 78

[15] Pontes F J, de Paiva A P, Belestrassi P P, Ferriera J R, de Silva MB 2012 Optimization of Radial Basis Function neural network employed for prediction of surface roughness in hard turning process using Taguchi's orthogonal arrays *Expert System with Applications* **39** 7776

IOP Conf. Series: Materials Science and Engineering **316** (2018) 012045 doi:10.1088/1757-899X/316/1/012045

[16] Hadiyat M A and Wahyudi R D 2013 Integrating Steepest Ascent for the Taguchi Experiment: A Simulation Study *Int. J. of Technology* **4** 280

[17] Hadiyat M A 2012 Penerapan Optimasi Multirespon Menggunakan Hybrid Principal Component Analysis - Taguchi pada Proses Turning Material Polyacetal *Proc. of Seminar Nasional Teknologi Industri* (Jakarta: Universitas Trisakti) p 100

[18] Wiryawan I 2014 *Penerapan Integrasi Kano, Kansei dan Servqual ke dalam QFD untuk Meningkatkan Kualitas Layanan di BCA KCP Wlingi Blitar* (Surabaya: Undergraduate Thesis University of Surabaya)

IOP Conf. Series: Materials Science and Engineering **316** (2018) 012046 doi:10.1088/1757-899X/316/1/012046

Segmentation of Natural Gas Customers in Industrial Sector Using Self-Organizing Map (SOM) Method

A M Masbar Rus[1], **R Pramudita**[1], **I Surjandari**[1]

[1]Industrial Engineering Department, Faculty of Engineering, Universitas Indonesia, 16424, Depok, Indonesia.

annisamarlin@ui.ac.id, redianipramudita@gmail.com, isti@ie.ui.ac.id

Abstract. The usage of the natural gas which is non-renewable energy, needs to be more efficient. Therefore, customer segmentation becomes necessary to set up a marketing strategy to be right on target or to determine an appropriate fee. This research was conducted at PT PGN using one of data mining method, i.e. Self-Organizing Map (SOM). The clustering process is based on the characteristic of its customers as a reference to create the customer segmentation of natural gas customers. The input variables of this research are variable of area, type of customer, the industrial sector, the average usage, standard deviation of the usage, and the total deviation. As a result, 37 cluster and 9 segment from 838 customer data are formed. These 9 segments then employed to illustrate the general characteristic of the natural gas customer of PT PGN.

1. Introduction

Nowadays the limitations of Indonesia's petroleum could not meet the demands of the country's needs, thus Indonesia need to find other applicable alternatives to substitute the petroleum. One of the alternatives which is considered to be the most suitable to bolster the country's energy needs is natural gas. In fact, Indonesia has abundant gas reserves to shift the energy needs from petroleum which the reserve is predicted to fulfil the demand until 41.2 years ahead [1].

Along with the transition of energy from oil to natural gas, gas consumption in Indonesia becomes increasing. This phenomenon could be seen in Figure 1., which depicts the significant rise of natural gas production and consumption in Indonesia [1].

The increase in the consumption of natural gas in Indonesia is influenced by some sector of gas customers. Based on Ministerial Regulation of Energy and Mineral Resources (ESDM) No. 03 of 2010 article 6, there is priority in gas usage for sector; a) oil gas lifting &, b) fertilizer industry, c) the provision of electric power and another industry. These 4 sectors are natural gas customers that have been classified by the Government.

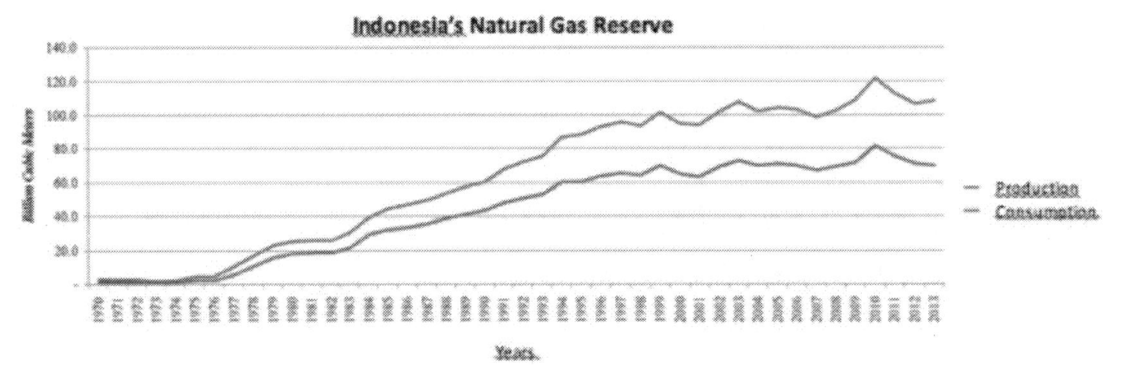

Figure 1. Graphic *of* Natural Gas *Consumption and Production* in Indonesia [1]

Natural gas which is used in Indonesia, commonly referred to as domestic gas, has already allocated towards each sector [2]. Based on Indonesia's natural gas balance 2014, Indonesia have allocated 59,45% of the total production of natural gas for the domestic needs. The detail allocation is shown in figure 2.

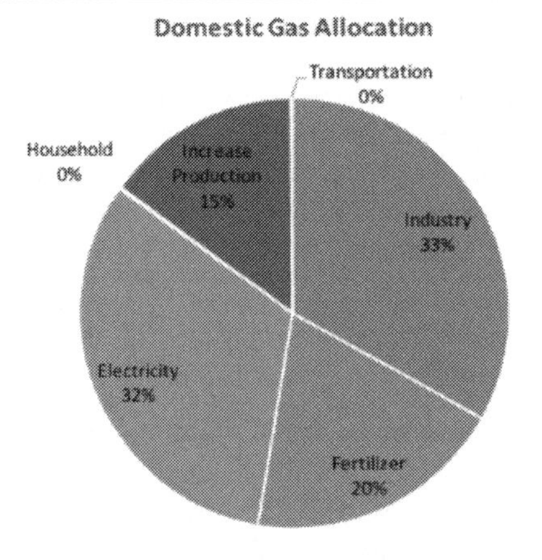

Figure 2. Domestic Gas Allocation [3]

Natural Gas is considered as more affordable in terms of price and have a lower level of pollution rather than other alternatives [4]. For this reason, industrial sector consumes more gas than other sectors as it is depicted in figure 2, thus industrial sector play an important role to the development of natural gas in Indonesia. Therefore, focus need to be made to the development of gas usage in industrial sector. Since natural gas is a non-renewable energy, thus the development is suggested to be supported with a robust and efficient plan. In order to do this, observation is made to the condition of the gas usage in Indonesia.

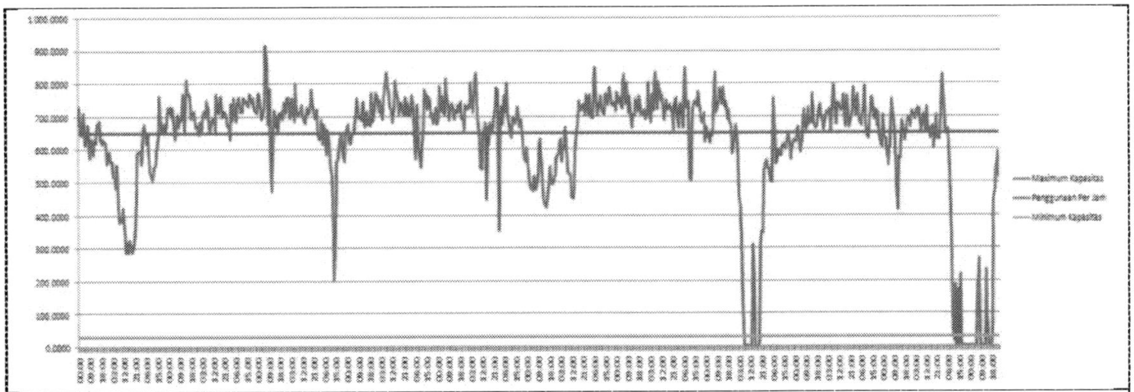

Figure 3. The example of consumption pattern of natural gas customer. [5]

Figure 3. shows a graphic of one of customer's natural gas usage in industrial sector of PT PGN. The graphic shows the costumer usage per hour and its maximum and minimum limits which determined according to the customer's contract. Based on the graphic, the customer's usage which exceed the maximum and minimum limit occurs very often. This condition happened on almost 80% of gas usage patterns sample in the industrial sector which influence the normal gas distribution. However, regardless of these deviation behavior, instead of penalizing the deviate customer, the producer still applying a same rate for all customers even though some of them exceed the maximum and minimum limit. This irregularity shows urgency in streamline natural gas usage for the future benefit.

One of the preliminary measure which could be done to manage natural gas usage effectively is by improving current segmentation process of gas customers in industrial sectors. This segmentation improvement could help the company to design an appropriate marketing mix for segment, thus the marketing strategies could be directed to the desired target [6].

In a nutshell, to address the challenge of enhancing natural gas usage efficiency, segmentation of gas users is proposed as preliminary measure. This research will employ Self-Organizing Map (SOM) for improving the segmentation process and employ gas usage data from customer of PT. PGN to represent the condition of gas usage in Indonesia.

2. Methodology

In this research, Self-Organizing Map (SOM) method will be used to conduct the segmentation by considering the homogeneity of the input data which has been determined for every sample. The sample which will be observed are the gas customers of a stated-owned natural gas enterprise, PT. Perusahaan Gas Negara (PGN) (Persero) Tbk, who is the biggest gas distributor in Indonesia.

The data collected is secondary data, i.e. real-time data of gas usage obtained from the Gas Management System (GMS) of PT PGN. Moreover, the information covered in this data is the identity of the customer and the results of temperature, pressure, and volume of usage on each of customer's gauge measurement. Furthermore, real-time data used in this research is the usage data per hour of 1172 gas customers in the industrial sector in Sub-Business Unit (SBU) I, encompass the southern part of Sumatra area and the western part of Java, within January to December 2014.

2.1. Data Preprocessing

In data pre-processing process, the previous real-time data will be pre-processed to produce variable data set as shown in Table 1. Later, this data will be used as attributes in the process of data processing.

Table 1. Variables and Values

IOP Conf. Series: Materials Science and Engineering **316** (2018) 012046 doi:10.1088/1757-899X/316/1/012046

Variable	Values
Area	Jakarta
	Bekasi
	Bogor
	Cilegon
	Cirebon
	Karawang
	Tangerang
	Palembang
Type of Customers	IMP 1
	IMP 2
	IMP 3
	IJK 1
	IJK 2
	IJK 3
Industrial Sector	Glass
	Wood
	Ceramic
	Paper
	Chemical
	Basic Metal
	Food
	Fabricated Metal
	Cement
	Textile
	Power Plant
	Others
Average Usage	Gas usage standard volume per time unit (Nm^3)
Standard Deviation of Usage	The standard deviation of the gas usage standard volume
Total Deviations of Usage	The total volume of deviate usage from maximum and minimum limits (Nm^3)

After that, data preprocessing is conducted by checking the data completion and the calculation of the average value, standard deviation, as well as the total deviation of gas usage as shown in table 1. Checking the completeness of the data is conducted by looking at whether each object is included in the whole database from January-December 2014. As a result, from the initial data set of 1172, there are 838 data set which is concluded as complete. Later, the average, standard deviation, and the total usage deviation of 838 data set is calculated using *Microsoft Excel.*

2.2. Data processing

In this process, the variable inputs which is used encompass variable area, type of customer, the industrial sector, the total deviation of use, average (per hour / day / week/ month) and standard deviation (units of hours / days / weeks / months). The *STATISTICA* software is employed for the calculation process.

At the beginning of the process, initialization of number of clusters, the coefficient learning rate (η) and neighbourhood coefficient (σ) is determined. The initial determination of the coefficient does not have any specific rules, thus the coefficients which is used in this study is a default number from *STATISTICA* software, namely $\eta = 0.1$ and $\sigma = 3$. As for the determination of the number of clusters, experiments are conducted for different number of clusters until all certain number of clusters are occupied by at least 1 type of customer and has the lowest error validity.

3. Results and Discussion

In the previous section, Self-Organizing Map (SOM) has been employed to produce clusters based on their homogeneity. In this research, the result of the cluster is differentiated based on time unit as shown in table 2.

Table 2. The processing error (daily)

Time unit	Training error	Test error	Validation error
Hour	0.515248	0.566476	0.540686
Day	0.486721	0.622279	0.537399
Week	0.514967	0.615034	0.537586
Month	0.729907	0.829682	0.762875

The table shows that grouping based on day unit demonstrating better results from other measure. This conclusion can be inferred from the error validation value that reaches 0,537399, which is lower than other group time. The error shows the distance between objects in the cluster and the distance between each cluster. The smaller the error then the closer the distance between objects in the cluster, and the further the distance between its clusters. The cluster result based on day-time unit is depicted in figure 4.

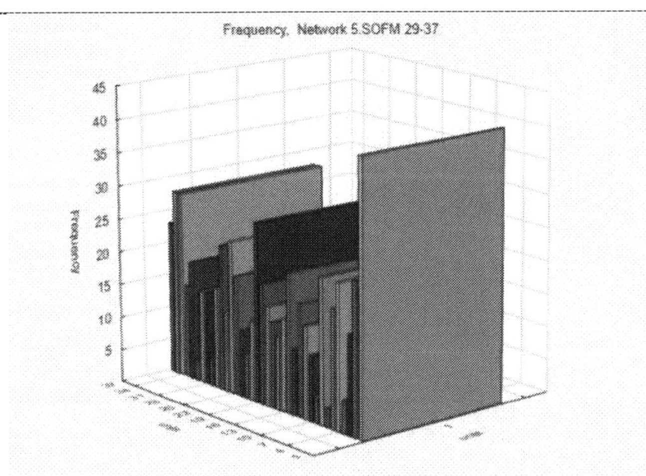

Figure 4. Frequency graphic of daily clustering results.

Objects of standard deviation on each cluster will interpret the usage distribution, which will be compared to its average. While, total deviation will interpret whether most of the usage pass the

IOP Conf. Series: Materials Science and Engineering **316** (2018) 012046 doi:10.1088/1757-899X/316/1/012046

maximum and minimum limit. Interpretation based on the obtained label will also be adapted with the type of customers on each cluster. Afterwards, the result of these labelling is used for the purpose of descriptive analysis.

Table 3. Interpretations of Each Cluster

Cluster	Distribution of The Usage	Deviation of The Usage	Type of Customers
1	very high	high	IMP3
2	very high	very high	IMP3
3	very high	high	IMP3
4	-	-	IMP1 IMP2 IMP3
5	low	high	IJK3
6	high	high	IJK1
7	-	-	IJK2 IMP1 IMP3
8	high	very high	IMP3
9	-	-	IMP3
10	very high	high	IMP3
11	very high	high	IMP3
12	high	very high	IMP3
13	high	very high	IJK2 IMP3
14	high	very high	IMP3
15	high	higher	IMP3
16	very high	high	IMP3
17	high	very high	IMP3
18	high	higher	IMP1 IMP3
19	high	higher	IMP3
20	high	higher	IMP3
21	high	very high	IMP3
22	high	very high	IMP3
23	very high	high	IMP2 IMP3
24	very high	high	IMP3
25	high	higher	IMP3
26	high	high	IMP2
27	high	high	IMP2
28	very high	high	IMP2
29	high	high enough	IMP1
30	high	high enough	IMP1
31	high	high enough	IMP1
32	high	high	IMP1
33	very high	high	IMP2
34	very high	high	IMP2
35	high	high	IMP2

Cluster	Distribution of The Usage	Deviation of The Usage	Type of Customers
36	very high	high	IMP2
37	high	high	IMP2

The results of cluster analysis on the previous sub-section shows that each cluster has a characteristic that can be used as a reference in the creation of customer segmentation.

1) Segment 1 (consisting of cluster #2) is the 3rd type of customer in Tangerang area, and diverse sectors. This segment is dominated by chemical industry sector. Distribution and deviation characteristics of the usage of this segment is very high. Therefore, this paper recommends PT PGN to pay more attention to this segment because of the high fluctuation in the usage which could cause problem in gas distribution. One of the proposed solution to control the distribution is by managing higher rates for this segment.

2) Segment 2 (consisting of cluster #8, #12, #13, #14, #15, #16, #22 and #23) is the 3rd type of customer located in Jakarta and Bogor. This segment is dominated by food industrial sectors. Characteristic deviation of this segment usage is very high. In other words, user in this segment often pass the limit even though its fluctuation is not as high as in Segment 1. Accordingly, appropriate supervision need to be assign to users in this segment as the previous segment. Moreover, if it is necessary, proper sanction could also be applied to reduce the occurrence of the deviation.

3) Segment 3 (consisting of cluster #1, #3, #10, #11, #16, #23 and #24) is a 3rd type of customers located in Tangerang and Karawangan area, which is one of industrial area in Indonesia. This segment has the most members and the usage distribution characteristics of this segment is very high or more likely to be unstable. Based on the information, this paper recommends PT PGN to determine the rate or the contract based on daily usage.

4) Segment 4 (consisting of cluster #18, #19, #20 and #25) is the 3rd type of customer located in Jakarta and Bekasi. Distributions and deviations of their usage are not too high. Accordingly, it is suggested to give standard supervision to maintain the distribution and the deviation.

5) Segment 5 (consisting of cluster #5) is the customer type of IJK3 or Service Industry with Big Capacity, which located in Jakarta area. Usage distribution level for this segment is relatively small but it has large deviations. This happened because the service industry operates continuously, thus its usage is considered high because it passes the maximum limit every day. Therefore, it is suggested that the contract is reviewed because it seems that the assign rates or contract is not suitable with the customer needs. Moreover, as the characteristic of this segment is different from other segments, different contract plan could help PT PGN to meet their customer needs.

6) Segment 6 (made up of #28, #33, #34 and #36) is IMP2 customer type which located in Tangerang and Jakarta. The usage distribution is high while its deviation is not too high. In other words, customer in this segment has big capacity with fluctuate usage. As a suggestion, implementation of daily rates could help the company to meet the customer needs.

7) Segment 7 (consisting of cluster #26, #27, #35 and #37) is the kind of customer in IMP2, in Bekasi and Bogor. Industrial sector in this segment is vary but mostly dominated by metal fabrication industry. Characteristics of the usage distribution and deviations are high. For this reason, it is suggested that higher rates policy for this segment is implemented as in segment 1.

8) Segment 8 (made up of cluster #6) consists of 2 IJK customers and a bit of IJK 1, in Jakarta. The spread of its usage is large but still below the limit. This segment could be considered as a good user because its usage is fluctuated but it does not interfere the distribution.

9) Segment 9 (consisting of cluster #29, #30, #31 and #32) is a type of IMP1 in Bogor area, with metal fabrication industry and textile. The usage distribution and deviations are high for customers with small capacity. Moreover, the usage is also fluctuated and often pass the minimum and maximum limit. Thus, review regarding rate, penalty policy and maximum and minimum limit could help the company to meet the customer needs.

4. Conclusion

By using the six variables (area, type of customer, industrial sector, daily average, standard deviation, and total deviation), 37 clusters are formed with lowest value of error validation. The clusters are based on similar standard deviation and its total deviation value. Moreover, each cluster is identified by the distribution and the deviation of its usage, which eventually produce nine customer segmentations. Furthermore, aside from the nine segments which are proposed, PT PGN is also expected to make the fares strategy that can focus on the criteria of each cluster. Thus, the distribution of natural gas could be managed effectively.

References

[1] British Petroleum. (2014). BP Statistical Review of World Energy. London: British Petroleum.
[2] Kementerian ESDM. (2014). Peta Jalan Kebijakan Gas Bumi Nasional. Jakarta: Kementerian Energi dan Sumber Daya Mineral
[3] Badan Pengkajian Penerapan Teknologi. (2014). Outlook Energi Indonesia 2014. Jakarta: Pusat Teknologi Pengembangan Sumberdaya Energi, BPPT
[4] Bakar, W. A., & Ali, R. (2010). Natural Gas. In P. Potocnik, Natural Gas (pp. 1-37). Johor: Sciyo.
[5] Perusahaan Gas Negara. (2012). Laporan Tahunan PGN. Jakarta: PGN.
[6] Kiang, M. Y., Hu, M. Y., & Fisher, D. M. (2006). The effect of sample size on the extended self-oranizing map network—A market segmentation application. Computational Statistics & Data Analysis , 51, 5940 – 5948.
[7] Badan Pusat Statistik. (2014). Distribusi Persentase Produk Domestik Bruto Triwulanan Atas Dasar Harga Berlaku Menurut Lapangan Usaha, 2000-2014 (Persen). Retrieved 2015 Februari from Badan Pusat Statistik.
[8] Direktorat Jenderal Minyak dan Gas Bumi. (2010). Artikel Mengenal Jenis-jenis Gas Bumi. Retrieved 2015 Februari from Kementerian Energi dan Sumber Daya Mineral: http://mobile.migas.esdm.go.id/?op=Artikel&id=4
[9] Fayyad, U., Piatetsky-Shapiro, G., & Smyth, P. (1996). From Data Mining to Knowledge Discovery in Databases. American Association for Artificial Intelligence , 37-54.
[10] Gullo, F. (2015). From Patterns in Data to Knowledge Discovery: What Data Mining Can Do. Physics Procedia , 62, 18-22.
[11] Han, J., & Kamber, M. (2000). Data Mining: Concepts and Techniques. San Fransisco: Morgan Kaufmann.
[12] Han, J., & RC., D. (1988). Algorithms for Clustering Data. New Jersey: Prentice Hall Inc.
[13] Hanafizadeh, P., & Mirzazadeh, M. (2011). Visualizing market segmentation using self-organizing maps and Fuzzy Delphi method – ADSL market of a telecommunication company. Expert Systems with Applications , 198–205.
[14] Harsya, T. R. (2013). Analisis Keekonomian Pemanfaatan Kilang LNG Arun Sebagai Receiving

Gas Terminal di Provinsi Aceh. Universitas Indonesia, Departemen Teknik Kimia. Jakarta: Fakultas Teknik Kimia Program Pasca Sarjana Universitas Indonesia.
[15] Hung, C., & Tsai, C.-F. (2008). Market Segmentation Based on Hierarchical Self-Organizing Map. Expert Systems with Applications , 780 - 787.
[16] Kaski, S. (1997). Data Exploration Using Self-Organizing Maps. Finlandia: Laboratory of Computer and Information Science, Department of Computer Science and Engineering, Helsinki University of Technology.
[17] Kementerian ESDM, Dirjen Migas. (2014). Kebijakan Alokasi Gas Bumi Untuk Dalam Negeri. Jakarta: Kementerian Energi dan Sumber Daya Mineral.
[18] Kementerian Pendidikan dan Budaya. (2015). Kamus Besar Bahasa Indonesia Versi Online. Retrieved 2015 April from Kamus Besar Bahasa Indonesia Versi Online: http://kbbi.web.id/segmentasi
[19] Kohonen, T. (2013). Essentials of the self-organizing map. Neural Networks , 52-65.
[20] Kohonen, T. (2001). Self-Organizing Maps. New York: Springer.
[21] Larose, D. T. (2004). Discovering Knowledge in Data: An Introduction to Data Mining. New Jersey: John Wiley & Sons Inc.
[22] Pratte, J. M. (2004). Fossil Fuels: Natural Gas. Kennesaw: Enviromental Science Activities for the 21st Century.
[23] Ra¨sa¨nen, T., Ruuskanen, J., & Kolehmainen, M. (2008). Reducing energy consumption by using self-organizing maps to create more personalized electricity use information. Applied Energy , 830 - 840.
[24] Roojiers, F., Wielders, L., Schepers, B., Croezen, H., & de Bruyn, S. (2009). Natural Gas as Transitional Fuel. AD Groningen: Castel International.
[25] Salazar, G., Veles, A., Parra, M., & Ortega. (2002). A Cluster Validity Index For Comparing Non-hierarchial Clustering Methods. Medellin: Engineering Faculty, University of Antioquia.
[26] Tamberi, F. (2007). Anomaly Detection Data Mining Techniques. Pisa: Department of Computer Science University of Pisa.
[27] Tan, P.-N., Steinbach, M., & Kumar, V. (2005). Introduction to Data Mining. Boston: Addison-Wesley.
[28] Vondracek, J., Pelikan, E., Konar, O., Cermakova, J., Eben, K., Maly, M., et al. (2008). A Statistical Model for the Estimation of Natural Gas Consumption. Applied Energy , 362–370.
[29] Zaki, M. J., & Meira Jr., W. (2014). Data Mining and Analysis: Fundamental Concepts and Algorithms . New York.

IOP Conf. Series: Materials Science and Engineering **316** (2018) 012047 doi:10.1088/1757-899X/316/1/012047

The Effect Of Local Coal And Smelting Sponge Iron On Iron Content Of Pig Iron

Soesaptri Oediyani[1], Pramita Juwita Sari[2], Djoko Hadi P[3*]

[1,2] Metallurgical Engineering Department, Faculty of Engineering, Sultan Ageng Tirtayasa University, Cilegon 42435, Indonesia
[3] Badan Tenaga Atom Nasional, Bandung 40112, Indonesia

*Corresponding Address: s_oediyani@untirta.ac.id

Abstract. The new regulation on mineral resources was announced by Ministry of Energy and Mineral resources (ESDM) of Indonesia at 2014 which it called Permen ESDM No 1/2014. Therefore, this research was conducted to add the value of local iron ores by using smelting technology. The objective of the research is to produce pig iron that meet the requirement of the new regulation of mineral resources such as 90% Fe. First, iron ores and coal mixed together with lime as a flux, then smelted in a Electric Arc Furnace at 1800°C. The process variables are (1; 1.25; 1.5 ; 1.75 ; 2.0) and the composition of coal (0.8%, 1.6%, 3.0%). The type of coal that used in this research was bituminous coal from Kalimantan and also the iron ores from Kalimantan. The products of the smelting technology are Pig iron and slag. Both pig iron and slag then analyzed by SEM-EDS to measure the iron content. The result shows that the maximum iron content on pig iron is about 95.04% meanwhile the minimum iron content on slag is about 3.66%. This result achieved at 1.6% coal and 2.0.

Keywords: Mineral Resources, Pig Iron, Slag, Smelting Technology

1. Introduction

Since 2014, mining products such as minerals or ores cannot be exported directly but must be processed first in the Indonesia. This is a Government Regulation (PP) No. 1 of 2014 and the Regulation of the minister of energy and mineral resources No. 1 in 2014 , which is derived from the Law of Mineral and Coal Mining No. 4 in 2009. One of the objectives of the government regulation is to give value adding to the minerals and to support the industrial development program which is being launched by the government. According to the regulation of minister of energy and mineral resources No. 1 of 2014 pig iron produced from iron ore processing must contain a minimum iron content of 90% (Permen ESDN, 2014).

In general, reducing agent used to produce pig iron is natural gas. However, nowadays, the natural gas is becoming increasingly scarce and expensive (7-8 dollars per MMBtu) [2] Therefore, in this research coal is used as an alternative reductant to replace natural gas. Utilization of local iron ore and coal as raw material for steel making to produce pig iron is expected to provide the added value of local mineral resources and meet the provisions of the regulation of minister of energy and mineral resources No. 1 of 2014 such as equal and more than 90% Fe. In this research, hematite iron ore and sub-bituminous coal from Kalimantan with reserves 562.98 million tons (Statistil Geologi dan ESDM, 2013) and 55.363 trillion tons (PSDG, 2013), respectively, were used.

Manufacture of pig iron from the raw material such as iron sponge, flux (CaO) and coal carried in Electric Arc Furnace at a temperature of 1800 °C. This process names smelting of iron ores and basicity is one of the main problems in the processing of pig iron. Basicity is a ratio of a mineral formed from alkaline oxides and a mineral formed from acidic. Flux is used to collect impurities from the raw materials and form a slag phase. The slag phase affects the iron content of pig iron. Therefore, the basicity must be maintained and controlled to achieve the highest iron content of pig iron. In addition, basicity also can affect the viscosity of slag such as basicity increases, the viscosity decreases. If the viscosity decreases, the diffusion of atoms in the slag toward the metal phase will be easier as a result the iron content of pig iron increases. On the other hand, coal is used as a reducing agent for reducing iron oxide from iron ores. The reduced iron is formed metal phase separately with slag phase (Suharto, 2013).

This research is a development of previous research. In previous research, iron ore hematite from Merangin Jambi was used. The value of basicity in previous research was low (0.24 ; 0.3 : 0.36) the result shows that the basicity affected the iron content of pig iron such as basicity increases, iron content of pig iron increases too. Therefore, in this research the value of the is higher (1 ; 1.25 ; 1.5 ; 1.75 ; 2) than previous research. Furthermore, according to ternary diagram of CaO-SiO_2-FeO if the value of basicity between 1-2, the affinity of SiO_2 and CaO is expected higher than the affinity of SiO_2 and FeO. In a result, SiO_2 and CaO will form slag phase but FeO remains in metal phase.

2. Methodology/ Experimental

Methodology of this research was motivated by the regulation of minister of energy and mineral resources on added value of local resources and the result of previous research that mentioned in introduction section. There are two type resource of iron such as iron ores which calls lateritic ores and iron sands. Lateritic ores can be divided into two type i.e. hematite and magnetite. Generally iron ores contain 40-65%Fe and the value must be added by using preparation process such as mineral dressing (comminution, sizing and concentration process) or others process such as agglomeration, roasting and calcinations. The result of preparation process usually contains 65% Fe that suitable for ironmaking process.

In this research, the preparation processes include ore beneficiation (crushing, grinding, sizing and magnetic concentration) and direct reduction of ores was used. Firstly, the iron ores from mining was crushed and grinded. Then, the product was screened to achieve the suitable size and it added the value by using magnetic concentrator. The product of magnetic concentrator names a concentrate which usually content 65% Fe. The concentrate then mixed with coal and binder to form a pellet.

Secondly, direct reduction of pellet is done after ore beneficiation conducted. The product of direct reduction is called sponge iron and became the raw material of this research. Before the research started, the composition of sponge iron, flux and coal were analyzed by using the proper testing such as XRF (for sponge iron and flux), proximate and ultimate analysis for coals. Figure1 and Figure2 show the sponge iron and sub-bituminous coal used in this research respectively.

This research was conducted in laboratory scale. The objective of the research is to produce pig iron that meet the requirement of the new regulation of mineral resources such as 90% Fe. Basicity is very important factor in the smelting process because it can affect iron content of pig iron. To obtain the desired basicity, it need the addition of CaO (lime). The variation values of basicity in this research were 1 ; 1.25 ; 1.5 ; 1.75 ; and 2. In addition, the variation values of coal were 0.8 %, 1.6 % and 2.4 %.

IOP Conf. Series: Materials Science and Engineering **316** (2018) 012047 doi:10.1088/1757-899X/316/1/012047

Figure 1. Sponge Iron

Figure 2. Sub-bituminous Coal

Meanwhile, the smelting temperature is fixed ($1800^{O}C$) and Electric Arc Furnace used is in laboratory scale. The product of smelting name is pig iron then analyzed by using SEM-EDS to determine the chemical composition and microstructure of pig iron. Figure 3, 4 and 5 show the Electric Arc Furnace as a laboratory scale, pig iron and slag that produced in this research, respectively.

Figure 3. Electric Arc Furnace

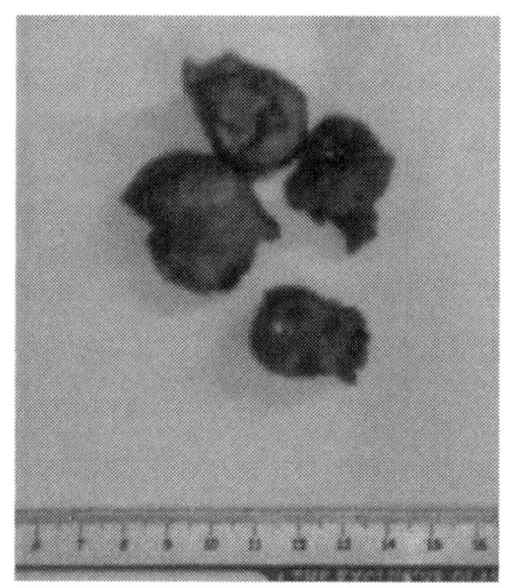

Figure 4. Pig iron

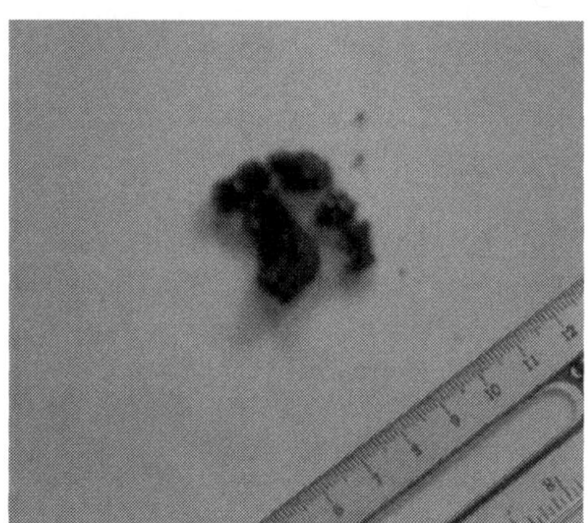

Figure 5. Slag

IOP Conf. Series: Materials Science and Engineering **316** (2018) 012047 doi:10.1088/1757-899X/316/1/012047

3. Results and Discussion

Tables 1, 2 and 3 show the composition of limestone, coal and sponge iron, respectively, as a raw materials that used in this research.

Table 1. Composition Of Limestone (CaO)

Substitute	%
CaO	53.20
MgO	0.28
SiO_2	0.70
LOI	41.80

Table 2. Proximate Analysis Of Coal

Substitute	Fixed Carbon	Moisture	Volatile Matter	Ash
%	37.98	7.70	41.95	12.37

Table 3. Chemical Composition of Sponge Iron

Substitute	%
Fe total	72.80
Fe Metal	65.60
Al_2O_3	1.03
SiO_2	4.60
MgO	0.36
CaO	0.14

3.1. Effect Of Basicity On Iron Content Of Pig Iron

Figure 6 shows the influence of basicity on iron content of pig iron. It could be seen that basicity affects the iron content of pig iron but the effect varies with the variation value of coal addition. At 0.8% coal, the highest value of iron content (90.73% Fe) is achieved at basicity equals to one but the iron content decrease slightly when the basicity increase. We can see that the iron content become 89.85% when the basicity increase to two. According to the previous research that conduct by Budi Syafari, if the basicity too high, it could increase the viscosity of the slag (Syafari, Budi, 2009). Therefore, the iron content of pig iron decreases slightly because diffusion of atom at interphase between slag and metal phase is inhibited. As equation (1) mentioned below, diffusion of atom will decrease if the viscosity increase.

$$D = \frac{Kb.T}{6\pi r\eta} \tag{1}$$

D: coefficient of diffusion, Kb: Boltzmann constant 91.38 x 10^{-23} J/mol.K), T: temperature (K), R: radius of particle (cm) and η: viscosity (poise).

Meanwhile, at 1.6% and 2.4% coal, the increasing of iron content depends on the increasing of basicity. As can be seen from reaction (2) that CaO will replace FeO in slag and form calcium silicate. Therefore, if basicity increases, the iron content of pig iron increases too. The smallest values of iron content (90.67 %Fe) is achieved at 1.6% coal while the highest iron content (95.04%) at 2.4% coal.

$$2FeO.SiO_2 + 2CaO \rightarrow 2CaO.SiO_2 + 2FeO \ \Delta G^0_{1923} = -113\ KJ \tag{2}$$

IOP Conf. Series: Materials Science and Engineering **316** (2018) 012047 doi:10.1088/1757-899X/316/1/012047

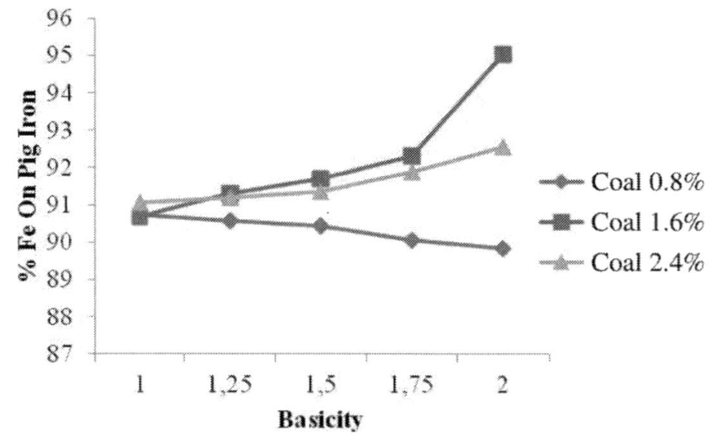

Figure 6. Effect Of Basicity On Iron content Of Fe

3.2. Effect Of Addition Of Coal On Iron content Of Pig Iron

The addition of limestone without accompanied by the addition of coal, the iron content of Fe in pig iron will not rise. Coal is a source of carbon used as a reducing agent for reducing iron oxides to increase the percentage of iron content of iron. In the process of smelting using EAF, limestone as a flux is added to react with impurities in sponge iron such as SiO_2. Moreover, it is also added coal as a reducing agent. If the content of C increases, the reduction process will increase too but there is an additional restrictions coal because excess coal can affect the viscosity. The reduction reaction of iron oxide by coal which occurs in the melting process can be seen as follows: (Rosenqvist, 1983)

$$<Fe> \rightarrow \{Fe\} \qquad \Delta G^{o}_{1823} = -\ 1.6 \text{ KJ} \tag{3}$$

$$<FeO> + \{C\} \rightarrow <Fe> + (CO) \qquad \Delta G^{o}_{1823} = -\ 139.86 \text{ KJ} \tag{4}$$

$$<FeO> + \{C\} \rightarrow \{Fe\} + (CO) \qquad \Delta G^{o}_{1823} = -\ 141.46 \text{ KJ} \tag{5}$$

Based on the research that has been done is seen that the additional of coal tends to increase the percentage of iron content of iron but after it reach a maximum point, the percentage of iron content is decreased. For more details can be seen in Figure 7. The decline in iron content caused by excess of coal could increase the viscosity. High viscosity means hot metal becomes condensed therefore atoms difficult to diffuse. Furthermore, the decline of iron content of iron is affected by the content of SiO_2 in coal. Table 4 shows the composition of coal ash by ultimate analysis (Amaliyah et.all, 2011).

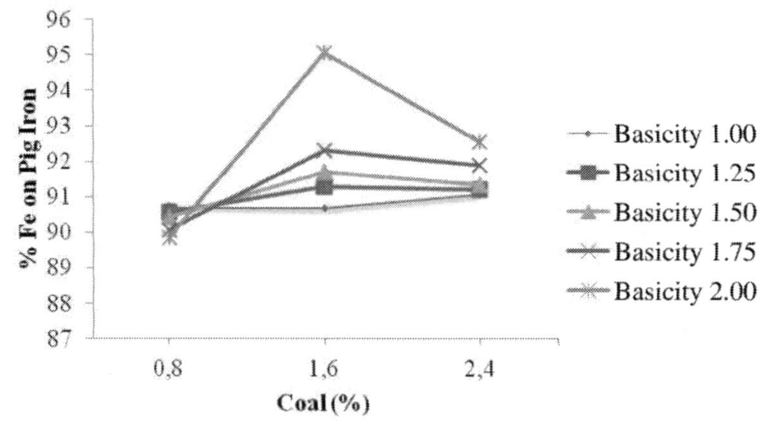

Figure 7. Effect of additional Coal on Iron content of Iron

Table 4. The Composition of Coal Ash

Substitute	%
SiO_2	48.96
Al_2O_3	29.26
CaO	6.08
MgO	1.13
SO_3	4.17
Na_2O	0.25
Fe_2O_3	6.83
TiO_2	2.27

3.3. Effect of Basicity on percentage of Iron in Slag

Figure 8 shows the effect of basicity on percentage of iron in slag. It can be seen that initially the percentage of iron in slag is high (8%) then is decrease slightly by increasing basicity. It means, CaO replace FeO in forms of fayalite because the affinity of CaO and SiO_2 higher then affinity of CaO and FeO that mention at equation (2). In addition, the additional of coal can also affect the content of iron in slag such as the increasing carbon content in the smelting process, the percentage of iron in slag is decreased.

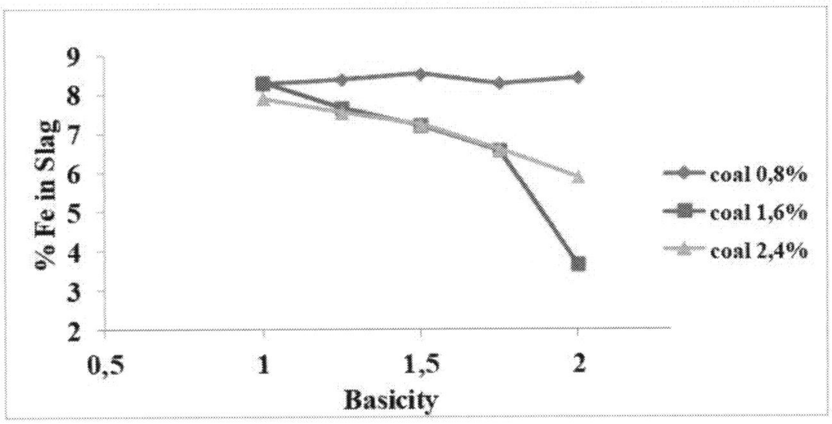

Figure 8. Effect of Basicity On Iron Content in Slag

4. Conclusion

Based on this research, it can be concluded that:

- Basicity affects the iron content of pig iron such as if basicity increases, the iron content increases too. The highest iron content (95.04%) achieved at basicity equals to 2.
- The addition of coal also can affect the iron content of pig iron. The improvement of iron content of pig iron has the same pattern as the effect of basicity i. e. iron content increase if the addition of coal increase as well. The highest iron content (95.04%) achieved at 1.6% of coal.
- Iron content in slag decrease when the basicity and the addition of coal increase. The highest iron content in slag (8.52%) is achieved at 1.5 basicity and 0.8% addition of coal.

5. Acknowledgement

This research was conducted by research collaboration between the Metallurgical Engineering Department of Engineering Faculty of Sultan Ageng Tirtayasa University with BATAN, Bandung.

Special thanks is given to my colleague, Mr. Djoko H.P and also to Paramita Juwita Sari, Ing Sakti, Agis Priyatna and Budi Syafari, students of Metallurgical Engineering Department.

6. References

[1] Peraturan Menteri Energi dan Sumber Daya Mineral Republik Indonesia, (2014), Jakarta. Statistik direktori geologi dan sumber daya mineral, (2013), http://www.dim.esdm.go.id.
[2] Data Sumber Daya dan Cadangan Berdasarkan data Pusat Sumber Daya Geologi (PSDG), (2013).
[3] Suharto.dkk, (2013), *Peleburan Sponge Iron Pada Tungku Kupola Menjadi Hot Metal*, UPT. Balai Pengolahan Mineral Lampung – LIPI.
[4] Syafari, Budi, (2009), *Pengaruh Temperatur, Waktu dan Fraksi Ukuran terhadap Persen Metalisasi Besi Spons Konsentrat Bijih Besi Laterit Kotawaringin Timur*, Universitas Sultan Ageng Tirtayasa, Cilegon-Banten.
[5] Rosenqvist, Terkel, (1983), *Principles of Extractive Metallurgy*, New York : McGraw-Hill.
[6] Amaliyah, Novriany., Muhammad Fachry, (2011), *Analisis Komposisi Batu Bara Mutu Rendah Terhadap Pembentukan Slagging Fouling Pada Boiler*, Universitas Hasanuddin, Tamalanrea-Makasar.

THE 15TH INTERNATONAL CONFERENCE QUALITY IN RESARCH (QIR) 2017

PREPARATION AND IONIC CONDUCTIVITY OF $Li_{3.9}Ca_{0.1}Ti_5O_{12}$ USING WASTE CHICKEN EGGSHELLS AS Ca SOURCE FOR ANODE MATERIAL OF LITHIUM-ION BATTERIES

Achmad Subhan[a], Dedy Setiawan[b], Sitti Ahmiatri Saptari[b]

[a]*Indonesian Institute of Sciences-Research Center for Physics (LIPI), Serpong, Indonesia*
[b]*Physics Department UIN Syarif Hidayatullah Jakarta, Ciputat, Indonesia*

ABSTRACT

$Li_{3.9}Ca_{0.1}Ti_5O_{12}$ has been synthesized as anode material for lithium-ion batteries parallel with $Li_4Ti_5O_{12}$ anode material using solid state reaction method in an air atmosphere. $LiOH.H_2O$, TiO_2, and waste chicken eggshells in the form of $CaCO_3$ were chosen as sources of Li, Ti, and Ca respectively and prepared using stoichiometric. The phase structure, morphology, and electrochemical impedance of as-prepared samples were characterized using XRD, SEM, and EIS. The XRD characterization revealed that in $Li_{3.9}Ca_{0.1}Ti_5O_{12}$ sample, all amount of dopant had entered the lattice structure of $Li_4Ti_5O_{12}$. The EDX image also detect the existence of Ca in the structure of $Li_{3.9}Ca_{0.1}Ti_5O_{12}$. The EIS characterization revealed that the $Li_{3.9}Ca_{0.1}Ti_5O_{12}$ sample had lower electrochemical impedance compared to the $Li_4Ti_5O_{12}$ sample. The diffusion coefficient were obtained by Faraday's method, and exhibited that the $Li_{3.9}Ca_{0.1}Ti_5O_{12}$ sample (1.46986×10^{-12} cm^2/s) had higher ionic conductivity than the $Li_4Ti_5O_{12}$ sample (4.40995×10^{-16} cm^2/s). According to the cycle performance test, the $Li_{3.9}Ca_{0.1}Ti_5O_{12}$ sample also had higher charge-discharge capacity and stability compared to the $Li_4Ti_5O_{12}$ sample.

Keywords: $Li_{3.9}Ca_{0.1}Ti_5O_{12}$; Anode Material; Lithium-ion Batteries; Waste Chicken Eggshell; Ionic Conductivity;

1. INTRODUCTION

Nowdays, lithium-ion batteries have been widely used in several technology applications, such as Electric Vehicle (EV) and Hybrid Electric Vehicle (HEV) because of its advantages. In terms of life cycle and redox potential, LiB have excellency in both characteristics compared to another batteries such as NiCd and NiMH. [1] In fact, the development of Li-ion batteries in its power and energy density is the key of successful application. [2]

However, the characteristics of lithium-ion batteries depend on its material components, including the anode. [3] One of the most developed anode material for lithium ion batteries is lithium titanium oxide (LTO). The researches take place in improving specific capacity and ionic conductivity. Several researches have been done including carbon coating and doping with Mg,

Nd, Gd, and Al atoms. [4-15] These treatment is believed could improve the structure of LTO and enhance both specific capacity and ionic conductivity.

Lithium titanium oxide (LTO) is a material with cubic spinel structure and some spaces in the lattice which give possibility of lithium ion to intercalate as shown in figure 1. This material is also well known as zero strain material, as the change of lattice parameter in its structure is almost negligible when lithium ions intercalate. The unit cell of $Li_4Ti_5O_{12}$ is built of 8 unit formula of $Li_{(8a)}[Li_{1/3}Ti_{5/3}]_{(16c)}O_{4(32e)}$ and the 16d sites is occupied by 1/6 Li atoms and 5/6 Ti atoms. [9] The only change is migration of lithium ion from 8a tetrahedral lattice into 16c octahedral lattice which forms $Li_7Ti_5O_{12}$. (Andreas Laumann, 2010) However, LTO has lower theoretical capacity, 175 mAh/g, and higher redox potential, 1.55 V Li^+/Li, compared to graphite anodes (372 mAh/g and ~100 mV Li^+/Li). This have been an advantage of the material compared to others. (Qianyu, 2015)

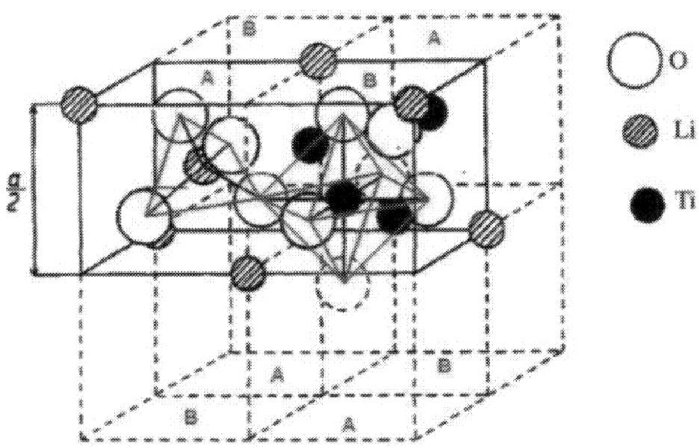

Figure 1 The cubic spinel structure of lithium titanium oxide [2]

However, chicken eggshells is known as the most unutilized kitchen waste which contain rich amount of $CaCO_3$. The $CaCO_3$ is believed could be a good Ca source for the dopant of LTO. The outer membrane of the eggshells can be taken and utilized for a natural $CaCO_3$ which is also believed having finer structure than the commercial one. [16]

2. EXPERIMENT

$Li_4Ti_5O_{12}$and $Li_{3.9}Ca_{0.1}Ti_5O_{12}$ were both synthesized by a simple solid state reaction method using stoichiometric calculation which the formula is shown in equation 1. $LiOH.H_2O$ and TiO_2 were chosen as Li and Ti sources, respectively. In this experiment, eggshell powder in the form of $CaCO_3$ was utilized as Ca source. Waste eggshell was cleaned, ground, and dried at 100^{0}C for 12h to get nature-based $CaCO_3$. Raw materials were mixed and pounded in a bowl for 8 h at room temperature. Powders were calcined at 350^{0}C for 1 h, and 700^{0}C for 2 h, before it was sintered at 800^{0}C for 4 h, with temperature rate of 5^{0}C per minute. Both processes are done in a muffle furnace KSL-1700X.

$$3.9LiOH.H_2O + TiO_2 + 0.1CaCO_3 \rightarrow Li_{3.9}Ca_{0.1}Ti_5O_{12} \quad (1)$$

X-ray Diffractometer(XRD) was used in characterizing powders crystalline phase and structure of waste eggshells, $Li_4Ti_5O_{12}$ and $Li_{3.9}Ca_{0.1}Ti_5O_{12}$ powderswith CuKα radiation. The XRD identified the phase with scan rate of 8.00 deg/min and scan range of 10.00 – 90.00 deg. Besides, SEM-EDX was used in characterizing morphology of $Li_{3.9}Ca_{0.1}Ti_5O_{12}$ and detecting the particle distribution of Ca in its structure.

Battery samples were prepared by mixing active materials with Acetylene Black (AB) as conductive material, and Polyvinylidene Flouride (PVDF) as binder in 80 : 10 : 10 weight ration in DMAC solution. The mixtures were spread onto Cu-foil in Automatic Thick Film Coating and dried at 81^{0}C in a drying box for 3 hours.Battery cells were prepared by assembling coin-type samples (d = 16mm) as working electrodes and coin-type Li-metal (d = 16mm) as reference electrode, while Celgard polymers were used as separator in a glove box which both moisture and oxygen content were at least condition and Ar gas was used as inert gas.

Electrochemical Impedance Spectroscopy was used in characterizing the electrochemical impedance of the anodes. The output of characterization was used to calculate Warburg coefficient and Diffusion coefficient of the anodes, which could determine the ionic conductivity.

Warburg coefficient was calculated with the following equation,

$$Z_{re} = R_e + R_{ct} + \sigma_w \omega^{-1/2} \tag{2}$$

Where, Z_{re} is total impedance, R_e is electrolyte resistance, R_{ct} is charge transfer resistance, σ_w is Warburg coefficient, and ω is angular frequency.

Diffusion coefficient which determine the ionic conductivity of anodes was calculated with the following equation,

$$D = \frac{R^2T^2}{2n^4A^2F^2\sigma^2C^2} \tag{3}$$

where, R is gass constant, 8.314 J/mol.K, T is absolute temperature, 298.15 K, F is Faraday constant, 96500 C/mol, A is area of samples, 0,000201 m^2, C is lithium ion concentration, 4370 mol/m^3, n is the number of electron transfered during lithium ion intercalation, 3 for $Li_4Ti_5O_{12}$, and σ Warburg coefficient (Ω.m^2.s$^{-1/2}$).

3. RESULT AND DISCUSSION

The XRD patterns of waste eggshell is shown by figure 2 in the form of $CaCO_3$ in accordance with the standard diffraction pattern of $CaCO_3$which [104] peak appeared as primary peak. [104] peak revealed that waste eggshell contains high amount of $CaCO_3$, indicates that waste eggshells as nature-based $CaCO_3$ can be used as source of Ca.

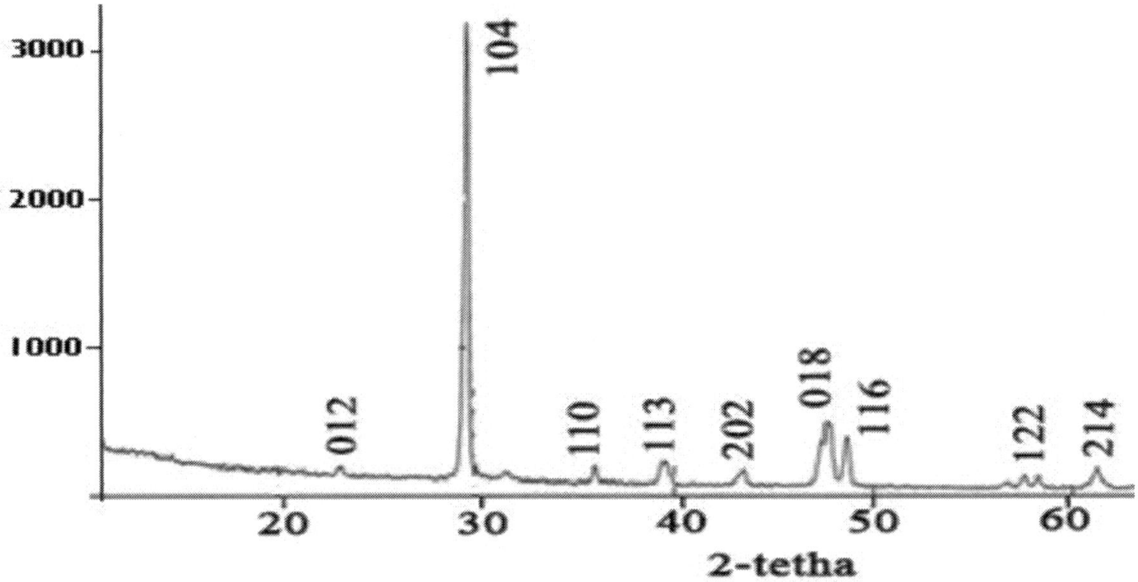

Figure 2 The XRD pattern of waste eggshell in the form of $CaCO_3$

The XRD patterns of $Li_4Ti_5O_{12}$ and $Li_{3.9}Ca_{0.1}Ti_5O_{12}$ is shown by figure 3. In accordance with the standard diffraction pattern of LTO, both samples can be indexed to a cubic spinel structure with the space group of Fd-3m. The XRD peak also revealed that $CaCO_3$ phase was not detected in $Li_{3.9}Ca_{0.1}Ti_5O_{12}$, indicates that Ca^{2+} ions have successfully entered into the 8a lattice structure of LTO and substitutes the amount Li ion which can be noted as $Li_{0.13}Ca_{0.03(8a)}[Li_{1/3}Ti_{5/3}]_{(16c)}O_{4(32e)}$. The morphology of Ca distribution was also detected in SEM-EDX images which is shown in figure 4. This evidence has proven that nature-based $CaCO_3$ is efficient to be used as a source of Ca as the dopant of LTO.

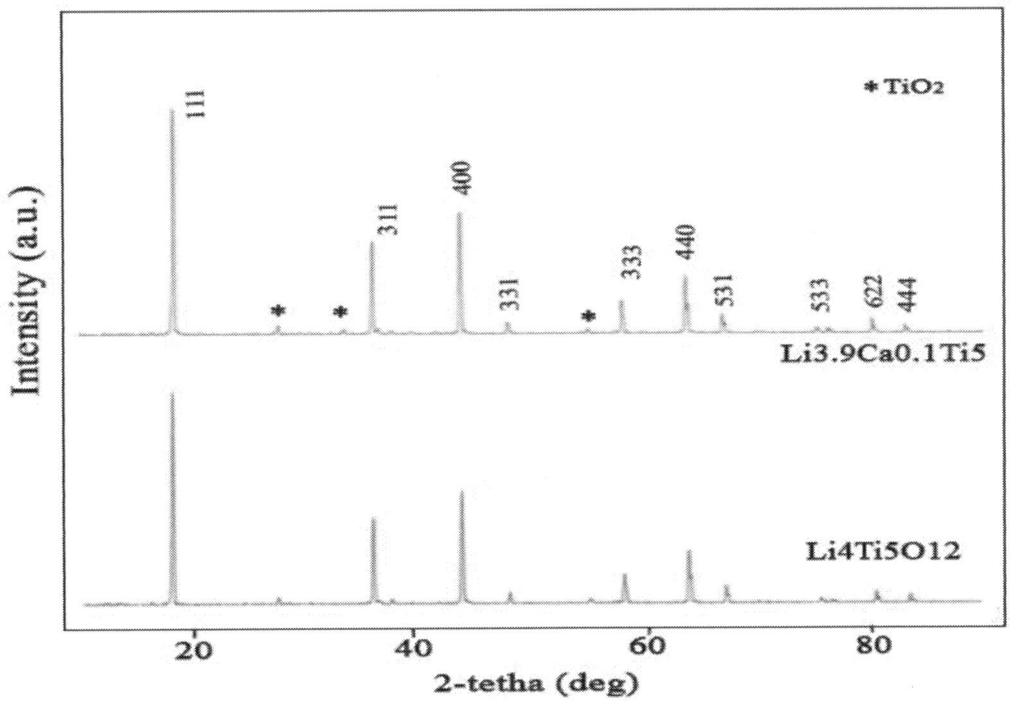

Figure 3 XRD pattern of $Li_4Ti_5O_{12}$ and $Li_{3.9}Ca_{0.1}Ti_5O_{12}$ sample

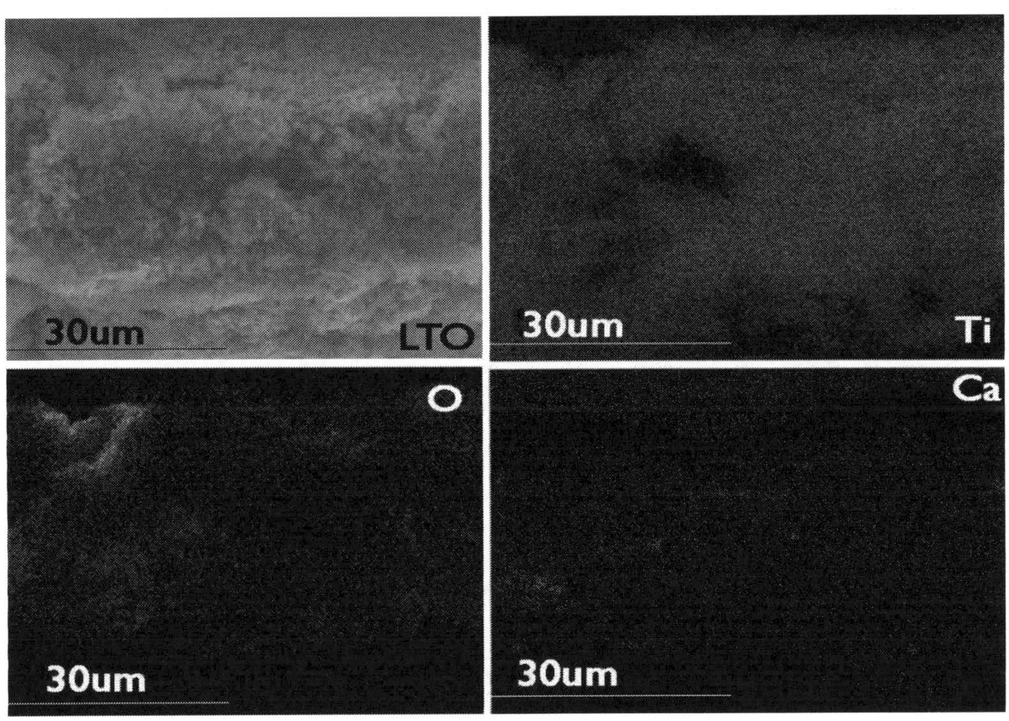

Figure 4 EDX mappings of Calcium (atomic % = 0.2) in the structure of $Li_{3.9}Ca_{0.1}Ti_5O_{12}$

The lattice parameter and composition of both samples are shown in table 1 which can be seen that there is no significant difference between $Li_4Ti_5O_{12}$ and $Li_{3.9}Ca_{0.1}Ti_5O_{12}$. This result indicates that small Ca doping will not effect on the lattice parameter of LTO.

Table 1. Lattice parameter and composition of $Li_4Ti_5O_{12}$ and $Li_{3.9}Ca_{0.1}Ti_5O_{12}$ sample

Sample	a (Å)	%LTO	%TiO_2 rutil
$Li_4Ti_5O_{12}$	8.3596(8)	97.2(16)	2.8(4)
$Li_{3.9}Ca_{0.1}Ti_5O_{12}$	8.3573(9)	95.1(16)	4.9(6)

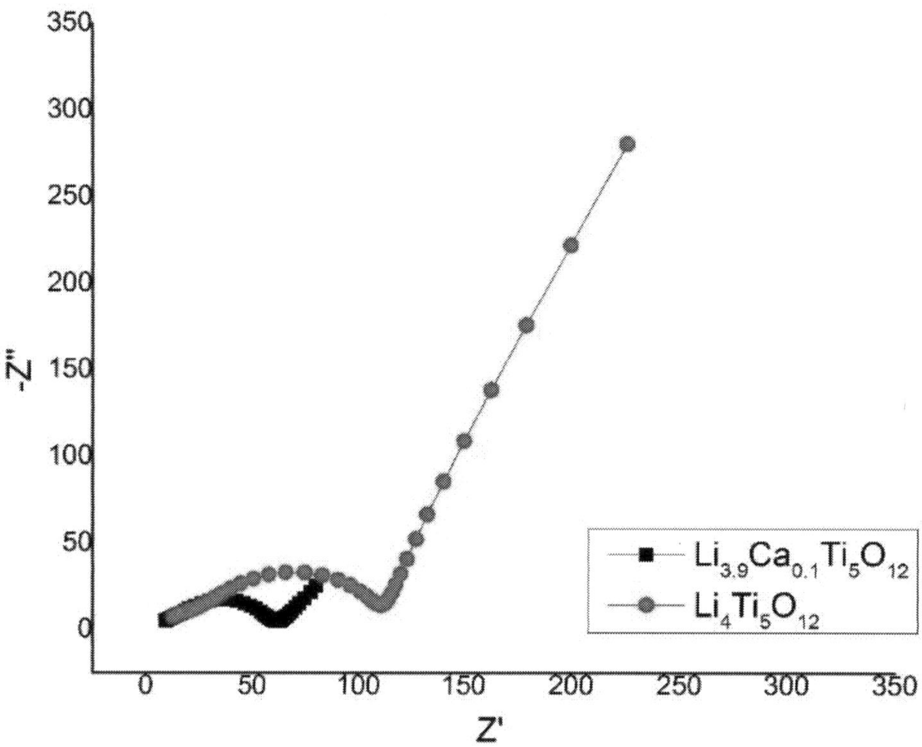

Figure 5 Electrochemical impedance of $Li_4Ti_5O_{12}$ and $Li_{3.9}Ca_{0.1}Ti_5O_{12}$

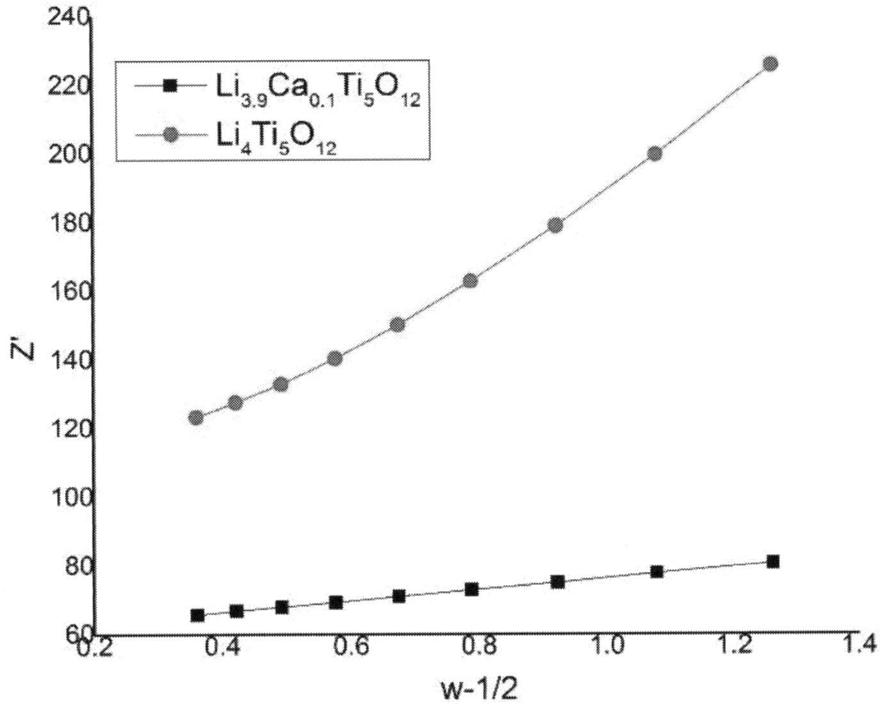

Figure 6 The plot of $\omega^{-1/2}$ and Z'

Figure 5 shows the electrochemical impedance of $Li_4Ti_5O_{12}$ and $Li_{3.9}Ca_{0.1}Ti_5O_{12}$. It can be seen that $Li_{3.9}Ca_{0.1}Ti_5O_{12}$ has lower electrochemical impedance, indicates that it will have higher ionic conductivity. Figure 6 shows the plot of $\omega^{-1/2}$ and Z'. The slope of the graph is called as Warburg coefficient. The diffusion coefficient was then calculated after knowing the Warburg coefficient.

Table 2. The Warburg coefficient and the diffusion coefficient of $Li_4Ti_5O_{12}$ and $Li_{3.9}Ca_{0.1}Ti_5O_{12}$

Sample	σ ($\Omega.m^2.s^{-1/2}$)	D (cm^2/s)
$Li_4Ti_5O_{12}$	113.300	4.40995 x 10^{-16}
$Li_{3.9}Ca_{0.1}Ti_5O_{12}$	16.529	2.07205 x 10^{-14}

From table 2, it can be seen that $Li_{3.9}Ca_{0.1}Ti_5O_{12}$ has higher diffusion coefficient than $Li_4Ti_5O_{12}$ sample, with the value of 2.07205 x 10^{-14} cm^2/s and 4.40995 x 10^{-16} cm^2/s respectively. It indicates that $Li_{3.9}Ca_{0.1}Ti_5O_{12}$ anode material has an optimum structure to have high ionic conductivity. The higher ionic conductivity indicates the easier lithium ion to intercalate.

A charge-discharge characterization also had been done to determine the electrochemical performance of both samples, cycle rates at 0.2C. Figure 7 shows that the $Li_{3.9}Ca_{0.1}Ti_5O_{12}$ anode material has higher charge-discharge capacity, with the value of 158.3 mAh/g and 150 mAh/g respectively, than the $Li_4Ti_5O_{12}$ anode material, with the value of 142.7 mAh/g and 127 mAh/g respectively. This has proven that the $Li_{3.9}Ca_{0.1}Ti_5O_{12}$ anode material has also more optimum structure to receive charges than the $Li_4Ti_5O_{12}$ anode material.

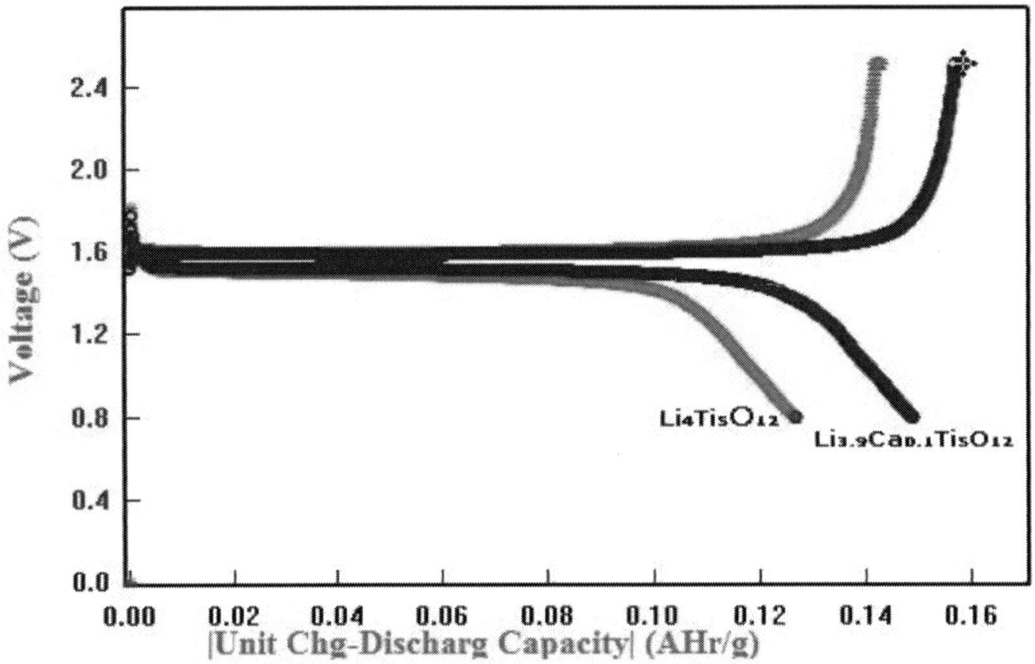

Figure 7 Charge-discharge (at 0.2C) characterization shows the $Li_{3.9}Ca_{0.1}Ti_5O_{12}$ anode material has higher charge-discharge capacity than the $Li_4Ti_5O_{12}$ anode material

In addition to that, figure 8 shows cycle performance of both anode material with cycle rate of 2C for 60 cycles, which LTO represents the $Li_4Ti_5O_{12}$ anode material, while LCaTO represents the $Li_{3.9}Ca_{0.1}Ti_5O_{12}$ anode material. An outstanding performance has been exhibited by the $Li_{3.9}Ca_{0.1}Ti_5O_{12}$ anode material which still has charge-discharge capacity of 104.1 mAh/g and 49.3mAh/g respectively after 61 cycle. These values are still higher than the $Li_4Ti_5O_{12}$ anode material charge-discharge capacity drop, with the value of 76.2 mAh/g and 30.3 mAh/g. The Coloumbic efficiency stabilizes at 100% after 61 cycle, which is due to the excellency of Li^+ intercalation during charge-discharge process. This stability of cycle can be donated to the enhancement of ionic conductivity of the $Li_{3.9}Ca_{0.1}Ti_5O_{12}$ anode material.

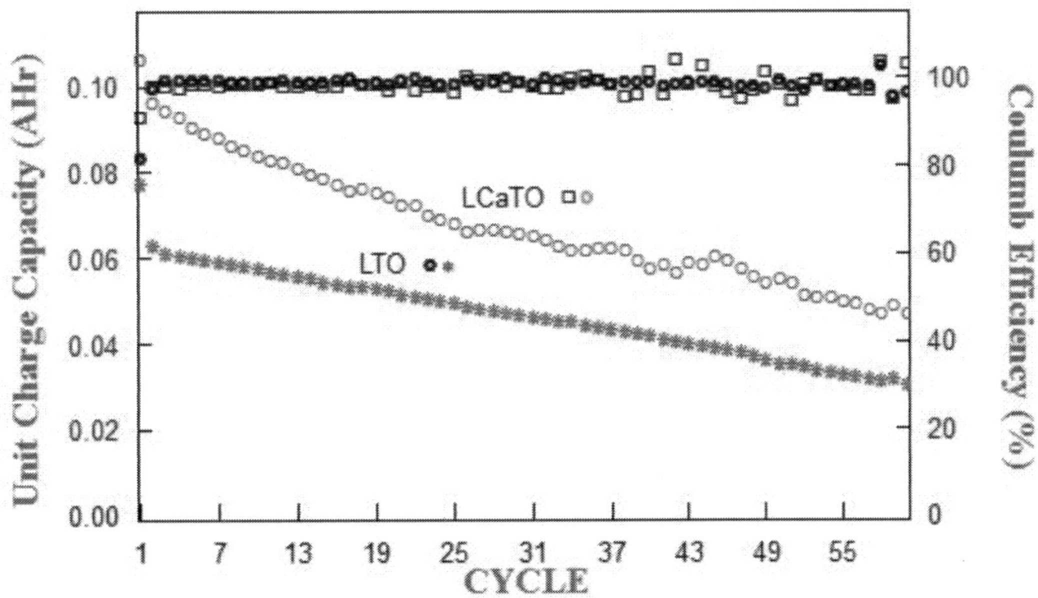

Figure 8 Cycle performance of the $Li_{3.9}Ca_{0.1}Ti_5O_{12}$and the $Li_4Ti_5O_{12}$ anode material, at 2C.

4. CONCLUSION

$Li_{3.9}Ca_{0.1}Ti_5O_{12}$ had been synthesis parallel with $Li_4Ti_5O_{12}$ using simple solid state method. Waste chicken eggshell was chosen to be utilized as Ca source. XRD pattern shows that waste chicken eggshells is in the form of $CaCO_3$ and can be utilized as Ca source for the dopant of LTO. XRD and SEM-EDX characterization on $Li_4Ti_5O_{12}$ and $Li_{3.9}Ca_{0.1}Ti_5O_{12}$ powder shows that both sample is match with the standard peak of LTO and Ca had been successfully entered to the structure of LTO and substitutes the amount Li ion which can be noted as $Li_{0.13}Ca_{0.03(8a)}[Li_{1/3}Ti_{5/3}]_{(16c)}O_{4(32e)}$in $Li_{3.9}Ca_{0.1}Ti_5O_{12}$ sample. EIS characterization shows that $Li_{3.9}Ca_{0.1}Ti_5O_{12}$ has higher diffusion coefficient than $Li_4Ti_5O_{12}$ sample, with the value of 2.07205 x 10^{-14} cm^2/s and 4.40995 x 10^{-16} cm^2/s respectively. The charge-discharge test exhibits that $Li_{3.9}Ca_{0.1}Ti_5O_{12}$sample (158.3 mAh/g and 150 mAh/g) has higher charge-discharge capacity and stability than $Li_4Ti_5O_{12}$ sample (142.7 mAh/g and 127 mAh/g). According to the cycle performance test, the $Li_{3.9}Ca_{0.1}Ti_5O_{12}$sample also has good stability with charge-discharge capacity of 104.1 mAh/g and 49.3mAh/g, still higher than $Li_4Ti_5O_{12}$ sample with charge-discharge capacity of 76.2 mAh/g and 30.3 mAh/g after 61 cycle, indicates that the $Li_{3.9}Ca_{0.1}Ti_5O_{12}$ anode material has high ionic conductivity and good for anode material candidate of lithium-ion battery.

5. ACKNOWLEDGEMENT

This research is fully funded by Indonesian Institute of Science-Research Center for Physics (P2F LIPI).

IOP Conf. Series: Materials Science and Engineering **316** (2018) 012048 doi:10.1088/1757-899X/316/1/012048

6. REFFERENCES

1. Oswal, Mehul. *A comparative Study of Li-ion Batteries.* University of Southern California. 2010.
2. Subhan, Achmad. *Fabrikasi dan Karakterisasi $Li_4Ti_5O_{12}$ untuk Bahan Anoda Baterai Lithium Keramik.* FTUI. Thesis. 2011
3. Laumann, Andreas. *Nouvel Routes to Li4Ti5O12 : Characterization and Phase Relation.* Ludwig-Maximilians Universitat Muenchen. 2010
4. Q. Zhang, M. G. Verde, J. K. Seo, X. Li, and Y. S. Meng, "Structural and Electrochemical Properties of Gd-Doped Li4Ti5O12 as Anode Material with Improved Rate Capability for Lithium-Ion Battery," Journal of Power Source 280, 355-362 (2015).
5. X. Li, J. Xu, P. Huang, W. Yang, Z. Wang, M. Wang, Y. Huang, Y. Zhou, M. Qu, Z. Yu, and Y. Lin, "Insitu Carbon Coating to Enhance Rate Capability of Li4Ti5O12 Anode Material and Suppress The Electrolyte Reduction Decomposition on Electrode," Electrochimica Acta 190, 69-75 (2016).
6. H. P. Liu, G. Wen, S. Bi, C. Wang, J. Hao, and P. Gao, "High Rate Cyclic Performance of Nanosized $Li_4Ti_5O_{12}$/graphene composite for Lithium Ion Battery," Electrochimica Acta 192, 38-44 (2016).
7. P. Wang, P. Li, T. F. Yi, X. Lin, Y. R. Zhu, L. Shao, M. Shui, N. Long, and J. Shu, "Improved Lithium Storage Performance Lithium Sodium Titanate Anode by Titanium Site Substitution with Aluminum," Journal of Power Source 293, 33-41 (2015).
8. F. Li, M. Zeng, J. Li, and H. Xu, "Preparation and Electrochemical Performance of Mg-doped $Li_4Ti_5O_{12}$ Nanoparticle as Anode Materials for Lithium-ion Battery," Int. J. Electrochem. Sci. 10, 10445-10453 (2015).
9. W. Wang, W. Xiong, and Z. Wang, "A Nanoparticle of Mg-doped $Li_4Ti_5O_{12}$ for High Rate Lithium-ion Battery," Electrochimica Acta 114, 198-204 (2013).
10. K. Ding, J. Zhao, J. Zhou, Y. Zhao, Y. Chen, Y. Zhang, B. Wei, L. Wang, and X. Hei, "Preparation and Characterization of Dy-doped Lithium Titanate ($Li_4Ti_5O_{12}$)," Int. J. Electrochem. Sci. 11, 446 – 458 (2016).
11. Q. Zhang, and X. Li, "High Rate Capability of Nd-Doped Li4Ti5O12 as Effective Anode Material for Lithium-ion Battery," Int. J. Electrochem. Sci. 8, 7816 – 7824 (2013).
12. Y. B. He, M. Liu, Z. D. Huang, B. Zhang, Y. Yu, B. Li, F. Kang, and J. K. Kim, "Effect of Solid Electrolyte Interface (SEI) Film on Cyclic Performance of Li4Ti5O12 Anodes for Li-ion Battery," Journal of Power Source 239, 269-276 (2013).
13. P. Wang, P. Li, T. F. Yi, X. Lin, H. Yu, Y. R. Zhu, S. Qian, M. Shui, and J. Shu, "Enhanced Lithium Storage Capability of Sodium Lithium Titanate via Lithium Site Doping," Journal of Power Source 297, 383-294 (2015).
14. Y. Li, Z. Wang, D. Zhao, and L. Zhang, "Gd-doped Single Crystalline Li4Ti5O12/TiO2 Nanosheets Composite as Superior Anode Material in Lithium Ion Battery," Electrochimica Acta 182, 368-375 (2015).
15. Q. Zhang, C. Zhang, B. Li, S. Kang, X. Li, and Y. Wang, "Preparation and Electrochemical Properties of Ca-doped $Li_4Ti_5O_{12}$ as Anode Material in Lithium-ion Battery", Electrochimica Acta 98, 146-152 (2013).
16. *Calcium Carbonate Content of Egg Shell.* Oxford : New York.

The tensile strength of mechanical joint prototype of lontar fiber composite

Jefri Bale[1*], Dominggus G.H. Adoe[2], Kristomus Boimau[3], Thomas Sakera[4]

[1,2,3,4]Department of Mechanical Engineering, Universitas Nusa Cendana, 85001 Kupang, NTT, Indonesia

[*]Email: jefri_bale@staf.undana.ac.id

Abstract. In the present study, an experimental activity has been programmed to investigate the effect of joint prototype configuration on tensile strength of lontar (Borassus Flabellifer) fiber composite. To do so, a series of tests were conducted to establish the tensile strength of different joint prototype configuration specimen of lontar fiber composite. In addition, post observation of macroscope was used to map damage behavior. The analysis of lontar fiber composite is a challenge since the material has limited information than others natural fiber composites materials. The results shown that, under static tensile loading, the tensile strength of 13 MPa produced by single lap joint of lontar fiber composite is highest compare to 11 MPa of tensile strength generated by step lap joint and double lap joint where produced the lowest tensile strength of 6 MPa. It is concluded that the differences of tensile strength depend on the geometric dimensions of the cross-sectional area and stress distribution of each joint prototype configuration.

1. Introduction

In a climate with increasing focus and need for environmentally sustainable material for automotive and aeronautics industry, also the economic impact of natural material and its benefits are recognized as a key driver for growth in the agricultural industry, especially for developing countries [1]. Researcher will be expected to wonder wheather mechanical company can do more to support automotive and aeronautic industry or other transportation industry in going green while also opening new market and sales channels for themselves. In this regard, there are six main issues that are driving product development of composite material in the transportation industry (Figure. 1). The first three issues are the development and use of lighter weight material which correlates with the issue of fuel efficiency, reduce CO2 emissions and save non renewable resources. Three other issues are driving the automotive and aeronautic industry to seek new materials are cost effectiveness, biodegradability and the capability of material.

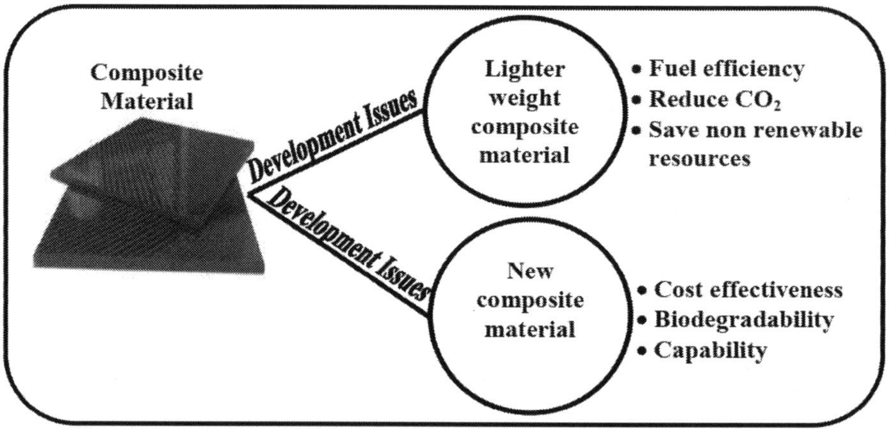

Figure 1. Driving product development of composite material

From the point of view of materials engineering, natural fiber is considered one of the new era environmentally friendly materials which have good properties to replace the conventional synthetic fiber as a reinforcement of composite materials in in all industries in coming centuries [2]. Natural fibers are used in various applications as explained by previous work [3]. Table 1 shows the comparison between natural and synthetic fibers.

Table 1.Comparison between Natural fibers and Synthetic fibers [4]

Aspect	Property	Natural Fibers	Synthetic Fibers
Technical	Mechanical Properties	Moderate	High
	Moisture Sensitivity	High	Low
	Thermal Sensitivity	High	Low
Environmental	Resource	Infinite	Limited
	Production	Low	High
	Recyclability	Good	Moderate

Apart from their biodegradability, natural fiber are becoming increasingly popular due to their abundance in tropical country, low cost, and recyclable nature in the field of composites for automotive and aeronautic application. The benefits of using natural fiber composites increase result from the fact that they are made from a renewable and sustainable plant fiber source, they release no net carbon dioxide, are 40% lighter than fiberglass and have a better energy management characteristics than glass fiber in their respective composite structures. For example, it takes 3.1 MJ of energy to produce 1 kilogram of kenaf, whereas it takes almost four times the same energy (~12 MJ) to produce 1 kilogram of glass fiber [5].

Previous works [6-8] in the field of lontar fiber composite investigated on the characteristics of the physical properties of lontar fiber, lontar fiber composite mechanical strength and influence of chemical treatment. Generally, their results shown that the tensile strength of lontar fiber composite with NaOH treatment had greater tensile strength than obtained from non treatment specimens. Their works also concluded that 30 % of lontar fiber volume fraction resulting optimal tensile and bending strength compared with others volume fraction.

The presence of geometrical discontinuities like cutouts and holes are an important machining operations to facilitate the assembly of composite component such as joining of riveted and bolted joints because composites cannot be welded directly like metal materials to ascertain the structural integrity or connections of complex composite products. Mechanical connections remain the key means for the transfer of loads between structural elements made of composite materials [9]. Several

IOP Conf. Series: Materials Science and Engineering **316** (2018) 012049 doi:10.1088/1757-899X/316/1/012049

important factors such as structural and architectural requirements, technological process, nature of materials, application speed, environment advantages and disadvantages, as well as the total cost, need to be considered when the type of connection is chosen [10]. An experimental study of bolt-hole clearance effects in double-lap, multi bolt composite joints indicated that variable clearances in multi-bolt joints significantly influence the load distribution. At the higher loads the distribution tends to even out, but this process may be interrupted by failure [11].

A comparative study on different drill points geometries and feed rate for composite laminates drillings concluded that low feed rate seemed appropriate for laminate drilling and the most adequate tool for higher feed rates is the twist drill with a 120° point angle for minimal delamination [12]. Early results about the effect of hole fabrication and hole diameter on the tensile strength of lontar fiber composite have been generated as preliminary study of the author [13]. The results shown that tensile strength of moulded hole lontar fiber composite specimens is higher compared to the drilled hole specimens in all diameter and width (d/w) ratio. It has been noted that tensile strength of drilled hole specimen has been mostly affected by the presence of initial defect due to drilling operations.

On the other hand, the use of lontar fiber composite in structure design of mechanical joint is a challenging task since very limited work has been done on the application of lontar fiber composite for mechanical fastened joint design. Against this background, the research plan has been undertaken, with an objective to investigate the variation of joint prototype and its effect on mechanical properties of lontar fiber. The present work thus aims to understand the tensile behavior of lontar fiber composites on different type of mechanical joint prototype by experimental method. This study is part of a project at Universitas Nusa Cendana as a center study of archipelagic dryland field to invent a variety of potential natural fibers in East Nusa Tenggara - Indonesia as an reinforcement for biodegradable plastic matrix composite materials for transportation industry.

2. Methodology

The composite specimen that used in this study was reinforced by lontar fiber with 5 cm of fiber length. The composite produced by pressure molding process that contains 0.32 of nominal lontar fiber volume as the reinforcement and polyester was employed as the matrix throughout 4 mm of specimen thickness. The orientation of fiber random against each other. In order to investigate mechanical joint prototype tensile strength, three types of joint prototype were studied, i.e single lap joint, double lap joint and step lap joint as seen in Figure. 2. Test was carried out with a servo-hydraulic testing machine with a capacity of 100 kN. The machine was equipped with a standard load cell and mechanical grips. Specimens were aligned and mounted first in the lower and afterwards in the upper grips of the test station. After mounting the specimens, any loading due to the gripping was minimized through controller panel. A unit of dial indicator was used to record all the test results such as load and displacement. Then, the tests were considered to begin. Static tensile tests were performed under a constant cross-head speed of 1 mm/min.

IOP Conf. Series: Materials Science and Engineering **316** (2018) 012049 doi:10.1088/1757-899X/316/1/012049

Step Lap Joint

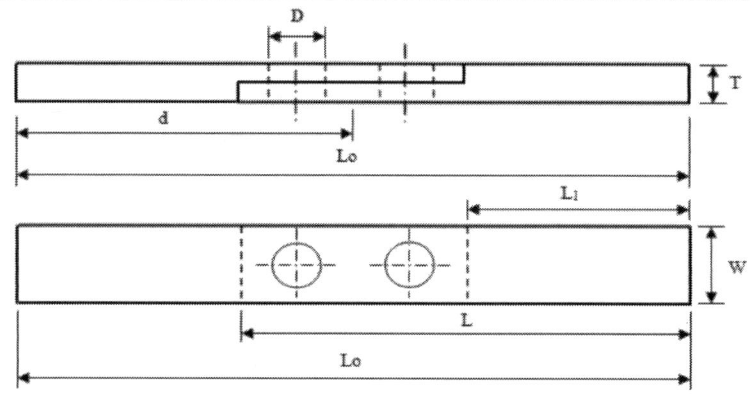

Double Lap Joint

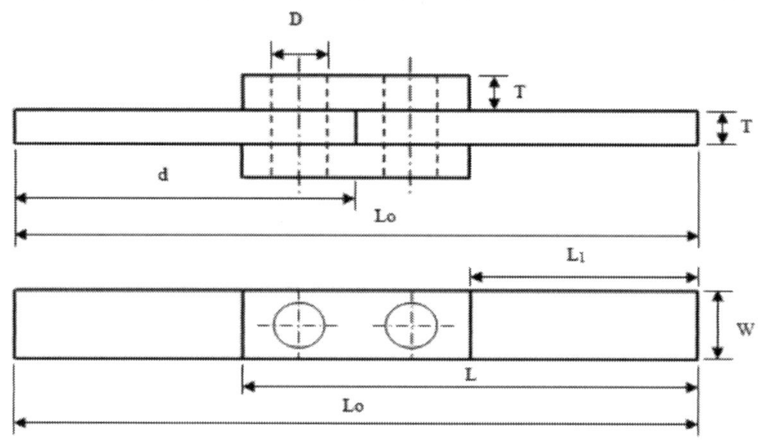

Single Lap Joint

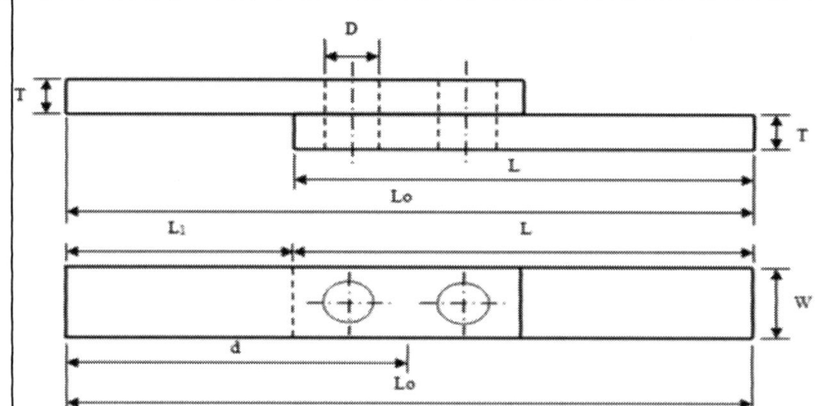

Lo : 250 mm L : 150 mm L_1 : 100 mm T : 4 mm
D : 8 mm d : 125 mm W : 50 mm

Figure 2. Specimen Geometry

IOP Conf. Series: Materials Science and Engineering **316** (2018) 012049 doi:10.1088/1757-899X/316/1/012049

The test procedure was based on standard ASTM D5961 which consisting of five specimens of each test for respectively each type of joint configuration. Figure. 3 shows the test set up and all type of specimens.

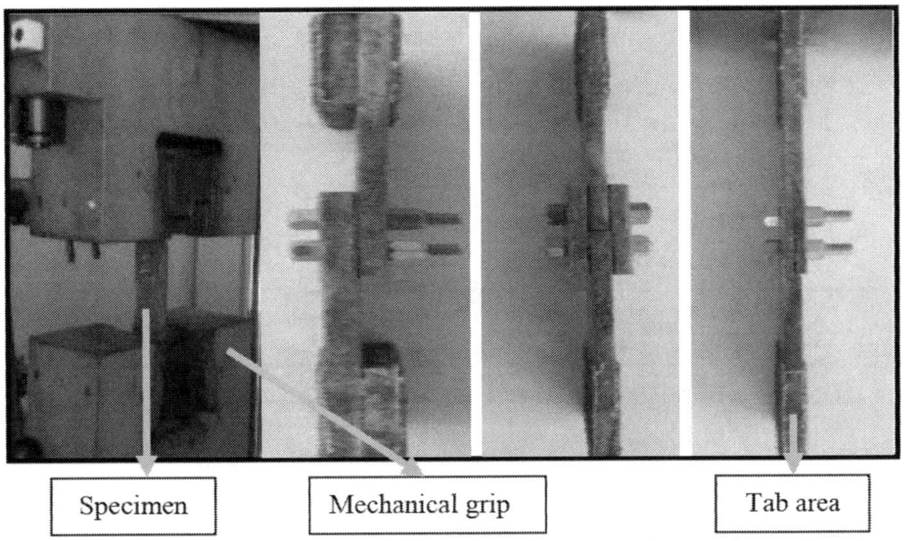

Figure 3. Test Set up and lontar fiber composite specimens

3. Results and Discussion

The tensile strength of mechanical joint prototype of lontar fiber composite was analyzed. Figure. 4 shows the load versus global displacement curves in static condition for the test specimens.

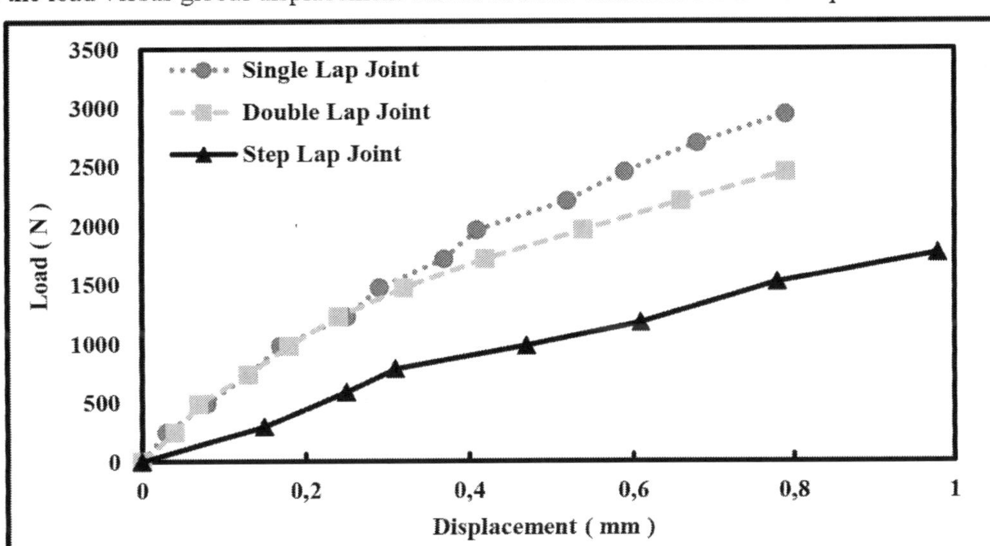

Figure 4. Load vs displacement of joint prototype of lontar fiber composite

Each specimen exhibits a typical of brittle behavior with suddenly failure where the highest ultimate tensile load of 2900 N resulting by single lap joint specimen which is 50 % greater than load value of step lap joint which is about 1500 N. Figure. 5 shows the typical stress and strain behavior of joint prototype of long random lontar fiber composite under tensile load.

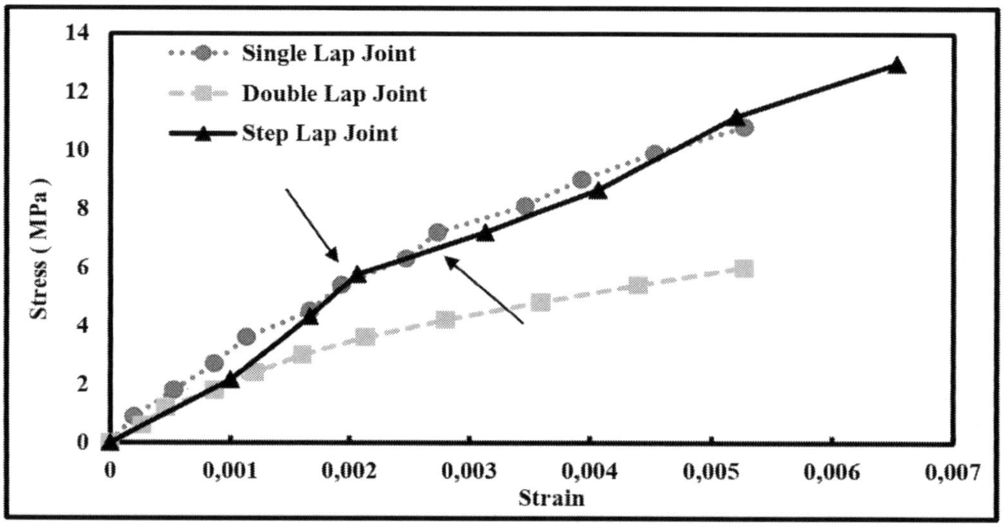

Figure 5. Stress vs strain of joint prototype of lontar fiber composite

The curve gives the relationship between stress and strain where stress is calculating from given load and strain obtained from global displacement. The stress - strain curve behavior shows the brittle behavior of composite material with the sudden failure of composite material under tensile loading in static condition. It was found that the highest stress of 13 MPa produced by step lap joint specimen and followed by single lap joint specimen where generated 11 MPa of tensile stress. Furthermore, the lowest tensile stress was produced by double lap joint specimen. At the beginning of the loading, a non linear behavior corresponding to the realignment of fibers and mechanical grips movement is observed. After this first phase, a linear behavior is obtained. From Figure. 5, the different slope can be observed on some part of curves during the second phase (shown by black arrow), could be associated to the changing of damage mechanism from matrix cracking into debonding and fiber breakage. This kind of behavior also explained by Arnautov et.al, 2015 in their study about carbon/epoxy composites joint.

Locally, the variation of the cross-sectional area of joint prototype produces the stress calculation and resulting local stress level between areas with different joint configuration. Joint prototype of Step lap joint has smaller cross-sectional area compare to other joint prototype configuration. Therefore, step lap joint generates the highest stress where indicate in Figure.5. This result is certainly different from Figure.4 due to the presence of cross-sectional area which is inversely proportional to the stress and load on strength calculation. The local strain of the specimen under the tensile load also entirely depends upon the thickness of the cross sectional area. The step lap joint specimen tends to show more strain compared to single and double lap joint specimen. It is quite evident that strain increases with decrease in thickness of cross-sectional area of specimen. In other words, the joint geometry affects the strength of mechanical joint prototype of lontar fiber composite.

The highest UTS of joint prototype of lontar fiber composite specimen is around 13 MPa resulting highest modulus measurement of 2.2 GPa as seen in Figure. 6.

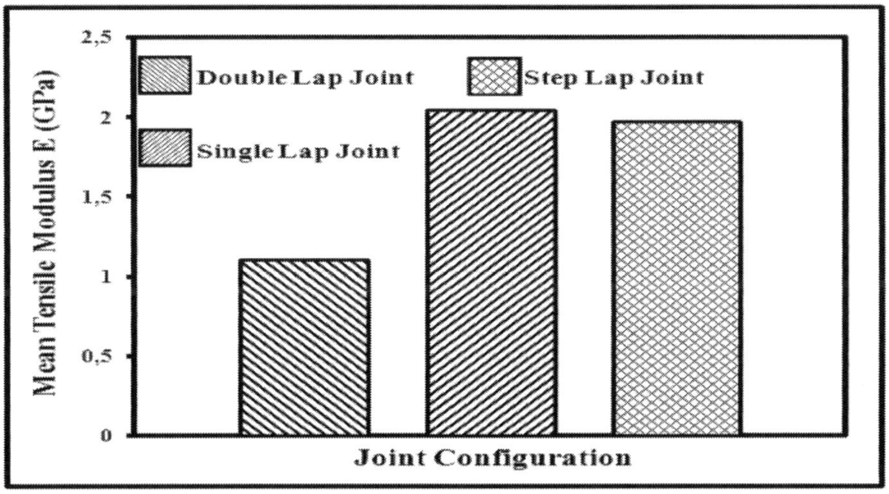

Figure 6. Tensile modulus of joint prototype of lontar fiber composite

The stiffness of single lap joint and step lap joint was found to be around 2 times higher compared to double lap joint. The likely reason for this is that the configuration of double lap joint contains cut off in the middle part then generates less effective stiffness of the bridging mechanism. Lower stiffer bridging generates lower ability to sustain the given load. Consequently, double lap joint produces lower tensile stiffness modulus compared to step and single lap joint. The macroscopic images of final failure of joint prototype can be seen in Figure. 7.

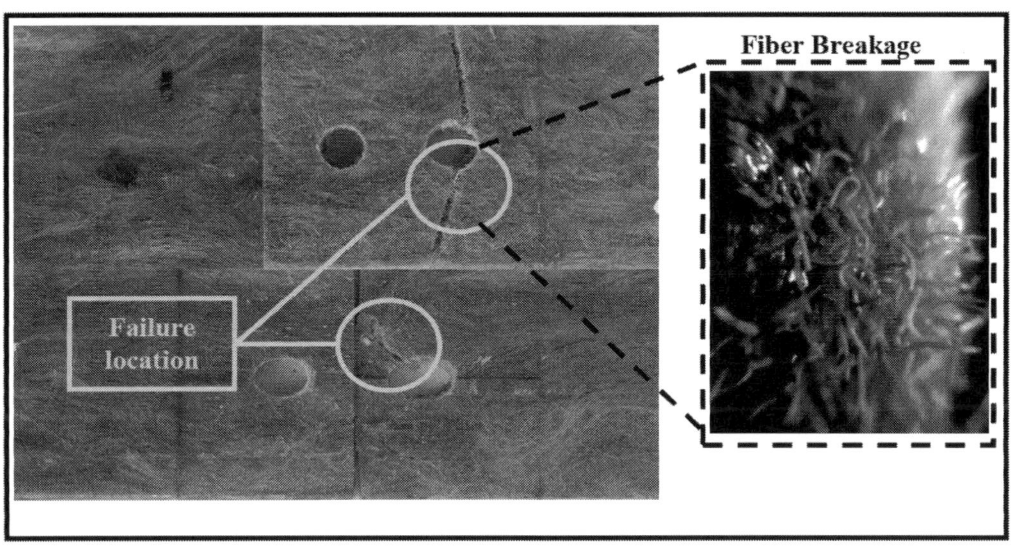

Figure 7. Final failure of joint prototype of lontar fiber composite

The damage of joint prototype of lontar fiber composite in tension is fairly transverse to the loading axis. The presence of the hole increases highest stress concentration that leads to strength degradation and initiate damage during the test. It was found that damage starts in form of cracking emanating due to matrix cracking around the hole and propagated along the width. The damage area propagated by the debonding of the fiber and the matrix at the interface located around the hole. After initial damage, the specimens continued to sustain the load under increasing displacements. When the critical point was reached, damage area reaches to the end of width side of the specimen and specimen ultimately failed mainly due to the fiber breakage. It means that, at final stage of specimen failure, fibers no longer able to carry the increasing of given load and damage has occurred completely.

4. Conclusion

This study presents experimental results of different mechanical joint prototype configuration of long random lontar fiber composite under static tensile loading. The results show that the differences of joint prototype configuration of lontar fiber composite have an effect on its tensile strength. It was found that the highest stress of 13 MPa produced by step lap joint specimen and followed by single lap joint specimen where generated 11 MPa of tensile stress. Furthermore, the lowest tensile stress of 6 MPa was produced by double lap joint specimen. The differences in tensile strength and strain possibly originated by different thickness of cross sectional area at of each joint prototype. The double lap joint experiences the lowest tensile modulus of 1.2 GPa due to less effective stiffness of the bridging mechanism on tensile strength of lontar fiber composite. The tensile damage of lontar fiber specimen with different joint prototype configurations occurs in the form of matrix cracking and ultimately failed mainly due to the fiber breakage.

5. Acknowledgment

The authors gratefully acknowledge the support of the Ministry of Research, Technology and Higher Education of Indonesia (KEMENRISTEKDIKTI) as a Grant of Fundamental Research 2017.

References

[1] Anandjiwala R D 2006 The Role of Research and Development in Global Competitiveness of Natural Fiber Products (Natural Fiber Vision 2020) p 1

[2] Ramesh M, Palanikumar K and Hemachandra Reddy K 2017 Plant Fibre Based Bio-Composites: Sustainable and Renewable Green Materials. *Renewable and Sustainable Energy Reviews* **79** 558

[3] Yousif B F, Shalwan A, Chin C W and Ming K C 2012 Flexural Properties of Treated and Untreated Kenaf/Epoxy Composites. *Materials and Design* **40** 378

[4] Saravana Bavan D and Mohan Kumar D 2010 Potential Use of Natural Fiber Composite Materials in India. *Journal of Reinforced Plastics and Composites* **29** 3600

[5] Khan M A, Hinrichsen G, Drzal L T 2001 Influence of Novel Coupling Agents on Mechanical Properties of Jute Reinforced Polypropylene Composite. *Compos Interf* **20** 1711

[6] Boimau K 2009 Characterization of Tension Behavior and Surface Topograph of Lontar Fiber Alkali Treatment. *Proceedings of the 8th Regional Conference on Annual Mechanical Engineering*, Semarang, Indonesia

[7] Boimau K and Limbong I 2010 The Effect of Volume Fraction on Tensile Strength of Composite Polyester Reinforced by Lontar Fiber. *Proceedings of Regional Conference on Environment Friendly in Sustainable*, Malang, Indonesia

[8] Boimau K , Bale J, Lagan M, Limbong I 2011 The Effect of Volume Fraction of Bending Strength of Composite Polyester Reinforced by Lontar Fiber. *Proceedings of the 11th Regional Conference on Annual Mechanical Engineering*, Yogyakarta, Indonesia

[9] Arnautov A, Nasibullins A, Gribniak V, Blumbergs I and Hauka M 2015 Experimental Characterization of The Properties of Double-Lap Needled and Hybrid Joints of Carbon/Epoxy Composites. *Materials* **8** 7578

[10] Sergiu P, Nicolae T, Paul C, and Dragos B 2012 Experimental Program Regarding The Behaviour of Composite Materials Joints. *Buletinul Institutului Politehnic Din Iaşi Secţia Construcţii. Arhitectură, Publicat de Universitatea Tehnică , Gheorghe Asachi din Iaşi Tomul LVIII (LXII)*, Fasc **4**

[11] Lawlor V P, McCarthy M A and Stanley W F 2005 An Experimental Study of Bolt-Hole Clearance Effects in Double-Lap, Multi-Bolt Composite Joints. *Comp. Struct* **71** 176

[12] Durão L M P, Gonçalves D J S, Tavares J M R S , de Albuquerque V H C , Aguiar Vieira A and Torres Marques A 2010 Drilling Tool Geometry Evaluation for Reinforced Composite Laminates, *Comp. Struct* **92** 1545
[13] Bale J, Boimau K and Tokoh R 2016 The Effect of Fabrication Process and Hole Diameter on Tensile Strength of Lontar Fiber Composite. *Sainstek Undana* **3** B-58

A preliminary study on the reduction of limonite ore by using rice husk as a reducing agent

Ahmad Maksum[a,b], Michael Kelvin E. Husein[a], Sulaksana Permana[a], Andi Rustandi[a], Johny Wahyuadi Soedarsono[a,*]

[a]Center of Minerals Processing and Corrosion Research, Department of Metallurgical and Materials Engineering,Universitas Indonesia, Depok 16424, Indonesia

[b]Department of Mechanical Engineering, Politeknik Negeri Jakarta, Depok 16425, Indonesia

Corresponding author : jwsono@metal.ui.ac.id

Abstract. The abundant of rice husk in Indonesia has encouraged researchers to study the feasibility of rice husk for substituting material that is more expensive or dangerous. In previous study, silica with a purity of 99.9% has been obtained from rice husk with calcinations process. Nevertheless, the gases resulting from the process were not used and left useless. Therefore, in this study, those gases derived from rice husk calcinations process were used as reducing agents during the ferronickel (Fe-Ni) production through a direct reduction process. The objective of this study was to investigate the effect of the amount of rice husk in the pellets on the increase of nickel content in the limonite reduction process. The limonite ore were crushed to the size of less than 150 mesh using disc-mill, and then were mixed with rice husk powder (10, 20, 30 in wt % mass) before being pelletized using bentonite as a binder. The resulted pellets were roasted at 500°C for 60 minutes and then quenched in water media. After drying process, the reduction process of 40g pellets was conducted at 1000°C for 90 minutes with 20g rice husk in furnace. The effects of additional rice husk on the direct reduction of limonite ore pellets were qualitatively analyzed by using X-ray powder diffraction (XRD) and quantitatively by Atomic absorption spectroscopy (AAS). Both analysis results showed that the reduction process followed the reaction scheme: $Fe_2O_3 \rightarrow Fe_3O_4 \rightarrow FeO$ and NiO phase was detected in the sample with 20% rice husk addition. The optimum concentration of Ni 1.23% was obtained for 20% rice husk addition.

Keywords: Direct reduction, Ferronickel, Limonite, Quenching, Rice husk.

1. Introduction

Nickel, as one of the commercially important metals in modern life, is widely used in the industrial manufacture of stainless steel, both for steel construction and household purposes, and the battery industries as well. In addition, nickel is also used in a wide variety of metal alloys and nickel plating [1]. At this time, the nickel was obtained from primary sources in the form of nickel sulfide ore (nickel associated with copper) and lateritic ore (nickel associated with iron and/or magnesium silicate) [2]. Approximately 60% of the total nickel production in the world uses a nickel sulfide ore as raw material. Its existence, however, is only about 30% of total world reserves of nickel [3]. With the depletion of nickel sulfide ore reserves caused by continuous exploration, the world industries are forced to find other sources that can be used in nickel production industries as an alternative raw material to substitute nickel sulfide.

For the last few years there has been an increase in the investigation of low-grade lateritic ore to substitute nickel sulfide as the nickel resource [4,5]. Lateritic ore is more difficult to be processed due to the lower nickel content and more complexity of its mineralogy structure as a result of their isomorphism crystal structure goethite and serpentine [6,7].

In the metallurgical extraction methods, nickel lateritic ore is generally processed through hydrometallurgy [8–10], pyrometallurgy [11], or a combination of both pyro- and hydrometallurgy [12]. Until now, the hydrometallurgical process, especially High Pressure Acid Leaching (HPAL), has been effective for dissolving nickel lateritic ore with the high recovery of nickel and cobalt [13]. However, this method requires higher investment costs mainly for the autoclave procurement. In addition, this method is also less effective when the content of magnesium in the lateritic ore is higher than 15%, and sometimes can cause problems for the environment and health as well.

The widely used nickel ore in Indonesia is the one with high nickel content above 1.5%, which is then used to produce ferronickel or nickel matte [14]. Various studies to increase levels of nickel from lateritic ore have been done, ranging from thermodynamic calculations for the reduction [15], the characterization of the microstructure and phase changes during the process of reduction and leaching [16], the mechanism of its reduction with coal as a reducing agent [17], and the effect of sodium sulfate [18] or sodium chloride [19] addition in the nickel lateritic ore reduction roasting process, to further increase the nickel content of the result of the reduction, magnetic separation and microwave [20].

Recently, many researchers have been using coal as reducing agent on the reduction process of nickel lateritic ore in order to produce ferronickel or nickel matte. Based on the above discussion, this study investigated the effect of the amount of rice husk in the pellets on the increase of nickel content in the limonite reduction process. The characteristics of limonitic ore and reduction product were analyzed by using X-ray diffraction (XRD) and Atomic Adsorption Spectroscopy (AAS). The results of this preliminary study may provide a reference for further investigation in related reduction process, and furthermore, it can answer the feasibility of rice husk as a sustainable alternative reducing agent.

2. Material and method

2.1. Materials

The selected reducing agent is the rice husk from Kerawang, West Java, Indonesia. The proximate analysis of the rice husk is listed in Table 1. The rice husk is a good reducing agent because of high volatile matter, and relatively low harmful element (S, P) content.

Table 1. Proximate and Ultimate Analysis of Dry Rice Husk.

Analysis Parameters	Value	Unit	Basis
Proximate			
Moisture	8.70	%	adb
Ash	23.85	%	adb
Volatile Matter	54.10	%	adb
Fixed Carbon	13.35	%	adb
Ultimate			
Carbon	33.14	%	adb
Hydrogen	5.14	%	adb
Nitrogen	0.55	%	adb
Total Sulfur	0.12	%	adb
Oxygen	37.20	%	adb

A limonite ore from North Sulawesi Province in Indonesia was selected for this study. The chemical composition of the limonite ore has silica and alumina content as seen in Table 2.

Table 2. Main chemical compositions of the low-grade nickel lateritic ore sample (mass fraction, %).

Fe_2O_3	NiO	Co_2O_3	SiO_2	Al_2O_3	Cr_2O_3	MnO	MgO	ZrO_2	Others
75.3	1.69	0.315	10.8	6.04	2.59	1.25	0.94	0.701	0.374

2.2. Roasting reduction method
The limonite ore was crushed and sieved to 80 wt%, passing 0.074 mm, while rice husk was crushed and screened to a size range of 0.5–2. The sieved limonite ore sample was thoroughly mixed with bentonite (2 wt.%) and sieved rice husk (the amount of rice husk addition were varied from 0 wt.% to 30 wt.%) for preparing pellets with a diameter of 12–16 mms, as shown in Table 3.

Table 3. The mixture composition of pellets.

Code	Limonite	Bentonite	Rice Husk
Sample 1	98%	2%	0%
Sample 2	88%	2%	10%
Sample 3	78%	2%	20%
Sample 4	68%	2%	30%

2.3. Characterization
The content of nickel and iron in the magnetic concentrate samples were determined by atomic absorption spectrometry (AAS, HITACHI 180/80, Japan). Meanwhile, the XRD patterns were recorded using a diffractometer (RIGAKU D/Max 2500, Japan) under the conditions of radiation: Cu Ka, tube current and voltage: 250 mA, 40 kV, scanning range: 10–80° 2θ, step size: 0.02° 2θ and scanning speed: 8°/minute.

3. Result and discussion

3.1. Initial characterizations
Before undergoing the roasting reduction, the chemical composition of the limonite ore was analyzed by XRF is given in Table 2.

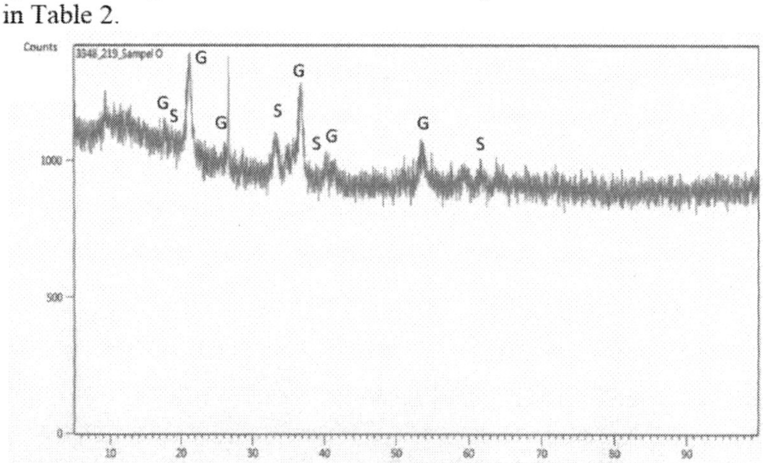

Figure 1. The XRD Pattern of Limonite ore with (S) phases as Spinel ($Al_{1.835}$ $Cr_{0.079}$ $Fe_{0.221}$ $Mg_{0.775}$ $Ni_{0.014}O_4$ $Zn_{0.007}$) and (G) phases as Goethite.

Based on that chemical composition, the XRD pattern resulted from XRD machine was analyzed by using Expert High Score Software to predict the phases that are presented in the limonite ore. And the result (Figure 1) showed that peaks from limonite ore sample matched with the goethite and spinel (Al1.835 Cr0.079 Fe0.221 Mg0.775 Ni0.014O4 Zn0.007) phase patterns standard. So, it was confirmed that goethite and spinel were the main essential minerals in the limonite sample ore. Meanwhile, the invisibility of crystalline structure of nickel was suggested that nickel was associated with other elements and formed as spinel.

3.2. Effect of rice husk addition on phase transformation during reduction

The effect of rice husk addition on mineral transformation of limonite ores after roasting reduction at 1000°C is shown in Figure 2.

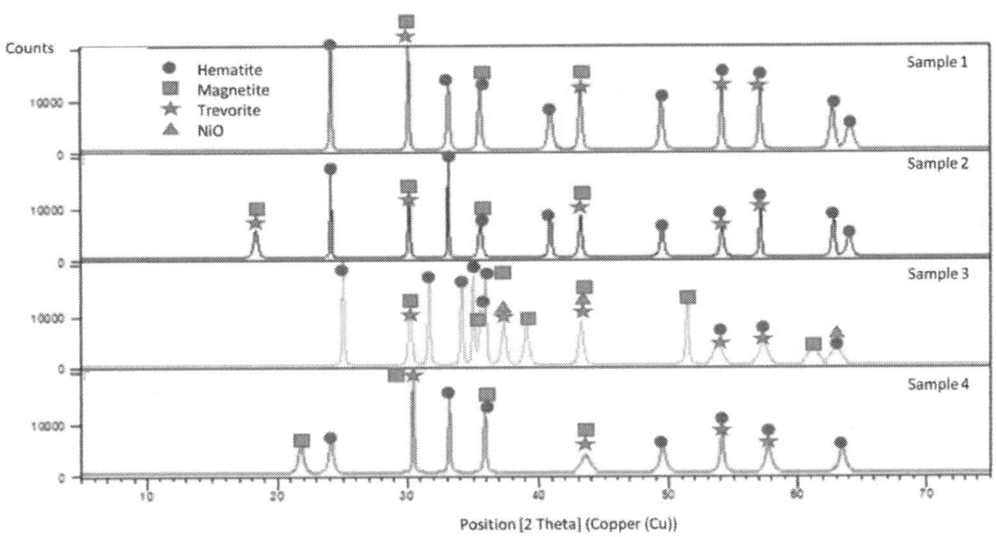

Figure 2. XRD patterns of magnetic concentrates obtained from the pellets reduced at 1000°C for 60 minutes.

In reference to Figure 2, the major phases that were present in sample 1 were magnetite, hematite, and trevorite. Meanwhile, NiO phase was not detected in the XRD pattern. It might happen due to the amount that was too small or existed in the trevorite form. Thermodynamically, trevorite phase, which might be formed due to the reduction of NiO by fixed carbon, was able to occur at low temperatures, namely 438 °C, as shown in Figure 3. In addition, NiO was easily reduced by CO. This was indicated by a negative value of the standard Gibbs free energy of reaction (3), and the curve of forming NiO was located above the forming of CO_2, as shown in Figure 3.

$$C + CO_2 = 2CO \qquad \Delta G^{\theta} = 166550 - 171TJ/mol \qquad (1)$$

$$NiO + C = Ni + CO \qquad \Delta G^{\theta} = 124800 - 175TJ/mol \qquad (2)$$

$$NiO + CO = Ni + CO_2 \qquad \Delta G^{\theta} = -40590 - 0.42TJ/mol \qquad (3)$$

IOP Conf. Series: Materials Science and Engineering **316** (2018) 012050 doi:10.1088/1757-899X/316/1/012050

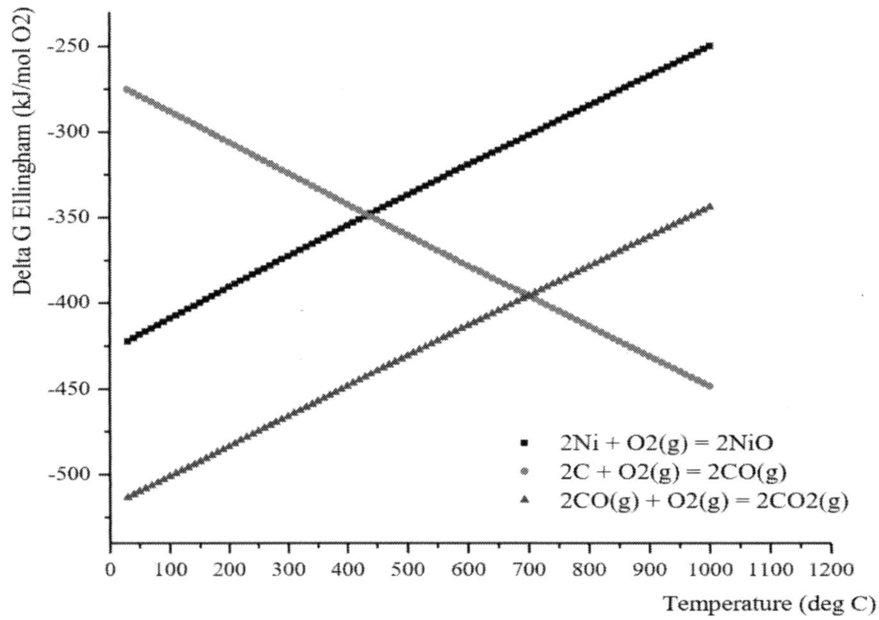

Figure 3. Standard Gibbs free energy (ΔG^{θ}) of carbon oxide and nickel oxide.

From the XRD pattern of sample 2 in Figure 2, it could be observed that the presence of magnetite phase in sample 2 was noticeably increased if compared to the first sample. This was suspected due to the utilization of rice husk in the limonite ore during the reduction process. The rice husk helped to produce more CO gas used in the reduction process. The utilization of rice husk in the pellet assured that the limonite ore would be in contact with the rice husk and when the rice husk in the pellet was starting to produce CO gas, it came in contact with the limonite ore optimizing the process working in tandem with the rice husk used as the reduction agent.

The increased amount of the rice husk has impacted significantly with the XRD pattern of sample 3 compared with the XRD pattern of sample 1 and sample 2. Although the major phases that were present in sample 3 were hematite, magnetite, and trevorite, NiO started to appear clearly. The magnetite phase that existed in sample 3 was also more intense and existed in more angles than the magnetite phase in the XRD pattern of the previous samples. This was due to the trevorite phases that had been further reduced to NiO and magnetite, triggered by the increased amount of CO that was caused by the utilization of higher amount of rice husk in the limonite ore.

In Figure 2, the number of trevorite phase peaks (from sample 1 to sample 3) was increased with the increasing of rice husk amount; and the NiO phase had even been detected in sample 3. However, the trend was not observed in the XRD pattern sample 4 in which the amount was the largest rice husk. It was suspected that the amount of rice husk was too much. This caused NiO trapped in premature melt of silica husks (rice husk ash) and then separated as a tail in magnetic separation process [21].

3.3. Effect of rice husk addition on the Ni and Fe contents of magnetic concentrates after reduction

The Ni and Fe content of magnetic concentrate in each sample was detected by using AAS, as given in Table 4. As shown in the Table 4, the highest Ni content was obtained from samples 3, which was prepared by using 20% additional RH. In the range of 0-20%, the increasing of additional RH amount would lead to higher iron contents in magnetic concentrate. These results were similar to the effects of sawdust [22–24], where the addition of sawdust as reducing agents would increase the amount of magnetic iron ore concentrate in the form of magnetite phase.

Table 4. Nickel and iron contents in magnetic fraction as a function of rice husk addition.

Code	Fe (wt.%)	Ni (wt.%)
Limonite Ore	39.68	1.19
Concentrate sample 1 (without RH)	9.48	0.55
Concentrate sample 2 (10% RH)	8.28	0.61
Concentrate sample 3 (20% RH)	19.81	1.23
Concentrate sample 4 (30% RH)	7.47	0.27

As shown in the Table 4, the highest Ni content was obtained from samples 3, which was prepared by using 20% additional RH. In the range of 0-20%, the increasing of additional RH amount would lead to higher iron contents in magnetic concentrate. These results were similar to the effects of sawdust [22–24], where the addition of sawdust as reducing agents would increase the amount of magnetic iron ore concentrate in the form of magnetite phase.

The amount of nickel content corresponding to the amount of iron content in various amounts of rice husk indicates a strong bond between nickel and iron in a compound. So it is more convinced that the Ni is incorporated in magnetite substance of one Fe atoms to form trevorite ($NiFe_2O_4$), as seen in the XRD pattern. Meanwhile, the relatively small nickel content in all concentrates indicates that most of the nickel content is in the tailings, so it is not attracted by magnets because of their smaller magnetic properties.

The interesting phenomenon was the decreasing Ni content in sample 4. This low amount of the Ni content in sample 4 was suspected due to the excess amount of the rice husk used in the process, which generated an excess tar and silica which act as a barrier for CO to reduce hematite. So, the fraction of magnetite and trevorite become low and affect to Ni recovery.

4. Conclusion

In this study, beneficiation of nickel from limonite ore was carried out by roasting reduction and magnetic separation. The increasing of Ni content was a result of the transformation of hematite into magnetite phase during roasting reduction by using rice husk as pellet addition and reducing gas agent. A small amount of nickel detected in the concentrate may be due to too strong a magnet being used. This causes a small amount of trevorite that should not be attracted by the magnet getting into the concentrate bin.

The optimum result, 1.23 % Ni, was reached in magnetic concentrate sample which was obtained from reduction of pellet with 20% additional RH. Nevertheless, it is still necessary to analyze the composition and the phase of the obtained tailings fraction to ascertain where the largest concentration of nickel, whether in concentrate or in tailings.

5. Acknowledgement

Special thanks are due to Mr. Agus Budi Prasetyo (Research Centre for Metallurgy, Indonesian Institute of Sciences - LIPI, Indonesia) and Mr. Delfiandra (Center for Testing and Identification of Goods, Directorate General of Customs and Excise, Jakarta, Indonesia) for helpful supports. The authors also wish to thank the Directorate of Research and Community Engagement, Universitas Indonesia under Postgraduate research grants for the financial support of this research.

References

[1] Bunjaku A, Kekkonen M, Pietilä K and Taskinen P 2012 Effect of mineralogy and reducing agent on reduction of saprolitic nickel ores *Miner. Process. Extr. Metall.* **121** 156–65

[2] Liu Y and Lee M 2015 Separation of Co and Ni from a chloride leach solutions of laterite ore by solvent extraction with extractant mixtures *J. Ind. Eng. Chem.* **28** 322–7

[3] Mudd G M 2010 Global trends and environmental issues in nickel mining : Sulfides versus laterites *Ore Geol. Rev.* **38** 9–26

[4] Senanayake G, Childs J, Akerstrom B D and Pugaev D 2011 Reductive acid leaching of

laterite and metal oxides - A review with new data for Fe(Ni,Co)OOH and a limonitic ore *Hydrometallurgy* **110** 13–32

[5] Guo Q, Qu J, Han B, Zhang P, Song Y and Qi T 2015 Innovative technology for processing saprolitic laterite ores by hydrochloric acid atmospheric pressure leaching *Miner. Eng.* **71** 1–6

[6] Guo Y, Dai H and Yang J 2013 Beneficiation Test Research on High Grade Hematite-Limonite Ore *Appl. Mech. Mater.* **298** 3029–33

[7] Tang X, Liu R, Yao L, Ji Z, Zhang Y and Li S 2014 Ferronickel enrichment by fine particle reduction and magnetic separation from nickel laterite ore *Int. J. Miner. Metall. Mater.* **21** 955–61

[8] Fan C, Zhai X, Fu Y, Chang Y, Li B and Zhang T A 2012 Leaching behavior of metals from chlorinated limonitic nickel laterite *Int. J. Miner. Process.* **110–111** 117–20

[9] Agacayak T and Koseler M 2015 Effect of Microwave Heating on the Leaching of Lateritic Nickel Ore in Perchloric Acid *J. Chem. Soc. Pakistan* **37** 230–5

[10] Astuti W, Hirajima T, Sasaki K and Okibe N 2016 Comparison of effectiveness of citric acid and other acids in leaching of low-grade Indonesian saprolitic ores *Miner. Eng.* **85** 1–16

[11] Li G, Shi T, Rao M, Jiang T and Zhang Y 2012 Beneficiation of nickeliferous laterite by reduction roasting in the presence of sodium sulfate *Miner. Eng.* **32** 19–26

[12] Oxley A and Barcza N 2013 Hydro – pyro integration in the processing of nickel laterites *Miner. Eng.* **54** 2–13

[13] Kyle J 2010 Nickel laterite processing technologies – where to next? *ALTA 2010 Nickel/Cobalt/Copper Conference, 24 - 27 May* (Perth, Western Australia)

[14] Soedarsono J W, Kawigraha A, Sulamet-ariobimo R D, Asy M A, Yosi A and Putra E M 2013 The Influence of Coal and Reduction Process Parameters in Producing Iron Nugget *Adv. Mater. Res.* **789** 517–21

[15] Swinbourne D R 2014 Understanding ferronickel smelting from laterites through computational thermodynamics modelling *Miner. Process. Extr. Metall.* **123** 127–40

[16] Rhamdhani M A, Hayes P C and Jak E 2009 Nickel laterite part 1 - Microstructure and phase characterisations during reduction roasting and leaching *Trans. Institutions Min. Metall. Sect. C Miner. Process. Extr. Metall.* **118** 129–45

[17] Li Y, Sun Y, Han Y and Gao P 2013 Coal-based reduction mechanism of low-grade laterite ore *Trans. Nonferrous Met. Soc. China* **23** 3428–33

[18] Jiang M, Sun T, Liu Z, Kou J, Liu N and Zhang S 2013 Mechanism of sodium sulfate in promoting selective reduction of nickel laterite ore during reduction roasting process *Int. J. Miner. Process.* **123** 32–8

[19] Zhou S, Wei Y, Li B, Wang H, Ma B and Wang C 2016 Mechanism of sodium chloride in promoting reduction of high-magnesium low-nickel oxide ore *Sci. Rep.* **6** 1–12

[20] Samouhos M, Taxiarchou M, Hutcheon R and Devlin E 2012 Microwave reduction of a nickeliferous laterite ore *Miner. Eng.* **34** 19–29

[21] Maksum A, Rustandi A, Permana S and Soedarsono J W 2017 Influence of roasting-quenching pretreatment on the rice husk silica prepared by calcination method *AIP Conf. Proc.* **1823** 20009-1-020009–8

[22] Luo S, Yi C and Zhou Y 2011 Direct reduction of mixed biomass-Fe2O3 briquettes using biomass-generated syngas *Renew. Energy* **36** 3332–6

[23] Guo D, Hu M, Pu C, Xiao B, Hu Z, Liu S, Wang X and Zhu X 2015 Kinetics and mechanisms of direct reduction of iron ore-biomass composite pellets with hydrogen gas *Int. J. Hydrogen Energy* **40** 4733–40

[24] Guo D, Zhu L, Guo S, Cui B, Luo S, Laghari M, Chen Z, Ma C, Zhou Y, Chen J, Xiao B, Hu M and Luo S 2016 Direct reduction of oxidized iron ore pellets using biomass syngas as the reducer *Fuel Process. Technol.* **148** 276–81

Microstructural and Mechanical Properties of Welded High Strength Steel Plate Using SMAW and SAW Method for LPG Storage Tanks

Winarto Winarto[1,a] Rini Riastuti [1], and Nur Kumeidi

[1] Department of Metallurgical and Materials Engineering, Universitas Indonesia, Depok - 16424, Indonesia

[a)] Corresponding author : winarto@metal.ui.ac.id

Abstract. Indonesian government policy to convert energy consumption for domestic household from kerosene to liquefied petroleum gas (LPG) may lead to the increasing demand for LPG storage tank. LPG storage tank with a large capacity generally used the HSLA steel material of ASTM A516 Grade 70 joined by SMAW or combination between SMAW and SAW method. The heat input can affect the microstructure and mechanical properties of the weld area. The input heat is proportional to the welding current and the arc voltage, but inversely proportional to its welding speed. The result shows that the combination of SMAW-SAW process yield the lower hardness in the HAZ and the fusion zone compared to the singe SMAW process. PWHT mainly applied to reduce residual stress of welded joint. The result shows that PWHT can reduce the hardness in the HAZ and the fusion zone in comparing with the singe SMAW process. The microstructure of weld joint shows a coarser structure in the combined welding process (SMAW-SAW) comparing with the single welding process (SMAW).

1. Introduction

Indonesian government policy to convert energy consumption for domestic household from kerosene to liquefied petroleum gas (LPG) may lead to demand for LPG storage tank. LPG storage tank with a large capacity generally used a low-carbon steel material ASTM A516 Grade 70.

In the fabrication process, Shielded Metal Arc Welding (SMAW), Submerged Arc Welding (SAW) and combination of both can be used for joining the LPG tank as can be seen in Figure 1.

The major variables that affect the weld involve heat input and include the welding current, arc voltage, and travel speed. Welding current is the most important. The higher the current, the deeper the penetration. The welding current should be selected based on the electrode size. The higher the welding current, the greater the melt-off rate (deposition rate). [1]

In general, deposition rate of SAW process is higher that of SMAW process. Hence, in welding a low carbon steel, SAW process will produce a deeper penetration & wider HAZ compared to SMAW Process. [1,2]

(a) (b)

Figure 1. Spherical (a) and ordinary (b) tanks for LPG storage

This paper purposed to study the effect of heat input on the microstructure and mechanical properties of the weld area. In addition, PWHT performed to the weldment is due to the plate thickness of 35 mm, in order to evaluate the effect of PWHT on the microstructure and mechanical properties of the weldment after welding.

2. Research Methodology

The research was carried out using the high strength steel plates with 35 mm thickness. The base material is specified by ASTM A516 Grade 70 having the chemical composition as shown in Table 1.

Table 1. Chemical composition of base metal – steel plate ASTM A516 Gr.70 (in weight %).

C	Si	Mn	P	S	Cr	V
0.21	0.34	1.10	0.02	0.01	0.01	0.01

The plate was prepared by V-grooved design as seen in Figure 2. The sample was then welded by SMAW and the combination of SMAW-SAW method as shown in Figure 3 with welding parameter which refers to ASME section IX are as follow:

Method	: SMAW & Combination of SMAW-SAW
Joint Design	: Butt joint with 70° V- groove.
Filler type	: E7016, E7018 & F7P8 – EH12K
Filler size	: 3.2 mm in diameter
Preheat Temp	: 65 – 95 °C (gas burner)
Interpass temp	: 250 °C max
Current	: 130 Ampere (SMAW) & 550 Ampere (SAW)
Voltage range	: 24 Volts
Travel speed	: 23 mm/min (SMAW) & 50 mm/min (SAW)
Heat Input	: 7.78 – 13.2 kJ/cm

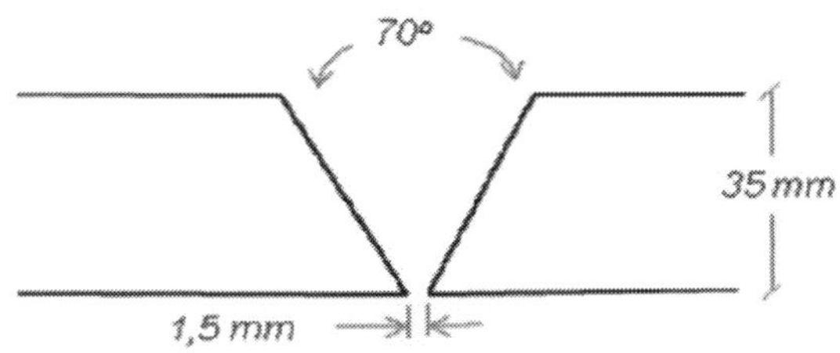

Figure 2. Welding Design with V-Goove

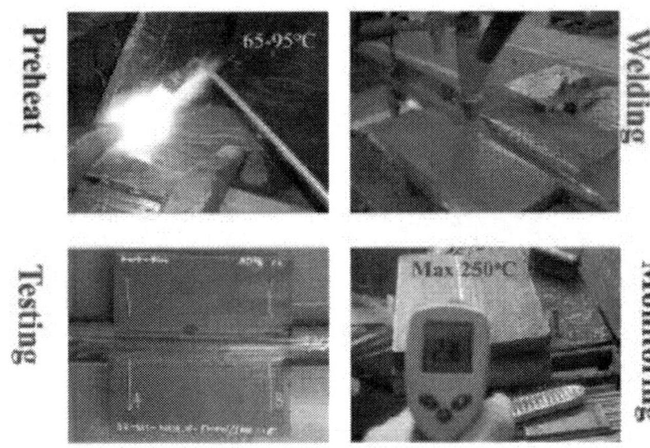

Figure 3. Stages of manufacturing process, from welding, testing and monitoring of joining steel plates

After welding, the specimen was post heat treated in the furnace by hating rate of 222 °C per hour and then holding at temperature of 650 °C within 1 hour 15 minutes and then cooling in furnace with the cooling speed of 280 °C per hour [3,4] as can be seen in Figure 4.

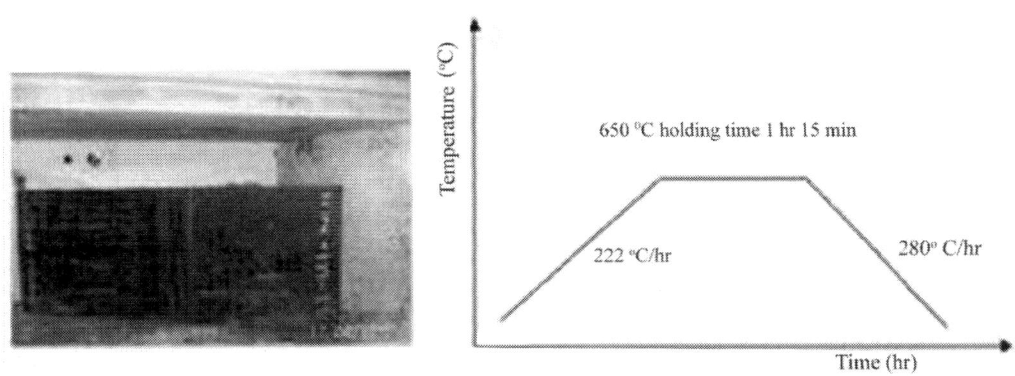

Figure 4. Stages of post weld heat treatment (PWHT).

IOP Conf. Series: Materials Science and Engineering **316** (2018) 012051 doi:10.1088/1757-899X/316/1/012051

3. Results And Discussion

The calculation of heat input is carried out to determine the amount of heat input received by the test specimen. The amount of heat input can determine changes in the microstructure [4,5]. The average heat input on the SMAW welding test specimen (P1) is 0.77 KJ/mm, while for SAW welding is 2,30 KJ/mm. The heat input received by the test specimen is directly proportional to the magnitude of the applied arc voltage and welding current. The greater the arc voltage and the welding current, the greater the heat input received by the test specimen. The speed of welding (travel speed), inversely proportional to the amount of heat input [4,6].

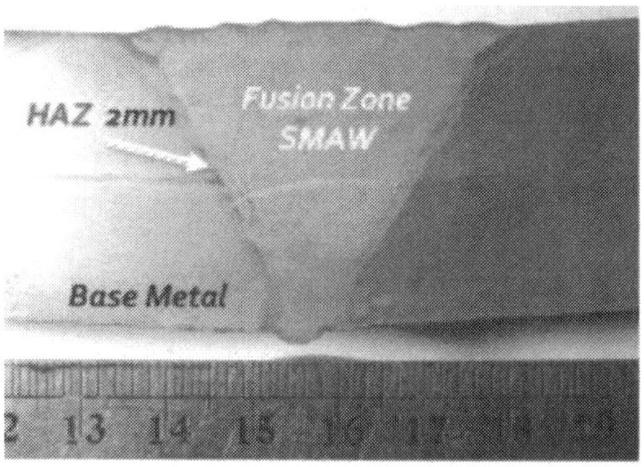

(a)

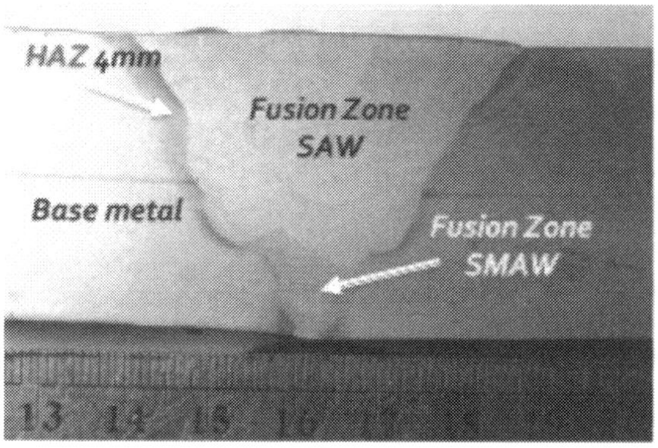

(b)

Figure 5. The macro structure of the welding of test specimens (a) welding P1 and (b) welding P2

Figure 5 is a macrostructure of the SMAW weld specimen (P1) and SMAW-SAW combination weld specimen (P2). P1 welding was done as much as 59 layers, while P2 was as much 30 layers. The previous welding area are exposed to high temperatures, so that the hardness decreases and the material's toughness increases [1,5,6]. The size of HAZ in P1 test specimen is about 2 mm wide, while the size of HAZ in P2 test specimen is about 4 mm wide. The size of HAZ depends on the amount of heat input received by the material. In another study it was also stated that the increase the heat input, the wider size of HAZ would occur [5].

(i)

(ii)

Figure 6. The base metal microstructures (i) before PWHT (ii) after PWHT; A = ferrite phase; B = perlite phase

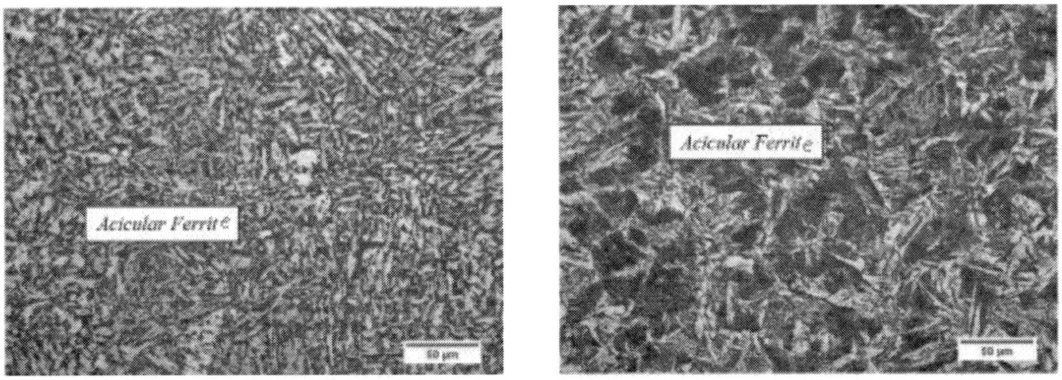

(i) **(ii)**

Figure 7. HAZ microstructure on P1 sample (i) before PWHT (ii) after PWHT; Grain growth occurs in acicular ferrite

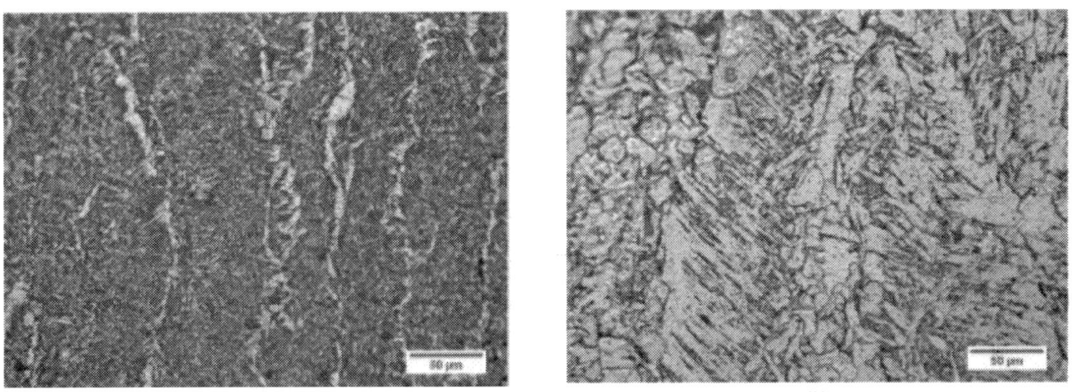

(i) **(ii)**

Figure 8. Fusion Zone microstructure on P1 sample (i) before PWHT (ii) after PWHT; A, acicular ferrite; B, grain boundary ferrite; C, polygonal ferrite; D. Widmanstaten Ferrite

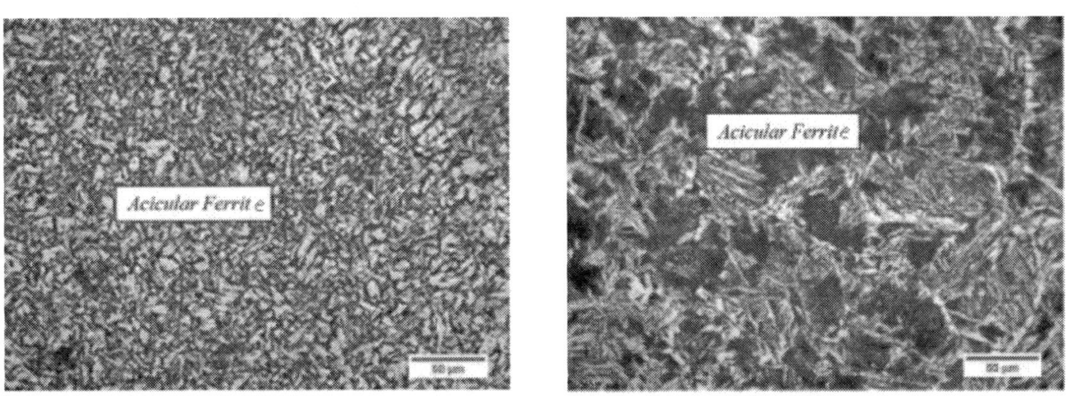

(i) **(ii)**

Figure 9. HAZ microstructure in sample P2 (i) before PWHT (ii) after PWHT; Grain growth occurs in acicular ferrite

IOP Conf. Series: Materials Science and Engineering **316** (2018) 012051 doi:10.1088/1757-899X/316/1/012051

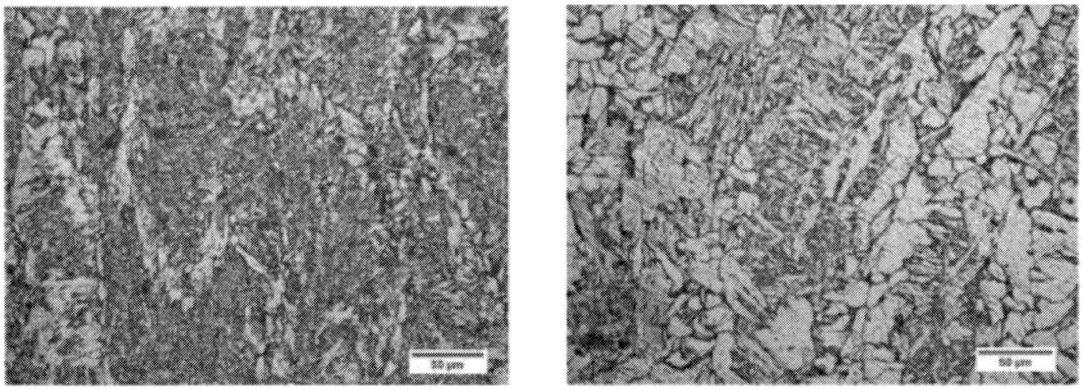

(i) (ii)

Figure 10. Fusion Zone microstructure in P2 sample (i) before PWHT (ii) after PWHT; A, acicular ferrite; B, grain boundary ferrite; C, polygonal ferrite; D, Widmanstaten Ferrite

In Figure 6 to Figure 10 shows the differences in the microstructure between P1 and P2 (before and after PWHT), respectively, ranging from HAZ P1, fusion zone P1, HAZ P2 and fusion zone P2. In each of these differences of microstructures can be seen the growth of ferrite grains. In another study mentioned that the existence of this ferrite growth will cause material hardness decreased and material toughness increased [5,6].

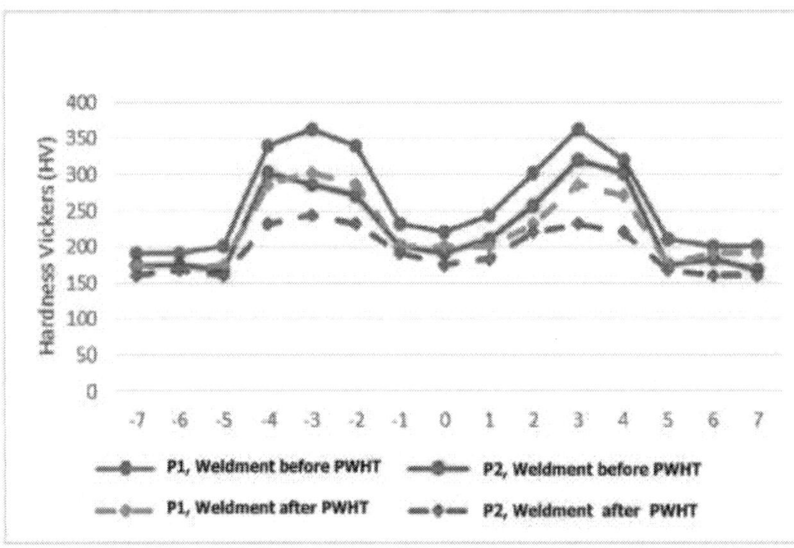

Figure 11. Graph of hardness distribution testing on P1 and P2 specimens before and after PWHT; 0 indicates the centre line of fusion zone

Figure 11 shows the hardness distribution of the test specimens on P1 and P2 either with before PWHT or after PWHT. In both specimen tests, there was a decrease in hardness after PWHT. According to research by Markku Pirinen and Bin Hu, the decrease of hardness in the material after welding can increase the toughness [7]. The high hardness in the HAZ area is due to the residual stresses occurring in the weld area. This residual stress is due to by tensile and compressive stress resulting the melting zone and solidifying zone of the filler material. At the time the filler metal is still in the liquid phase, it will be compressive force in HAZ and it will change the microstructure of the HAZ. While at the time of solidifying of filler metal, there will be tensile force in HAZ. The residual stress caused by the welding process can be reduced by reheating treatment after welding (PWHT).

4. Conclusion

The following conclusions are as follows:

1. Weldment produced by the combination process (SMAW and SAW) gives the lower hardness in both of HAZ and weld metal compared with weldment produced by the single SMAW process.
2. The purpose of PWHT are to reduce residual stress of welded joint and it shows that PWHT can reduce the hardness of weldment in both either in single SMAW process or in combination process (SMAW and SAW).
3. The microstructure for all area of weld joint (HAZ and weld metal) produced by the combination processes is seems a bit coarser comparing with weld produced by the single SMAW process.

5. Acknowledgement

The authors would like to acknowledge the Directorate of Research and Public Services, Universitas Indonesia for the financial support as part of PITTA Research Grant in 2017.

6. References

[1] Hu, J., Du, L.X, Wang, J.J, & Gao, C.R. Effect of welding heat input on microstructures and toughness in simulated CGHAZ of V–N high strength steel. Materials Science & Eng. A577, 2013: 161–168.

[2] Bubbico, Roberto & Marchini, Mauro. "Assessment of an explosive LPG release accident: A case study". Journal of Hazardous Materials Volume 155, Issue 3, 2007: 558–565.

[3] Funderburk, R. Scott. Key Concepts in Welding Engineering: Post welds Heat Treatment. Welding Innovation Vol. XV, No. 2, 1998.

[4] Desai, Pramathesh. Monitoring Heat Treatment to Improve Weld Quality: Controlled pre- and postweld heat treatments can dramatically affect weld metal characteristics. Welding Journal, 2010: 109-110.

[5] Pang, W., Ahmed, N. & Dunne, D. Hardness and microstructural gradients in the heat affected zone of welded low-carbon quenched & tempered steels. Australasian Welding Journal, 56 (2), 2011: 36-48.

[6] Popović, Olivera, et al. The effect of heat input on the weld-metal toughness of surface welded joint. 14th Int'l Research/Expert Conference, TMT 2010, Mediterranean Cruise, 2010: 61-64.

[7] Pirinen, Markku. The effects of welding heat input on the usability of high strength steels in welded structures. Lappeenranta University of Technology, Lappeenranta, Finland, 2013.

Upgrading tantalum and niobium oxides content in Bangka tin slag with double leaching

J W Soedarsono[1], S Permana[1], J K Hutauruk[1], R Adhyputra[1], A Rustandi[1], A Maksum[1,2], K S Widana[3], K Trinopiawan[3] and M Anggraini[3]

[1]Center of Minerals Processing and Corrosion Research, Department of Metallurgical and Materials Engineering,Universitas Indonesia, Depok 16424, Indonesia
[2]Department of Mechanical Engineering, Politeknik Negeri Jakarta, Depok 16425, Indonesia
[3]Nuclear Energy Agency Indonesia, Pasar Jum'at, Jakarta 12440, Indonesia

E-mail : jwsono@metal.ui.ac.id

Abstract. Tantalum has become one of the 14 types of critical materials where the level of its availability is assumed as the midterm critical metal. Benefits of the element tantalum in the electronics field increased the deficit balance of supply / demand, as more variations of electronic products developed. The tantalum experts calculated the level of availability until 2020. Base on the previous studies, tin slag is a secondary source of tantalum and niobium. This study uses tin slag from Bangka, Indonesia, abbreviated, Bangka Tin Slag (BTS). BTS was roasted, water quenched and sieved, abbreviated BTS-RQS.BTS was roasted, water quenched and sieved, abbreviated BTS-RQS.BTS-RQS was roasted at a temperature 700□C given sample code BTS-R700QS, while roasted at 800°C given sample code BTS-R800QS.A variable leaching experiment on BTS-R700QS was solvent concentration variable and on BTS-R800QS was time variable. The entire residue was characterized by X-Ray Fluorescence (XRF), and the optimum results are on the BTS-R800QS leaching into 5 M NaOH for 20 min followed by 5M HCl for 50 min, with content of Ta2O5 and Nb2O51.56% and 1.11%, respectively. The result of XRF measurement showed was the increasing of TNO content due to the increasing solvent concentration and time of acid leaching. The discussion of thermodynamics this study used was HSC Chemistry 6 as a supporting data.

Keywords: Bangka tin slag, leaching, niobium pentoxide, tantalum pentoxide, X-Ray Fluorescence.

1. Introduction

Currently, tantalum and niobium are the two of 14 critical minerals [1], counted based on tantalum midterm deposit availability until year 2020 [2]. Demand and supply in 2013 forecast of Ta_2O_5 in 2016 is minus 212,000 lbs. and in in 2015 is minus 79,000 lbs. [3]. Demand for tantalum is increasing in line with the increasing demand of high technology gadgets, as tantalum is a key element in electronics, automotive, and medicals industries [4].

Tantalite is the main mineral that contains tantalum [2], tantalite is recovered from mining and its availability in earth crust is about 2 ppm [5]. In nature, tantalum pentoxide (Ta_2O_5) is always accompanied by niobium pentoxide (Nb_2O_5). Both oxides minerals have chemical properties similarities [6]. Due to the tantalum low availability, I have done many researches to recovery tantalum from tin slags [7–14]. Table 1 shows sources of tin slags complete with their tantalum and niobium oxides.

Previous researches have shown that Indonesian tin slags have $(TaNb)_2O_5$ at about 2.7% [8]. At recent research, Indonesian tin slags, that have $(TaNb)_2O_5$ at about 0.97%, dissolved with HF, can be increased to 2.0% of $(TaNb)_2O_5$, while dissolved with HCl followed by NaOH, can be increased to 1.71% of $(TaNb)_2O_5$[15].

Table 1. Tantalum and niobium pentoxide-containing tin slags

No.	Country	Ta_2O_5 (%)	Nb_2O_5 (%)
1.	Nigeria[7]	20.8	14.1
2.	Australia[16]	10	5.5
3.	Zaire[8]	9.8	7.8
4.	South Africa[16]	5	7
5.	Brasil[16]	1-2	2-4
6.	Thailand[17]	0.9	1.4
7.	Malasyia[18]	0.55~8.16	1.33~5.57
8.	Indonesia[15]	0.33	0.64

Methods used to increase TNO contents of tin slags in previous researches are: electromagnetic separation [19], dissolution in hydrochloric acid and hydrofluoric acid followed by decalcination at 800°C, then by final melting with carbon [9], alkaline and acid washing, then concentrates is applied by selective carbochlorination at temperature lower than 500°C [10], dissolved with HF-H_2SO_4[12], Chlorination (Cl_2+N_2 or Cl_2+CO+N_2) at 500-1,000°C[11], in situ carbothermic reduction using coarse concentrates powder with collie coal reductor and metallurgical cokes at 1,200-1,400°C[13], chlorination with chlorine gas in front of hydrochloric acid leaching and chlorination with carbon tetrachloride vapor at a relatively low temperature[14], separation with magnetic properties is choosed in increasing tantalume and niobium contents due to their low magnetic properties[20].

The aim of this research is to upgrading TNO contents in tin slags with different roasting temperatures, leaching concentration and time variations. This research will see the effects of roasting at 700 and 800°C, both followed by water quenching - sieving; BTS-R700QS leaching, in an alkaline followed by acid with concentrations variations in constant leaching time; also the effects of BTS-R800QS, in an alkaline followed by acid leaching with time variations, to TNO contents.

2. Experiment

2.1. Materials

Tin slag samples in this research were from PT. Timah Tbk., Indonesia. Characterization result of beginning BTS by X-Ray Fluorescent (XRF) Spectro Xepos Ametek was presented on previous research [15]. This study used two abbreviations namely MOO and EMO. MOO is short for major other oxides, MOO are SiO_2, CaO, TiO_2, Al_2O_3, Fe_2O_3 and ZrO_2. EMO stands for element and minor other oxides, EMO are oxides and elements not present in MOO, Ta_2O_5 and Nb_2O_5. This research uses leaching solutions of 1M, 4M and 5M HCl, also 1M, 4M and 5M NaOH (technical solution). The leaching equipment used was magnetic stirrer IKAMAG RH motor with adjustable motor and speed (at scale 3, displayed dial are 0-6, the speed range 100-200 rpm). This research discussion uses HSC Chemistry 6.0 software [21] as thermodynamic analysis. Fig. 1 shows this research scheme.

IOP Conf. Series: Materials Science and Engineering **316** (2018) 012052 doi:10.1088/1757-899X/316/1/012052

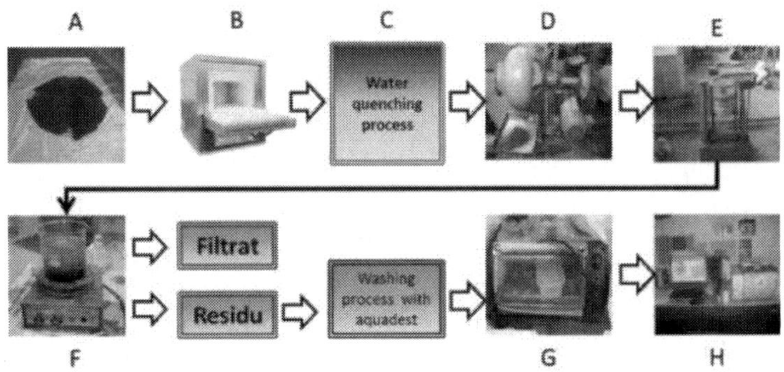

Fig. 1. Research scheme : (A) Bangka Tin Slag (B) Roasting Process (C) Water Quenching Process (D) Milling Process (E) Sieving Process (F) Leaching Process (G) Calcination Process (H) X-Ray Flouresncent analysis

2.2. Roasting Procedure

BTS was roasted at 700°C for 1 hour, water quenched and heated in oven, to vaporized its water content (BTS-R700QS). Then, sample was sieved with +100, -100+150,-150+200, -200+250 and -250 mesh and each sample was characterized with XRF.

The other BTS sample was roasted at 800°C for 1 hour, water quenched and heated in oven, to vaporize its water content (BTS-R800QS). Then, sample was sieved with +100, -100+150 and -150+200 mesh and each sample was characterized with XRF. Table 2. shows sample codes.

2.3. Leaching procedure at BTS-R700QS sample with solvent concentration variable

Sample BTS-R700QS leaching with particle size of +100 was conducted at room temperature for 30 min. The variable was leaching solution concentration.

The leaching solution used was 1 M NaOH (sample code A), while other samples with 4M NaOH (sample code D). Sample weight was 50 g and solvent volume was 500 ml. Leaching filtrate by NaOH was washed with distilled water to eliminate remaining NaOH, then, followed by calcinations in oven to remove moisture. Both calcinated samples divided into two, these four samples then used for second leaching.

Table 2. Sample codes

Sampel code	Roasting Temperature	1st Leaching	2nd Leaching
A	700°c	s : NaOH 1 M	
B		s : NaOH 1 M	s : HCl 1 M
C		s : NaOH 1 M	s : HCl 4 M
D		s : NaOH 4 M	
E		s : NaOH 4 M	s : HCl 1 M
F		s : NaOH 4 M	s : HCl 4 M
G	800°c	s : NaOH 5 M; t : 20 m	
H		s : NaOH 5 M; t : 20 m	s : HCl 5 M; t : 20 m
I		s : NaOH 5 M; t : 20 m	s : HCl 5 M; t : 20 m
J		s : NaOH 5 M; t : 50 m	
K		s : NaOH 5 M; t : 50 m	s : HCl 5 M; t : 50 m
L		s : NaOH 5 M; t : 50 m	s : HCl 5 M; t : 50 m

Note : s = solvent, t = time, M = Molar, m = minutes

Second leaching was applied to two samples (sample code A) with 1 M HCl and 4 M HCl (sample code B and C), and also to the other two samples (sample code D)with 1 M HCl and 4M HCl (sample code E and F). This leaching uses 400 ml solvent for each sample.

2.4. Leaching procedure at BTS-R800QS sample with time of leaching variable

BTS-R800QS sample leaching with particle size of -150+200 mesh was conducted at room temperature with leaching time variable.

First leaching was done with alkaline solution where the solid to solution ratio was 1:20. Sample BTS-R800QS of 12 g was leached with 5 M NaOH for 20 min (sample code G) and the other sample leaching were of 10 g with 5M NaOH for 50 min(sample code J). Filtrates of first leaching were washed with distilled water, then calcined in oven to remove moisture. Both calcinated samples divided into two, these four samples then used for second leaching.

Second leaching was done with solid to solution ratio of 1:30. Leaching applied to two samples G used 5 M HCl with leaching time of 20 and 50 min (samples code H and I). The other two samples J used 5M HCl with leaching time of 20 and 50 min (samples code K and L).

3. Results and discussion

On comparing the other oxide content between BTS-R700QS, particle size +100 mesh, and BTS-R800QS, particle size +150-200 mesh, SiO_2 content is decreasing while EMO content is increasing.

On previous research [15]BTS-RQS did not show content changes of SiO_2 and EMO, on the other hand, DTA-TGA characterization result showed weight decreasing starting 100°C. Table 3 shows composition of BTS, BTS-R700QS particle size +100 mesh and BTS-R800QS particle size +150-200 mesh.

Table 3. Chemical composition of BTS, BTS-R700QS (+100 mesh) and BTS-R800QS (+150-200 mesh)

Sampel	Ta_2O_5 (%)	Nb_2O_5(%)	SiO_2(%)	CaO(%)	TiO_2(%)	Al_2O_3(%)	Fe_2O_3(%)	ZrO_2(%)	EMO (%)
BTS	0.33	0.64	34.26	15.44	11.92	11.7	8.84	4.78	12.06
BTS-R700QS (+100 Mesh)	0.4	0.88	18.91	13.23	10.86	6.11	9.36	6.33	33.92
BTS-R800QS (+150-100 Mesh)	0.39	0.99	25.22	16.38	11.47	8.24	9.69	7.11	20.51

3.1. Roasting Temperature

If we look at illustration of relation between valuable mineral particles and host rock particles [22], anisothropy thermal expansion, breakage-strain and residual stress will affect liberation of valuable mineral on partially or fully locked condition. Roasting and water quenching will make valuable mineral and host rock particles bonding separation easier. Separation process is possible due to specific gravity, magnetism and solubility[23][24][25].

Roasting and quenching result on 700°C and 800°C shows that there are significant weight fraction increment of particle size -100+150, -150+200 and -200+250 mesh. At particle size +100, weight fraction is decreasing at about 50%. Previous research [15] shows the same phenomenon, so it can be concluded that decreasing particle size is effective on particle size less than 100 mesh. Fig. 2 shows relation of weight distribution to particle size of sample BTS, BTS-R700RQS and BTS-R800QS.

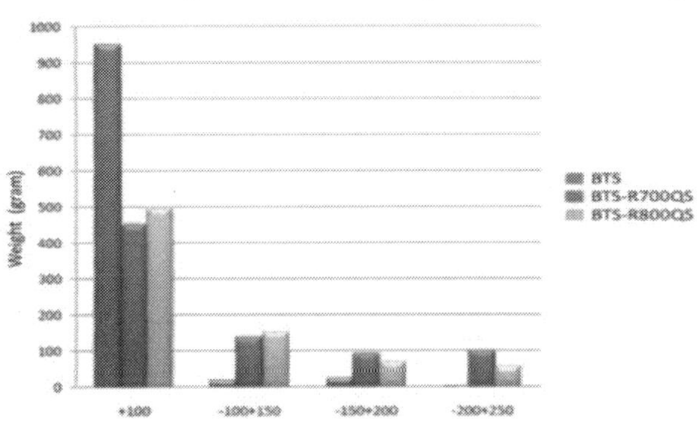

Fig.2. Weight and particle size distribution relation of BTS, BTS- R700QS and BTS-R800QS.

IOP Conf. Series: Materials Science and Engineering **316** (2018) 012052 doi:10.1088/1757-899X/316/1/012052

3.2. Leaching with concentration variation and constant time

Leaching products show maximum value, at sample code C, Ta_2O_5 and Nb_2O_5 content 1.51 and 1.03% respectively, while sample code F, Ta_2O_5 and Nb_2O_5 content 1.42 and 1.12% respectively. Alkaline leaching of sample code A shows increment of TNO content when the second leaching using higher acid concentration. Leaching at sample code D, when the first was using 1M HCl, Ta_2O_5 was decreasing to 0.614% while on the second was using acid of 4M HCl; $Nb_2O_5$was decreasing to 1.12%. Fig. 3 shows the increment of TNO at sample code A after leaching with higher concentration. This can be explaining with formula (1) where solution concentration will increase reaction rate.

$$r = \frac{dC}{dt} = k.C^n \quad (1)$$

where : r = reaction rate, k= constanta, C= concentration and n = reaction order

Irregularities shown up at sample code D, after leached with increased concentration from 1M HCl to 4M HCl, Nb_2O_5 contents decreasing. Leaching conditions is more efficient at sample code A. Gibbs energy (ΔG), calculated using software[21], of sample dissolving with NaOH shows Ta_2O_5 reaction has the smallest ΔG, however it is reversible also for SiO_2, TiO_2, Al_2O_3, Fe_2O_3 and ZrO_2. Table 4 shows the reaction with NaOH followed with HCl.

Table 4. Reaction with NaOH and followed with HCl[21]

Reaction with Alkaline	Δ G (kcal)	Reaction with Acid	Δ G (kcal)
$Ta_2O_5 + 2NaOH = 2\ NaTaO_3 + H_2O$	-45.235	$2NaTaO_3 + 2HCl = Ta_2O_5 + 2NaCl + H_2O$	**)
$Nb_2O_5 + NaOH =$*)			
$SiO_2 + 4NaOH = Na_4SiO_4 + 2H_2O$	**)	$Na_4SiO_4 + 4HCl = SiO_2 + 4NaCl + 2H_2O$	**)
$CaO + NaOH =$*)			
$TiO_2 + 2NaOH = Na_2TiO_3 + H_2O$	**)	Na2TiO3+HCl= *)	
$Al_2O_3 + 2NaOH = 2NaAlO_2 + H_2O$	-8.165	$NaAlO_2 + 4HCl = NaCl + AlCl_3 + 2H_2O$	**)
$Fe_2O_3 + 2NaOH = 2NaFeO_2 + H_2O$	-4.262	$NaFeO_2 + 4HCl = FeCl_3 + NaCl + 2H_2O$	**)
$ZrO_2 + 2NaOH = Na_2ZrO_3 + H_2O$	-5.954	$Na_2ZrO_3 + HCl =$*)	

*) data unavailable **) data cannot be calculated

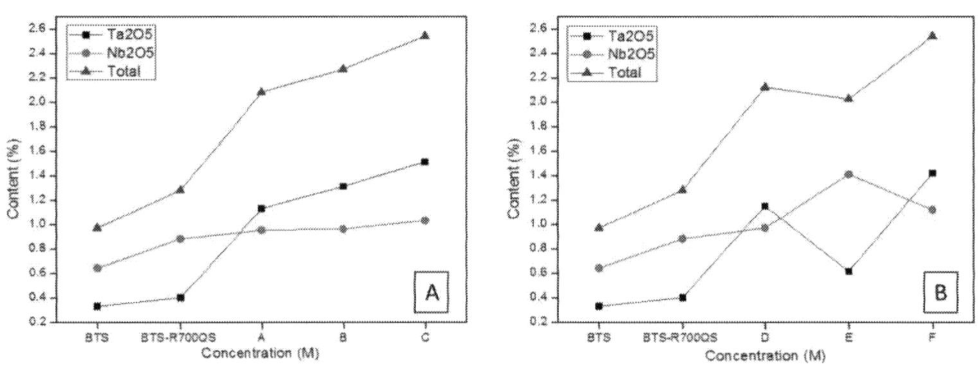

Fig. 3. Ta_2O_5,Nb_2O5 and total content, leaching products of first leaching with (a) 1M NaOH (b) 4M NaOH

Generally, SiO_2 and Fe_2O_3 contents are decreasing as per sample code A and D when second leaching was with1M HCl. TiO_2 content resulting increament in all leaching conditions. At leaching sample code A and D, leaching will be effective if second leaching using HCl 1 M for SiO_2 and Fe_2O_3. Fig. 4 shows SiO_2, TiO_2 and Fe_2O_3 content of leaching products with first leaching were (a) 1M NaOH (b) 4 MNaOH.

IOP Conf. Series: Materials Science and Engineering **316** (2018) 012052 doi:10.1088/1757-899X/316/1/012052

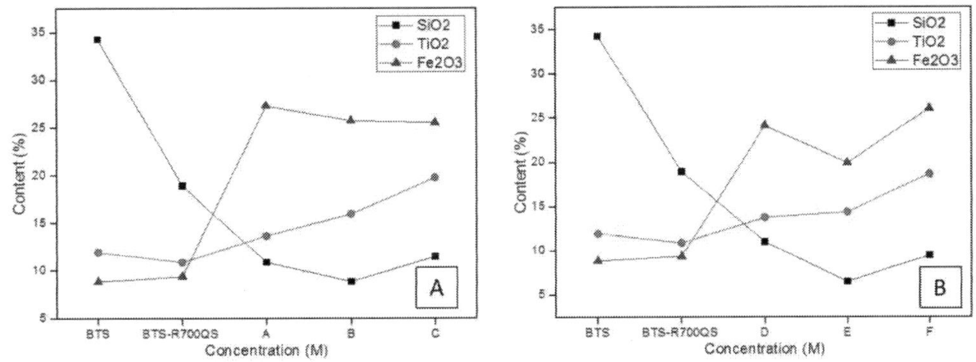

Fig. 4. SiO_2, TiO_2 and Fe_2O_3 content of leaching products with first leaching were (a) 1M NaOH (b) 4 MNaOH

CaO and Al_2O_3 content decreasing were on sample code A and D at second leaching with 4 M HCl, while ZrO_2 decreasing was on sample code D at second leaching with4 M HCl. At leaching on sample code A and D will be effective if second leaching was with 4 M HCl for CaO and Al_2O_3 oxides. Fig. 5 shows CaO, Al_2O_3 and ZrO_2 content of leaching products with first leaching were (a) 1M NaOH (b) 4M NaOH.

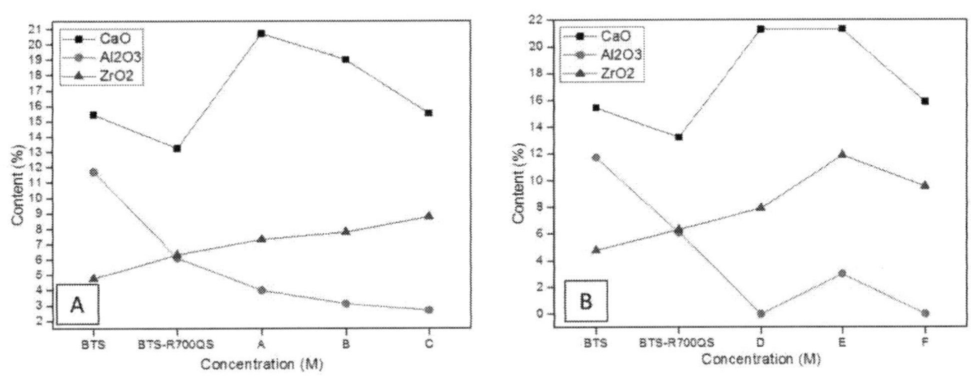

Fig. 5. CaO, Al_2O_3 and ZrO_2 content of leaching products with first leaching were (a) 1M NaOH (b) 4M NaOH

Seeing from content increament of TNO, disolving of SiO_2, Fe_2O_3 and TiO_2, and dissolving of CaO, Al_2O_3 and ZrO_2, Dissolving with sample A and second leaching with4 M HCl gave more efficient leaching. End product of both leaching have the same TNO value, that is 2.54%.

3.3. Leaching with time and concentration variation

Leaching products reached its maximum at sample code I with Ta_2O_5 and Nb_2O_5 content are 1.56 and 1.11 % respectively. Alkaline leaching of sample code G when followed by 5M HCl leaching in 20min was resulting a decreasing Ta_2O_5 content to be 0.8%, sample code G when followed by 5M HCl leaching in 50min was resulting Nb_2O_5 content to be 1.11%. Alkaline leaching sample code J when followed with HCl 5 M leaching in 50min was resulting decreasing Nb_2O_5 content to 0.89%. Fig. 6 shows Ta_2O_5, Nb_2O_5 and total content of leaching products with 5M NaOH for time variations of (a) 20 min (b) 50 min.

IOP Conf. Series: Materials Science and Engineering **316** (2018) 012052 doi:10.1088/1757-899X/316/1/012052

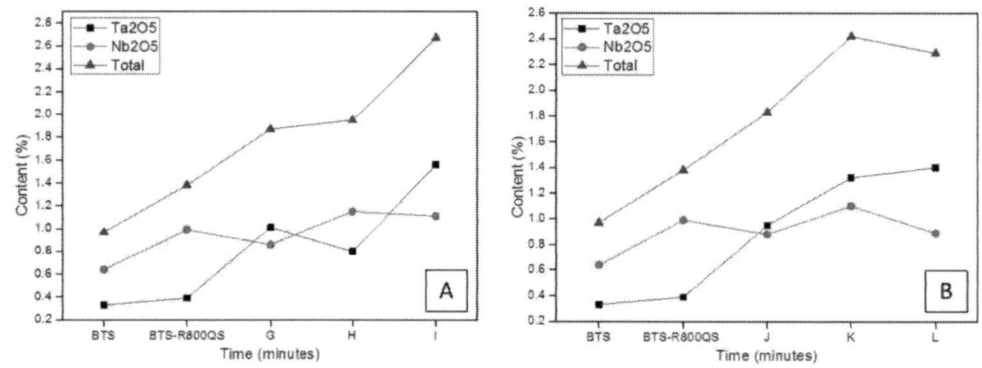

Fig. 6. Ta_2O_5, Nb_2O_5 and total content of leaching products with 5M NaOH for time variations of (a) 20 min (b) 50 min.

In favorable conditions for sample code J, second leaching with 5M HCl in 50 min resulting reduction in CaO content only. For this sample code and leaching condition, all of them did not show any reduction significantly of SiO2, TiO_2 and Fe_2O_3 oxides. Fig. 7 shows SiO_2, TiO_2 and Fe_2O_3 content of leaching products with 5M NaOH for time variations of (a) 20 min (b) 50 min.

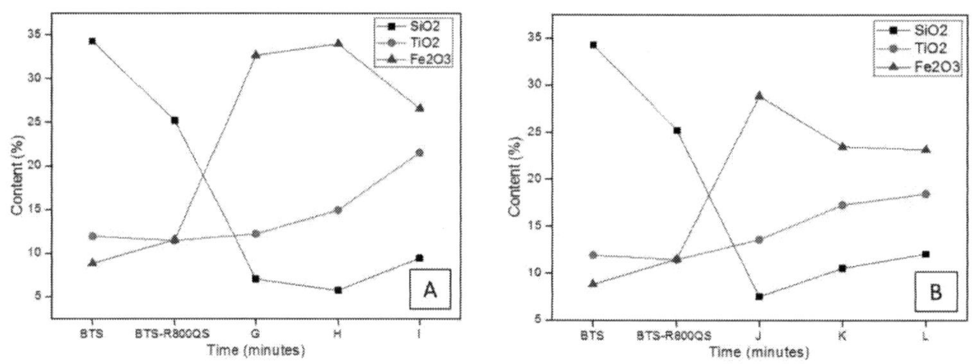

Fig. 7. SiO_2, TiO_2 and Fe_2O_3 content of leaching products with 5M NaOH for time variations of (a) 20 min (b) 50 min.

In favorable conditions for sample code G and J, second leaching with 5M HCl in 50 min resulting reduction in CaO content only. Fig. 8 shows CaO, Al2O3and ZrO2content of leaching products with NaOH 5M for time variations of (a) 20 min (b) 50 min.

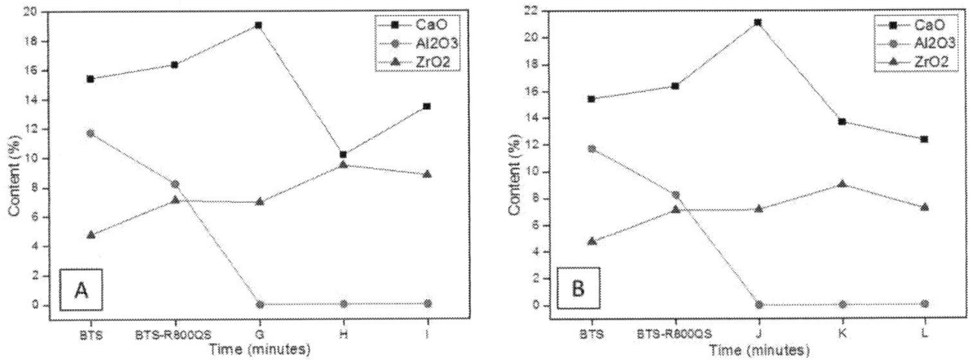

Fig. 8.CaO, Al_2O_3 and ZrO_2 content of leaching products with 5M NaOH for time variations of (a) 20 min (b) 50 min.

Seeing from content increament of TNO, dissolving of SiO2, Fe2O3 and TiO2, and dissolving of CaO, Al2O3and ZrO2, dissolving with sample G and second leaching with 5M HCl in 50 mingave more efficient leaching. TNO content at sample code I was 2.67%.

The final conclusion of this study, the roasting sample code BTS-R800QS with increasing leaching time (total time of 60 min for sample code C and F to of 70 min for sample code I) and increasing leaching solvent concentration producing more optimal results. However, when leaching variable increased at sample code BTS-R800QS in condition 5M NaOH in 50 min followed 5M HCl in 50 min, TNO content becomes 2.29%.

4. Consclusions

Based on the study above, upgrading TNO content in Bangka tin slag with double leaching have result as follows:

1. The particle size reduction with roasting and water quenching was effective for particle size below than 100 mesh.
2. Leaching experiments on BTS-R800QS into 5M NaOH for 20 minfollowed by 5M HCl for 50 min gave an optimum TNO grade of 2.67%.
3. Up to 5M NaOH solvent concentration concentration for 20 min followed by 5M HCl for 50 min was seen with increasing solvent concentration and leaching time of TNO content increased.
4. However, when the leaching variable was increased to a 5M NaOH solvent concentration variable for 50 min and followed by 5M HCl for 50 min, the leaching result of TNO content decreased, only 2.29%.

Acknowledgements

The author would like to thanks Ang Kok Djiang, Wahyu Kartika M.Si. for support this paper and providing valuable comment. We thanks Dr. rer. nat. Agustino Zulys M.Sc. for thoughtful discussions. The support of National Nuclear Energy Agency Indonesia is gratefully acknowledged.

References

[1] Wellmer F and Dalheimer M 2012 The feedback control cycle as regulator of past and future mineral supply Min. Depos. 47 713–29
[2] Laura Talens Peiro, Gara Villalba Mendez R U A 2011 Rare and Critical Metals as By-products and The Implications for Future Supply
[3] Stratton P 2013 Outlook for The Global Tantalum Market (New York)
[4] Edmond E H and February O 2007 Commerce Resources Corporation and Tantalum
[5] Bose D K and Gupta C K 2001 Extractive Metallurgy of Tantalum Miner. Process. Extr. Metall.

Rev. 22 389–412
[6] Ayanda O S and Adekola F A 2011 A Review of Niobium-Tantalum Separation in Hydrometallurgy J. Miner. Mater. Charact. Eng. 10 245–56
[7] Odo J.U., Okafor WC, Ekpe SO and CC. and N 2014 Extraction of Niobium from Tin Slag Int. J. Sci. Res. Publ. 4 1–7
[8] Gaballah I, Allain E, Meyer-Joly M-C and Malau K 1992 A Possible Method for the Characterization of Amorphous Slags Recovery of Refractory Metal Oxides from Tin Slags Metall. Trans. 23 249–59
[9] Cenerazzo J A, Mosheim C E, Zionsville and Corrodo E. Marvisa 1971 US Patent 3.585.024-Up Grading The Tantalum and Columbium Contents of Tin Slag 1–4
[10] Gaballah I and Allain E 1994 Recycling of Strategic Metals from Industrial Slag by a Hydro- and Pyrometallurgical Process Resour. Conserv. Recycl. 10 75–85
[11] Gaballah I, Allain E and Djona M 1997 Extraction of Tantalum and Niobium from Tin Slags by Chlorination and Carbochlorination Metall. Mater. Trans. 28
[12] Subramanian C, Suri A K and Atomic B 1998 Recovery of Niobium and Tantalum from Low Grade Tin Slag - A Hydrometallurgical Approach Environ. Waste Manag. 100–7
[13] Mudzanapabwe N T, Chinyamakobvu O S and Simbi D J 2004 In Situ Carbothermic Reduction of a Ferro-Columbite Concentrate in The Recovery of Nb and Ta as Metal Matrix Composite from Tin Smelting Slag Waste Dump Mater. Des. 25 297–302
[14] Brocchi E A and Moura F J 2008 Chlorination Methods Applied to Recover Refractory Metals from Tin Slags Miner. Eng. 21 150–6
[15] Permana S, Soedarsono J W, Rustandi A and et.al 2016 Other Oxides Pre-removed from Bangka Tin Slag to Produce a High Grade Tantalum and Niobium Oxides Concentrate IOP Conference Series : Materials Science and Engineering vol 012006 p 10
[16] Köck W and Paschen P 1989 Tantalum-processing, properties and applications Jom 41 33–9
[17] Bunnakkha C and Jarupisitthorn C 2012 Extraction of Tin from Hardhead by Oxidation and Fusion with Sodium Hydroxide 22 1–6
[18] Sulaiman M Y 1993 Simultaneous determination of thorium and uranium in tin slag Sci. Total Environ. 131 187–95
[19] Sidney M. Heins 1964 US Patent 3.129.896-Separation of Heavy Metal Contituents of Tin Slag 0–1
[20] Bulatovic S M 2010 Handbook of Flotation Reagents Vol.2 (Amsterdam: Elsevier Science & Technology Books)
[21] Roine A, Kotiranta T and et.al 2006 HSC Chemistry, Version 6.0 03 5641
[22] L. Free M 2013 Hydrometallurgy Fundamentals and Applications (United States of America: John Wiley & Sons. Inc., Hoboken, New Jersey)
[23] Callister W D and Rethwisch D G 2014 Materials Science and Engineering Materials Science and Engineering ed D Sayre (John Wiley & Sons, Inc.)
[24] Saylor D M, Rohrer G S, Carter W C, Langer S A and Jr E R F 2001 Residual-Stress Predictions in Polycrystalline Alumina J. Am. Ceram. Soc. 54
[25] Liu S, Guo Y, Qiu G and Jiang T 2013 Preparation of Ti-rich material from titanium slag by activation roasting followed by acid leaching Trans. Nonferrous Met. Soc. China 23 1174–8

IOP Conf. Series: Materials Science and Engineering **316** (2018) 012053 doi:10.1088/1757-899X/316/1/012053

Preparation of Fe_3O_4/Bentonite Nanocomposite from Natural Iron Sand by Co-precipitation Method for Adsorbents Materials

Perdamean Sebayang[1,2,*], Candra Kurniawan[1], Didik Aryanto[1], Eko Arief Setiadi[1], Konni Tamba[3], Djuhana[2], and Toto Sudiro[1]

[1]Research Center for Physics – Lembaga Ilmu Pengetahuan Indonesia, Bld. 440-442 Puspiptek Office Area, Tangerang Selatan 15314, Indonesia
[2]Departement of Mechanical Engineering, Universitas Pamulang, Jl. Surya Kencana No. 1, Tangerang Selatan 15417, Indonesia
[3]Department of Physics, Universitas Negeri Medan, Jl. W. Iskandar Pasar V Medan Estate, Medan 20221, Indonesia

*E-mail: pard003@lipi.go.id

Abstract. An adsorption method is one of the effective ways to filter the heavy metals wastes in aqueous system. In this paper, the Fe_3O_4/bentonite nanocomposites were successfully prepared from natural iron sand by co-precipitation method. The chemical process was carried out by dissolving and hot stirring the milled iron sand and bentonite in acid solution and precipitating it by NH_4OH. The sediment was then washed using distilled water to neutralize pH and dried at 100 °C for 5 hours to produce Fe_3O_4/bentonite powders. The samples were characterized by XRD, FTIR, BET, TEM, VSM and AAS. All samples were composed by Fe_3O_4 single phase with a spinnel structure and lattice parameter of 8.373 Å. The transmittance peak of FTIR curve proved that the Fe_3O_4 particles and bentonite had a molecular bonding. The addition of bentonite to Fe_3O_4 nanoparticles generally reduced the magnetic properties of Fe_3O_4/bentonite nanocomposites. The optimum condition of 30 wt% bentonite resulted 105.9 m^2/g in surface area, 14 nm in an average particle size and 3.2 nm in pore size. It can be used as Cu and Pb adsorbent materials.

1. Introduction

Natural iron-sand was an abundant material in Indonesia and not optimally harnessed nowadays, especially in the form of magnetic nanoparticles [1]. The processing of magnetite (Fe_3O_4) nanoparticles from natural iron-sand will enrich its economical value and increase its potential applications such as for biosensor, magnetic hyperthermia, drug delivery systems, microwave absorber and adsorbent materials [2–5]. There are some common methods which can be used to produce magnetic nanoparticles, such as sol-gel, co-precipitation, and hydrothermal processes [6–8]. Among them, the chemical co-precipitation is a simplest method [8].

Due to the non-biodegradable properties of metal waste, the effective and efficient methods are required to escape the heavy metal ions from an environment. The adsorption method is one of the effective ways to filter the heavy metals wastes in aqueous system because of its high efficiency, relatively stable properties and non-harmful by-products [9]. The adsorption capability of magnetic nanoparticle was considered to be one of the effective ways to adsorb pollutants and heavy metals ions from aqueous and gaseous systems [10]. In addition, bentonite is a functional material which widely used to adsorb the radioactive wastes. The adsorption capacity of adsorbent materials is resulted from a relatively high surface area and a net negative charge on their structure. This can attract and hold the heavy metals cation. The magnetic adsorbent can separated the heavy metal ions from the medium by a simple magnetic process. Hashemian et al. studied the adsorption of Co (II) by Fe_3O_4/bentonite nanocomposite. They found that the addition of Fe_3O_4 to bentonite increases the adsorption capacity. In that paper, Fe_3O_4 was prepared by using chemical co-precipitation method of 2 mmol Fe(II) chloride and 4 mmol Fe(III) chloride [10]. However, the utilization of natural iron sand as starting material for producing Fe_3O_4 in the preparation of adsorbent material of Fe_3O_4/bentonite nanocomposite has not been carried out yet.

In this paper, we discussed the preparation of Fe_3O_4/bentonite nanocomposite from natural iron-sand. The samples were prepared by co-precipitation method and its performance was evaluated to absorb Cu and Pb heavy metal wastes. The samples characterization was carried out by XRD, FTIR, TEM and VSM.

2. Experimental Method

The Iron-sand from Bingei River, North Sumatera was used as starting material to form Fe_3O_4. The material was milled by a planetary ball milling for 15 hours. The 10 grams of milled powder was mixed with 63 mL of HCl (37%, 12 M) in a beaker glass. The solution was hot stirred at 70 °C and 450 rpm for 60 minutes. The chemical reaction that take placed during stirring is as follows [11],

$$Fe_3O_{4(s)} + 8HCl_{(l)} \rightarrow 2FeCl_{3(l)} + FeCl_{2(l)} + 4H_2O_{(l)} \quad (1)$$

To remove the impurities, the solution was filtered by filter paper No. #40 which equals to 8 micron in pore size. The remaining $FeCl_3$ and $FeCl_2$ mixed solution was then hot stirred again at 70 °C for 60 minutes. The 6.5 M NH_4OH was added to form Fe_3O_4 sediment according to the following reaction [12],

$$2FeCl_{3(l)} + FeCl_{2(l)} + 8NH_4OH_{(l)} \rightarrow Fe_3O_{4(s)} + 8NH_4Cl_{(l)} + 4H_2O_{(l)} \quad (2)$$

The Fe_3O_4 sediment was washed with distilled water repeatedly to obtain normal pH (~7). The Fe_3O_4 slurry was then dried at 70 °C for 24 hours to obtain Fe_3O_4 powder.

The 30 g of commercial Bentonite powder from Adrich was activated by hot stirring in 250 mL HCl at 80 °C for 60 minutes. The bentonite solution was washed with aquades to obtain normal pH. The Bentonite slurry was then dried at 70 °C for 5 hours to obtain an activated bentonite powder.

The Fe_3O_4/bentonite nanocomposite was made by mixing 10 gram of Fe_3O_4 and varied composition of an activated Bentonite powder from 30 to 70 wt.% in the distilled water as a medium. The solution was stirred at 80 °C for 60 minutes with stirring speed of 300 rpm. The 5 M NaOH solution was dropped slowly until 100 mL. The sediment product was washed with distilled water repeatedly to obtain normal pH (~7) and then dried at 100 °C for 5 hours to obtain Fe_3O_4/bentonite nanocomposite powder. The heavy metal adsorption test was carried out using artificial waste water containing 706 ppm Pb and 277 ppm Cu ions. The 50 mg nanoparticle samples were mixed with 25 mL of waste water and mixed for 30 minutes by Shaker Mill.

The sample characterizations such as elemental, atomic bonding, surface area and magnetic properties analysis were performed using Torontech X-Ray Fluorescence, Simadzu Fourier Transform Infrared, Quantracom BET and Dexing Vibrating Sample Magnetometer, respectively. The structure analysis was performed by Rigaku XRD and Tecnai TEM. The heavy metal adsorbent performance test of Cu and Pb elements were conducted by AAS analysis.

3. Results and Discussions

The results of XRF elemental analysis of milled iron sand and co-precipitated product are shown in Table 1. It can be seen that the concentration of iron element after chemical treatment increases while another elements such as Ti and Si concentration decrease.

Table 1. XRF elemental analysis of milled iron sand and co-precipitated product.

No.	Element	Milled iron sand (wt%)	Co-precipitated product (wt%)
1	Fe	91.82	93.35
2	Ti	4.11	3.83
3	Si	1.83	0.61
4	Others	2.24	2.21
	Total	100.00	100.00

The correlation between transmittance percentage (%) and wavenumber of Fe_3O_4 after co-precipitation, pure bentonite and composites containing 30, 50, and 70 wt.% bentonite are shown in Figure 1. According to transmittance curve, the formation of Fe_3O_4/bentonite nanocomposites can be confirmed. It is indicated by the transmittance peak shifting at different wavenumber. As shown in Table 2, the peak of nanocomposite samples is shifted from the original peaks of Fe_3O_4 and Bentonite powders. The transmittance peaks at wavenumber (υ) = 401 and 570 cm^{-1} correspond to Fe-O stretch vibration bond of magnetic Fe_3O_4 nanoparticles. The transmittance peaks at υ = 462.9 and 1041.5 cm^{-1} correspond to Si-O-Si bend and stretch vibration bond of bentonite, respectively. The formation of Fe_3O_4/bentonite nanocomposite leads to transmittance peak shifting of Si-O-Si bond. This is due to an energy changes and interaction between bentonite and Fe_3O_4. The functional group of Si-O-Si with bending vibration at υ = 447.49 and 1026.13 cm^{-1} indicates that the bentonite coexist in the nanocomposite. The functional groups of bentonite are the C-H and O-H with stretching vibration at υ = 794.67 and 3626.17 cm^{-1}, respectively [13], and also (Al-Al-OH) with bending vibration at υ = 918.12 cm^{-1}. The peak of O-H functional groups in the samples C, D and E with stretching vibrations at υ = 3410.15 cm^{-1} was shifted due to the effect of bentonite addition. The summarizing results of peaks and the corresponding vibration types are showed in Table 2.

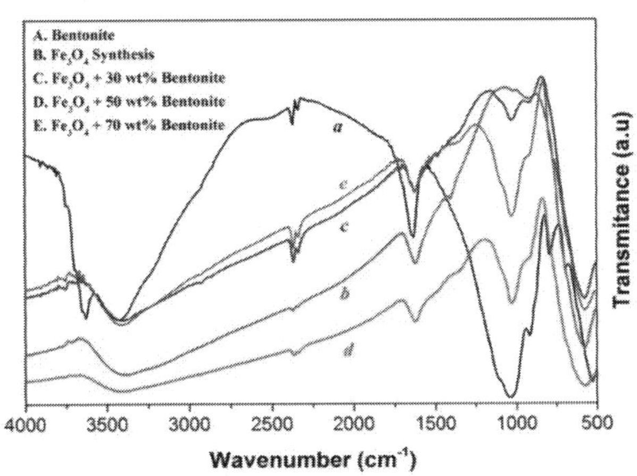

Figure 1. FTIR analysis of Fe_3O_4/bentonite nanocomposite: A) bentonite, B) Fe_3O_4 after co-precipitation (synthesis), C) Fe_3O_4/30 wt% bentonite, D) Fe_3O_4/50 wt% bentonite, and E) Fe_3O_4/70 wt% bentonite.

IOP Conf. Series: Materials Science and Engineering **316** (2018) 012053 doi:10.1088/1757-899X/316/1/012053

Table 2. Functional bonding analysis performed by FTIR measurement.

No.	Func. Group	Sample A (bentonite)	Sample B (Fe_3O_4-synthesis)	Sample C (Fe_3O_4/30 wt% bentonite)	Sample D (Fe_3O_4/50 wt% bentonite)	Sample E (Fe_3O_4/70 wt% bentonite)	Vibration type
1.	Fe-O	-	401.19	-	-	-	*Stretching*
2.	Si-O-Si	462.92	-	447.49	466.2	462.92	*Bending*
3.	Al-O-Si	524.64	-	-	-	-	*Bending*
4.	Fe-O	-	570.93	570.93	570.93	570.93	*Stretching*
5.	C-H	794.67	-	-	-	-	*Stretching*
6.	Al-Al-OH	918.12	-	918.12	-	-	*Bending*
7.	Si-O-Si	1041.56	-	1026.13	1033.85	1033.85	*Stretching*
8.	O-H	1635.64	1627.9	1627.92	1627.92	1627.92	*Bending*
9.	C-H	2368.59	-	2337.72	2337.72	2337.72	*Stretching*
10.	O-H	3425.58	3387.2	3410.15	3410.15	3410.15	*Stretching*
11.	O-H	3626.17	-	-	-	-	*Streching*

The XRD analysis of bentonite, iron sand after milling for 15 hours, Fe_3O_4 after co-precipitation and Fe_3O_4/30 wt% bentonite nanocomposite are shown in Figure 2. The results show that bentonite is composed of quartz (Q) and montmorillonite (M) phases. While, milled iron sand, co-precipitated Fe_3O_4, and Fe_3O_4/30 wt% bentonite samples consist of a single phase of magnetite (Fe_3O_4) with spinel cubic structures and lattice parameters of 8.373Å.

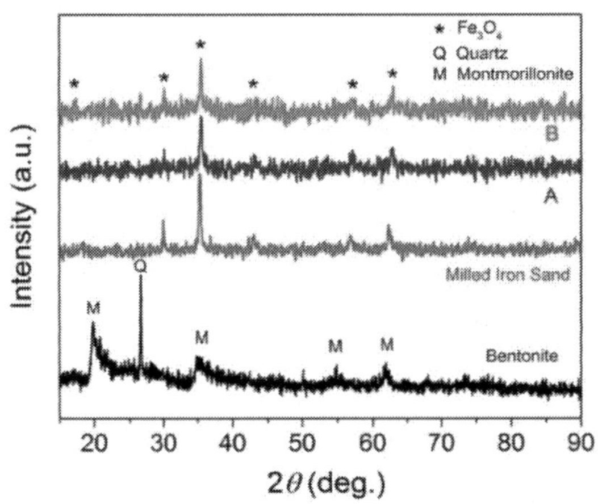

Figure 2. XRD patterns of bentonite, milled iron-sand (15 h), co-precipitated Fe_3O_4 (**A**) and Fe_3O_4/30 wt% bentonite (**B**).

The hysteresis curves of pure Fe_3O_4 and Fe_3O_4/bentonite nanocomposites measured by Vibrating Sample Magnetometer (VSM) are shown in Figure 3. It can be seen that the addition of bentonite decreases the remanence magnetization (M_r), saturation magnetization (M_s) and increases the coercivity field (H_c). This is due to the paramagnetic properties of bentonite. In general, a high magnetization is needed for good performance of adsorbent materials. The optimum magnetization properties of Fe_3O_4/bentonite nanocomposites is achieved by 30 wt.% bentonite addition that results saturation M_s = 22 emu/g, remanence M_r = 3.74 emu/g, and coercivity H_c = 119 Oe. Based on the results of magnetic properties characterization, an adsorbent performance analysis was carried out for the aforesaid sample.

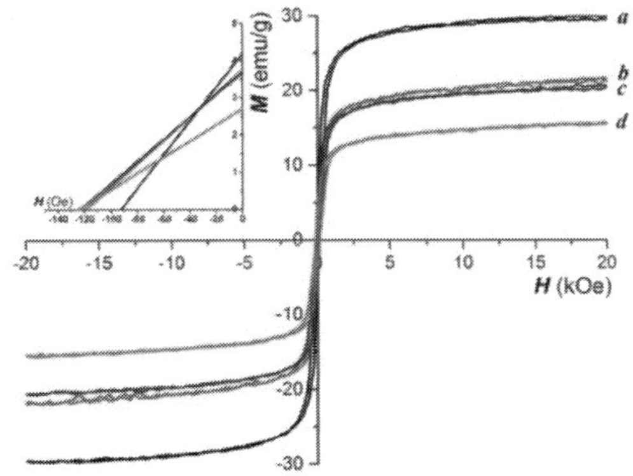

Figure 3. Hysteresis curves of Fe_3O_4/bentonite nanocomposites: (a) pure Fe_3O_4, (b) 30% bentonite, (c) 50% bentonite, and (d) 70% bentonite.

The surface area characterization was measured by BET method (Brunauer-Emmet-Teller) at liquid N_2 temperature (77.35K). The measurement results show that the surface area (SA) of pure Fe_3O_4 and Fe_3O_4/30 wt% bentonite in dry condition are 129.82 and 105.92 m^2/g, respectively. In comparison to the results of previous studies, the surface area of present result is higher [14,15]. Dongbei et al. reported that the surface area of Fe_3O_4/bentonite with ratio 10:1 and 1.5:1 is 19.6 and 26.4 m^2/g, respectively [14]. Dong Wan et al. using the same materials of Fe_3O_4/bentonite obtained 53.02 m^2/g [15]. It is well known that the surface adsorption raises with the increase of surface area. Following the $6/(\rho \times SA)$ formula [16], the average particles sizes of pure Fe_3O_4 and Fe_3O_4/bentonite 30% are 12 and 14 nm, respectively.

The Figure 4 shows the TEM images of pure Fe_3O_4. The average particle size of Fe_3O_4 nanoparticle is 14.5 nm. This result is almost the same as particle size obtained by BET analysis. Combining the results of TEM and BET, it can be concluded that the particle size is around < 20 nm.

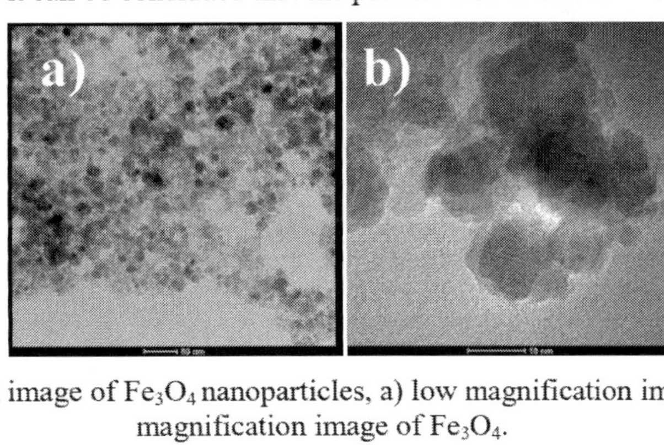

Figure 4. TEM image of Fe_3O_4 nanoparticles, a) low magnification image, and b) high magnification image of Fe_3O_4.

In this study, the adsorption performance of Fe_3O_4/30 wt% bentonite to the heavy metal ions is compared to that of pure Fe_3O_4. The Cu and Pb heavy metals adsorption test were carried out using an Atomic Absorption Spectrophotometer (AAS). According to the results, the concentration of Cu and Pb metal ions decreases to 46.3 and 509 for pure Fe_3O_4. While for Fe_3O_4/30 wt% bentonite nanocomposite, the concentration of Cu and Pb metal ions are 49 and 15 ppm, respectively. There is no significant difference in the adsorption performances of both materials to the Cu ion. But, the addition of bentonite significantly increases the adsorption performance of Fe_3O_4 to the Pb ion. When bentonite in contact with water, it will expand and increases the surface area of sample. The summarizing results of adsorption performance of Fe_3O_4/bentonite nanocomposite to Cu and Pb ions are showed in Table 3.

Table 3. Adsorption performance of Fe_3O_4/bentonite nanocomposite to Cu and Pb ions.

Pollute Element	Concentration (ppm)			Adsorption (%)	
	Initial Value	After adsorption			
		Pure Fe_3O_4	Fe_3O_4/30 wt% bentonite	Pure Fe_3O_4	Fe_3O_4/30 wt% bentonite
Cu	277	46.3	49	83.28	82.31
Pb	706	509	15	27.90	97.84

4. Conclusion

The Fe_3O_4/bentonite nanocomposites from natural iron-sand have been successfully prepared by co-precipitation method. All samples are composed of a single phase of Fe_3O_4 with the spinel cubic structure, lattice parameter of 8.373 Å and average particles size of around < 20 nm. The addition of bentonite into Fe_3O_4 nanoparticles generally reduces the magnetic properties of Fe_3O_4/bentonite nanocomposite. The optimum condition is achieved in Fe_3O_4/30 wt% bentonite composition with the saturation magnetization M_s = 22 emu/g, remanence magnetization M_r = 3.74 emu/g, and coercivity H_c = 119 Oe. In the aforesaid condition, the nanocomposite sample has a surface area of 105.9 m^2/g and pore size of 3.2 nm. It can absorb the 82.3 % Cu and 97.8 % Pb ions. Accordingly, this material can be considered as heavy metal adsorbent materials.

Acknowledgement

This work is financially supported by Program Riset Mandiri from Indonesian Institute of Sciences Tahun 2017.

References

[1] Ridwan and Manaf A 2009 Riset dan Pengembangan Nanopartikel Magnetik untuk Pengolahan Limbah Cair *Indones. J. Mater. Sci.* **Special Edition** 1–8
[2] Kodama R . 1999 Magnetic nanoparticles *J. Magn. Magn. Mater.* **200** 359–72
[3] Akbarzadeh A, Samiei M and Davaran S 2012 Magnetic nanoparticles: preparation, physical properties, and applications in biomedicine *Nanoscale Res. Lett.* **7** 144
[4] Gubin S P, Koksharov Y A, Khomutov G B and Yurkov G Y 2005 Magnetic nanoparticles: preparation, structure and properties *Russ. Chem. Rev.* **74** 489–520
[5] Lu A-H, Salabas E L and Schüth F 2007 Magnetic Nanoparticles: Synthesis, Protection, Functionalization, and Application *Angew. Chem. Int. Ed.* **46** 1222–44
[6] Suh S K, Yuet K, Hwang D K, Bong K W, Doyle P S and Hatton T A 2012 Synthesis of Nonspherical Superparamagnetic Particles: *In Situ* Coprecipitation of Magnetic Nanoparticles in Microgels Prepared by Stop-Flow Lithography *J. Am. Chem. Soc.* **134** 7337–43
[7] Junejo Y, Baykal A and Sözeri H 2013 Simple hydrothermal synthesis of Fe_3O_4-PEG nanocomposite *Open Chem.* **11**
[8] Meng H, Zhang Z, Zhao F, Qiu T and Yang J 2013 Orthogonal optimization design for preparation of Fe_3O_4 nanoparticles via chemical coprecipitation *Appl. Surf. Sci.* **280** 679–85
[9] Liu J, Zhao Z and Jiang G 2008 Coating Fe_3O_4 Magnetic Nanoparticles with Humic Acid for High Efficient Removal of Heavy Metals in Water *Environ. Sci. Technol.* **42** 6949–54
[10] Hashemian S, Saffari H and Ragabion S 2015 Adsorption of Cobalt(II) from Aqueous Solutions by Fe_3O_4/Bentonite Nanocomposite *Water. Air. Soil Pollut.* **226**
[11] Baqiya M A, Taufiq A, Sunaryono, Ayun K, Zainuri M, Pratapa S, Triwikantoro and Darminto 2017 Spinel-Structured Nanoparticles for Magnetic and Mechanical Applications *Magnetic Spinels - Synthesis, Properties and Applications* ed M S Seehra (InTech)

[12] Bajpai A K and Likhitkar S 2013 Investigation of magnetically enhanced swelling behaviour of superparamagnetic starch nanoparticles *Bull. Mater. Sci.* **36** 15–24
[13] Wu L, Ye Y, Liu F, Tan C, Liu H, Wang S, Wang J, Yi W and Wu W 2013 Organo-bentonite-Fe_3O_4 poly(sodium acrylate) magnetic superabsorbent nanocomposite: Synthesis, characterization, and Thorium(IV) adsorption *Appl. Clay Sci.* **83-84** 405–14
[14] Wu D, Zhu C, Chen Y, Zhu B, Yang Y, Wang Q and Ye W 2012 Preparation, characterization and adsorptive study of rare earth ions using magnetic GMZ bentonite *Appl. Clay Sci.* **62-63** 87–93
[15] Wan D, Li W, Wang G, Chen K, Lu L and Hu Q 2015 Adsorption and heterogeneous degradation of rhodamine B on the surface of magnetic bentonite material *Appl. Surf. Sci.* **349** 988–96
[16] Rouquerol F, Rouquerol J, Sing K S W, Llewellyn P and Maurin G 2014 *Adsorption by powders and porous solids: principles, methodology and applications* (Academic Press, Inc.)

IOP Conf. Series: Materials Science and Engineering **316** (2018) 012054 doi:10.1088/1757-899X/316/1/012054

Properties of carbon composite paper derived from coconut coir as a function of polytetrafluoroethylene content

Fredina Destyorini[1], Indriyati[2], Nanik Indayaningsih[1], Bambang Prihandoko[1] and Anne Zulfia Syahrial[3]

[1]Research Center for Physics, Indonesian Institute of Sciences (LIPI), South Tangerang, Indonesia
[2]Research Unit for Clean Technology, Indonesian Institute of Sciences (LIPI), Bandung, Indonesia
[3]Department of Metallurgical and Materials Engineering, Faculty of Engineering, Universitas Indonesia, Depok, Indonesia

E-mail: fred007@lipi.go.id; fredinadestyorini@gmail.com

Abstract. The carbon composite papers were produced by utilizing carbon materials from coconut coir. In the present work, carbon composite papers (CCP) were prepared by mixing carbon materials in the form of powder and fibre with polymer (ethylene vinyl acetate and polyethylene glycol) in xylene at 100°C. Then, polytetrafluoroethylene (PTFE) with different content was used to treat the surface of CCP. The properties of PTFE-coated CCP were analysed by means of contact angle measurement, tensile testing, porosity, density, and electrical conductivity measurements. As expected, all CCP's surfaces treated with PTFE were found to be hydrophobic with contact angle >120° and relatively constant during 60 minutes measurement. Furthermore, water contact angle, density, and mechanical properties of CCP generally increase with increasing PTFE content. However, the porosity and electrical conductivity of CCP decrease slightly as the PTFE content increased from 0 wt% to 30 wt%. Based on the observation and analysis, the optimum PTFE content on CCP was 20 %, in which the mechanical properties and hydrophobicity behaviour were improved significantly, but it was only caused a very small drop in porosity and electrical conductivity

1. Introduction

Coconut coir is an abundant agricultural waste in Indonesia but not yet optimal utilizations. It contains hydrocarbon such as cellulose, hemi-cellulose and lignin as major composition [1]. The carbonization processes of hydrocarbon in inert gas to temperature of 1300°C produces charcoal with high carbon content and conductive [2]. Due to this abundance and advantages, several researchers have initiated to use the potential of coconut coir for raw material of carbon paper for gas diffusion layer (GDL) in proton exchange membrane fuel cells (PEMFC) [3,4]. Previous study in our lab revealed that coconut carbon fibre has internal porous (in the range of 1-5 μm) to enhance reactant gases reach active sites in the catalyst layer [2].

GDLs are generally available in the form of carbon paper and carbon cloth. At present, they are composed of carbon materials, which are produced from fossil fuels such as coal and petroleum, raising concerns about sustainability. The utilization of coconut coir as carbon source for the manufacture of carbon paper for GDLs could resolve these issues.

As one of the critical components in PEMFC, GDL main functions are to supply reactant gases, remove product water, and conduct electron between adjacent components, and provide mechanical support for membrane electrode assembly (MEA) [5]. In order to perform these functions, carbon paper as GDL should be porous (50%-90% of porosity), hydrophobic (contact angle >90°), conductive, and strong enough [6]. Hydrophobic property controlled by hydrophobic treatment is needed to remove liquid water saturation in the cathode effectively. Although carbon substrates used as GDL are usually hydrophobic, they are typically treated with a hydrophobic agent such as polytetrafluoroethylene (PTFE), which is used for the commercial carbon paper [7,8]. PTFE treatment applies to both, anode and cathode sides. This feature reduces the flooding of water in the GDL and provides the reactant access to the catalyst layer [9]. However, its impact on other parameters such as electrical conductivity and porosity should be considered for an appropriate cell design. It has been reported that usage PTFE with too high content could decrease porosity and conductivity of carbon paper [8,10]. On the other hand, too low PTFE content would make water generated in the cell cannot be removed effectively. Therefore, PTFE content in the GDL should be balanced and optimized to achieve the best performance of fuel cell.

In this study, carbon paper or GDL was prepared from coconut coir. The coconut coir carbon paper (CCP) was then treated with various concentration of PTFE to improve its hydrophobic property. The optimum PTFE content was determined based on investigation of their physical and mechanical properties. Relationship among those parameters will be revealed as an effect of PTFE content.

2. Experiment

2.1. Materials
Coconut coir obtained from local market, was cut into a length of ±2 mm. Ethylene vinyl acetate (EVA) and poly ethylene glycol (PEG) purchasedfrom Aldrich Chemical Co., Inc. (St. Lois, MO, USA) were used as binder and dispersant, respectively. Xylene (Brataco Chemica) was used as solvent for making the CCP. Polytetrafluoroethylene (PTFE) emulsion obtained from MTI Corporation was used as hydrophobic agent of CCP.

2.2. Production of carbon materials
Coconut coir was processed by two stages, namely: carbonization and pyrolysis. In the carbonization process, coconut coir was heated at 500°C for an hour under N_2 atmosphere and cooled down to room temperature to produce charcoal which had high carbon content. Subsequently, the charcoal was processed through pyrolysis at 1300°C for an hour under N_2 atmosphere to improve the electrical conductivity, eliminate impurities, and improve other properties of carbon. The resulting carbon material was named as carbon fibre. Some of carbon fibre was grinded into a powder of ±74 μm sizes (200 mesh).

2.3. Preparation of CCP
Carbon fibre (70 wt%) and carbon powder (10 wt%) were mixed with EVA and PEG in xylene at 100°C for 2 hours to make slurry. To form a paper, the slurry was cast on molded glass, rolled, and then dried at room temperature for 24 hours to evaporate the solvent. To provide the hydrophobic property, the CCPs were treated with different quantities of PTFE for 30 minutes, varied from 10 to 30 wt%, and dried at room temperature for 24 hours, and finally sintered at 350°C.

2.4. Characterization of CCP
Characterization and analysis were carried out on untreated and PTFE-treated CCP samples. For evaluating the hydrophobic properties of a CCP, surface contact angles between water and the carbon

paper were measured using sessile drop method. For each measurement, a 50 μL water droplet was placed onto the surface of CCP and images were captured every 20 minutes for 1 hour after the droplet attached to the sample surface. Furthermore, the droplet shape was analyzed using Bashforth and Adams tables [11] to determine the contact angle of the samples. Large contact angles imply high hydrophobia. Porosity and density were examined by the kerosene density method using Archimedes principle in accordance with BS 1902: Part 1A standard. Through-plane electrical conductivity was measured using HIOKI 3522-50 HITESTER LCR-meter. The mechanical properties of the carbon paper samples were measured by using a universal testing machine (Orientec Co. Ltd, Model UCT-5T) according to ASTM D828-97. For each sample, 7 rectangular specimens (70×10 mm) were tested and an average tensile strength and Modulus Young were determined.

3. Results and discussion

3.1. Hydrophobicity (water contact angle) of CCP

The hydrophobic properties of a solid surface could be determined by the surface contact angle. When liquid water was used as a wetting agent, hydrophilic carbon papers will have contact angle smaller than 90°, while hydrophobic carbon papers have contact angle between 90 and 180° [12]. Fig. 1 shows the contact angle of CCP with different content of PTFE. The contact angle of untreated CCP (0 wt% PTFE) changes during 60 minutes of measurement process. At the first minute when the water droplet attached to the surface of untreated CCP, it showed the hydrophobicity with contact angle greater then 90°, indicating that the substrate is resistant to wetting. However, water droplet started to penetrate into the porosity of CCP, then untreated CCP became wet and the contact angle dropped to 0° (40 to 60 min). This indicates that the untreated CCP is not able to resist the water penetration for a long period. When the paper is impregnated with PTFE, contact angles increase from approximately 115° up to 135-145°. All the PTFE-treated CCPs exhibit high contact angles and are relatively constant during the measurement process as can be seen in Fig. 1. It is noted that the droplets show relatively similar contact angle on CCPs with different PTFE content (10 to 30 wt%). This elucidates that PTFE treatment significantly affects the hydrophobicity of carbon paper, but the contact angle does not change too much for different PTFE content.

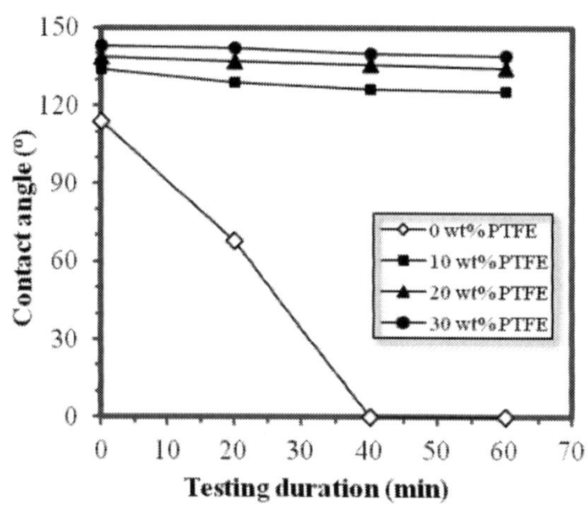

Figure 1. Contact angle of CCP with different PTFE content

3.2. Porosity and density of CCP

Porosity is needed to diffuse and transport reactant/product in the electrode. The porosity and density measurements using kerosene density method were only performed on PTFE-treated CCPs. These parameters cannot be measured by this method for the untreated CCP because EVA as a binder will dissolve in kerosene [13]. The porosity and density value of PTFE-treated CCPs are shown in Fig. 2 and Fig. 3, respectively. It is proved that increasing PTFE content from 10 to 30% decrease the porosity of CCPs for about 8 %.

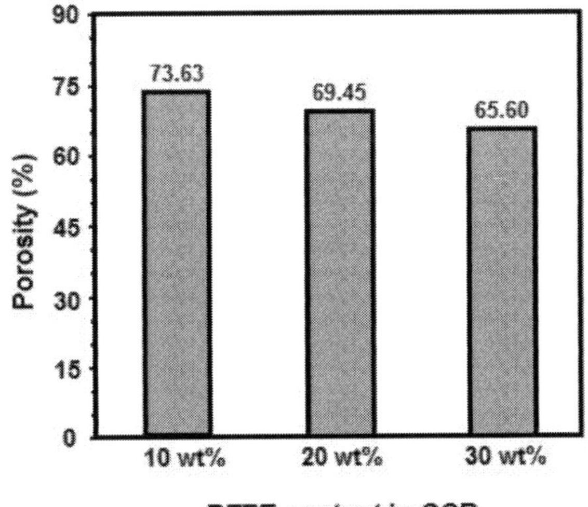

Figure 2. Porosity of CCP with different content of PTFE

This could be explained that PTFE particles penetrate through the pores of CCP and deposit onto the carbon fibres or accumulate at the fibre crossing. As a consequence, it reduces the overall porosity of CCPs. Yoon and Park [6] reported that carbon paper should have high porosity of 50-90% to effectively function as GDL, and the porosity of all treated CCPs is in this range. It should be noted that higher GDL porosity improves mass transport, leading to higher current density produced by PEMFC, but it will decrease the electrical conductivity as well as the mechanical properties [5] that will be shown in the next section. Fig. 3 shows the density of CCP as a function of PTFE content. The density of CCP gradually increased as PTFE content increased. This is a reasonable result since PTFE penetrate into the CCP's pores and accumulate at the fibre crossing as mentioned above. The densities increase from 0.42 to 0.55 gram/cm^3 for PTFE 10 to 30%, respectively.

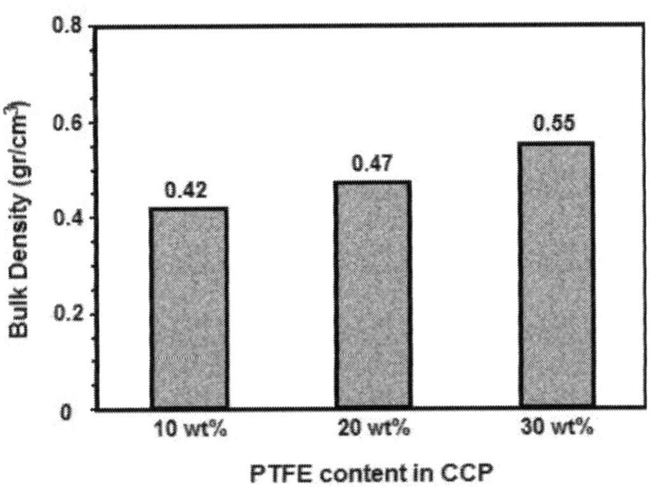

Figure 3. Density of CCP with different content of PTFE

IOP Conf. Series: Materials Science and Engineering **316** (2018) 012054 doi:10.1088/1757-899X/316/1/012054

3.3. Electrical conductivity of CCP

The values of through-plane electrical conductivity of all samples with different PTFE content are represented in Figure 4. As predicted that an increase of PTFE content up to 30% leads to decrease in the electrical conductivity for almost 50% compared to untreated CCP. It occurs because of the increasing numbers of non-conductive PTFE particles that isolated the conductive carbon materials and disconnected the conduction path in the CCP [14].

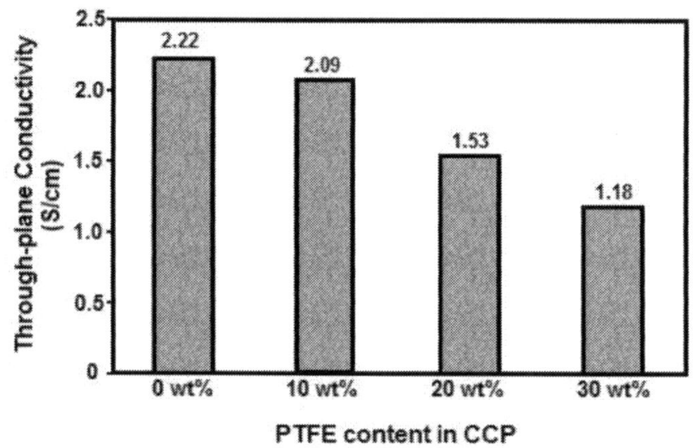

Figure 4. Through-plane electrical conductivity of CCP with different content of PTFE

3.4. Mechanical properties of CCP

Fig. 5 and Fig. 6 show the mechanical properties of untreated CCP and PTFE-treated CCPs with varying PTFE content. It is observed that PTFE treatment could improve the tensile strength and Modulus Young of CCP. In this case, PTFE also acts as the matrix in the composites that bonds the carbon fibres together and transfers loads between them. A decrease in porosity of CCP because of the addition of PTFE content may also cause an improvement of tensile properties [15] of CCP. It is noticed that increasing PTFE content from 10 to 20% gives significant improvement to the mechanical properties of CCP compared to increasing PTFE content from 20 to 30%.

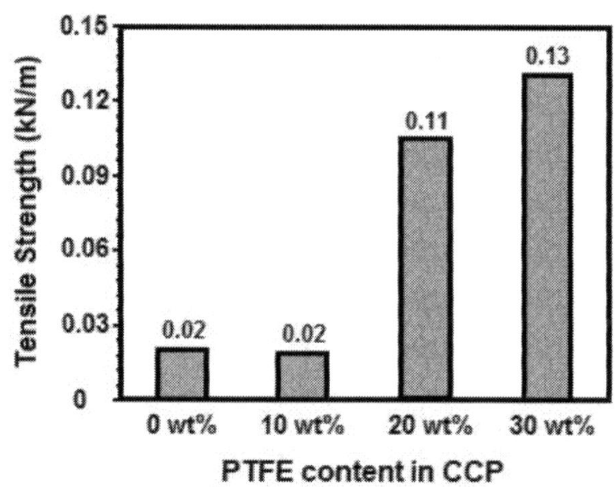

Figure 5. Tensile strength of CCP with different content of PTFE

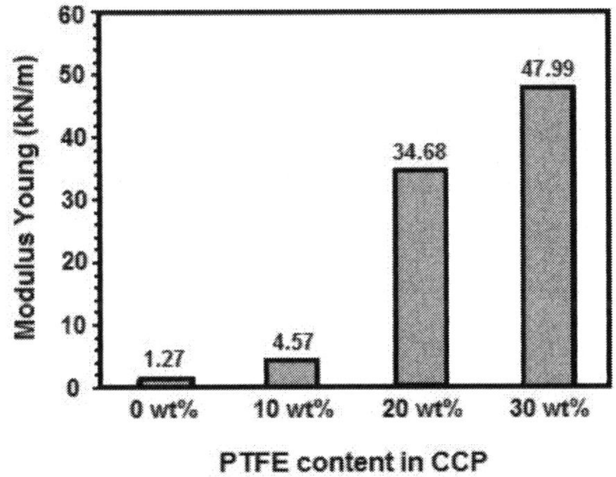

Figure 6. Modulus Young of CCP with different content of PTFE

4. Conclusion

The effects of PTFE hydrophobic polymer content on the physical and mechanical properties of carbon paper from coconut coir have been investigated. The porosity and electrical conductivity are found decreasing slightly, while the hydrophobicity and mechanical properties of carbon paper increase with the increasing of PTFE content in carbon paper. The use of 20 wt% PTFE in carbon paper from coconut coir is the optimum content because this content could improve CCP's tensile properties significantly and maintain the hydrophobicity of CCP, but it is only caused a small drop in porosity and electrical conductivity.

Acknowledgement

This work was supported by Research Center for Physics - Indonesian Institute of Sciences (LIPI), and funded by the Indonesian Institute of Sciences under Unggulan Programme No. 3406.004.001.012

References

[1] Ali M 2011 Coconut fibre: A versatile material and its applications in engineering (Review) *J. Civ. Eng. Constr. Technol.* **2** 189–97

[2] Indayaningsih N, Priadi D, Zulfia A and Suprapedi 2011 Analysis Of Coconut Carbon Fibers For Gas Diffusion Layer Material *Key Eng. Mater.* **462–463** 937–42

[3] Indayaningsih N, Zulfia A, Priadi D and Hendrana S 2016 Preparation of carbon composite from coconut fiber for gas diffusion layer *Ionics (Kiel).* **22** 1445–1449

[4] Kinumoto T, Matsumura T, Yamaguchi K, Matsuoka M, Tsumura T and Toyoda M 2015 Material Processing of Bamboo for Use as a Gas Diffusion Layer in Proton Exchange Membrane Fuel Cells *ACS Sustain. Chem. Eng.* **3** 1374–80

[5] Kandlikar S G, Garofalo M L and Lu Z 2011 Water management in a PEMFC: Water transport mechanism and material degradation in gas diffusion layers *Fuel Cells* **11** 814–823

[6] Yoon G H and Park Y il 2012 Enhanced hydrophobicity of GDL by a novel coating process in PEM fuel cells *Int. J. Precis. Eng. Manuf.* **13** 1153–9

[7] Song W, Yu H, Shao Z, Yi B, Lin J and Liu N 2014 Effect of polytetrafluoroethylene distribution in the gas diffusion layer on water flooding in proton exchange membrane fuel cells *Chinese Journal of Catalysis* **35** 468–473

[8] Lobato J, Cañizares P, Rodrigo M A, Ruiz-López C and Linares J J 2008 Influence of the Teflon loading in the gas diffusion layer of PBI-based PEM fuel cells *J. Appl. Electrochem.* **38** 793–802

[9] Daino M M and Kandlikar S G 2012 3D phase-differentiated GDL microstructure generation with binder and PTFE distributions *Int. J. Hydrogen Energy* **37** 5180–5189
[10] Fishman Z and Bazylak A 2011 Heterogeneous Through-Plane Porosity Distributions for Treated PEMFC GDLs. II. Effect of MPL Cracks *J. Electrochem. Soc.* **158** B841–5
[11] Bashforth F and Adams J C 1883 *An attempt to test the theories of capillary action by comparing the theoretical and measured forms of drops of fluid. With an explanation of the method of integration employed in constucting the tables which give the theoretical forms of such drops* (Cambridge: Cambridge [Eng.] University Press)
[12] Park S and Popov B N 2009 Effect of cathode GDL characteristics on mass transport in PEM fuel cells *Fuel* **88** 2068–73
[13] Sangwal K 2007 *Additives and crystallization processes: from fundamentals to applications* (Chichester: John Wiley & Sons, Inc.)
[14] Hung T F, Huang J, Yangandy F L W and Chen-Yang 2007 Physical and Electric Properties of a PTFE-Containing Gas Diffusion Layer for Proton Exchange Membrane Fuel Cell *Polym. Prepr.* **48** 477–8
[15] Wang X, Chi Y and Cao Y 2014 The effect of PTFE on the mechanical, friction, and wear properties of CF/PMMA composites *J. Thermoplast. Compos. Mater.* **27** 1232–1238

IOP Conf. Series: Materials Science and Engineering **316** (2018) 012055 doi:10.1088/1757-899X/316/1/012055

Preparation of anatase TiO_2 nanoparticles using low hydrothermal temperature for dye-sensitized solar cell

N Sofyan[1,2], A Ridhova[1], A H Yuwono[1,2] and A Udhiarto[3]

[1] Department of Metallurgical and Materials Engineering, Faculty of Engineering, Universitas Indonesia, Depok 16424, Indonesia
[2] Tropical renewable Energy Center, Faculty of Engineering, Universitas Indonesia, Depok 16424, Indonesia
[3] Department of Electrical Engineering, Faculty of Engineering, Universitas Indonesia, Depok 16424, Indonesia

E-mail: nofrijon.sofyan@ui.ac.id

Abstract. One device being developed as an alternative source of renewable energy by utilizing solar energy source is dye-sensitized solar cells (DSSC). This device works using simple photosynthetic-electrochemical principle in the molecular level. In this device, the inorganic oxide semiconductor of titanium dioxide (TiO_2) has a great potential for the absorption of the photon energy from the solar energy source, especially in the form of TiO_2 nanoparticle structure. This nanoparticle structure is expected to improve the performance of DSSC because the surface area to weight ratio of this nanostructures is very large. In this study, the synthesis of TiO_2 nanoparticle from its precursors has been performed along with the fabrication of the DSSC device. Effort to improve the size of nanocrystalline anatase TiO_2 was accomplished by low hydrothermal treatment at various temperatures whereas the crystallinity of the anatase phase in the structure was performed by calcination process. Characterization of the materials was performed using X-ray Diffraction (XRD) and scanning electron microscope (SEM), while the DSSC performance was examined through a high precision current versus voltage (I-V) curve analyzer. The results showed that pure anatase TiO_2 nanoparticles could be obtained at low hydrothermal of 100, 125, and 150 °C followed by calcination at 450 °C. The best performance of photocurrent-voltage characteristic was given by TiO_2 hydrothermally synthesized at 150 °C with power conversion efficiency (PCE) of 4.40 %, whereas the standard TiO_2 nanoparticles has PCE only 4.02 %. This result is very promising in terms low temperature and thus low cost of anatase TiO_2 semiconductor preparation for DSSC application.

1. Introduction

Titanium dioxide (TiO_2) has long been the subject of in-depth research because of its wide range applications such as sensors [1], optical emission, photonic crystals, catalysts, photocatalysts [2], purification of environment [3], photovoltaic cells, batteries and electrochemical devices [4-6]. This material is one of the inorganic oxide metals known as transitional metal oxides (TMO) [7]. The discovery that crystalline TiO_2 can be used as electrode in the dye-sensitized solar cell (DSSC) has driven even more intensive research due to its high efficiency and low fabrication cost [6].

The presence of oxide semiconductor layer plays a very important role in the DSSC for converting light energy into electrical energy. Moreover, the interaction between the dye and semiconductor oxide particles will be very instrumental in determining the process of converting light energy into electrical energy [8].

In the field of nanomaterials, it has been well known that with the small size down to the nanometer scale, the ratio between the surface area to volume ratio (surface-area-to-volume-ratio) of the material will be greater [9]. In this aspect, the possibility of surface interactions with the environment will be higher, and hence, the use of TiO_2 nanomaterial in the form of nanoparticles, nanotubes and or mesoporous as semiconductor oxide layers in a DSSC is expected to improve the system performance [10].

There are many routes for synthesizing TiO_2 nanopartcile, one of which through the hydrolysis of titanium alkoxide followed by a sol-gel reaction in acidic environments and hydrothermal growth [11]. However, the hydrothermal temperature treatment for this method is usually above 190 °C, which is still quite high. Because of that, in this paper, preparation of TiO_2 nanoparticles was optimized at relatively low hydrothermal temperature that can be applied for DSSC using titanium tetra isopropoxide precursor by the sol-gel technique in a tetra methyl ammonium hydroxide basic environment. Monodisperse TiO_2 nanoparticles of homogeneous morphology were obtained by optimizing the synthesis method. The results obtained are reported and discussed.

2. Experimental Setup

2.1. Preparation of TiO_2

The precursors were titanium tetra isopropoxide (TTIP) with chemical formula $Ti\{OCH(CH_3)_2\}_4$ and tetra methyl ammonium hydroxide (TMAH) with chemical formula $N(CH_3)_4^+OH^-$, both were purchased from Sigma Aldrich. These precursors were used during the sol-gel followed by hydrothermal process. The preparation of TiO_2 is in according to the literature [11] with a slight modification to lower the hydrothermal temperature process and the method is explained as the following: A total of 40.5 mL TTIP was added into 112.5 ml of distilled water with a ratio of 1: 50 at 15 °C under a magnetic stirrer for 1 hour. White precipitates resulting from the mixing process were filtered and washed three times using 50 ml of distilled water. The precipitates were transferred into a solution of TMAH and were refluxed at 100 °C for 4 hours. TiO_2 content was set at 10 % by mass. The colloids produced from peptizing were then treated hydrothermally at three different low temperature variations 100, 125, and 150 °C for 12 hours in a Teflon autoclave. The colloidal suspension was then centrifuged at 5,000 rpm and the obtained precipitates were washed with dry ethanol (Merck KGaA, Germany). At the last stage, TiO_2 nanoparticles were calcined at 450 °C for two hours before being characterized using X-ray diffraction (XRD, ARL OPTX-2050) and scanning electron microscope (SEM, FEI Quanta-650).

2.2. DSSC Fabrication

The TiO_2 nanoparticle paste was prepared by adding an amount of methanol into TiO_2 nanoparticle powder along with two drops of TRITON X-100 (Sigma Aldrich). The blend was mixed until a homogeneous phase was obtained. Two fluorine-doped tin oxide (FTO, Solaronix) conductive glass substrates were prepared. One of the glass substrates was drilled to create two tiny perforations at one side and dry cleaned by using methanol. The other FTO glass was also dry cleaned by using methanol and attached to a flat layer. This substrate was tape masked and coated with TiO_2 nanoparticle paste using a doctor-blade method under an area of about 1 cm^2. The paste was dried at 450 °C for an hour and allowed to cool down before soaking it in the commercial organic dye solution (RK1, Solaronix) and air dried. As a comparison, the commercial TiO_2 nanoparticle (Degussa P25) was also prepared under the same condition. The other perforated FTO glass substrate was coated with a platinum paste (Platisol, Solaronix) and dried at 400 °C for an hour. Both the coated glass substrates were then attached to one another separated by a spacer and sealed to avoid electrolyte leakage. Once dried, the electrolyte (Iodolyte, Solaronix) was then injected through the two tiny holes drilled at one side of the substrate. The holes were then sealed and the cell was ready for characterization. Cell activity was measured by using a simple ammeter, while the cell performance was tested by using a Semiconductor Parameter Analyzer (SPA, Agilent 4155A) with a standard illumination of about 100 mW/cm^2.

The power conversion efficiency (PCE) of the device was calculated in accordance with the following equation [12]:

$$\eta = \frac{FF \times J_{SC} \times V_{OC}}{I_{in}} \times 100 \qquad (1)$$

where J_{sc} is the short-circuit photocurrent density (mA cm^{-2}), V_{oc} is the open-circuit voltage (volts), I_{in} is the intensity of the incident light (mW cm^{-2}) and FF is the fill factor defined as:

$$FF = \frac{i_{max} V_{max}}{i_{OC} V_{OC}} \qquad (2)$$

where i_{max} is the maximum photocurrent, V_{max} is the maximum voltage, respectively, and i_{oc} is the open-circuit current (mA). The values of i_{max} and V_{max} can be extracted from the maximum power of the I-V characteristics.

3. Results and Discussion

The samples prepared from three different hydrothermal temperature variations and calcined at 450 °C were characterized by using X-ray diffraction. As a comparison, diffraction pattern from the commercially available TiO_2 nanoparticle (Degussa P25) was also taken. The diffraction patterns are shown in figure 1.

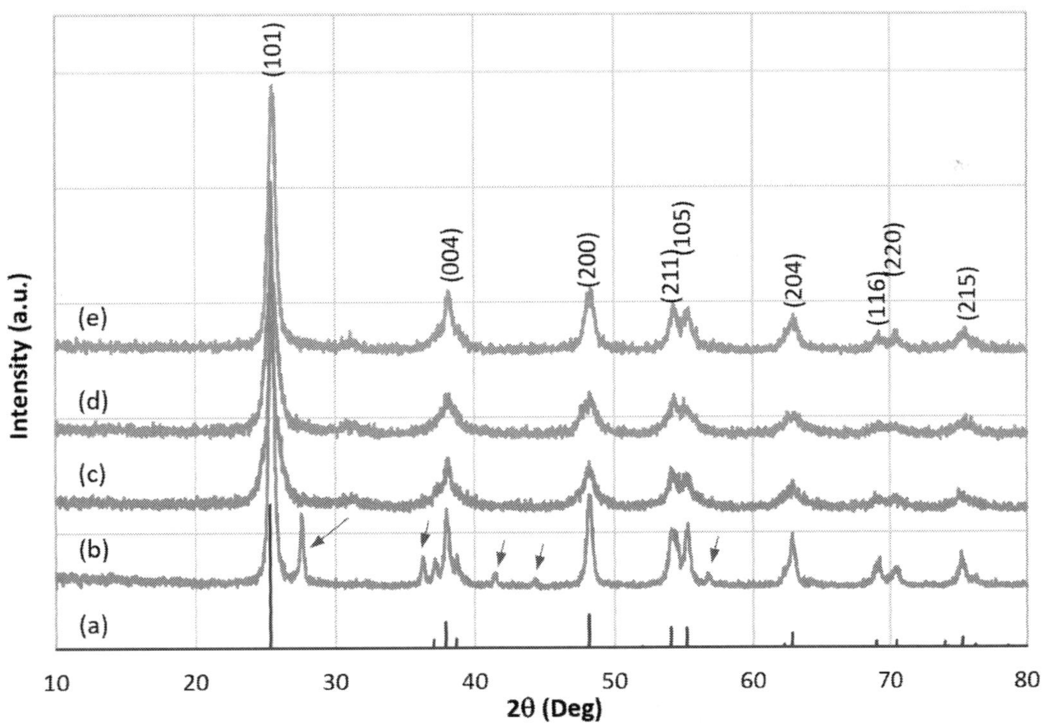

Figure 1. X-Ray diffraction patterns of TiO_2 (a) reference (JCPDS No. 73-1764), (b) Degussa P25, and hydrothermally synthesized at (c) 100 °C, (d) 125 °C, and (e) at 150 °C.

As seen from figure 1, all of the samples show the same trend and confirmed the pure anatase crystal structure (JCPDS file No. 73-1764). The three distinct diffraction peaks observed at 2θ value of 25.31°,

IOP Conf. Series: Materials Science and Engineering **316** (2018) 012055 doi:10.1088/1757-899X/316/1/012055

37.79°, and 48.04° are corresponding to (101), (004), and (200) anatase crystal planes, respectively [13]. No other characteristic peaks of rutile phase are detected in the diffractograms of the as-synthesized TiO_2. Other peaks of rutile, however, are observed in the commercial TiO_2 nanoparticles (Degussa P25) as indicated by the arrows in figure 1(b).

Based on the diffractograms, the crystallite size of the as-synthesized TiO_2 nanoparticles was calculated from the Debye–Scherrer's equation using the XRD line broadening [14]:

$$c_s = \frac{k\lambda}{B\cos\theta} \qquad (3)$$

where c_s is the crystallite size, λ is the wavelength of the X-ray radiation with Cu Kα = 0.15406 nm, k is a constant taken as 0.94, θ is the diffraction angle, and B is the full width at half maximum (FWHM) peak in radians.

All of the diffraction peaks were taken into account from which the average crystallite size of each temperature variation was obtained, i.e. 5.9 nm, 6.3 nm, and 8.6 nm at temperature variation 100 °C, 125 °C, and 150 °C, respectively. It seems that with the increase of hydrothermal temperature resulted in increase of crystallite size. This can be understood since increasing the hydrothermal temperature will result in more energy for the crystal to grow.

Scanning electron microscope was used to study the surface morphological features of the TiO_2 nanoparticle and the results are given in figure 2. Figure 2(a) is the secondary electron image of the commercial TiO_2 nanoparticle (Degussa P25), whereas the hydrothermally synthesized TiO_2 at 100, 125, and 150 °C is given in figure 2(b), 2(c), and 2(d), respectively.

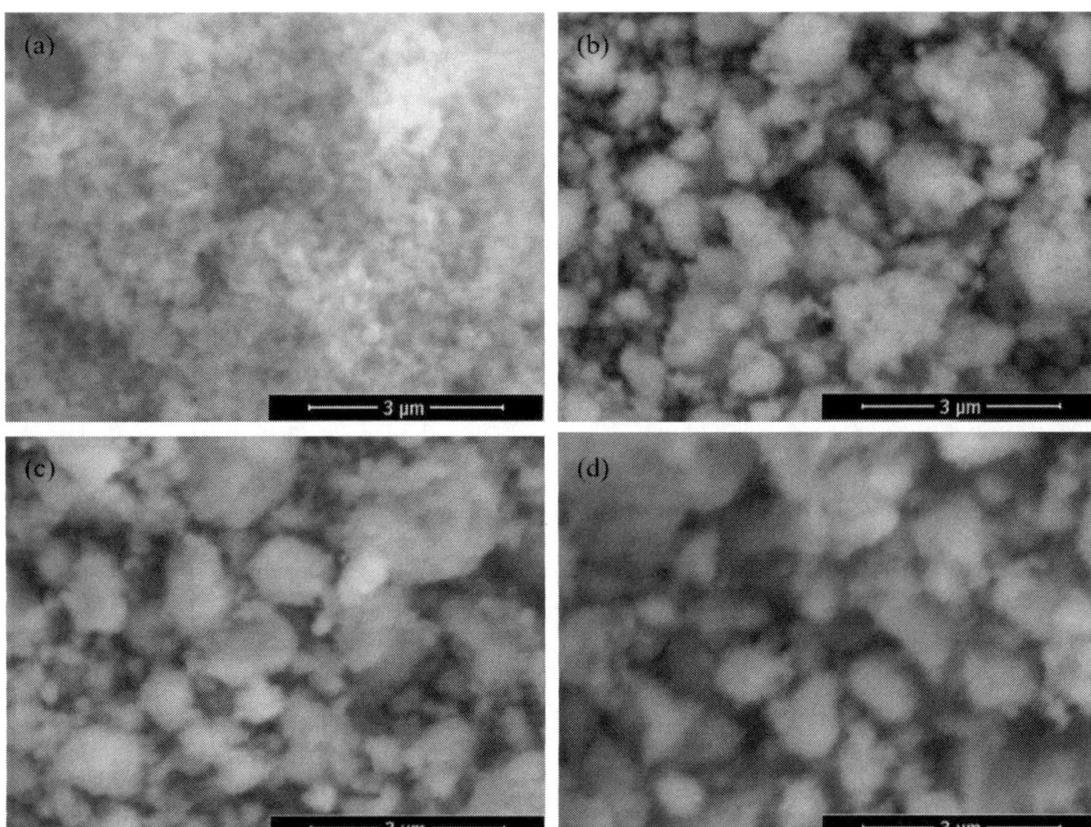

Figure 2. Secondary electron images of (a) commercial TiO_2 nanoparticle and hydrothermally synthesized TiO_2 at (b) 100 °C, (c) 125 °C, and (d) at 150 °C.

As seen in figure 2, the as-synthesized TiO_2 nanoparticles are still quite diverse in size and some of them are still clumped and agglomerated as compared to those of commercial nanoparticles TiO_2. Nonetheless, image processing shows that the average particle size is 450, 500, and 590 nm for temperature variation 100, 125, and 150 °C, respectively. This result is in agreement with that of the crystallite size confirmed in the X-ray diffraction analysis mentioned previously. Compared to the commercial TiO_2 nanoparticles with an average particle size < 80 nm, the obtained particle sizes from the as-synthesized TiO_2 nanoparticles are still quite large. Nonetheless, since all of the as-synthesized TiO_2 nanoparticles have shown the characteristics of pure anatase, the material thus could be used as semiconductor photoactive layer in dye-sensitized solar cell (DSSC).

Photocurrent-voltage characteristics of the DSSC from commercial TiO_2 (Degussa P25) and as-synthesized TiO_2 at various hydrothermal temperature are given in figure 3. As seen in figure 3, photocurrent-voltage characteristic of both as-synthesized TiO_2 nanoparticle hydrothermally synthesized at various temperatures and commercial TiO_2 nanoparticle show a current density of around 10 mA/cm^2. Power conversion efficiency (PCE) of the DSSC are given in table 1.

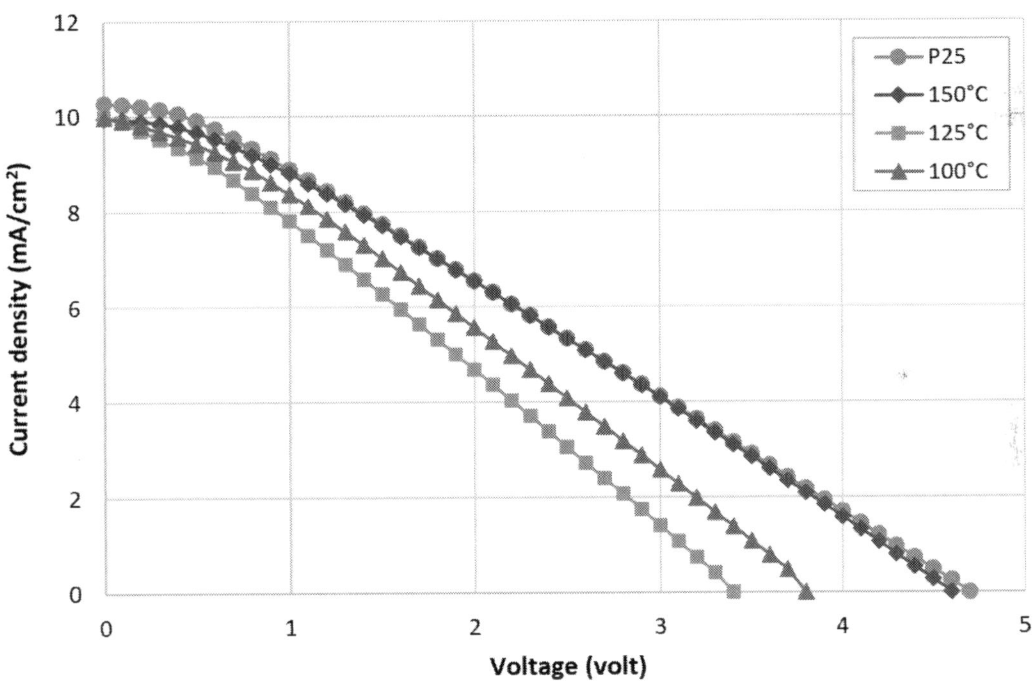

Figure 3. Photocurrent-voltage characteristics of the DSSC from the as-synthesized TiO_2 at various hydrothermal temperatures and commercial TiO_2 nanoparticle.

As seen in table 1, the PCE from all of the samples are above 4 %. For the as-synthesized TiO_2 nanoparticle, it seems that there is an anomaly on the PCE of hydrothermal at 125 °C. In according to the crystallite and particle size, the increase in hydrothermal should also increase the PCE, which is not the case. This could result from particle inhomogeneity and or agglomeration; however, the exact reason has yet to be confirmed. The interesting point is that PCE from the as-synthesized TiO_2 hydrothermally treated at various temperatures is larger than that of commercial TiO_2. This can be understood since the commercial TiO_2 nanoparticle contains rutile phase. This rutile phase can reduce the power efficiency due to different interparticle connectivity associated with the particle packing density results in a slow electron transport [15, 16].

Table 1. Power conversion efficiency of the DSSC from the as-synthesized TiO_2 at various hydrothermal temperatures and commercial TiO_2 nanoparticle.

	100 °C	125 °C	150 °C	P25
V_{oc} (volt)	3.90	3.60	4.70	4.80
I_{oc} (mA)	9.99	9.99	9.96	10.28
V_{max} (volt)	0.45	0.45	0.45	0.40
I_{max} (mA)	9.42	9.14	9.78	10.06
FF	0.11	0.11	0.09	0.08
PCE (%)	4.24	4.11	4.40	4.02

4. Conclusion

In this work, low hydrothermal temperature method for synthesizing TiO_2 nanoparticles has been successfully carried out at 100, 125, and 150 °C followed by a calcination at 450 °C. The method was able to produce a pure anatase crystalline structure. The smallest crystallite size was achieved at hydrothermal temperature of 125 °C, however, the best performance for DSSC application was achieved at hydrothermal temperature of 150 °C with power conversion efficiency 4.40 %. This result is very promising in terms low temperature and thus low cost of anatase TiO_2 semiconductor preparation for DSSC application

Acknowledgments

The authors would like to express their gratitude for the funding from the Directorate of Research and Community Services (DRPM), Universitas Indonesia, through Hibah PITTA No. 823/UN2.R3.1/ HKP.05.00/2017

References

[1] Varghese O K, Gong D, Paulose M, Ong K G, Dickey E C, Grimes C A, 2003 *Adv. Mater.* **15** 624
[2] Livraghi S, Votta A, Paganini M C, Giamello E, 2005 *Chem. Commun.* **4** 498
[3] Ikezawa S, Homyara H, Kubota T, Suzuki R, Koh S, Mutuga F, Yoshioka T, Nishikawi A, Ninomiya Y, Takahashi M, Baba K, Kida K, Hara T, Famankinwa T, 2001 *Thin Solid Films*, **386** 173
[4] Hagfeldt A and Grätzel M, 1995 *Chem. Rev.* **95** 49
[5] Kavana L, Attiaa A, Lenzmann F, Elder S H, Grätzel M, 2000 *J. Electrochem. Soc.* **147** 2897
[6] Grätzel M, 2001 *Nature* **414** 338
[7] Liu L, Ji Z, Zou W, Gu X, Deng Y, Gao F, Tang C, Dong L, 2013 *ACS Catalysis* **3** 2052
[8] Narayan M R, 2012 *Renew. Sust. Energ. Rev.* **16** 208
[9] Yuwono A H, Ramahdita G, Sofyan N, 2012 *Adv. Mater. Research* **557-559** 468
[10] O'Regan B and Grätzel M, 1991 *Nature* **353** 737
[11] Burnside S D, Shklover V, Barbé C, Comte P, Arendse F, Brooks K, Grätzel M, 1998 *Chem. Mater.* **10** 2419
[12] Fernando J M R C and Senadeera G K R, 2008 *Curr. Sci.* **95** 663
[13] Lekphet W, KeT-C, Su C, Sireesha P, Kathirvel S, LinY-F, Chen B-R, Li W-R., 2017 *Solar Energy* **142** 1
[14] Cullity B D, 1978 *Elements of X-ray Diffraction* (Massachusetts: Addison-Wesley).
[15] Koelsch M, Cassaignon S, Minh C T T, Guillemoles J F, Jolivet J P, 2004 *Thin Solid Films* **451–452** 86
[16] Park N G, van de Lagemaat J, Frank A J, 2000 *J. Phys. Chem. B* **104** 8989

Optimization of the Water Volume in the Buckets of Pico Hydro Overshot Waterwheel by Analytical Method

Budiarso[1], Dendy Adanta[1]*, Warjito[1], A I Siswantara, Pradhana Saputra[1] and Reza Dianofitra[1]

[1]*Department of Mechanical Engineering, Faculty of Engineering, Universitas Indonesia, Kampus Baru UI Depok 16424, Indonesia*

E-mail: dendyadanta@ymail.com

Abstract. Rapid economic and population growth in Indonesia lead to increased energy consumption, including electricity needs. Pico hydro is considered as the right solution because the cost of investment and operational cost are fairly low. Additionally, Indonesia has many remote areas with high hydro-energy potential. The overshot waterwheel is one of technology that is suitable to be applied in remote areas due to ease of operation and maintenance. This study attempts to optimize bucket dimensions with the available conditions. In addition, the optimization also has a good impact on the amount of generated power because all available energy is utilized maximally. Analytical method is used to evaluate the volume of water contained in bucket overshot waterwheel. In general, there are two stages performed. First, calculation of the volume of water contained in each active bucket is done. If the amount total of water contained is less than the available discharge in active bucket, recalculation at the width of the wheel is done. Second, calculation of the torque of each active bucket is done to determine the power output. As the result, the mechanical power generated from the waterwheel is 305 Watts with the efficiency value of 28%.

1. Introduction

Rapid economic and population growth in Indonesia lead to increased energy consumption, including electricity needs. Electrical energy development that is more efficient at a lower cost becomes vital since energy resource's price tend to increase whilst its availability was limited. Meanwhile, Indonesia is a developing country where electricity supply is not evenly distributed in the entire region even though electricity is the primary need of everyone [1].

Construction of hydroelectric power (hydropower) is one of the potential electricity supply solution for developing countries such as Indonesia in which some areas, including remote areas, have rich water energy potential. The remote areas in Indonesia have several characteristics, such as the lack of infrastructure, clean water, inadequate electricity, irrigation, education, economy, and low productivity [2]. In designing hydroelectric power generation, the selection of water turbine technology should be considered as each turbine has its specifications. There are at least two parameters that are usually used in determining the type of turbine to be used namely head and flow rate [4].

[1] dendyadanta@ymail.com

Selection of the type of water turbine can be seen on the graph that describes the characteristics of relation between flow rate (m^3/s) and high net fall (m) in order to get the type of turbine that is appropriate and suitable to the site conditions of turbines as shown in fig. 1 [4]. The experimental site used is at Salam Lake UI with head is 2,7 m and discharge 0,041 m^3/s.

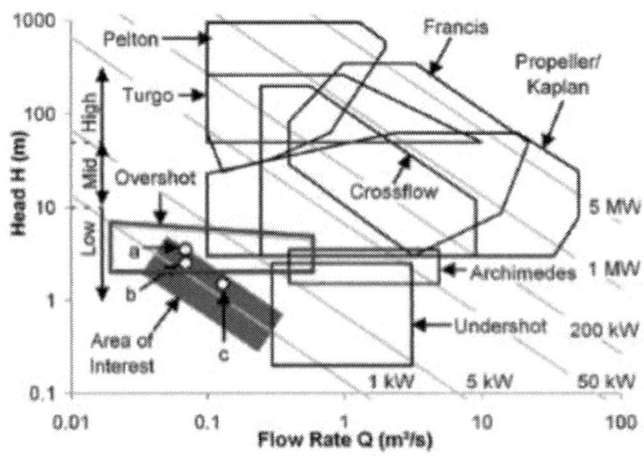

Figure 1. Flow rate vs Head [4]

From the information obtained in figure 1 the type of turbine that is suitable to be used is in Salam Lake is overshot waterwheel. The overshot waterwheel is a hydroelectric power plant that utilizes the gravity of water to rotate the shaft [5].

Although overshot is a fairly old technology, very few studies that examine the overshot waterwheel are available. There are at least four studies that discuss overshot turbines. Euren and Weyer (2005) identified suitable irrigation for overshot and undershot waterwheel [7]. Aryo (2012) used overshot to generate electricity at very low head condition, 25 cm [8]. Capecchi (2013) tells the origin of the overshot waterwheel used in the 18th century [6]. Wibawa et.al. (2014) designed an overshot turbine that will be used for electrification in the village of Bendosari Malang Regency Indonesia [9].

There are several advantages of using overshot waterwheel as a power plant in remote areas in Indonesia such as ease of operation, maintenance and maintenance and the low cost of investment costs are not too high. This is because the river conditions in Indonesia contain a lot of garbage in the form of dried leaves from trees, household waste and others that may interfere with the turbine's operation. To preventsuch interference, this study attempts to optimize bucket dimensions with the conditions available so that garbaes do not have a significant effect on the tubine. In addition, the optimization also has a good impact in incrasing the amount of generated power.

2. Methodology

The method that is used to design the overshot waterwheel is the analytical and numerical method, by specifying the dimensions in accordance with the provisions of the flow rate and the height. Analytical method also includes stating the mass and the volume of water contained in each active blade. Then, ascertaining the rpm, torque, and the power of the waterwheel is done to know the center mass point on each active bucket.

2.1. Waterwheel design analytic

Water condition at Salam Lake Universitas Indonesia with head (H) is 1.15 m and discharge (Q) is 0,041 m^3/s. So, available hydro dynamic power = 1090 N

2.1.1. Flow velocity

$$V = \sqrt{2.g.h} \tag{1}$$

2.1.2. Peripheral speed of the waterwheel

$$U_1 = \frac{V_1 \cos \alpha_1}{2} \tag{2}$$

2.1.3. Waterwheel rotation

$$n = \frac{60\ U_1}{\pi\ D_1} \tag{3}$$

2.1.4. Tip speed channel

The tip speed channel can be obtained using the open channel hydraulics equation

$$h_1 + \frac{\alpha\ V_1}{2g} + z_1 = h_2 + \frac{\alpha\ V_2}{2g} + z_2 \tag{4}$$

$$R = \frac{A}{p} \tag{5}$$

$$V = \frac{1}{n} R^{\frac{2}{3}} S^{\frac{1}{2}} \tag{6}$$

flowrate on channel

$$Q = A.V \tag{7}$$

2.1.5. Dimension

Outer diameter

$$D = \frac{1000\ N\ 2}{\rho.g.Q.\eta} \tag{8}$$

On the outer diameter, the obtained results is divided by 7,so that the waterwheel diameter will be smaller and can be installed at Salam lake UI.

Inner diameter

Planned $\mathrm{t/R_0}\ =\ 0{,}15\ (0{,}05 < t/\mathrm{R_0} < 0{,}25)$,with:

$$\mathrm{t = R_0 - R_0} \tag{9}$$

and,

$$0{,}15R_0 = R_0 - R_1$$
$$R_1 = R_0 - 0{,}15\ R_0$$
$$0{,}15R_0 = R_0 - R_1$$
$$R_1 = 0{,}85\ R_0$$
$$R_0 + 0{,}85\ R_0 = \frac{1{,}03}{2}$$
$$R_0 = 0{,}515m$$

Thus:

$$R_1 = 0{,}85R_0$$
$$R_1 = 0{,}43m$$

Then the value taken is

$$R_1 = 0{,}43m;\ D_1 = 0{,}875m$$

2.1.6. The bucket design

It is known that the optimum blade shape is the shape of flat bottomed, where the bottom of the blade-shaped is radial to the axis of the shaft with 1/4th length from annulus width [10].

$$\frac{R_0}{\sin\beta}=\frac{h+R_1}{\sin\alpha}=\frac{h'}{\sin\theta} \quad (10)$$

$$\sin\alpha=\frac{\sin\beta\,(h+R_1)}{R_0} \quad (11)$$

$$\theta=180^0-\beta-\alpha \quad (12)$$

$$h'=\frac{D_0\sin\theta}{\sin 115^0} \quad (13)$$

Overall length

$$s=h'+h \quad (14)$$

Design of the bucket can be seen on figure 2:

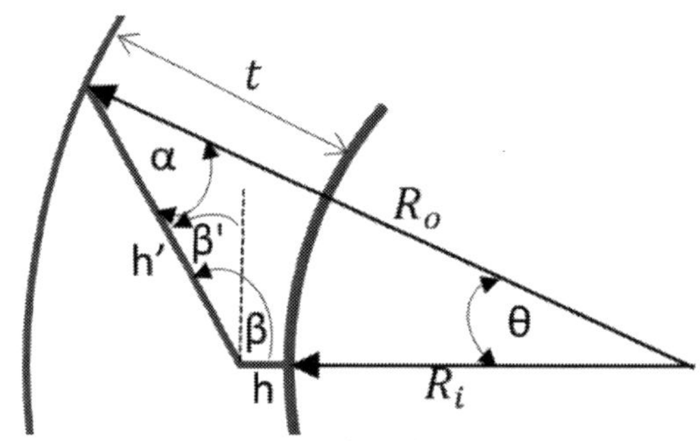

Figure 2. Bucket design [10]

Number of buckets:

$$z=\frac{\pi\,D_0\sin\beta'}{a.s+t'} \quad (15)$$

Active buckets:

$$i=\frac{N(rps)}{rpm}\,z \quad (16)$$

Distance between the bucket on outer side:

$$z_o{}'=\frac{D_0\,\pi}{z} \quad (17)$$

Distance between the bucket on inner side:

$$z_i{}'=\frac{D_i\,\pi}{z} \quad (18)$$

The result of all the calculation above can be seen on table 1.

Table 1. Design summary

Description	Value	Unit
V	4,75	m/s
U_1	1,43	m/s
n	28	RPM
Q	0,19	m^3/s
D_0	1,03	m
D_1	0,87	m
b	0,80	m

a	53,1	degree (^{0})
θ	11,87	degree (^{0})
h'	11,43	cm
s	11,30	cm
z	25	-
i	11,67	-
z_o'	13	cm
z_i'	11	cm

After obtaining the design results of the calculation, the visualization of 2D and 3D can be seen in fig. 3.

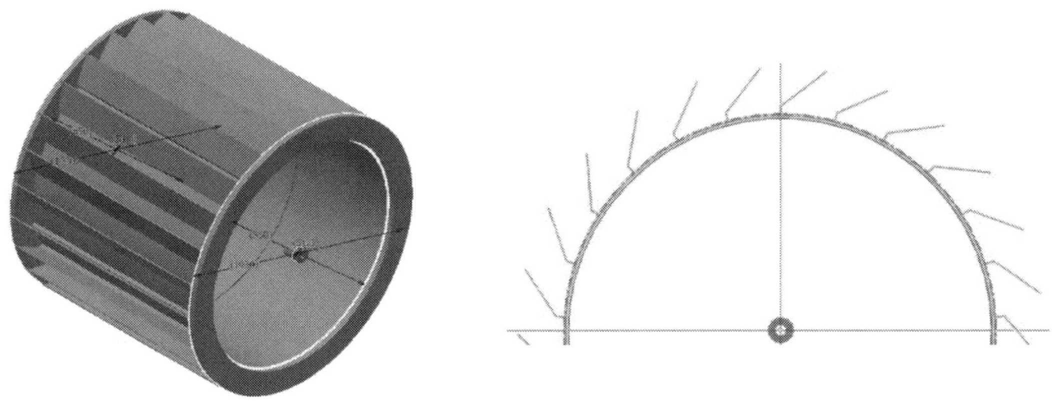

a b

Figure 3. a. overshot waterwheel design, b. side view of the 2D overshot waterwheel

2.2 Optimization of volume

To get the value of power and torque, the calculation is done by determining the buckets volume and the centroid points for each bucket.

2.2.1. Volume of each bucket

To get the volume and mass on a bucket, the calculation is performed by dividing the area into several areas. For example the bucket number 1 on fig. 3. It is divided into three areas, namely there are two triangles and one rectangle. After the surface of each component is obtained, they are multiplied by the width of the bucket itself. Therefore we obtained the volume and mass of water in the bucket number 1. Overshot waterwheel should be able to accommodate the discharge of produced water per second. Then we got 80 cm width to accommodate a flow of 0,041 m^3/s. with assuming the initial calculation of 40 cm to get an area of each bucket.

IOP Conf. Series: Materials Science and Engineering **316** (2018) 012056 doi:10.1088/1757-899X/316/1/012056

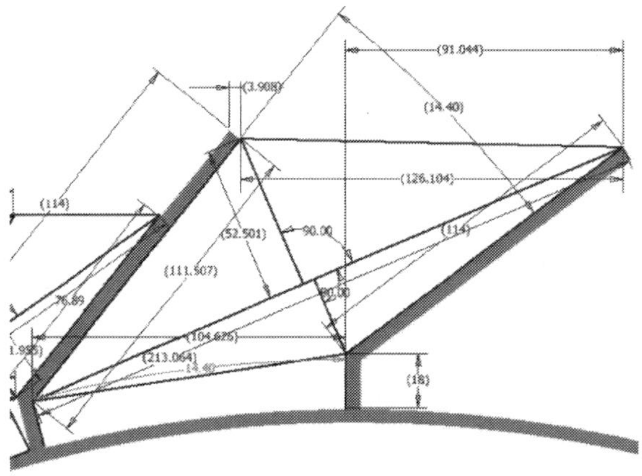

Figure 4. Dimension of bucket number 1

Table 2. Area of triangle 1 of bucket number 1

Description	Value	Unit
½ x base x height	55,38	cm^3
area x width of bucket	4430,4	cm^3
convert cm^3 to m^3	0,0044304	m^3
Convert to mass	4,4304	kg

Table 3. Area of triangle 2 of bucket number 1

Description	Value	Unit
½ x base x height	27,69	cm^3
area x width of bucket	2215,2	cm^3
convert cm^3 to m^3	0,0022152	m^3
Convert to mass	2,2152	kg

Table 4. Area of rectangle of bucket number 1

Description	Value	Unit
½ x base x height	18,9	cm^3
area x width of bucket	1512	cm^3
convert cm^3 to m^3	0,001512	m^3
Convert to mass	1,512	kg

Total Mass on bucket 1 = 8,1576 kg.
The total area of bucket 1 = 8157,6 cm^3/s. If the amount total of water contained is less than the available discharge in active bucket, recalculation is done at the width of the wheel.

2.2.2. Center point of mass

Center point of mass is calculated to determine the torque of each active bucket. After the total torque of each active bucket is known, then we may get the value of the power output. This

calculation applies to all active bucket. For the example on bucket number 1:
Weight percentage on bucket number 1:

$$I_1, I_2, I_3 = A_{triangle\ 1}\ x\ total\ area \quad (19)$$

Mass each surface on bucket number 1:

$$w_1, w_2, w_3 = Total\ volume\ (kg) x\ weight\ presentage \quad (20)$$

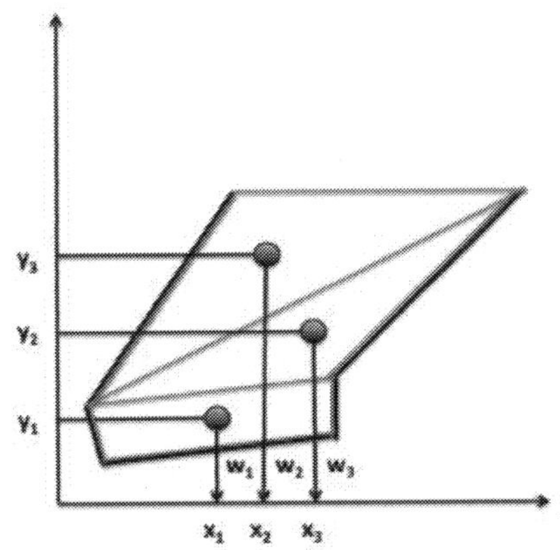

Figure 5. Each center point of surfaces on bucket number 1 [11]

$$x_G = \frac{\sum w_i x_i}{\sum w_i} = \frac{w_1x_1 + w_2x_2 + w_3x_3 + \cdots}{w_i + w_2 + w_3 + \cdots}$$

$$x_G = \frac{g(m_1x_1 + m_2x_2 + m_3x_3 + \cdots)}{g(m_i + m_2 + m_3 + \cdots)}$$

$$x_G = \frac{m_1x_1 + m_2x_2 + m_3x_3 + \cdots}{m_i + m_2 + m_3 + \cdots} = \frac{\sum m_i x_i}{\sum m_i} \quad (21)$$

The same calculation applies to y axis:

$$y_G = \frac{m_1y_1 + m_2y_2 + m_3y_3 + \cdots}{m_i + m_2 + m_3 + \cdots} = \frac{\sum m_i y_i}{\sum m_i} \quad (22)$$

After the x_G and y_G is known, then we can get the center point of mass on the bucket 1. Which is the β and radius from center point of waterwheel to center point of the bucket and is known with 86.64^0 and 468 mm distance.

3. Results and Discussion

On the center of mass calculation, we can get point coordinates (*x,y*) on the bucket 1 itself (x_G = 79,8 and y_G = 54,8), and with a length of 468 mm radius and at an angle of $86{,}64^0$ ($\cos 86{,}64^0$ = 0,0586) from center point of the wheel to the center point of bucket 1.
Torque on bucket 1 can be determine by:

$$\tau = \cos\beta\ x\ length\ of\ mass\ x\ g\ x\ total\ mass \quad (23)$$

$$\tau = 0{,}0586\ x\ 0{,}468\ x\ 9{,}81\ x\ 8{,}1576 = 2{,}194 \text{ Nm}$$

The same calculation applies to all other active buckets and the optimum result is can be seen in table 5:

Table 5. Torque each bucket

Description	Value	Unit

Bucket 2	9,61	Nm
Bucket 3	15,10	Nm
Bucket 4	18,03	Nm
Bucket 5	18,89	Nm
Bucket 6	17,26	Nm
Bucket 7	14,24	Nm
Bucket 8	7,75	Nm
Bucket 9	0,93	Nm

Total Torque = 104 Nm.
After all the torque from each bucket are known, the power output can be calculated by:

$$P_{output} = \tau_{total}\ x\ rotation \tag{24}$$

$$P = 104\ Nm\ x\ 2{,}93\ rad/s = 305\ Watts$$

From the result of the power output and torque that is known, then the efficiency can be obtained by:

$$\eta = \frac{P_{output}}{P_{hydro}} \tag{25}$$

$$\eta = \frac{305}{1090} = 28\%$$

Table 6. Analytical result

Description	Value	Unit
Torque	104	Nm
Power	305	Watts
Efficiency	28	%
Bucket active	8	-
Bucket volume	40,28	cm^3

4. Conclusion

The result of investigation by using analytical method got maximum performance of overshot water wheel at head 1,15 meters and discharge 0,041 m^3/s, which resulted in 305 Watt with wheel width 0,8 meters and has mechanical efficiency equal to 28%. The volume of water in the bucket is the most in bucket number 1 because the bucket arm leads to the vertical axis of the wheel while the largest torque is on bucket number 5 because it has the longest force arm.

References

[1] B. Ho-Yan, "Design of a low head pico hydro turbine for rural electrification in Cameroon," 2012.
[2] P. Adhikari, U. Budhathoki, S. R. Timilsina, S. Manandhar, and T. R. Bajracharya, "A Study on Developing Pico Propeller Turbine for Low Head Micro Hydropower Plants in Nepal," *J. Inst. Eng.*, vol. 9, no. 1, pp. 36–53, 2014.
[3] B. Thomas, "Pico-Hydropower Franchising in Rural Honduras," *Int. J. Serv. Learn. Eng. Humanit. Eng. Soc. Entrep.*, vol. 6, no. 1, pp. 46–63, 2011.
[4] A. A. Williams and R. Simpson, "Pico hydro–Reducing technical risks for rural

electrification," *Renew. Energy*, vol. 34, no. 8, pp. 1986–1991, 2009.
[5] M. Pigaht and R. J. van der Plas, "Innovative private micro-hydro power development in Rwanda," *Energy Policy*, vol. 37, no. 11, pp. 4753–4760, 2009.
[6] S. Vicente and H. Bludszuweit, "Flexible design of a pico-hydropower system for Laos communities," *Renew. Energy*, vol. 44, pp. 406–413, 2012.
[7] Pradhana Adhika Saputra, "Perancangan Volume Bucket Kincir Air Piko Hidro Overshot dengan Metode Analitik dan Numerik," Universitas Indonesia, 2017.
[8] R. Pradito, "Perancangan Kincir Air Piko Hidro Undershot dengan Metode Analitik dan Numerik," Universitas Indonesia, 2017.
[9] H. Helmizar, "Turbine wheel-a hydropower converter for head differences between 2.5 and 5 m." University of Southampton, 2016.
[10] G. Muller and C. Wolter, "The breastshot waterwheel: design and model tests," *ICE Proceedings-Engineering Sustain.*, pp. 203–211, 2004.
[11] A. Tevata and C. Inprasit, "The effect of paddle number and immersed radius ratio on water wheel performance," *Energy Procedia*, vol. 9, pp. 359–365, 2011.
[12] S. Paudel, N. Linton, U. C. E. Zanke, and N. Saenger, "Experimental investigation on the effect of channel width on flexible rubber blade water wheel performance," *Renew. energy*, vol. 52, pp. 1–7, 2013.
[13] A. B. Praful Yelguntwar, Oranay Bhange, Yogesh Lilhare, "Design, Fabrication & Testing Of A Waterwheel For Power Generation In An Open Channel Flow," *Int. J. Res. Eng. Adv. Technol.*, vol. 2, no. 1, 2014.
[14] E. Quaranta and R. Revelli, "Optimization of breastshot water wheels performance using different inflow configurations," *Renew. Energy*, vol. 97, pp. 243–251, 2016.
[15] E. Quaranta and R. Revelli, "CFD simulations to optimize the blade design of water wheels," *Drink. Water Eng. Sci.*, vol. 10, no. 1, p. 27, 2017.
[16] Dendy Adanta, Budiarso, Warjito, A.I. Siswantara, "STUDY OF TURBULENCE MODELLING ON PICO-HYDRO TURBINE," in *Quality in research*, 2017.
[17] Kris Biantoro, "Kincir Air Langkah Atas dengan Daya Keluaran 1.4 kW," Depok, 1989.
[18] G. Müller and K. Kauppert, "Performance characteristics of water wheels," *J. Hydraul. Res.*, vol. 42, no. 5, pp. 451–460, 2004.
[19] E. Popov, "Mekanika Teknik, Erlangga." Jakarta, 1984.
[20] J. Senior, P. Wieman, and G. Muller, "The Rotary Hydraulic Pressure Machine for Very Low Head Hydropower sites," in *Hidroenergia conference*, 2008, pp. 1–8.
[21] W. G. S. Hugh W. Coleman, *Experimentation and Uncertainty Analysis for Engineering*. Canada: Wiley Inter Science, 1999.
[22] Isti Sujandari, "Design of Experiments: Blocking in One-Factor ANOVA," Depok.
[23] E. Quaranta and R. Revelli, "Performance characteristics, power losses and mechanical power estimation for a breastshot water wheel," *Energy*, vol. 87, pp. 315–325, 2015.

IOP Conf. Series: Materials Science and Engineering 316 (2018) 012057 doi:10.1088/1757-899X/316/1/012057

The Design of Connection Solid Oxide Fuel Cell (SOFC) Integrated Grid with Three-Phase Inverter

Darjat[1,2,3], Sulistyo[2], Aris Triwiyatno[1], Humaid Thalib[1]
[1]Department of Electrical Engineering, Diponegoro University, Semarang, 50275, Indonesia
[2]Department of Mechanical Engineering, Diponegoro University, Semarang, 50275,Indonesia
[3]Corresponding author, e-mail: dr.darjat@gmail.com

Abstract. Fuel cell technology is a relatively new energy-saving technology that has the potential to replace conventional energy technologies. Among the different types of generation technologies, fuel cells is the generation technologies considered as a potential source of power generation because it is flexible and can be placed anywhere based distribution system. Modeling of SOFC is done by using Nernst equation. The output power of the fuel cell can be controlled by controlling the flow rate of the fuels used in the process. Three-phase PWM inverter is used to get the form of three-phase voltage which same with the grid. In this paper, the planning and design of the SOFC are connected to the grid.

1. Introduction

Energy is one of the major needs in human life. Growing energy needs can be used as indicators of human prosperity, but along with it will cause problems in the provision of business. Most people still rely on fossil fuels to comply energy needs. So, the longer the existing fossil energy will be dwindling. During 2000-2011, the amount of energy consumption increased by an average of 3% per year [1].

The amount of energy consumption continues to increase by increasing population growth. Excessive consumption of fossil fuels as a primary energy source has contributed to a variety of environmental problems, which in turn speeds up the process of global climate change. The issue of climate change was triggered by the use of fossil fuels has prompted various parties, especially energy users to perform a variety of efforts, including by using renewable energy to reduce the use of fossil energy that can reduce greenhouse gas emissions.

SOFC is a renewable energy that produces direct current output power so it takes inverter to convert the direct current becomes alternating current power. The output voltage and frequency of the inverter should be equal to the voltage and frequency on the grid, so the system can work [2], [3].

Solid oxide fuel cell (Solid Oxide Fuel Cell / SOFC) is an electrochemical cell that converts chemical energy into electrical energy by way of oxidizing fuel. Solid oxide fuel cell operates at a temperature of 600°C-1000°C and has a high efficiency that can reach 65%, and the rest of the clean combustion is water (H2O) without pollutants [4].

System cells solid oxide fuel generally consists of a tank of fuel / hydrogen and oxygen tanks, pressure sensors and temperature, measuring current and voltage, the valve actuator to regulate the mass flow (mass flow) of the second tank, heating (heater), and canals disposal.

Research on solid oxide fuel cells SOFC modeling is very interesting and has also been developed by some researchers previously and this paper uses a model which is already developed in principle and almost the same with each other. Padulles and his team did modeling of solid oxide fuel cells for electric energy system simulation [5]. Zhu and Tomsovic have developed models of previous studies and analyzing the event of load changes [6]. Wang was modeling of solid oxide fuel cells by observing the types of tubular thermodynamic systems [4].

IOP Conf. Series: Materials Science and Engineering **316** (2018) 012057 doi:10.1088/1757-899X/316/1/012057

2. SOFC System Description

2.1 Nernst Equation

Nernst equation is used to determine the output voltage of the fuel cell is as follows:

$$E_{cell} = E_o + \frac{RT}{nF} \ln \frac{p_{H_2} p_{O_2}^{-\frac{1}{2}}}{p_{H_2O}} \quad (1)$$

Parameter Eo is ideal voltage of the cell SOFC (volts), R is the universal gas constant (atm / mol K), T is the operational temperature (K), p_{H_2}, p_{H_2O}, and p_{O_2} is the pressure hydrogen, water, and oxygen (atm).

2.2 Characteristics of Orifice Valve

If the proportional partial pressure with the molar flow of each gas that passes through the canal, it can be written:

$$\frac{q_{H_2}}{p_{H_2}} = \frac{K_{an}}{\sqrt{M_{H_2}}} = K_{H_2} \quad (2)$$

$$\frac{q_{H_2O}}{p_{H_2O}} = \frac{K_{an}}{\sqrt{M_{H_2O}}} = K_{H_2O} \quad (3)$$

Parameter q_{H_2} and q_{H_2O} is the molar flow of hydrogen and water through anode valve (mol / s), p_{H_2} and p_{H_2O} is the partial pressure of hydrogen and water. K_{H_2} and K_{H_2O} are molar constant of hydrogen valves and water (mol / s atm).

2.3 Partial Pressure

Solid oxide fuel cells (SOFC) use reaction of hydrogen gas, oxygen gas and water as products. Each gas that is used has partial pressure [12]. Partial pressure will be used to calculate Nernst voltage. Partial pressure of hydrogen and oxygen are:

$$p_{H_2} = \frac{1/K_{H_2}}{1+\tau_{H_2}s}\left(\frac{1}{1+\tau_f s} q_f^{in} - 2K_r I\right) \quad (4)$$

$$p_{O_2} = \frac{1/K_{O_2}}{1+\tau_{O_2}s}\left(\frac{1}{r_{H-O}}\frac{1}{1+\tau_f s} q_f^{in} - K_r I\right) \quad (5)$$

2.4. Use of Fuel

The use of fuel or fuel utilization is the ratio between the flow rate of the fuel that reacts ($q_{H_2}^r$) with the flow rate of fuel coming ($q_{H_2}^r$) [13]. Constraint in modeling SOFC is a fuel utilization whose value must be maintained in the range of 0.7 to 0.9. Fuel utilization is important indicator for the purposes of security and efficiency. The relationship expressed in the equation:

$$U_f = q_{H_2}^r / q_{H_2}^{in} \quad (6)$$

$$q_{H_2}^{in} = \frac{2K_r I}{U_f} \quad (7)$$

Parameter U_f is fuel utilization and I is the input current SOFC stack.

2.5 Output Voltage and Voltage Drop

The output voltage of the fuel cell can be written into the equation:

$$V_{dc} = V_o - V_{ohm} - V_{act} - V_{conc} \quad (8)$$

IOP Conf. Series: Materials Science and Engineering **316** (2018) 012057 doi:10.1088/1757-899X/316/1/012057

V_{dc}= SOFC output voltage (volts), V_o= Nernst voltage (volts), V_{ohm} = ohmic voltage drop, V_{act} = activation of voltage drop, and V_{conc} = concentration of voltage drop.

Ohmic voltage drop caused by the resistance on the electrolyte and the interconnection between the fuel cell.

$$V_{ohm} = I\ R_{ohm} \tag{9}$$

V_{ohm}ohmic = voltage drop,I=load current, and R_{ohm}=ohmic losses.
Activation voltage drop caused by the activation energy barrier before the chemical reaction occurs.

$$V_{act} = \ \alpha + \beta \ln I \tag{10}$$

V_{act}=activation of voltage drop,I=load current, α= tafel constant, and β=tafel slope.
Concentration voltage drop caused byt he differences ion between the reactant so it forms concentration gradient. The formation of the concentration gradient effected voltage reduction in the cell.

$$V_{conc} = \frac{-RT}{2F} ln \frac{I}{I_L} \tag{11}$$

I= input current or load current, I_L= limiting current density, R = universal gas constants (atm / mol K), T = operating temperature (K), and F the Faraday constant (C / mol).

3. Inverter System

The inverter is power electronics series that is used to convert or change the direct voltage (DC) to an alternating voltage (AC). There are several types of inverter [7] that divided based on the criteria, such as:

a. Based on the amount of phase: the single-phase inverter and multi-phase, for example, three-phase which has been the type of inverter used to interconnect the output to grid of electrical power system (grid).
b. Based on DC input sources: the voltage source inverter (VSI-Voltage Source Inverter), and current source inverter (CSI-Current Source Inverter).
c. Based on the method of setting signal shape: square wave, pulse amplitude modulation (PAM) and pulse width modulation (PWM).
d. Based on the output wave form: that is a square wave, square terraces and sinusoidal.

Equipment usage is selected based on the type of application. Inverter usually wears pulse width modulation (PWM) to produce alternating output voltage. An inverter is categorized as type of voltage source inverter (Voltage fed Inverter) if the input in form of voltage source and can e categorized as a current source inverter (Current fed inverter) if resources in the form of a current source. If the input voltage is the voltage that can be regulated, then this inverter can be called variable dc inverter linked.

4. Design System

4.1 SOFC Stack

Simulation design of solid oxide fuel cells (SOFC stack) requires the value of the parameters to calculate the output plant, as shown in Table 1.

IOP Conf. Series: Materials Science and Engineering **316** (2018) 012057 doi:10.1088/1757-899X/316/1/012057

Table 1. *SOFC Parameters*

Parameter	Explanation	value	unit
T	Absolute temperature	1273	K
F	Faraday Constant	96487	C/mol
R	Gas Universal Constanta	8314	J/(kmol K)
Io	current Initial	100	A
E_o	Ideal Voltage	1,18	V
N_o	The Number of cell	384	
U_f	Fuel usage	0,85	
KH_2	Molar constant of hydrogen valve	$8{,}32\times10^{-6}$	kmol/(s atm)
KH_2O	Molar constant of water valve	$2{,}77\times10^{-6}$	kmol/(s atm)
KO_2	Molar constant of oxygen valve	$2{,}49\times10^{-5}$	kmol/(s atm)
τH_2	Time response of hydrogen flow	26,1	S
τH_2O	Time response of water flow	78,3	S
τO_2	Time response of oxygen flow	2,91	S
R	Ohmic loss	0,126	Ω
rH-O	Hydrogen-oxygen flow ratio	1,145	
τ_f	Time response of fuel prossesor	5	S

Block series for SOFC stack system is shown in Figure 3.2. Block simulink stack SOFC is also consist of Nernst voltage calculation, the partial pressure (PH2, pH2O, and pO2), fuel flow (qf), and voltage drop (Vohm).

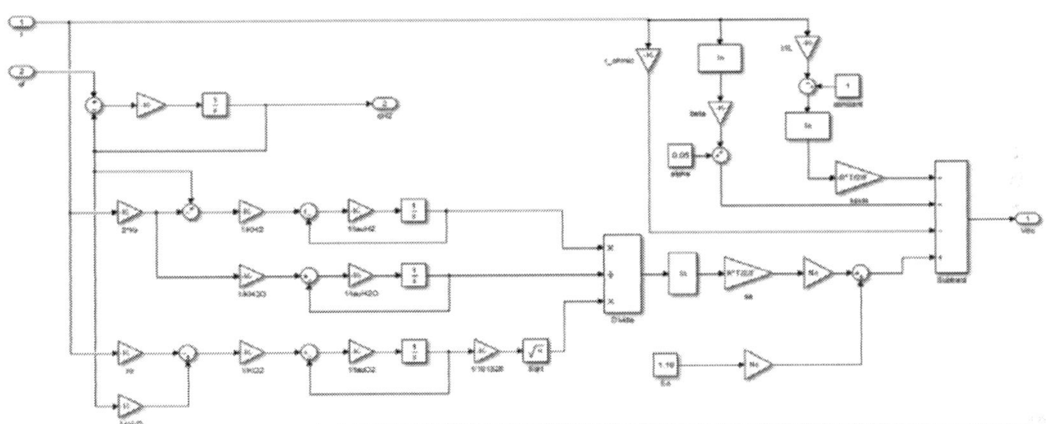

Figure 1. SOFC Stack

4.2 The series of Grid-side Controller

Grid-side controller series consists of several blocks which are PLL, dq-abc transformation, PWM hysteresis, and active power controlling. Image of grid-side controller series is shown in Figure 2.

IOP Conf. Series: Materials Science and Engineering **316** (2018) 012057 doi:10.1088/1757-899X/316/1/012057

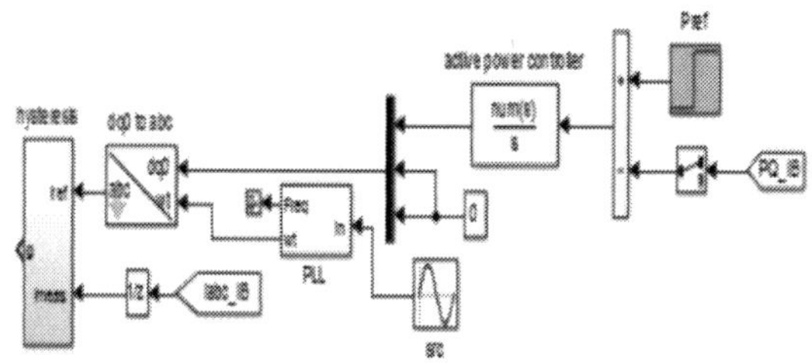

Figure 2. Grid-side Controller Series.

Inverter is supplied voltage from the dc voltage source and inverter releases voltage and ac current that flows to L filter and to the grid.

4.3 Inverter series connected to the Grid

Inverter series is used to convert from DC voltage to AC voltage. Inverter input voltage obtained from a series of SOFC. While, output inverter is in form of voltage and AC current of three-phase that is connected to a three-phase transformer with Y-Y system. Transformer on the series of this system is used for protection. This system needs 0,001 mH LCL filter connected to the grid which has voltage of 440 Volt.

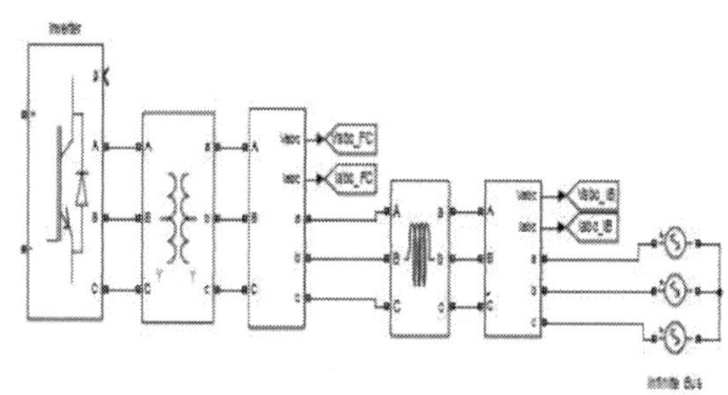

Figure 3. The series Inverters connected to the Grid

5. Results and Discussion

Research result of Grid Integrated SOFC shows some chart and waves with two variations, namely temperature variations and variations in the amount of the cell.

5.1 Temperature Variations

The temperature is 873 K with amount of cell is 450

IOP Conf. Series: Materials Science and Engineering **316** (2018) 012057 doi:10.1088/1757-899X/316/1/012057

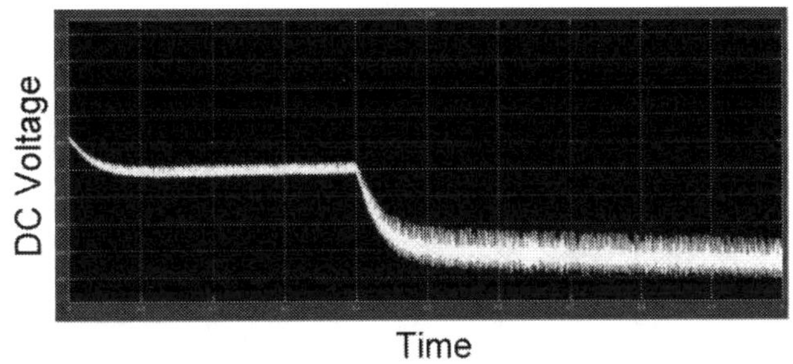

Figure 4. DC Voltage Graph SOFC T = 873 K

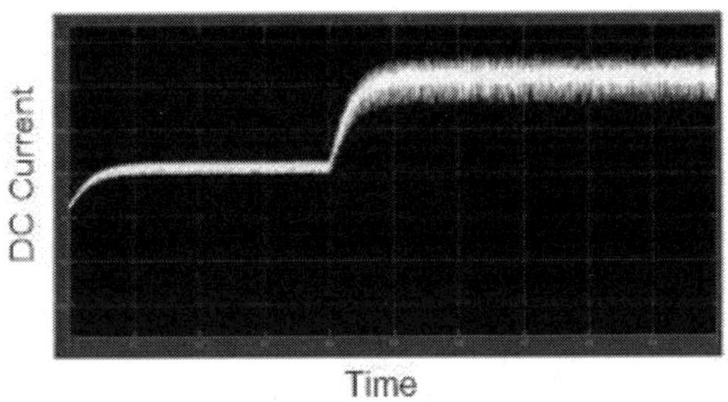

Figure 5. DC current Graph SOFC T = 873 K

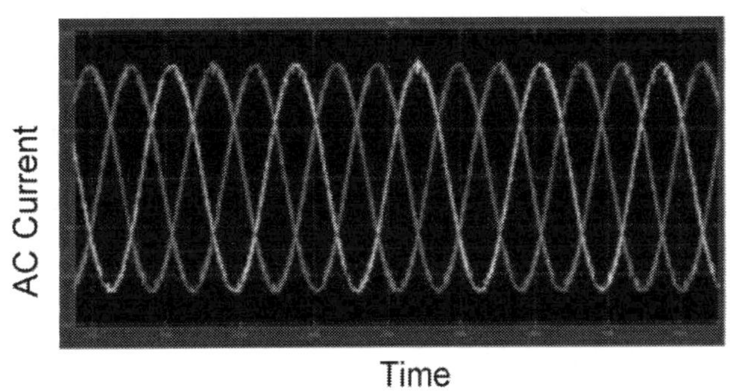

Figure 6. AC Current Wave of Three Phase Output Inverter T = 873 K

Figure 4 shows SOFC dc voltage graph with the voltage value 435.6 volts at the time 0.6 seconds. While Figure 5 is a graph of the dc output current SOFC which has current value 151.2 Ampere at the time 0.6 seconds.

In Figure 6, it can be seen that the AC current wave of three-phase on the grid with temperature 873 K in form of sinusoidal wave

The temperature is1273 K with cell amount 450

IOP Conf. Series: Materials Science and Engineering **316** (2018) 012057 doi:10.1088/1757-899X/316/1/012057

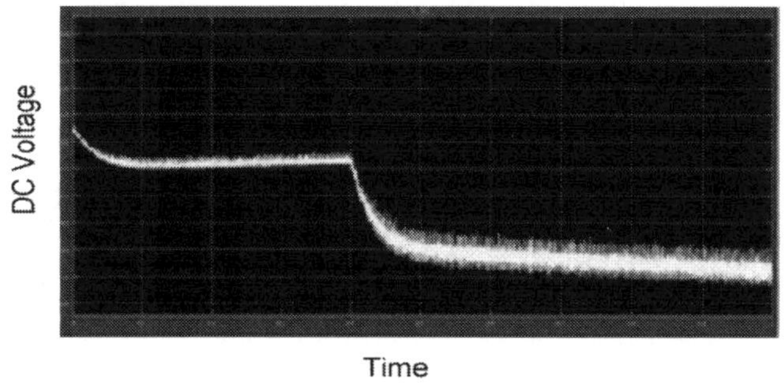

Figure 7. Graph of DC Voltage SOFC T = 1273 K

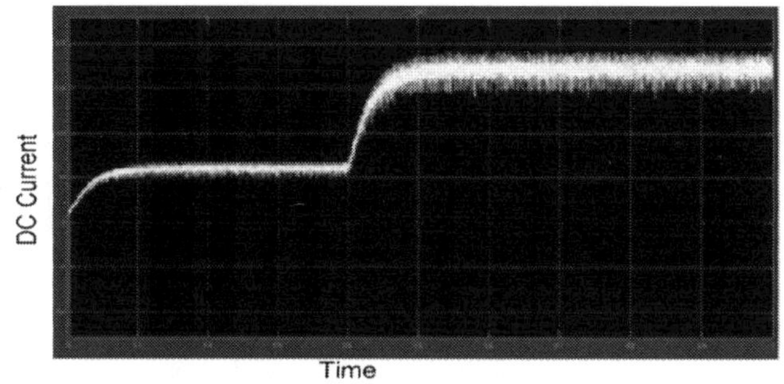

Figure 8. DC Current Graph SOFC T = 1273 K

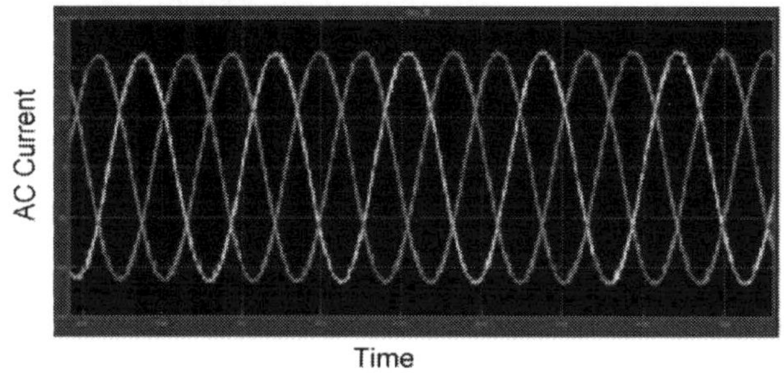

Figure 9. AC Current Wave of Three Phase Inverter Output T = 1273 K

Figure 7 shows graph of dc voltage SOFC with the voltage value 398.8 volts at the time 0.6 seconds. While, Figure 8 is a graph of the dc output current SOFC which has current value 171.8 Ampere at the time 0.6 seconds.

In Figure 9, it can be seen that AC current wave of three-phase on the grid with the temperature 1273 K in form of sinusoidal wave.

IOP Conf. Series: Materials Science and Engineering **316** (2018) 012057 doi:10.1088/1757-899X/316/1/012057

5.2 Variation of Cell amount

Cell amount is 350 with temperature 1273 K

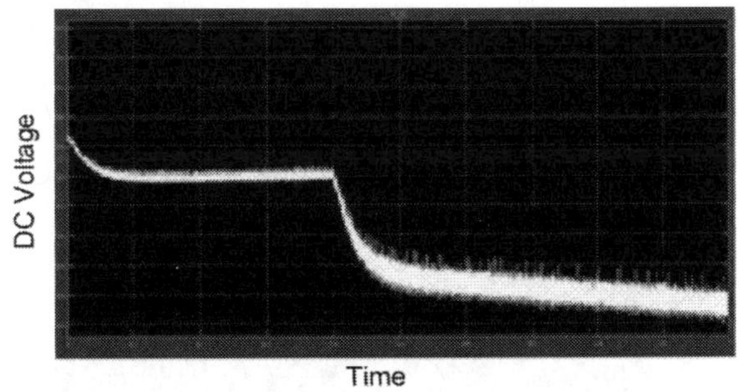

Figure 10. DC Voltage Graph SOFC with 350 cells

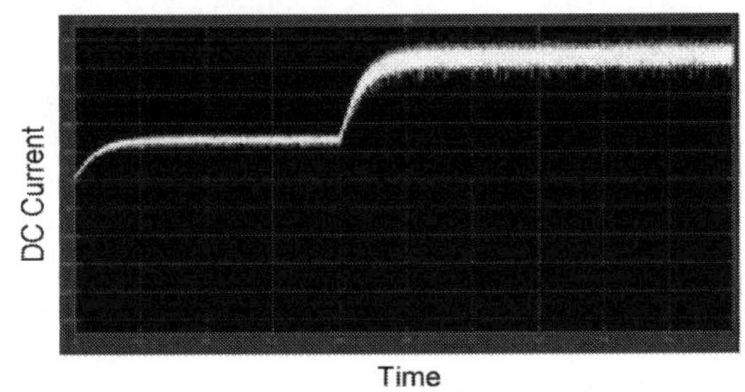

Figure 11. Graph DC current SOFC with 350 cells

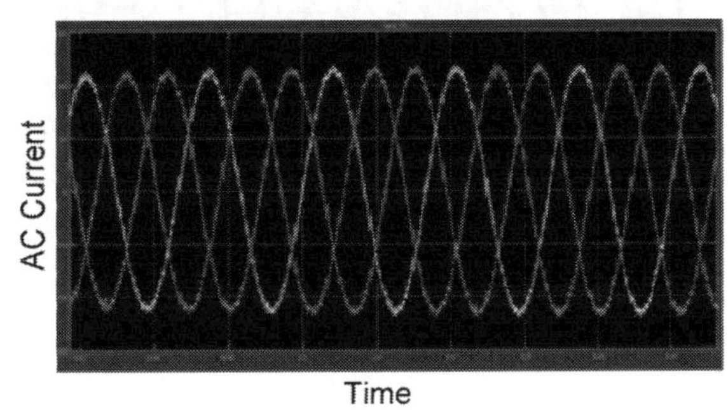

Figure 12. AC Current Wave of Three Phase Output Inverter with 350 Cells

Figure 10 shows a graph of voltage dc SOFC with voltage value 302.1 volts at the time 0.6 seconds. While, Figure 11 is graph of the dc output current SOFC which has current value 214.1 Ampere at the time 0.6 seconds.

In Figure 12, it can be seen that AC current wave of three-phase on grid with 350 cells in form of sinusoidal wave

Amount of cell is 750 with temperature 1273 K

IOP Conf. Series: Materials Science and Engineering **316** (2018) 012057 doi:10.1088/1757-899X/316/1/012057

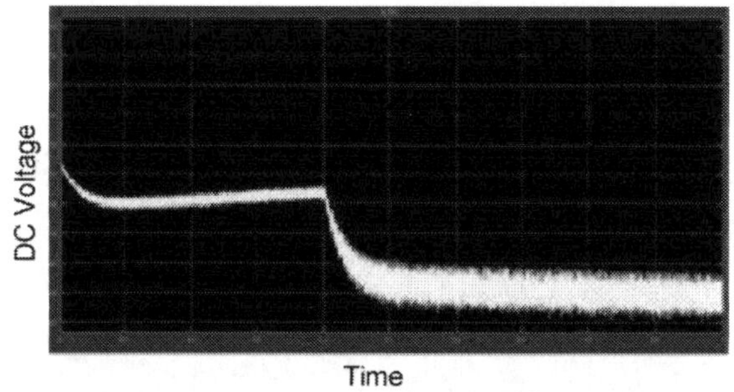

Figure 13. The Graph of DC Voltage SOFC with 750 cells

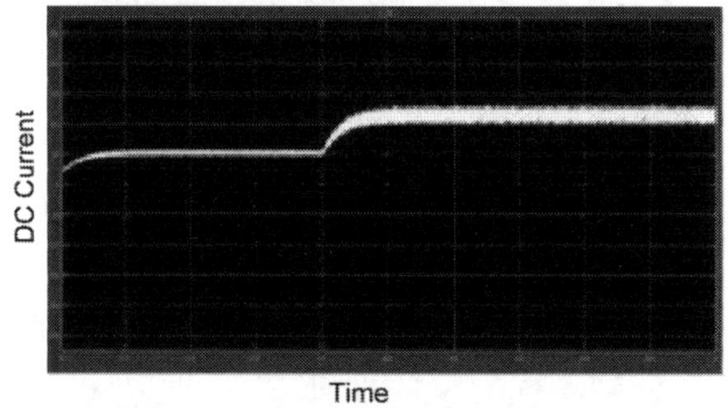

Figure 14. The Graph of DC current SOFC with 750 cells

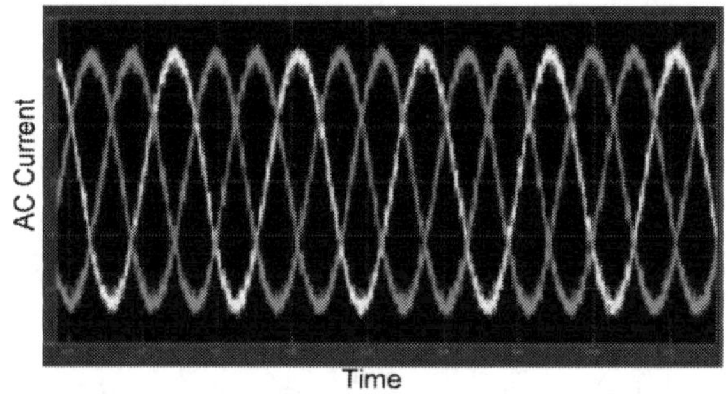

Figure 15. AC Current Wave of Three Phase Output Inverter with 750 Cells

Figure 13 shows graph of dc voltage SOFC with voltage value 692.2 volts at the time of 0.6 seconds. While, Figure 14 is graph of the dc output current SOFC which has current value 110.1 Ampere at the time of 0.6 seconds.

In Figure 15, it can be seen that AC currents wave of three-phase on grid with 750 cells in form of sinusoidal wave

6. Conclusion

SOFC research result is integrated into the grid showed that SOFC sends voltage and current to the grid and then the inverter sends AC current wave of three-phase to the grid. On the temperature variation, higher temperature value causes the value of the output dc voltage lower SOFC by reduction 36.8 volts. Then, the higher temperature causes the higher value of the output dc current SOFC with enhancement

20.6 Ampere. In the variations of cell amount, the higher the voltage value causes enhancement 390 Volts. Then, the greater number of cells causes the value of the output dc current SOFC lower with reduction 104 Amperes.

Reference

[1] K. Energi *et al.*, "INDONESIA ENERGY."
[2] G. Inverters, A. Abdalrahman, and A. Alshazly, "Simulation and Implementation of," vol. 60, no. 4, pp. 41–49, 2012.
[3] I. Singh, V. Kumar, and D. Parmar, "Design and Simulations of Solid Oxide Fuel Cell Power System using Three Phase Inverter for Grid Applications," vol. 2, no. 7, pp. 399–402, 2013.
[4] C. Wang, "MODELING AND CONTROL OF HYBRID," no. July, 2006.
[5] J. Padulles, "An integrated SOFC plant dynamic model for power systems simulation," pp. 495–500, 2000.
[6] Y. Zhu and K. Tomso, "De v elopment of models for analyzing the load-following performance of microturbines and fuel cells," vol. 62, 2002.

IOP Conf. Series: Materials Science and Engineering **316** (2018) 012058 doi:10.1088/1757-899X/316/1/012058

Two-phase Frictional Pressure Drop of Propane with Prediction Methods of Viscosity and Density in 500 μm Diameter Tube

S Novianto[1,2], A S Pamitran[1], R Koestoer[1], K Saito[3]

[1] Department of Mechanical Engineering, Faculty of Engineering, Universitas Indonesia, Kampus UI Depok 16424, Indonesia

[2] Department of Research, Sekolah Tinggi Teknologi Texmaco, Pabuaran-Subang, Indonesia

[3] Department of Applied Mechanics and Aerospace Engineering, Waseda University,1-104, Totsuka-machi, Shinjuku-ku, Tokyo 169-8050, Japan

E-mail: pamitran@eng.ui.ac.id

Abstract. The experimental study of frictional pressure drop of two-phase flow with propane on the microchannel has been done. The aim of the present research is to characterize pressure drop of evaporative propane in microchannel with 500 μm diameter and 0.5 m length. The experimental apparatus used heating process in the test section with closed loop process system. Variable of research are mass flux of 360 to 915 kg/m^{2}s and vapor quality of 0 to unity. The homogeneous and separated model used to determine the two-phase flow frictional pressure drop. Some existing correlations of two-phase flow viscosity and density used to predict frictional pressure drop, including parameter C that used non-dimensional parameter of two-phase viscosity number $N_{\mu tp}$ which developed as a function of two-phase viscosity and density. The comparison of prediction frictional pressure drop showed that the Hibiki, Xuejiao and Pamitran correlation from separated model.

1. Introduction

The natural refrigerant is became popular and more intensively discussed since the issue of environment. Propane is one of friendly refrigerant with zero ODP and its GWP is less than four [1]. Moreover, the natural refrigerant (R-290) has indicated to replace R-22 in terms of its hydrodynamic performance [2].

Two-phase flow in microchannels application at industries have more popular recently. Microchannel technology appears to be a very good opportunity to minimize the charge without performance loss. [3] research the performance of a refrigeration system with a cooling capacity between 1 kW and 2 kW provided with air-to-propane minichannel heat exchanger. [4] reported the experimental performance heat pump with the power 100 kW using propane as working fluids. The experimental apparatus used shell and tube heat exchanger using minichannels.

Pressure drop on heat exchanger are mainly importance topic in saving energy issue. Therefore, research on pressure drop in microchannels is very important and more conducted. Homogeneous pressure drop models is the first model which assumed the equal vapor and liquid velocities. The attainment of thermodynamic equilibrium between the phase and the use of a suitably defined single phase friction factor for two-phase. Separated pressure drop models assumed constant but not necessarily equal velocity for the vapor and liquid phases. The use of empirical correlations or simplified concepts to relate the two-phase friction multiplier ($\emptyset^2$) and the void fraction (α) to the independent variables of the flow.

[5] have showed flow regime, void fraction, rising velocity slug bubble and drop friction. The accumulation data on flow characteristics to produce newly C as function of inner diameter. The value of C parameter decrease with decreasing inner diameter. [6] used R-134a, R-410A, R-290 and CO_2 as working fluids with tube diameter 1,5 and 0,5 mm. Pamitran et al. develop *C* parameter as function Reynolds number and Weber number and used viscosity two phase from [7] equation. [8] develop C parameter with non-dimensional parameter. The analysis indicates that the two-phase Reynolds number, Re_{tp}, the gas quality, x , and the two-phase viscosity number, $N_{\mu_{tp}}$, to be the key parameters for all flow conditions: liquid turbulence-vapor turbulence, liquid turbulence-vapor laminar, liquid laminar-vapor turbulence and liquid laminar-vapor laminar.

The aim of this experiment is prediction of frictional pressure drop two-phase flow with some viscosity and density parameter in horizontal diameter 0.5 mm tube.

2. Methodhology

2.1. Experiment set up

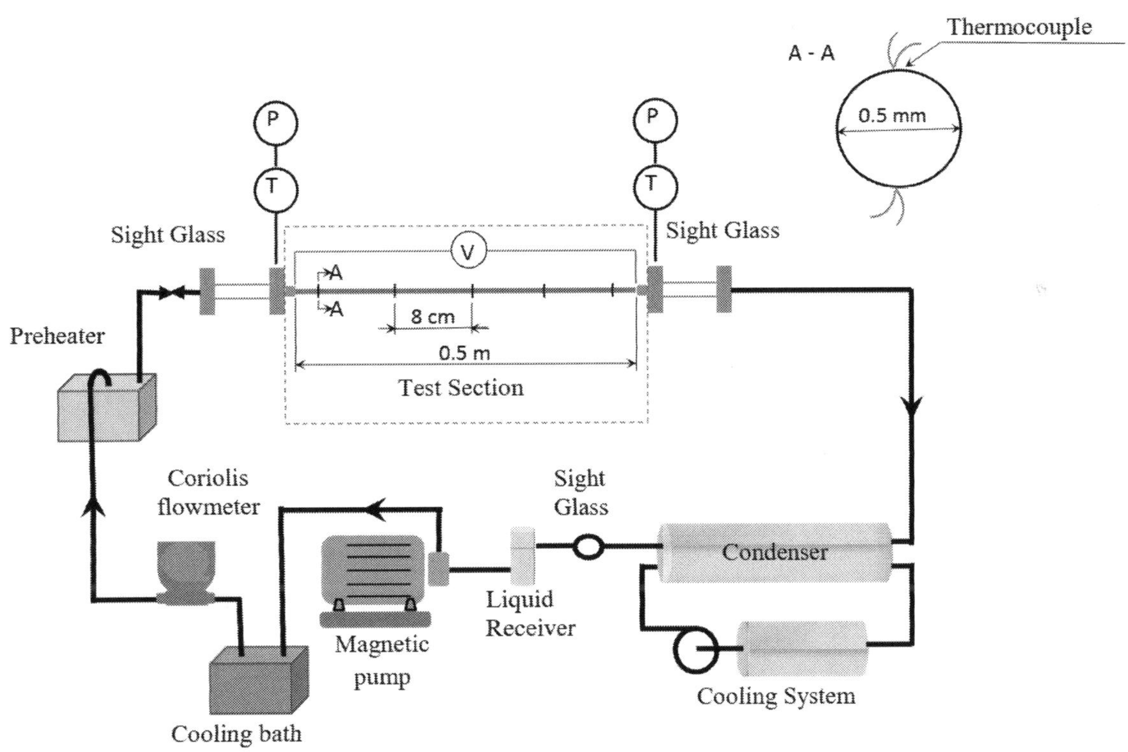

Figure 1. Experimental apparatus

Figure 1 describes the experimental apparatus. The main observation is on the test section heated by electrical heater. The test section is a horizontal tube with diameter of 500 μm and length of 0.5 m. Surface temperature in the test section were measured by attaching K-type thermocouples at the top and bottom of the test section. There are five temperature measurement locations on the test section. Moreover, inserted thermocouple and pressure transmitter are installed at the inlet and outlet of the test section. A condensing unit is used to condense the vapor of working fluids. After the condensation process, the liquid of working fluids is pumped by a magnetic pump. A cooling bath is placed after the magnetic pump to maintain the working fluids in liquid phase condition. A Coriolis flow meter is used for measuring flow rate of the working fluids. Before the working fluid enters the test section, it is flowed in a preheater to adjust the inlet temperature of working fluids. There are sight glasses at the inlet and outlet of the test section for observation the phase of working fluids.

2.2. Data reduction

[9] introduced confinement number (Co) as a ratio of capillary length and hydraulic diameter.

$$Co = \frac{\left[\frac{\sigma}{g(\rho_l-\rho_v)}\right]^{1/2}}{d_h} \quad (1)$$

Where $\sigma, g, \rho_l, \rho_v$ and d_h represent surface tension, acceleration due to gravity, liquid density, vapor density, and hydraulic diameter, respectively. Kew and Cornwell defined that the microchannel flow occurred when Co > 0.5. The average calculated Co for the present experimental data is 2.38. Based on the classification proposed by Kew and Cornwell, it can be classified as a microchannel flow.

The measured temperature and pressure is recorded by a Data Acquisition. Vapor quality is calculated using the following equation.

$$x_o = \frac{\Delta i + i_{fi} - i_f}{i_{fg}} \quad (2)$$

Where $x_o, \Delta i, i_{fi}, i_f,$ and i_{fg} represent outlet quality, increasing enthalpy, fluid inlet enthalpy, enthalpy of saturated liquid, and latent heat of vaporization, respectively.

Pressure drop friction from homogeneous equation is given as:

$$\left(\frac{dP}{dz}F\right) = \frac{2f_{tp}G^2}{D\bar{\rho}} \quad (3)$$

Where$\left(\frac{dP}{dz}F\right)$,f_{tp}, G and $\bar{\rho}$ represent pressure drop friction, two-phase friction factor, mass flux and mean density, respectively.

Assuming that the friction factor (f_{tp}) may be expressed in terms of the Reynolds number by Blasius equation:

$$f_{tp} = 0.079\left[GD/\mu_{tp}\right]^{-0,25} \quad (4)$$

The equation of mean density used Adams and Cicchiti. The equation is given as:

$$\bar{\rho} = \left[\frac{x}{\rho_v} + \frac{(1-x)}{\rho_l}\right]^{-1} \quad (5)$$

$$\bar{\rho} = x\rho_v + (1-x)\rho_l \quad (6)$$

Where ρ_v and ρ_l represent vapor density and liquid density, respectively.

Pressure drop from separated models. These models calculate the frictional pressure gradient by the frictional pressure gradient of a single-phase flow. Depending on the chosen approach, this flow can be either for the liquid (or vapor) phase considered as total flow in the channel, or for the liquid (or vapor) considered to flow alone in the channel, both corrected by the corresponding two-phase friction multiplier parameter [10]. The letter is given by the following this equation:

$$\left(\frac{dp}{dz}F\right) = \left(\frac{dp}{dz}F\right)_f \emptyset_f^2 \quad (7)$$

Pressure drop is correlated as a function of Reynolds number. Reynolds number of two phase flow is a function of viscosity, mass flux, and hydraulic diameter. This present study uses some two phase viscosity correlations, viz. [11], [12], [13], and [7] as shown in the following equations.

$$\mu_{tp} = \left[\frac{x}{\mu_v} + \frac{(1-x)}{\mu_l}\right]^{-1} \tag{8}$$

$$\mu_{tp} = x\mu_v + (1-x)\mu_l \tag{9}$$

$$\mu_{tp} = \bar{\rho}[xv_v\mu_v + (1-x)v_l\mu_l] \tag{10}$$

$$\mu_{tp} = \mu_l(1-\alpha)(1+2.5\alpha) + \mu_v\alpha \tag{11}$$

Where $\mu_{tp}, \mu_v, \mu_l, v_l, v_v, \alpha$, and x represent two-phase viscosity, vapor viscosity, liquid viscosity, liquid specific volume, vapor specific volume, void fraction and quality, respectively.

3. Result and discussion

Experiment two-phase flow boiling on microchannel have done successfully. The data result showed mass flux between 360 to 915 kg/m²s. Heat flux between 30 to 43 kW/m². Saturation temperature between 38.3 to 48.7°C. Flow regime experiment on micro channel found as liquid turbulence-vapor turbulence.

[8] develop new C parameter based on the data experiment. The parameter has correlation with viscosity number. The equation viscosity number is given as:

$$N_{\mu_{tp}} = \frac{\mu_{tp}}{\left(\rho_{tp}\sigma\sqrt{\frac{\sigma}{g\Delta\rho}}\right)^{0,5}} \tag{12}$$

The common logarithm of the two-phase viscosity number data experiment have bigger range number. Xuejiao experiment reported the common logarithm of the two-phase viscosity number between – 3 to – 2 at the same regime flow, liquid turbulence and vapor turbulence.

Table 1. Data experiment with parameter measurement quality, logarithm viscosity number, C parameter and regime flow regime.

x_o	$\log(N_{\mu_{tp}})$	C Parameter			Flow Regime
		Xuejiao	Pamitran	Hibiki	
0,35 – 0,97	(- 3.40) – (- 2,65)	6,67 – 19,2	10,27 – 58,3	3.09	Turbulence – Turbulence

Table 1. Describing data experiment with parameter measurement i.e. quality, logarithm viscosity number, C parameter and regime flow regime. Frictional pressure drop from separated models is effected by C parameter. [5] develop C parameter as function of hydraulic diameter. For the tube with 0.5 mm, The C parameter has constant 3.09 value. [8] develop C parameter for flow regime turbulence-turbulence as function of viscosity number, Reynolds two-phase and quality. The equation from Li and Hibiki is given as:

$$C = 6.28N_{\mu_{tp}}^{0.78}Re_{tp}^{0.67}x^{0.32} \tag{13}$$

Data experiment showed the C parameter from Xuejiao on flow regime turbulence-turbulence lower than C from Chisholm. [6] develop C parameter as function of Weber number and Reynolds number. The equation is given as:

$$C = 3\ 10^{-3}We_{tp}^{-0.433}Re_{tp}^{1.23} \tag{14}$$

Data experiment showed the C parameter from Pamitran 10.27 to 58.3 value. Some of data experiment over than 20 for turbulence-turbulence.

Characteristic of pressure drop from separated models more predominated by C parameter. Flow regime is effected to two-phase frictional multi player. This data experiment showed on turbulence-turbulence condition.

Figure 2. Describing pressure drop prediction on the all data experiment with various parameter i.e. mass flux, heat flux and saturation temperature. For round type is frictional pressure drop from homogeneous models. Calculating frictional pressure drop used varying mean density from Adams and Cicchiti. Calculating two-phase friction factor used varying two-phase viscosity from Adams, Cicchiti and Dukler.

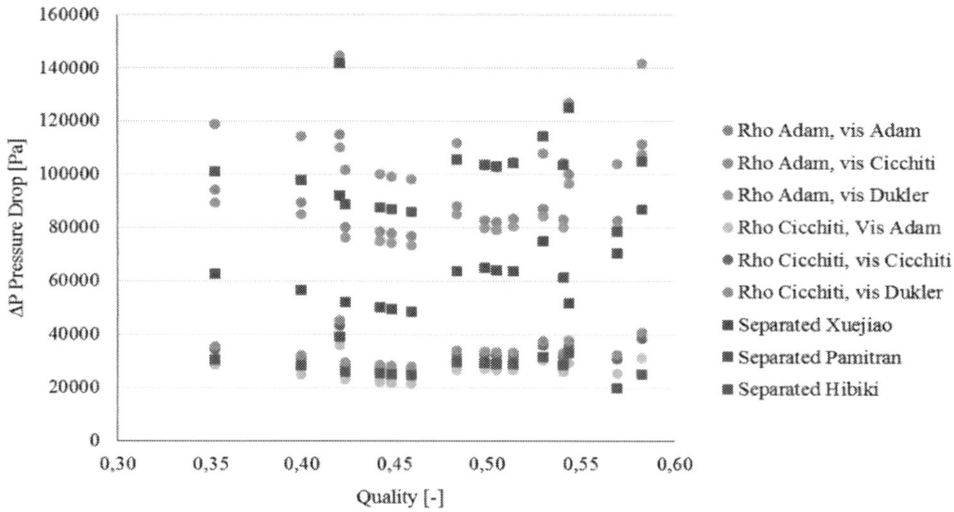

Figure 2. Homogenous and separated frictional pressure drop calculation. Round type is homogeneous models, square type is separated models.

The figure showed frictional pressure drop with density from Adams and two-phase viscosity from Cicchiti has higher pressure drop. Vice versa frictional pressure drop with density from Cicchiti and two-phase viscosity from Adams has lower pressure drop.

For square type is frictional pressure drop from separated models. There are three correlation pressure drop i.e. Xuejiao, Pamitran and Hibiki. The figure showed frictional pressure drop from Pamitran correlation has higher pressure drop. The lower frictional pressure drop occur on Hibiki correlation.

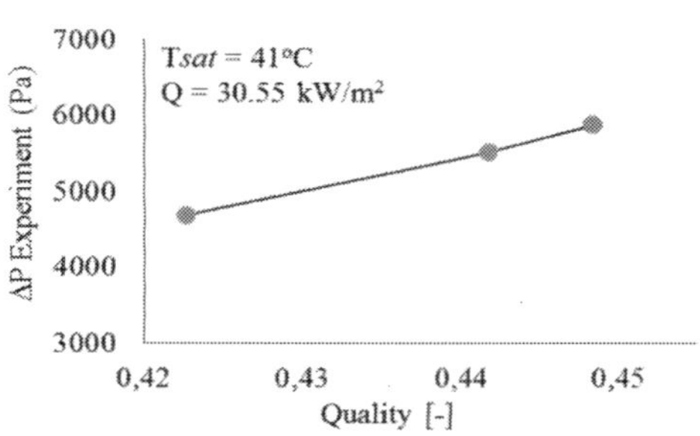

Figure 3. Effect of quality on pressure drop

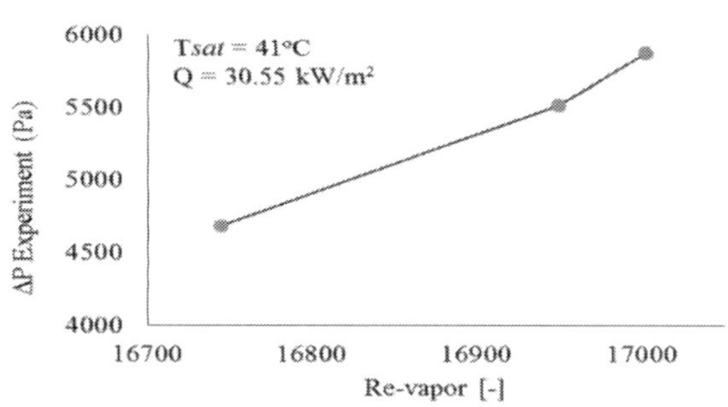

Figure 4. Effect of Reynolds vapor on pressure drop

Figure 3 showed the increasing quality will effect on increasing pressure drop. Then figure 4 showed the increasing Reynolds vapor will effect on increasing pressure drop. The condition experiment on 41°C saturation temperature and 30.55 kW/m^2 heat flux.

On quality 0.42 showed the pressure drop 4682 Pa, and quality 0.45 showed the pressure drop 5873 Pa. Both of figure have inter connecting. The vapor Reynolds number increase if quality increase. The figure showed increasing quality or vapor Reynolds number will increase frictional pressure drop. Differences of quality occur due to mass flux experiment is different.

4. Conclusion

Experimental frictional pressure drop on the microchannel has done successfully. Frictional pressure drop from homogeneous models showed the frictional pressure drop with density from Adams and two-phase viscosity from Cicchiti has higher pressure drop. Vice versa frictional pressure drop with density from Cicchiti and two-phase viscosity from Adams has lower pressure drop.

Quality and vapor Reynolds number predominant to effect on frictional pressure drop in microchannel. Heat transfer on microchannel have big effect to contribute evaporating fluid to be vapor.

5. Acknowledgement

This study is funded by Hibah Ristekdikti PUPT 2017.

6. References

[1] K.-I. Choi, A. Pamitran, J.-T. Oh, and K. Saito, "Pressure drop and heat transfer during two-phase flow vaporization of propane in horizontal smooth minichannels," *International Journal of Refrigeration,* vol. 32, pp. 837-845, 2009.

[2] N. Ghazali, Q. A. Yousif, A. S. Pamitran, S. Novianto, and R. Ahmad, "OPTIMIZATION OF THE FRICTION FACTOR AND FRICTIONAL PRESSURE DROP OF R22 AND R290," *INTERNATIONAL JOURNAL OF TECHNOLOGY,* vol. 7, pp. 227-234, 2016.

[3] M. R. Hoehne and P. Hrnjak, "Charge minimization in systems and components using hydrocarbons as a refrigerant," *Urbana,* vol. 51, p. 61801, 2004.

[4] A. Cavallini, E. Da Riva, and D. Del Col, "Performance of a large capacity propane heat pump with low charge heat exchangers," *International Journal of Refrigeration,* vol. 33, pp. 242-250, 2010.

[5] K. Mishima and T. Hibiki, "Some characteristics of air-water two-phase flow in small diameter vertical tubes," *International journal of multiphase flow,* vol. 22, pp. 703-712, 1996.

[6] A. Pamitran, K.-I. Choi, J.-T. Oh, and P. Hrnjak, "Characteristics of two-phase flow pattern transitions and pressure drop of five refrigerants in horizontal circular small tubes," *International Journal of Refrigeration,* vol. 33, pp. 578-588, 2010.

[7] D. Beattie and P. Whalley, "A simple two-phase frictional pressure drop calculation method," *International Journal of Multiphase Flow,* vol. 8, pp. 83-87, 1982.

[8] X. Li and T. Hibiki, "Frictional Pressure Drop Correlation for Two-Phase Flows in Mini and Micro Single-Channels," *International Journal of Multiphase Flow,* 2016.

[9] P. A. Kew and K. Cornwell, "Correlations for the prediction of boiling heat transfer in small-diameter channels," *Applied Thermal Engineering,* vol. 17, pp. 705-715, 1997.

[10] J. G. Collier and J. R. Thome, *Convective boiling and condensation*: Clarendon Press, 1994.

[11] W. McAdams, W. Woods, and L. Heroman, "Vaporization inside horizontal tubes-II-benzene-oil mixtures," *Trans. ASME,* vol. 64, pp. 193-200, 1942.

[12] A. Cicchitti, C. Lombardi, M. Silvestri, G. Soldaini, and R. Zavattarelli, "Two-phase cooling experiments: pressure drop, heat transfer and burnout measurements," Centro Informazioni Studi Esperienze, Milan1959.

[13] A. Dukler, M. Wicks, and R. Cleveland, "Frictional pressure drop in two-phase flow: A. A comparison of existing correlations for pressure loss and holdup," *AIChE Journal,* vol. 10, pp. 38-43, 1964.

IOP Conf. Series: Materials Science and Engineering **316** (2018) 012059 doi:10.1088/1757-899X/316/1/012059

CFD investigation of pentamaran ship model with chine hull form on the resistance characteristics

Yanuar[1] and W Sulistyawati[2]
[1] Department of Mechanical Engineering, Universitas Indonesia, Depok 16424, Indonesia
[2] Graduate Student Department of Mechanical Engineering, Universitas Indonesia, Depok 16424, Indonesia

E-mail: yanuar@eng.ui.ac.id

Abstract. This paper presents an investigation of pentamaran hull form with chine hull form to the effects of outriggers position, asymmetry, and deadrise angles on the resistance characteristics. The investigation to the resistance characteristics by modelling pentamaran hull form using chine with symmetrical main hull and asymmetric outboard on the variation deadrise angles: 25°, 30°, 35° and Froude number 0,1 to 0,7. On calm water resistance characteristics of six pentamaran models with chine-hull form examined by variation of deadrise angles by using CFD. Comparation with Wigley hull form, the maximum resistance drag reduction of the chine hull form was reduced by 15.81% on deadrise 25^0, 13.8% on deadrise 30^0, and 20.38% on deadrise 35^0. While the smallest value of total resistance coefficient was generated from chine 35^0 at R/L:1/14 and R/L:1/7. Optimum hull form for minimum resistance has been obtained, so it is interesting to continue with angle of entrance and stem angle of hull for further research.

1. Introduction

Many studies established that pentamaran better than monohull, i.e.: large volume and deck, low resistance at high speed, good stability, and good sea keeping performance. The factors affected mutual on multihull were viscosity resistance due to the wide variation of the wetted surface on the changes of pressure and speed, and wave making resistance as effect engagement of the cross section along the ship and spreads of the wave. Outriggers position to the main hull contributed to influence on frictional resistance and stability [1]. The Asymmetric outrigger configuration and separation distance resulted change in resistance and there is no single configuration that consistently outperforms the other configuration across the entire speed range [2]. Frictional resistance on pentamaran increased as a consequence on enhancement of wetted surface area hull form, but wave making resistance can be reduced with a slender hull form. [3] has defined wave making resistance reduction and enhancement of frictional resistance were affected by the hull length to width ratio (L/B).

Minimization resistance of ship there are several ways such as modeling hull form or other ways, e.g.: polymer, microbubbles [4]. Hull form very significant impact on the ship, then modeling with the precise configuration on pentamaran was indispensable. Researches on pentamaran generally use Wigley hull form, this study use chine hull form to investigate and compare with Wigley. The specialty of this chine hull form in some studies shown advantages such as: reducing resistance, easier and faster in the building process. Experiment [5] shown symmetrical chine hull form on catamarans could reduce wave resistance, interference between hull due to resistance will tend to fall on Fr > 0,5. [6] found that chine more profitable on Fr > 0.75 compared with NPL series, while in sea keeping performance; rounded models are superior in displacement conditions. [7] had compared chine for catamarans and Wigley model by a computer program SHIPFLOW. Wigley hull form in deep waters have a higher value Cw (wave coefficient) at 0.3 < Fr <0.6, and in shallow waters the highest Cw value at Fr 0.4. While the chine hull form, Cw tend to fall at 0.6 <Fr <0.7 in line with the decrease in depth, and the highest value Cw in shallow waters occurred at Fr 0.4.

Ship performance is determined by parameters: L/B, $A_P / V^{2/3}$, LCG, deadrise angle, angle variation along the hull, and shape of chine [8]. [9] had used multihull with chine get hydrodynamics of the catamaran on deadrise angle 0°-20° lift coefficient will improve with decreasing spacing of outriggers and high Froude numbers in line increase of deadrise angle. But at larger deadrise angles, high Froude numbers and the gap of outriggers more widen the complex effects occur where the waves flow direction resulting in pressure on the rear area of hull, and also resulted in lift coefficient. [10] were using configuration pentamaran Wigley hull form with numerical method to analyse hydrodynamic characteristics at speed variation. Specifically of this study is investigating the calm-water total resistance, frictional and residuary characteristic of a pentamaran with chine hull form at variation of deadrise angles on 25^0, 30^0, 35^0. The numerical analyses based on computational fluid dynamics (CFD) modelling and validation with last experimental data.

2. Methodology

Experimental and numerical modelling techniques are very important to get hull form with good hydrodynamic, performance and safety. Experimental for multihull ships were first performed by [11] on catamarans to obtain optimal hull with minimum resistance. [12] with numerical computational approach based on [13] determined resistance and motions on catamaran, trimaran, quadrimaran and pentamaran using Wigley hull form on several configurations. He mentioned that the waves making resistance on multihull always the biggest component as the speed or value of Fr increase.

Based on the hypothesis of Froude that the total resistance consists of two components, i.e.: frictional resistance (R_F) and the residuary resistance (R_R).

$$R_T = R_F + R_R \tag{1}$$

The analysis data experiment using ITTC'57 methodology for calculate the total resistance coefficient (C_T):

$$C_T = C_F + C_R \tag{2}$$

$$C_F = \frac{0.075}{(\log Rn - 2)^2} \tag{3}$$

Pentamaran consists of one main hull, two inners and two outers hull, with different Reynold number, then the frictional resistance coefficient calculated by:

$$C_F = C_{Fmain}\frac{S_{main}}{S_T} + C_{F1}\frac{2side_1}{S_T} + C_{F2}\frac{2Side_2}{S_T} \tag{4}$$

Then the residual coefficient was given by:

$$C_R = C_T - C_T \tag{5}$$

2.1. CFD Method

This study was investigation a K-ε model transformed the outer boundary layer and the free stream [14]. In a large variety of different flow, K-ε model has been used successfully. The K-ε model [15] is numerical integrated equation of viscous sub layer used in Computational Fluid Dynamics (CFD) to simulate characteristics of turbulent flow. It was generally used to mixing-length model, to find algebraically turbulent length scales in high complexity flows. K is turbulent kinetic energy determines in the turbulence, ε. A nonlinear boundary condition on ε allows using of a homogeneous boundary condition.

The mathematical model for free surface flow was based on the homogenous multiphase Eulerian–Eulerian fluid approach. This approach shares the same velocity field and other relevant fields such as temperature, turbulence, etc. Numerical method was derived from Navier Stokes' equations where using a Volume of Fluid (VOF) approach. Incompressible Reynolds averaged Navier-Stokes (RANS) was expressed by [16].

$$\rho\frac{\vartheta U_i}{\vartheta t}+\rho U_j\frac{\vartheta U_i}{\vartheta x_j}=-\frac{\vartheta P}{\vartheta x_i}+\frac{\vartheta T_{ij}}{\vartheta x_j} \tag{6}$$

$$\left(\frac{\vartheta \rho}{\vartheta t}+U_j\frac{\vartheta \rho}{\vartheta x_j}\right)+\rho\frac{\vartheta U}{\vartheta x_i}=0 \tag{7}$$

Equation (6) as momentum equation (second Newtonian law for fluids) and equation (7) as continuity equation. Where, ρ: fluid's density, U_i: velocity components, P: pressure, and T_{ij}: components of viscous stress tensor as follows:

$$T_{ij}=(2\mu S_{ij}-\rho\overline{U_i U_j}) \tag{8}$$

$$S_{ij}=\frac{1}{2}\left(\frac{\vartheta U_i}{\vartheta x_j}+\frac{\vartheta U_j}{\vartheta x_i}\right) \tag{9}$$

where ${}^{\vartheta U_i}/_{\vartheta x_i}=0$ as effect of turbulence on the flow by Reynolds Stress (RS) $\rho\overline{u_i u_j}$. And the equation of volume fraction transport [17]:

$$\frac{\partial c}{\partial t}+\frac{\partial (cU_j)}{\partial x_j}=0 \tag{10}$$

c; volume fraction $=\frac{V_{air}}{V_{total}}$, ρ; fluid density and μ; viscosity is calculated as:

$$\mu=\mu_{air}c+\mu_{water}(1-c) \tag{11}$$

$$\rho=\rho_{air}c+\rho_{water}(1-c) \tag{12}$$

The turbulence model wall laws have restrictions on the y+ value at the wall. The wall y^+, a non-dimensional distance from the wall to the first grid point was similar to local Reynolds number represented in Equation 13. It's determining whether the influences in the wall-adjacent cells are laminar or turbulent. Turbulence models deal with the flow in the boundary layer can be further subdivided into a laminar (viscous) sublayer and a fully turbulent region.

$$y^+=\frac{U_\tau y}{v} \tag{13}$$

where, Uτ; frictional velocity as comparison between shear stress at the wall (τ_w) and density (ρ), y; distance from wall surface, v; kinematic viscosity. The value y^+ of hull form was depending the wall surface adjusting to the flow in the boundary layer.

2.2. Last Experimental Result

The results of the experimental data by [18] was used to validate the CFD analysis on Wigley pentamaran hull form with asymmetric inner side-hulls. The experiment had done to 9 configurations at variations Fr 0.1 - 0.7. On model IIA with configuration S/L (separation/ length of main hull) 3/16 and R/L (stagger/ length of main hull) 1/ 20 at Fr 0.35 the resistance was reduced by 25%. At Froude numbers ranging from 0.2 to 0.7, the resistance coefficients have a downwards trend as shown in figure 1.

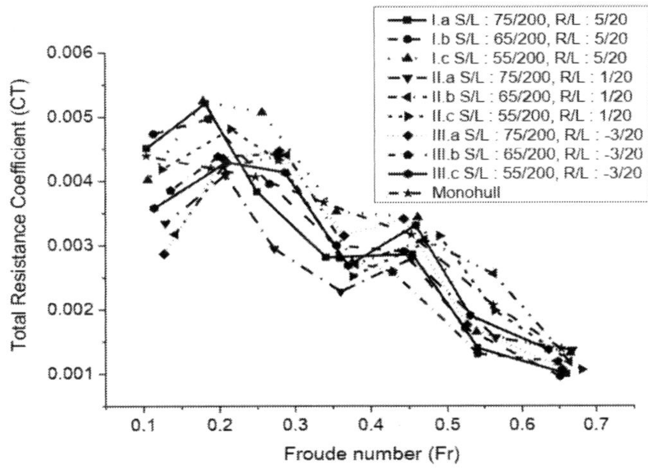

Figure 1. Total resistance coefficient [18]

Table 1. Model hull specification

Main Dimension	Wigley			Chine 25^0		
	Main	Inner side	Outer side	Main	Inner side	Outer side
LOA	2.00 m	1.00 m	0.75 m	2.40 m	1.02 m	0.93 m
B	0.20 m	0.10 m	0.07 m	0.20 m	0.10 m	0.07 m
T	0.07 m	0.07 m	0.06 m	0.07 m	0.07 m	0.06 m
H	0.15 m	0.15 m	0.14 m	0.15 m	0.15 m	0.14 m
Cb	0.57	0.59	0.58	0.59	0.67	0.55
WSA	0.39 m^2	0.18 m^2	0.11 m^2	0.55 m^2	0.30 m^2	0.21 m^2
Δtot		24.5 kg			≈24.5 kg	
Main Dimension	Chine 30^0			Chine 35^0		
	Main	Inner side	Outer side	Main	Inner side	Outer side
LOA	2.75	1.08 m	0.93 m	3.30 m	1.12 m	0.93 m
B	0.20 m	0.10 m	0.07 m	0.20 m	0.10 m	0.07 m
T	0.07 m	0.07 m	0.06 m	0.07 m	0.07 m	0.06 m
H	0.15 m	0.15 m	0.4 m	0.15 m	0,15 m	0.14 m
Cb	0.52	0.65	0.55	0.44	0.62	0.55
WSA	0.61 m^2	0.31 m^2	0.21 m^2	0.69 m^2	0.32 m^2	0.21 m^2
Δtot		≈24.5 kg			≈24.5 kg	

The results of IIa configuration as initial modelling then performed with pentaraman chine hull form at variation of deadrise angles: 25°, 30°, 35°. The first step was modelling an initial pentamaran which used Wigley linesplan of Yanuar experimental. The second, modified the Wigley hull form with chine form in the same S/L position. The specifications of both hull form was shown in table 1. Midship section of Wigley pentamaran and chine hull forms were designed as shown in figure 2 and 3. Transformation Wigley to chine form with some differences main hull L/B: 12; 13.75; 16.5 in constant B/T: 2.9 and H/T: 2.1, there are 6 new configurations from the initial configuration. In each deadrise angle of chine hull form there are two configurations as shown in figure 4. Each configuration at each chine hull form is a change in the position of inner and outer side hull from its original position.

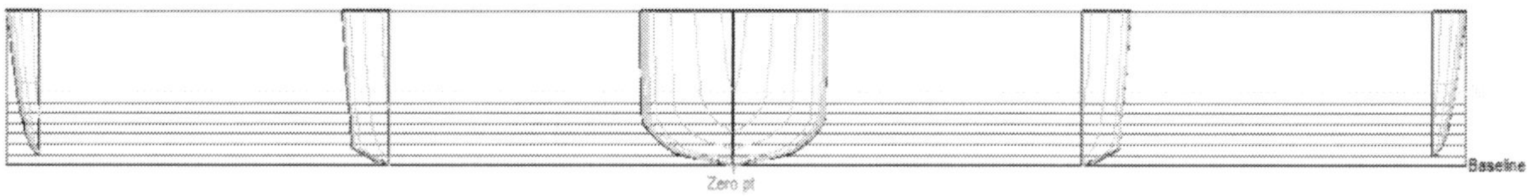

Figure 2. Pentamaran Wigley

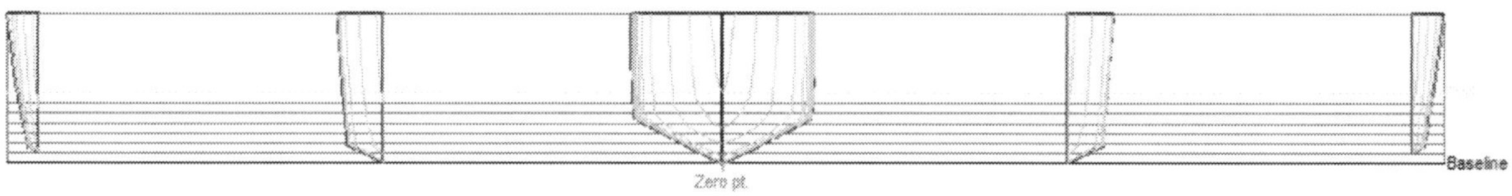

Figure 3. Pentamaran chine

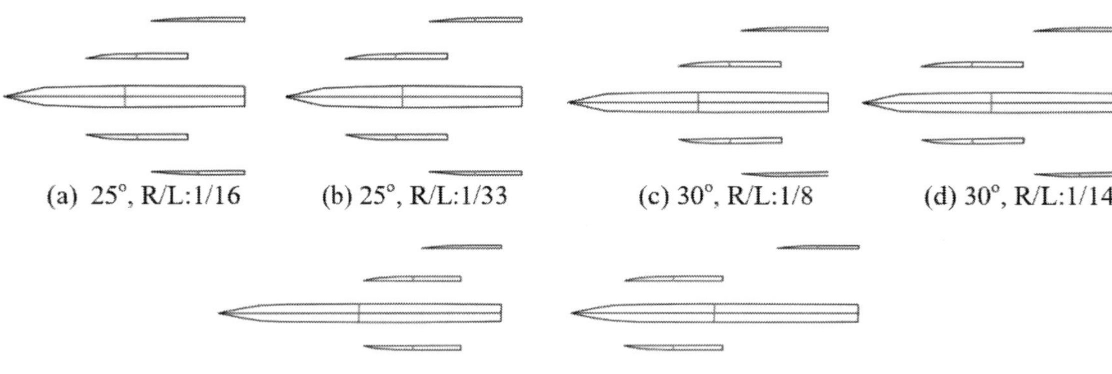

(a) 25°, R/L:1/16 (b) 25°, R/L:1/33 (c) 30°, R/L:1/8 (d) 30°, R/L:1/14

(d) 35°, R/L:1/5 (f) 35°, R/L:1/7

Figure 4. Six new configurations of chine hull form

2.3. Meshing Technique

Table 2. CFD simulation properties

Property	Fine Mesh
Type of mesh	Tetrahedral, mixer
Domain Physics	Homogeneous water/air multiphase, K-ε, automatic wall function, buoyancy model –density difference, standard free surface model
Boundary physics:	
Inlet	Inlet, volume fraction, turbulence intensity 0.05
Outlet	Outlet, relative pressure
side wall	Wall, free slip condition
Top	Opening, relative pressure 0 Pa
Hull/ Bottom	Wall, no slip condition
Symmetry plane	Along centerline of the hull
Solver settings:	
Convergence criteria	Residuary type: RMS, Target: 10-e4
Multiphase control	Volume fraction coupling

[19] and [17] revealed that mesh quality is very important to perform a reliable CFD. Total element number in the mesh should be large enough to represent the geometry and flow phenomena in the domain. Physical domains at free surface model for domain boundaries in detailed are provided in table 2. The mesh settings for element size was repeated 5 times (4.5M, 7M, 9M, 10M and 11M) to evaluate convergence. For verification, i.e. convergence and the assessment of numerical uncertainty, processing mesh has been carried out to the total resistance (force) which element size of boundary 0.1 and the hull of 0.005 to 0.003. The meshing results on the model shown in figure 5, which produced about 10 million elements. ICEM meshing method was selected to produce the mesh automatically. The mesh generated in ICEM using a low transition options to refine the mesh gradually. Unstructured tetrahedral elements were selected in the entire domain and hull.

(a) Meshing domain (b) Hull meshing detail

Figure 5. CFD mesh analysis results

Total number of elements, mesh size as well as value of Y^+ was considered for several variations of mesh. The initial mesh created 4.5M elements then increased until to 11M elements with Y^+ on the hull ~5. In multiphase flow, volume fraction is considered to prevent a large residual. Repeated convergence was assumed through normal residuals. The residual RMS of domain required on 1E -04. The control time step for steady state problems, CFX uses a robust implicit formulation allowing the large time step can be determined, and accelerate convergence. Time step permits fixed time step size used for the entire flow domain. The flow was dominated by advection (horizontal mass movement which resulted in the change) the size of the time step should be scale length divided by the speed scale [20]. A steady state calculation usually requires between fifty to a hundred-time steps to achieve convergence. But overall a great time step makes an unstable convergence, otherwise if time step too small would be very slow convergence.

3. Results and Discussion

Transformation Wigley pentamaran into chine hull form on constant displacement causes the change of Lwl so the domain changed, particularly, the stern and bow hull forms adjust to the ship length. In addition to domain size, mesh settings and CFD setup follow the previously described arrangement, i.e.: mesh size, boundary conditions, time step and convergence controls. From the results of the CFD analysis and comparison with experimental data was shown in figure 6, which both graphs of total resistance coefficient (C_T) show similar trend. The difference between the experimental result and CFD was obtained by the equation (14). The difference average result of the cases was 3.6%, which shows CFD setup well-defined.

$$\Delta C_T = \frac{C_{T.exp} - C_{T.CFD}}{C_{T.exp}} x100\% \quad (14)$$

IOP Conf. Series: Materials Science and Engineering **316** (2018) 012059 doi:10.1088/1757-899X/316/1/012059

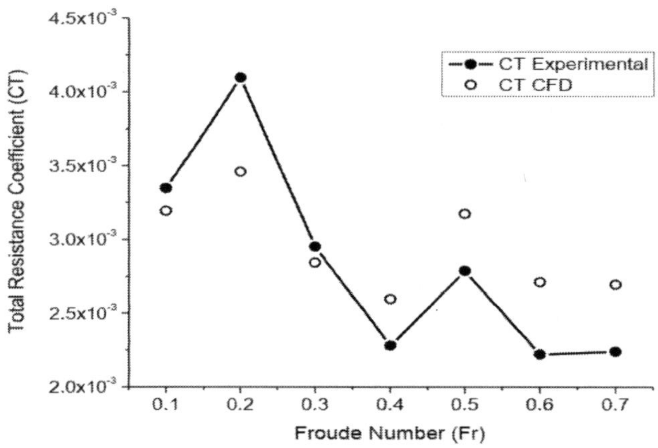

Figure 6. Comparison of C_T between experimental and CFD of Wigley hull form

The resistance coefficient results of 6 chine hull forms C_T, C_F and C_R were shown in figure 7 – 9, respectively. The results of C_T showed the smallest value was generated from chine 35^0 R/L:1/14 and R/L:1/7 with similar trend. The highest average of all chine hull form for C_T occur in 0.2-0.4 Fn. Then, by increase of Fr the C_T values gradually began to fall. At deadrise angle 35^0, it significantly has the smallest C_T compared to deadrise 25^0 and 30^0. It indicates that increased deadrise angle caused decrease the C_T value. The average reduction of C_T value as shown in figure 7 with chine 30^0 to 35^0 was about 9.75%, 25^0 to 30^0 was about 2.9%, and 25^0 to 35^0 was about 12.3%. The frictional resistance acting on a hull form for experimental refers to equation 4, while in CFX-Post could be calculated by performing an area integral of the wall shear in the x-direction. In figure 9 shows C_F graph has similar trend to C_T, where the lowest C_F value generated by chine 35^0 as well. While the greatest value of residual resistance (C_R) was shown by chine hull form 35^0 at R/L: 1/14 and R/L: 1/7. It means this model has bigger wave resistance component than the other forms.

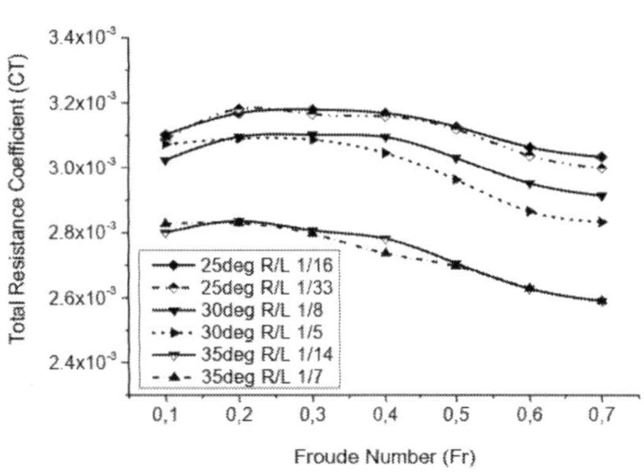

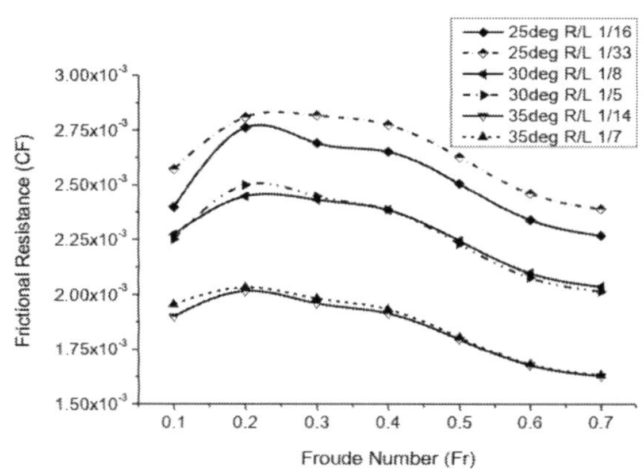

Figure 7. Result on total resistance coefficient (C_T) **Figure 8**. Result on friction coefficient (C_F)

The C_T graph of Wigley hull form appeared hump and hollow phenomenon, while the chine hull form there is no significantly increase or decrease of C_T in Fr variation. The drag reduction of Wigley pentamaran to chine hull form was represented by the equation (15). Percentage of drag reduction (D_R) results of 6 chine hull forms was shown in figure 10. While the maximum drag reduction successively to chine form set out in table 3.

$$DR(\%) = \left|\frac{CT_{wigley} - CT_{chine}}{CT_{wigley}}\right| \tag{15}$$

Table 3. Maximum drag reduction of chine hull form

chine	25deg R/L 1/16	25deg R/L 1/33	30deg R/L 1/8	30deg R/L 1/5	35deg R/L 1/14	35deg R/L 1/7
DR (%)	15,81	15,55	13,80	10,65	20,21	20,38

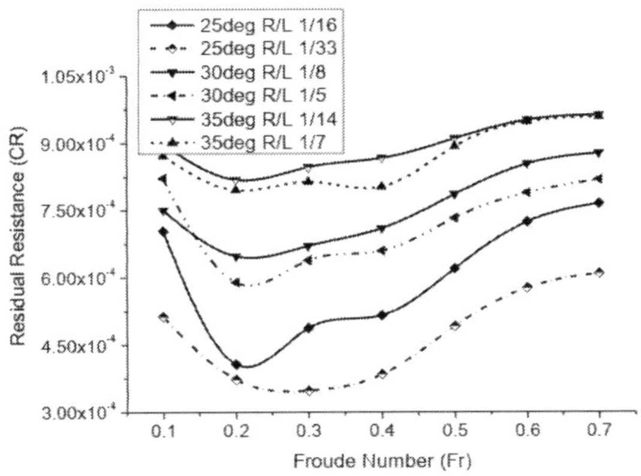

Figure 9. Result on residual resist. coeff. (C_R)

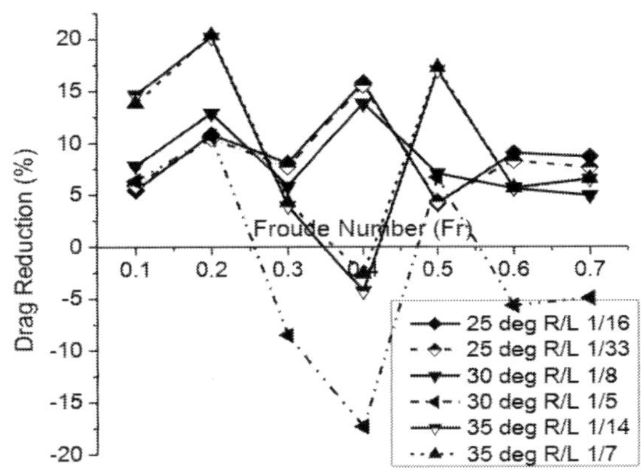

Figure 10. Percentage of drag reduction (D_R)

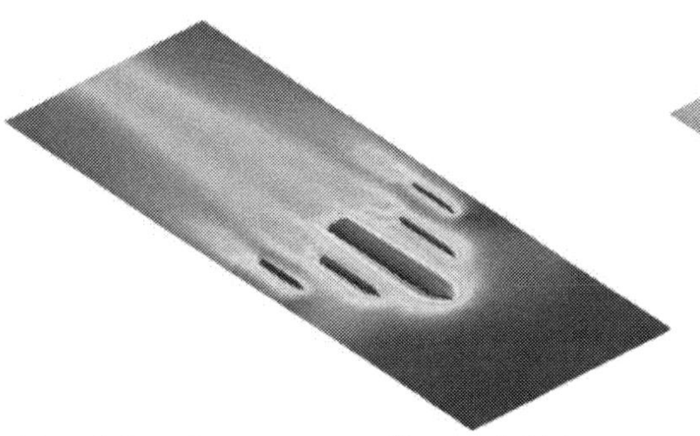

(a) Wigley Contours on z/Lpp = 0.035

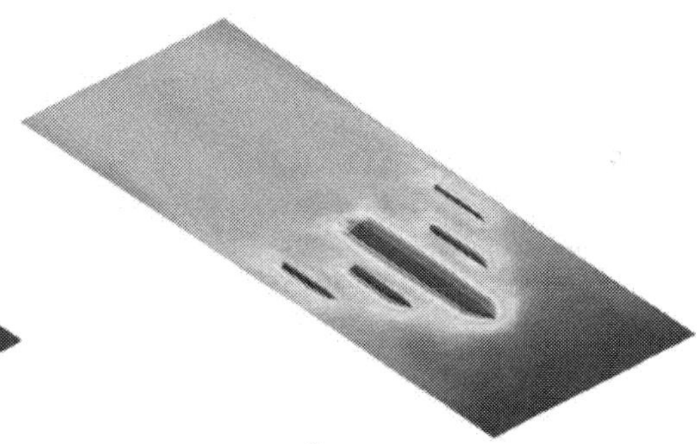

(b) Chine 25^0 Contours on z/Lpp = 0.03

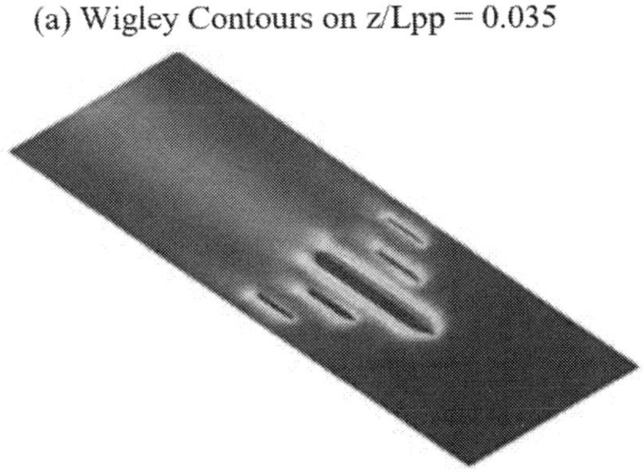

(c) Chine 30^0 Contours on z/Lpp = 0.025

(d) Chine 35^0 Contours on z/Lpp = 0.02

Figure 11. Free surface elevation (reverse side), z/Lpp, of global wave pattern, Fn = 0.7

The contour of wave volume fraction from CFD simulation at Fr 0.7 for Wigley and chine forms were shown in figure 11. There are differences contour of wave volume fraction for Wigley and each chine, which blue colour (a) and (b) shows bigger wave than the dark blue, (c) and (d). It seems that the hull form of Wigley generates bigger wave effect than chine.

4. Conclusion

The comparison of CFD analysis between Wigley to chine form with inner and outer asymmetric hull showed the effect of increasing deadrise angle can improve drag reduction. The highest drag reduction of chine 35^0 due to the big deadrise angle would be causing unconstructed waves angle between each hull. Some advantages chine hull form compared to Wigley are:

- C_T value tends to be smaller on higher L/B with constant B/T and H/T;
- Increasing C_T value not fluctuate with increasing Fr;
- The greater deadrise made the smaller C_T;
- Drug reduction could reach 20.38 % at the largest angle 35^0 on Fr 0.2.

The implication for further research acknowledged needs to investigate the effect of the angle of entrance and stem angle of hull form to the resistance characteristic.

5. Acknowledgement

This work has been financially supported by Directorate of Research and Community Services, Universitas Indonesia (Hibah Ristek Dikti 2017). The authors would like to acknowledge the support of pentamaran team of Department of Mechanical Engineering.

6. References

[1] Oller, Erik, Vasilios Nikou, and Konstantinos Psallidas 2003 Focused Mission High Speed Combatant *DTIC Doc.*

[2] Yanuar, Ibadurrahman, Kurniawan T Waskito, S Karim, and M Ichsan 2017 Interference Resistance of Pentamaran Ship Model with Asymmetric Outrigger Configuration *J. of Marine Sci. and Appl.* **16**:42-47

[3] Ikeda, Yoshiho, Emiko Nakabayashi, and Ai Ito 2005 Concept design of a pentamaran type fast RoRo ship *J. of the Japan Soc. of Nav. Archit. and Oc. Eng* **1**:35-42

[4] Yanuar, Gunawan, Sunaryo, and A Jamaluddin 2012 Micro-bubble drag reduction on a high-speed vessel model *J. of Mar. Sci. and Appl.* Vol **11** Issue 3 Pages 301-304

[5] Chengyi, and Wang 1994 Resistance charicteristic of high-speed catamaran and its application *Shipbld. of China* **3**:003

[6] Blount, Donald L 1995 Factors Influencing the Selection of a Hard Chine or Round-Bilge Hull for High Froude Numbers *FAST '95* Travemunde Germany

[7] Moraes, HB, JM Vasconcellos, and RG Latorre 2004 Wave resistance for high-speed catamarans *Oc. Eng.* **31**(17):2253-2282

[8] Begovic, E, et al 2004 Pentamaran hull for medium size fast ferries *Hydrodynamics VI* (Eds L Cheng and K Yeow):**23-28**

[9] Bari, Ghazi S, and Konstantin I Matveev 2016 Hydrodynamic modeling of planing catamarans with symmetric hulls *Oc. Eng.* **115**:60-66

[10] Tarafder, Md Shahjada, Mir Tareque Ali, and Md Shahriar Nizam 2013 Numerical prediction of wave-making resistance of pentamaran in unbounded water using a surface panel method *Proc. Eng.* **56**:287-296

[11] Hsiung CC and Xu H 1988 Determining Optimal Forms of a Catamaran for Minimum Resistance by the Mathematical Programming *Meth.' Schiffstech* Bd **35**

[12] Peng 2001 Numerical computation of multi-hull ship resistance and motion *Dalhousie Univ. Halifax* Canada

[13] Michell JH 1898 The Wave-Resistance of a Ship *Phil. Magazine* Series **5** Vol 45 Xo 272 London January pp 105-1 23
[14] Menter FR 1994 Two-equation eddy-viscosity turbulence models for engineering applications *AIAA J.* **32**(8):1598 –605
[15] Versteeg, Henk Kaarle Malalasekera, Weeratunge 2007 An introduction to Computational Fluid Dynamics *The Finite Vol. Meth.* Pearson
[16] Wilcox DC 2006 Turbulence Modelling for CFD *3rd Ed. DCW Industr.*
[17] Ferziger JH and Peric M 2002 Computational Methods for Fluid Dynamics *3rd ed. Spring. Verlag* pp **157–206**
[18] Yanuar, Gunawan, Kurniawan T and A Jamaluddin 2015 Experimental Study Resistances of Asymmetrical Pentamaran Model with Separation and Staggered Hull Variation of Inner Side-Hulls *Inter. J. of Fluid Mech. Research* Vol **42** No 1
[19] Blazek J 2001 Computational Fluid Dynamics: Principles and Applications *Elsev. Sci. Ltd* Oxford England
[20] Seo JW, Seol DM, Lee JH and Rhee SH 2010 Flexible CFD meshing strategy for prediction of ship resistance and propulsion performance *Inter J. Nav. Arch. Oc. Eng.* **2** pp 139-14

IOP Conf. Series: Materials Science and Engineering **316** (2018) 012060 doi:10.1088/1757-899X/316/1/012060

Mathematical Modeling – The Impact of Cooling Water Temperature Upsurge on Combined Cycle Power Plant Performance and Operation

Ahmad Indra Siswantara[a], Hariyotejo Pujowidodo[b], Asyari Darius[c], Gun Gun Ramdlan Gunadi[d]

[a]Department of Mechanical Engineering, Universitas Indonesia, Depok 16424, Indonesia
[b]Centre for Thermodynamics, Engine, and Propulsion, BPP Teknologi, Serpong 15314, Indonesia
[c]Mechanical Engineering Department, University of Darma Husada, Jakarta, Indonesia
[d]Mechanical Engineering Department, State University of Jakarta, Depok 16424, Indonesia

*E-mail address:*a_indra@eng.ui.ac.id

Abstract. This paper presents the mathematical modeling analysis on cooling system in a combined cycle power plant. The objective of this study is to get the impact of cooling water upsurge on plant performance and operation, using Engineering Equation Solver (*EES™*) tools. Power plant installed with total power capacity of block#1 is 505.95 *MWe* and block#2 is 720.8 *MWe*, where sea water consumed as cooling media at two unit condensers. Basic principle of analysis is heat balance calculation from steam turbine and condenser, concern to vacuum condition and heat rate values. Based on the result shown graphically, there were impact the upsurge of cooling water to increase plant heat rate and vacuum pressure in condenser so ensued decreasing plant efficiency and causing possibility steam turbine trip as back pressure raised from condenser.

Keywords : heat rate, condenser vacuum, power, cooling temperature

1. Introduction

In combined cycle power plant, energy produced from the conversion gas heat and steam expansion through turbine into mechanical shaft coupled to electric generator. The magnitude of generator power depends on expansion process where gas, steam and its condensate release their potential energy (named enthalpy). Heat and steam enthalpy would change into blades rotation (mechanical energy), exhaust heat and cooling medium (sensible heat). As the energy sources generally come from natural gas and oil, then its combustion exhaust gas recovered into steam in steam generator equipment. Cooling water plays important role to give the condenser the optimum vacuum pressure or enthalpy drop in turbine system.

With principle applying heat balance in turbine and condenser, it could be known and calculated the amount cooling water and its outlet temperature for the determined output power and turbine inlet steam. When demand of output power increased, thus condenser must be able to achieve the suitable vacuum condition to expand steam optimally. Conversely if inlet temperature of cooling water increased, it could reduce the latent heat for condensation and for compensating this, steam consumption to be magnified.

The cooling water temperature upsurge is the important parameter that effects plant performance as well its reliability. Anozie, A. N. and O. J. Odejobi (2011). developed calculation model for simulation of a thermal plant at various circulation water flowrate and saturation pressure , to determine the optimum condenser cooling water flowrate for the process. The influence of circulation cooling water on the efficiency thermonuclear plant also conducted by Gañán, J., et al. (2005). to improve plant performance with appropriate cooling system. The important role of cooling system for power plant as well was concerned by Wu, J., et al. (2001) to study simulation of cooling water discharge from power plant to give contribution in feasibility studies of engineering options that increase the cooling capacity of the waterbody. This parameter is one of the most influential that affect operation variables such as condenser pressure, heat rate, fuel consumption, and cycle efficiency which had interested Sanathara,M.V., et.al (2013) and Haldkar, V., et al. (2013) to study the optimum cooling water mass flow at Steam Power Plant 120 MW. Cooling water inlet temperature and flow are the potential parameter which influence vacuum condition in condenser, as was observed and evaluated by Singh, S.P., et.al. (2014). Ibrahim, M., and Badawy, M.R., (2014) evaluated the effect of cooling water site condition parameters to heat transfer performance of condenser. Meanwhile Ali, S. H., et.al., (2014). and Boi, Rakesh, Mehta, Nirvesh, & Bojak, Khedar. (2015). have studied the back pressure of condenser would impact heat rate and steam power plant performance.

Basically in combined cycle power plant, there are two sections of energy conversion i.e. combustion heat and steam power to mechanical power. Combustion convert chemical energy (contained in fuel) into combustion heat and its high pressure utilized to rotate the turbine blades. Meanwhile at the second one, the exhaust gas from the gas turbine recovered in heat steam generator to convert into steam that used to produce power through steam turbine.

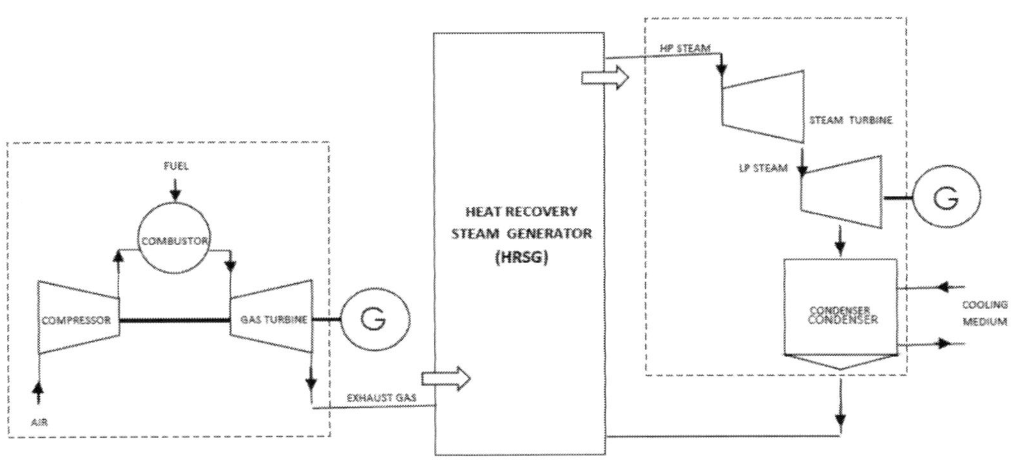

Figure 1. Principle of Energy Conversion in Combined Cycle Power Plant Modeled

System cycle efficiency indicated by the parameter of heat rate that is the ratio of the amount input energy to output power achieved. Output power depends on the turbine enthalpy drop, which determined by condenser performance for rejecting heat to cooling medium. This paper presents the study of the cooling water upsurge to power plant performance and operation using mathematical modeling heat balance on turbine and condenser system

IOP Conf. Series: Materials Science and Engineering **316** (2018) 012060 doi:10.1088/1757-899X/316/1/012060

2. Methods

Mathematical Modeling for analyzing the impact of cooling water temperature rise to plant performance and its operation, was built through principle of mass and heat balance calculation in control volume bounded at condenser and steam turbine. For this modeling, assumptions taken are steady state, heat transfer parameters at condenser as surface area, global heat transfer coefficient is constant. As the variables of analysis are operation parameters such as pressure P, temperature T, flowrate of fuel G_F , steam G_s and cooling water G_{CW} also generator power from gas and steam turbine W_{GT} and W_{ST}, those collected from the plant design, performance test/and or operation data record. Firstly it had to be identified the cycle process of gas system and steam flow to turbine and condenser as well power generator to understand the thermodynamic state and cycle of power generated.

Output power generator W_{ST} is the result of energy balance upon steam turbine and condenser. Enthalpy drop (ΔH) defined by the difference of total enthalpy inlet turbine H_{inturb} and condensation enthalpy H_{cond}. These functions are defined in the following equations based on the energy balance. And W_{GT} obtained from measured value from performance test/and or operation record.

As the performance value determined by the Net Plant Heat Rate (NPHR) based on the amount of total heat from combustion (HHV and LHV basis) and steam divided to the net total output power generator at gas and steam turbine after subtracted auxiliary power. For combustion heat calculated by variables fuel consumption and calorific value.

$$W_{ST} = (H_{inturb} - H_{cond}).\eta_{ST} \quad (1)$$

$$H_{inturb} = H_{HP} + H_{LP} \quad (2)$$

$$H_{HP} = H(P_{HP}, T_{HP}) \quad (3)$$

$$H_{LP} = H(P_{LP}, T_{LP}) \quad (4)$$

$$H_{cond} = G_{CW}.h_{cond} \quad (5)$$

$$h_{cond} = h_{cond}.sat\ (P_{cond}) \quad (6)$$

$$H_{condwtr} = G_{CW}.cp.\ (T_{hot} - T_{cold}) \quad (7)$$

$$Q_{HHV} = G_F.H_{HHV} \quad (8)$$

$$Q_{LHV} = G_F.H_{LHV} \quad (9)$$

$$NPHR_{HHV} = (Q_{HHV} + H_{inturb})/(W_{GT} + W_{ST} - P_{aux}) \quad (10)$$

$$NPHR_{LHV} = (Q_{LHV} + H_{inturb})/(W_{GT} + W_{ST} - P_{aux}) \quad (11)$$

Where cp is specific heat of cooling water in KJ/kg.°C ; all enthalpies in KJ and W in kW units. Subscripts HP and LP indicates high pressure and low pressure state conditions whereas hot and cold explained temperature for outlet and inlet from and into condenser of cooling medium. Notation η_{ST} is turbine efficiency. For h_{cond} is specific enthalpy for condensation at saturated pressure P_{cond}.

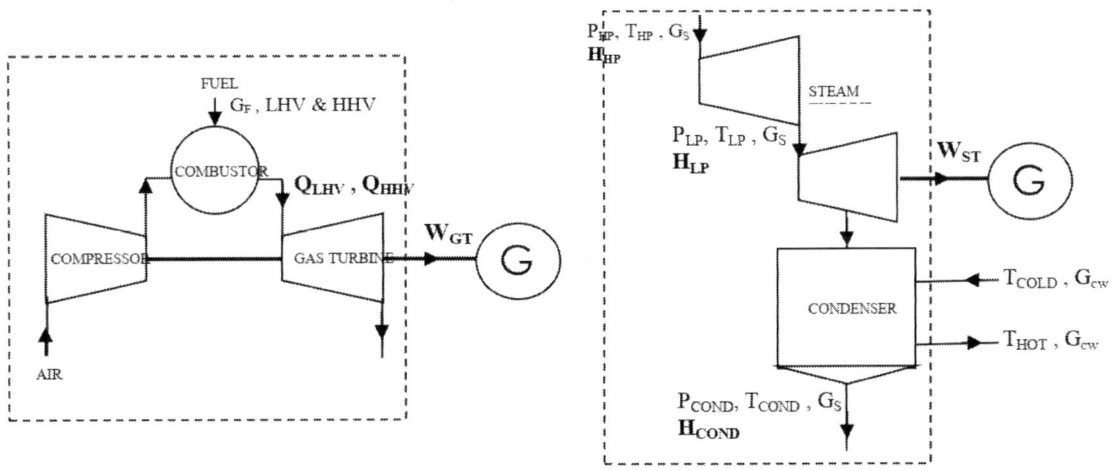

(a)

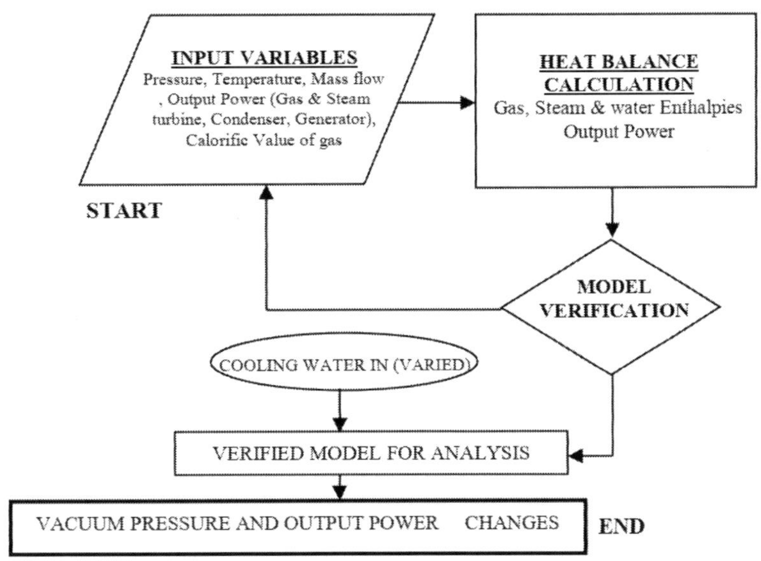

(b)

Figure 2. Volume Control analysis of heat balance (a) and flow diagram of modeling (b)

Calculation model need to be verified to ensure or know the bias (error), compared to the performance test/and or operation record. As the parameters for verification are the input variables and calculated variables. Model that has been verified, then for the analysis variable is the rise of cooling water temperature concerning to the vacuum pressure condenser and output power changes.

3. Result and Discussion

The combined cycle power plant comprises 2 blocks (block#1 and block#2) with gas turbines, HRSGs, and steam turbines system that identified from plant design/process flow diagram. Each

IOP Conf. Series: Materials Science and Engineering **316** (2018) 012060 doi:10.1088/1757-899X/316/1/012060

steam turbine has one condenser for rejecting heat from High Pressure (HP) and Low Pressure (LP) stages. Mathematical modeling verified with performance test of block#2 which input data given in table 1 below.

Table 1. Performance Test Input Data for Model Verificatio

No	Parameter	Symbol	Value	Unit
1	Gas fuel Flowrate GT1	V_{gas1}	47018	kg/hr
2	Gas fuel Flowrate GT2	V_{gas2}	47186	kg/hr
3	Higher Heating Value Gas	HHVgas	13048	kCal/kg
4	Lower Heating Value Gas	LHVgas	11834	kCal/kg
5	Output Power Generator GT1	wgen1	230131	kW
6	Output Power Generator GT2	wgen2	229402	kW
7	HP Steam massflow ST21	GHP1	247.7	ton/hr
8	HP Steam massflow ST22	GHP2	252.3	ton/hr
9	HP Steam massflow ST23	GHP3	251.8	ton/hr
10	HP Steam inlet Pressure ST21	P13	6810	kPa
11	HP Steam inlet Pressure ST22	P23	6870	kPa
12	HP Steam inlet Pressure ST23	P33	6830	kPa
13	Condenser Pressure ST21	P12	10.09	kPa
14	Condenser Pressure ST22	P22	9.684	kPa
15	Condenser Pressure ST23	P32	9.684	kPa
16	LP Steam massflow ST21	GLP1	4.531	ton/hr
17	LP Steam massflow ST22	GLP2	4.616	ton/hr
18	LP Steam massflow ST23	GLP3	4.607	ton/hr
19	LP Steam inlet Pressure ST21	P13	360	kPa
20	LP Steam inlet Pressure ST22	P23	360	kPa
21	LP Steam inlet Pressure ST23	P33	350	kPa
22	In/out cooling water Cond#1	T_{cold21}/T_{hot21}	31.9/41	°C
23	In/out cooling water Cond#2	T_{cold22}/T_{hot22}	31.7/40.8	°C
24	In/out cooling water Cond#3	T_{cold23}/T_{hot23}	32/40.5	°C
25	Spesific heat cap. water	cp	4.183	kJ/°C.kg
26	Thermal Efficiency ST	η	79-89	%
27	Output Power Generator ST21	W_{gen21}	70400	kW
28	Output Power Generator ST22	W_{gen22}	71300	kW
29	Output Power Generator ST23	W_{gen23}	70100	kW

Table 2. Performance Test Input Data for Model Verification

No	Variables	Units	Performance Test	Calculation Model	Error (%)
1	Heat Rate HHV	kCal/kWh	1830.77	1831	-0,01256
2	Heat Rate LHV	kCal/kWh	1660,45	1661	-0,03312
3	Output Power ST21	kW	70.4333	70.274	0,226171
4	Output Power ST22	kW	71.3333	70.964	0,517711
5	Output Power ST23	kW	70.1000	69.979	0,172611

IOP Conf. Series: Materials Science and Engineering **316** (2018) 012060 doi:10.1088/1757-899X/316/1/012060

To clarify the hypothesis that the raised cooling water would decrease the plant performance and operation, previously it was important to analyze the operation data. Figure 3 (from model recalculation) shows that with referring to plant design there are the most influential parameters of condenser pressure and cooling water, which values subsequently are 0.083 bar.abs and 29 °C for cold water-in for mass flow 16300 ton/hr.

But from observation when steam turbine pressure, temperature and steam flow operated at 62.99 bar; 510 °C; 276.48 kg/s (HP) and 4.72 bar; 313 °C; 69.55 kg/s (LP), the condenser actually operates at average values of pressure 10 kPa.abs and temperature below 30 °C thus output power turbine produces more than 120 MWe. Conversely when temperature cold-in higher than 30 °C corresponds to the output power below 100 MWe.

It is clearly shown that the cooling water upsurge has the impact upon output power (related to plant performance) and possibility the change of condenser pressure.

For inlet cooling temperature changing from 30 ÷ 35 °C, the condenser pressure and output power will vary with corresponding to condenser inlet heat. Inlet heat of condenser depends on thermodynamic state of steam, correlating to condensation pressure or temperature at condition of steam out from turbine. It could be seen that the higher cooling water temperature causing output power more higher, this is because latent heat of condensation at correspond pressure saturated more higher. This modeling shows averagely that the condenser pressure is still far from the limit design pressure -520 mmHg, nevertheless this bring the potential problem turbine could be tripped by the raised back pressure.

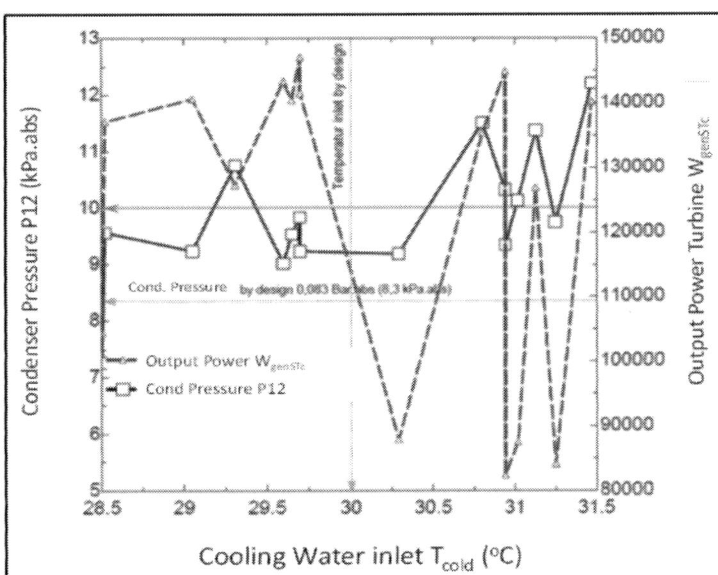

Figure 3 The Effect of Raised Cooling Water upon Condenser Pressure and Output Power

IOP Conf. Series: Materials Science and Engineering **316** (2018) 012060 doi:10.1088/1757-899X/316/1/012060

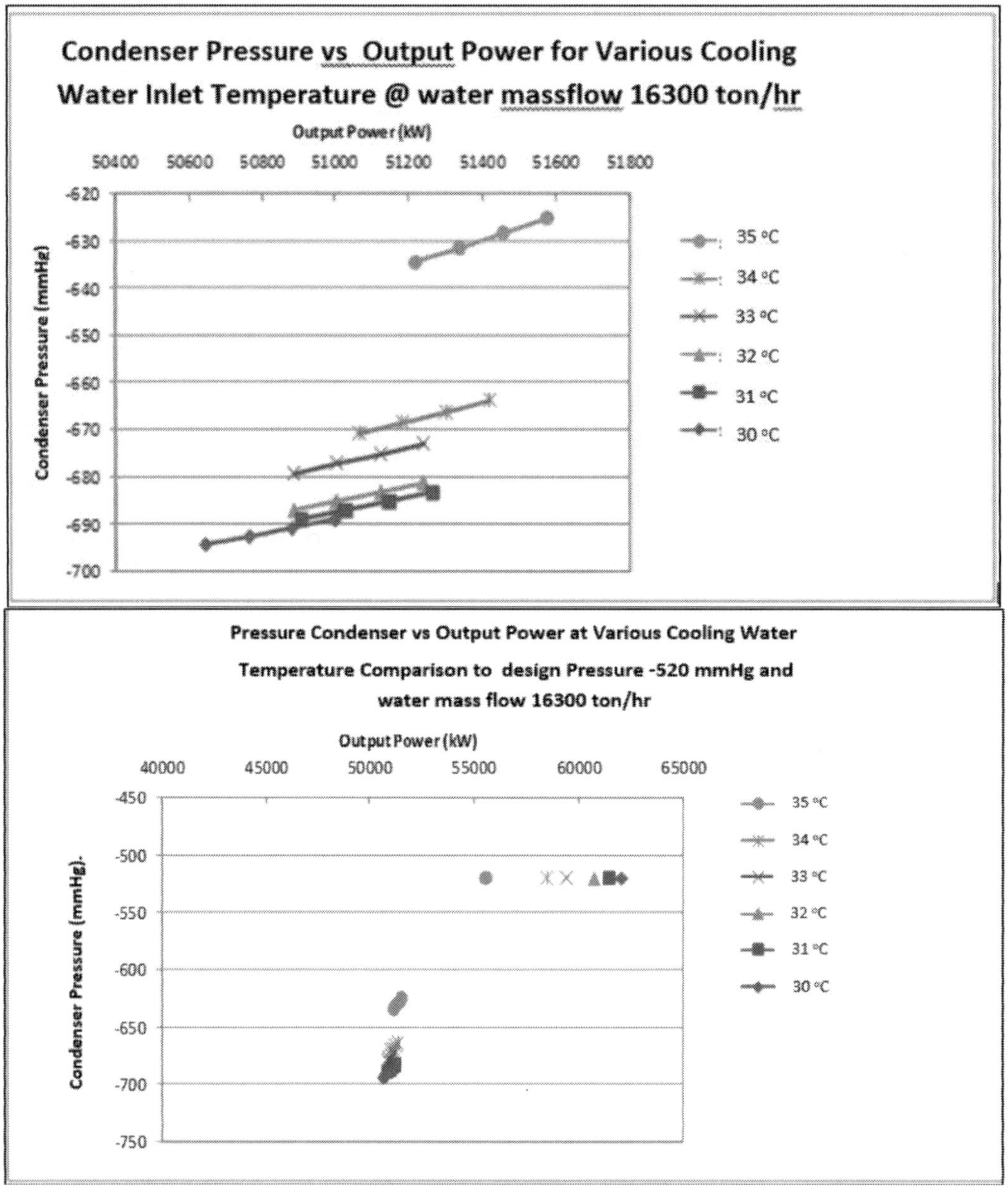

Figure 4 The Change of Condenser Pressure and Output Power for Various Cooling Water Temperature

4. Coclusion

Effect from the cooling water temperature upsurge has increased the condenser pressure as well the output power would decrease. For the change of cooling water temperature from 30 ÷ 35 °C, the condenser pressure is above 620 mmHg. abs that is still safely operated from limit pressure 520 mmHg.abs. At the lower cooling water temperature, the condenser operated at higher pressure in condenser. Otherwise, the pressure would be lower when the cooling water temperature increases. This modeling has given the good result with bias less than 1% compared to operation data.

Acknowledgements
The authors would like to thanks DRPM Universitas Indonesia for funding this research through "Hibah Publikasi Internasional Terindeks untuk Tugas Akhir Mahasiswa UI 2017" and to PT. CCIT Group Indonesia for EES software license

References

[1] Anozie, A. N., & Odejobi, O. J. (2011). The search for optimum condenser cooling water flow rate in a thermal power plant. *Applied Thermal Engineering, 31*(17–18), 4083-4090. doi: http://doi.org/10.1016/j.applthermaleng.2011.08.014

[2] Chantasiriwan, Somchart. (2015). Effects of Cooling Water Flow Rate and Temperature on the Performance of a Multiple-Effect Evaporator. *Chemical Engineering Communications, 202*(5), 622-628. doi: 10.1080/00986445.2013.858040

[3] Gañán, J., Rahman Al-Kassir, A., González, J. F., Macías, A., & Diaz, M. A. (2005). Influence of the cooling circulation water on the efficiency of a thermonuclear plant. *Applied Thermal Engineering, 25*(4), 485-494. doi: http://doi.org/10.1016/j.applthermaleng.2004.07.001

[4] Haldkar, Vikram, Sharma, Abhay Kumar, Ranjan, RK, & Bajpai, VK. (2013). Parametric analysis of surface condenser for thermal power plant. *Int. J. Therm. Technol, 3*, 155-159.

[5] Sanathara, Milan V, Oza, Ritesh P, & Gupta, Rakesh S. (2013). Parametric analysis of surface condenser for 120 MW thermal power plant. *International Journal of Engineering Research & Technology, 2*(3).

[6] Wu, J., Buchak, E. M., Edinger, J. E., & Kolluru, V. S. (2001). Simulation of cooling-water discharges from power plants. *Journal of Environmental Management, 61*(1), 77-92. doi: http://dx.doi.org/10.1006/jema.2000.0396

IOP Conf. Series: Materials Science and Engineering **316** (2018) 012061 doi:10.1088/1757-899X/316/1/012061

Model Predictive Control Based on System Re-Identification (MPC-SRI) to Control Bio-H_2 Production from Biomass

A Wahid[1] **and H M H Taqwallah**[1]
[1]Chemical Engineering Department, Universitas Indonesia, Depok, Indonesia

E-mail: wahid@che.ui.ac.id

Abstract. Compressors and a steam reformer are the important units in biohydrogen from biomass plant. The compressors are useful for achieving high-pressure operating conditions while the steam reformer is the main process to produce H_2 gas. To control them, in this research used a model predictive control (MPC) expected to have better controller performance than conventional controllers. Because of the explicit model empowerment in MPC, obtaining a better model is the main objective before employing MPC. The common way to get the empirical model is through the identification system, so that obtained a first-order plus dead-time (FOPDT) model. This study has already improved that way since used the system re-identification (SRI) based on closed loop mode. Based on this method the results of the compressor pressure control and temperature control of steam reformer were that MPC based on system re-identification (MPC-SRI) has better performance than MPC without system re-identification (MPCWSRI) and the proportional-integral (PI) controller, by % improvement of 73% against MPCWSRI and 75% against the PI controller.

1. Introduction

A biohydrogen from biomass process consists of raw material processing unit, gasification unit, char decomposer unit, compression unit, steam reforming, char combustors units, unit coolers, H_2S removal units, and pressure swing absorber unit. Among these processes, the compression and the steam reformer units are the very important units in generating biohydrogen effectively since they influence the subsequent performance of the units and will determine the amount of the next product. Compression unit tasked to raise the pressure coming out of the gasification unit in order to reach high pressure to operate. The high pressure needed for H_2S removal operation conditions. This material has been eliminated, because it contains toxic substance that can be harmful for humankind.

In order to achieve an optimal condition for the operation of the plant, it is necessary to use a controller on the units. Therefore, if there is any kind of disturbance outside, it can be rejected to maintain the stability for the operation of the plant. Research on controling these units have been carried out [1] using the PI controllers. This research used a MPC which is one of the advanced control. The MPC is a model-based control and a multi variable control [2-3]. Therefore, the application of MPC controllers is expected to be better than the PI controller. To develop a plant model, this research employs a process simulator [4-5]. Since the factors which influenced the MPC is the used of the model, so the best models of the systems need to find out. In this case, we used a new approach called the system re-identifications [6]. System re-identification turns commonly used in image processing world [7-8].

2. Methodology

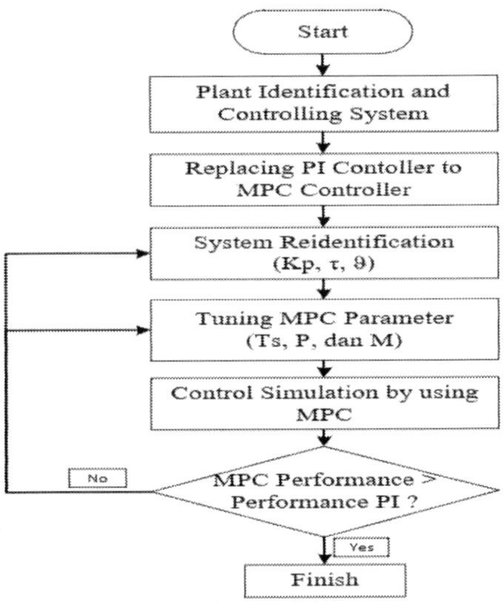

Figure 1 Flowchart of the research

2.1 Flowchart of the Research

The methodology of this research outline consists of several steps. These steps can be done through performing a simulation using the software for process simulation. Flowchart of research methods can be seen in Fig. 1. This research used a process simulator (UniSim$^{(R)}$) operated in dynamic mode [5]. While the plant used is the result of the pre-design [9].

2.2 Simulation Study

Figure 2 shows the control system of compressor and steam reformer at the bio-hydrogen from biomass plant, which consists of four pressure controls and two temperature controls. Because of the compressor used is a centifugal type, each compressor must be equipped with anti-surge control to avoid the surge (back flow) due to the inlet flow rate to compressors below the surge limit.

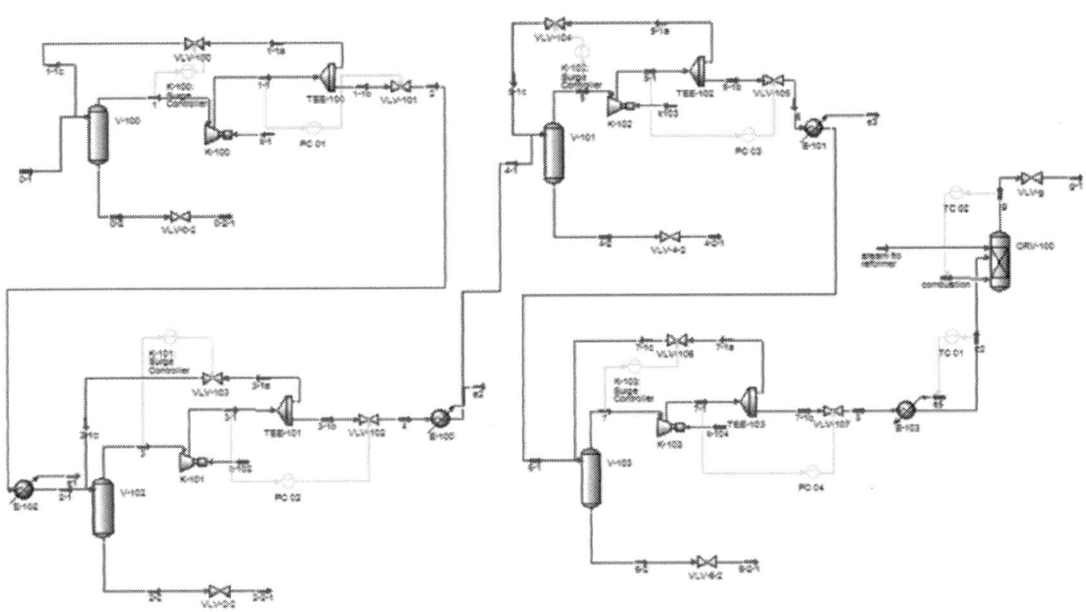

Figure 2 Compressor and steam reformer controls using UniSim$^{(R)}$

Table 1 shows the controlled variables (CV), process variable source, and output target or manipulated variables (MV). When using PI controller, the closed mode system identification carried out by model testing. The system will be given set point change so that it can obtain empirical models

IOP Conf. Series: Materials Science and Engineering **316** (2018) 012061 doi:10.1088/1757-899X/316/1/012061

of FOPDT using the method II [10-11]. MPC using the FOPDT parameters: K_p (process gain), θ (dead-time), and τ (time constant), directly in the control. The FOPDT model is expressed in transfer function equation as follows:

$$G(s) = \frac{K_p e^{-\theta s}}{\tau s + 1} \tag{1}$$

where $G(s)$ is transfer function, and s is Laplace Transform variable (complex number).

Table 1 Controlled Process Variable

Unit	Controlled Variable	Process Variable Source	Output Target
Compressor	Stream Pressure that will come to H2S Removal unit	Capacity Control : motor speed	Capacity Control : Control valve
Steam Reformer	Inlet Stream Temperature of Steam Reformer	Heat of Heat Exchanger after Compressor Unit	Control Valve
	Outlet Stream Temperature of Steam Reformer	Combustion of Steam Reformer	Control Valve

Empirical model parameters obtained from model testing at PI controller is the initial model of a system. This model will be identified back to get a new model. If the new model has a control performance better than the early models, the integral of absolute error (IAE) as a measure of performance, it will be re-identification to get IAE worse (larger) than before, which means that it is the limit of re-identification. The last model which has the smallest IAE is the best model to be used in the MPC with system re-identification (MPC-SRI), while the MPC without system re-identification (MPCWSRI) used as the comparison, beside the PI controller [12]. To tune the MPC parameters (sampling time, *T*, prediction horizon, *P*, and control horizon, *M*) used the Dogherty-Cooper method [13].

3. Results and Discussion

3.1 System Re-Identification using MPC

Modeling the MPC controller is done through the same way with the PI controllers with each controller setpoint change to see the change of reaction generated by the controller to achieve stability. In this research, there are 4 compressors and 1 steam reformer. In a steam reformer there are 2 pieces of control (heater temperature control and combustor temperature control). Thus, in this study there are 4 pressure controllers and 2 temperature controllers. To control the pressure in the compressor unit, the setpoint will be changed by +7 and -7 kPa, while the temperature control for steam reformer unit will be changing the setpoint of +10°C and-10°C. Each controller is done changing the setpoint until it reaches the lowest IAE. It is aimed to obtain the maximum and the precise models. The parameters model obtained will be continued to the modification of MPC (*T*, *P* and *M*).

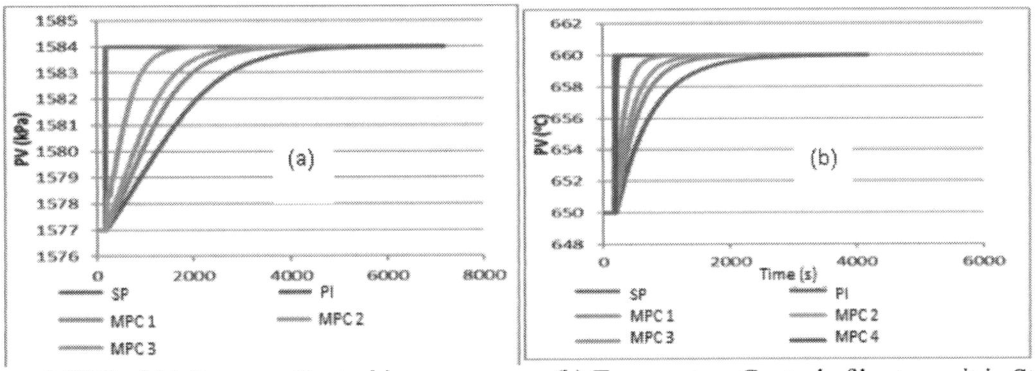

Figure 3 PRC of (a) Pressure Control in compressor, (b) Temperature Control of heater unit in Steam Reformer

In reidentifying the system we will calculate the empirical value of the system, the value of empirical can be seen from Figure 3. The results of this modeling will be used to see the response of the controller when disorders given. The result of modeling the latter on each unit is the best models, so that it will do the adjustment. Model Testing performed to obtain the minimum IAE values for each controller to get more accurate results. Results of tuning on each controller can be seen in Table 2, where PC1 – PC4 are the pressure controls of the compressor 1 – 4 and M1 – M4 are models result from the re-identifications. Table 3 shows the results of tuning in temperature control of heater and combustor.

From modeling conducted on each controller, the modeling results are used to do further adjustment MPC which is modeling the latter. The results of the empirical best setting for each controller are appreared in Table 4.

Table 2 Tuning of PC in compressors

FOPDT	PC1				PC2				PC3			PC4	
	M1	M2	M3	M4	M1	M2	M3	M4	M1	M2	M3	M1	M2
K_p	-0,65	-0,51	-0,23	-0,15	-0,71	-0,57	-0,30	-0,16	-0,42	-0,28	-0,70	-0,70	-0,70
τ	0,50	3,42	7,25	6,26	16,93	29,08	42,83	6,97	29,31	23,42	1,56	5,25	1,00
θ	1,30	0,30	1,52	1,09	2,31	8,04	9,40	1,27	5,07	7,09	0,82	0,09	1,17

Table 3 Tuning of TC in the heater and combustor

	TC1				TC2
	M1	M2	M3	M4	M1
K_p	0,22	0,22	0,60	3,40	0,31
τ	36,00	9,48	3,50	0,51	2,00
θ	0,33	0,52	0,67	0,33	0,17

Table 4 Result of FOPDT PRC of each Control

No.	Controller	Kp	τp	Θ
1	PC 1	-0,15	6,26	1,09
2	PC 2	-0,16	6,97	1,27
3	PC 3	-0,70	1,56	0,82
4	PC 4	-0,70	1,00	1,17
5	TC 1	3,40	0,51	0,33
6	TC 2	0,31	2,00	0,17

3.2 MPC Controller Performance

The results of the PI controller tuning compressor units - 3 were setup using Ziegler Nichols method, where the best results obtained with a value of Kc = 0.7187 and Ti = 0.3556. As for the MPC controller, empirical values used are Kp = -0.7006, τp = 1.5563, and θ = 0.8168, parameter value adjustment for fine tuning the MPC outcome of this unit is

Step Respon Length (SRL) = 50
Prediction Horizon (P) = 50
Control Horizon (M) = 3
Control Interval (Ts) = 2

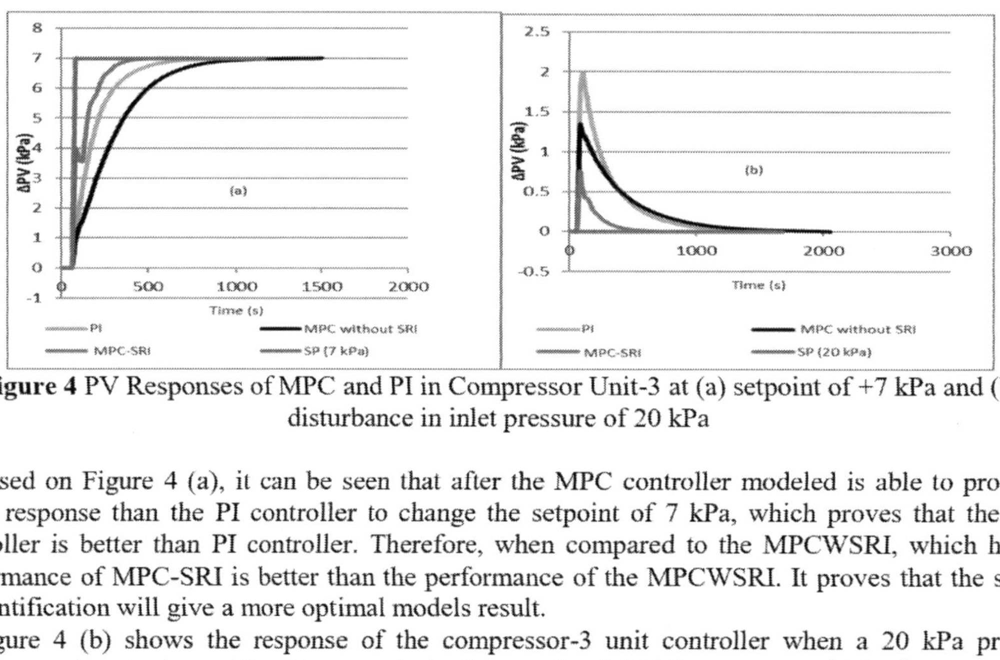

Figure 4 PV Responses of MPC and PI in Compressor Unit-3 at (a) setpoint of +7 kPa and (b) disturbance in inlet pressure of 20 kPa

Based on Figure 4 (a), it can be seen that after the MPC controller modeled is able to provide a faster response than the PI controller to change the setpoint of 7 kPa, which proves that the MPC controller is better than PI controller. Therefore, when compared to the MPCWSRI, which has the performance of MPC-SRI is better than the performance of the MPCWSRI. It proves that the system re-identification will give a more optimal models result.

Figure 4 (b) shows the response of the compressor-3 unit controller when a 20 kPa pressure disturbance input given. The blue graph is the result of MPC response which shows a better performance than the PI controller performance results. Both controllers managed to control the system, but MPC to control and restore the initial conditions more quickly than the PI controller. Therefore, when compared to the results of the MPC performance without Re-identification it comes to a proof that the MPC needs to be re-identified first.

Thus, based on both of these pictures it can be calculated from the value of the parameter controlling the success of each method to be compared quantitatively. The results of the calculations are presented in Table 5.

Table 5 The Result of Performance of Control in Compressor Unit-3

Parameter	PI	MPC	MPC without Reidentification
Offset	0	0	0
Rise Time (s)	1160	801	1500
Peak Time	0	0	0
Overshoot	0	0	0
Settling Time (s)	780	440	1090
Decay Ratio	0	0	0
Period	0	0	0
Maximum Deviation	0	0	0
IAE	42	20	76
ISE	129	47	264

Ten parameters are tested on Compressor Unit-3 shows that the performance of the MPC controller is much better than the PI controller. PI controller requires a longer time to reach a steady state. Similar results were also obtained from the MPC without Re-identification which requires a longer time to reach a steady state.

The same thing will be generated in Compressor Unit 1, Compressor Unit 2, Compressor Unit 4, Unit Heater and Combustor in the Steam Reformer. Each MPC controller produces better performance than the performance of PI controller. MPC has proved its ability in controlling the pressure and the temperature. MPC tuning parameters in this research performed using a non-adaptive adjustment based on calculations developed by Dougherty Cooper [13] and used the method of fine tuning also [14-16]. Tuning results using a non-adaptive method showed no response to setpoint changes given to the controller. This suggests that the non-adaptive MPC controller cannot function properly. As for the result of fine tuning the MPC controller shows a very maximum result. it is an evidence from the comparison between MPC controller and PI controller are very significant. Therefore, the fine tuning method used is very precise.

Non-adaptive MPC showed no changes can be caused by the range of prediction horizon (P) is too big, based on the calculations in which the process of this system does not require high P. In addition, the control interval (T) required by the results of the calculation are also quite large, whereas in doing the fine tuning control method which takes the value of T is small enough to obtain optimum results. Otherwise, if it is done through the sufficiently high control interval, then the process becomes quickly unstable or excessive response to the setpoint influence itself. Thus, the adjustment using a non-adaptive method is not recommended for use in MPC. MPC is a control that can optimize a control. The downside of this is that if the MPC is given a high interference, the response will be very bad then it tends to be unstable and always does the oscillations. Unlike the PI controller, it is capable of controlling a given system even though the disturbance is high. MPC will perform much better than the PI when a small disturbance given, but does not apply if the MPC is given a high interference, then the MPC will tend to be unstable. Hence, the model is referred to as the controller of MPC works for optimization.

Based on the above research, a measurable percentage increases in performance of MPC after the modeling and the tuning. Results improved performance of MPC can be seen in Table 6. Table 6 shows that the performance of MPC is much better than the performance of the PI controller by % improvement of 75%. This means that the modeling which done is very useful to increase the performance of a controller. This can be seen when the MPC controller does not do the re-identification in this research used the empirical model, the MPC controller performance will not be better and tend to be bad, that is why it is very necessary to do the re-identification.

Table 6 Improvement for each control performance

Controller	IAE PI	IAE MPC-SRI	IAE MPC Without Reidentification	% Improvement (To MPCWSRI)	% Improvement (To PI)
PC 1	84	11	58	81	87
PC 2	39	15	289	95	62
PC 3	42	20	76	74	52
PC 4	50	18	142	87	64
TC 1	11	0,4	3,9	90	96
TC 2	249	27	31	13	89

4. Conclussions

1. Controlling the pressure and the temperature steam reformer compressor using the MPC has better results than the PI controller. MPC shows that it can optimize the system, unlike the case with PI controller that takes a long time to optimize the system.
2. Empirical models used to perform tuning PI on previous studies have not produced an optimum performance when used in the MPC.
3. MPC parameter modification cannot be done with non-adaptive tuning strategy, so it must be done by fine tuning strategies.
4. System Re-identification and fine tuning of MPC is able to improve the performance of pressure control of compressor and temperature in steam reformer, resulting in a better performance by % improvement of 73% against MPCWSRI and 75% against the PI controller.
5. MPC controller successfully improves the performance of the pressure control to the compressor 1 as much as 7.4 times better, compressor 2 as much as 2.6 times better, compressor 3 2 times better, Compressor 4 as much as 2.7 times better, the heater temperature steam reformers as much as 27 times better, and the column temperature on the combustor steam reformer as much as 9 times better.

Acknowledgments

We express our gratitude to the Universitas Indonesia which has funded this research through the scheme of Hibah Publikasi Internasional Terindeks untuk Tugas Akhir Mahasiswa (PITTA) No.2068/UN2.R3.1/PPM.00/2017

References

[1] A. Wahid and M. Iqbal. 2016. Compressor and steam reformer control in the biohydrogen from biomass design plant using PI controller. *Sinergi* 20 (2): 81-90

[2] Paul Şerban Agachi, Zoltán K. Nagy, Mircea Vasile Cristea, and Árpád Imre-Lucaci. (2006). *Model Based Control: Case Studies in Process Engineering* (Weinheim: WILEY-VCH Verlag GmbH & Co. KgaA)

[3] E. F. Camacho and Carlos Bordons. (1998). *Model Predictive Control* (Great Britain: Springer-Verlag London Limited)
[4] Aspen. (2012). *Aspen Hysys Operation Guide : Reference Guide*
[5] Honeywell. (2009). *UniSim® Design Dynamic Modeling: Reference Guide*
[6] A. Wahid. and R. N. Nararya. (2015). Optimation of gassification and char combustor unit controls in the biohydrogen from biomass design plant using reidentification model predictive control. *Proceeding of Seminar Nasional Teknik Kimia Indonesia V.* Yogyakarta, 12 – 13 Oktober
[7] George A. Cushen. A person re-identification system for mobile devices. *Proceeding of 2015 11th International Conference on Signal-Image Technology & Internet-Based Systems (SITIS),* 23-27 Nov. 2015
[8] S. Gong., M. Cristani, S. Yan, C. C. Loy. (2013). *Person Re-identification* (London: Springer-Verlag).
[9] I. A. Budianta, F. Abqari. (2011). Plant design of biohydrogen from biomass. *Report of Plant Design.* Depok: Departemen Teknik Kimia Universitas Indonesia
[10] T. E. Marlin. (2000). *Process Control: Designing Processes and Control Systems for Dynamic Performance* (United States: McGraw-Hill Higher Education)
[11] C. A. Smith & A. B. Corripio (1985). *Principles and Practice of Automatic Process Control,* (United States: John Wiley & Sons Inc.)
[12] A. Wahid, A. Ahmad, and V. Cyntia. (2013). Distillation column control using multiple model predictive control based on representative model predictive control. *Proceeding's The 13th International Conference on QiR (Quality in Research).* Yogyakarta, Indonesia, 25-28 June 2013
[13] Dougherty, D., Cooper, D. (2003). A practical multiple model adaptive strategy for single loop MPC. *Control Engineering Practice*, 11, 141-159
[14] Ahmad, A. and Wahid, A. (2007). Application of Model Predictive Control (MPC) Tuning Strategy in Multivariable Control of Distillation Column. *Reaktor* 11(2): 66-70
[15] A. Wahid and A. Ahmad (2015). Min-max controller output configuration to improve multi-model predictive control when dealing with disturbance rejection. *International Journal of Technology.* 6 (3): 504-515
[16] A. Wahid and A. Ahmad (2016). Improved multi-model predictive control to reject very large disturbances on a distillation column. *International Journal of Technology.* 6: 962-971

IOP Conf. Series: Materials Science and Engineering **316** (2018) 012062 doi:10.1088/1757-899X/316/1/012062

Producing armyworm (*spodoptera sp.*) Bioinsecticide based on cysteine protease of red ginger (*zingiber officinale var. Rubrum*)

N T Afnan[1], D F Nur[1], T S Utami[1], M Sahlan[1], A Wijanarko[1], and H Hermansyah[1,2]
[1]Bioprocess Engineering, Engineering Faculty, University of Indonesia, Depok 16424, Indonesia
[2]DRPM Universitas Indonesia, University of Indonesia, Depok 16424, Indonesia

Abstract. Armyworm (*Spodoptera sp.*) is highly polyphagous defoliator on various horticulture and grain plants. Various chemical insecticides have been created to control it. There is a need to create an eco-friendly and specific insecticide which only affect armyworm's nervous system. This research investigates cysteine-protease's enzyme activity of red ginger (*Zingiber officinale var. Rubrum*) which is called zingibain. Its catalytic site matches with residue site in armyworm's body so it can be used as bioinsecticide raw material which meets the criterias above. Fresh red ginger rhizomes were washed and extracted. The juice was then deposited in low temperature and centrifuged to get rid of its starch content. It was filtrated to remove large contaminants and poured into Potassium Phospate buffer. The liquid was then centrifuged again for 30 minutes before collecting the supernatant. Fresh leaves were then dipped into crude ginger protease extract and fed to fourth instar-armyworms. Leaves dipped into non-diluted extract were barely eaten by armyworm while the 50% and 25% dilution was half eaten and most eaten. The crude red ginger extract was not strong enough to kill them although the research showed its enzymatic activity reaches up to 169 PU. It still needs improvement to be produced as commercial bioinsecticide.

Keywords: Acetylcholinesterase (AChE) Inhibitor; Armyworm (*Spodoptera sp.*); Bioinsecticide; Cysteine Protease; Red Ginger (*Zingiber officinale var. Rubrum*)

1. INTRODUCTION

Indonesia is an agricultural country that has 8 million hectares agricultural land and will be expanded 500 thousand hectares due to national food self-sufficiency plan [1]. Beside alocating some space for agricultural land, food self-sufficiency needs supporting factor such as insecticide. In the other hand, chemical insecticide usage can polute environment, mutates insects become resistant or tolerant [2], degrades the quality of surface water [3[, accumulated in mammals body, and exposes human fetuce through pregnancy period [4]. Those side effects can be avoided by replacing chemical insecticide with bioinsecticide which specifically affects certain insect. So that, it less harm to environment, under water-living organisms, mammals, and human.

Armyworms become the object of this research because its polyphagous characteristic [5]. It destroy plants by eating every part of leaves included its tissue, bone (midrib), and vein. Armyworm's life cycle starts from egg which laid down in groups which consist of up to 300 eggs each group and only needs at least 30 days to repeat its cycle. Armyworms active at night while hiding during the day. So that, its activity barely detected by farmer. A plant that attacked by armyworms can lose most of its leaves in a night and gradually wilt and die.

Red ginger (*Zingiber officinale var. Rubrum*) is known for its powerful proteolytic enzyme called zingibain. Its enzyme is included as cysteine proteases and match perfectly with armyworm's (*Spodoptera sp.*) catalytic site in nervous system [4]. In addition, red ginger can be found relatively easy in Indonesia. That makes red ginger's enzyme is potential to be used as the main material of bioinsecticide [6] to attack armyworm.

The test result shows that dipped-leaves were less favored than the negative control (leaves which wasn't dipped into red ginger extract). The leaves was dipped into red ginger extract with three various concentration such as 25%, 50%, and 100%. Leaves which was dipped into 25% concentration remained the least leaves, the 50% concentration remained half leaves, and the 100% concentration barely eaten by armyworms.

2. MATERIAL AND METHODS

This research was conducted at Bioprocess Laboratory, Chemical Engineering Department, Engineering Faculty, University of Indonesia, Depok. Fresh red ginger rhizomes were purchased from local market in Depok, armyworms were purchased from Bogor Agricultural University, tyrosin, caseine, and trichloroacetic acid (TCA) (technical grade) were purchased from PT. Pasifik Kimia Indonesia.

2.1. Extraction of crude red ginger cystein protease (zingibain)

800 g fresh red ginger rhizomes were cleaned and wind dried then be shredded and filtered to obtain its 100% concentration [6]. Then Potassium phospate buffer 0.1M, pH 7, was added to crude extract with 1:1 ratio to keep the pH neutral. Crude extract was centrifuged at 4600 rpm for 30 minutes to get rid of its starch [7]. The concentration was then diluted with demineral water to a concentration of 50% and 25%.

2.2. Ginger protease activity determination

The definition of protease activity is under a certain temperature the concentration of protease which is used to produce 1μg tyrosine by hydrolyze casein. The steps of determining the activity of ginger protease is as following:

Mix 1 ml of crude extract and 1 ml of 0.5% casein then the solution is incubated for 10 min at 37^{0}C. The Assay is stopped by adding 1 ml of 0.4 M TCA [7]. The color reaction was observed by pouring 2 ml of extract-casein-TCA solution with 6 ml of biuret and 0.3 ml Folin-Ciocaltaeu reagent. Then using the solution of blank group to adjust the extinction, and measured the absorbance at 750 nm. The absorbances were compared with free tyrosin standard curve which was made by checking the absorbance of free tyrosin in 20, 40, 60, 80, and 100 ppm. The formula of protease activity calculation is:

$$C_t = \frac{A_t}{A_{std}} \times C_{std} \tag{1}$$

Where C_t: Concentration of free tyrosin in crude extract (ppm)
C_{std}: Concentration of free tyrosin in standard curve (ppm)
A_t: Crude enzyme absorbance
A_{std}: Tyrosin standard absorbance

2.2.1. Incubation time determination
Incubation time was considered to obtain the highest enzyme activity of crude red ginger extract. The mixed solution of extract-casein was incubated in 37°C for 10, 20, and 30 minutes. These solutions were measured the absorbance with UV-Vis spectrofotometer.

2.3. Organoleptic and lifetime test
Organoleptic and lifetime test was conducted to observe if there's any physical changes during the storage time and condition. The organoleptic test was done everyday for two weeks while lifetime test was done everyday for 4 days.

Crude extracts without dilution was placed in different temperature storage (10°C, 27°C, and 37°C) and observed for its smell and colour changes. To make sure that the crude extracts still had protease activity, the solutions (25%, 50%, and 100% concentration) were tested by colour reaction and absorbance test which had been mentioned above.

2.4. Efficacy test
Efficacy test was the major test of this research which observed the correlation between dipped-leaves with remaining leaves uneaten. The test conducted by dipping 5 leaves into each concentration of crude extract (25%, 50%, and 100%) plus negative control. Each day for 3 days, the leaves were observed and replaced by the new one while armyworm's movement was observed.

3. RESULTS AND DISCUSSION
3.1. Protease activity assay
3.1.1. Incubation Time Determination
Incubation time determination was done before doing protease activity assay. The determination of optimum incubation time was the least time with the highest activity. The result shown on the Table 1 below:

Table 1. Incubation Time Determination

No.	Incubation Time (minutes)	Protease Activity (ppm)
1.	10	82.7159
2.	20	81.233
3.	30	81.5897

The result shows that there's only slight different of protease activity among the variables (10, 20, and 30 minutes). It seems that incubation time didn't really affect protease activity as long as it was incubated in enzyme's optimum temperature (37°C). Various amount of protease activity among the variables might be caused by the solutions (crude extract-casein) didn't homogenize the exact same way. But since 10 minutes incubation showed the highest protease activity with the least time, there's no need to incubate the extract for longer time.

3.1.2 Protease Activity
Protease activity for each concentration and each storage condition is provided in Table 2 below:

Table 2. Protease Activity

No.	Storage time	Protease Activity (ppm)								
		37°C			27°C			10°C		
		25%	50%	100%	25%	50%	100%	25%	50%	100%
1.	Day 1	32.19	57.88	168.4	12.82	58.36	151.9	21.61	33.47	67.35

IOP Conf. Series: Materials Science and Engineering **316** (2018) 012062 doi:10.1088/1757-899X/316/1/012062

2.	Day 2	28.35	41.27	101.7	28.84	54.29	84.83	20.39	31.94	54.87
3.	Day 3	25.63	43.36	109.8	29.68	58.40	142.9	24.33	32.01	78.95
4.	Day 4	27.99	42.37	109.7	31.27	53.28	169.1	24.12	39.51	81.53

Those data are visualized in graphics for better comparison. These are the graphics:

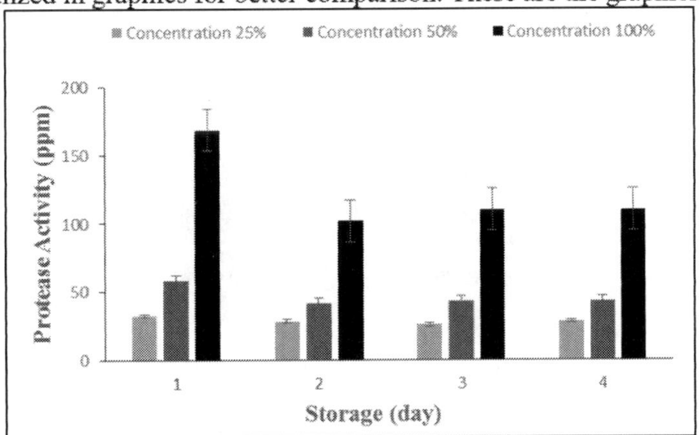

Figure 1. Protease Activity that Stored in 37^{0}C

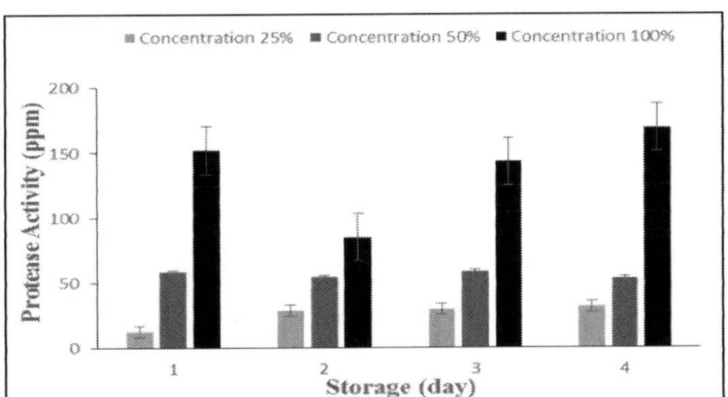

Figure 2. Protease Activity that Stored in 27^{0}C

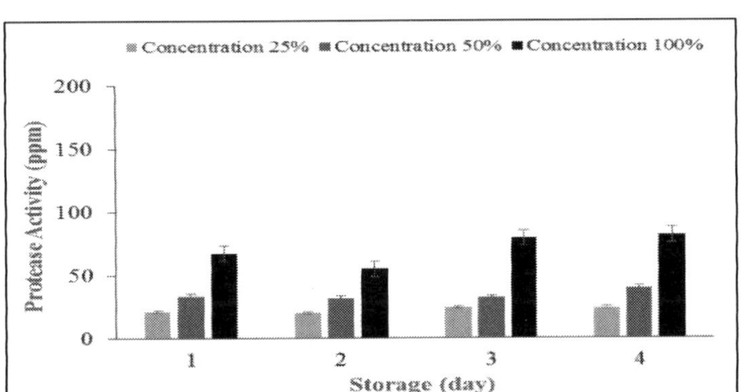

Figure 3. Protease Activity that Stored in 10^{0}C

From data above, there were inconsistency of protease activity measured. The data should be steady or decreasing since there must be denaturation within the crude extract. But since each sample was analyzed by its absorbance through UV-Vis spectrofotometer the denaturation was not detected.

IOP Conf. Series: Materials Science and Engineering **316** (2018) 012062 doi:10.1088/1757-899X/316/1/012062

Meanwhile, colour reaction test by Lowry method gave better understanding of what happened with those samples.

Theoretically, peptide bonds of protein react with copper under alkaline conditions to produce Cu^{+} which reacts with Folin-Ciocalteau reagent [8]. The reactions result in a blue colour. In first day of measurement, every sample turned into blue due to colour reaction test. But since the third day, samples which stored in 27^{0} and $37^{0}C$ turned into greenish. Proved that there's chemical changes of enzyme within the crude extract. The changes of colour reaction test in first day and third day is shown in Figure 4 below:

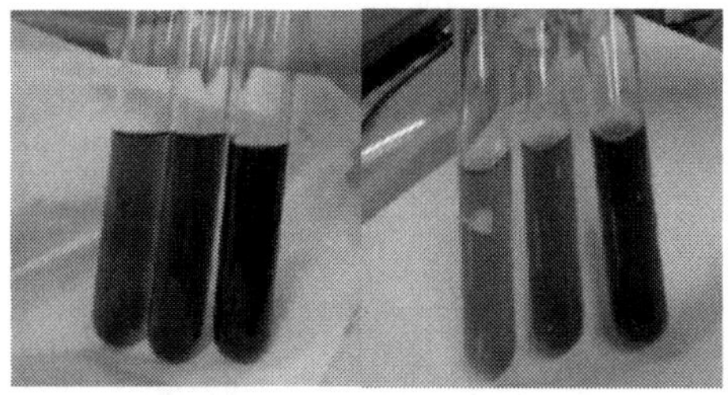

Figure 4 Colour reaction test, colour changes in: (a) first day (b) third day

The colour reaction test result doesn't neglect protease activity assay through UV-Vis spectrofotometer but completes the understanding that the samples still had protease activity. But the activity was not as good as it was first which strongly indicated denaturation process.

3.2 Organoleptic test

Organoleptic test was a qualitative test which measured subjectively by author's part of body such as nose and eyes. Parameters were made so that the qualitative data can be quantified. The parameters are shown in Table 3 below:

Table 3 Parameters for Organoleptic test

No.	Value	Smell Parameters	Colour Parameters
1.	5	Fresh and Strong	Brown and Homogen
2.	4	Fresh, less strong	Brown, little sediment
3.	3	Fresh	Brown, much sediment
4.	2	Acidic	Blackish brown
5.	1	Highly acidic	Blackish brown, much sediment

Result of organoleptic test is shown on the Figure 5 below:

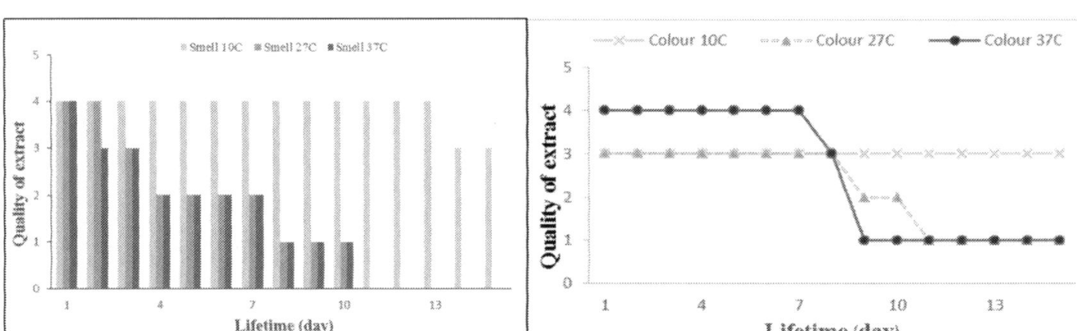

Figure 5 Organoleptic Test: Smell & Colour changes

Organoleptic test was conducted to know the lifetime of crude red ginger extract. The result shows that the extract which stored in 10^0C had the longest lifetime (longer than 15 days). Crude extract that stored in 27^0C and 37^0C only last for 3 days before its smell and colour got worse. Smell and colour changes in this test shows the declining quality of protein content in each extract. These data strengthen the understanding of denaturated enzyme which estimated in the discussion of protease activity assay.

Extract which stored in 10^0C had the longest lifetime because denaturation process was slowed down. In the other hand, denaturation process gets faster in room temperature (27^0C) and enzyme's optimum temperature (37^0C). From the organoleptic test, lifetime of crude red ginger extract in various temperature can be assumed and the storage condition can be determined.

3.3 Efficacy test
Result of efficacy test is shown on the Figure 6 below:

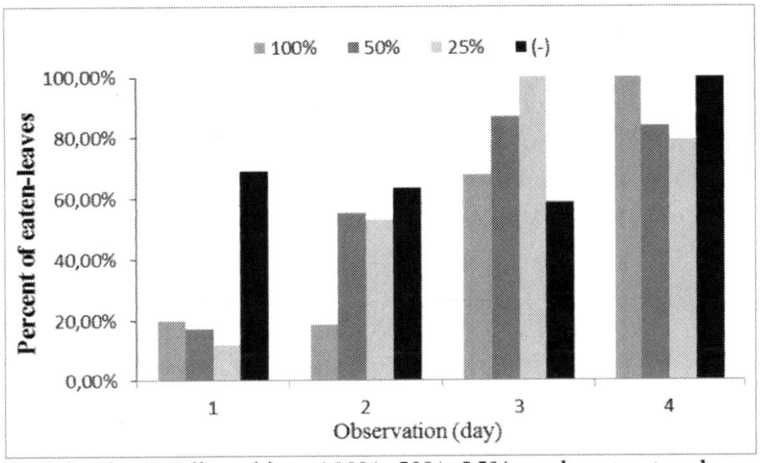

Figure 6 Remaining leaves dipped into 100%, 50%, 25% crude extract and negative control

Table 4 Amount of armyworms alive during the experiment

No.	*Crude extract concentration*	*Amount of died-larvae*	*Mortality*
1	Negative control (-)	0	0%
2	25%	1	20%
3	50%	1	20%
4	100%	0	0%

Based on the observation done, leaves that dipped into various concentration of crude red ginger extract are eaten and affect the armyworm. It shows that leaves which were dipped into 25%, 50%, and 100% concentration gradually eaten out. From the third day, most of armyworms entered the fourth instar so that there were significant area of eaten-leaves. Only 25% and 50% concentration can kill the worm with 20% mortality. The efficacy test needs to be done to third instar-armyworm or below which armyworm's immune system still develop and not complete yet.

4. Conclusion
Crude red ginger extract's protease activity assay was best conducted with 10 minutes incubation time. Spectrofotometer absorbance shows that extract stored in 27^0-37^0C had more protease activity than the

one stored in 10°C. Protease activity of crude red ginger extract reaches up to 169 ppm (in 27°C). In the other hand, organoleptic test shows that crude red ginger extract which stored in 10°C hadn't spoiled for 14 days or more. So, crude red ginger extract is best to be kept in 10°C and pretreated 10 minutes in 27° or 37°C before used. The efficacy test shows that the dipped-leaves affected the armyworms with 20% mortality. Crude red ginger's low purity level can be the cause of the low mortality.

5. Acknowledgement

Authors acknowledge the research grant supported by DRPM Universitas Indonesia that has resulted this article. This also funded partially by USAID, Indonesian Ministry Of Research, Technology, and Higher Education. Authors also wish to thank Ms. Rhennie in assisting with UV-Vis spectrofotometer analysis.

6. References

[1] Kementerian Pertanian 2015 *Rencana Kerja Kementerian Pertanian Tahun 2016,* Jakarta: s.n.

[2] Moekasan T 2000 *Penerapan PHT pada Sistem Tumpang Gilir Bawang Merah dan Cabai.* Jakarta: Badan Penelitian dan Pengembangan Pertanian.

[3] Gumilang A P 2010 *Menuju Pertanian Organik*, Last modified October 2010: news.detik.com/opini-anda/1464155/menuju-pertanian-organik

[4] Potera C 2007 *Environmental Health Perspective,* V.115, p. A 78, Cysteine Assistance

[5] Mursyahadah N, Hartani N, and Hendra M 2015 *Prosiding Seminar Sains dan Teknologi FMIPA Unmul* Uji Efektifitas Ekstrak Daun Tigaron (Crateva religiosa g. Forst.) terhadap Mortalitas Ulat Grayak (Spodoptera litura F.)(Lepidoptera: Noctuidae) di Laboratorium. Samarinda

[6] Asfi S H, Rahayu Y S, and Yuliani 2015 *LenteraBio,* 4(1) Uji Bioaktifitas filtrat Rimpang Jahe Merah (Zingiber officinale) Terhadap Tingkat Mortalitas dan Penghambatan Aktifitas Makan Larva Plutella xylostella secara In-vitro..

[7] Qiao Y y, et al. 2009 *Computer and Computing Technologies in Agriculture II,* V.3, pp. 1619-1628, Economic Methods of Ginger protease's Extraction and Purification..

[8] Waterborg J H 2002 *The Protein Protocols Handbook.* s.l.:Humana Press, pp. 7-9 The Lowry Method for Protein Quantitation

IOP Conf. Series: Materials Science and Engineering **316** (2018) 012063 doi:10.1088/1757-899X/316/1/012063

Anaerobic digestion of citrus waste using two-stage membrane bioreactor

Ria Millati[1]*, Lukitawesa[2], Ervina Dwi Permanasari[1], Kartika Wulan Sari[1], Muhammad Nur Cahyanto[1], Claes Niklasson[3], Mohammad J. Taherzadeh[2]

[1]Department of Food and Agricultural Product Technology, Faculty of Agricultural Technology, Universitas Gadjah Mada, Jl. Flora Bulaksumur, Yogyakarta 55281, Indonesia

[2]Swedish Centre for Resource Recovery, University of Borås, Allégatan 1, 50190 Borås, Sweden

[3]Department Chemical and Biological Engineering, Chalmers University of Technology, 41296 Gothenburg, Sweden

*Corresponding author. E-mail address: ria_millati@ugm.ac.id

Abstract. Anaerobic digestion is a promising method to treat citrus waste. However, the presence of limonene in citrus waste inhibits anaerobic digestion process. Limonene is an antimicrobial compound and could inhibit methane forming bacteria that takes a longer time to recover than the injured acid forming bacteria. Hence, volatile fatty acids will be accumulated and methane production will be decreased. One way to solve this problem is by conducting anaerobic digestion process into two stages. The first step is aimed for hydrolysis, acidogenesis, and acetogenesis reactions and the second stage is aimed for methanogenesis reaction. The separation of the system would further allow each stage in their optimum conditions making the process more stable. In this research, anaerobic digestion was carried out in batch operations using 120 ml-glass bottle bioreactors in 2 stages. The first stage was performed in free-cells bioreactor, whereas the second stage was performed in both bioreactor of free cells and membrane bioreactor. In the first stage, the reactor was set into 'anaerobic' and 'semi-aerobic' conditions to examine the effect of oxygen on facultative anaerobic bacteria in acid production. In the second stage, the protection of membrane towards the cells against limonene was tested. For the first stage, the basal medium was prepared with 1.5 g VS of inoculum and 4.5 g VS of citrus waste. The digestion process was carried out at 55°C for four days. For the second stage, the membrane bioreactor was prepared with 3 g of cells that were encased and sealed in a 3×6 cm^2 polyvinylidene fluoride membrane. The medium contained 40 ml basal medium and 10 ml liquid from the first stage. The bioreactors were incubated at 55°C for 2 days under anaerobic condition. The results from the first stage showed that the maximum total sugar under 'anaerobic' and 'semi-aerobic' conditions was 294.3 g/l and 244.7 g/l, respectively. The corresponding values for total volatile fatty acids were 3.8 g/l and 2.9 g/l, respectively. Methane production of citrus waste taken from the first stage under 'anaerobic' condition in membrane and free-cells bioreactors was 11.2 Nml and 7.2 Nml, respectively. Whereas, methane production of citrus waste taken from the first stage under 'semi- aerobic' condition in membrane and free-cells bioreactors was 8.8 Nml and 5.7 Nml, respectively. It can be seen from the results of the first stage that volatile fatty acids from 'anaerobic' condition was higher than that of 'semi-aerobic' condition. The absence of oxygen provides the optimal condition for growth and metabolism of facultative and obligatorily anaerobic bacteria in the first stage. Furthermore, polyvinylidene fluoride membrane was able to protect the cells from antimicrobial compounds.

1. Introduction

In 2015, juice and marmalade production created around 30-36 million tons of citrus [1][2]. Without proper treatment, this waste causes health and environmental problems. Fortunately, citrus waste consists mostly of organic matter. This makes it suitable for treatment by anaerobic digestion, which produces methane that can be used as an energy source.

Digestion is a complex process. When a feedstock is placed in an anaerobic digester, the facultative bacteria (hydrolytic bacteria) break down the complex molecules, using up the oxygen in the feed. They will also use the oxygen in the air trapped inside the digester. In the second and third steps, the acidogen and acetogen bacteria form low molecular weight volatile fatty acids (VFA), as well as carbon dioxide and hydrogen. In the last step, the methanogenic bacteria convert volatile fatty acids into water, carbon dioxide, and methane. However, the hydrolytic and acetogenic bacteria are generally not as sensitive as methanogens and grow much faster. Hence, there would be accumulation of VFA that results in low pH and could kill methanogens. To further complicate the problem, citrus contains a flavour compound, limonene, that has an antimicrobial effect. Limonene can cause inhibition of methane production because the injured methanogens take a longer time to recover than the injured acidogen and acetogen bacteria. As a result, methane production will be decreased and VFA will be accumulated.

One way of solving the methanogen acidification problem is by applying two-stage anaerobic digestion [3]. The purpose of two-stage digestion is to separate the conditions for each stage, *i.e.* acid formation (1st stage) and methane formation (2nd stage). Two-stage anaerobic digestion provides optimal conditions for each stage, such as pH, temperature, and alkalinity, to achieve optimal growth of digesting bacteria. It is also suggested [4] that digestion of citrus waste be conducted using a membrane bioreactor in which cells are encased in a hydrophilic membrane. Hence, limonene, being a hydrophobic compound, will be prevented from accessing the cells. In this work, a membrane bioreactor was applied in the 2nd stage of digestion to protect the cells from limonene in the citrus waste.

2. Methodology

2.1. Inoculum

The inoculum was obtained from a thermophilic biogas plant (55°C) (Borås Energy and Environment AB, Borås, Sweden). It was kept at 55°C for three days to acclimatize the cells and to settle the sludge. The acclimated sludge was homogenized, filtered through a sieve with a pore size of 1.0 mm in order to remove the remaining large particles. It was then centrifuged (Heraus Megafuge 8, Thermo Scientific, Germany) at 1649g for 20 minutes to separate the supernatant and the suspended sludge. The suspended sludge was later used as an inoculum either as free cells or cell containment in membrane sachets.

2.2. Preparation of membrane cell

Membrane cell was prepared by inserting 3 g of the suspended sludge of inoculum into a sachet made from a hydrophilic polyvinylidene fluoride (PVDF) membrane (Durapore®, Thermo Fisher Scientific Inc., Sweden) [4]. The membrane sachet was prepared by cutting the membrane into a size of 3 cm x 6 cm and sealed on three sides of the membrane with heating. After the insertion of the cells, the sachet was sealed to close. The membrane cells were immediately used for the experiment.

2.3. Citrus wastes

The citrus wastes were supplied by Brämhult juice AB (Borås, Sweden) and stored at - 20°C before use. The wastes were from orange juice process and contained in (%db): total solid (TS) (23.9), volatile solid (VS) (23.1), carbohydrate (18.5), protein (1.28), and fat (0.23). The citrus wastes were thawed and grounded to achieve a size of around 20 mesh before use.

2.4. Medium

The basal medium was added to supply necessary nutrients for the microorganism. The composition of the basal medium was as follows (in mg/l): NH_4Cl (280), $K_2HPO_4.3H_2O$ (330), $MgSO_4.7H_2O$ (100) and $CaCl_2.2H_2O$ (10). One millilitre of trace nutrient solution was added per litre of basal medium. The trace nutrient solution contained (in mg/l): $FeCl_2.4H_2O$ (2000), H_3BO_3 (50), $ZnCl_2$ (50), $MnCl_2.4H_2O$ (500), $CuCl_2.2H_2O$ (38), $(NH_4)_6$ $Mo_7O_{24}.4H_2O$ (50), $CoCl_2.6H_2O$ (2000), $NiCl_2.6H_2O$ (142), $Na_2SeO_3.5H_2O$ (164).

2.5. Anaerobic Digestion

2.5.1 First Stage. A 500-ml glass bottle bioreactor was used as a batch digester with a total working volume of 250 ml. Each bioreactor contained 4.5 g VS of ground citrus waste and 1.5 g VS of free cells of inoculum. The basal medium was added until the working volume was achieved. There was no pH adjustment for the basal medium in the 1st stage. The reactors were then closed and incubated under either 'anaerobic' or 'semi-aerobic' conditions at 55°C for 4 days. To obtain 'anaerobic condition', the bioreactor was flushed with a gas mixture of 80% nitrogen (N_2) and 20% carbon dioxide (CO_2). To obtain 'semi-aerobic condition', the bioreactor was injected with 100 ml air/day. The bioreactors were closed with rubber septum and green cap (GL 45). The digesters were shaken manually each day. Control sample containing only citrus waste and basal medium was prepared in order to investigate whether citrus waste produced VFA and methane without the inoculum. Moreover, inoculum and basal medium were applied as a blank in order to measure gas production from the inoculum alone.

2.5.2 Second Stage. A 120-ml glass bottle bioreactor was used as a batch digester with a total working volume of 50 ml. The glass digester was filled with 10 ml of liquid from the first stage of digestion, 40 ml of basal medium, and 3 g of free cells or encased cells. For the 2nd stage, the pH of the basal medium was adjusted to 7. A glass digester containing only 3 g of inoculum and 50 ml basal medium was prepared as a blank. The glass digester was tightly sealed with 20 mm aluminium crimp cap with inserted grey PTFE/butyl rubber septum. The glass digesters were incubated under 'anaerobic condition' at 55°C for 2 days in the incubator and shaken manually each a day. The experimental set up are shown in Figure 1.

All the experiments of the 1st stage and the 2nd stage were run in triplicate and the results were shown as an average.

IOP Conf. Series: Materials Science and Engineering **316** (2018) 012063 doi:10.1088/1757-899X/316/1/012063

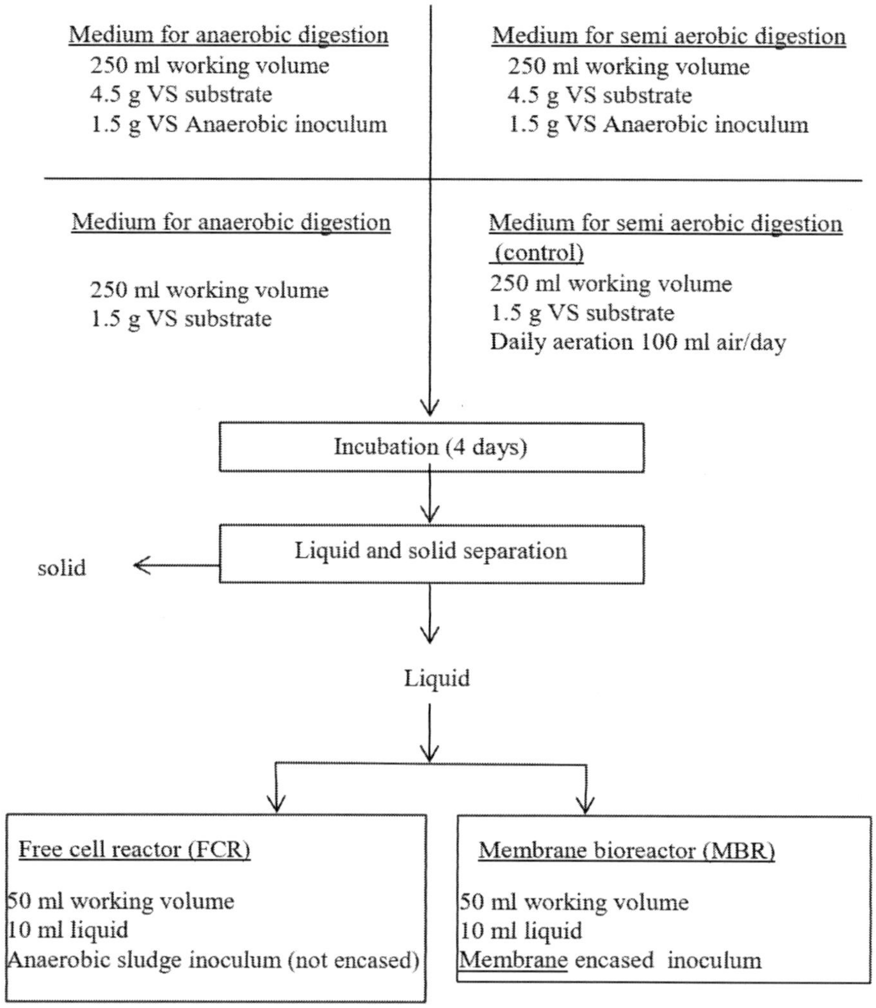

Figure 1. Experimental set up of first and second stage digestion

2.6. Analysis

Volatile fatty acids (VFA) including acetic acid, propionic acid, isobutyric acid, butyric acid, isovaleric acid and total sugar including glucose, sugar mix were analyzed using a High Performance Liquid Chromatography (HPLC) (Waters, USA) that was equipped with an ion-exchange column (Aminex HPX-87H, Biorad, Richmond, CA, USA). It was operated at 50°C, with a UV-VIS detector (Waters 486, Milipore, Milford, MA) at 210 nm with 5 mM sulfuric acid solution (Sigma Aldrich) as an eluent at a flow rate of 0.6 ml/min. Prior to analysis, samples were first filtered using syringe filter (Acrodisc 32 mm with 0.2 µm support membrane).

VS dissolution of the substrate was determined by subtracting VS of the substrate in the end experiment with VS of the substrate in the 1st-day experiment. Volatile solid content in the beginning and in the end experiment were determined [5].

Methane production was measured using a Gas Chromatograph (Varian 450 GC, U.S.A.) with a capillary column equipped with a thermal conductivity detector (TCD). N_2 was used as a carrier gas, and the instrument was operated with injector and oven temperature at 75°C and 50°C, nitrogen column flow 2.0 ml/min and detector temperature at 250°C. A 0.25 ml pressure lock syringe (VICI, U.S.A.) was used for the gas sampling.

3. Results and discussion

In this work, two-stage anaerobic digestion of citrus waste was applied in order to overcome acidification problem that is potential to inhibit methane production. Several parameters including sugar production, volatile fatty acid (VFA) production, methane production, pH value, and volatile solid (VS) dissolution were observed.

3.1. Production of total sugar and volatile fatty acids in the 1st stage of anaerobic digestion of citrus waste

Sugar is one of the hydrolysis step products. In the hydrolysis step, the complex organic matters, such as carbohydrate, protein, and fats are hydrolyzed into their monomers such as sugars, amino acids, and fatty acids. In this study, only total sugar production that was observed because the main chemical composition of citrus waste was carbohydrate. The total sugar production of anaerobic digestion of citrus waste is presented in Figure 2. Total sugar production from the digestion of citrus waste in all conditions continued to increase but started to decrease after the 1st day and reached a plateau since the 2nd day of digestion. The reason for this could be due to the facts that all the substrate was already converted to sugars and the sugars were converted to volatile fatty acids. It can be observed that the highest concentration of total sugar from the digestion under 'anaerobic condition' was bigger than that of under 'semi-aerobic condition'. The concentration of total sugar under 'anaerobic' and 'semi-aerobic' conditions was 294.3 g/l and 244.7 g/l, respectively. Microorganisms of many different genera are responsible for hydrolysis, most of them are facultative anaerobic bacteria and obligatory anaerobic bacteria [6]. The results of this work showed the presence of oxygen in 'semi- aerobic' condition decreased production of sugar.

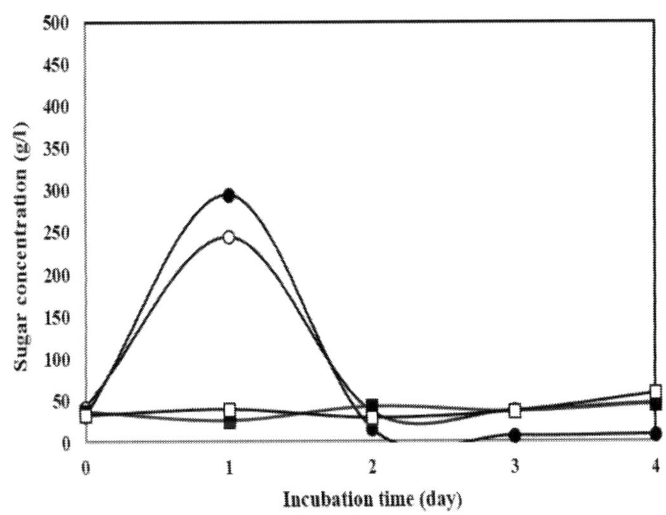

Figure 2. Sugar production of citrus waste anaerobic condition (●), semi-aerobic condition (○), control anaerobic condition (■) and control semi-aerobic condition (□) for 4 days.

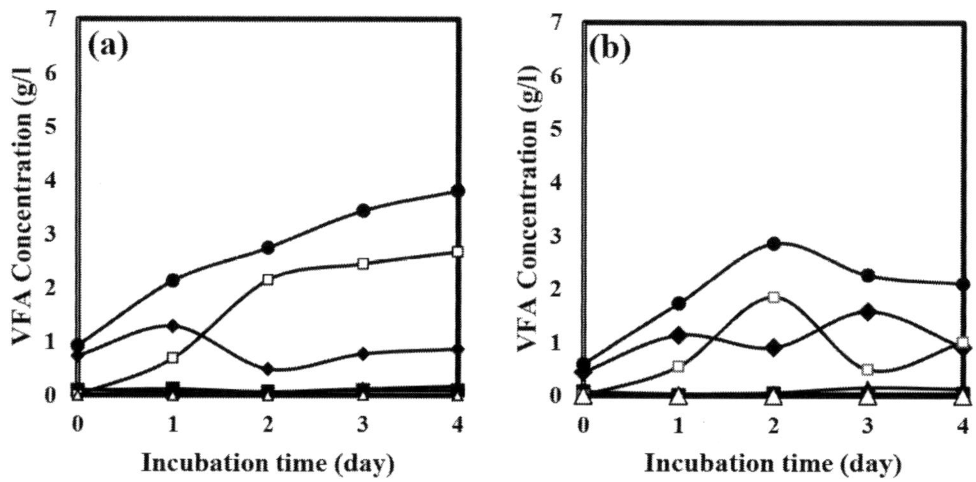

Figure 3. Volatile fatty acids (VFA) profile of citrus waste added with inoculum in anaerobic condition (a) and semi-aerobic condition (b). Concentration of total VFA (●), acetic acid (♦), propionic acid (■), isobutyric acid (▲), butyric acid (□) and isovaleric acid (△) for 4 days of incubation.

Figures 3(a) and 3(b) show the profile of concentration of total VFA as well as the individual VFA including acetic acid, propionic acid, isobutyric acid, butyric acid, and isovaleric acid. Volatile fatty acids are precursors of methane, therefore the production of VFA is necessary to be observed. Under 'anaerobic' condition, the production of VFA continued to increase until the last day of digestion. Whereas under 'semi-aerobic condition', VFA production already reached nearly a constant level after day-2. Concentration of VFA that has a longer chain than butyric acid was constantly low. In the 'semi-aerobic' condition, some of the butyric acid was converted to acetic acid at early stage. This is probably due to some of the acetogenic bacteria that are facultatively anaerobic. The concentration of butyric acid decreased and the concentration of acetic acid increased after the 1st day. As a result, the concentration of total VFA was almost constant. If Figure 3(a) and Figure 3(b) were compared, production of VFA from the digestion of citrus waste under 'anaerobic condition' was higher than that of 'semi-aerobic condition'. The production of VFA results in the decrease of pH of the medium. Initially, the pH of the medium was 6.6-7.3. After digestion for 4 days, the pH dropped to 4.7-4.9.

The volatile solid (VS) dissolution by microorganisms was also measured to investigate the amount of organic materials consumed to produce sugar and VFA by the digesting bacteria. The VS dissolution of citrus waste in 'anaerobic' condition and 'semi-aerobic' condition was 63.46%, 59.29%, respectively. This result is relevant to the VFA production. The highest VFA production during four days of digestion of citrus waste in 'anaerobic' and 'semi-aerobic' was 3.8 g/l, 2.9 g/l, respectively. The greater the dissolution of volatile solid, the higher the production of VFA. After anaerobic digestion was run for 4 days, the process was stopped. The liquid and the solid part was separated and the liquid part was used as substrate for the 2nd stage to produce methane.

3.2. Production of methane in the 2nd stage of anaerobic digestion of citrus waste

The 2nd phase of anaerobic decomposition is dominated by a group of methanogenic bacteria, which grow in neutral condition (pH 6.7-7.5) [7]. Thus, the pH of the 2nd stage medium was adjusted to 7 as the substrate (the liquid part) from the 1st stage had a low pH. One of the indicators of digestion process is pH value. A drop in pH value is an indication of a disturbance of the fermentation process [6]. The measurement of the pH value in the medium in all experiments on the last day of digestion showed the value of around 7. This means that methanogens were growing in their appropriate condition.

Figure 3(a) and Figure 3(b) show that methane was produced from the liquid taken from the 1st stage both under 'anaerobic' and 'semi-aerobic' conditions, respectively. To evaluate the performance of MBR, the methane production was compared with bioreactor using only free cells as inoculum (FCR). Anaerobic digestion of liquid from the 1st stage under 'anaerobic' condition produced higher methane compared to that of 'semi-aerobic' condition. As the VFA in the liquid from semi-aerobic condition was lower than anaerobic condition, accordingly, it can be understood that the methane production of the liquid from semi-aerobic condition was also lower.

Furthermore, the methane was produced higher by using MBR system than FCR system. The accumulated methane production in MBR and FCR systems under 'anaerobic' condition were 11.2 Nml and 7.2 Nml, respectively. The corresponding values under 'semi-aerobic' were 8.8 Nml and 5.7 Nml, respectively. The result indicates that the methane-producing bacteria encased in PVDF membrane were better protected against any harmful compound present in the medium. It is well known that citrus waste contains 0.5% peel oil and more than 90% of it is an antibacterial component, named-limonene. Limonene at concentration of 400 μL/L was reported to inhibit the anaerobic digestion process and caused ultimate failure during continuous operation at mesophilic conditions [8]. According to the biochemical reaction series of anaerobic digestion, methane is formed from acetate (by acetogenic methanogens) and H2/CO2 (by hydrogenotrophic methanogens) [6]. Limonene was found to be inhibitory more for acetogenic methanogens rather than hydrogenotrophic methanogens [9]. Acetic acid consumption was inhibited in the presence of a limonene concentration of 9.10-4 M. Limonene being hydrophobic was prevented diffusing into the hydrophilic membrane. As a result, the cells capacity against inhibition effect is enhanced. Previous study reported that no methane was produced on the first 10 days of anaerobic digestion of orange peel using single stage digestion with FCR system [10]. The methane started to be produced on day-20. The result indicated that digestion of citrus waste on single stage FCR system faced a long lag phase. On the contrary, application of two-stage membrane bioreactor in this work resulted in 8.8-11.2 mL of methane on day-2 (in the 2nd stage) and no lag phase was observed.

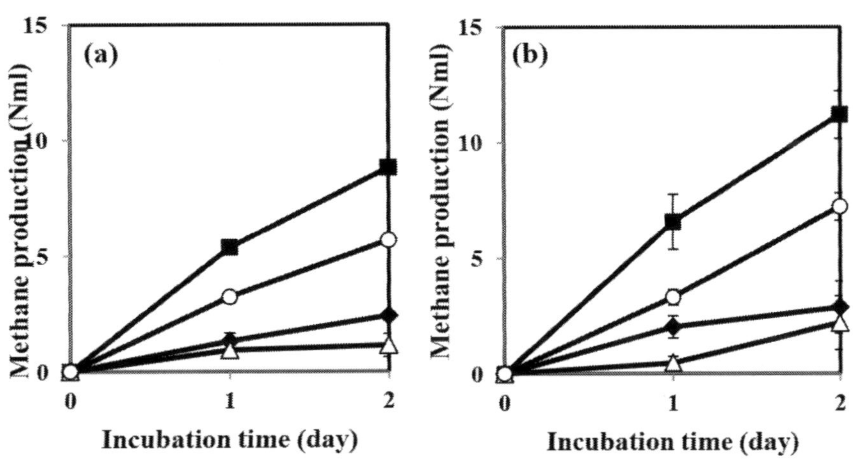

Figure 4. Methane production in citrus waste 'semi-aerobic' condition (a), 'anaerobic' condition (b) for blank MBR (♦), control FCR (△), citrus waste FCR (○) and citrus waste MBR (■) for two days.

4. Conclusion

- Volatile fatty acids (VFA) produced from the digestion of citrus waste under 'anaerobic condition' was higher than that of under 'semi-aerobic', The higher metabolite products production from the digestion under anaerobic condition indicates that the presence of oxygen in 'semi-aerobic' inhibited metabolism of the microorganisms in the 1st stage.
- Two-stage anaerobic digestion could minimize lag phase of methane production from the digestion of citrus waste

References

[1] FAO., Food and Agriculture Organization., 2014, http://www.faostat.org/.
[2] Wilkins, M.R., Widmer, W.W., Grohmann, K. 2007 Simultaneous saccharification and fermentation of citrus peel waste by Saccharomyces cerevisiaeto produce ethanol *Process Biochem*. **42** 1614-19.
[3] Bouallagui, H., Ben Cheikh, R., Marouani, L., Hamdi, M. 2005 Mesophilic biogas production from fruit and vegetable waste in a tubular digester *Bioresour. Technol.* **86** 85–9.
[4] Youngsukkasem, S., Barghi, H., Rakshit, S. K., &Taherzadeh, M. J. 2013 Rapid Biogas Production by Compact Multi-Layer Membrane Bioreactor: Efficiency of Synthetic Polymeric Membranes, *Energies* 6211-24.
[5] Sluiter, A., Hames, B., Ruiz, R., Scarlata, C., Sluiter, J., Templeton, D. 2008 *Determination of ash in biomass*, National Renewable Energy Laboratory, Colorado.
[6] Deublein, D., and A, Steinhauser 2008 *Biogas from waste and renewable resources* Wiley-VCH, Weinheim.
[7] Micolucci, F., Gottardo, M., Bolzonella, D., Pavan, P. 2014 Automatic process control for stable bio-hythane production in two-phase thermophilic anaerobic digestion of food waste *International Journal of Hydrogen Energy* **39**(31) 17563-72.
[8] Mizuki, E., Akao, T., & Saruwatari, T. 1990 Inhibitory effect of Citrus *unshu* peel on anaerobic digestion *Biol, Wastes* **33**(3) pp. 161–8.
[9] Crane, A., Nelson, W.O., Brown, R.E. 1957 Effects of D-limonene and a-D-pinene on in vitro carbohydrate dissimilation and methane formation by rumen bacteria *J. Dairy Sci.* **40** 1317–23.
[10] Sanjaya, Adhitya P., Cahyanto, Muhammad N., Millati, R. 2016 Mesophilic batch anaerobic digestion from fruit fragments *Renewable Energy* **98** 135– 41.

IOP Conf. Series: Materials Science and Engineering **316** (2018) 012064 doi:10.1088/1757-899X/316/1/012064

Techno-economic analysis of lipase enzyme production from agro-industry waste with solid state fermentation method

I M Hidayatullah[1], R Arbianti[1], T S Utami[1], M Suci[1], M Sahlan[1], A Wijanarko[1], M Gozan[1], H Hermansyah[1]

[1]Department of Chemical Engineering, Faculty of Engineering, Universitas Indonesia, Depok, West Java 16424, Indonesia

E-mail: heri.hermansyah@ui.ac.id

Abstract. Needs for this kind of catalyst derived from biological raw materials (biocatalysts) has increased along with development of products based on eco-friendly. To achieve the needs of biocatalyst (enzyme), large production is necessary. This study aimed to get the best conditions and design equipment to produce lipase enzyme based on solid state fermentation using SuperPro Designer v9.0. Several equipment such as *Tray Bioreactor*, *Mixing Tank* 1, *Filter Press*, centrifuge, *Mixing Tank* 2, and a dryer have been improved during the simulation. Economic analysis in the form of NPV, IRR, Payback Period, and the Benefit Cost Ratio was evaluated respectively. The result showed that production of 10 kg enzyme with NPV Rp112.796.147.423,00; IRR 54.20%; Payback Period 1.95 years; and Benefit Cost Ratio of 3.36 was more advantageous.

1. Introduction

The usage of enzymes for various industrial needs has increasing every year. In 2015, national needs of enzyme reached 2,500 tons while enzyme consumption growing up to 4-6% each year which is still import up to 99% usage [1]. Data showed that the enzyme market value is quite promising. However, it is necessary to build a system of enzyme production that are capable to produce independently because its market share is high (minimum competitor) and its development still limited only for laboratory scale. The process of making the enzyme, in terms of biomass cultivation medium used, was divided into 2 types fermentation: fermentation with solid medium (solid state fermentation) and fermentation with liquid medium (submerged fermentation). The advantages of solid state fermentation compare to the submerged fermentation include the minimization of contamination due to the lack of the water, the needs of a few fermentation media, the high yield of products, and the requirement of simple equipment [2].

2. Methods

2.1. Determination of mass balance, simulation, and equipment sizing

Determination of mass balance conducted by using secondary data from related researches about production of enzyme that has been done before [3,4,5]. The material needed including agro-industry waste; *Aspergillus sp*.; nutrients; olive oil as inducer; and skim milk powder as additives that needs to be added so that the enzyme would not denatured by heat drying. After calculating the mass balance of each processes, then sizing the main equipment [6,7] such as Tray Bioreactor, Mixing Tank 1, Filter Press, Mixing Tank 2, and Dryer. Simulation of lipase enzyme production then will be performed using SuperPro Designer v9.0. The results will be used as reference to analyse its economical related to the profitability of enzyme production.

2.2. Determination of profitability analysis (NPV, IRR, PBP, B/C)

To determine the profitability analysis of lipase production, studies carried out related to the the parameters that will affect the profitability of production [8,9,10]. The Total Purchase Equipment (TPE) is calculated based on the main equipment. The calculation is based on literature and vendor equipment purchased estimation with scale (six-tenths-factor-rules). Then the Total Capital Investment (TCI) will be calculated using delivered equipment cost percentage method [11]. After the calculation of TCI completed, the calculation of operating costs is followed by the assumption that the plant life is around 20 years. The operating costs including raw material costs, labor costs, maintenance costs, insurance costs, electricity and water utilities costs, as well as the costs of distribution for raw material and finish product. After determining the operating costs, the calculation of capital investment was done using the loan assumption from the bank as much as 40% with BI (Bank Indonesia) rate is about 7.5% and the remaining amount is 60% from capital investment. MARR value was calculated to determine the annual capital repayment that has to be paid. Afterall, the determination of product's prices was carried out based on production batch number and annual production capacity. In the end, the prices will be used to determine the value of NPV, IRR, payback period, and Benefit Cost Ratio.

2.3. Variation of substrate price, product price, exchange rate of rupiah, and production capacity toward profitability analysis

The variation of sensitivity level is done by using independent variables such as the price of agro-industry waste, in this study is rice bran. The rice bran deviation of variation started from -30 until +30 with the range of variation is about ±10. The price of lipase enzyme deviation of variation started from -10 to +10 with range of variation is about ±2. The exchange rate of rupiah against dollar with the deviation of variation started from -20 to +20 with range of variation is about ±5.

3. Results and discussion

3.1. Mass balance 1 kg and 10 kg basis

The calculation of mass balance 1 kg and 10 kg basis is used to determine the amount of raw material that needed to produce lipase enzyme with 1 kg and 10 kg production capacity. The following table below are the detail of raw material composition.

Table 1. Raw material composition in enzyme lipase production 1 kg and 10 kg basis

	Usage Amount			
Raw Material	1 Kg Basis [Kg/h]	10 Kg Basis [Kg/h]	1 Kg Basis [Kg/Batch/day]	10 Kg Basis [Kg/Batch/day]
Agro-industry waste	1.70	16.96	8.48	84.80
Olive oil	0.03	0.34	0.17	1.70
Aspergillus sp.	0.17	1.70	0.85	8.50
$MgSO_4.7H_2O$	0.00076	0.00763	0.0038	0.0382
Sodium phosphate	0.031	0.305	0.15	1.53
Ammonium sulphate	0.013	0.127	0.064	0.636
Calcium chloride	0.00064	0.00636	0.003	0.032
Distilled water	18.06	180.62	90.31	903.12
Glucose	0.025	0.254	0.13	1.27

	Usage Amount			
Raw Material	1 Kg Basis [Kg/h]	10 Kg Basis [Kg/h]	1 Kg Basis [Kg/Batch/day]	10 Kg Basis [Kg/Batch/day]
Potassium phosphate	0.005	0.051	0.03	0.25
Skim milk powder	2.143	21.430	10.71	107.15
Urea	0.0058	0.0577	0.03	0.29
Total	22.19	221.86	110.93	1109.31

3.2. Process simulation of enzyme lipase production 1 kg and 10 kg basis

The process simulation of lipase enzyme production is using SuperPro Designer v9.0 software. From the simulation, it was obtained that the production is feasible for lipase enzyme product 1 kg/batch (with 1 kg basis) and 9.92 kg/batch (with 10 kg basis).

3.3. Profitability analysis of enzyme production

Profitability analysis is determined by following a few methods that are accepted as parameters of the economic feasibility of production. The methods including Net Present Value, Internal Rate of Return, Payback Period, and Benefit Cost Ratio. Profitability analysis is related to the level of sensitivity for some items which is define as the variation of research.

Table 2. Economical parameters of lipase enzyme production 1 kg and 10 kg basis

		Amount	
Parameters	Unit	1 kg basis	10 kg basis
Capital expenditure (CAPEX)	Rp	2,974,435,697	25,219,549,876
Operational expenditure (OPEX)	Rp/years	1,668,331,785	8,594,329,744
Maintenance	Rp/years	14,880,000	126,100,000
MARR/WACC	%	10	10
Enzyme production capacity	kg/years	288	2880
Enzyme's price	Rp/kg	9,947,300	9,947,300
Tax rate	%	25	25
Loan period	years	6	6
Internal rate of return (IRR)	%	22.32	54.20
Net present value (NPV)	Rp	5,264,974,339	112,796,132,099
Payback period	years	7.50	1.95
Benefit cost ratio (B/C)	-	1.16	3.34

After specify the economical parameters that needed to measure the sensitivity, the sensitivity graph will be obtained as shown below.

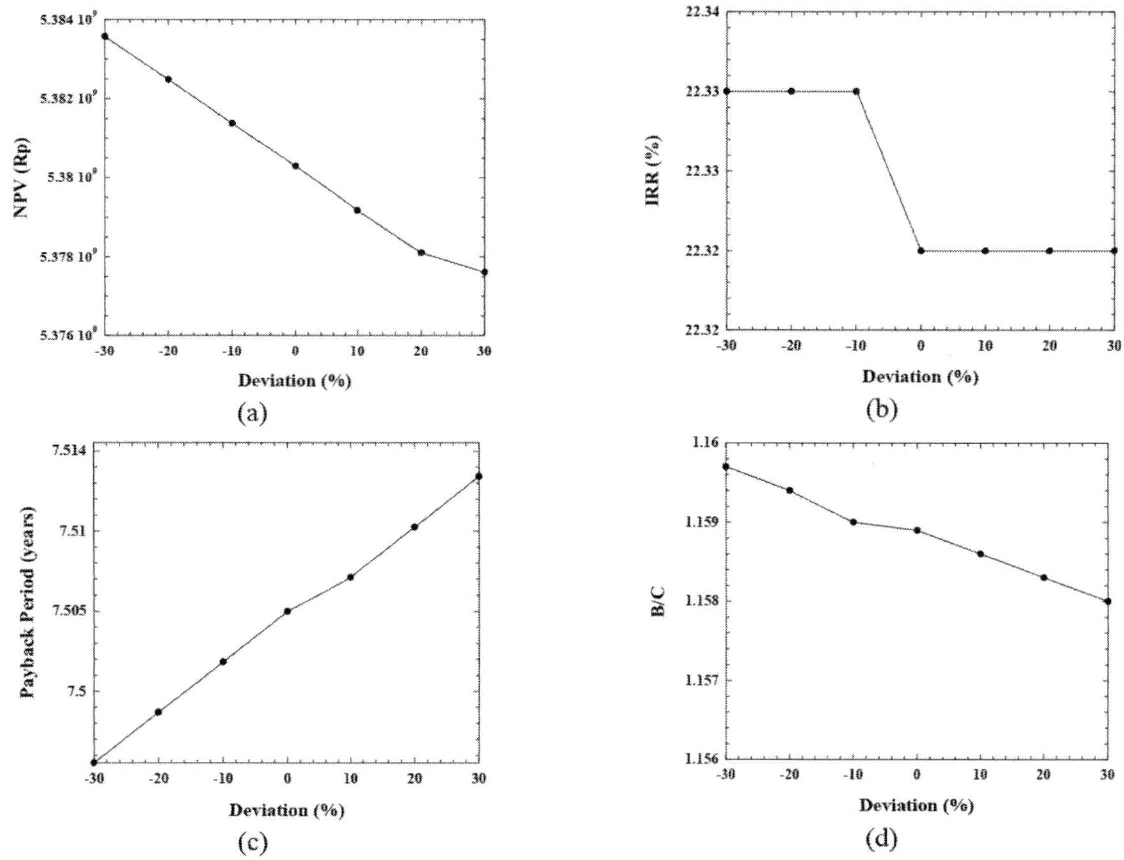

Figure 1. Sensitivity analysis of 1 kg basis substrate prices deviation (a) toward NPV (b) toward IRR (c) toward Payback Period (d) toward B/C

The results showed that the price of substrate which has been varied by the deviation, rises against payback period and declines toward NPV, IRR, and also B/C. On the other hand, IRR and B/C graphs are less sensitive than the NPV and Payback Period graphs. The highest result of NPV, IRR, and B/C with lowest Payback Period in the amount of substrate price is Rp 139.00/kg, and Rp 5,383,574,903 of NPV; 22.33% of IRR; 7.5 years of Payback Period; and 1.1597 of B/C.

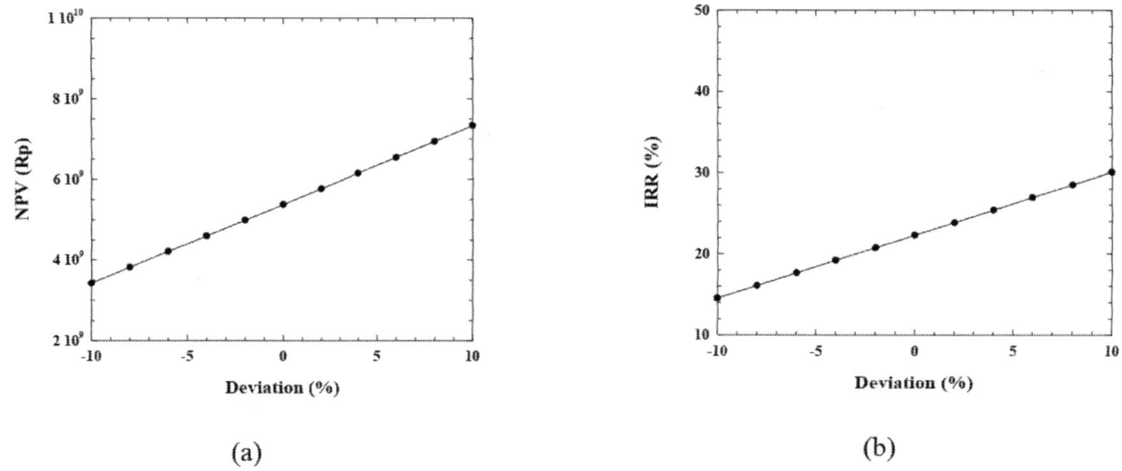

IOP Conf. Series: Materials Science and Engineering **316** (2018) 012064 doi:10.1088/1757-899X/316/1/012064

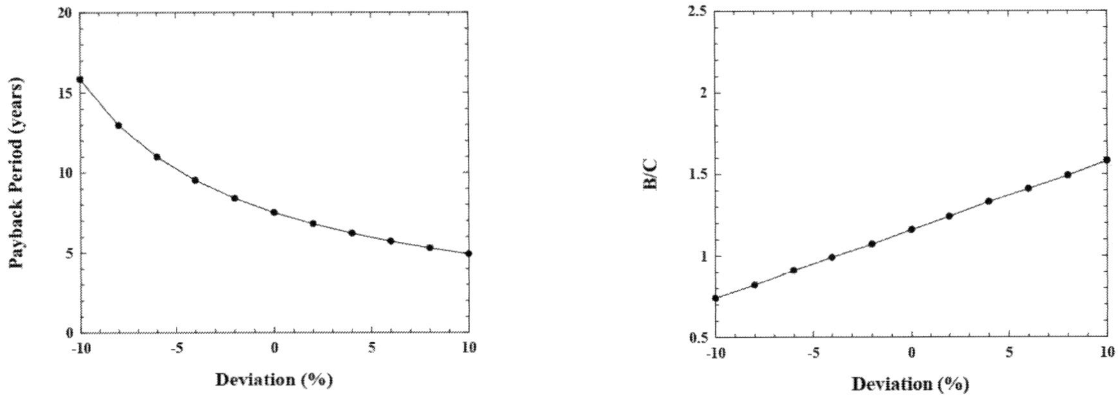

(c)

(d)

Figure 2. Sensitivity analysis of 1 kg basis enzyme prices deviation (a) toward NPV (b) toward IRR (c) toward Payback Period (d) toward B/C

Based on the graphic above, the best variation was obtained at 10th deviation with enzyme prices is about Rp 10,924,035.00 and Rp 7,327,689,990.00 of NPV; 30.07% of IRR; 4.92 years of Payback Period; and 1.58 of B/C.

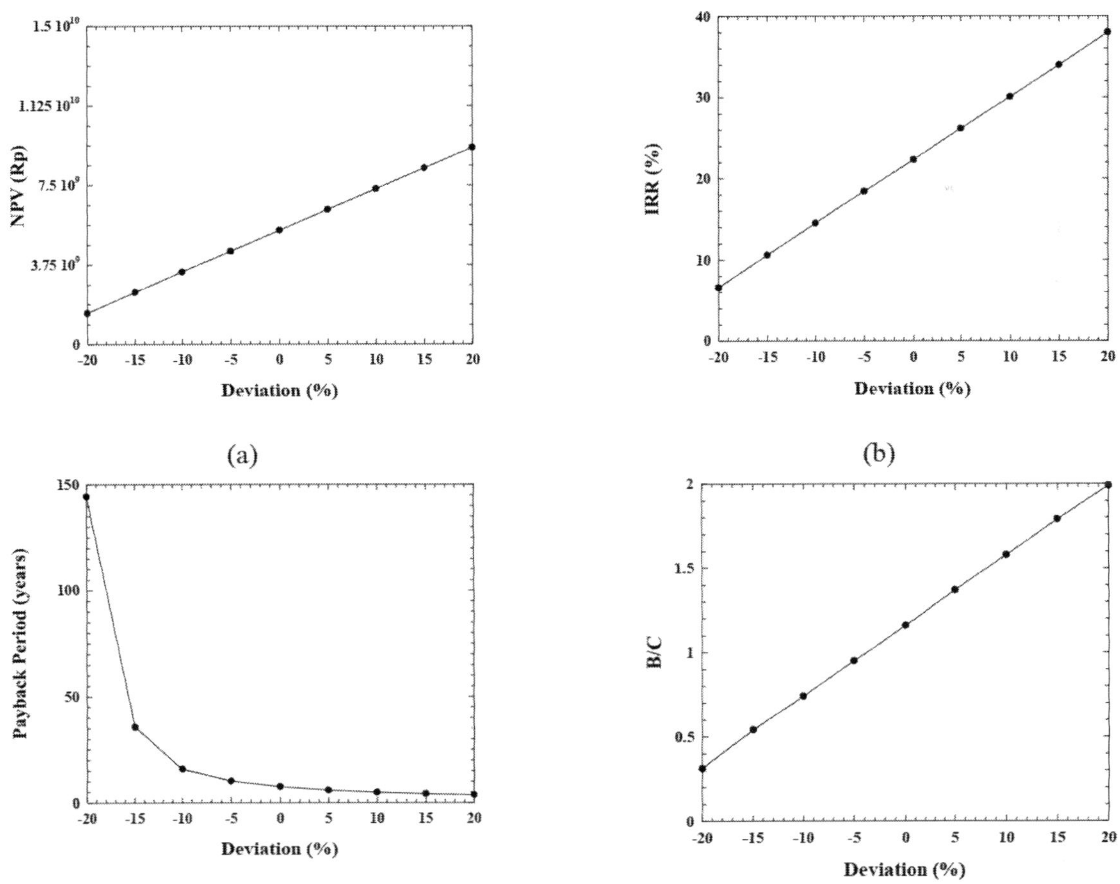

(a)

(b)

(c)

IOP Conf. Series: Materials Science and Engineering **316** (2018) 012064 doi:10.1088/1757-899X/316/1/012064

(d)

Figure 3. Figure 3 Sensitivity analysis of 1 kg basis rupiah's value deviation (a) toward NPV (b) toward IRR (c) toward Payback Period (d) toward B/C

Based on the graphic above, the best variation was obtained at 20th deviation with the exchange rate of rupiah upon dollar is Rp 15,923.00 and Rp 9,278,365,765.00 of NPV; 37.98% of IRR; 3.66 years of Payback Period; and 1.99 of B/C.

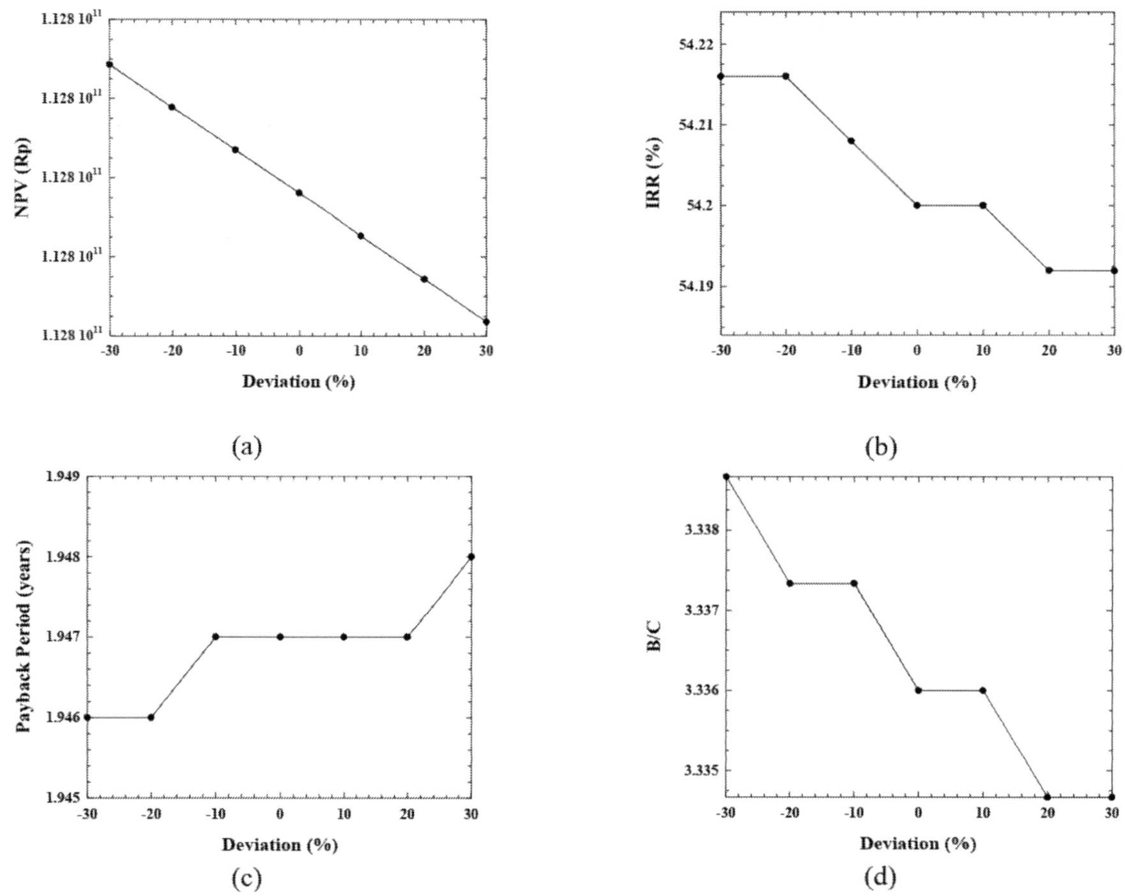

Figure 4. Sensitivity analysis of 10 kg basis substrate prices deviation (a) toward NPV (b) toward IRR (c) toward Payback Period (d) toward B/C

Based on the graphic above, the best variation was obtained at 30th deviation with the price of substrate is about Rp 139.00 and Rp 9,278,365,765.00 of NPV; 37.98% of IRR; 3.66 years of Payback Period; and 1.99 of B/C.

IOP Conf. Series: Materials Science and Engineering **316** (2018) 012064 doi:10.1088/1757-899X/316/1/012064

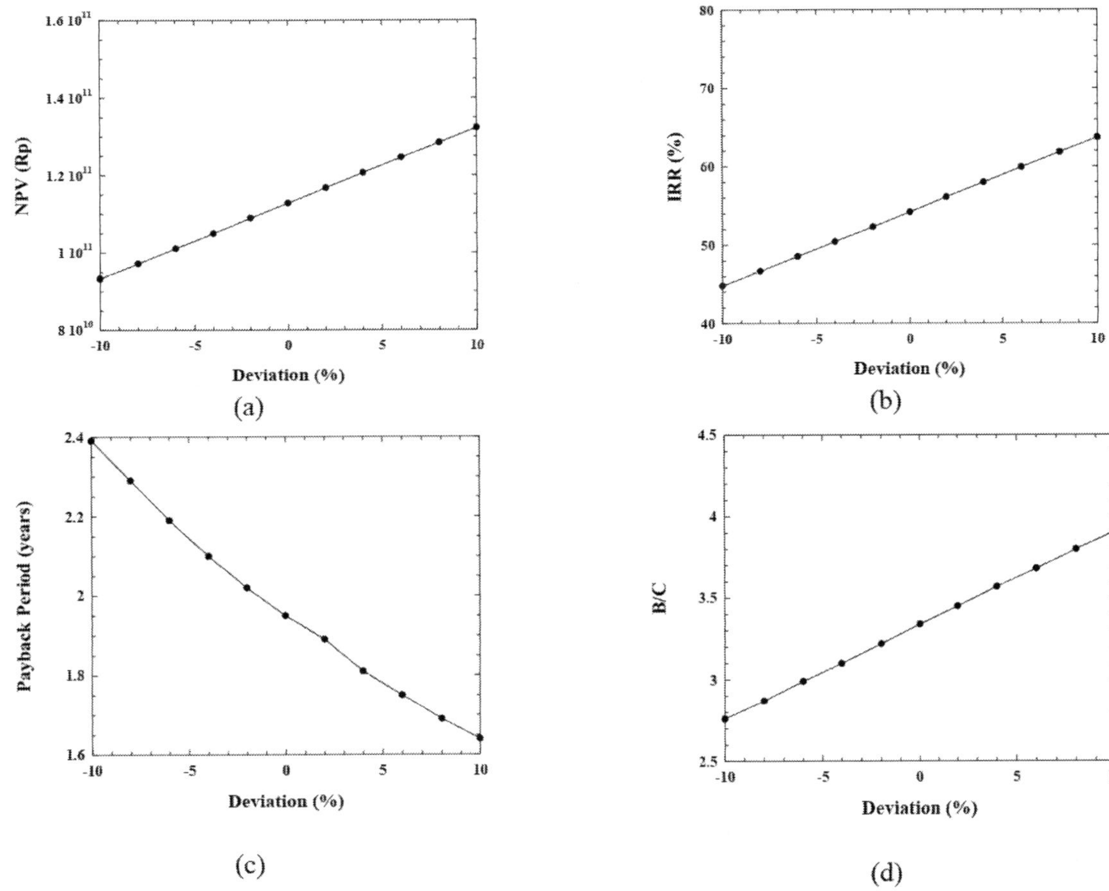

Figure 5. Sensitivity analysis of 10 kg basis enzyme prices deviation (a) toward NPV (b) toward IRR (c) toward Payback Period (d) toward B/C

Based on the graphic above, the best variation was obtained at 10th deviation with the price of enzyme is about Rp 10,924,035.00 and Rp 132,313,586,823.00 of NPV; 63.72% of IRR; 1.64 years of Payback Period; and 3.91 of B/C.

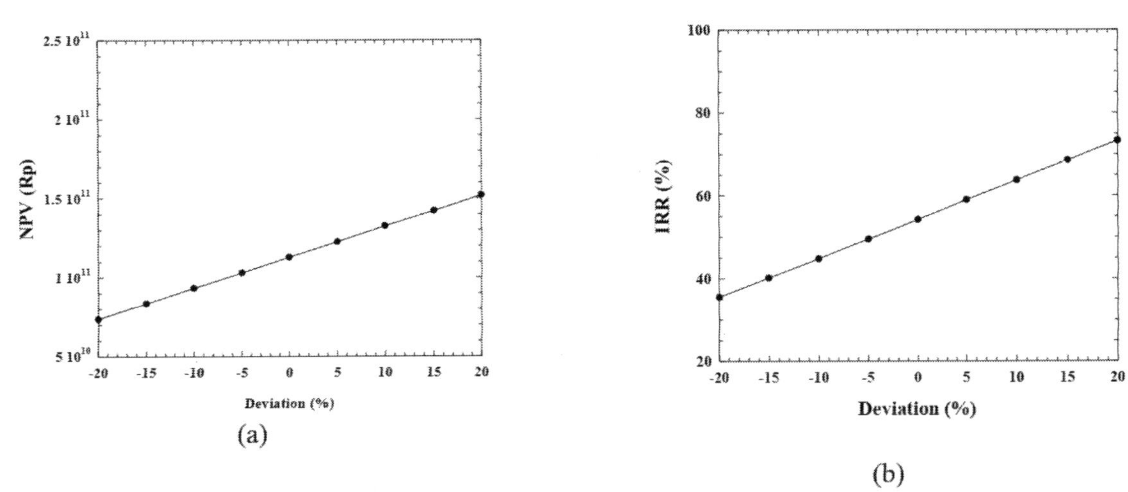

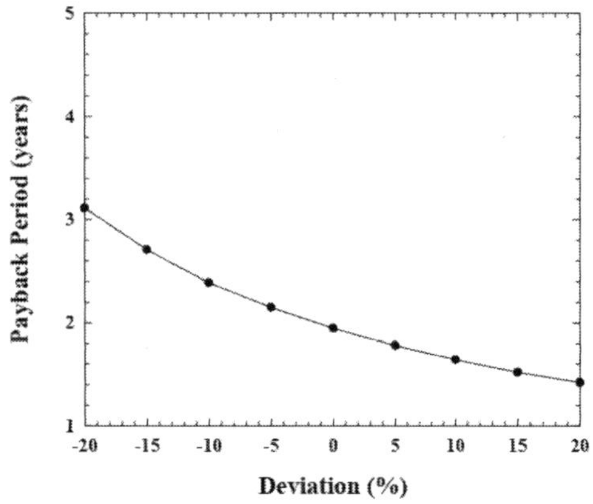

(c)

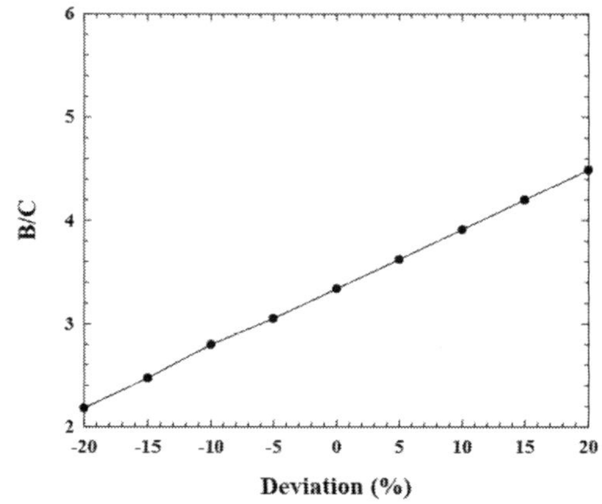

(d)

Figure 6. Sensitivity analysis of 10 kg basis rupiah's value deviation (a) toward NPV (b) toward IRR (c) toward Payback Period (d) toward B/C

Based on the graphic above, the best variation was obtained at 20th deviation with the exchange rate of rupiah upon dollar is Rp 15,923.00 and Rp 151,823,286,083.00 of NPV; 73.32% of IRR; 1.42 years of Payback Period; and 4.49 of B/C.

4. Conclusion

To obtain the lipase enzyme product of 1 kg and 10 kg production capacity, requires 1.7 kg and 16.96 kg of agro-industry waste as the substrate. The results of the simulation using SuperPro Designer is 1 kg of lipase product with 1.7 kg substrate and 9.92 kg of lipase product with 16.96 kg substrate. While, the highest result of NPV, IRR, Payback Period, and B/C on each 1 kg and 10 kg basis is Rp 15,923.00 exchange rate. By Rp 199.00/kg of substrate's price; Rp 9,947,305.00 of enzyme's price; and Rp 13,269.00 of rupiah's value upon dollar; the results for 1 kg basis is Rp 5,377,014,216.00 of NPV; 22.31% of IRR; 7.51 years of Payback Period; and 1.16 of Benefit Cost Ratio. On the other hand, the results for 10 kg basis is Rp 112,796,147.00 of NPV; 54.20% of IRR; 1.95 years of Payback Period; and 3.36 of Benefit Cost Ratio. Therefore, the 10 kg of production capacity is more profitable than 1 kg basis in terms of profitability method of NPV, IRR, Payback Period, and Benefit Cost Ratio.

5. References

[1] Navis 2014 Available online at: http://www.bppt.go.id/teknologi-agroindustri-dan-bioteknologi/1987-potensi-dan-tantangan-menuju-kemandirian-enzim-nasional, Accessed on 31 March 2016
[2] Farahbakhsh A, Ghasemi M and Ataie S E 2013 *Int. Conf. on Environment, Agriculture and Food Sciences* (Kuala Lumpur, Malaysia)
[3] Aliyah A, Edelweiss E, Sahlan M, Wijanarko A and Hermansyah H 2016 *Inter.l J. Technol.* **7** 1392
[4] Utami T S, Hariyani I, Alamsyah G and Hermansyah H 2017 *Energy Proceedia* **136** 41–46
[5] Lehninger AL 1997 *Dasar-Dasar Biokimia* (Jakarta: Penerbit Erlangga)
[6] Brownell L and Young E 1959.*Process Equipment Design: Vessel Design 1st ed* (Michigan: John Wiley & Sons, Inc.)
[7] McCabe W, Smith J and Harriott P 2005 *Unit Operation of Chemical Engineering 7th ed* (New York: McGraw-Hill Education)
[8] Castilho L, Polato C, Baruque E, Sant'Anna G and Freire D 2000 *Biochem. Eng. Journal* **4** 239–247
[9] Giatman, M 2006 *Ekonomi Teknik 1st ed* (Jakarta: RajaGrafindo Persada)
[10] Oleskowicz-Popiel P, Klein-Marcuschamer D, Simmons B and Blanch H 2014 *Bio resource Technol.* **158** 294–299

[11] Silla H 2003 *Chemical Process Engineering 1st ed* (New York: Marcel Dekker)

6. Acknowledgement

The authors gratefully acknowledge KEMENRISTEK DIKTI, University of Indonesia, and USAID SHERA for supporting this research.

IOP Conf. Series: Materials Science and Engineering **316** (2018) 012065 doi:10.1088/1757-899X/316/1/012065

CO_2 Absorption from Biogas by Glycerol: Conducted in Semi-Batch Bubble Column

Pratiwi puji lestari[1], Aswati Mindaryani[1], S. K. Wirawan[1]

[1]*Department of Chemical Engineering Universitas Gadjah Mada, Yogyakarta 55281, Indonesia*

E-mail: amindaryani@ugm.ac.id and pratiwi.1309@gmail.com

Abstract. Biogas is a renewable energy source that has been developed recently. The main contents of Biogas itself are Methane and carbon dioxide (CO_2) where Methane is the main component of biogas with CO_2 as the highest impurities. The quality of biogas depends on the CO_2 content, the lower CO_2 levels, the higher biogas quality. Absorption is one of the methods to reduce CO_2 level. The selections of absorbent and appropriate operating parameters are important factors in the CO_2 absorption from biogas. This study aimed to find out the design parameters for CO_2 absorption using glycerol that represented by the overall mass transfer coefficient (K_La) and Henry's constant (H). This study was conducted in semi-batch bubble column. Mixed gas was contacted with glycerol in a bubble column. The concentration of CO_2 in the feed gas inlet and outlet columns were analysed by Gas Chromatograph. The variables observed in this study were superficial gas velocity and temperatures. The results showed that higher superficial gas velocity and lower temperature increased the rate of absorption process and the amount of CO_2 absorbed.

Keywords: Biogas; CO_2 absorption; Glycerol; Henry constant; The overall mass transfer coefficient

1. Introduction

The development of science and technology will be based on the increasing energy needs. This is demonstrated by the many development of renewable energy used to meet human needs. One of the renewable energy source that has a potential to be developed is biogas. At this time, the development of biogas utilization has been done by KESDM in 2015 through three schemes, namely through the state budget, semi-commercial biogas program ("*Program Biru*") and development of biogas-based power generation program by private sector [1].

The main content of the biogas are methane (CH_4) and carbon dioxide (CO_2). Some research on combustion characteristics has been proven that the presence of CO_2 in fuel combustion can reduce the rate of reaction. High CO_2 concentration in fuels can reduce its heating value and reduce the quality of biogas [2]. Common method that has been used in the refining process of biogas is absorption using amine-based compound. However, this method has some constraints which are costly, corrosive, and high energy requirements for regeneration [3].

Based on the CO_2 gas separation scheme provided by Wahono in his journal namely "*Pemanfaatan Zeolit Lokal Gunung Kidul-Yogyakarta untuk Optimasi Sistem Biogas*", the most appropriate method for biogas purification process by physical absorption [4]. Physical absorption is the absorption process of dissolved gas (solute) by physical interaction, diffusion of gasses into the water, or gas dissolving into the liquid phase without chemical reaction. Physical absorption process is primarily used when the concentration of dissolved CO_2 is high [5]. The main problem is high potential of solvent loss [3];[6].

One of the compounds that can be used as physical absorbents is glycerol. Glycerol has low vapor pressure, stable, non-toxic, cheap and available in vast quantities as a by-product of biodiesel production. According to Aschenbrenner and Peter in his Journal of Comparative Study of Solvent Properties for Carbon Dioxide Absorption, the terminal –OH have a higher affinity for CO_2 molecules that increase CO_2 absorption [7]. Dissolution of CO_2 in the glycerol caused by Van der Waals and electrostatic forces. The bond between the absorbent and CO_2 in the physical absorption is lower than the bond formed by chemical absorption [8]. Therefore, glycerol is possible to be used as an absorbent.

The aim of this study was to investigate CO_2 absorption by glycerol in semi-batch bubble column. Overall mass transfer and Henry's constant are measured under various conditions. Parameters such as superficial velocity and temperature are varied.

2. Methodology/Experimental

2.1. Setup

The schematic diagram of CO_2 absorption in glycerol was performed in a semi-batch bubble column as shown in Figure1.Glycerol (96%) was purchased from PT. Brataco. The simulated biogas (mixed gas) consists of 40% CO_2 containing N_2 was supplied by PT.Samator Gas. The liquid volume in the column was 250 ml. The column was immersed in a water bath to control the temperature of the liquid. Mixed gas supplied from cylinder and the gas flow rates were controlled by volume flow meters. For injection and uniform distribution of the gas phase, a gas sparger (sintered sparger) was installed at the center of bottom column. The concentration of outlet gas from the absorber was analysed with gas chromatography (GC). Range of gas superficial velocity from 0,56 cm/s to 0,73 cm/s and temperature from 293 to 333 K have been investigated.

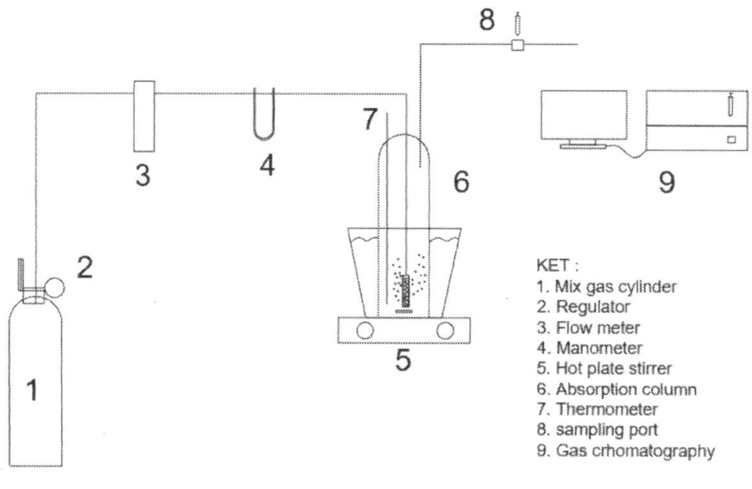

Figure 1.CO_2Absorption Process Schematic Experimental Setup

2.2. Determination of K_La and H

Assuming that the liquid and gas are perfectly mixed, the carbon dioxide absorption process can be described by applying mass balance:
Mass balance of CO_2 in gas phase:

$$\frac{dC_{AG}}{dt} = \frac{1}{V.\varepsilon_G}(C_{AG}in - C_{AG}) - \frac{K_La}{\varepsilon_G}(\frac{C_{AG}}{H} - C_{AL}) \quad (1)$$

Mass balance of CO_2 in liquid phase:

$$\frac{dC_{AL}}{dt} = K_La(\frac{C_{AG}}{H} - C_{AL}) \quad (2)$$

Where C_{AL} is the carbon dioxide concentration in the liquid phase, C_{AG} is carbon dioxide concentration in the gas phase and H is the Henry's constant.

For K_La and H value, the solution of equations 1 and 2 resulted C_{AL} and C_{AG} as a function of time. C_{AG} value calculation result is compared with C_{AG} research data. The difference between the calculation of C_{AG} and research data expressed by SSE (Sum Square of Error), Optimum K_La and H value has minimum SSE value. For superficial gas velocity variations and other temperature, the same calculation is carried out.

3. Results and Discussion

3.1. Effect of Temperature

To study and compare the effect of temperature, CO_2 concentration was measured at three different temperatures (293, 313 and 333K), all operation pressure is atmospheric pressure and inlet gas superficial velocity is 0.73 cm/s. Figure 2 shows that the increasing of temperature in the absorber lead to lower C_{AL} value in the solution. The lower C_{AL} value indicates the decrease of gas solubility in glycerol. At higher temperatures, the kinetic energy of molecules and the strength of their collision are larger [9]. The negative impact of temperature to solubility can be given a similar explanation [10]. CO_2 capacity of glycerol is optimum at 293K that is equal to 1.511 mol_{CO2} / $L_{glycerol}$.

IOP Conf. Series: Materials Science and Engineering **316** (2018) 012065 doi:10.1088/1757-899X/316/1/012065

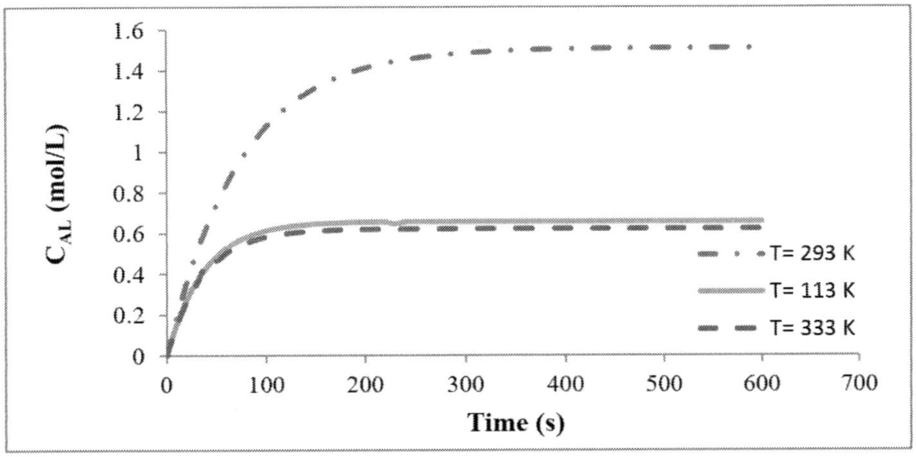

Figure 2. CO_2 Concentration in liquid phase (C_{AL})Profile in Various Temperature (v_s=0.0073 m/s)

3.1.1. Henry's Constant. Table 1 shows the Henry's Constant for constant for CO_2 in glycerol at studied temperatures. As expected, the H value for CO2 in glycerol increase along with the increasing temperature, due to the temperature is inversely proportional to gas solubility. The higher Henry's constant value means that the release of gas molecules from the liquid is easier [9].

Table 1. Henry's Constant Value of CO_2-Glycerol in Various Temperature

Temperature (K)	H (dimensionless)
293	0.0165
313	0.0289
333	0.0340

Henry's constant of experimental data at low pressure can estimated as a solubility in weight-fraction Henry's constant by Iwai [11].

$$H^* = \frac{P.y_{CO2}}{w_{CO2}} \quad (3)$$

Table 2 shows the comparison of Henry's constant obtained both of calculation and the values obtained inthe other absorbents.

Table 2. Henry's constant in various absorbent at different temperatures

H*(MPa)								
T(K)	This Work	T(K)	Selexol[12],[13]	sulfolane[14]	PEGME250[15]	EGMEE[15]	Genosorb[16]
293	4.47						
303	7.1	298	3.6	8.7	9.5	10.3	10.3
313	8.77	313	4.7	11.1	10.8	10.8	10.8
333	11.85	333	6.5	15.8	13.3	13.9	13.9

IOP Conf. Series: Materials Science and Engineering **316** (2018) 012065 doi:10.1088/1757-899X/316/1/012065

3.1.2. Heat of Absorption CO_2-Glycerol. Heat of absorption is directly associated with the energy required to regenerate absorbent. Heat of absorption is also a significant factor to recycle the CO_2 in glycerol [17]. The enthalpy change that occurs when CO_2 gas absorbed into glycerol can be described using the Van't Hoff equation [18].

$$\boldsymbol{ln(H) = ln(H^0) - \frac{\Delta H_{Abs}}{R}\frac{1}{T}} \tag{4}$$

The influence of temperature on Henry's law constant is present in Figure 3. that plot the natural logarithm of Henry's law constant to inverse temperature. Linearized equation of Equation 3 is $\ln(H) = \mathbf{16,289 - 2278\frac{1}{T}}$. Based on the linearized equation, ΔH_{abs} value obtained from the slope of the Van't Hoff equation is -14.12 kJ/mol. Negative result shows that the process occurs is exothermic [16]. Based on the theory of Jaycock and Parfitt, the value of ΔH_{abs} (energy requirements) range in the physical processes is -20 kJ/mol to -0 kJ/mol. According to Jaycock and Parfitt theory, the process of CO_2 absorption by glycerol is physical absorption [19].

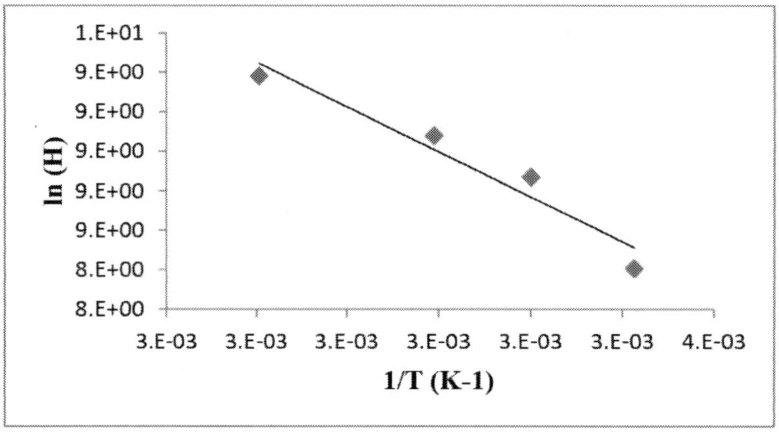

Figure 3. Dependence of Henry's law constant with temperature

The calculated values of heat of absorption for Glycerol-CO_2 system and other absorbent are given in table 3.

Table 3. Heat of Absorption CO_2 in various absorbent

No	Komponen	ΔH_{abs} (kJ/mol)
1	CO_2-Gliserol	-14.12
2	CO_2-MEG (Monoethyleneglycol)[9]	-11.77
3	CO_2-Propilenecarbonat[13]	-12.2
4	CO_2-selexol[13]	-14.3
5	CO_2-Genosorb 1753[16]	-14.48
6	CO_2-water[20]	-12.72

3.2. Effect of Superficial Gas Velocity

To study and compare the effect of gas superficial velocity, CO_2 concentration was measured at three different gas superficial velocity (0.0056; 0.0062; and 0.0073 m/s), all operation pressure is atmospheric pressure and temperature is 303 K. Figure 4. shows that the concentration of CO_2 in the gas out in various superficial velocity increase throughout the absorption process. In superficial velocity of 0.0073 m/s, the constant concentration of CO_2 out from bubble column can be achieved more quickly compared to superficialvelocity of 0.0056 m/s and 0.0062 m/s. This is possibly due to the amount of CO_2 gas that flows in the absorption column, so higher superficial velocity results high amount of CO_2 flowing and contact with the glycerol. Higher gas superficial velocity will increase the amount of CO_2 gas bubble dissolved into the glycerol. However, increasing of gas superficial velocity can increase the size of gas bubbles [21]. The Large size of gas bubbles will increase turbulence in the liquid phase, and also increase the flotation velocity [22].

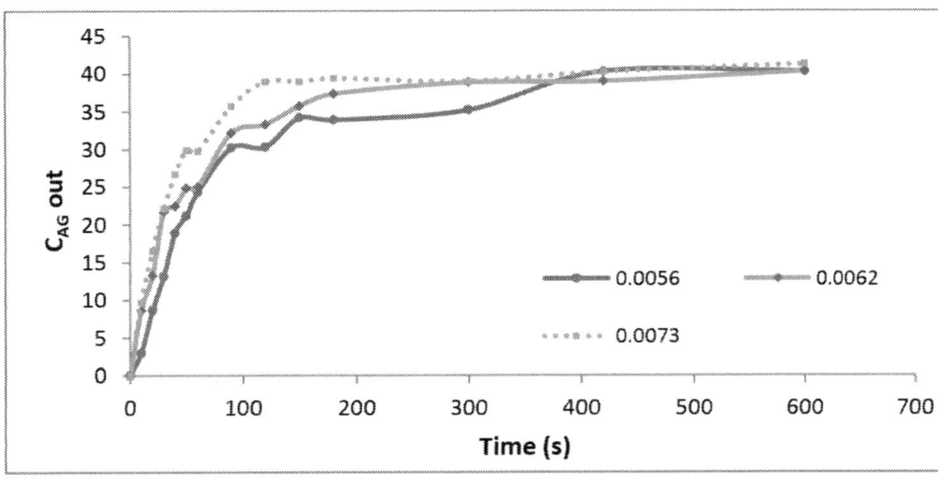

Figure 4. CO_2 Concentration Profile in Various Superficial Velocity (T=303K)

3.2.1. Mass Transfer Coefficient (K_La). The gas-liquid mass transfer coefficient (K_La) obtained in this study shows the magnitude of the mass transfer from the gas phase into the liquid in the absorption column. Figure 5. shows that the greater value of superficial velocitywill increase the value of K_La. According to Bouaifi, the increase of superficial velocity may increase the contact area between the gas and liquid and also K_La value [23]. For high superficial velocity, turbulence in the absorption column expressed by the Reynolds number (Re) will be even greater. High turbulence causes higher contact that occurs between the gas bubbles and fluid [24].

IOP Conf. Series: Materials Science and Engineering **316** (2018) 012065 doi:10.1088/1757-899X/316/1/012065

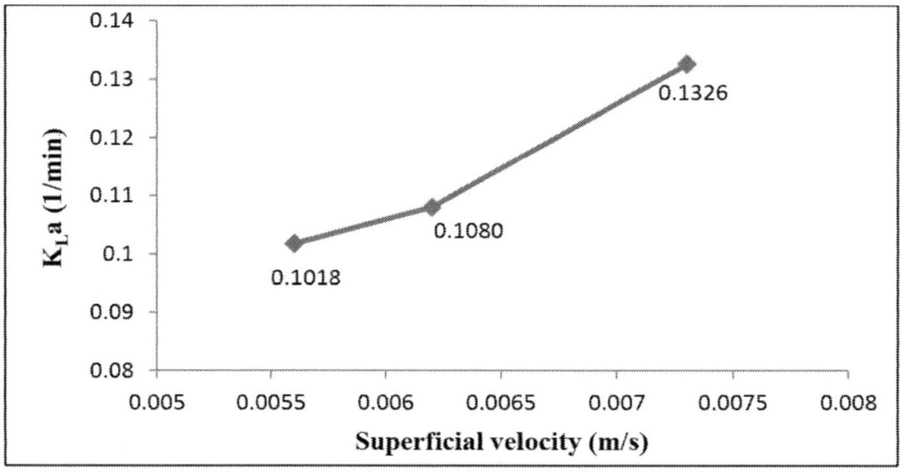

Figure 5. K_La Value to Gas Superficial Velocity (T= 303 K)

The relations between Re numbers and K_La can be explained by the empirical approach in the form of dimensionless number [25]

$$Sh_L = \frac{K_L a.d_p}{D} = 2 + b' Re_G^{0,779} Sc_L^{0,546} \left(\frac{d_p.g^{\frac{1}{3}}}{D^{\frac{2}{3}}}\right) \tag{5}$$

The suitability of theoretical and calculations in this research is determined by calculating the relative error. Comparison K_La in this research and theoretical calculations are shown in Table

Table 4. K_La and K_La Hungmark Equation Value Comparation

Vs	K_La (1/min)		Relative Error
(m/s)	Research Data	Hungmark, 1967	(%)
0.0056	0.102	0.121	18.52
0.0062	0.108	0.130	20.78
0.0073	0.133	0.149	12.29

In Table 4 shows that the theoretical calculation error compared to data research superficial velocity range from 12% - 20%. These error values could give justification that optimization for parameter determination in mathematical model is completely acceptable.

4. Conclusion

In this study, glycerol as a new absorbent for CO_2 absorption bubble column has been studied. CO_2 absorption capacity of glycerol is 1.5 $mol_{CO2}/L_{glycerol}$. Higher superficial gas velocity and lower temperature increased the rate of absorption process and the amount of CO_2 absorption. The relationship between the value of Henry's law constant toward the time followed the equation ln(H)=16.289 – (2278)1/T.

5. Acknowledgement

The authors are grateful for the financial support of Chemical Engineering Department Universitas Gadjah Mada.

IOP Conf. Series: Materials Science and Engineering **316** (2018) 012065 doi:10.1088/1757-899X/316/1/012065

6. References

[1] Kementerian Energi dan Sumber Daya Mineral 2015 *RENSTRA KESDM* (Jakarta)

[2] Deublein D and Angelika S 2008 *Biogas from Waste and Renewable Resources: An Introduction* (Weinhem: WILEY-VCH Verlag GmbH & Co. kGaA)

[3] Seo M W *et al* 2014 Methanol absorption characteristic for the removal of H_2S, COS, and CO_2 in pilot-scale biomass-to-liquid process *J. Energy* **66** 56-62

[4] Wahono S K 2008 *Pemanfaatan Zeolit Lokal Gunung Kidul-Yogyakarta untuk Optimasi Sistem Biogas, Seminar Nasional Fundamental dan Aplikasi Teknik Kimia.*

[5] Dindore V, Brilman D, Geuzerbrock F, Versteeg G 2004 Membrane-solvent selection for CO_2 removal using membrane gas-liquid contactors *J. Sep. Purif. Technol.* **40** 133-145

[6] Wellinger A and Lindeberg A 1999 *Biogas Upgrading and Utilization. Task 24: Energy from Biological Conversion of Organic Wastes.*

[7] Aschenbrenner O and Peter S 2010 Comparative study of solvent properties for carbon dioxide absorption *J. Energy Environ. Sci.* **3** 1106-13

[8] Plasynski S I and Chen Z Y 2008 Review of CO_2 capture technologies and some improvement opportunities, J. Energy Fuel **45** 644-6

[9] Serpa F *et al* 2013 *Solubility and Thermodynamic Properties of Carbon Dioxide in MEG/Water Mixture, III Iberoamerican Conf. on Supercritical Fluids Cartagena de Indias* (Colombia)

[10] Shamiri A, Shafeeyan M S, Tee H C, Leo C Y, Aroua M K, Anghamohammadi N 2016 *J. of Nat. Gas Sci. Eng.* **35** 605-13

[11] Iwai Y, Anai Y, Arai Y 1981 Prediction of solubilities for volatile hydrocarbons in low density polyethylene using *UNIFAC-FV* model *J. Polym. Eng. Sci.* **21** 1015-18

[12] Jou F Y, Deshmurkh R D, Otto F D, Mather A E 1990 solubility of H_2S, CO_2, CH_4 and C_2H_6 in sulfolane *J. Fluid Phase Equilib.* **56** 313-24.

[13] Xu Y 1990 *Thermodinamic Investigation (1) Gases and Vapor in Liquid (2) Calorimetric Studies (Ph.D. Thesis)* (Canada: University of Alberta)

[14] Schmidt K A G 1997 *Solubility of Sulphur Dioxide in Mixed Polyethylene Glycol Ethers (M. Sc. Thesis Chemical Engineering)* (Canada: University of Alberta)

[15] Henni, A.,Tontiwachwuthikul, P., Chakma A., (2005), Solubility of Carbon Dioxide in Polyethylene Glycol Ethers, The Canadian Journal of Chemical Engineering, Volume 83, pp. 358-361.

[16] Rayer A V, Henni A, Tontiwachwuthikul 2011 High pressure physical solubility of carbon dioxide (CO_2) in mixed polyethylene glycol dimethyl ethers (Genosorb 1753) *Can. J. Chem. Eng.* **9999** 1-8

[17] Kim Y E, Moon S J, Yoon Y I, Jeong S K, Park K T, Bae S T, Nam S C 2014 Heat of absorption and absorption capacity of CO2 in aqueous solution of amine containing multiple amino groups J. Sep. Purif. Technol **122** 112-8

[18] Schüler N, Hecht K, Dittmeyer R 2012 On the solubility of carbon dioxide in binary water-methanol mixture *J. Chem. Eng. Data* **57** 2304-8

[19] Jaycock M J and Parfitt G D 1981 *Chemistry of Interfaces* (Chichester: Ellis Howood Limited Publisher)

[20] Carroll J J, Slupsky J D, Mather A E 1991 The solubility of carbon dioxide in water at low pressure *J. Phys. Chem. Ref. Data* **20** 1201-9

[21] Kazakis N A, Mouza A A, Paras S V 2008 Experimental study of bubble formation at metal potous sparger: effect of liquid properties and sparger characteristics on the inital bubble size distribution *Chem. Eng. J.* **137** 256-81

[22] Behkish A 2004 *Hydrodynamic and Mass Transfer Parameters in Large-scale Slurry Bubble Column Reactors (Dissertation)* (Pittsburg: Chemical Engineering University)

[23] Bouaifi M, Hebrard G, Bastoul D, Roustan M 2001 A comparative study of gas hold-up, bubble Size, interfacial area and mass transfer coefficient in stirred gas-liquid reactors and bubble columns J. Chem. Eng. Process. **40** 97-111.

IOP Conf. Series: Materials Science and Engineering **316** (2018) 012065 doi:10.1088/1757-899X/316/1/012065

[24] Eckenfelder W 2000 *Industrial Water Pollution Control 3rd ed.* (New York: Mc Graw Hill Companies Inc.)

[25] Hungmark G A 1967 Holdup and mass transfer in bubble columns *J. Ind. Eng. Chem. Process. Des. Dev.* **6** 218.

IOP Conf. Series: Materials Science and Engineering **316** (2018) 012066 doi:10.1088/1757-899X/316/1/012066

Effect of some variable in cellulase production by *Aspergillus niger* ITBCC L74 using solid state fermentation

B Abdullah, S Maftukhah, E Listyaningrum and F Faradhiba

Chemical Engineering Department, Engineering Faculty, Diponegoro University
Prof. Soedarto, SH Street, Kampus Undip Tembalang, Semarang, Indonesia 50275

Maftukhah-tuchah@yahoo.com

Abstract

Cellulase is a very important enzyme for ethanol production, food, papper, etc, from lignocellulose and others. Rice straw and corn cob are the largest agricultural waste in Indonesia, while the water hyacinth weed is a plant that has not been used optimally. The content of cellulose is high enough on rice straw, water hyacinth and corn corb so it can be used as a substrate in the production of cellulase to increase the economic value of the rice straw, hyacinth, and corncob. As for the purpose of this study is to use the rice straw, water hyacinth, and corn cob as substrates of cellulase enzyme, determine the effect type of substrates, moisture content and fermentation time in production of cellulase enzyme and also determining the optimum conditions for production of cellulase enzymes. The method is solid fermentation system and using fungi *Aspergillus niger ITBCC L74* as inoculum. The variable used were fermentation time is 2 , 4 , 6 , 8 and 10 days, moisture content is 50, 60, 70, and 80%, as well as the type of substrate is rice straw, water hyacinth, and corn cob. The results showed that the highest protein content in the crude enzyme of the rice straw, water hyacinth and corncobs @ is 0.0153 mg/ml, 0.0194 mg/ml and 0. 0146 mg/ml, respectively. The optimum enzyme activity were for the rice straw, water hyacinth and corn cobs @ 2.569 U/ml, 1.606 U/ml and 1.302 U/ml, respectively. The optimum moisture content were obtain for rice straw, water hyacinth and corn cob respectively 80%, 70% and 60%. And the optimum fermentation time for rice straw, corn cob, and water hyacinth is on the sixth day. In this study showed the highest enzyme activity on the type of rice straw substrate with a water content of 80% and fermentation time 6 day.

Keywords: Cellulase enzyme; *Aspergillus niger* ITBCC L74; rice straw; water hyacinth; corn cob

1. Introduction

Rice straw is the largest agricultural waste in Indonesia. There are around 20 million per year. According to BPS data in 2006, the paddle field area in Indonesia is 11.9 million ha. The tilth of each hectare of rice fields can produce till 12-15 tons of dry materail, it depends on location and varieties. So far, only 31-39% of rice straw use as fodder while another is about 36-62% used as fertilizer or be burned and about 7-16% used to indutrial requiement [1].

Water hyacinth (*Eichhornia crassipes*) is a free-floating aquatic plants. Water hyacinth can grow and spread easily through the water supply to other water bodies. Water hyacinth has vegetative breeding speed rapidly, especially in the tropics and subtropics [2]. Because of the proliferation of water hyacinth which so quickly, this water hyacinth become weed for some areas in Indonesia. The amount

IOP Conf. Series: Materials Science and Engineering **316** (2018) 012066 doi:10.1088/1757-899X/316/1/012066

of water hyacinth has not been used optimally so many of researchers trying to develop the potential of water hyacinth into something that has high economic value.

According to BPS data on 2014, the production of dry-cornseed is about 19.03 million tons. The huge corn seed production create waste that called corncobs. Corncob usually dont used anymore or have low economic value.

Cellulase is one of the enzymes produced by microorganisms. Cellulase enzyme has an important role in the process of bioconversion of organic waste into glucose, single cell protein, animal feed, ethanol and others [3]. Cellulase is an enzyme that can hydrolyzes β (1-4) in the cellulose. Cellulase enzymes can be obtained from various sources such as plants, insects, and microorganisms. Cellulase produced by fungi, bacteria, plants, and ruminants. Fungi are the main microorganisms that produce cellulase enzymes. Examples of filamentous fungi is *Trichoderma* and *Aspergillus* [4].

The purpose of this study is to use the straw, water hyacinth, and corn cobs as substrates for cellulase enzyme production, determine the effect type of substrates, moisture content and fermentation time in production of cellulase enzyme and also determine the optimum conditions for production of cellulase enzymes.

2. Material and method

2.1.Cellulosic materials

The raw material consists of water hyacinth which obtained from the river in Tirto Husodo Timur street Tembalang while corn cobs and straw obtained from Kecapi village Jepara. Before using, each substrat is chopped and sieved to a homogeneous size. While *Aspergillus niger* ITBCC L74 was obtained from the Laboratory of Microbiology and Bioprocess Bandung Institute of Technology.

2.2. Organism and culture condition

Culture of *Aspergillus niger* was maintained on PDA and incubated at 30 ° C for 7 days [5,6,1].

2.3. Media preparation and enzym production

Medium composition described by Mandles [22] and Weber was used for fermentation. The media contained (per liter of distiled water) : urea 0,3 g, $(NH_4)_2SO_4$ 1,4 g, $CaCl_2$ 0,002 g, $MgSO_4.7H_2O$ 0,3 g, KH_2PO_4 2 g, $FeSO_4.7H_2O$ 0,005 g, $MnSO_4.7H_2O$ 0,0016 g, $ZnSO_4.7H_2O$ 0,0014 g, tween-80 2 ml, and polypeptone 1 g. Production of cellulose enzym started by mixing each substrat and nutrient solution based on variable of moisture content. Adjusting the pH of fermentation and then sterilized by autoclaving 121°C for 15 min. This was cooled and inoculated with 1x1 cm og organism from PDA culture plates. Then incubated at ±30 °C and fermentated with a predetermined time variables [5,6,1].

Fermented enzyme is extracted with distilled water so that would be obtained filtrate and solids. After that, the filtrate and solids are separated using a centrifuge. The filtrate obtained is used to test the protein and the activity of the enzyme

2.4. Cellulase assay

The first analysis is the analysis of proteins by Lowry method. This analysis to indicate the levels of enzymes in the sample. Then, analysis for determine the activity of the enzyme produced is used CMC 1% is added to the enzyme extract, and glucose levels were tested by the method of DNS (3,5-dinitrosalycylic acid) [7].

3. Results and discussion

3.1. Raw material analysis for production of cellulase enzyme

This research was conducted on various types of substrates, namely rice straw, water hyacinth, and corn cobs. Characterization of the raw material of each substrate is presented in Table 1.

Table.1 Results of Raw Materials Characterization for Production of Cellulase Enzyme

Parameter	Rice Straw		Water Hyacinth		Corn Cob	
	References	Analysis Results	References	Analysis Results	References	Analysis Results
Cellulose (%)	35-45 (a)	38.26	17.8-31 (c)	29.54	32.3-45.46 (d)	39.68
Hemicellulose (%)	30-40 (a)	35.73	22-43.4 (c)	35.71	39-49 (d)	40.58
Lignin (%)	5-20 (a)	10.28	7-26.36 (c)	7.46	6.7-13.9 (d)	13.43
Protein (%)	3-5 (b)	4.81	11.9-20 (c)	12.18	3.3-5.32 (e)	4.61
Ash (%)	7.8-15.6 (b)	10.92	15-25.7 (c)	15.21	2.77-2.99 (e)	2.70

(a) [8]; (b) [1]; (c) [9]; (d) [10];(e) [11]

Based on Table 1, it can be seen that there is a difference between the raw material content of the substrates used with that obtained from the literature. Such differences may be caused due to several factors such as climate, soil, pest / disease, as well as the type of the plant it self [12]. These factors certainly can't be controlled, even though the same plant species, but other factors also depend on where they are grown also different and can cause the content of the substrate is different from the literature.

From Table 1, it can be seen that the corn cob contains cellulose which is the highest of 39.68%, followed by rice straw 38.26%, and 29.54% of water hyacinth. Corn cob hemicellulose also contains the highest of 40.58%, followed by rice straw 35.73%, and 35.71% of water hyacinth. Cellulose and hemicellulose in the raw materials are carbon source for the growth of *Aspergillus niger* for production of cellulase enzyme [13]. Water hyacinth has a lower cellulose content than straw or corn cobs, but also had the lowest lignin content is 7.46%, followed by rice straw 10.28% and 13.43% corn cobs. Lignin is a very complex molecules composed of units fenilpropan are connected in three-dimensional form that is difficult to be degraded. The higher proportion of lignin, the higher resistance to enzymatic degradation [14]. Because the carbon content is relatively high compared with cellulose and hemicellulose, lignin has a high energy content. Lignin naturally brown. If the substrate turns a little white color, it means there are some lignin missing. Lignin makes the substrate becomes hard and clay. If the substrate becomes softer, meaning its protector lignin already damaged [1]. Water hyacinth has the highest protein content is 12.18%, followed by rice straw 4.81%, and 4.61% corn cob. Protein is an amino acid that has the elements C, H, O, N as nutrients that get the highest enzyme [15]. Carbon acts as an enzyme cofactor [23], whereas nitrogen required in the fermentation process because it can affect the activity of *Aspergillus niger* [1]. In growth, carbon and nitrogen are used for the components of fungal cell of the body. This process occurs because during the fermentation of *Aspergillus niger* using nutrients for growth [16]. Water hyacinth also has the highest ash content is 15.21%, followed by rice straw 10.92%, and 2.7% of corn cob. The ash content in a material reflects mineral content, although the value of the ash can't be used as an index to determine the number of elements. Fermentation requires an adequate intake of minerals to support the growth of mold (*Aspergillus niger*) so as to obtain optimal results [17].

3.2. Analysis of protein content

Based on the research that has been done by varying the variable type substrate, water content, as well as the long fermentation time, results for analysis of protein content are presented in Table 2.

Table 2 Results for Analysis of Protein Content on a Variety of Variables

Substrate	*Moisture Content*	**Protein Content (mg/ml)** 2 days	**4 days**	**6 days**	**8 days**	**10 days**
Rice Straw	50%	0.0068	0.0056	0.0122	0.0117	0.0086
	60%	0.0063	0.0055	0.0112	0.0109	0.0078
	70%	0.0108	0.0076	0.0129	0.0124	0.0104
	80%	0.0144	0.0116	0.0153	0.0147	0.0142
Water Hyacinth	50%	0.0102	0.0083	0.0140	0.0133	0.0097
	60%	0.0171	0.0109	0.0161	0.0141	0.0124
	70%	0.0183	0.0139	0.0194	0.0156	0.0146
	80%	0.0038	0.0019	0.0089	0.0081	0.0073
Corn Cob	50%	0.0121	0.0087	0.0144	0.0136	0.0114
	60%	0.0132	0.0095	0.0146	0.0140	0.0115
	70%	0.0102	0.0074	0.0118	0.0119	0.0093
	80%	0.0049	0.0046	0.0097	0.0092	0.0075

3.2.1. Effect of fermentation time on protein content. According Ariyani et al (2014) [16], increased protein content in line with the growth of molds (fungi) due to the fungi body consists of elements containing nitrogen. Besides enzymes produced by the fungus is also a protein. This is also supported by Garraway and Evans (1984) [18], which states fungal cell wall contains 6.3% protein, while fungal cell membranes with hyphae contains 25-45% protein and 25-30% carbohydrates. In its growth, fungi using carbon and nitrogen for the fungus body cell components. This process occurs because during the fermentation of *Aspergillus niger* ITBCC L74 using nutrients (especially carbohydrates) for growth and increased protein content.

From Table 2 it can be seen that time of the fermentation affects the amount of protein content in the production of cellulase enzymes. On the third substrates highest protein content are produced on average are on the sixth day fermentation period. The highest protein content using substrates water hyacinth is 0.0194 mg/ml, followed by rice straw 0.0153 mg/ml, and corn cob 0.0146 mg/ml. After the sixth day, then the protein content decreased. It is because on the eighth day onwards *Aspergillus niger* ITBCC L74 are used currently on death phase, where the number of dead cells more than cells that grow [19].

3.2.2. Effect of water content on protein content. According Idiawati et al (2014) [5], the solid fermentation water content have an important role in the biosynthesis and secretion of many enzymes, especially cellulases. The fermentation process is closely related to microorganisms. Water has a very important function in the metabolism of organisms, including as a source of the element hydrogen and oxygen required for the biosynthesis of cell components, play an important role in the process of enzymatic hydrolysis and very important role in the transport of nutrients and products of metabolites through the cell membrane , Fermentation is a solid substrate fermentation using insoluble substrate but contain enough water for growth and development of microorganisms were inoculated into the substrate itself. Therefore, the presence of water affects the growth of *Aspergillus niger* ITBCC L74 to produce cellulase enzymes.

In Table 2 shows the water content can affect the protein content in the prtoduction of cellulase enzymes. Water content is too low leads to reduced solubility of nutrients in the substrate [1]. However, the optimum water content also depends on the type of substrate used. The highest protein content produced by using water hyacinth substrate with water content of 70% 0.0194 mg/ml, followed by rice straw with water content 80% 0.0153 mg/ml, and corn cob with water content 60% 0.0146 mg/ml.

IOP Conf. Series: Materials Science and Engineering **316** (2018) 012066 doi:10.1088/1757-899X/316/1/012066

3.2.3. Effect of Substrate Type on Protein Content. Based on Table 2, the type of substrate can affect the production of cellulase enzymes. It can be seen from Table 1 on the content of each substrate. Water hyacinth has a lower cellulose content than straw or corn cobs, but also has a low lignin content. Lignin is a very complex molecules composed of units phenyl propane are connected in three-dimensional form that is difficult to be degraded. The higher the proportion of lignin, the higher the resistance to chemical and enzymatic degradation [14]. Because the carbon content is relatively high compared with cellulose and hemicellulose, lignin has a high energy content. Lignin naturally brown. If the substrate turns a little white color, it means there are some lignin damaged. Lignin makes the substrate becomes hard and clay. If the substrate becomes softer, it means the protector lignin already damaged [1]. In Table 1 shows that the corn cobs have a structure tougher than straw and water hyacinth for the lignin content most among the three substrate. Instead, the water hyacinth has a softer structure because the lignin content least of the third substrate. The highest protein content produced by using a substrate of water hyacinth 0.0194 mg/ml, followed by rice straw 0.0153 mg/ml, and corn cobs 0.0146 mg/ml.

3.3. Analysis of enzyme activity

The enzyme activity is expressed in U/ml. 1 unit is the amount of enzyme needed to break down cellulose into 1 mole of reducing sugar per minute. Based on the research that has been done by varying the variable type substrate, water content, as well as the long fermentation time, the enzyme activity analysis results are presented in Table 3.

3.3.1. Effect of fermentation time on enzyme activity. Based on Table 3, the fermentation time relationship with the activity of the enzyme can be expressed in Figure 1

Table 3 Results for Analysis of Enzyme Activity on Various Variables

Substrate	*Moisture Content*	Enzyme Activity (U/ml) 2 days	4 days	6 days	8 days	10 days
Rice Straw	50%	0.517	0.694	1.267	0.643	0.491
	60%	0.967	1.312	1.673	0.782	0.394
	70%	1.226	1.353	1.783	1.117	0.536
	80%	1.092	1.581	2.569	2.214	1.226
Water Hyacinth	50%	0.591	0.679	1.073	1.055	0.269
	60%	0.517	0.973	1.125	0.794	0.205
	70%	0.001	1.555	1.631	1.606	0.730
	80%	0.617	0.657	0.986	0.637	0.442
Corn Cob	50%	0.086	0.238	0.364	0.111	0.061
	60%	0.694	1.023	1.302	1.251	0.719
	70%	0.415	0.651	1.023	0.466	0.086
	80%	0.035	0.086	0.466	0.339	0.010

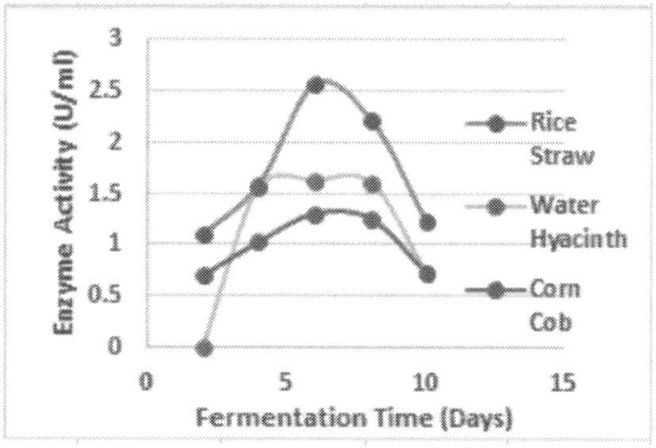

Figure 1 Effect of Fermentation Time on Enzyme Activity in Various Substrate

Time of fermentation to produce cellulase enzyme is one of the essential aspects for microbial growth affect the production of enzymes. In Table 3 and Figure 1 shows the effect of variations in fermentation time on the activity of enzyme cellulase. The enzyme activity will increase along with increasing fermentation time. However, the enzyme activity decreased on the eighth day and the tenth day. It is caused by the interaction between the enzyme cellulase to cellulose began to decline due to the accumulation of the product formed on the previous day, causing inhibition of the enzyme cellulase [20]. Besides the growth of mold is also influenced by the availability of nutrients [5]. If the availability of nutrients many lifetimes, the microorganisms are also getting longer. Conversely, if the availability of nutrients less than the lifetime of microorganisms will be more brief. The enzyme activity highest of rice straw 2.569 U/ml, followed by water hyacinth 1.631 U/ ml, and corn cob 1.302 U/ml.

3.3.2.Effect of water content on enzme activity. Based on Table 3, effect of water content on activity of the enzyme can be expressed in Figure 2.

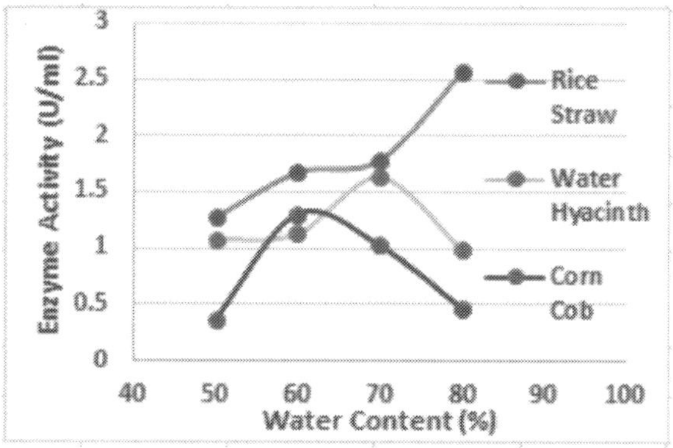

Figure 2. Effect of Water Content on Enzyme Activity in Various Substrates

The production of cellulase enzymes is influenced by factors such as keeping the water content. On Solid State Fermentation (SSF) water content has an important role in the biosynthesis and secretion of enzymes, especially cellulases [5]. The fermentation process is closely related to microorganisms. Water has an important function in microorganisms of which play a role in the metabolism of an organism as well as in the transport of nutrients. Solid State Fermentation (SSF) is fermented using substrates that are not soluble but contain enough water for growth and development of microorganisms were inoculated into the substrate. Therefore, the presence of water affects the growth of *Aspergillus niger* ITBCC L74 to produce cellulase enzymes. In Solid State Fermentation (SSF), the water content maximum is around 50% - 80% [16].

The growth of microorganisms is affected by the water content in the solid substrate. If the water content in low numbers it will reduce nutrient diffusion and solubility of the solid substance. Causing insufficient supply of nutrients to the growth of microorganisms and enzymes to be disrupted as well as lower production of the enzyme [20]. However, if the moisture content high, it will cause a decrease in porosity and gas volume, but it will increase the contamination causing a decrease in the production of the enzyme [21]. In the solid fermentation requires sufficient water conditions for mold growth, but not too wet because it will encourage the growth of bacteria.

Based on Table 3 and Figure 2 are shows that the water content effect on enzyme activity. The highest enzyme activity was obtained at the straw with water content of 80% followed by water hyacinth of 70% and corn cob of 60% with enzyme activity of 2.569, 1.631 , 1.302 U/ml, respectively.

3.3.3. Effect of substrate type on enzyme activity. Based on Table 1 and 3, Figure 1 and 2 show that the enzyme activity was effected mainly cellulose content in substrate due to as a carbon source for growth of *Aspergillus niger* ITBCC L74 to produce cellulase enzymes. The cellulose content of rice straw and corn cob almost similar (38.26 and 39.68%) with enzyme activity of 2.569 and 1.302 U/ml, Although the water hyacinth has lower cellulose content (29.54%) than corn cob, but the enzyme activity was higher (1.631 U/ml), it might be due to protein and mineral content more higher than other was needed for growth as macro and micro element. Lignin content in substrate also affect to the enzyme activity.

4. Conclusion

The optimum fermentation time to produce cellulase enzymes obtained was six days. *Based* on the analysis of enzyme activity in the to the long fermentation time, obtained. The highest enzyme activity obtained was 2.569 U/ml by using rice straw as substrate with a water content of 80%. Followed by water hyacinth and corn cob with enzyme activity of 1.631 and 1.302 U/ml at water content of 70% and 60%, respectively.

References

[1] Sa'adah Z, Ika N and Abdullah 2008 Produksi enzim selulase oleh *Aspergillus niger* menggunakan substrat jerami dengan sistem fermentasi padat *Skripsi* Diponegoro University

[2] Pinnata D and Damayanti A 2013 Pemanfaatan selulosa asetat eceng gondok sebagai bahan baku pembuatan membran untuk desalinasi *Skripsi* Institut of Sepuluh November

[3] Chalal D 1983 Solid state fermentation with *Trichoderma reesei* for cellulase production *Appl. Env. Micro* **vol 49** p 205-10

[4] Ikramulhaq, Javed M, Saleem T and Siddiq Z 2005 Cotton saccharifying activity of celluloses produced by co-culture of *Aspergillus niger* and *Trichoderma viride J. Agric & Biol. Sci* **vol 1(3)** p 241-45

[5] Idiawati N, Harfinda E and Arianie L 2014 Produksi enzim selulase oleh *Aspergillus niger* pada ampas sagu *J. Natur Indonesia* **vol 16(1)** p 1-9

[6] Safaria S, Idiawati N and Zaharah T 2013 Efektivitas campuran enzim selulase dari *Aspergillus niger* dan *Trichoderma reesei* dalam menghidrolisis substrat sabut kelapa *JKK* **vol 2(1)** p 46-51

[7] Acharya P, Acharya D and Modi A 2008 Optimization for cellulase production by *Aspergillus niger* using saw dust as substrate *African J. Biotechnology* **vol 7(22)** p 4147-52

[8] Kawee-ai A, Srisuwun A, Tantiwa N, Nontaman N, Boonchuay P, Kuntiya A, Chaiyaso T and Seesuriyachan P 2016 Eco-friendly processing in enzymatic xylooligosaccharides production from corncob: influence of pretreatment with sonocatalytic–synergistic fenton reaction and its antioxidant potentials *Ultrasonics Sonochemistry* **vol 31** p 184–192

[9] Gunnarson C and Petersen C 2007 Waterhyacinths as a resource in agriculture and energy production: A literature review *Waste Management* **vol** *27* p 117– 29

[10] Baadhe R, Potumarthi R and Mekala N 2014 Influence of dilute acid and alkali pretreatment on reducing sugar production from corncobs by crude enzymatic method: A comparative study *Bioresource Tech.* **vol 162** p 213–17

[11] Pointner M, Kuttner P, Obrlik T, Jager A and Kahr H 2014 Composition of corncobs as a substrate for fermentation of biofuels *Agronomy Research* **vol 12(2)** p 391–6

[12] AAK 2009 Petunjuk praktis bertanam sayuran (*Kanisius* Yogyakarta)

[13] Saropah D, Jannah A and Maunatin A 2012 Kinetika reaksi enzimatis ekstrak kasar enzim selulase bakteri selulolitik hasil isolasi dari bekatul *Jurnal Alchemy* **vol 2(1)** p 34-35

IOP Conf. Series: Materials Science and Engineering **316** (2018) 012066 doi:10.1088/1757-899X/316/1/012066

[14] Anggarawati D 2012 Aktivitas enzim selulase isolat SGS 2609 BBP4B-KP menggunakan substrat limbah pengolahan rumput laut yang dipretreatment dengan aam *Skripsi*
[15] Ariyanti Y 2015 Kandungan bahan organik dan protein kasar tongkol jagung (*zea mays*) yang diinokulasi dengan fungi *Trichoderma sp.* pada lama inkubasi yang berbeda *Skripsi* Hasanudin University Indonesia University
[16] Ariyani S, Asmawit and Utomo P 2014 Optimasi waktu inkubasi produksi enzim selulase oleh *Aspergillus niger* menggunakan fermentasi substrat padat *J.B Industri* **vol 5(2)** p 61-7
[17] Purwanti F, 2012 Kualitas nutrien onggok yang difermentasi *Aspergillus niger* dengan penambahan level urea dan zeolit yang berbeda *Skripsi* Institut of Bogor Agriculture
[18] Garraway M and Evans R 1984 Fungal nutrition physiology (New York: John Wiley and Sons)
[19] Hura H 2015 Analisis keberadaan *Candida albicans* dan *Aspergillus spp*, serta keluhan kesehatan dan perilaku penjual tentang bahaya kesehatan pada pakaian bekas di pasar Melati kelurahan Tanjung Selamat kecamatan Medan Tuntungan, kota Medan tahun 2015 *Skripsi* Sumatra Utara University
[20] Amriani F 2013 Praperlakuan fisik dan biologi terhadap biomasa eceng gondok untuk produksi enzim selulase oleh *Aspergillus niger* dan *Tricoderma reseei Skripsi* Sumatra Utara University
[21] Sohail M, Sidiqi R, Ahmad A and Khan S 2009 Cellulase production from *Aspergillus niger* MS82 : effect of temperature and pH *New Biotech.* Vol **25(6)** p 437-42
[22] Raza F, Raza N, Hameed U and Mariam L 2011 Solid state fermentation for the production of B-glucosiade by co-culture of *Aspergillus niger* and *A.oryzae Pak. J. Bot* **vol 43(1)** p 75-83
[23] Purkan, Purnama H and Sumarsih S 2015 Produksi enzim selulase dari *Aspergillus niger* menggunakan sekam padi dan ampas tebu sebagai induser *J. Imu Dasar* **vol 16(2)** p 95-102

Ozone Disinfection of *Vibrio vulnificus* in Shrimp Pond Water

Wara Dyah Pita Rengga *[a], Echa Cahya Julyta Putri[a], Ria Wulansarie[a], Agus Suryanto[b]
[a]Chemical Engineering Department, Engineering Faculty, Universitas Negeri Semarang, Semarang, Indonesia, 50229
[b]Electrical Engineering Department, Universitas Negeri Semarang, Kampus Sekaran Gunungpati, 50229, Indonesia
E-mail address: wdpitar@mail.unnes.ac.id

Abstract. One variety of shrimp, L.Vanamei, often uses brackish water during the operation in the shrimp pond. Chlorination and ultraviolet are usually used for disinfection of brackish water. However, it is ineffective and forms sediment in the water distribution. It can be a negative impact on the water quality cause a contamination on the shrimp, so the farmers might have loss of profit because Vibrio vulnificus causes infection and dead on the shrimp. It affects the safety of consumers and should be minimized. The purpose of this study is to reduce the number of V. vulnificus bacteria in the pond water. The water was put in the storage tanks then pumped to filter out the impurities of the water. Furthermore, the water set the flow rate in 1 LPM, 2 LPM, and 3 LPM. After that, the ozone was injected to the water flow to sterilize the V. vulnificus bacteria. Finally, the water was returned to the original tank. The water from the tank was taken through a valve and analyzed in 0, 3, 7, 12, 18, 24, 30 minutes. The sample was analyzed immediately using a Total Plate Count method to determine the number of V. vulnificus bacteria in the shrimp pond water. The flow rate shows that the longer time of ozone made a lower amount of Vibrio v. bacteria. In 2 LPM water, it shows the optimum results of V. vulnificus. bacteria reduction for 88.1% compared to the flow rate of 1 LPM and 3 LPM with the bacteria reduction of 68,8% and 70.6%. This study shows that the ozone with a flow rate of 2 LPM circulation is the most effective method to help reducing the number of V. vulnificus in brackish water distribution system in the shrimp environment and potentially as a disinfectant.

Keywords: Flow rate; Ozone; Shrimp pond water; Vibrio vulnificus

1. Introduction

Indonesia is a country of *L. vannamei* shrimp exporter in the world. State competitors China, Vietnam, Thailand have been out of business in exporting shrimp *L. vannamei*. Main export destination countries of Indonesia *L. vannamei* is America and Japan. This causes the shrimp production was decreased till 50% of production and the impact on exports. The production of shrimp in Indonesia and provide greater foreign exchange for the country. Many aquatic organisms are cultured experienced the mass death causing high economic losses caused by being infected with pathogenic microorganisms [1]. Various factors that cause crop failures that occur in shrimp ponds in Indonesia and resulted in the production of shrimp needs to be followed up. Problems often occur shrimp farm is an attack infectious *Vibrio vulnificus*. Vibriosis disease caused by the bacterium *Vibrio sp* cause of death in the shrimp in a short time. A pathogenic of bacteria comes from families Vibrionaceae with curved rod shape such as a comma. These bacteria has the properties gram negative with the size of 2-3 μm [2]. The shrimp were infected with these bacteria generally provide some clinical symptoms that can be seen with the naked eye, namely: shrimp looks weak, dark red or pale, and on the antennae and legs of shrimp swimming in red.

Based on previous research, ozone has been demonstrated to exterminate a wide range of viruses, including Venezuelan equine encephalomyelitis, hepatitis, influenza, vesicular stomatitis, infectious bovine rhinotracheitis virus [3]. Ozone is used to treat the human pathogenic bacteria such as *Escherichia coli*, *Staphylococcus aureus*, *Pseudomonas fluorescens*, *Salmonella typ*hi, and *Klebsiella pneumoniae* by using ozonizer [4,5]. Ozone is also used to analyze the ability of the virus to defend on the fungus Alternaria solani conidia Sorauer [6]. The results obtained proved lethal activity of ozone and provide precise data on ozone concentration times exposure. The research has been give that ozone technology can be applied to areas that have a virus, bacteria, also fungi. Because of its high oxidation potential, ozone oxidizes cell components of the bacterial cell wall to reduce the number of microorganisms become inactive. This is because ozone technology disrupts the integrity of the bacterial cell envelope through oxidation of lipoproteins and phospholipids.

Several methods to reduce the number of bacteria *Vibrio sp* has been carried out, among others, is a filtration method that has been used to reduce the number of bacteria *Vibrio sp* [7] and obtained the resulting decline in the number of *Vibrio sp* to 103 CFU / mL. Disinfection of Escherichia coli bacteria in water using ozone technology has also been done and prove that the technology can dilaplikasikan ozone in water treatment processes to inactivate bacteria. Further research to increase shrimp production again can be done by reducing the effect of Infection Myo Necrosis Virus (IMNV) using ozone technology. Based on these studies used ozone from the ozone generating devices are commonly referred to as ozone generator (ozonator). The success of ozone technology is further adopted to reduce the number of bacteria *Vibrio sp*. The ability to reduce the number of both bacterial or viral microorganisms using ozone is quite good, so in this study will be used the same technology that ozone technology. Ozone technology can be used as a disinfectant to inactivate bacteria, fungi, and other microorganisms. This research is used to reduce *V. vulnificus* artificial shrimp ponds of water, because it will determine a decrease in the number of theses bacteria alone. The system used in this study is continuous. Artificial pond water is passed on ozone by varying the contact time.

The aim of research conducted with using a variation of flow rate as well as the pond water disinfection time progresses. Observations were made on the number of *V. vulnificus* before and after disinfection and analyzed using the methods Total Plate Count (TPC). It is expected ozone technology can be a solution to overcome the problems of crop failures in shrimp aquaculture caused by pathogenic bacteria is *V. vulnificus*. In this research, the number of bacteria Vibrio sp contained in shrimp pond water vibriosis lowered to reduce diseases that attack the shrimp so that it can increase the harvest of shrimp.

2. Methodology/Experimental

Ozone technology is applied for disinfection of pathogenic bacteria in shrimp pond water. The *Vibrio vulnificus* was provided by the Balai Besar Perikanan Budidaya Air Payau (BBPBAP) in Jepara Central Java. The bacterial isolate was inoculated into the nutrient broth using a sterilized loop and was allowed to grow for 24 hours at 37°C in an incubator.

The disinfection process is performed to determine the effectiveness of ozone technology used in the disinfection of *V. vulnificus* bacteria. In this study, water samples containing *V. vulnificus sp* will be analysed in amounts of before and after the disinfection process. The experimental setup consisted of 1 L gas washing bottle with a tap for sampling at different time intervals. Disinfection is done using ozone technology. The process used continuously until the number of *V. vulnificus sp* reached 1000 CFU/mL. The laboratory scale installation used to apply the ozone treatment is shown in Figure 1.

IOP Conf. Series: Materials Science and Engineering **316** (2018) 012067 doi:10.1088/1757-899X/316/1/012067

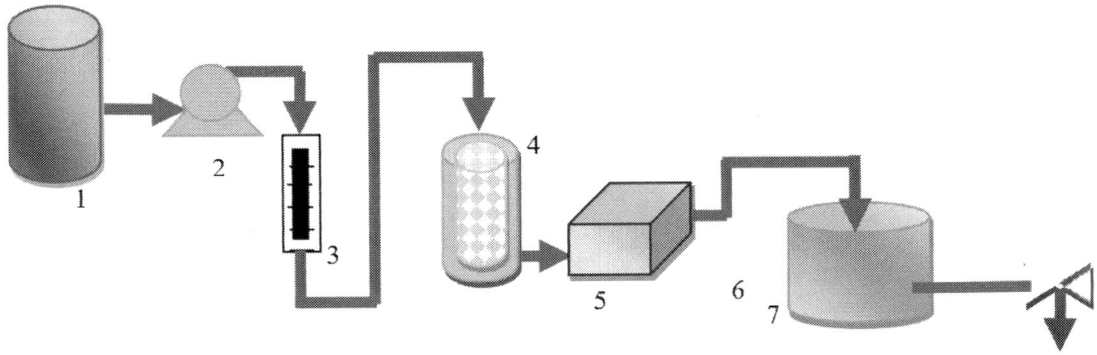

Figure 1 Research equipment system for *V. vulnificus*. Disinfection (1) Shrimp pond water tank, (2) pump, (3) flow meter, (4) micro filter, (5) ozonator, (6) sample valve (7) water tank after disinfection

In the research, oxygen from air is generated to ozone by ozonator (hanaco TSH-278) with flow around 0.0325 g/h. The tank is filled as much as 4 L of shrimp pond water. For the first step was take the initial sample before the disinfection process is carried out and inserted into the sample bottles that have been sterilized, then the ozone pass through to the pump was turned on at a total flow rate (ozone and water) of 1, 2, and 3 Liter/min (LPM) interchangeably. The flow rate of incoming oxygen stream ranged from 1 LPM to 3 LPM. The next step was turned on the stopwatch. Ozone output after disinfection process for taking sample after 3, 7, 18, 24, and 30 minutes and put in a sterile sample bottle.

Shrimp Pond water samples before and after (valve sample) the disinfection processes in sample bottle were tested using the Total Plate Count (TPC) method to determine the number of bacteria contained. Sterilization of tools used for TPC method and used Thiosulfate-Citrate-Bile Salts-Sucrose (TCBS) agar for the media. The decrease in the number of bacteria in the shrimp pond can be calculated by the percentage of bacterial amount after disinfection process to the number of early bacteria present in the shrimp pond.

3. Results and Discussion

The disinfection process are performed on samples of shrimp pond water. The disinfection process performed using ozone generated by the ozone generator with the time variation of 0, 3, 7, 12, 18, 24, and 30 minutes. In this research using of pH 5. This study also uses a variation of the flow rate of the pump output which amounted to 1, 2, and 3 LPM. Table 1 shows that the total vibrio bacteria on variation of flow rate for time. The shrimp pond water from the tank flows along with the ozone at each flow rate indicates that there is a decrease in the number of bacteria within 3 minutes of the initial bacterial count in the shrimp pond water tank is 3200 CFU/mL. The highest decline occurred in 7 minutes and after that the decline was lower than before. Within 30 minutes the decrease of bacterial count reaches 1000 CFU/mL for the flow rate of 1 LPM, 380 CFU/mL for flow rate of 2 LPM, and 940 CFU/mL for flow rate of 3 LPM.

Flow rate increased with the increase of flow rate from 1 LPM to 2 LPM but decreased at 3 LPM flow rate. This is due to the decrease in ozone used at the LPM flow rate 3 and the ozone residence time in the reactor. However at 1 LPM flow rate of high ozone level but the reaction damaging bacterial cell wall is still low. Although, dissolved ozone is more effective in reacting with bacteria, the ozone gas is less interactive because less time and much is wasted. Thus, at higher ozone concentrations, ozone may not participate proportionally in the inactivation process due to the constraints of the gas-ozone interaction with the bacteria.

Table 1 Decrease of *Vibrio sp* bacteria on variation of flow rate

Time (Minute)	Total Bacteria (CFU/mL) flow rate 1 LPM	flow rate 2 LPM	flow rate 3 LPM
0	3,2 x 10^3	3,2 x 10^3	3,2 x 10^3
3	3,1 x 10^3	3,0 x 10^3	3,1 x 10^3
7	2,0 x 10^3	8,7 x 10^2	1,9 x 10^3
12	1,4 x 10^3	8,5 x 10^2	1,2 x 10^3
18	1,2 x 10^3	6,7 x 10^2	1,1 x 10^3
24	1,1 x 10^3	6,6 x 10^2	1,0 x 10^3
30	1,0 x 10^3	3,8 x 10^2	9,4 x 10^2

Decrease of bacteria at flow rate 1, 2, and 3 LPM can be seen in Figure 2. Ability of ozone to disinfect water shrimp ozone pond in water at solute condition at pH = 7 indicates that at flow rate of 2 LPM, decrease of bacteria count reach 88,1% compared to 1 LPM reach 68,8% and for flow rate 3 LPM reach 70,6%. Ozone disinfectant power includes oxidation processes to kill bacterial protoplasm, membrane oxidation, and ozonolysis of unsaturated fatty acids which are external membranes. Low ozone concentrations and short contact times are sufficient to disinfect shrimp pond water. Extrapolation of fixed yields should also take into account the weather conditions preferring the adaptation and development of various strains of the most diverse genera. *Vibrio v.* causes red tailed disease in vanamei shrimp which is a severe acute and severe illness that has been detected as the cause of death problems because the shrimp are not resistant to attack from the vibrio bacteria. The dose administered at 2 LPM streams is 0.542 ppm ozone for 30 minutes can decrease the number of *vibrio v.* bacteria up to 88.1% according to the minimal minimum amount of *Vibrio v.* in shrimp ponds is 1000 CFU/mL.

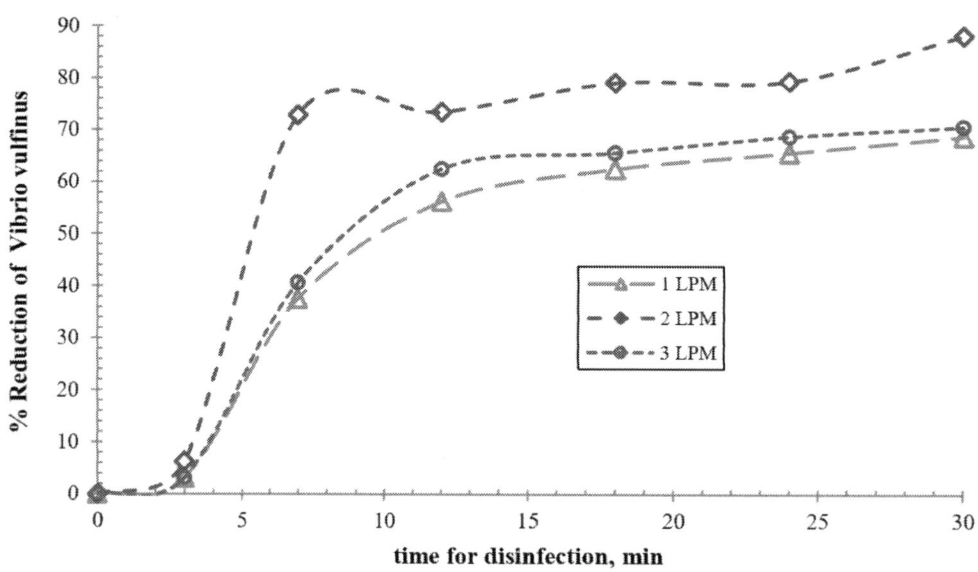

Figure 2. Reduction *Vibrio Vulnifucus* at variation flow rate

In Figure 3 shows the effect of ozone on the TPC results for the LPM flow rate of 2 LPM over a certain time interval. In the TPC test it appears that the number of bacteria still alive after the infectious process is evidenced by the presence of white round sign as a bacterium *Vibrio vulnificus*. At increasing time, the number of bacteria on the TPC is less and less. This is because the longer time the meeting between the bacteria and ozone during ozone is not turned into oxygen because the dissolved ozone can turn into oxygen back in 15 minutes [8].

The method for killing *Vibrio v.* bacteria can be done by determining the flow rate of the infected shrimp pond water flow appropriately. The mechanisms were predicted that the bacterial body was known to be protected by relatively dense cell membranes. If the ozone starts to come into contact with the cell wall as it flows with a certain flow rate it will disrupt the metabolism of bacterial cells. The cell wall is very important for bacteria because it is guaranteed the organism can maintain its shape. A sufficient amount of ozone permeates the cell membrane during this flow rate because the continuous system has a contact time between ozone and bacteria during the flow. This can cause bacterial damage. As the ozone molecule makes contact with the cell wall, the oxidative reaction that gives the structure of a small hole in the cell wall. At the same time there is inhibition and blocking the operation of the enzymatic control system. A newly created hole in the cell wall has injured bacteria and bacteria begin to lose its shape while the ozone molecules continue to make holes in the cell wall. If it has several times the ozone collision in a few seconds then the bacteria wall can no longer maintain its shape and eventually the bacteria cells die. The longer the contact time between the ozone produced then the bacteria will be damaged in the cell wall and then will die and found the result that the effective flow rate is 2 LPM. Chlorine is the decontamination agent is the most commonly used in the industry for frozen shrimp to kill potential pathogens. Chlorine completely eliminates this organism so that many dead organisms die. However, long contact with chlorine causes severe respiratory tract damage. There is research that serves to replace chlorine as a disinfectant natural materials chitosan. However, the ability of chitosan as a disinfectant is lower than shrimp pond treatment by using ozone, because with chitosan decrease the bacteria only reach 60% [9]. These results suggest the possibility of using ozone depat decontamination of pathogenic bacteria in seafood plants, a change that will reduce the health problems of workers.

E. coli bacteria have high sensitivity to ozone treatment and compare with other bacterial strains of *Salmonella typhi*, *Klebsiella pneumoniae*, and *Pseudomonas fluorescens*. The decrease in the number of *E.coli* bacteria reaches 88% for 15 minutes with ozone [5]. Direct count method indicates that after disinfection, the bacterial cell loses its ability to grow. This is almost the same as the bacterial sensitivity V. *vulfinicus* against ozone.

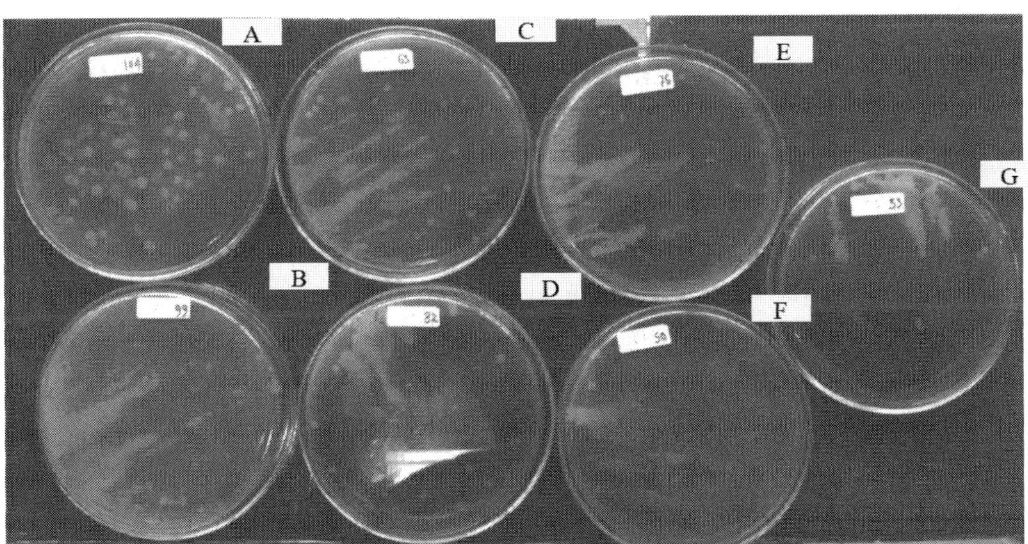

Figure 3. Effect of ozone concentration for *Vibrio vulnificus* population on a nutrient agar plate. [Ozone]=0,542 ppm; Flow rate=2 LPM; ozonisation duration=A: 0 minute, B: 3 minutes, C: 7 minutes,
D: 12 minutes E: 18 minutes, F: 24 minutes, G: 30 minutes.

4. Conclusion

Disinfection experiments were conducted with ozone treatment used shrimp pond water with *Vibrio vulnificus* bacteria. Decrease of a mount of bacteria is effective method for pathogenic bacteria of shrimp for flow rate of 2 LPM compared to 1 and 3 LPM.

5. Acknowledgments

Authors thankfully acknowledge for funding support received from the Chemical Engineering Department, Universitas Negeri Semarang, Ministry of Research, Technology and Higher Education of the Republic of Indonesia.

6. References

[1] Elvis, A.M. and Ekta, J.S. (2011). *Ozone therapy: A clinical review. Journal of Natural Science, Biology, and Medicine*, Volume 2, Number 1, pp. 66–70.
[2] Austin, B.D. and A. Austin. 1988. *Bacterial Fish Pathogen: Disease and Farmdes and Wild Fish.* Ellis Horwood, Chicester. pp. 364.
[3] Guzel-Seydim, Z.B., Greene A.K., and Seydim A.C. (2004). *Use of Ozone in the Food Industry*.Lebensm.-Wiss. u.-Technol. Technol. Volume 37: 453–460.
[4] Kowalski, W.J., Bahnfleth, W.P., Striebig, B.A., and Whittam, T.S.(2003). *Demonstration of a Hermetic Airborne Ozone Disinfection System'': Studies on E. coli*, AIHA Journal, Volume 64, pp. 222–227.
[5] Prabakaran M., Tamil Selvi S., Merinal, S., and Panneerselvam A. (2012). *Effect of ozonation on pathogenic bacteria. Advances in Applied Science Research*, Volume 3, Number 1, pp. 299-302.
[6] He, H., Zheng, L., Li, Y., Song, W. (2015). *Research on the Feasibility of Spraying Micro/Nano Bubble Ozonated Water for Airborne Disease Prevention.* Ozone: Science & Engineering, 37, pp. 78–84.
[7] Huq, B. Xu, M. A. R. Chowdhury, M. S. Islam, R. Montilla and R. R. Colwell. 1996. *A Simple Filtration Method To Remove Plankton-Associated Vibrio cholerae in Raw Water Supplies in Developing Countries.*
[8] Viera, M.R., Guiamet, P.S., de Mele, M.F.L., Videla, H.A. 1999. *Use of dissolved ozone for controlling planktonic and sessile bacteria in industrial cooling systems.* International Biodeterioration and Biodegradation Vol 44, Issue 4, pg.201-207.
[9] Chaiyakosa, S., Charernjiratragul, W., Umsakul, K, Vuddhaku, V. (2007) *Comparing the efficiency of chitosan with chlorine for reducing Vibrio parahaemolyticus in shrimp*, Food Control, volume 18, number 9, pp.1031–1035.

IOP Conf. Series: Materials Science and Engineering 316 (2018) 012068 doi:10.1088/1757-899X/316/1/012068

15th International Conference on Quality in Research (QiR 2017)

Ozone Technology for Pathogenic Bacteria of Shrimp (Vibrio sp.) Disinfection

Ria Wulansarie[a]*, Wara Dyah Pita Rengga[a], Rustamadji[a]

[a]*Chemical Engineering Department, Engineering Faculty, Universitas Negeri Semarang, Semarang, Indonesia, 50229*

Abstract

One of important marine commodities in Indonesia, shrimps are susceptible with Vibrio sp bacteria infection. That infection must be cleared. One of the technologies for disinfecting Vibrio sp. is ozone technology. In this research, Vibrio sp. is a pathogenic bacterium which infects Penaeus vannamei. Ozone technology is applied for threatening Vibrio sp. In this research, ozonation was performed in different pH. Those are neutral, acid (pH=4), and base (pH=9). The sample was water from shrimp embankment from Balai Besar Perikanan Budidaya Air Payau (BBPBAP) located in Jepara. That water was the habitat of Penaeus vannamei shrimp. The brand of ozonator used in this research was "AQUATIC". The used ozonator in this research had 0,0325 g/hour concentration. The flow rate of sample used in this research was 2 L/minute. The ozonation process was performed in continuous system. A tank, pipe, pump, which was connected with microfilter, flowmeter and ozone generator were the main tools in this research. It used flowmeter and valve to set the flow rate scalable as desired. The first step was the insert of 5 L sample into the receptacle. Then, by using a pump, a sample supplied to the microfilter to be filtered and passed into the flow meter. The flow rate was set to 2 LPM. Furthermore, gas from ozonator passed to the flow for the disinfection of bacteria and then was recycled to the tank and the process run continuously. Samples of the results of ozonation were taken periodically from time 0, 3, 7, 12, 18, 24 to 30 minutes. The samples of the research were analyzed using Total Plate Count (TPC) test in BBPBAP Jepara to determine the number of Vibrio sp. bacteria. The result of this research was the optimal condition for pathogenic bacteria of shrimp (Vibrio sp.) ozonation was in neutral condition.

Keywords: pH, ozone, vibrio sp, shrimp, disinfection, shrimp embankment, pathogenicic bacteria;

1. INTRODUCTION

Indonesia is a maritime country which has abundant marine commodities. One of the important export commodities in Indonesia is shrimp. There are two species of shrimp which are cultivated in Indonesia; they are Penaeus vannamei and Penaeus monodon (the popular name in Indonesia is Windu shrimp). Some regions sizable shrimp producer in Indonesia are Lampung, Maluku, Jepara, and Kendal. Those shrimps are susceptible with bacteria infection. Bacteria that attack is Vibrio sp, and viruse that attack shrimp is Myo. The cause of shrimp diseases discussed in this study is the Vibrio sp bacteria. The bacteria are pathogenic bacteria, comes from the family Vibrionaceae with curved rod shape such as a comma. The bacteria are gram-negative has the properties to the size of 2-3 μm [1]. The shrimp were infected with these bacteria generally can be identified by some of the symptoms that can be seen with the naked eye, namely: shrimp looks weak, dark red or pale, and on the antennae and legs of shrimp swimming in red. Pathogenesis of vibriosis, bacteria enter through the blood and circulation of tissue causes damage and inflammation of the blood vessels, skin, and base of the fins and followed by infection of the heart [2].

In the shrimp farm in the Sidorukun Village, Kendal found the Vibrio sp bacteria of 3.5 x 10^2 CFU/mL in 2014 (Report on Laboratory Test Results BBPBAP Jepara 2014), and increased in 2016 with the number of bacteria 7.8 x 10^3 CFU/mL (Report of Laboratory Test Results BBPBAP Jepara in 2016) caused by embankment water contaminated with Vibrio sp bacteria. The maximum of Vibrio sp. in embankment water is 1x10^2 CFU/mL [3]. The Vibrio sp. bacteria arises because water quality in embankments that are not good.

The presence of bacteria and viruses on shrimp make shrimp die and will have an impact on the harvest of shrimp. That causes Indonesia's per capita income will be reduced and will reduce the level of the Indonesian economy. Hence the need for a technology to disinfect shrimp embankments water systems

* Corresponding author.
E-mail address: ria.wulansarie@mail.unnes.ac.id

containing bacteria and viruses so that the water quality of embankment where shrimp live to be more qualified. This causes the shrimp survival and better quality so that more abundant harvests shrimp and the Indonesian economy will increase. In this research, microorganism that cause disease in shrimp is Vibrio sp. bacteria.

Several methods to reduce the number of Vibrio sp bacteria has been carried out, among others, is a filtration method that has been used to reduce the number of Vibrio sp bacteria [4] and obtained the resulting decline in the number of Vibrio sp from 10^4 to 10^3 CFU/mL. Disinfection of Escherichia coli bacteria in water using ozone technology has also been done by Wulansarie [5] and prove that the technology can be applied in water treatment processes to inactivate bacteria by 10^6 to 10^1 CFU/mL. Additionally, Mukti [6] has also been conducting research to increase shrimp production return 100% of the previous began to fall due to virus Infection Myo Necrosis Virus (IMNV) technology using ozone. Based on background of study, ozone technology will be applied to reduce the number of bacteria Vibrio sp. Ozone technology acts as a disinfectant to inactivate bacteria, fungi, and other microorganisms. The sample was water from shrimp embankment from Balai Besar Perikanan Budidaya Air Payau (BBPBAP) located in Jepara.. The system used in this study is continuous. The study was conducted using a variety of disinfection and pH conditions of embankment water. Observations were made on the number of Vibrio sp before and after disinfection and analyzed using the methods Total Plate Count (TPC), which was conducted in the BBPBAP Laboratory, Jepara. It is expected ozone technology can be a solution to overcome the problems of crop failures in shrimp aquaculture caused by pathogenic bacteria is Vibrio sp.

2. METHODOLOGY/ EXPERIMENTAL

Ozone technology is applied for threatening pathogenic bacteria of shrimp, Vibrio sp. In this research, ozonation was performed in different pH. Those pH are neutral (water shrimp embankment without treatment), acid (pH=4), and base (pH=9). The sample was water from shrimp embankment from Balai Besar Perikanan Budidaya Air Payau (BBPBAP) located in Jepara. That water was the habitat of Penaeus vannamei shrimp. The brand of ozonator used in this research was "AQUATIC". The used ozonator in this research had 0,0325 g/hour concentration. The flow rate of sample used in this research was 2 L/minute. The ozonation process was performed in continuous system. A tank, pipe, pump, which was connected with microfilter, flowmeter and Ozone generator were the main tools in this research. Microfilter containing membrane 10 micron used to filter pollutant. This research used a form of variable pH of sample. It used flowmeter and valve to set the flow rate scalable as desired. The first step was the insert of 5 L sample into the receptacle. Then, by using a pump, a sample supplied to the microfilter to be filtered and passed into the flow meter. The flow rate was set to 2 LPM. Furthermore, gas from ozonator passed to the flow for the disinfection of bacteria and then was recycled to the tank and the process run continuously. Samples of the results of ozonation were taken periodically from time 0, 3, 7, 12, 18, 24 to 30 minutes. In this case the samples tested at acidic and alkaline conditions, to determine the optimal pH of embankment water for disinfection using ozone. Condition of pH was 4 for acidic and 9 for alkaline. The research equipment system can be seen in Picture 1.

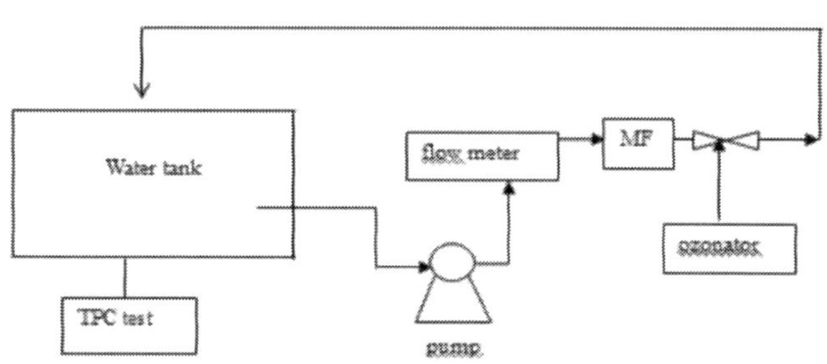

Picture 1. Equipment system for Vibrio sp. disinfection

After disinfected by ozone, the sample will be analyzed. The samples of the research were analyzed using Totat Plate Count (TPC) test in BBPBAP Jepara to determine the number of Vibrio sp. bacteria in the samples.

3. RESULTS AND DISCUSSION

The result of this research can be seen in Picture 2. From the picture we can see the effect of ozone for Vibrio sp.

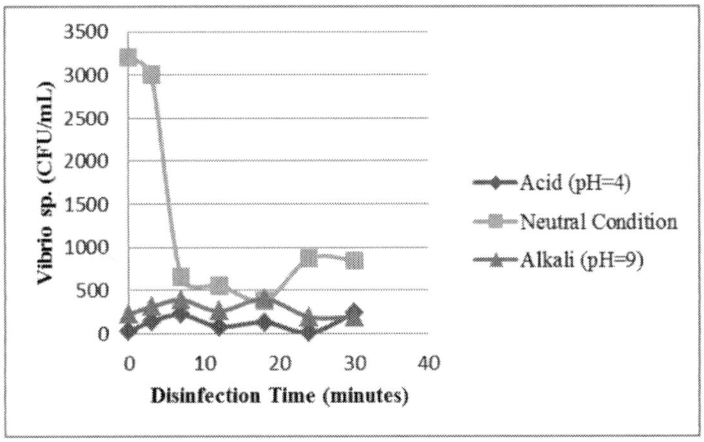

Picture 2. The result of disinfection Vibrio sp. using ozone

From Picture 2 we can see the optimum degradation of Vibrio sp. bacteria. That condition was in neutral condition, the condition that there wasn't no treatment in the samples. That neutral condition has pH=6,7. In that condition degradation of Vibrio sp. reach until 73.4%. The number of Vibrio sp. in final disinfection process was 850 CFU/mL. That number of Vibrio sp. still exceed the maximum limit [3]. According Peratitus [7] there are three ozonation; direct ozonation (in acid condition/pH<4), ozonation by ozone and OH radical (in pH= 4 – 9), and ozonation by OH radical (p>9). According Manley and Niegowski [8], ozonation by OH radical is more powerfull rather than ozonation by ozone.

Table 1. Reduction and Oxidation Potential [8]

Number	Active species	Redox Potential (mV)
1	Fluorin	+3,06
2	Radikal hidroksil ($OH^\bullet$)	+2,80
3	Ozon (O_3)	+2,07
4	Hidrogen peroksida (H_2O_2)	+1,77
5	Permanganat	+1,67
6	Klor dioksida	+1,50
7	Asam Hipoklorit	+1,49
8	Gas Klorin	+1,36

The reduction and oxidation potential of ozone and OH radical can be seen in Table 1. So from that table we can see that the ozonation process will be better in neutral condition (pH= 4-9) because OH radicals and ozone had synergy for degradation of Vibrio sp. bacteria. After disinfection the number of bacteria was decrease so that it is suitable with Dahuri [9] "water quality is one of the keys to success in shrimp farming". The research result also suitable with Graham [10] that say "ozone, UV rays, microfiltration is effective for removal of E. coli bacteria and all coliforms. Ozone can be produced by corona discharge [11]. Ozone can act as a disinfectant against pathogens, reducing taste and odor and the ability to oxidize compounds [12] which suitable with the research.

4. CONCLUSION

From this research can be conclude that the optimal condition for pathogenic bacteria of shrimp (Vibrio sp.) ozonation was in neutral condition of samples (pH=6,7).

5. ACKNOWLEDGEMENT

The authors thank to BBPBAP Laboratory, Jepara for their support and their contribution in our research.

6. REFERENCE

[1] Austin, B.D. and A. Austin. 1988. *Bacterial Fish Pathogen: Disease and Farmdes and Wild Fish.* Ellis Horwood, Chicester. 364 p.

[2] Irianto, A. 2005. Patologi Ikan Teleostei Gadjah Mada University Press. Yogyakarta.

[3] Bintari, N.W.D., Kawuri, R., Dalem, A.A.G. 2016. *Identifikasi Bakteri Vibrio Penyebab Vibriosis pada Larva Udang Galah (Macrobrachium rosenbergii (de Man))*. Jurnal Biologi 20 [2]: 53 – 63.

[4] Huq, A., and R. R, Cholwee. *Vibrios in the marine and estuarine environment: tracking Vibrio cholerae.* J. Ecosyst. Health, in press.

[5] Wulansarie, R. 2015. *Sinergy of Ozone Technology and UV rays in the drinking water supply as a Breakthrough Prevention of Diarrhea Diseases in Indonesia.* Waste Technology, Vol 3 (2), 55-57.

[6] Mukti, I. 2012. *Aplikasi Teknologi Oksidasi Lanjut (Ozon dan UV) Untuk Disinfeksi Infectious Myo Necrosis Virus (IMNV) Pada Tambak Udang di Kabupaten Pesawaran,Provinsi Lampung.*

[7] Peratitus (ed.). 2003. *Ozone Reaction Kinetics for Water and Wastewater System*, London: A CRC-Press.

[8] Manley, T. C. and Niegowski, S. J. 1967. *Ozone.* Encyclopedia of Chemical Technology, Vol. 14, 2nd ed., pp. 410–432. New York, US: Wiley.

[9] Dahuri, R., J. Rais., S.P. Ginting dan M.J. Sitepu. 2004. *Pengelolaan Sumberdaya Wilayah Pesisir dan Lautan Secara Terpadu*. PT. Pradnya Paramita, Jakarta : 220.

[10] Graham, P. P. N. 2005. *"Treatment of a secondary municipal effluent by ozone, UV and microfiltration: microbial reduction and effect on effluent quality."* Journal of Desalination 186 47-56.

[11] Alsheyaba, A. H. M, Mohammad A.T.. 2007. *"Optimisation of ozone production for water and wastewater treatment."* Journal of Desalination 217 1–7.

[12] Suslow, T. V. 2004. *Ozone Applications for Postharest Disinfection of Edible Horticultural Crops, ANR Publication.*

IOP Conf. Series: Materials Science and Engineering **316** (2018) 012069 doi:10.1088/1757-899X/316/1/012069

Influence of Nutrient Impregnated into Zeolite Addition on Anaerobic Digestion of Palm Oil Mill Effluent (POME)

M Mellyanawaty[1], F M A Chusna[2], H Sudibyo[2], N Nurjanah[1], W Budhijanto[2,*]
[1] Department of Environmental Engineering, Faculty of Engineering, Universitas Muhammadiyah Tasikmalaya, Tasikmalaya, 46196 Indonesia
[2] Department of Chemical Engineering, Faculty of Engineering, Universitas Gadjah Mada, Yogyakarta 55281 Indonesia

[*]Corresponding author: wiratni@ugm.ac.id

Abstract. Palm oil mill effluent (POME) was wastewater generated from palm oil milling activities which was brownish liquid, acidic with pH 3-4, and contained soluble materials which were hazardous to the environment. It was characterized by high organic loading (COD 40,000–60,000 mg/L). According to its characteristics, POME was identified as a potential source to generate renewable energy through anaerobic digestion. In other words, a combination of wastewater treatment and renewable energy production would be an additional advantage to the palm oil industries. Methanogenesis was the rate limiting step in anaerobic digestion. In the conventional anaerobic digester, it required large reactors and long retention time. The addition of microbial immobilization media was to improve anaerobic reactor performance in term of higher organic removal and methane production. Additionally, better performance could lead to reduction of reactor volume and shorter retention time in high rate anaerobic digester. The loading of essential microorganism nutrient into the media might increase the affinity of bacteria to attach and grow on the media surface.
Activating or inhibition effects of natural and modified zeolite addition in anaerobic digestion of POME was studied in batch reactors using erlenmeyer of 1,000 mL at COD concentrations of about 8,000 mg/L. Zeolite was impregnated with nickel and magnesium at concentrations of 0.0561 mg Ni/g zeolite and 0.0108 mg Mg/g zeolite. The effect of the different zeolite addition was determined by the measurement of soluble COD (sCOD), Volatile Fatty Acids (VFAs) and biogas production. Greater effect of modified zeolite was observed in zeolite impregnated with nickel with a 54% increase of biogas production. Meanwhile, the modified zeolite impregnated with magnesium had no positive impact to the methanogenic bacteria activities.

1. Introduction

Indonesia is one of the largest palm-oil-producing countries in the world. Domestic production of palm oil had increased in the last few years. Thereby, the palm oil waste which was the by-product of palm milling process also increased. Huge quantities of waste were generated from palm oil industry. One of which was palm oil mill effluent (POME). The fresh POME is a hot, acidic (with pH between 1–5), and brownish colloidal suspension with a high amount of total solids (40,500 mg/L). It also contained high organic material with the level of Chemical Oxygen Demand (COD) was about 50,000 mg/L and Biochemical Oxygen Demand (BOD) was about 25,000 mg/L [1]. As the consequence, the generation of POME had created environmental issue for the palm oil industry. According to its high organic loading, POME could become a promising source for biogas production and potentially boosted up the renewable energy sector.

One preferred process to utilize POME as the energy source was anaerobic digestion. Anaerobic digestion has considerable advantages during the process such as minimum sludge production, less energy demand, no unpleasant odor and low microorganism nutrient requirements [2]. It implied that the level of COD and BOD in POME was also reduced as the biogas was produced. During the process of converting POME into methane, carbon dioxide, and water, there were three steps comprising of hydrolysis, acidogenesis (including acetogenesis) and methanogenesis [3]. Hydrolysis as the first step was the reaction where complex organic matters (i.e. carbohydrates, proteins, lipids) were degraded into sugar, amino acid, and fatty acids. In the second step, acidogenic bacteria converted these sugar, fatty acids, and amino acids into organic acids which mainly consisted of acetic acid. The product from acetogenesis step would be converted by methanogenic bacteria into methane and carbon dioxide as final products.

In anaerobic digestion process, one common problem was the slow growth of methanogenic bacteria which made the methanogenesis be the rate limiting step. Accordingly, in the conventional anaerobic digesters, it required longer retention time and followed by larger reactor dimension [4]. Nevertheless, several methods were developed to optimize the anaerobic digestion process by preventing the loss of bacterial biomass by using a support, a material to which the microbes attached [5]. Natural zeolite was selected as the immobilization media because it had large surface area and contained essential mineral for the anaerobic microbe's growth such as potassium, iron, etc. [6]. In this study, the impregnated minerals were nickel and magnesium. They would be impregnated into zeolite as immobilization medium. Hopefully, the impregnation could enhance and stimulate the microorganism activities. The aim of this study is to investigate the effect of the modified zeolite to the anaerobic digestion of POME using the effluent of biodiesel digester as inoculum. This study was preliminary research to evaluate the effect of mineral/nutrient addition into zeolite as immobilization media.

2. Methodology/experimental

2.1. POME and Inoculum Characterization

The palm oil mill effluent (POME) as substrate was collected from PT. Perkebunan Nusantara VII, Lampung. The table 1 showed the characteristics of POME used in this study. The inoculum used in this anaerobic process was digested biodiesel waste produced by the biodiesel industry operating in East Java with the COD of 1,400 mg/L, the VFA concentration of 300 mg/L, and pH of 8.17.

Table 1. Characteristics of Lampung Palm Oil Mill Effluent (POME)

Parameter	Value	Unit
pH	4.09	-
Chemical Oxygen Demand	10,000 – 16,000	mg/L
Protein	0.15	%
Nitrogen	0.0295	%
Fe	0.258	mg/L
Phosphate	37.175	mg/L
Phenol	≤ 0.0001	mg/L
Oil and grease	115	mg/L
Potassium	1,459.86	mg/L
Sulphate	1,032.93	mg/L
Ammonia	125	mg/L

2.2. Modified and Natural Zeolite

The natural zeolite with particles size of 1–2 mm was obtained from Tasikmalaya, West Java, Indonesia. Nickel and magnesium cations were impregnated to the natural zeolite surface. Zeolite was homo-ionized with NaCl 1 N for 24 hours before washed with aquadest and was soaked with NiSO4 and $MgCl_2$ solution for 24 hours. The initial concentrations of $NiSO_4$ and $MgCl_2$ were 200 mg/L for each. Thereafter, the modified zeolite was used as the immobilization media in anaerobic reactors.

The nickel and magnesium level inside the zeolite was determined by mass balance from the Ni^{2+} and Mg^{2+} level reduction of the solution before and after impregnation using ICP method (Perkin Elmer Optima 8300). The calculated Ni^{2+} and Mg^{2+} concentrations inside the solid were 0.0561 mg Ni/g zeolite and 0.0108 mg Mg/g zeolite respectively.

2.3. Anaerobic Batch Reactor

One liter Erlenmeyer flasks were modified to be used as batch anaerobic reactors equipped with a rubber cap with a glass pipe for measuring gas pressure and a drain pipe for liquid sampling port in the bottom part. It was connected through gasometer for gas measurement (figure 1). The gasometer was filled with acidic salt water to reduce biogas solubility [7]. A mixture of diluted POME (sCOD 8,000 mg/L) and inoculum with volumetric ratio of 3:1 was added. The anaerobic digestion process was conducted for 30 days.

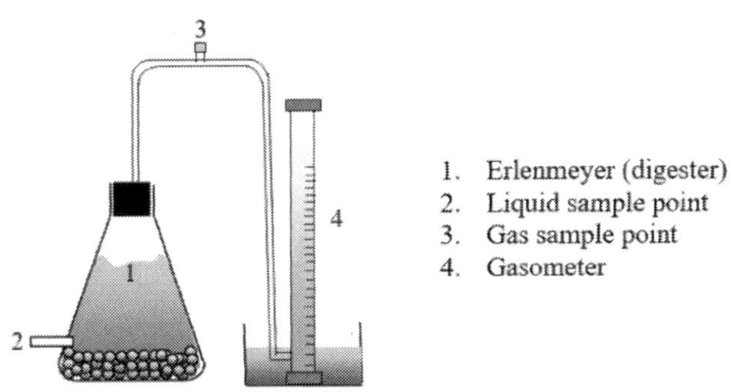

Figure 1. Anaerobic Digestion Batch Reactor

2.4 Analytical Methods

The volume of biogas was calculated following the gasometer equation from study by Walker et al [7]. The sCOD was used as a measure of the oxygen equivalent of the organic content inherently of a sample that is susceptible to oxidation by a strong chemical oxidant. In this study, the closed reflux method of COD analysis and VFA analysis method used procedures explained by APHA [8]. The procedures for COD analysis consisted of equipment preparation, sample and reagent preparation and reading method. For equipment preparation, the COD analysis required the reflux instrument which consisted of Digital Reactor Block heater HI 839800 and Hach DR 900 colorimeter. The pH was measured using Lutron pH meter PH-208. Meanwhile, the biogas composition was determined by gas chromatography Shimadzu GC 8A, Japan.

2.5 Batch Anaerobic Digestion

Three experimental sets of batch reactors were conducted in anaerobic digestion in which each contained different immobilization medium. The label of each reactor following the sCOD concentration was used and the nutrient was impregnated into the zeolite. The table 2 showed the label of the entire reactors in this experiment.

Table 2. The Label of the Anaerobic Digester (Reactor)

Reactor No.	$sCOD_{initial}$ (mg/L)	Cation	Type of immobilization media
1	8,000	-	8Z (Control)
2	8,000	Ni	8Z-Ni
3	8,000	Mg	8Z-Mg

3. Results and discussion

3.1 Results

The figure 2 shows the results obtained by the sCOD measurements in the process of anaerobic digestion of POME using zeolite as immobilization media. The significant sCOD removal was occurred from day 0 to day 8. At day 8 onwards, the sCOD concentration levels decreased slightly. The sCOD removal efficiency for the reactor 8Z, 8Z-Ni, and 8Z-Mg were 43.69%, 53.40%, and 60.52% respectively. The high sCOD efficiency removals indicated that the complex organic matter was decomposed and converted by microorganisms into methane and carbon dioxide as the final product of anaerobic digestion. Acidogenic microorganisms degrade the organic materials only until the production of VFAs which were also organic matter. Generally, the addition of nutrient which was used in this study i.e: Ni^{2+} and Mg^{2+}, provided the sCOD removal higher than control.

The effect of the addition of nickel and magnesium ion into natural zeolite as immobilization medium over the production of VFAs during the anaerobic process could be seen in figure 3. The production of VFAs significantly occurred in first 4 days. At day 4, the concentration of VFAs were between 3,400–4,260 mg/L. These values significantly increased from 2,143 mg/L. Reactor 8Z-Ni produced VFAs higher than 8Z indicating that nickel in the zeolite could stimulate the acidogenic microorganisms activities. However, the reactor 8Z-Mg produced VFAs lower than 8Z indicating that the activity was not improved after magnesium ion was added into the zeolite pores.

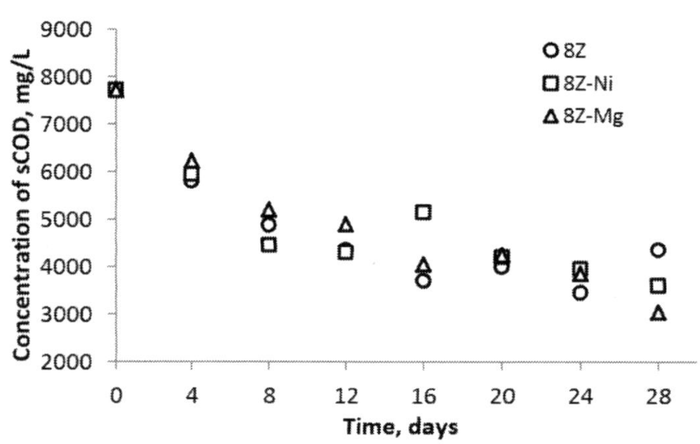

Figure 2. Concentration of sCOD during the Anaerobic Digestion of POME Using Immobilization Medium

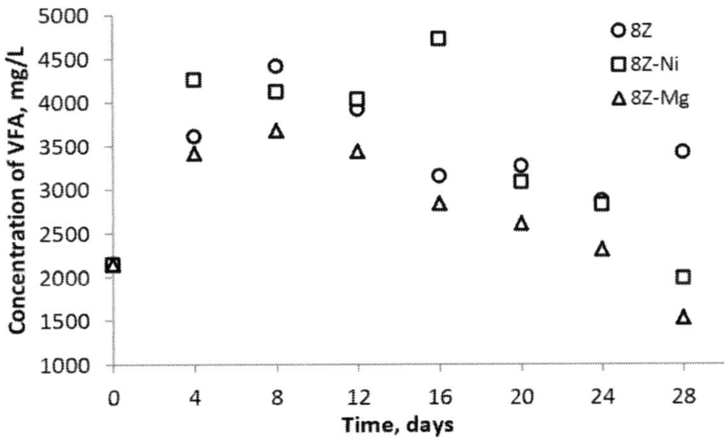

Figure 3. Production of VFAs during the Anaerobic Digestion of POME using Immobilization Medium

The biogas production was evaluated by measuring the total biogas and methane production. The anaerobic digestion batch trials with nickel in zeolite provided the highest biogas production as shown in figure. 4. However, reactor 8Z-Mg produced biogas lower than control. During the process, reactor 8Z, 8Z-Ni, and 8Z-Mg obtained the cumulative biogas production as many as 550.05 mL, 850.07 mL, and 256.74 mL respectively.

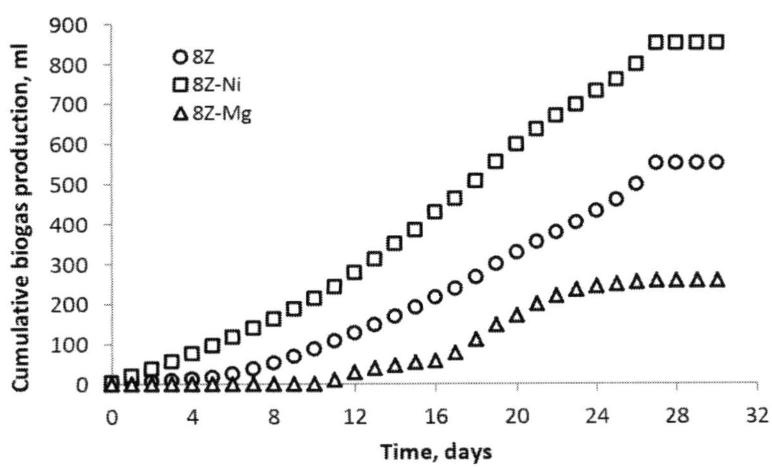

Figure 4. Cumulative Biogas during the Anaerobic Digestion of POME using Immobilization Medium

3.2 Discussion

The presence of nickel in the zeolite, in this study, enhanced the production of biogas as large as 54%. The presence of nickel in anaerobic digestion processes could increase acetate utilization rate of methanogenic bacteria. Similarly, Milán et al. [9] observed the modified zeolite with nickel increase the methanogenic specific activity by 2.8 times in anaerobic digestion of piggery waste. Nickel was a unique micronutrient for methanogenic bacteria. It stimulated the conversion of acetic acid into methane and carbon dioxide by a mixed methanogenic population [3]. The addition of material with ion exchange capacity (e.g. bentonite, phosphorite, and zeolites) or inorganic absorbent material (e.g. clay, manganese oxides) to anaerobic digester had shown good results [10–13].

Meanwhile, the presence of magnesium into the zeolite decreased the production of biogas by 53%. Similarly, Uludag-Demirer et al. [12] had investigated the effect of $MgCl_2$, $Mg(OH)_2$, and Na_2HPO_4. The result showed no improvement in the methane production yield of the anaerobic digester. The physical phenomenon observed during and after the anaerobic digestion was the formation of precipitate. The precipitate was ammonium magnesium phosphate ($MgNH_4PO_4.6H_2O$) or struvite. Study by Miles and Ellis [14] explained that struvite was a mineral which often precipitated from wastewater during anaerobic digestion when ammonium, phosphate, and magnesium ions were released. In high concentrations of dissolved magnesium, orthophosphates, and ammonia ions, there was strong potential for formation of struvite in the biological treatment of organic wastes. A high magnesium concentration was found in potential substrates for anaerobic (co-)digestion olive oil mill wastewater. In high concentration, magnesium became inhibitory on the anaerobic digestion process [15].

Research by Romero-Guiza et al [15] and Perez Rodriguez et al [16] reported that the bentonite support impregnated with Mg^{2+} caused struvite crystal appeared whose size remains similar with time. This attributed to the fact that the precipitates were formed from magnesium exchange ions transferred to the solution which took place over a very short period of time. It could be interpreted that the presence of Mg^{2+} on the zeolite as immobilization medium in anaerobic digestion of POME did not affect the biogas production. It likely promoted the struvite precipitation. Further study related to the struvite formation during and after anaerobic digestion of POME could be exciting.

4.Conclusion

A major stimulation to microorganism in anaerobic reactor was observed after the addition of immobilization medium. The impregnation of nickel ion into natural zeolite improved the anaerobic reactor performance. On the other hand, magnesium-ion-impregnated natural zeolite only slightly influenced the AFBR performance. To sum up, the application of nickel-ion-impregnated natural zeolite opened the possibility to allow AFBR to perform as efficient as possible and might lead to the considerable reduction of volume reactor for the future design of AFBR.

References

[1] Ayu E D, Halim L, Mellyanawaty M, Sudibyo H and Budhijanto W 2017 The effect of natural zeolite as microbial immobilization media in anaerobic digestion at various concentrations of palm oil mill effluent (POME) *AIP Conf. Proc.* **1840**

[2] Prasetyo E, Sudibyo H and Budhijanto W 2017 Determination of the Optimum Hydraulic Retention Time in Two Stages Anaerobic Fluidized Bed Bioreactor for Landfill Leachate Treatment **49** 1–11

[3] Gerardi M H 2003 *The Microbiology of an Anaerobic Digesters* ed M H Gerardi (Canada: John Willey $ Sons, Inc.)

[4] Sudibyo H, Guntama D and Budhijanto W 2017 Assessment of Start-Up Mechanisms for Anaerobic Fluidized Bed Reactor in Series Based on Mathematical Simulation **110007** 1–10

[5] Bucke C and Brown D E 2014 Immobilized Cells *Philos. Trans. R. Soc. Lond. B. Biol. Sci.* **300** 369–89

[6] Wirawan S ., Sudibyo H, Setiaji M ., Warmada I . and Wahyuni E . 2015 Development of natural zeolite adsorbent: chemical analysis and preliminary TPD adsorption study *J. Eng. Sci. Technol.* 87–95

[7] Walker M, Zhang Y, Heaven S and Banks C 2009 Potential errors in the quantitative evaluation

of biogas production in anaerobic digestion processes *Bioresour. Technol.* **100** 6339–46

[8] APHA 1998 *Standard Methods for the Examination of Water and Wastewater, 20th Edition* (APHA American Public Health Association)

[9] Milán Z, Villa P, Sánchez E, Montalvo S, Borja R, Ilangovan K and Briones R 2003 Effect of natural and modified zeolite addition on anaerobic digestion of piggery waste. *Water Sci. Technol.* **48** 263–9

[10] Montalvo S, Díaz F, Guerrero L, Sánchez E and Borja R 2005 Effect of particle size and doses of zeolite addition on anaerobic digestion processes of synthetic and piggery wastes *Process Biochem.* **40** 1475–81

[11] Tada C, Yang Y, Hanaoka T, Sonoda A, Ooi K and Sawayama S 2005 Effect of natural zeolite on methane production for anaerobic digestion of ammonium rich organic sludge. *Bioresour. Technol.* **96** 459–64

[12] Uludag-Demirer S, Demirer G N, Frear C and Chen S 2008 Anaerobic digestion of dairy manure with enhanced ammonia removal *J. Environ. Manage.* **86** 193–200

[13] Halim L, Mellyanawaty M, Cahyono R B, Sudibyo H and Budhijanto W 2017 Anaerobic digestion of palm oil mill effluent with lampung natural zeolite as microbe immobilization medium and digested cow manure as starter *AIP Conference Proceedings*

[14] Miles A and Ellis T G 2001 STRUVITE PRECIPITATION POTENTIAL FOR NUTRIENT RECOVERY FROM ANAEROBICALLY TREATED WASTES *Water Sci. Technol. a J. Int. Assoc. Water Pollut. Res. Tech* **43** 259–66

[15] Romero-Güiza M S, Mata-Alvarez J, Chimenos J M and Astals S 2016 The effect of magnesium as activator and inhibitor of anaerobic digestion *Waste Manag.* **56** 137–42

[16] Pérez Rodríguez J L, Carretero M I and Maqueda C 1989 Behaviour of sepiolite, vermiculite and montmorillonite as supports in anaerobic digesters *Appl. Clay Sci.* **4** 69–82

Acknowledgement

This work was funded by Penelitian Kerja Sama Antar Perguruan Tinggi (PEKERTI) research grant from Ministry of Research, Technology and Higher Education of the Republic of Indonesia (KEMENRISTEKDIKTI) 2017.

Abbreviations

POME	Palm Oil Mill Effluent
AFBR	Anaerobic Fluidized Bed Reactor
sCOD	Soluble Chemical Oxygen Demand
VFA	Volatile Fatty Acid

IOP Conf. Series: Materials Science and Engineering **316** (2018) 012070 doi:10.1088/1757-899X/316/1/012070

CHARACTERIZATION OF MODIFIED ZEOLITE AS MICROBIAL IMMOBILIZATION MEDIA ON POME ANAEROBIC DIGESTION

Rochim B Cahyono[a,*], Sri Ismiyati[b], Simparmin Br Ginting[b], Melly Mellyanawaty[c], Wiratni Budhijanto[a]
[a]*Department of Chemical Engineering, Faculty of Engineering, Universitas Gadjah Mada, Yogyakarta, Indonesia*
[b]*Department of Chemical Engineering, Faculty of Engineering, Lampung University, Bandar Lampung, Indonesia.*
[c] *Department of Environmental Engineering, Faculty of Engineering, Universitas Muhammadiyah Tasikmalaya, Tasikmalaya, Indonesia*
**Corresponding Author's: rochimbakti@ugm.ac.id*

ABSTRACT

As the world's biggest palm oil producer, Indonesia generates also huge amount of Palm Oil Mill Effluent (POME) wastewater and causes serious problem in environment. In conventional method, POME was converted into biogas using lagoon system which required extensive land area. Anaerobic Fluidized Bed Reactor (AFBR) proposes more effective biogas producing with smaller land area. In the proposed system, a immobilization media would be main factor for enhancing productivity. This research studied on characterization of Lampung natural zeolite as immobilization media in the AFBR system for POME treatment. Various activation method such as physical and chemical were attempted to create more suitable material which has larger surface area, pore size distribution as well as excellent surface structures. The physical method was applied by heating up the material till 400°C while HCl was used on the chemical activation. Based on the result, the chemical activation increased the surface area significantly into 71 m^2/g compared to physical as well as original zeolite. The strong acid material was quite effective to enforce the impurities within zeolite pore structure compared to heating up the material. According to distribution data, the Lampung zeolite owned the pore size with the range of 3 – 5 µm which was mesopore material. The pore size was appropriate for immobilization media as it was smaller than size of biogas microbial. The XRD patterns verified that chemical activation could maintain the zeolite structure as the original. Obviously, the SEM photograph showed apparent structure and pore size on the modified zeolite using chemical method. The testing of modified zeolite on the batch system was done to evaluate the characterization process. The modified zeolite using chemical process resulted fast reduction of COD and stabilized the volatile fatty acid as the intermediate product of anaerobic digestion, especially in the beginning of the process. Therefore, the chemical activation process was most suitable to produce the immobilization media from Lampung natural zeolite for POME waste treatment

Keywords: Anaerobic digestion; Modified natural zeolite; Palm Oil Mill Effluent (POME); Anaerobic Fluidized Bed Reactor (AFBR); immobilization media

IOP Conf. Series: Materials Science and Engineering **316** (2018) 012070 doi:10.1088/1757-899X/316/1/012070

1. INTRODUCTION

The world palm oil industry is growing up continuously to fulfill the human needs related to food and hygiene products. Indonesia produced 32.5 million tons of palm oil this year. It renders Indonesia as largest palm oil producer in the world that supply around 45% of total global demand. During palm oil production, this industry generates huge liquid waste, called Palm Oil Mill Effluent (POME) which has high polluting properties such as 25000 mg/l Biological Oxygen Demand (BOD), 55000 mg/l Chemical Oxygen Demand (COD) and 19610 mg/l of Suspended Solid (SS) [1]. Thus, direct releasing of POME into the surroundings would cause pollution and environment degradation. Conventional anaerobic digestion using pond system has been utilized as common and primary method for processing of POME in Indonesia which needs large land and small gas production. In addition, the method has also limitation to handle POME as substrates with high content of inhibitor. In order to resolve the problem, the development of anaerobic digestion using immobilize cell is proposed which own some advantages such as contain more amount of bacteria, prolong the residence time as well as reduction of reactor volume [2]. Cell immobilization is a treatment using limited motion of the bacteria within the reactor using specific space or porous material. By this method, the combination of high cell concentration and flow rate would improve the productivity. The porous material would enhance the contact between substrate and bacteria to increase the biogas production.

Commonly, the material contains several cation such as K^+, Na^+, Ca^{2+} and Mg^{2+} and possess the minimum pore diameter of 10 Å, high surface area and large capacity of ion exchange. The zeolite could control ammonia concentration and increase the growth rate of bacteria in the waste water treatment by exchange the ammonium ion with specific ion of zeolite [3-5]. In addition, Fernandez et al reports that zeolite has excellent characteristic to attract the microorganism and cation exchanger in the industrial wastewater treatment [6]. However, the addition of high amount of zeolite in the substrate would increase the viscosity of wastewater which disturbs the mass transfer of substrate and microorganism.
In order to enhance the activity and remove the impurities, the natural zeolite should be prepared and activated using several methods such as chemical, physic as well as combination of chemical and physic. Beside the impurities, the activation process could also modify the cation type, Si/Al ratio and other characteristic which is suitable for specific wastewater effluent. Physic activation which only remove impurities is proceed trough heating up the zeolite until temperature of 200-400°C for a couple hours [7]. Contrary, the chemical activation is done using base or acid solution not only to remove the impurities but also alter the Si/Al ratio. The reduction of Si/Al ratio is achieved by base activation in the high pH [8]. Chemical activation using acid solution (HCl) on the Turkey zeolite causes de-alumination process which increase Si/Al ratio [9]. Based on the facts above, the activation process on the natural zeolite could improve the performance of that material on the wastewater treatment. Therefore, the purposes of this research was to find the best activation method and theirs characterization of natural zeolite from Lampung in order to utilize on anaerobic immobilization process of POME.

IOP Conf. Series: Materials Science and Engineering **316** (2018) 012070 doi:10.1088/1757-899X/316/1/012070

2. METHODOLOGY/ EXPERIMENTAL

2.1. Material

Natural zeolite ((klinoptilolit) was obtained from CV. Minatama Lampung. In order to avoid effect of particle size on the experiment, the zeolite was crushed and sieved to similar size. That zeolite was washed using aquadest and dried in an oven at 110°C for 3 h. The main properties of the zeolite were listed in the Table 1. Raw material of POME was obtained from ponds in a palm oil mill located in Lampung, Indonesia which the characteristic as shown in the Table 2.

Table 1. The main properties of natural zeolite from Lampung

Chemical composition (%mass)	Si	K	Ca	Fe	Al	Mg	Mn	Sr	Rb
	60.7	11.5	10.0	8.48	8.21	0.42	0.38	0.20	0.09
Specific surface area (m^2/g)	42.16								
Total pore volume (cc/g)	0.147								
Average pore radius (Å)	194.63								

Table 2. Characterization of POME as substrate in the anaerobic process.

Parameter	Amount	Unit
pH	3.97	-
Chemical Oxygen Demand	11,995	mg/L
Protein	0.15	%
Nitrogen	0.0295	%
Total sugar	0.314	%
Fe	0.258	mg/L
Sulfide	5.8	mg/L
Phosphate	37.175	mg/L
Phenol	$\leq$ 0.0001	mg/L
Nickel	0.143	mg/L
Oil and grease	115	mg/L
Potassium	1,459.86	mg/L
Sulphate	1,032.93	mg/L
Ammonia	125	mg/L

2.2. Experiment procedure

a. Zeolite activation process

In the physical activation, the natural zeolite was washed and heated inside the furnace with air atmosphere and temperature of 400°C for 3 hours. While the chemical activation was done by using HCl 2M to wash and soak the natural zeolite in the room temperature for 3 hours. Subsequently, the zeolite was sieved and washed using aquadest until the pH 7 and then dried inside the oven on temperature of 110°C for 3 hours. The cooling process was proceed using desiccators for both physical and chemical activation. Several characterizations were applied to evaluate the activation process such as surface area, pore volume, XRD.

a. Anaerobic digester using POME as substrate

The activated zeolite was tested on the anaerobic digestion using POME as substrate based on equipment which was shown in the Figure 1. The anaerobic digester was equipped with digester, bubbler and gasometer. The digester was constructed from acrylic with the diameter of 9.4 cm and total volume of 2.8 liter. The gas production was indicated by bubbler which was made by Erlenmeyer 250 mL and measured by gasometer. The gasometer was filled with acidified salt water to reduce biogas solubility [13].

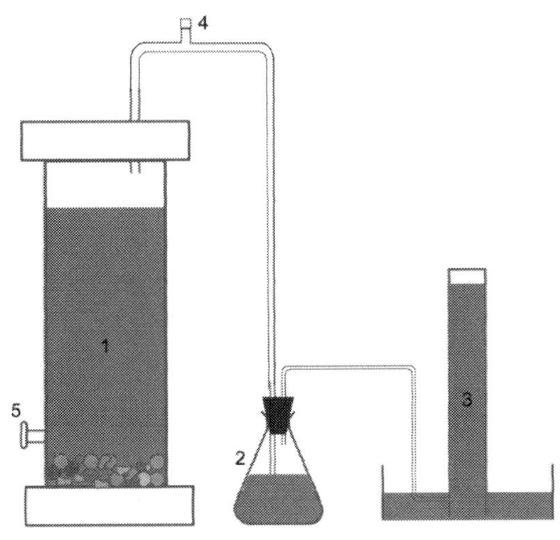

1. Digester biogas
2. *Bubbler*
3. Gasometer
4. *Rubber septum*
5. *Nipple* for liquid sampling

Figure 1. The experiment apparatus for batch anaerobic process

The experiment was carried out in duplicate for 40 days in room temperature. Diluted POME (SCOD 2,000 mg/L) and digested cow manure were mixed with ratio 3:1 (v/v) and filled into the digester without pH adjustment. Different masses of zeolite were added, i.e. 280, 560, and 840 g zeolite/ g SCOD digester liquid, and one digester was set as control without zeolite addition. The nitrogen flushing was done for five minutes before the observation.

Biogas production was calculated using the through gasometer equation. The determination of soluble chemical oxygen demand was carried out by closed reflux colorimetric method (APHA, 2005) by means of COD heater HI 839800 and Hach DR 900 colorimeter. The analysis of volatile fatty acid was conducted by distillation method, resulting VFA concentration as acetic acid. The pH was measured by Lutron pH meter PH-208, calibrated with 7.0 buffer solution. Methane concentration was measured by Gas Chromatography Shimadzu GC 8A, Japan. The number of total anaerobic bacteria was analyzed by total plate count procedure

3. RESULTS AND DISCUSSION

3.1. Zeolite activation and characterization

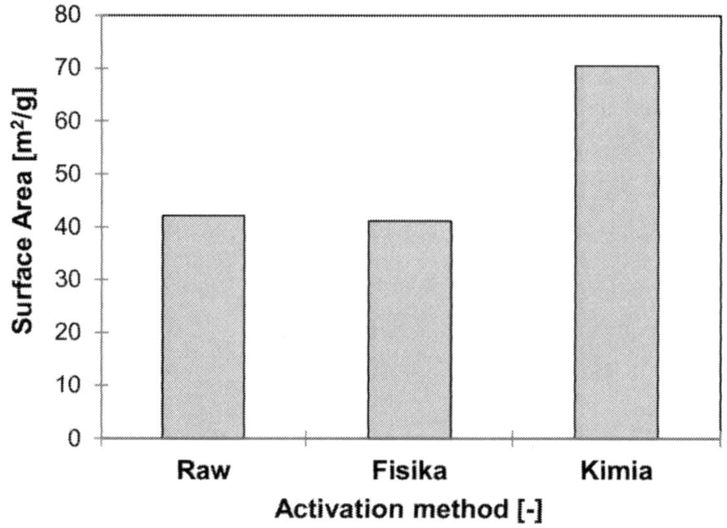

Figure 2. Effect of activation method on the zeolite surface area

As mentioned before that surface area of immobilization medium was key parameter on this anaerobic process. Figure 2 shows the effect of activation method on zeolite surface area. Cleary that physical activation resulted no significant effect on the surface area while the chemical activation generated more surface area, around twice within the zeolite material. The physical activation using heating up process was ineffective to remove the impurities which attach strongly within zeolite surface. It was different with chemical activation which used HCl to dissolve the impurities. Those activated zeolite owned high surface area, around 71 m^2/g. Therefore, the chemical activation was more effective to remove the impurities and enhance the surface area compared to physical activation. However, the activation process should maintain the original structure of zeolite which was effective in the immobilization process.

IOP Conf. Series: Materials Science and Engineering **316** (2018) 012070 doi:10.1088/1757-899X/316/1/012070

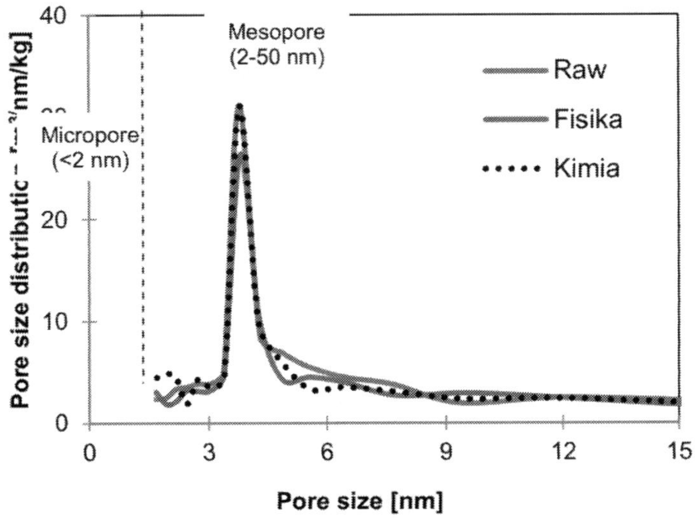

Figure 3. Pore side distribution of zeolite in each activation process

Beside the surface area, the pore size distribution as a effect of activation method was evaluated and shown in the Figure 3. Generally, the activation method was not change the pore size distribution significantly since the profile of distribution line was quite dense. The activated zeolite holds pore size of 4 nm, predominantly. In detail evaluation, it was shown that the chemical activation create more pore in the size of 4 nm compared to physical activation. This result confirmed the surface area analysis that the chemical activation was more effective to produce activated zeolite as immobilization agent in viewpoint of available space for microorganism. In addition, the pore size of Lampung zeolite was categorized as mesopore region which was more than 2 nm. This condition offer more advantages when it used as immobilization agent because the biogas microorganism could not penetrate into pore area. The microorganism would lodge in the surface of zeolite material and have more contact with substrate to improve the degradation process. Therefore, the surface morphology of zeolite was key parameter in this anaerobic process.

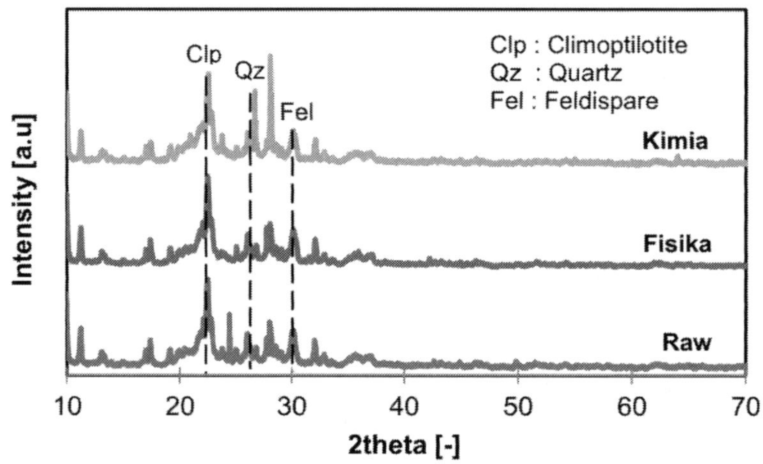

Figure 4. X-Ray Diffaraction (XRD) pattern of Lampung zeolite in each activation process.

Figure 4 shows the crystalline structure of Lampung zeolite in each experimental treatment. The natural zeolite had some crystalline such as climoptilotite, quartz, and feldispare. Obviously, this structure was maintained stable in each activation process. It confirmed that the chemical activation could generate more surface area, pore size with stable crystalline structure. This activation process removed only impurities without break the structure of zeolite material. It was important result to have the immobilization agent with high space for microorganism and good surface structure.

Figure 5. SEM Photo of Lampung zeolite in various activation process.

Figure 5 shows SEM photo of the zeolite in each experiment treatment with similar enlargement. This photo could be used to analyze the effect activation process on the surface of the zeolite. Clearly, the activation step resulted the clean and bright zeolite material. Chemical activation created brighter material compared to physical activation which indicated the cleanness of the material. Again, the SEM photo confirmed that the chemical activation produced better material for immobilization agent in the anaerobic process.

3.2. Testing on Immobilization-anaerobic process

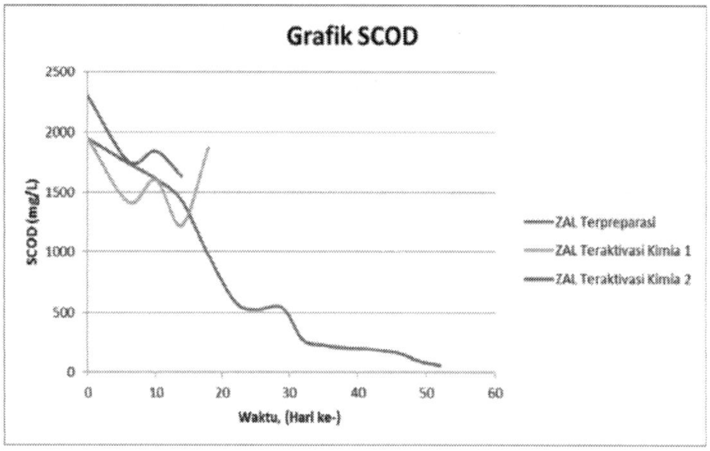

Figure 6. Effect zeolite as immobilization agent on the anaerobic process

Figure 6 shows the average sCOD concentration during the batch anaerobic process using POME as substrate and activated zeolite as immobilization agent. Generally, the anaerobic process using zeolite demonstrated normal process which was indicated by decreasing of sCOD during the experiment. In beginning reaction, the chemical activated zeolite shows better performance compared to natural zeolite which was indicated by high decreasing rate of sCOD. Even the testing was limited by time, this result offer great possibility to use natural zeolite as immobilization agent in the anaerobic process using POME as substrates.

4. CONCLUSION

Chemical activation on Lampung zeolite offered promising result for anaerobic-immobilization process in term of high surface area as well as pore size distribution. In addition, the crystalline structure was stable and maintain in place during the activation process. Anaerobic testing using activated zeolite showed that the material had high possibility as immobilization agent.

5. REFERENCES

[1] Khanal.S. Kumar., (2008), *Biotechnology for Bioenergy Production : Principles and Application*, University of Hawai'i Manoa, Wiley-Blackwell, John Wiley and Sons Publication

[2] Shuler, M. and Kargi, F., 2002, *Bioprocess Engineering Basic Concepts 2nd Ed.*, Prentice Hall, New Jersey

[3] Rouquerol, F., J. and K. S. W. Sing, (1999), *Adsorption by Powders and Porous Solids: Principles, Methodology and Applications*, Academic Press, San Diego, CA

[4] Montalvo, S., Guerrero, L., Borja, R., Sánchez, E., Milán, Z., Cortés, I., Rubia, MA., (2012), *Application of natural zeolits in anaerobic digestion processes: A review, Applied Clay Science* 58: 125–133
[5] Borja, R., Sanchez, E., Weiland, P., Travieso, L., dan Martin, A., (1994), *Kinetics of Anaerobic Digestion of Cow Manure with Biomass Immobilized on Zeolite,* Chemical Engineering Journal, 54, B9-B14.
[6] Fernández, N., Montalvo, S., Polanco, F., Guerrero, L., Cortés, I., Borja, R., Sánchéz, E, dan Travieso, L., (2007), *Real Evidence About Zeolite as Microorganisms Immobilizer in Anaerobic Fluidized Bed Reactor*, Journal of Process Biochemistry, 42: 721-728
[7] Rosita, N., Erawati, T., Moegihardjo, M., (2004), *Effect of activation method on zeolite effectiveness as adsorbent (in Bahasa Indonesia).* Majalah Farmasi Airlangga, 4 No. 1, 20-25.
[8] Jozefaciuk, G., Bowanko, G., (2002), *Effect of Acid and Alkali Treatments on Surface Areas and Adsorption Energies of Selected Minerals*, Journal Clays and Clay Minerals, 50 No. 6, 771- 783
[9] Ozkan, F.C., Ulku, S., (2005), *The Effect of HCl Treatment on Water Vapor Adsorption Characteristics of Clinoptilolite Rich Natural Zeolite*, Journal Microporous and Mesoporous Materials 77, 47-53.

IOP Conf. Series: Materials Science and Engineering **316** (2018) 012071 doi:10.1088/1757-899X/316/1/012071

Kinetics of Anaerobic Digestion of Palm Oil Mill Effluent (POME) in Double-Stage Batch Bioreactor with Recirculation and Fluidization of Microbial Immobilization Media

L I Ramadhani[1], S I Damayanti[1], H Sudibyo[1,2], W Budhijanto[1,2,*]
[1] Chemical Engineering Department, Universitas Gadjah Mada, Yogyakarta 55281 Indonesia
[2]Center for Energy Studies, Universitas Gadjah Mada, Yogyakarta,55281, Indonesia

[*]Corresponding author: wiratni@ugm.ac.id

Abstract. Palm Oil Mill Effluent (POME) becomes big problem for palm oil industries, especially for Crude Palm Oil (CPO) industry since it produces 3 tons of POME for every ton of CPO production.The high amount of organic loading in POME makes it potential as a substrate in anaerobic digestion to generate biogas as renewable energy source. The most common but conventional method by using open lagoon is still preferred for most CPO industry in Indonesia to treat POME because of its simplicity and easiness. However, this method creates new major problem for the water bodies since it has no significant chemical oxygen demand (COD) removal and needs wide area. Besides, greenhouse gas (CH_4) is also released during the process. An innovation was made in this study by designing vertical column process equipment to run an anaerobic digestion of POME. The vertical column was functioned as anaerobic fluidized bed reactor (AFBR). To enhance the digestion rate in AFBR, natural zeolite was used as the immobilization media and the inoculum was taken from digested biodiesel waste. This research aimed to determine the kinetic constants of double-stage anaerobic POME digestion for COD removal and biogas production. To get close to the real condition, the POME used in this experiment had 8,000 mg/L of sCOD (the real sCOD was ±16,000 mg/L). The experiment was conducted under room temperature with up-flow velocity between 1.75 and 2.3 cm/s for optimum fluidization of immobilization media.

1. Introduction

POME is a highly polluting wastewater causing serious problem to the environment due to its highly pollutant content. For every ton fresh fruit bunches processed in Palm Oil Mill, there will be 120-200 kg crude palm oil (CPO), 230-250 kg empty fruit bunches, 130-150 kg fiber, 60-65 kg nuts and 0.7 m^3 wastewater effluent [1]. Almost every palm oil mill in Indonesia uses opened lagoon method to treat the wastewater namely POME. However, the effectiveness of this method is not reliable due to the high retention time and wide area required. Besides, greenhouse gas, for instance, CO_2 and CH_4 are released directly to the atmosphere. On the other hand, the existence of CH_4 in the gas released during POME treatment represents the potential to be used as the source for biofuel production. What made POME different from any other wastewater is its high volatile fatty acid (VFA) content. The presence of abundant VFA in POME makes it an attractive substrate for anaerobic digestion due to the higher methane yield obtained than proteins or carbohydrates. In this case, high VFA content waste can be regarded as a large potential for renewable energy source [2].

The solution of the aforementioned problem related to the ineffectiveness of open lagoon method is anaerobic digestion. By this process, the biogas produced during treatment can be controlled and space can be saved by using compact vertical process equipment. In general, anaerobic digestion process consisted of acidogenesis and methanogenesis process [3]. The former focuses on removing the pollutant represented by Chemical Oxygen Demand (COD) parameter while the latter focuses on converting organic material represented by VFA parameter into biogas. Both processes can run simultaneously and sequentially. Because of different optimum pH level for both processes, the optimum operating can be achieved if anaerobic digestion process is conducted separately and sequentially. Two-phase anaerobic digestion system had been recommended for the treatment of wastewater containing high VFA content such as dairy waste [4], ice cream factory effluents [5], fish meal processing waste [6], slaughterhouse waste [7], and olive mill solid waste [8].

In this work, POME was treated sequentially by separating acidogenesis and methanogenesis processes into two anaerobic fluidized bed reactors (AFBR). Acidogenesis process ran on the first reactor and methanogenesis process ran on the second reactor. The vertical column AFBR was used due to its effectiveness to process high organic loading rate (OLR) wastewater with lower hydraulic retention time (HRT). To prevent wash-out phenomenon, natural zeolite was used in this study as the microbial immobilization media. The immobilization media was then fluidized to maximize the contact area between substrate and microorganism [9] so that the microorganism growth could be enhanced. The aim of this research was to study the kinetics of anaerobic digestion of POME in double-stage anaerobic fluidized bed reactor by synthesizing mathematical modelling for batch process.

2. Methodology/experimental

2.1 Materials

The materials needed in this research were Palm Oil Mill Effluent (POME), natural zeolite as the immobilization media and digested biodiesel waste as microorganism isolate source. The POME was obtained from PT. Perkebunan Nusantara VII, Lampung, Indonesia with Chemical Oxygen Demand (COD) as much as 13,580 mg/L and pH 4.48. Digested biodiesel waste was taken from the biodiesel industry operating in East Java with the COD of 1,980 mg/L and pH of 7.53. The natural zeolite with diameter range of 2-2.38 mm came from Tasikmalaya, West Java.

2.2 Anaerobic digestion of POME

Anaerobic digestion was conducted in two bioreactor columns made of acrylic equipped with close loop recirculation system for fluidization (figure 1). The first AFBR was a 15 L acidogenic bioreactor and the second one was a 10 L methanogenic bioreactor with 150 gram of immobilization media from natural zeolite in each bioreactor. This system was supported by one dosing pump and two centrifugal pumps working 24 hours a day with on-off system for each bioreactor.

To ensure that in the first AFBR there was only acidogenesis process and in the second AFBR there was only methanogenesis process, the pH level was set at acidic condition (pH 5-5.5) and neutral condition (pH 7) respectively. The pH inside the second AFBR was adjusted by adding NaOH to the methanogenic bioreactor. These batch bioreactors ran for 14 days for acidogenic bioreactor and 21 days for methanogenic bioreactor.

IOP Conf. Series: Materials Science and Engineering **316** (2018) 012071 doi:10.1088/1757-899X/316/1/012071

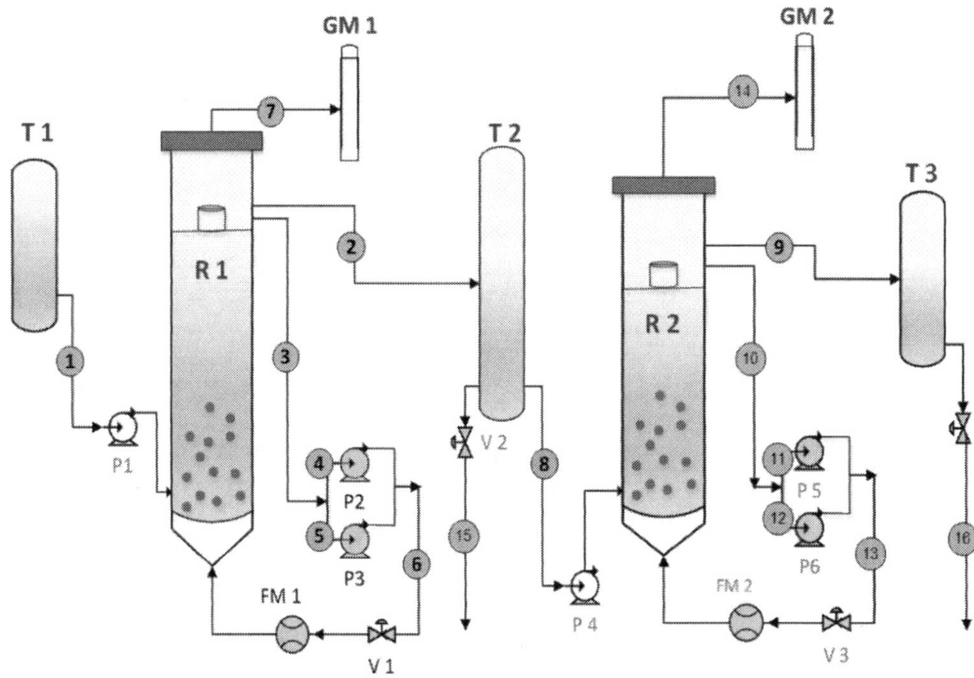

Annotations:

R1	:	Acidogenic AFBR	GM 1,2	:	Gas meter
R2	:	Methanogenic AFBR	FM 1,2	:	Flow meter
Tank 1	:	Influent tank	Stream 1,8	:	Influent stream to R1 and R2
Tank 2	:	Intermediate tank	Stream 2	:	Effluent stream from R1
Tank 3	:	Effluent tank	Stream 3,9	:	Suction stream for recirculation
P1,P4	:	Dossing pump	Stream 4,14	:	Influent stream into cirulating pump 1
P2,P5	:	Circulating pump 1	Stream 5,12	:	Influent stream into cirulating pump 2
P3,P6	:	Circulating pump 2	Stream 9	:	Effluent stream from R2
V1,V3	:	Controlling valve for circulating stream	Stream 6, 13	:	Influent stream to AFBR for fluidization
V2, V4	:	Controlling valve for effluent stream	Stream 7, 14		Effluent stream of biogas
			Stream 15, 16	:	Effluent stream

Figure 1. Experimental set up

2.3 Analysis of sCOD, VFA and CH_4

In this work, parameter used to represent organic loading of the POME was soluble Chemical Oxygen Demand (sCOD) while to represent available substrate for biogas production, it was used Volatile Fatty Acid (VFA). The analysis of sCOD and Volatile Fatty Acid (VFA) during the experiment followed the standard procedure by APHA [10]. The sCOD analysis was conducted with closed reflux colorimetric method. The VFA analysis used the titrimetric method. The gas volume was measured by using the gasometer method outlined by Walker et al [11] while the methane content was analyzed by using Gas Chromatography (GC) Shimadzu GC 8A.

2.4 Mathematical modelling for Batch AFBR

The growth rate for both acidogenic and methanogenic cells were defined from cell mass balance combined with specific growth rate represented by Contois equation [12,13,14] due to its slow rate process. The Contois equation for the growth rate of acidogenic and methanogenic microbes were described by equation (1) and equation (2).

$$\frac{dX_1}{dt} = \left(\frac{\mu_{m1}.C_{sCOD}}{K_{SX1}.X_1 + C_{sCOD}} - k_{d1} \right).X_1 \tag{1}$$

$$\frac{dX_2}{dt} = \left(\frac{\mu_{m2}.C_{VFA}}{K_{SX2}.X_2 + C_{VFA}} - k_{d2} \right).X_2 \tag{2}$$

Batch mathematical modelling for AFBR was developed based on the mass balance of sCOD, VFA and CH_4. The developed model still assumed that both acidogenesis and methanogenesis existed together. From the models, it was expected that the different pH application could lead to the conclusion in which constants were affected by pH adjustment after the kinetic constants were all determined. The determination was conducted by fitting the experimental data using MATLAB R2016b by means of *ODE15S* and *FMINCON* solver. The simultaneous ordinary differential equations were shown by equation (3) – equation (7).

$$\frac{d(C_{sCOD})}{dt} = -\frac{1}{Y'_{X1/CsCOD}} \left(\frac{\mu_{m1}.C_{sCOD}}{K_{SX1}.X_1 + C_{sCOD}} - k_{d1} \right).X_1 \tag{3}$$

$$\frac{1}{Y'_{X1/CsCOD}} = \frac{1}{Y_{CVFA/CsCOD}}.Y_{CVFA/X1} + \frac{1}{Y_{X1/CsCOD}} \tag{4}$$

$$\frac{dC_{VFA}}{dt} = \left[\frac{1}{Y_{CVFA/CsCOD}}.Y_{CVFA/X1} \right] \left(\frac{\mu_{m1}.C_{sCOD}}{K_{SX1}.X_1 + C_{sCOD}} - k_{d1} \right).X_1 - \frac{1}{Y'_{X2/CVFA}} \left(\frac{\mu_{m2}.C_{VFA}}{K_{SX2}.X_2 + C_{VFA}} - k_{d2} \right).X_2 \tag{5}$$

$$\frac{1}{Y'_{X2/CVFA}} = \frac{1}{Y_{X2/CVFA}} + Y_{CCH4/X2} \tag{6}$$

$$\frac{d(C_{CH4})}{dt} = Y_{CCH4/X2} \left(\frac{\mu_{m2}.C_{VFA}}{K_{SX2}.X_2 + C_{VFA}} - k_{d2} \right).X_2 \tag{7}$$

3. Result and discussion

3.1 sCOD removal and VFA formation in acidogenic AFBR

Main function of anaerobic fluidized bed reactor on the first stage as the acidogenic reactor was to convert sCOD into Volatile Fatty Acid (VFA). Acidogenic AFBR ran on acidic condition (pH 5-5.5) with starting values of sCOD and VFA consentration as much as 1,925 mg/L and 4,518 mg/L. In this work, sCOD measurement was all the soluble organic matter besides VFA. Acidic condition was preferred on this stage to maximize the growth of acidogenic microbes and to prevent the growth of methanogenic microbes which is too sensitive toward pH decreasing. The figure 2 (a) showed that sCOD concentration decreased significantly until day 12. After 12 days of experiment, the sCOD was relatively stable. This phenomenon indicated that first, the acidogenic microorganism could possibly grow well (must be confirmed with the VFA profile) and the second, the system was ready to be switched into the continuous operation if removal of sCOD was the main concern.

VFA formation is the most important aspect that indicates acidogenesis process ran well. The VFA concentration profile within 13 days of process is shown on figure 2 (b). The decrease of VFA at the beginning (until day 3) indicated that the microorganism was still in the adaptation period by consuming substrate. However, it started to increase from day 4 to day 10. This indicated that acidogenic microorganism started dominating after 3 days of anaerobic digestion under acidic condition. However, the increase of VFA content inside the reactor has its maximum value. After day 10, the VFA content decreased again and was followed by significant biogas production as shown by figure 2 (c). By this situation, it could be seen that the acidogenic microorganism had maximum tolerance to the VFA concentration of the substrate.

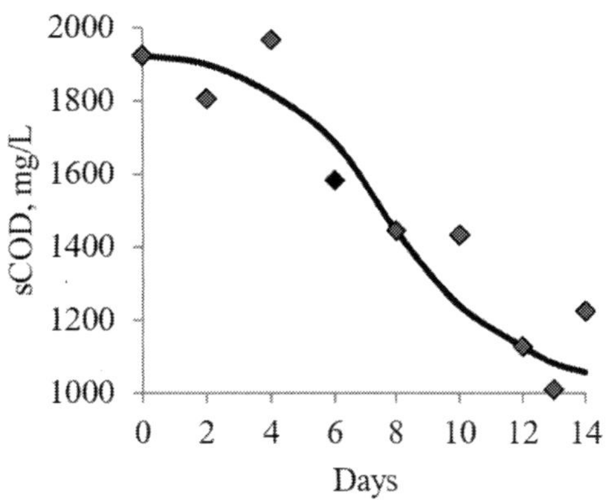

(a)

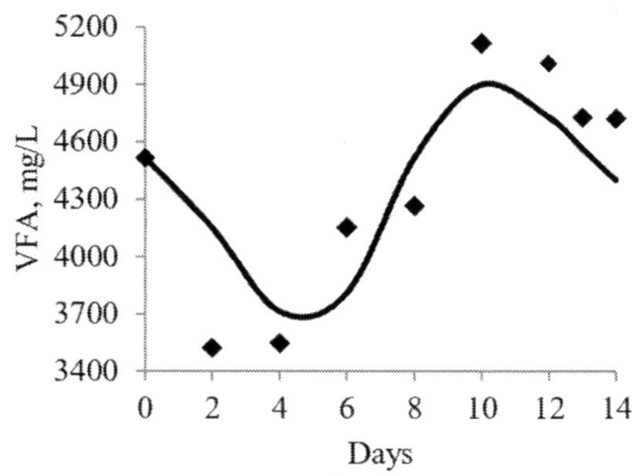

(b)

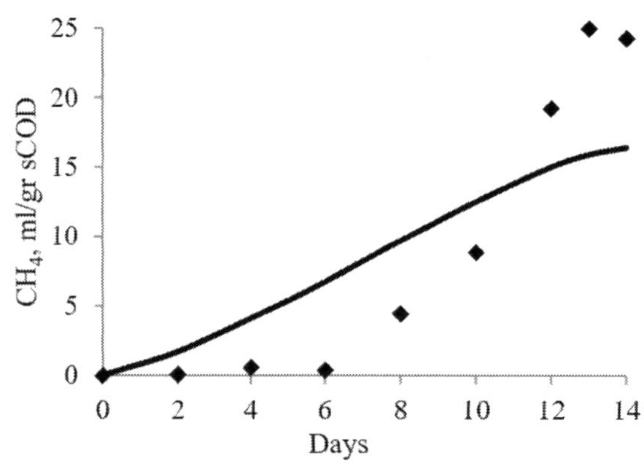

(c)

Figure 2. Profile of a). sCOD change against time b). VFA formation against time c). CH_4 production against time in acidogenic AFBR

3.2 Profile concentration of COD and VFA at methanogenic AFBR

Batch phase of methanogenic AFBR was observed for 21 days. The sCOD profile insignificantly decreased in this reactor. This condition proved that neutral condition had an important role to minimize acidogenesis process even this process was still existed (see figure 3). The VFA increase in the first 5 day of digestion, as shown by figure 3 (b), showed the adaptation phase of the microorganism. On the next day, decreasing VFA profile indicated that methanogenic microorganism started dominating in the anaerobic digestion under neutral condition. It was confirmed by CH_4 formation shown by figure 2 (c). Methane as methanogenesis process result started to be formed on day 4 and significantly increased on day 15. The decrease of CH_4 formation from day 15 indicated that for biogas production, there was minimum concentration of VFA between 1,000 – 1,500 mg/L.

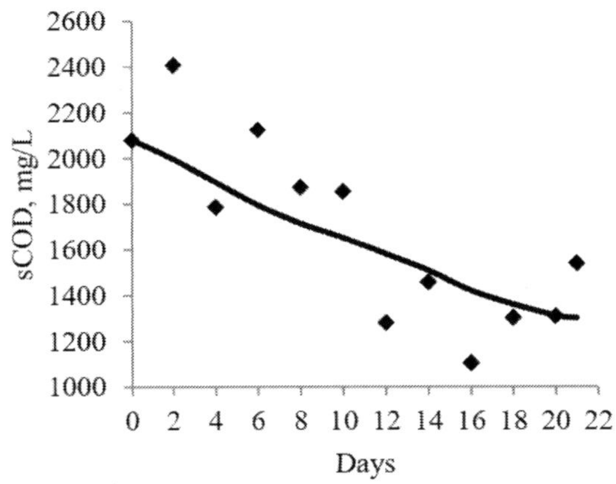

(a)

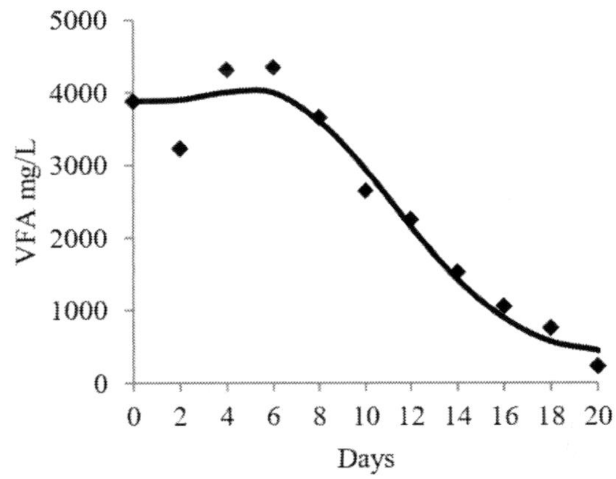

(b)

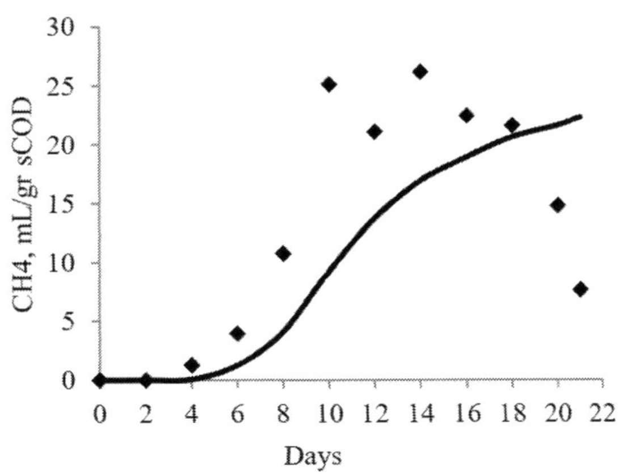

(c)

Figure 3. Profile of a). sCOD change against time b). VFA formation against time c). CH_4 production against time in methanogenic AFBR

The highest methane volume was 1,948.76 mL on day 14 and the highest purity achieved on day 18 was 63.16 % v/v methane. Methane purity aimed in this research was higher than batch anaerobic digestion of POME by using natural zeolite as immobilization media in the packed bed bioreactor done by Ayu et al. [1] and Purnomo et al. [15] which was about 54%. Fluidization system used in this work had been proven as an effective way to optimize anaerobic process based on the previous study by Ayu et al. [1].

3.3 Kinetics constants

The table 1 showed the kinetics constant of the kinetics equations. On the first stage AFBR of POME digestion, acidogenesis was observed as the dominant process. For acidogenic and methanogenic cells, the growth rate (μ_m) and the death rate (k_d) in acidogenic AFBR showed relatively the same values. Nevertheless, sCOD reduction for both VFA conversion and acidogenic cell maintenance represented by $Y'_{X1/sCOD}$ as large as 235.4989 mg acidogenic cell/mg sCOD showed that the process dominating in this AFBR was acidogenic. It was confirmed by low yield of cell formation per mg VFA reduction ($Y'_{X2/VFA}$) as a parameter of methanogenic process.

Table 1. Kinetics constants of batch AFBR

Constants	Value		Unit
	Acidogenic AFBR	Methanogenic AFBR	
μ_{m1}	2.1321	0.000	day^{-1}
μ_{m2}	2.1321	0.6735	day^{-1}
K_{SX1}	0.0066	175.3444	mg COD/mg acidogenic cell
K_{SX2}	173.3997	0.7618	mg VFA/mg methanogenic cell
$Y_{sCOD/VFA}.Y_{VFA/X1}$	235.4989	0.000	mg VFA/mg acidogenic cell
$Y_{X2/VFA}$	0.0283	3.3251	mg methanogenic cell/mg VFA
$Y_{CH4/X2}$	0.2285	0.0016	[mg CH_4/L]/[mg methanogenic cell/L]
$Y_{X1/sCOD}$	0.000	0.5792	mg acidogenic cell/mg sCOD
$Y'_{X1/sCOD}$	235.4989	0.0000	mg acidogenic cell/mg sCOD
$Y'_{X2/VFA}$	0.0281	3.3251	mg methanogenic cell/mg VFA
k_{d1}	1.000	0.0000	day^{-1}
k_{d2}	0.9441	0.0000	day^{-1}

In the second stage AFBR (the methanogenic one), methanogenesis was observed as the main process existed proven by the dominant growth of methanogenic microbes. The growth rate of methanogenic microbes (μ_{m2}) was 0.6735/days while μ_{m1} was equal to zero. It indicated that there were only methanogenic microbes could grow well in the methanogenic AFBR. The very low opportunity for acidogenic microbes to survive was shown by the values of sCOD reduction both for VFA conversion ($Y_{sCOD/VFA}.Y_{VFA/X1}$) and for cell maintenance ($Y_{X1/sCOD}$) which were very low. Although methanogenesis succesfully dominated the process, methane conversion was not satisfying yet. The yield of CH_4 formation ($Y_{CH4/X2}$) was only 0.0016 [mg CH_4/L]/[mg methanogenic cell/L] compared to the yield of methanogenic cell growth from VFA consumption ($Y_{X2/VFA}$) which was 3.3215 mg methanogenic cell/mg VFA. Optimization could be made by shifting the process configuration to continuous process and applying diferent retention time to the methanogenic AFBR.

4. Conclusion

In the acidic condition of acidogenic AFBR, acidogenic dominated the process although methanogenic microbes could still survive. The pH level had an important impact to separate acidogenesis and methanogenesis process on double-stage system. Under neutral condition, it could be claimed that methanogenic microbes were successfully dominating the second stage of AFBR. Further study could focus on identifying and measuring the real concentration of both acidogenic and methanogenic microbes together with trying new process configuration for biogas production optimization.

References

[1] Ayu, E D, Halim L, Mellyanawaty M, Sudibyo H and Budhijanto W, 2017, The Effect of Natural Zeolite as Microbial Immobilization Media in Anaerobic Digestion at Various Concentrations of Palm Oil Mill Effluent (POME) *AIP Conference & Proceedings* **1840** 110005

[2] Mobarak-Qamsari E, Kasra-Kermanshahi R, Nosrati M and Amani T 2012 Enzymatic Pre-hydrolysis of High Fat Content Dairy Wastewater as a Pretreatment for Anaerobic Digestion *Int. J. Environ. Res.* **6** (2) 475-480

[3] Prastyo E, Sudibyo H and Budhijanto, W, 2017 Determination of the Optimum Hydraulic Retention Time in Double Stage Anaerobic Fluidized Bed Bioreactor for Landfill Leachate Treatment *J. Eng. Technol. Sci.* **49** (3) 388-399

[4] Demirer G N and Chen S 2005 Two-Phase Anaerobic Digestion of Unscreen Dairy Manure *Process Biochem.* **40** 3542-3549

[5] Borja R, Banks CJ 1995 Anaerobic digestion of Ice-cream Wastewater: A Comparison of Single and Two-Phase Reactor System *Bull. Environ. Contamin. Toxicol.* **54** (3), 466-471

[6] Guerrero L, Omil F, Mendez R, and Lema JM 1999 Anaerobic Hydrolysis and Acidogenesis of Wastewater from Food Industries with High Content of Organic Solids and Protein *Water Res.* **33** (15) 3281-3290.

[7] Wang Z and Banks C J 2003 Evaluation of A Two-Stage AnaerobicDigester for The Treatment of Mixed Abattoir Wastes *Process Biochem* **38** (9)1267–1273

[8] Travieso L, Dominguez J R, Rinco´n B, Sa´nchez E, Benı´tez F, Borja R, Raposo F, Colmenarejo M F 2008 Batch Culture Growth of Chlorella Zofingiensis on Effluent Derived from Two-Stage Anaerobic Digestion of Two-Phase Olive Mill Solid Waste *Electron J Biotechnol* **11** (2) 1–8

[9] Nicolella C, Van Loosdrecht M C M and Heijnen J J 2000 Wastewater Treatment with Particulate Biofilm Reactor *J. Biotechnol.* **80** 1-33

[10] American Public Health Association (APHA) 2005 *Standard Methods for the Examination of Water and Wastewater* (New York: American Public Health Association)

[11] Walker M, Zhang Y, Heaven S and Banks C 2009 Potential Errors in the Quantitative Evaluation of Biogas Production in Anaerobic Digestion Processes *Bioresour. Technol.* **100** 6339-6346

[12] Echigu E A and Ghally A E 2014 Kinetic Modelling of Continous-Mix Anaerobic Reactors Operating Under Diunarlly Cyclic Temperatur Environment *Am. J. Biochem. Biotechnol* **10** (2) 130-142

[13] Hussain, A., Dubey, S.K. and Kumar V 2015 Kinetic study for aerobic treatment of phenolic wastewater, *Water Resour. Ind.* **11** 81-90

[14] Sudibyo H, Shabrina Z L, Halim L and Budhijanto W 2017 Mathematical Modelling and Statistical Approach to Assess the Performance of Anaerobic Fixed Bed Reactor for Biogas Production from Piyungan Sanitary Landfill Leachate *Energy Procedia* **105** 256-262.

[15] Purnomo C W, Mellyanawaty M and Budhijanto W 2017 Simulation and Experimental Study on Iron Impregnated Microbial Immobilization in Zeolite for Production of Biogas *Waste Biomass Valorization* **8** (7) 2413–2421

Acknowledgement

The study was funded through the Competitive Research Grant "Tim Pasca Sarjana" from Ministry of Research Technology and Higher Education (KEMENRISTEKDIKTI) 2017.

Nomenclature

C_{sCOD}	soluble chemical oxygen demand (mg/L)
C_{VFA}	volatile fatty acid concentration (mg/L)
C_{CH4}	methane concentration (mg/L)
X_1	acidogenic cell concentration (mg/L)

IOP Conf. Series: Materials Science and Engineering **316** (2018) 012071 doi:10.1088/1757-899X/316/1/012071

X_2	methanogenic cell concentration (mg/L)
μ_{m1}	maximum specific growth rate of acidogenic cell (day^{-1})
μ_{m2}	maximum specific growth rate of methanogenic cell (day^{-1})
K_{SX1}	half-saturation constant associated with sCOD (mg sCOD/mg acidogenic cell)
K_{SX2}	half-saturation constant associated with VFA (mg VFA/mg methanogenic cell)
$Y_{X1/CsCOD}$	yield of cell formation per mg sCOD reduction (mg acidogenic cell/mg sCOD)
$Y_{X2/C\,VFA}$	yield of cell formation per mg VFA reduction (mg methanogenic cell/mg VFA)
$Y'_{X1/CsCOD}$	total yield of cell formation per mg sCOD reduction (mg acidogenic cell/mg sCOD)
$Y'_{X2/CVFA}$	total yield of cell formation per mg VFA reduction (mg methanogenic cell/mg VFA)
$Y_{sCOD/VFA}.Y_{VFA/X1}$	yield of VFA formation per mg acidogenic cell (mg VFA/mg acidogenic cell)
$Y_{CCH4/X2}$	yield of CH_4 formation per mg methanogenic cell/L increase ($mg\ CH_4/L$)/[mg methanogenic cell/L)
k_{d1}	death rate constant of acidogenic cell
k_{d2}	death rate constant of methanogenic cell
Abbreviations	
POME	Palm Oil Mill Effluent
AFBR	Anaerobic Fluidized Bed Reactor
sCOD	Soluble Chemical Oxygen Demand
VFA	Volatile Fatty Acid

IOP Conf. Series: Materials Science and Engineering **316** (2018) 012072 doi:10.1088/1757-899X/316/1/012072

Modeling Lab-sized Anaerobic Fluidized Bed Reactor (AFBR) for Palm Oil Mill Effluent (POME) treatment: from Batch to Continuous Reactors

Muhammad Mufti Azis*, Hanifrahmawan Sudibyo and Wiratni Budhijanto

Chemical Engineering Department, Faculty of Engineering, Universitas Gadjah Mada Jalan Grafika 2, Kampus UGM Yogyakarta, Indonesia.

Corresponding author: muhammad.azis@ugm.ac.id

Abstract. Indonesia is aiming to produce 30 million tones/year of crude palm oil (CPO) by 2020. As a result, 90 million tones/year of POME will be produced. POME is highly polluting wastewater which may cause severe environmental problem due to its high chemical oxygen demand (COD) and biochemical oxygen demand (BOD). Due to the limitation of open pond treatment, the use of AFBR has been considered as a potential technology to treat POME. This study aims to develop mathematical models of lab-sized Anaerobic Fluidized Bed Reactor (AFBR) in batch and continuous processes. In addition, the AFBR also utilized natural zeolite as an immobilized media for microbes. To initiate the biomass growth, biodiesel waste has been used as an inoculum. In the first part of this study, a batch AFBR was operated to evaluate the COD, VFA, and CH_4 concentrations. By comparing the batch results with and without zeolite, it showed that the addition of 17 g/gSCOD zeolite gave larger COD decrease within 20 days of operation. In order to elucidate the mechanism, parameter estimations of 12 kinetic parameters were proposed to describe the batch reactor performance. The model in general could describe the batch experimental data well. In the second part of this study, the kinetic parameters obtained from batch reactor were used to simulate the performance of double column AFBR where the acidogenic and methanogenic biomass were separated. The simulation showed that a relatively long residence time (Hydraulic Residence Time, HRT) was required to treat POME using the proposed double column AFBR. Sensitivity analyses was conducted and revealed that μ_{m1} appeared to be the most sensitive parameter to reduce the HRT of double column AFBR.

Keywords: POME; AFBR; natural zeolite; modeling; batch reactor; continuous reactor; COD; VFA; CH_4

1. Introduction

Indonesia is aiming to produce 30 million tons of bulk palm oil (crude palm oil / CPO) per year by 2020. As a result, it is estimated that the amount of liquid waste namely palm oil mill effluent (POME) from all palm oil mills in Indonesia is expected to be as much as 90 million tons of POME per year. By utilizing appropriate technology, there is a huge potential energy from that amount of POME which may generate 900,000 tons of methane, which is equivalent to 1.3 million kilo liters of diesel oil, equivalent to 5 million MWh of electricity, and the power plant capacity (gas engine) 660 MW [1].

The current conventional practice of POME treatment is conducted using open pond system (lagoon). The pond system is simple and inexpensive, however it requires a large area and the amount of methane release could be enormous which resulted on large greenhouse gas emissions from the lagoons. According to Hanum [2], anaerobic treatment requires approximately 7 hectares of land for typical palm oil mill with a capacity of 30 tons of fresh fruit bunches (FFB)/hour. In addition, the lagoon treatment requires a long residence time, i.e. about 120 to 140 days. Although the current price of land in the surrounding area of palm oil mills and palm oil plantations is still considerably affordable, several palm oil mills are located near the outskirt of cities and far from plantations with high land price. Apparently, this would be the future trend in palm oil business. As a result, there is a need to develop new technology to process POME with low residence time.

Bioreactor process design requires at least three essential factors, namely: thermodynamics, kinetics of microbial growth, and transfer phenomena [3]. The process of acquiring parameters for kinetic growth is typically conducted on a small scale batch reactor. The kinetic parameters obtained from batch experiments will be very useful if one would like to increase the production capacity from batch to continuous operation.

In the bioprocess, the shift of operating conditions from batch to continuous operation certainly presents different characteristics. In terms of the availability of the substrate, batch process gives substrate concentrations change over the time and this allows the batch reactor to reach the lowest substrate concentration. As a result, the growing rate of microorganisms decelerates along with the reduced substrate or can even reach dead phase. Unlike batch reactor, continuous reactors maintains the availability of substrate concentration within the range of certain values that tends to be stable because there is always inlet feed substrate within the range of certain COD and VFA concentrations. Thus, the growth of microorganisms are more stable and more productive in terms of biogas production. However, various technical problems are often encountered during continuous operation such as infections and unwanted mutations of microorganisms [3].

The use of Anaerobic Fluidized Bed Reactors to treat POME has been investigated intensively in our group [4, 5, 6, 7]. The conversion of POME to methane follows two reaction steps as illustrated in Figure 1.

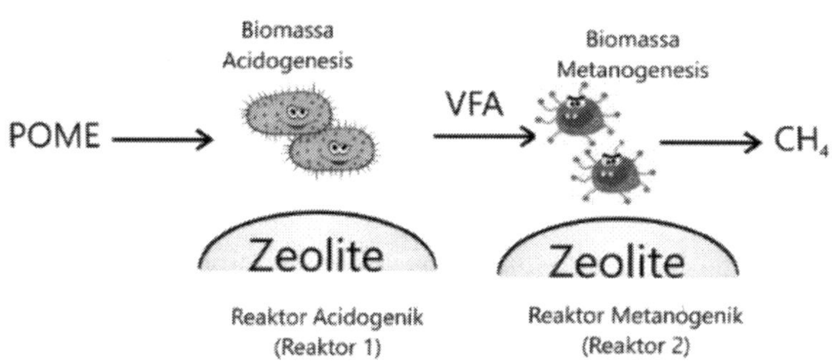

Figure 1 A simplified mechanism of POME conversion into methane

In batch operation, acidogenic and methanogenic microorganisms are present in the same reactor. During continuous operation, we proposed the use of double-stage AFBR to separate acidogenic and methanogenic microorganisms. In addition, the reactor utilizes natural zeolite from Lampung as an immobilized media for microbes. Zeolite is expected to allow microbes to grow and cover the surface of the zeolite so that the bioreactor may work well to break down POME with certain residence time. The experimental results from batch reactors of POME treatment have been presented in [5] using biodiesel waste as the inoculum (starter).

IOP Conf. Series: Materials Science and Engineering **316** (2018) 012072 doi:10.1088/1757-899X/316/1/012072

The aim of this study was to simulate the batch reactor to obtain a number of kinetic parameters which describes the microbial growth. Further, the kinetic parameters which is obtained from the batch reactor were used to assess the performance of double stage AFBR.

2. Experimental and Modeling Methods

2.1. Experimental Methods

The detail of experimental method for batch reactor has been described in [5]. Briefly explained here, batch reactor (2.8 L) was used to process 2 L of POME with an initial value of soluble COD of 8220 mg/L. The amount of COD, VFA, and CH_4 were monitored periodically. The amount of zeolite loading was varied and the experiment results used in the present study was 17 g zeolite/g sCOD. The values of soluble Chemical Oxygen Demand (sCOD) and Volatile Fatty Acid (VFA) concentration were measured according to the Standard Methods for Examination of Water and Wastewater. The pH was measured using a Lutron PH-208 pH meter. The analysis method of the gas volume was performed using gasometer and the amount of methane concentration was analyzed using Gas Chromatography (GC).

2.2. Simulation and parameter estimation of kinetic coefficients for batch reactor

The kinetic microbial growth was predicted using Contois equation in accordance with [7]. The mass balance of biomass and substrate are presented [8]:

$$\frac{dX_1}{dt} = \frac{\mu_{m1}.C_{sCOD}.X_1}{K_{SX1}.X_1 + C_{sCOD}} - k_{d1}.X_1 \quad (1)$$

$$\frac{dX_2}{dt} = \frac{\mu_{m2}.C_{VFA}.X_2}{K_{SX1}.X_2 + C_{VFA}} - k_{d2}.X_2 \quad (2)$$

$$\frac{dC_{sCOD}}{dt} = -\frac{1}{Y_{X1/sCOD}}.\frac{dX_1}{dt} - m_{s,X1}.X_1 \quad (3)$$

$$\frac{dC_{VFA}}{dt} = Y_{VFA/X1}.\frac{dX_1}{dt} - \frac{1}{Y_{X2/VFA}}.\frac{dX_2}{dt} - m_{s,X2}.X_2 \quad (4)$$

$$\frac{dC_{CH_4}}{dt} = Y_{CH_4/X2}.\frac{dX_2}{dt} \quad (5)$$

Simulations of batch reactor were conducted by solving equation (1) to (5) using an ODE15s function in a MATLAB commercial software. For parameter estimation, there were 12 parameters to estimate in equation (1) to (5). These parameters were estimated using a gradient search method in lsqnonlin function in MATLAB. The aim of the optimization process was to minimize the SSE and for this purpose "weighting factors" were introduced to obtain larger sensitivity for a certain parameter. The SSE as well as weighting factor are described in equation (6) and (7).

$$SSE = \sum \left(w_{C,i}.\sum \left([C_i]_{\exp} - [C_i]_{calc}\right)\right)^2 \quad (6)$$

$$w_{C,i} = \frac{1}{[C_i]\exp,\max} \quad (7)$$

2.3. Simulation of Double-Stage AFBR

Simulation of Double-Stage AFBR was conducted by steady state simulation of well-mixed reactors.

a. For acidogenic reactor, the equations used to conduct simulations are:

$$0=\frac{X_{1,in}-X_{1,out}}{\theta_1}+\frac{\mu_{m1}.C_{sCOD}.X_1}{K_{SX1}.X_1+C_{sCOD}}-k_{d1}.X_1 \quad (8)$$

$$0=\frac{C_{sCOD_{1,in}}-C_{sCOD_{1,out}}}{\theta_1}-\frac{1}{Y_{X1/sCOD}}.\frac{\mu_{m1}.C_{sCOD}.X_1}{K_{SX1}.X_1+C_{sCOD}}-m_{s,X1}.X_1 \quad (9)$$

$$0=\frac{C_{VFA_{1,in}}-C_{VFA_{1,out}}}{\theta_1}+Y_{VFA/X1}.\frac{\mu_{m1}.C_{sCOD}.X_1}{K_{SX1}.X_1+C_{sCOD}} \quad (10)$$

b. Similarly, for methanogenic reactor, the following equations are used:

$$0=\frac{X_{2,in}-X_{2,out}}{\theta_2}+\frac{\mu_{m2}.C_{VFA}.X_2}{K_{SX1}.X_2+C_{VFA}}-k_{d2}.X_2 \quad (11)$$

$$0=\frac{C_{VFA_{1,out}}-C_{VFA_{2,out}}}{\theta_2}-\frac{1}{Y_{X2/VFA}}.\frac{\mu_{m2}.C_{VFA}.X_2}{K_{SX1}.X_2+C_{VFA}}-m_{s,X2}.X_2 \quad (12)$$

$$0=\frac{0-C_{CH_4,out}}{\theta_2}+Y_{CH_4/X2}.\frac{\mu_{m2}.C_{VFA}.X_2}{K_{SX1}.X_2+C_{VFA}} \quad (13)$$

Simulations of equation (8) to (13) were conducted using fsolve function in MATLAB. Sensitivity analyses were conducted by substituting each parameter while keeping other parameters constant.

3. Results and Discussion

3.1. *Simulation of Batch AFBR*

Figure 2 shows the results of experiments and simulations of batch reactor in the presence and absence of zeolite. As time increases, it can be seen that generally the value of soluble COD decreases. In addition, the value of VFA (indicator of intermediate products) tends to decrease rather slowly as a function of time. Observation of methane accumulation, as an indicator of biogas formation, indicates that biogas production increased rapidly during the first 20 days before reaching a relatively stable values until the end of the experiment.

The results from experiments without zeolite (control) showed that after 21 days, a decrease of COD from 8220 to 995 mg/L can be observed or equivalent to the value of 88% COD removal. To study the effect of zeolite, addition of 17 g zeolite/g sCOD has been conducted. It could be seen that by adding 17 g zeolite/g sCOD, a larger drop of COD could be observed after 21 days where the value of COD dropped from 8220 to 615 mg/L, which is equivalent to 92% COD removal. Thus, the effect of adding zeolite is generally positive with respect to COD reduction over a span of 21 days.

Formation of VFA as a result of COD conversion into biogas is also presented in Figure 2. It appears that the dynamics of VFA is similar between the control condition and the 17 g/g sCOD. There was an increase during the first 10 days which was followed by a constant decline until the end of the experiment.

With the aid of methanogenic microorganisms, the conversion of organic matters in POME resulted in the production of methane gas. As seen in the third panel of Figure 2, methane production appeared to show the same characteristics with the presence and absence of zeolite. During the first 40 days, the concentration of methane increased steadily until it reached values ranging between 400-500 ml/gSCOD.

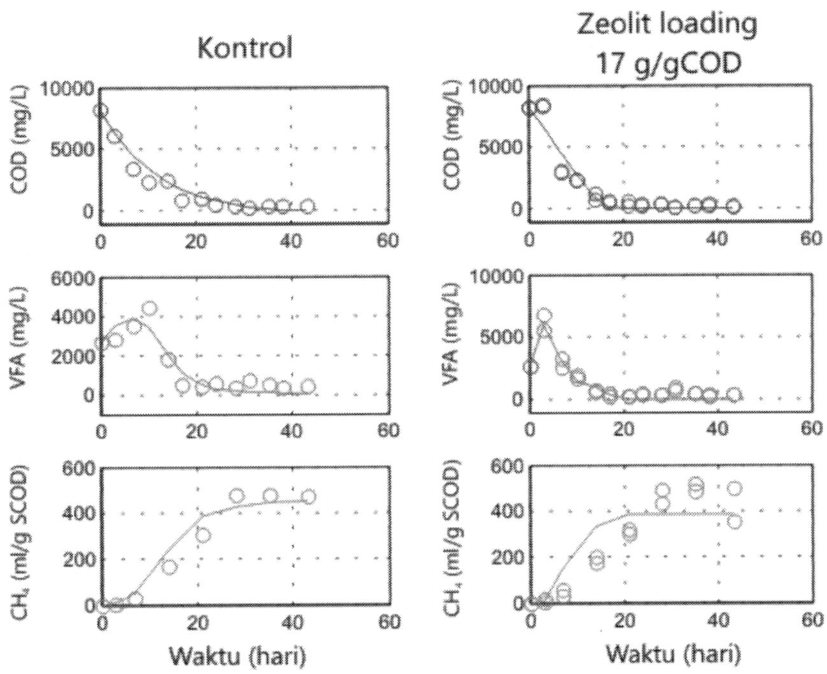

Figure 2. Simulation of batch reactor (sCOD, VFA, and CH_4 concentrations) as a function of *zeolit loading* with innoculum of biodiesel waste. Experimental data is denoted as (o) and simulation (–) .

The results of parameter estimations from batch reactor with biodiesel waste inoculum are presented in Table 1. The simulation results (Figure 1) generally show that the proposed model can explain the experiment results quite well. Based on the fitted parameters, the trend of parameters which are consistent with the decrease of COD along with increasing zeolite loading is μ_{m1}, μ_{m2} and $Y_{VFA/X1}$, $Y_{X2/VFA}$, and k_{d1}. Thus, these parameters can be used as indicators to enhance batch reactor performance although it is not trivial to understand their complex and simultaneous effects together.

Table 1 Result of parameter estimations from batch reactor with biodiesel waste as innoculum

Kinetic Parameters	control	17 g/gCOD
μ_{m1}	0.96867	1.3004
K_{SX1}	5.5449	4.6403
K_{SX2}	2.7683	1.3495
μ_{m2}	0.9614	1.0644
$Y_{CH4/X2}$	0.11728	0.02523
$Y_{VFA/X1}$	9.6437	23.464
$Y_{X1/sCOD}$	0.086438	0.16664
$Y_{X2/VFA}$	0.56066	0.64129
k_{d1}	1.0207	1.224
k_{d2}	0.59802	0.000296
$m_{s.X1}$	6.2736	3.7196
$m_{s.X2}$	0.009475	3.31E-11

3.2. Simulation of Double-Stage AFBR

Simulation of double-column AFBR is conducted by solving equation (8) to (13). Figure 3 gives the results of steady-state COD, VFA and CH_4 as a function of residence time in reactor 1 and reactor 2 of Double-Column AFBR. Figure (3) shows the simulation results obtained from kinetic parameters using biodiesel waste as inoculum and zeolite loading of 17 g/ g sCOD.
The main target of double-stage AFBR simulation is to obtain high COD removal rate with short residence time. The simulations showed that in order to achieve COD removal as much as 75%, the required residence time is still relatively high, i.e ca. 28 days.

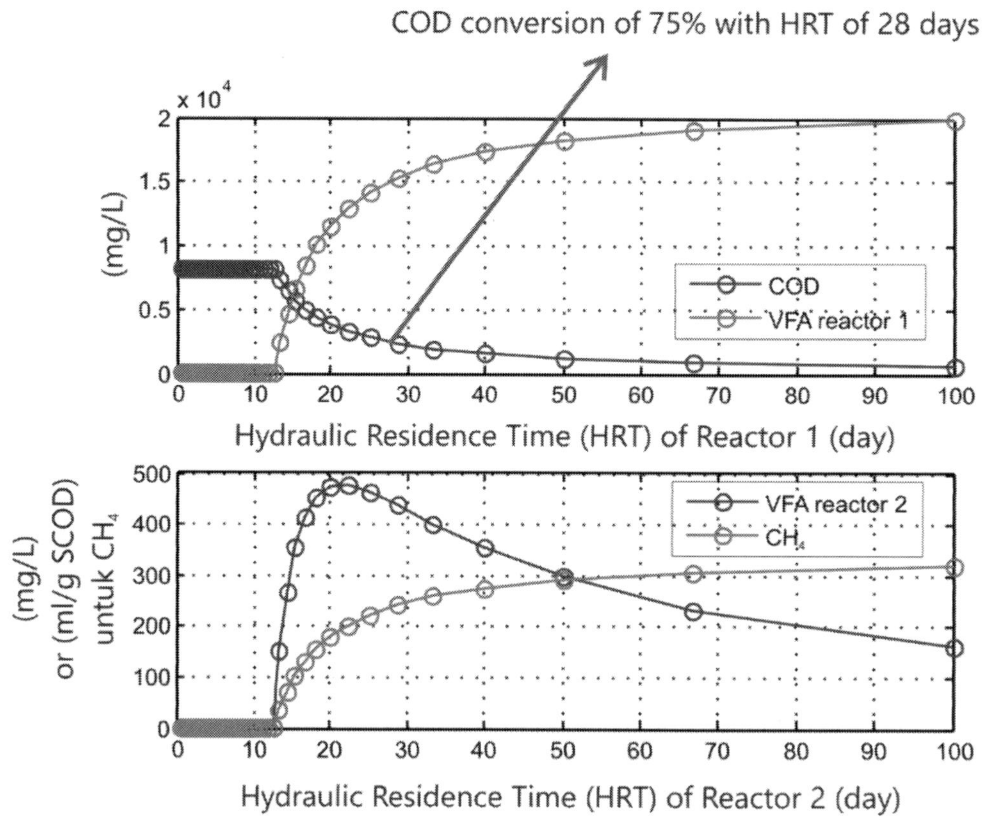

Figure 3. Simulation of Double-Stage AFBR. Operating conditions: biodiesel waste as inoculum and zeolite loading of 17 g zeolite/g sCOD.

3.3 Sensitivity Analyses of Double Stage AFBR simulations

Based on the results that we obtained from Figure 3, further analysis such as sensitivity analyses of kinetic parameters on the performance of Double-Column AFBR have been conducted. The sensitivity analysis was performed by changing one parameter value by keeping the values of other parameters constant in base case. The result showed that μ_{m1} appears to be the most sensitive parameter. By changing the value of μ_{m1} to 3 times greater ($3.\mu_{m1,base\ case}$), 75% conversion of COD removal could be obtained with a residence time ≈ 1 day. Other parameters appear to be less sensitive and give more or less similar results as displayed in Figure 3.

4. Conclusions

Mathematical models to simulate batch and double-stage AFBR have been proposed in the present study. Several conclusions can be withdrawn from the present work:

- The experiment results from batch reactor using biodiesel waste as the inoculum showed that the addition of zeolite gave a positive impact i.e. higher COD removal than the control condition (without addition of zeolite).
- The modeling of batch reactor with 12 kinetic parameters appears to be sufficiently well to explain the experiment data. Parameters μ_{m1}, μ_{m2} and $Y_{VFA/X1}$, $Y_{X2/VFA}$ and k_{d1} appears to increase with higher zeolite loading, in which h resulted in higher COD removal rate. Thus, these parameters can be used as indicators to enhance batch reactor performance although it is not trivial to understand their complex and simultaneous effects together.
- The result from the simulation of Double-Stage AFBR showed that the required residence time for high COD removal remains high (ca. 28 days). Sensitivity analysis showed that the increase of μ_{m1} to 3 times greater than the reference value gave a dramatic impact where 75% of COD removal could be obtained in $\approx$ 1 day.

Acknowledgement
We would like to thank BPDP Sawit for the financial support to this project. We would also like to acknowledge the fruitful collaboration with the Chemical Engineering Department of Universitas Lampung for their support in providing the POME and zeolite used in this study.

List of symbols:

sCOD	soluble chemical oxygen demand (mg/L)
VFA	volatile fatty acid (as acetic acid) concentration (mg/L)
CH_4	methane concentration in the liquid (mg/L)
X_1	acidogenic cell concentration (mg/L)
X_2	methanogenic cell concentration (mg/L)
μ_{m1}	maximum specific growth rate of acidogenic cell (day-1)
μ_{m2}	maximum specific growth rate of methanogenic cell (day-1)
KSX_1	half-saturation constant associated with sCOD (mg sCOD/mg acidogenic cell)
KSX_2	half-saturation constant associated with VFA (mg VFA/mg methanogenic cell)
$Y_{X1/COD}$	yield of cell formation per mg sCOD reduction (mg acidogenic cell/mg sCOD)
$Y_{X2/VFA}$	yield of cell formation per mg VFA reduction (mg methanogenic cell/mg VFA)
$Y_{CH4/X2}$	yield of CH_4 formation per mg methanogenic cell/L increase (mg CH_4/L]/[mg methanogenic cell/L)
$Y_{VFA/X1}$	yield of VFA formation per mg acidogenic cell (mg VFA/mg acidogenic cell)
kd_1	death rate constant of acidogenic cell
kd_2	death rate constant of methanogenic cell
θ	hydraulic retention time (HRT)

References

[1] Chin, M.J., Poh, P.E., Tey, B.T., Chan. E.S. and Chin, K.L., 2013. Renew. Sust. Energ. Rev., 26, pp.717–726.

[2] Hanum, F., 2009. Pengolahan Limbah Cair Pabrik Kelapa Sawit Dari Unit Deoiling Ponds Menggunakan Membran Mikrofilter. Tesis Sekolah Pascasarjana Universitas Sumatera Utara. Medan

[3] Nielsen, J. and Villadsen, J., 1994, *Bioreaction Engineering Principles*, pp. 415-439, Springer Science+Business Media New York

[4] Halim, L., Mellyanawati, M., Cahyono, R.B., Sudibyo, H. and Budhijanto, W., 2017, AIP Conf. Proc. Vol. 1840 110003

[5] Setyowati,P.A.H., Halim, L., Mellyanawati M., Sudibyo, H. and Budhijanto, W., 2017 AIP Conf. Proc. Vol. 1840 110004

[6] Ayu, E.D., Halim,L., Mellyanawaty, M., Sudibyo, Budhijanto, W., 2016, AIP Conf.

Proc. Vol. 1840 110005
[7] Sudibyo, H., Shabrina, Z.L., Halim, L. and Budhijanto, W., 2017, Energy Proc. Vol. 105, pp.256-262.
[8] Fogler, H.S., 2006, *Elements of Chemical Reaction Engineering 4th ed.*, pp.418-439, Pearson International Edition

IOP Conf. Series: Materials Science and Engineering **316** (2018) 012073 doi:10.1088/1757-899X/316/1/012073

Effect of biofilm and selective mixed culture on microbial fuel cell for the treatment of tempeh industrial wastewater

Rita Arbianti, Tania Surya Utami[1], Vifki Leondo, Elisabeth, Syafira Andyah Putri, Heri Hermansyah
[1]Department of Chemical Engineering, Faculty of Engineering, Universitas Indonesia, Kampus UI Depok 16424, Indonesia

Abstract. Microbial Fuel Cell (MFC) provides a new alternative in the treatment of organic waste. MFC produces 50-90% less sludge to be disposed than other methods. MFC technology can utilize existing microorganisms in the waste as a catalyst to generate electricity and simultaneously also serves as a wastewater treatment unit itself. Tempeh wastewater is one of the abundant industrial wastewater which can be processed using MFC. Research using the selective mixed culture is very likely to do due to the good result on COD removals by adding mixed culture. Microorganisms in tempeh wastewater consist of bacteria gram positive and gram negative. This study focused on the aspects of waste treatment which is determined by decreased levels of COD and BOD. Variations in this study are the formation time of biofilm and the addition of selective gram. MFC operated for 50 hours. For a variation of biofilm formation, experiments were performed after incubation by replacing incubation substrates used in the formation of biofilms. Biofilm formation time in this study was 3 days, 5 days, 7 days and 14 days. Gram positive and gram negative bacteria were used in selective mixed culture experiments. Selective mixed culture added to the reactor by 1 mL and 5 mL. Selection of gram-positive or gram-negative bacteria carried by growing mixed culture on selective media. COD and BOD levels were measured in the wastewater before and after the experiment conducted in each variation. Biofilm formation optimum time is 7 days which decrease COD and BOD levels by 18.2% and 35.9%. The addition of gram negative bacteria decreases COD and BOD levels by 29.32% and 51.32%. Further research is needed in order to get a better result on decreasing levels of COD and BOD.

Keywords: Biofilm; BOD; COD; Microbial fuel cell; Tempeh wastewater

1. Introduction

Industrial wastewater normally would be processed and send to the centralized waste treatment facility. The centralized facility mostly use aerobic waste treatment like activated sludge, trickling filter, or an oxidation ditch and pond aeration which consumes energy for the aeration process about 70% of the total energy required for waste treatment. The conventional method still produces considerable excess sludge (Du, 2007).

MFC (Microbial Fuel Cell) is a device that uses bacteria to generate electricity from organic and non-organic compounds. MFC provides a new alternative in the treatment of organic waste. MFC produces 50-90% less sludge to be disposed of than other methods (Holzman, 2005). MFC technology can utilize existing microorganisms in the waste as a catalyst to generate electricity and simultaneously also serves as a wastewater treatment unit itself (Liu et al., 2004).

Tempeh wastewater is one of the abundant industrial wastewater which can be processed using MFC. Tempeh industry generally is a household industry which disposes its wastewater directly into sewers or rivers without being processed first. Tempeh industrial wastewater is one of wastewater that causes a lot of problems to the environment. Whereas it still contain high BOD and COD levels (Wiryani, 2007). In utilization effort, tempeh industrial wastewater can be used as a substrate in the MFC system for the production of electricity.

The bacteria used in the MFC system can be either a pure culture of bacteria added to a medium or mixed culture of bacteria that is naturally contained in the material that has high organic matter content. Tempeh wastewater contains nutrients such as nitrogen, phosphorus, and organic materials such as carbohydrates, vitamins, and protein that act as a nutrient for bacteria. MFC operated using a mixed culture on the addition of 5 mL consortia to the reactor could decrease COD levels by of 27.34% (Mulyana, 2015). Previous research on MFC more emphasis to focus on optimizing the power and electricity generated. The focus of research in decreased levels of COD and BOD is necessary due to the electricity output of the MFC is still not as high as expected. This study examines the potential MFC as an efficient waste processing.

Research using the selective mixed culture is very likely to do due to the good result on COD removals by adding mixed culture. Microorganisms in tempeh wastewater consist of bacteria gram positive and gram negative (Nout et al., 1990). One of a group of bacteria will have a more dominant role in the treatment of tempeh wastewater compared to the other.

One of important factor in biological wastewater treatment is the ability to separate the biomass from the effluent in the process of clarification. At the sewage treatment using activated sludge, existing biomass in the form of a single cell (bacteria), micro colonies, and sets that form the floc (Lear and Lewis, 2012). Excess sludge in large amounts that need to be removed and separated. Disposal of sludge is quite expensive and spend considerable funds on waste management. Biofilm formation potentially reduces the amount of sediment cells.

Based on the above considerations, the present study will focus on waste treatment based on MFC using biofilm and addition of selective mixed culture. In this article, we study the effect of biofilm time formation and selective mixed culture on waste removal.

2. Material and methods

Tubular reactor which made of acrylic was used in this experiment. Volume of reactor is 650 mL. The anode and cathode that used was made from carbon in a certain size. Four cylindrical anodes were used. Each anode has diameter and length of $8x10^{-3}$ m $8x10^{-2}$ m. A plat cathodes are used which has length and width of $4x10^{-2}$ metre and $5x10^{-3}$ metre. Cathode contact directly with the air called air-cathode. The anode and cathode are connected with external circuit using copper wire and 820-ohm resistor.

Before used, the electrode was prepared. Graphite electrode immersed in a solution of 0.1 M HCl for 1 day and then rinsed with distilled water. It was soaked again into a solution of 0.1 M NaOH for 1 day and then rinsed again using distilled water. The reactor is filled with artificial tempeh wastewater. Artificial tempeh wastewater was made from the boiled soybeans using water at a ratio of 3:5 (Nout et al., 1990). Tempeh boiled water was incubated for 3 days. Each experiment was conducted using 292.5 mL tempeh wastewater, 292.5 mL of electrolyte solution (potassium persulfate 0.03 M), and 62.5 mL phosphate buffer pH 7.

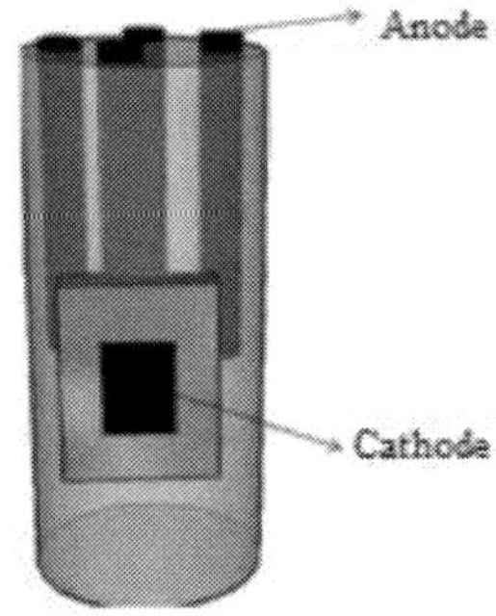

Figure 1. Reactor design

MFC operated for 50 hours. For a variation of biofilm formation, experiments were performed after incubation by replacing incubation substrates used in the formation of biofilms. Biofilm formation time in this study was 3 days, 5 days, 7 days and 14 days. Gram positive and gram negative bacteria were used in selective mixed culture experiments. Selective mixed culture added to the reactor by 1 mL and 5 mL. Selection of gram-positive or gram-negative bacteria carried by growing mixed culture on selective media. To cultivate gram-negative bacteria, MacConkey agar was used as growth media. To cultivate gram-positive bacteria, Mannitol Salt agar was used as growth media.

COD and BOD levels were measured in the wastewater before and after the experiment conducted in eachvariation.CODcontentmeasurementwasconducted by spectrophotometri method. Reagents used were HACH COD reagents Cat. 2125925 20-1500 mg/L. Before measured COD, calibration curve was made to determine the amount of COD in tempeh wastewater. Standard solution was made to determine the calibration curve. BOD measurement was conducted using BOD5(20) methods. The solubility of oxygen was measured using Dissolved Oxygen (DO) meter Hanna HI 7041s instrument.

3. Results

In Figure 2, 7 days' biofilm formation resulting the highest COD removal by 18.2%. COD removals on 14 days' biofilm formation was lower than 7 days' biofilm formation. BOD removals of waste have the same pattern with COD removals. 7 days' biofilm formation resulted highest BOD removals by 35.9% and followed by 14 days' biofilm formation by 34.26%.

IOP Conf. Series: Materials Science and Engineering **316** (2018) 012073 doi:10.1088/1757-899X/316/1/012073

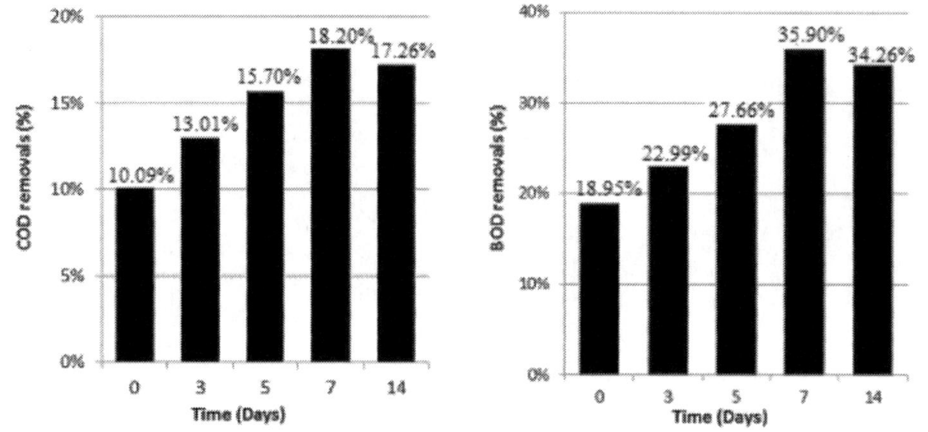

(a) (b)

Figure 2. Effect of biofilm formation time to (a) COD and (b) BOD removals

After the reactor operated for 50 hours, the largest reduction in COD levels based on Figure 3 was the addition of gram negative bacteria as much as 5 mL of 29.39%. The result was higher than previous research conducted by Mulyana (2014) and Citrasari (2015) with the addition of 5 mL mixed culture of bacteria which decreased COD levels by 26.19% and 27.35%.

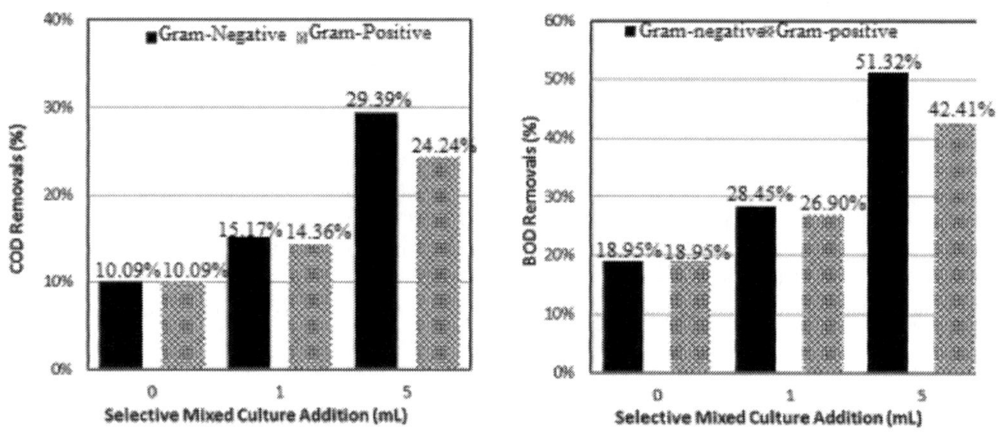

(a) (b)

Figure 3. Effect of addition of selective mixed culture on (a) COD and (b) BOD removals

4. Discussion

4.1. Effect of biofilm formation time to waste removals

At 14 days' biofilm formation, substrate transfer into biofilm more restricted than 7 days' biofilm. Substrate consumption rate increases as biofilm thickness and density increases. Number of bacteria which is catalyst in biochemical reaction increases as biofilm thickened (Seker et al., 1995). The thicker the biofilm the bacteria in biofilms have a better survival (Watnick and Kolter, 2000). So that the bacteria in the 14 days' biofilm has a better shelf life. However, at a certain thickness and density, the diffusion factor is more influential than the rate of reaction in substrate degradation of biofilm. The diffusion rate of the substrate to the bacteria in biofilm will decrease due to the decreasing diffusion coefficient (Seker et al., 1995). At 14 days the effect of biofilm diffusion rate is greater than the influence of the reaction rate of bacterial metabolism.

4.2. Effect of selective mixed culture addition to waste removals

On the addition of 1 mL selective mixed culture, waste removals of gram-negative just a bit larger than waste removals of gram-positive. However, the addition of 5 mL gram-negative resulted much greater waste removals than gram-positive bacteria. BOD removals of waste have the same pattern. Addition of gram-negative bacteria as much as 5 mL decreased BOD levels by 51.32%.

Based on the bacterial staining test which has been done, the most dominant bacteria in the tempeh wastewater are gram-negative bacteria. The high number of gram-negative showed that gram-negative bacteria more easily adapt to media tempeh wastewater. Tempeh wastewater contained diverse bacterial species such as *Citrobacter, Enterobacter agglomerans, Enterobacter cloacae, Klebsiella pneumoniae and Klozaenae* which is a gram-negative bacteria and *Lactobacillus casei, Enterococcus faecium, Staphylococcus epidermidis, and Streptococcus Dysgalactiae* which is gram-positive bacteria (Nout, 1990). Eventhough the addition of selective bacteria in the same amount, gram-negative bacteria result better COD removal. With the better adaptation ability, the growth of gram-negative bacteria was also better so its number will have increased in the reactor. The more bacteria, the more waste is degraded (Citrasari, 2014). Some species of a group of gram-negative bacteria in tempeh wastewater such as *Enterobacter cloacae* is resistant to Chromate (Fujie et al., 1996). Gram negative bacteria *Pantoea agglomerans* also commercially used in the paper mill waste treatment (Dewit et al., 2006).

5. Conclusion

Addition of gram-negative bacteria resulted greater waste removals than the addition of gram-positive bacteria. Addition of gram-negative bacteria as much as 5 mL decreased COD and BOD levels by 29.32% and 51.32%. The optimal Biofilm formation time for waste removal was 7 days' biofilm. It decreased COD and BOD levels by 18.2% and 35.9%.

6. Acknowledgement

The authors thank the Universitas Indonesia, which funded this research through the PUPT Research Grants Program, Universitas Indonesia, 2016.

7. References

[1] Citrasari, A. E., (2014), *Pengaruh jenis elektrolit dan penambahan jumlah bakteri konsorsium dari limbah cair tempe terhadap produksi listrik pada microbial fuel cell*, Depok, Universitas Indonesia.

[2] Dewit, Gerald, (2006), *Wastewater treatment compositions*, US 20060194303 A1.

[3] Du, Z., Li H., Gu T., (2007), *A state of the art review on microbial fuel cells : A promising technology for waste water treatment and bioenergy*, Biotechnol. Adv., **25** (5) : 464-482.

[4] Fujie, Koichi, et al., (1996), *Optimal operation of bioreactor system developed for the treatment of chromate wastewater using Enterobacter cloacae,* Wat. Sci. Tech., **33**: 173-182.

[5] Holzman, D. C., (2005), *Microbe power, Environmental Health Science*, **11**: 754-757.

[6] Lear, G. and Lewis G. D., (2012), *Microbial biofilm: Current research and applications*, New Zealand: Caister Academic Press.

[7] Liu, H., Ramnarayanan, Logan R., (2004), *Production of electricity during wastewater treatment using single chamber MFC*, Environ. Sci. Technol., **28**: 228-235.
[8] Mulyana, Guruh M., (2015), *Uji kinerja reaktor tubular membraneless microbial fuel cell (MFC) dengan substrat limbah cair Tempe*, Depok, Universitas Indonesia.
[9] Nout, M. J. R. and Rombouts F. M., (1990), *Recent developments in Tempe research, Journal of Applied Bacteriology,* **69**: 609-633.
[10] Seker, Sule, et al., (1994), *The effects of biofilm thickness on biofilm density and substrate consumption rate in a differential fluidized bed biofilm reactor*, Journal of Biotechnology, **41**: 39-47.
[11] Watnick, Paula and Kolter Roberto, (2000), *Biofilm, city of microbes*, Journal of Bacteriology, **182**: 2675-2679.
[12] Wiryani, Erry, (2007), *Analisis kandungan limbah cair pabrik tempe kedelai*, BIOMA, ISSN. 1410-8801.